Second Edition

PHYSICS
concepts and models

E. J. WENHAM

G. W. DORLING

J. A. N. SNELL

B. TAYLOR

Longman

E. J. Wenham, formerly Vice Principal of Worcester College of Higher Education, has been closely involved in curriculum development in physics both at home and overseas. He has acted as a consultant to UNESCO, OECD and the British Council and currently to the Nuffield Advanced Level Physics Project revision. He was an assistant organizer for the Nuffield Ordinary Level Physics Project and, later, the joint editor with Eric Rogers of the revised edition. Awarded the Bragg Medal and Prize of the Institute of Physics in 1977, he was made an M.B.E. for services to physics education in 1982.

G. W. Dorling is Head of Science at Wymondham College, Norfolk. Prior to this he was Senior Lecturer in the Physical Sciences Department at Worcester College of Higher Education, when he was also involved in in-service work with physics teachers both in the U.K. and abroad. He continues to maintain his links with teacher training in his role as teacher–tutor in the University of Cambridge Department of Education. At one time a member of the consultative committee of the Nuffield Combined Science Project, he is currently closely involved in the work of the Schools Council Integrated Science Project. He is an examiner in Advanced level physics for one of the examination boards.

J. A. N. Snell is a teacher of physics at St Edward's School, Oxford. During his time as head of the Physics Department at the school he has introduced the Nuffield Physics courses at both O and A level, project work in applied science and technology and has written courses in practical electronics. He is currently developing the teaching of microelectronics and computing, with particular reference to physics and technology, in the school.

B. Taylor, who is currently concerned with the preparation and operation of programmed learning systems used world-wide, has been Science Adviser to the Leicestershire Education Authority, a member of the Physics Department of Rugby School and Senior Physics Master at St. John's College, Southsea. He has organized many in-service courses both at home and abroad; he is an examiner for one of the school examination boards and represents the Association for Science Education on the Physics Education Committee of the Royal Society.

The cover photograph shows the Aurora Borealis photographed from Krunna, Sweden, October 7th 1981; photograph courtesy of M. J. Taylor, Physics Department, Southampton University.

LONGMAN GROUP UK LIMITED
Longman House, Burnt Mill, Harlow, Essex CM20 2JE, England and Associated Companies throughout the World

First published 1972
Second edition 1984

First published by Longman Group Limited 1985
Third impression 1987
ISBN 0 582 35580 X

Printed and bound in Great Britain by Scotprint Ltd, Musselburgh.

PREFACE

Since the publication of the first edition in 1972 certain trends in the teaching of physics have emerged. Discussions between the various Examination Boards and the universities in the United Kingdom have resulted in agreement on a core 'syllabus'; statistics suggest that fewer of those students who take physics are aiming at careers with a direct involvement in it; somewhat lower standards of sophistication have to be assumed both at entry to and exit from the study; the microprocessor and microcomputer have become familiar tools in education as well as commerce and industry. In addition, more emphasis is now placed on the applications of physics and their implications. We have therefore made substantial alterations in this second edition to reflect these changes.

Physics is, and will remain, the fundamental science. The main aim of this book is to provide students with a clear understanding of the principles of the science and an ability to relate them to the physical world. We feel this is essential not only for students who continue their study of physics after school but also for those who enter other fields which require good knowledge of the basic principles.

Physics is much more than a catalogue of facts and formulae. Here we have presented it as a *process* of developing concepts and models and a *structure* of closely interwoven facts and ideas. Together the structure and process provide a field of knowledge which raises questions about the nature of the physical world around us and offers methods and material for the resolution of these questions.

Whilst we may assume that many readers will have some background knowledge and understanding of simple mathematics, we have borne in mind the needs of those who are just beginning the subject. All important concepts are developed step-by-step from first principles. Surprisingly few mathematical skills are required at this level for the study of physics. Confidence in the handling of simple algebra, graphs and numbers will take the student a very long way. Where the use of a calculus method is essential, care has been taken over its presentation.

Students require practice in order to see the relevance of physical concepts to the natural world and to apply them to real problems. To provide this practice we have included a number of questions and problems at the end of the book. As with real-life problems, some are open-ended and without clear-cut solutions, but these provide good opportunity for discussion and speculation. Answers are supplied for the majority of the problems together with hints pointing towards a solution where this is appropriate. Worked problems are also given at relevant points throughout the book.

The selection of the material owes a great deal to the Nuffield Advanced Physics Project as well as to our own first edition. The proposal for a common core of study at the Advanced level of the British General Certificate of Education Examination has also been important. Our material is grouped into units but, as experience with the first edition has shown, it is not essential for teachers to follow the order in which these units appear. Even so, it is hoped that the first unit, which sets the scene and provides essential vocabulary for the study, will be taken first. Students studying independently would be well advised to take the units in the order given.

We have attempted to make the material tell a coherent story taking note of the past and present and looking towards the future. Social applications have been looked at where appropriate and the rapid development of the microprocessor and microcomputer is outlined in a new unit dealing with digital electronics.

Units of the Système International are used throughout with the addition, where expedient, of certain decimal fractions and multiples of those units to which special names have been given, e.g. the kilowatt hour, and of some units which are defined in terms of the best available experimental values of certain physical constants, e.g. the electron-volt.

Naturally this text has been strongly influenced by a series of very distinguished books on physics published in recent years. Special reference must be made to the publications of the Nuffield Foundation Science Teaching Project, the Physical Science Study Committee and to Project Physics. We also owe a debt to such other authors as Arons, Bennet, Rogers, Sears, Zemansky, Warren and Feynman.

The authors are particularly grateful to Professor E. F. W. Seymour (University of Warwick), Mr W. K. Mace (King Edward VI School, Sheffield), Mr P. Jordan (Highfields Schools, Wolverhampton) and Dr T. Martin (Worcester College of Higher Education) who have read the manuscript and whose comments and suggestions have been of the greatest value. Our thanks are also due to Mr E. Howard who read the unit on electronics and to all those who, finding errors or difficulties in the first edition, wrote and told us of them. This is greatly appreciated. Nevertheless, the authors must accept full responsibility for any errors, whether of fact or interpretation, which may still exist in their work.

January, 1984   E. J. Wenham
                               G. W. Dorling
                               J. A. N. Snell
                               B. Taylor

CONTENTS

Models and the Language of Physics

Chapter 1

INTRODUCTION

Many of the well-known laws of physics are associated with the names of scientists. Any study of gases soon introduces the laws of Boyle, Charles, and perhaps of Gay-Lussac. Planetary astronomy leads to the laws of Kepler at an early stage in its development. Much elementary optics is based on Snell's law. Electricity demonstrates its importance with any number of laws – those of Coulomb, Ohm, Kirchhoff, Ampère, Faraday, Lenz. It is an easy task to add to the list.

Some of these laws are quite remarkable for the degree of perception shown by their discoveries. Who but Kepler could have deduced his three laws of planetary motion from the mass of data supplied by Tycho Brahe? Even in our age of computers this still ranks as a magnificent achievement.

But, when one has marvelled at the skill of these scientists in reducing a mass of observations into a succinct law of behaviour; when one has admired their implicit belief in a physical world whose behaviour can be expressed in the regularities described by these laws, one may still be surprised at the dullness of so many of these laws. Can a relationship between electric current and potential difference have been the culmination and the driving force of the life of Georg Ohm? Can a relationship between the pressure and the volume of a gas sample have been the goal of Robert Boyle? Physics is, at times, presented as if it were.

1.1 Robert Boyle

Robert Boyle (1627–91), a contemporary of Christopher Wren and Samuel Pepys, wrote an account of many of his experiments in a book called *New Experiments, Physico-Mechanicall, Touching the Spring of the Air and its Effects.* Written in the style of a letter to his nephew it begins:

> 'Receiving in your last letter from Paris a desire that I would add some more experiments to those I formerly sent you over; I could not be so much of your servant as I am, without looking upon that desire as a command; and consequently, without thinking myself obliged to consider by what sort of experiments it might the most acceptably be obeyed. And at the same time, perceiving by letters from some other ingenious persons at Paris, that several of the Virtuosi there were very intent upon the examination of the interest of the air; in hindering the descent of the quicksilver in the famous experiment touching a vacuum; I thought I could not comply with your desires in a more fit and seasonable manner, than by prosecuting and endeavouring to promote that noble experiment of Torricellius; and by presenting your Lordship an account of my attempts to illustrate a subject, about which its being so much discoursed of where you are, together with your inbred curiosity and love of experimental learning, made me suppose you sufficiently inquisitive.'

Reading the account of Boyle's work on the air, it soon becomes clear that he was motivated not merely by a desire to satisfy the whims of his young nephew. The work was stimulated by the previous experiments of Torricelli to which Boyle refers.

1.2 Torricelli's experiment

Only a few years before Boyle's experiment with air, Torricelli, a pupil of Galileo, had shown that if a long tube, closed at one end, was filled with mercury and inverted with the open end immersed in a bowl of mercury, the column of mercury in the tube did not exceed about 760 mm in height (see Fig. 1.1). The space which appeared above the mercury appeared to be 'empty.'

The phenomenon could not be denied. But Torricelli's explanation of why the mercury

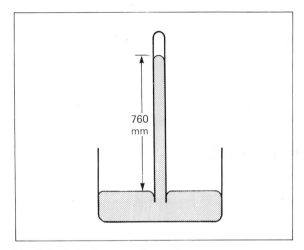

Fig. 1.1 A simple mercury barometer.

should behave in this way was at the centre of controversy during the middle of the seventeenth century. To quote Boyle, Torricelli believed that:

> 'the falling of the quicksilver in the tube to a determinate height, proceedeth from the aequilibrium, wherein it is at that height with the external air, the one gravitating, the other pressing down with equal force upon the adjacent mercury.'

This view was determined by Torricelli's *theory* (or *model*) of an earth surrounded by a 'sea of air' which exerts a pressure on us and on our surroundings, living, as we are, at the bottom of this 'sea.'

What had led Torricelli to perform the experiment with the mercury? No one knows for sure but there is some evidence to suggest that he had evolved his model of the atmosphere *before* doing his famous experiment. The fact that water could not be pumped up higher than about 10 m using a single lift pump was known to Galileo who mentions it as an unsolved problem in his book *The Dialogues concerning Two New Sciences.* Torricelli himself was a keen disciple of Galileo. It seems likely that his theory was proposed as a solution to this problem. His experiment with the mercury thus becomes an attempt to verify his solution by experiment.

Despite the experiment, the theory was not generally accepted. Many of the critics could not understand why, if the theory was acceptable, the mercury in a barometer tube did not fall when the

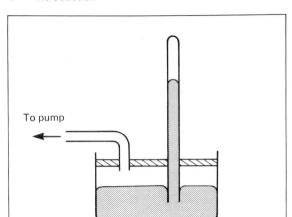

To pump

Fig. 1.2

barometer was sealed off from the surrounding air. However, Boyle was firmly convinced of the correctness of Torricelli's theory and set out, in a sequence of famous experiments, to demonstrate that a number of deductions which might be made from the theory could, in fact, be verified by experiment. For example, one of his first experiments was to use a pump, designed by Robert Hooke, to remove the air from the top of the bowl of mercury (Fig. 1.2 shows a modern version of the arrangement).

He reasoned that, if Torricelli's model of the atmosphere was satisfactory, the mercury in the tube should fall since there would be less air

pressure to support the column. This is a simple experiment to perform today. But Boyle found it so difficult that he felt called upon to give a detailed description of his technique.

So we see that Boyle was led towards his famous law, not by a desire to find how the pressure and volume of a gas varied but by a need to examine a theory or model of the behaviour of the air. His motive was to test some ideas then current as to *why* things behaved as they did, and not simply to find out *how* they behaved. But, as we shall see, further evidence about the behaviour of the physical world came to light as a result of his investigations.

Since this is not a book about the history of physics, some reason must be given for considering the work of Robert Boyle. Does the law by which Boyle's name is celebrated indicate the sort of questions he asked about the physical world? History strongly suggests that it is not. It suggests that Boyle was not so much interested in answering questions about how the world behaves but in why it behaves as it does. This is typical of the way in which most of science, and thus physics, evolved. While experiment, which occupies so much of a scientist's time, is concerned with answering 'How?', that question is itself posed by theories and models which have been proposed to answer 'Why?'.

Of course, no one is going to ask 'Why?' before there is some clear reason for doing so.

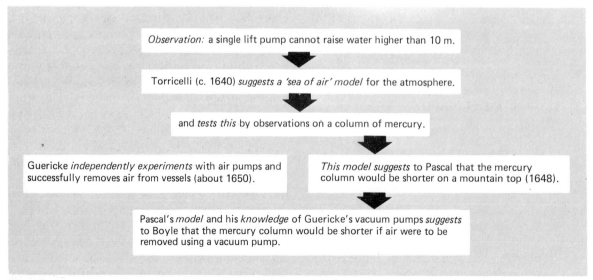

Fig. 1.3

What the history of physics seems to show is that the question 'Why?' usually arises at a very early stage in the investigation of new phenomena and that attempts to answer it are at once made by proposing theories or models which it is hoped represent the underlying causes. Such theories suggest new experiments which may both serve to test the theory and bring new facts to light. With this in mind, let us re-examine the work of Boyle (Fig. 1.3).

1.3   Laws

We are now in a position to consider the meaning of this word as it is used by scientists. At its simplest, a *law* is a summary of observed, measurable behaviour. Philosophers have discussed at length the impertinence of scientists who assume that what is true for a necessarily limited number of observations made in the past is true for all similar observations both now and in the future. But, in practice, scientists have disregarded the warnings of the philosophers on the grounds that their approach is justified by its success.

Kepler's laws of planetary motion, which will be discussed in Section 28.2, offer a good example. They summarize many observations of the positions of the planets relative to the background of the fixed stars. Their formulation depends solely on these observations and not on any preconceived theory as to why the planets behave as they do.

There are many laws in physics of this kind. They include:

The relationship between the angle of incidence which a light beam makes with a mirror and its consequent angle of reflection. This may be expressed in the simple law that the angle of incidence is equal to the angle of reflection.

The relationship between the extension of a loaded spring and the load itself (Hooke's law).

The relationship between the activity of a radioactive substance and time. This law is rather more complex than those cited above but, nevertheless, the relationship can be expressed by the simple exponential law.

In all these cases, the law describes a simple mathematical relationship between two measurable quantities. The laws do not depend upon any preconceived notions about causes. They merely summarize the answers to the appropriate 'How?' question. (For example, how does the extension of the spring change with load? How does the activity of a radioactive substance change with time?) Such laws imply that the physical world behaves in a coherent fashion. This implication is basic to the sciences.

1.4   Theories and models

If we believe that the physical world behaves in the coherent way suggested by such laws, we are driven to ask 'Why?'. It is man's attempts to answer this second question which have led to that body of knowledge which we know as science.

Our knowledge of the behaviour of the physical world can never be complete. It follows that our knowledge of the causes of that behaviour can never be certain. The best we can do is to guess (intelligently) some likely cause; to build imaginative pictures of the processes which give rise to this observed behaviour. These imaginative theories or *models*, as these pictures are called, have been the essential feature of scientific speculation throughout its recorded history. They are imaginative adventures of the mind and they bring the essential ingredient of creativity into science.

The wave theory of light is an excellent example of a modern theory. This suggests that light energy is transported from place to place in a wave-like manner. It can be shown that light, if regarded as a wave, can be reflected and refracted according to the laws which are found to apply to light. So a wave provides us with a satisfactory model for light – satisfactory in many respects; indeed, it might be preferable to talk of the wave *model* for light rather than of the wave *theory* of light. However, the two words are often used as though they were interchangeable.

It must by now be clear that the word *model* does not mean quite the same thing to a physicist as it does to the man in the street. When we talk of a *model ship* we usually mean a scaled-down copy of the real thing. The model ship is *like* the real ship. When physics took over the word *model*, the essential feature of the everyday usage which was preserved was in the word *like*. Very often, in physics we are dealing with processes which arise from underlying causes which are not directly perceived. This is true of the behaviour of light. No one can *see* a wave of light. The only sort of waves which we can see are those which might be generated in a stretched rope, a 'slinky' spring, or on the surface of a sheet of water, whether in a ripple tank or on the ocean itself. We can experiment with these waves and examine their characteristic properties. Even more important, we can write down mathematical expressions to describe their behaviour and make verifiable deductions from those expressions. When we talk of light energy travelling in waves, we mean that the energy is travelling *like* waves travelling down a rope or over water. We consider that a wave model is a good model of the way in which light behaves. It is in this sense that the scientist uses the word *model*. Let us consider two examples.

a) Newton's theory of gravitation

The theory of universal gravitation was proposed by Newton in an attempt to account for the observed motions of the planets. The essence of the theory is that all bodies attract all other bodies and, in their turn, are attracted by all other bodies because they possess mass. This force of attraction is called the force of gravity. It can account for the observed motions of the planets; for the attraction of all terrestrial bodies to the earth; for the behaviour of comets, of earth satellites, of space probes; it even provides a basis for a model for the formation of the galaxies. In so far as this force cannot be directly experienced, it becomes an act of speculation to say that the planets move round the sun because a force attracts them towards the sun just as (*like*) a force acts on a mass which is whirled round one's head on an attached string. In this sense, Newton's theory is a model for the behaviour of the planets (or of earth satellites, or whatever is being considered). But there is an important sense in which this model is different from its predecessors. In order to incorporate fully the experience summarized in the three laws of Kepler the force between a planet and the sun must vary inversely as the square of the separation, r, between the planet and the sun. The forces on masses which are attached to strings do not behave in this way. Furthermore, the force must be proportional to

the product of the mass of the planet, m_p, and the mass of the sun, m_s. In full

$$F \propto \frac{m_p m_s}{r^2}$$

All this is incorporated in Newton's theory of gravitation. The model is, to a considerable degree, a mathematical model. Certain features of this model can best be expressed in mathematical terms. This is a characteristic of a physical model and, as physics has developed, so its models have become more and more mathematical in nature.

But Newton took matters further than this. If the movement of the planets depends on some force of attraction between the planets and the sun – a force which depends on their mass – then this force ought to be evident between *all* bodies. This is what Newton proposed. He suggested that the force holding the planets in their orbits was the same force as that which attracted all earthly bodies to the earth. It is for this reason that Newton's theory is often referred to as the theory of universal gravitation. He had extended the familiar terrestrial force of gravity out to the edge of the solar system and beyond.

The history of the development of this theory has received very detailed attention from the historians of science. It is often used to illustrate the relationship between observation, theory, and law. By carefully disregarding some of the planetary theories which have not proved successful, it is possible to represent in Newton's work a clear-cut relationship between these entities. The argument is often propounded as follows: careful observations made by Tycho Brahe were summarized in laws of planetary motion by Kepler. This led Newton to propose a hypothesis (or tentative theory) to account for these laws. From this theory he made new deductions which, on being justified, led to his hypothesis attaining the status of a theory.

It is difficult to assess the usefulness of such an analysis. It almost certainly does not reflect the historical development of the theory – too much is left unsaid. However, with hindsight, it is often tempting to re-structure scientific developments into such a 'logical' order. It is doubtful whether such an ordering is justifiable. Theories and models are rarely propounded on a basis of purely objective fact. They must be seen against the current background of scientific thought.

We have tried to suggest in the previous discussion of the work of Boyle that theory (speculation) and the making of measurements proceed side by side, each acting on the other: a theory suggests some worth-while observations. The results modify the theory. The modified theory, in its turn, suggests new observations.

Certainly, Tycho Brahe and Kepler, influenced still by the scholasticism of the Middle Ages, had their own theories, and these were derived from Greek mathematical speculation. The theories did not prove fruitful, however, and it is a tribute to the work of these men and an object lesson for science, that their experimental work had value because their own theories were not allowed to influence the objectivity of their observations and analyses.

Newton's theory was successful, but not because he followed that pseudo-logical process which some commentators suggest. Newton's tremendous contribution was the insight into what is a fruitful theory. It is to Newton that we owe the view that a theory should be sufficient – sufficient to explain observed facts, and to propose new observations. The ability to know what is sufficient is one of the marks of genius.

If one wanted a story to show how Newton came to propose his theory of universal gravitation, one could do worse than consider the legend of Newton and the apple. Told in its usual form (how Newton sat under a tree and an apple fell to the ground nearby and so he discovered the law of gravity) it has often been laughed at. But think of it again. Newton is sitting in a Cambridge college garden, on a sunny day in late autumn, pondering new mathematical ideas he has just worked out to describe motion under a central force. This is the problem for which his laws of motion have been evolved. He has calculated that Kepler's laws must mean that the planets are acted upon by some force proportional to their masses. All this follows from his new dynamics, but it does not constitute a complete theory. An apple falls to the ground nearby. He stares at it, unheedingly for a moment. And then the flash of inspiration. Of course!

There is no more evidence to support this version of the story than there is to support the more commonly expressed view of Newton's careful and logical analysis of the problem. But as an allegory, it describes the process of the evolution of theories (when these have been more fully documented) much better than does the usual picture.

Before leaving this famous theory, let us consider one further aspect of the traditional story. It contains within it a suggestion that a theory is essentially different from a hypothesis erected on the basis of a single successful established deduction. While the status of the theory is clearly enhanced once a proposed consequence proves successful, one swallow does not make a summer. The status of a theory is further and further enhanced as it is shown experimentally to have further and further validity. But, no matter how many deductions are successfully tested, the theory can never be proved to be true in the absolute sense.

One needs only to show that one deduction is contrary to experience to invalidate, or, at least, to limit, the theory. This view has been expressed very cogently by the philosopher Karl Popper. In a discussion of the way in which a scientific theory can be distinguished from other types of theories or beliefs he has written:

> 'But I shall certainly admit a system as empirical or scientific only if it is capable of being *tested* by experience. These considerations suggest that not the verifiability but the falsifiability of a system is to be taken as a criterion of demarcation. In other words: I shall not require of a scientific system that it shall be capable of being singled out, once and for all, in a positive sense; but I shall require that its logical form shall be such that it can be singled out, by means of empirical tests, in a negative sense: it must be possible for an empirical scientific system to be refuted by experience.' (Popper, 1959.)

This is written with the careful choice of words common to philosophers, but the meaning is clear. Popper is attempting to distinguish scientific theories in the same way that Newton distinguished forces – by considering their unique properties. And just as we turn Newton's first law of motion round so that it tells us what is, or what is not, properly described as a force, so we can use this passage to tell us what is, and what is not, a meaningful scientific theory.

b) Rutherford's model of the atom

As another example of a theory in physics, let us consider the model of an atom whose origin is usually ascribed to Lord Rutherford. This is a specific case of an attempt to visualize something which is unseen and unseeable in terms with which we are familiar. We shall consider shortly some of the evidence which led to the firm establishment of a picture of matter as being built up of small, chemically indivisible particles (atoms). As further experimental evidence accrued, it became necessary to extend this theory and to visualise atoms as having some sort of structure.

On the basis of his work in radioactivity, Rutherford suggested that the atom was a system in which the greatest part of the mass of the atom was concentrated in a minute nucleus. He conceived that such an atom would have a positive central charge Ne and be surrounded by a compensating charge of N electrons (e is the electronic charge and N a positive integer). In making deductions from his theory the N electrons were considered as constituting a charge $-Ne$ supposed uniformly distributed throughout a sphere of radius R.

Here we have a purely pictorial representation of the atom. It allows both qualitative understanding of certain atomic events, and quantitative mathematical treatment. By 'understanding' we mean that acceptance of the model allows us to offer an explanation of certain observed behaviour: in this case, the scattering of alpha particles.

The important thing to remember about a pictorial model such as this is that it is no more than a model. It provides an interpretation of the way atoms behave. Too much must not be read into the picture. This is a model of the atom seen 'with the eyes half-closed.' For example, you cannot ask of this model 'what is the structure of the surrounding sphere of electric charge?' The model was not proposed to answer that question and, as Rutherford shows, it is irrelevant to the problem in hand.

Eventually, the detail in the model may be improved and a structure may be suggested both for the sphere of the electric charge and for the central nucleus. But this aspect of the model will remain unnecessary in interpreting Rutherford's experiments. We will return to this point later when we are considering models for the structure of matter.

1.5   The continuation of the work of Boyle

In order to illustrate the process by which physics builds models and theories of the physical world, we shall consider one model – that of the structure of matter – in detail. But before doing so let us look again at the work of Boyle as he continued his investigations into the behaviour of gases. These led to his celebrated law.

Soon after the publication of his book on *The Spring of the Air*, a number of attacks were launched against his advocacy of Torricelli's theory. One of these, by Linus, is particularly interesting because of the way in which Boyle dealt with the argument. Boyle's explanation of the behaviour of the mercury column in Torricelli's experiment involved the support of the column by the weight of the external air. Since the pressure of the air is applied directly to the surface of the mercury in the surrounding dish, the theory necessarily attributes to that air a certain resilience, or springiness, in resisting the crushing force of its own weight; hence the title of the book itself. Linus objected to this interpretation, saying that whilst he agreed that air had both weight and springiness, these were not sufficient to support the mercury column. He suggested that there was also some agency within the space above the mercury which held up the mercury column.

Boyle's reply to this appeared in an appendix to the book and was introduced thus:

> 'The other thing, that I would have considered touching our adversary's hypothesis is, that it is needless. For whereas he denies not, that air has some weight and spring, but affirms, that it is very insufficient to perform such great matters as the counterpoising of a mercurial cylinder of 29 inches as we teach that it may; we shall now endeavour to manifest by experiments purposely made, that the spring of the air is capable of doing far more than it is necessary for us to ascribe to it, to save the phenomena of the Torricellian experiment.'

The important point to note here is Boyle's statement that Linus' additional hypothesis is 'needless'. He intends to show that his own theory is sufficient. This is an important feature of all theories in physics since the time of Boyle. Such theories never, or should not, incorporate more elements than are sufficient to provide some explanation of the matter in hand. Especially has this to be borne in mind when dealing with those areas of speculation best described in terms of models. We saw in considering Rutherford's model of the atom how easy it is to ascribe more features to the model than can either be justified or needed for the immediate experimental evidence which Rutherford was considering.

1.6   The discovery of Boyle's Law

The apparatus which Boyle used to test his ideas was the same as that still to be seen in many teaching laboratories today (see Fig. 1.4). He poured just enough mercury into the J-shaped tube to trap a volume of air in the left-hand limb. The pressure of this trapped air was then very nearly atmospheric. Then he poured more mercury in at the open end until the difference in levels was about 760 mm. This meant that the compressed air in the short closed limb was supporting both the pressure of the atmosphere acting on the top of the mercury in the open limb and a column of mercury whose pressure was equal to that of the atmosphere. Boyle noted that the volume of the compressed air was about one half of its volume when the levels in the two columns were the same, i.e. when the air was under atmospheric pressure alone. In his discussion of the result of this experiment he points out that the smaller the space into which the air is compressed the greater its pressure (or *spring*). He goes on

> 'So that here our adversary may plainly see, that the spring of air, which he makes so light of, may not only be able to resist the weight of 29 inches (760 mm), but in some cases of above one hundred inches (250 cm) of quicksilver ...'

Boyle took a sequence of readings relating the compression of air to the pressure upon it, but seems to have been unable to see any mathematical relationship between the results. It was not in fact Boyle, but Richard Townley, who first detected in Boyle's readings, the relationship

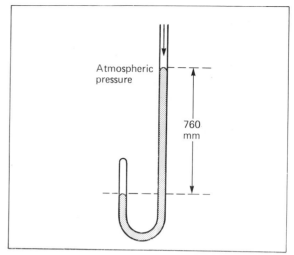

Fig. 1.4   J-tube for test of Boyle's law.

which we now know as Boyle's law, namely that the volume of a given mass of gas is inversely proportional to the pressure upon it.

This discovery so captured Boyle's imagination that the dispute which led to it seems to have been almost forgotten. Boyle himself, in further experiments, seems to have shown that the law remains true when the pressure on the gas sample is reduced below atmospheric. He also seems to have tested the effect of temperature and to have realized at least implicitly that to his law must be added the words 'at constant temperature.'

By his investigations into the 'spring of air' and his establishment, at least experimentally, of the law which bears his name, Boyle not only confirmed the Torricellian theory but considerably extended it. He tried to produce a model of the air which would account for its compressibility. In fact he produced two, saying he was not willing 'to declare peremptorily for either of them against the other.' His first model consisted of 'a heap of little bodies, lying one upon another, as may be resembled to a fleece of wool.'

His second was of particles whirled around in the 'subtle fluid' which was thought at that time to fill all space. Boyle said of this model:

> 'it imports very little, whether the particles of air have the structures requisite of springs, or be of any other form (how irregular soever) since their elastic power is not made to depend upon their shape or structure, but upon the vehement agitation.'

Here is yet another example of the way Boyle showed himself to be one of the forerunners of modern physical thinking. He would neither commit himself to one particular theory before he could produce experimental evidence to justify this comment nor would he ascribe to any one model more characteristics than he needed for the matter in hand (as when he said it matters little what the structure of the whirling particles is.)

This seems an appropriate point to take up the problem of a present-day model for the structure of matter, and to use this to show how models can be built up in physics.

1.7   A model for a gas

Experimental work with gases since Boyle's time has shown that gases share many of the same physical properties at atmospheric pressure.

They all conform to Boyle's law in the way their volumes change under applied pressure. They all expand as their temperature is raised at constant pressure and they all have (at low pressures) the same coefficient of expansion.

Since these properties seem to be largely independent of their chemical nature, we may expect that any model of a gas designed to explain these properties will be particularly simple. Let us try to build such a model.

The main criteria to be satisfied are:
a) the model should be as simple as possible;
b) the model should be consistent with well-established theories or models in related fields of study;
c) the model should lead to further predictions which can be verified by experiment.

Criterion (b) immediately gives us a starting point for the model – we shall imagine the gas to be an assemblage of separate (or discrete) particles.

This is a speculation of great antiquity which had been used as a qualitative way of thinking about natural phenomena by such men as Bacon, Boyle, Hooke and Newton. But it was John Dalton who, around 1803, was the first scientist to use the particulate model in a direct sense in order to provide a description of observed behaviour – in this case the laws of chemical combination. The development of the model in the hands of Gay-Lussac and Avogadro enabled the relative masses of the particles to be calculated. Furthermore, it was found necessary to classify these particles in terms of *atoms* and *molecules*.

The laws of chemical combination allow no other easy interpretation. So, in developing a model for the physical behaviour of gases, it would be unreasonable to assume anything other than a particulate nature for the gas.

The first feature of a gas that our model must explain is its high degree of compressibility and springiness. We have a choice between Boyle's two hypotheses. Either (a) the particles are in contact but themselves have an intrinsic springiness, like a ball of wool, or (b) the particles are far apart in a constant state of motion. This state of motion could account for the pressure exerted by a confined volume of gas by the collisions made by these particles with the side of the vessel.

Is there any other property of a gas which might guide one's choice?

1.8   Pressure and temperature

When the temperature of a sample of gas, confined to a given volume, is raised, its pressure also rises. The higher the temperature of the gas becomes, the greater the pressure it exerts on its container. The pressure of the gas is said to be temperature dependent. On the basis of the two proposed models, we must either assume that the intrinsic springiness or elasticity of the particles increases with temperature, or (model b) the particles move faster as the temperature rises.

It now seems possible to make a reasonable choice between the two models for an independent piece of physical theory suggests that heat is a measure of the energy transferred to or from a body by a gradient of temperature. A considerable body of experimental evidence can be interpreted in terms of this theory. A model of a gas as widely-spaced moving particles has much to commend it. It can be made consistent with this view of heat – for then the energy transferred to such a gas when it is heated becomes in the simplest model the kinetic energy of its moving particles.

This is an attractive possibility. The model we are proposing would not only serve to explain the properties of gases, but also extend the scope of two other major theories: the particulate nature of matter and the identification of heat with energy transfer.

1.9   Assumptions within the model

Before we seek to analyse this model mathematically, we must be clear about how much or
how little we shall assume about the nature of
these particles which are to be in a constant state
of agitation. The pressure is assumed to result
from the collisions between these particles and the
walls of the container. Such particles must have
mass, size and shape. They clearly attract each
other to some extent, or the existence of liquid and
solid forms of the same gaseous materials could
not be explained. If the particles are to bounce off
each other and the walls, there must also be some
sort of repulsive force when they get close to each
other, and to the molecules of the walls.

Let us consider the difference in volume
between gases and liquids. At atmospheric
pressure, every 1 cm^3 of water gives rise to 1600 cm^3
of steam. The space occupied by the gas is largely
'empty'. It should not be an unreasonable approximation to assume that the particles themselves
occupy a negligible proportion of this space.

And what of the inter-molecular forces? The
mathematical analysis will be considerably
simplified if we can assume that these are of very
short range compared with the average separation
of the particles of the gas and with the dimensions
of the vessel. Then, the forces will only act during
the actual process of collision. One outcome of
this is that the pressure exerted by a gas is independent of the exact nature of the material of the
walls of the container, since the forces between
wall and gas particles only act for a small proportion of the total time. This fact is justified by
experience.

The gas pressure is not only independent of the
nature of the container, but also of its shape. To
achieve this, the motion of the particles must be
evenly distributed over every direction on any
short time average. We may imagine that this
random distribution is maintained by internal
collisions within the gas and in the collisions with
the walls by particles which are moving in random
directions as they approach the walls. A random
distribution of directions must be taken to imply a
random distribution of speeds as well, from the
very high to the very low. However, we may
assume that there is some time-independent
average speed. The model itself has been partly
suggested by the energy theory of heat. If the
energy transferred to or from a gas as a result of a
temperature gradient is to be interpreted, at least
in part, as the kinetic energy of the motion of the
particles, then the total kinetic energy must presumably remain constant if the temperature of the
gas remains constant.

The constancy of the total kinetic energy of the
moving particles in the gas also means that either
all the collisions are perfectly elastic, or that over
any short time period, inelastic collisions are
equally balanced by hyper-elastic collisions.

So, summarizing, we are proposing a model of
a gas as an assemblage of particles in a state of
constant motion. The particles themselves have
mass but occupy a volume which is negligible
compared with the total volume of the gas. Their
velocities are distributed at random, but with a
well-defined average numerical value. Collisions
with the walls and between particles will all be
assumed to be elastic on average.

One further word about what the model does
not assume. It does not assume any structure for
the particles. It is a matter of complete indif-
ference as to whether they are what a chemist
would call *atoms* or *molecules*. For this reason, we
shall adhere to the term *particle*.

Further development of this model demands a
certain familiarity with the study of dynamics.
This will occupy Unit 2; moreover the considerations given to the ideas of pressure and of temperature deserve to be taken further. This will be
done in Unit 3.

Chapter 2

THE LANGUAGE OF PHYSICS

In common with the practioners of all other subjects, physicists tend to use some words in rather special ways, usually with quite restricted meanings. Obvious examples include 'weight', 'force', 'power', 'mass'. In all such cases the dictionary definition will be broader than the one implied in a physics text. For example the 1972 edition of Chambers' *Twentieth Century Dictionary* defines the noun *force* as 'strength, power, energy: efficacy: validity: influence: vehemence: violence: coercion . . .' and ends 'any cause which changes the direction or speed of the motion of a portion of matter'. In the sciences the noun *force* has that latter meaning. It is *not* a synonym for 'strength, power, energy' and the rest.

Careful, unambiguous definitions are essential for the precise descriptions which engineers and scientists expect of one another.

Physics is also a science which is heavily dependent upon the language of mathematics for it provides the tools with which physicists can operate.

In this chapter we shall explore some of the niceties of the languages which physicists use.

2.1 Physical Quantities

What is it that we do when we measure a length, or a current, or a mass? We compare the length (or the current, or the mass) in question with a *standard* length (or a current or a mass). For a length the standard is the metre and when we say that 'the bench is six metres long' we mean that the bench is six times *longer* than the standard metre. The value of the length is thus the product of a number and a unit – the unit being the size of the agreed standard. In this case the length is $6 \times (1)$ m. This is termed a *physical quantity*.

The *value* of a physical quantity is the product of a numerical value and a unit. For example, the physical quantity which we call the wavelength of a radio transmitter may have the value

$$\lambda = 450 \, \text{m}$$

where m is the symbol for the unit of length called the metre. It could just as well be written $\lambda/\text{m} = 450$, a way of writing often found on the axes of graphs.

Again, the density ϱ of hydrogen gas at s.t.p. has the value

$$\varrho = 0.09 \, \text{kg m}^{-3}$$

It is often helpful to adopt the convention of stating the units at the end of a line of calculation, like this

$$\varrho = m/V \ldots \varrho \text{ in kg m}^{-3}$$

It is regarded as wrong to write

$$\text{density} = \varrho \, \text{kg m}^{-3}$$

because ϱ, being a physical quantity, already includes both the numerical value and the unit. Numerical quantities must always have the units stated; symbols never.

2.2 Systems of units

The need to agree a system of units for trade and industry was realized long ago. As communications improved, so it became necessary for local systems of units to give way to national ones. New needs gave rise to new units. For example, during the second half of the eighteenth century James Watt, who drew fees based on the fuel saved by his improved steam engine, made careful measurements of the work done by his engines. This rating of performance led to the scientific notion of 'work.'

The international scientific community has adopted a number of conventions about physical quantities and the appropriate symbols for them as well as the International System of Units (SI) and prefixes. A number of non-SI units remain in use, however, even though they are discouraged. In this book we shall generally conform to the international system but we shall use non-SI units where these offer an advantage.

2.3 SI – The international system of units

This international system uses three kinds of unit: base, supplementary, and derived.

There are seven base units, one for each of the independent physical quantities: length, mass, time, electric current, temperature, luminous intensity, and amount of substance (Table 2a). The two supplementary units apply to plane and to solid angles (Table 2b). The derived unit for any other physical quantity is obtained from the combination of appropriate base units. For example, if a displacement is measured in metres and the time required for that displacement is measured in seconds, then the corresponding speed is measured in 'metre per second' (m s^{-1}). This is the derived unit for speed.

Table 2a Names and symbols for SI base units

Physical quantity	Name of SI unit	Symbol for SI unit
Length	metre	m
Mass	kilogram	kg
Time	second	s
Electric current	ampere	A
Thermodynamic temperature	kelvin	K
Luminous intensity	candela	cd
Amount of substance	mole	mol

If the rate of change of the displacement, which is the speed, is itself changing with time, then the acceleration must be measured in metre per second/second – that is, in metre per second squared (m s^{-2}).

A number of the derived units have been given specific names. For example, the unit of force, which derives from the equation $F = m a$ will have units $(\text{kg})(\text{m s}^{-2})$ or kg m s^{-2}. This unit is called the newton.

Not all units have convenient names. The unit of momentum, which derives from momentum $= m v$, is kg m s^{-1} and no shorthand form exists.

Table 2b Names and symbols for SI supplementary units

Physical quantity	Name of SI unit	Symbol for SI unit
Plane angle	radian	rad
Solid angle	steradian	sr

We shall not, however, offer here a list of all the derived units we shall use; we shall indicate how each one is constructed as the need for it arises.

The mole

All molecules or atoms of a pure substance have the same mass – which is, of course, very, very small. For example, the mass of 1 atom of the isotope of carbon known as carbon-12 is 19.92×10^{-27} kg.

The number of molecules in a sample of a pure substance of total mass M is M/m where m is the mass of a single molecule. This is usually quoted in *moles* which, in SI, are the base units of amount of substance.

■ The mole is the amount of a substance of a system which contains as many elementary entities as there are atoms in 0.012 kg of carbon-12.

The accepted abbreviation is *mol*. Note that the standard mass of the reference substance is 0.012 kg and not, as might have been expected, 12 kg.

Problem 2.1 How many atoms are there in 0.012 kg of carbon-12?

Each carbon-12 atom has a mass of 19.92×10^{-27} kg. Therefore the number of carbon-12 atoms in 0.012 kg is

$$\frac{0.012}{19.92 \times 10^{-27}} = 6.02 \times 10^{23}$$

That is the number of entities in a mole. A mole of electrons, say, will also contain 6.02×10^{23} electrons.

Background

We shall have reason frequently to refer to the mole and it is helpful to understand how the idea for the unit arose. We may start by referring to an empirical law first described by Gay-Lussac in 1808. He suggested that under conditions of equal pressure and temperature, gases which react chemically do so in simple volumetric proportions. So, for example, when oxygen and hydrogen react, 2 volumes of hydrogen combine with 1 volume of oxygen to produce 2 volumes of water (as vapour).

In 1811, Avogadro proposed as an hypothesis which could account for Gay-Lussac's law in terms of a molecular model that, under conditions of equal temperature and pressure, equal volumes of all gases contained the same number of molecules.

Avogadro was unable to find the exact value of that number but he realized that it must be very, very large. Since then many different determinations, all of them indirect, have been made and the currently accepted figure is 6.022×10^{23} mol^{-1}. This is the *Avogadro constant* (N_A). Measurements of the densities of gases indicate that, at s.t.p., this amount of a gas occupies 22.4×10^{-3} m^3 (22.4 litres or 22.4 dm^3).

Later it was realized that Avogadro's hypothesis could lead to a numerical scale of relative masses for molecules and atoms.

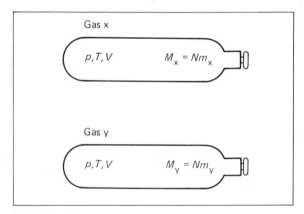

Fig. 2.1

Consider cylinders of two different gases, x and y, each containing the same volume V under the same conditions of pressure and temperature (Fig. 2.1). Since the volumes are equal, the numbers of molecules (N) in the two cylinders are also equal.

If the masses of the single molecules are m_x and m_y respectively the masses of the two gas samples are

$$M_x = N m_x \quad \text{and} \quad M_y = N m_y$$

and it follows that

$$\frac{M_x}{M_y} = \frac{m_x}{m_y}$$

The densities ϱ_x and ϱ_y must be in the same ratio.

The relative masses of gas molecules can be found by the simple expedient of weighing large samples in bottles at the same pressure and temperature. One particular particle was chosen as a standard against which all others could be compared. Today, as we have seen, the chosen particle is the atom of carbon-12.

Masses compared in this way were known as gram-molecular masses, the gram-molecule of a pure substance X being the amount of X whose mass *in grams* was equal to its relative molecular mass. Today this is known as the *mole* of the substance X which is defined as given above. That definition tells us that 1 mole of carbon-12 has a mass of 12 g. Its *molar mass* is 12 g mol^{-1} or 0.012 kg mol^{-1}. Similarly 1 mole of molecular oxygen has a mass of 32 g. Its molar mass is 0.032 kg mol^{-1}.

2.4 Dimensions

Measurement systems were at first based on three basic measures – those of length, of mass, and of time. As the sciences developed, other fundamental measures were added – those of current, temperature, luminous intensity and amount of substance. But, for the moment, we shall confine ourselves to the first three.

It is common in physics literature to find such statements as 'velocity has the dimensions L T^{-1}'. This peculiar sentence merely means that, independently of the system of units used, the measurement of velocity requires that the numerical value of a length be divided by the numerical value of a time. The statement may also appear in the form

$$[velocity] = L T^{-1}$$

Put in another way, velocity has dimensions $+1$ in length and -1 in time.

To take another example, volume requires that the numerical value of one length should be multiplied by the numerical values of another length and then by another.

$$[volume] = L^3$$

As the ratio of a mass to a volume, density must have dimensions given by

$$[density] = [mass]/[volume] = M L^{-3}$$

We know that $F = m\,a$ is the equation which defines force. Dimensionally we may write

$$[force] = [mass] \times [acceleration]$$
$$= M\,L\,T^{-2} \qquad (2.1)$$

Incidentally, this equation will provide us with the units of force. In SI the units of mass, of length, and of time are the kilogram, the metre, and the second. From Eq. 2.1 it is clear that the units of force must be kg m s^{-2} (which we now recognize as the newton). There are occasions when this is a useful technique.

We can employ 'the method of dimensions' to check the validity of equations and even to predict the form of equations. The ability to do this depends on the necessity that all the dimensions on each side of an equation must be the same.

Consider, for example, the student who derives an equation for the period of a simple pendulum: $T = 2\pi\sqrt{l/g}$.

The left-hand side has the dimension of time T.

The constant 2π is dimensionless – and we can learn nothing about that. We may now write the dimensional equation:

$$\left[\sqrt{\frac{length}{acceleration}} \right] = \frac{L^{\frac{1}{2}}}{(L\,T^{-2})^{\frac{1}{2}}}$$
$$= L^{\frac{1}{2}} L^{-\frac{1}{2}} T$$
$$= T$$

The student's equation *is* dimensionally consistent.

Let us now consider how the method can be applied to suggesting the form of an unknown equation. We shall take an example where we do know the form of the equation, but later in the book we shall use the method in cases where we do not.

For a stone whirled round in a horizontal circle on the end of a string we might expect the tension in the string to depend on the speed of the stone along its path, on the radius of that path and on the mass of the stone. So let us write

$$force = k\,m^x v^y r^z$$

where k is a dimensionless constant, m is the mass, v the speed, and r the radius; $x,\ y$ and z are the powers to which these parameters are raised – the dimensions.

The dimensional equation, in which k does not appear, will be

[force] = [mass]^x × [speed]^y × [radius]^z
M L T^{-2} = M^x × (L T^{-1})^y × L^z

Equating the indices of L: $1 = y + z$
M: $1 = x$
T: $-2 = -y$

Evidently $x = 1$, $y = 2$ and $z = -1$. So the equation must be

$$F = k \, m^1 \, v^2 \, r^{-1}$$

$$= k \, m \, \frac{v^2}{r}$$

It requires other arguments to find that k, the dimensionless constant, is in fact 1 (see Section 7.2).

2.5 Precision

We can never make measurements with absolute precision. Using an ordinary metre stick we are unlikely to measure the height of a table to anything better than a millimetre. We might express the length we find as 743 ± 1 mm, implying that the true value is unlikely to be less than 742 mm or more than 744 mm. This precision is quite high. 1 mm in 750 mm is about 0.13%.

This is another useful way in which to communicate information about the precision of a measurement. For example a resistor labelled '1000 ohm, 10%' is unlikely to have a resistance differing from 1000 ohm by more than 100 ohm. We can reply on it having a resistance between 900 and 1100 ohm. Another resistor labelled '1000 ohm, 2%' will have a resistance between 980 and 1020 ohm – and will cost perhaps twice as much as the other one!

All the measurements we make possess some degree of uncertainty and great care must be exercised when we process them arithmetically. Suppose we ask a student to verify the value of π by drawing a circle and measuring its diameter and its circumference. He does this to the nearest millimetre and reports his results as: diameter = 163 mm; circumference = 512 mm. He then punches these figures into his pocket calculator and announces that π is, 3.1411043. What can be said about this? In the first place we note that each of the two measurements he made involved three digits with meaning – these are *significant figures*. And even the last of each of these figures might

well have been a millimetre out either way. We can say at once that the last five digits of his value for π are meaningless. They come from the arithmetic and arithmetic is blind to meaning. That leaves him with 3.14. And that does agree with the accepted value of 3.142 (to four significant figures).

Remember that no arithmetical process can produce accuracy which is not in the measured values. If one of the values measured in an experiment is known only to, say, two significant figures whilst the others are known to three significant figures, the result of any multiplication or division with those values cannot be trusted to more than two figures. Generally speaking, when handling measured quantities arithmetically, note the first digit where uncertainty occurs and stop the calculation there.

2.6 Standard form and other matters

Physicists prefer to adopt a standard form when writing numerical values. Their aim is to write an intelligible number. For example a book of data will quote the expansivity of aluminium as 23×10^{-6} K^{-1} rather than print 0.000 023 K^{-1}.

There is too, a strong preference for the use of indices ± 3, ± 6, ± 9, etc., since these are directly related to the standard prefixes shown in Table 2c. But this is not always followed; for example, Planck's constant is usually quoted as 6.6×10^{-34} J s, and the speed of light as 2.998×10^8 m s^{-1}.

It is interesting to note that in medical practice, in engineering, in the construction industry and elsewhere, whole numbers are preferred. Then there can be no ambiguity about the position of

Table 2c Prefixes for SI units. These are used so that, normally, the numeric part lies between 0.1 and 1000.

Sub-multiple	Prefix	Symbol	Multiple	Prefix	Symbol
10^{-1}	deci	d	10^1	deca	da
10^{-2}	centi	c	10^2	hecto	h
10^{-3}	milli	m	10^3	kilo	k
10^{-6}	micro	μ	10^6	mega	M
10^{-9}	nano	n	10^9	giga	G
10^{-12}	pico	p	10^{12}	tera	T
10^{-15}	femto	f	10^{15}	peta	P
10^{-18}	atto	a	10^{18}	exa	E

the decimal point. Most of us are now familiar with the 5 ml dose of medicine or the 500 mm length of timber work top!

Problem 2.2   What can be said about a report by a student that he has measured the strength of the magnetic field and that it is $3\,425\,493\,649\,\mu$T?

Evidently this student was using a calculator with ten digits in the display and that it is highly unlikely that more than three of these figures are significant. So we prefer $3\,430\,000\,000\,\mu$T.

To express such a quantity in microtesla is manifestly absurd! So we change it to 3430 T. Which is, in standard form, 3.43×10^3 T.

2.7   Change and rate of change

The study of physics often involves the measurement of such quantities as length, time, temperature, energy, etc., and frequently we need to consider what happens when that quantity changes. Fortunately there is a universally understood shorthand ready for us to use. We use the operator Δ (delta) which, combined with the symbol for the quantity, means a *finite increment* in that quantity. For instance, if the symbol l represents the length of a spring, then Δl represents a finite increment in that length. The spring may change in length from 0.30 m to 0.35 m when the force acting on it is increased. In this case $\Delta l = 0.35$ m $- 0.30$ m $= 0.05$ m.

The rule for calculating changes is:

$$\text{change} = \text{final state} - \text{initial state} \quad (2.2)$$

If this rule is applied to the situation when the force is reduced and the length goes from 0.50 m to a final length of 0.45 m when $\Delta l = 0.45$ m $- 0.50$ m $= -0.05$ m. The significance of the sign is that it indicates that the change is a *decrease* of length.

When a change takes place it always takes some time to go from the initial to the final state. The temperature of the water in the cooling system of a car engine may change from 293 K to 353 K (i.e. from 20 °C to 80 °C) during the 5 min from starting up. The temperature rises at an average rate of 12 K per min since the change of 60 K occupied a time of 5 min. It is sufficient to write for

$$\frac{\text{change of temperature}}{\text{change of time}}$$

$$\frac{\Delta T}{\Delta t} = \frac{353 \text{ K} - 293 \text{ K}}{5 \text{ min} - 0 \text{ min}} = \frac{60 \text{ K}}{5 \text{ min}} = 12 \text{ K per min}$$

or better:

$$\frac{\Delta T}{\Delta t} = 0.2 \text{ K s}^{-1}$$

At the end of the journey, the car is put into its garage and it is found that the engine cools from 353 K to 323 K (i.e. from 80 °C to 50 °C) in 20 min (1200 s).

$$\frac{\Delta T}{\Delta t} = \frac{323 \text{ K} - 353 \text{ K}}{1200 \text{ s} - 0 \text{ s}} = -\frac{30 \text{ K}}{1200 \text{ s}} = -\frac{1}{40} \text{ K s}^{-1}$$

$$= -0.025 \text{ K s}^{-1}$$

In each of these examples, $\Delta T/\Delta t$ is a shorthand form for the average *rate of change of temperature with time* in the interval Δt and the sign indicates whether an increase or a decrease of temperature is involved.

The operator Δ is never used on its own but is always combined with a symbol for the physical quantity whose value is changing. This convenient shorthand is one we shall frequently meet.

The car journey itself provides us with another example which we shall analyse in more detail. As the car starts from rest (when $t = 0$), the distance s covered will change as the car accelerates. The speedometer will give readings which we may take to be *instantaneous* values of the speed while this is going on. Table 2d collects together some information about such an event:

Table 2d

Time, t/s	Distance from origin, s/m	Speedometer reading, v/m s^{-1}
0	0	0
1	1.5	3
2	6	6
3	13.5	9
4	24	12
5	37.5	15
6	54	18

These values, which are graphed in Fig 2.2 are in fact derived from the equation relating distance to speed which is $s = 1.5\,t^2$. We shall now use

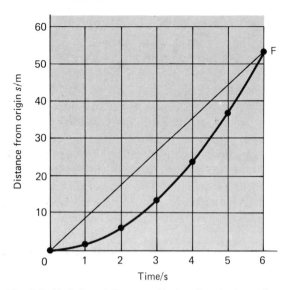

Fig. 2.2 Variation of distance with time for a body moving so that $s = 1.5\,t^2$.

these numbers and the equation to explore the ideas of average and instantaneous speed.

Asked for the *average speed* over the whole 6 seconds we write:

$$\text{average speed} = \frac{\text{final distance} - \text{initial distance}}{\text{time interval}}$$

$$= \frac{\Delta s}{\Delta t}$$

$$= \frac{(54 - 0)\,\text{m}}{6\,\text{s}} = 9\,\text{m s}^{-1}$$

This is the gradient of the line OF on the graph, and we can see that it tells us nothing at all about the events which were taking place. The car was accelerating so that its speed was increasing throughout the time interval and the gradient of the graph was continually changing.

If asked what the speed of the car was when $t = 4$ s, we might use the interval from 4 s to 5 s and say:

average speed between $t = 4$ s and $t = 5$ s

$$\frac{\Delta s}{\Delta t} = \frac{(37.5 - 24)\,\text{m}}{1\,\text{s}} = 13.5\,\text{m s}^{-1}$$

We would know this value of $\Delta s/\Delta t$ to be too high for the instantaneous speed at $t = 4$ s because the speed was increasing during this time interval. We could choose a shorter time interval.

Perhaps the 0.1 s between $t = 4$ s and $t = 4.1$ s. Then $\Delta t = 0.1$ s.

Using the equation to find s at $t = 4.1$ s, we have

$$s_{4.1} = 1.5(4.1)^2 = 25.215\,\text{m}$$

So

$$\Delta s = (25.215 - 24)\,\text{m} = 1.215\,\text{m}$$

And

$$\frac{\Delta s}{\Delta t} = \frac{1.215\,\text{m}}{0.1\,\text{s}} = 12.15\,\text{m s}^{-1}$$

That is much closer to the speedometer reading of $12\,\text{m s}^{-1}$. So let us try an even shorter time interval: say $\Delta t = 0.01$ s from $t = 4$ s to t $= 4.01$ s.

Using the equation to find s at $t = 4.01$ s, we have

$$s = 1.5(4.01)^2 = 1.5 \times 16.0801 = 24.120\,15\,\text{m}$$
$$\Delta s = (24.120\,15 - 24)\,\text{m} = 0.120\,15\,\text{m}$$

And

$$\frac{\Delta s}{\Delta t} = \frac{0.120\,15\,\text{m}}{0.01\,\text{s}} = 12.015\,\text{m s}^{-1}$$

It should now be clear that we can calculate speeds which can be brought closer and closer to the actual instantaneous speed by choosing smaller and smaller values of the time interval Δt.

The *limiting value* of $\Delta s/\Delta t$ as Δt tends towards zero is written $\mathrm{d}s/\mathrm{d}t$ and is known as the *first differential coefficient* of s with respect to t.

$$\frac{\mathrm{d}}{\mathrm{d}t}(s) = \lim_{\Delta t \to 0} \frac{\Delta s}{\Delta t}$$

$\mathrm{d}/\mathrm{d}t$ is an operator describing a process which, in this case, is being applied to the distance s.

$\Delta s/\Delta t$ gets closer and closer in value to $\mathrm{d}s/\mathrm{d}t$ *as* Δt gets smaller and smaller; and in the limit $\mathrm{d}s/\mathrm{d}t$ is the instantaneous speed.

This technique of choosing small values of a changing physical quantity will be used several times in this book (see, for example, the work on the discharge of a capacitor in Section 26.7). You may wish to work Questions 1.10 and 1.11 at the end of this book to gain practice with this important idea.

Where the relationship between two such variables as s and t in the example above can be expressed algebraically, it is usually possible to apply the techniques of a branch of mathematics called *differentiation* to find the instantaneous rate of change ds/dt.

2.8 Vector and scalar quantities

When the direction as well as the magnitude of a quantity must be quoted we are dealing with a *vector* quantity. The simplest example of a vector quantity is displacement. This involves two pieces of information – the distance between two points and also the direction of one point from the other. For example when a pilot of an aircraft states his position he is giving his displacement from the airport (or some other stated landmark). He might say, 'My position is 100 km due east of Rome.'

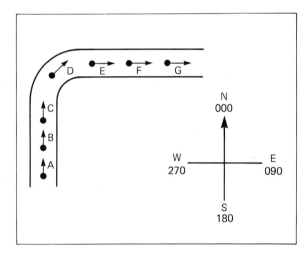

Fig. 2.3

Another example of a vector is velocity, which is concerned with a speed and a direction. If his navigation is to be successful, a pilot needs to know the wind velocity, i.e. both the magnitude of the wind speed and its direction.

Scientists make a distinction between speed and velocity. Speed is an example of a *scalar* quantity and it is usually adequate to describe the rate at which a car is moving along a road, as a reading from a speedometer without any reference to direction since this is fixed by the road. Figure 2.3 helps to make the distinction. Thus a car travelling along the road shown can do so with a constant speed of $50 \, km \, h^{-1}$ all the way from A to G. But although its speed is constant, its velocity is not, at any rate not all the time. At A, B, and C its

velocity is $50 \, km \, h^{-1}$ in a direction 000 (due north); at D its velocity has changed to $50 \, km \, h^{-1}$ heading 045 and it continues to change until it reaches E when it has become $50 \, km \, h^{-1}$ heading 090 and then continues to be constant through points F and G. This is a case in which the vector is changing, not by changing its magnitude but by changing its direction.

A vector is thus a quantity that has both a magnitude and a direction associated with it. It can be represented by an arrow drawn to scale so that its length gives the magnitude and its orientation gives the direction of the quantity.

2.9 The addition of vector quantities

Suppose an aircraft starts at point A and flies until it reaches B where it has a displacement of 100 km heading 090 from A. Figure 2.4 shows how it can be represented.

It then turns and flies until it is at C which has a displacement of 50 km, 045 from B. This part of the flight can be represented by Fig. 2.5.

These two flights can be added by means of a vector diagram drawn to scale. Note that the head of one vector is joined to the tail of the next (Fig. 2.6).

Now this aircraft could have reached C by flying along the route AC (Fig. 2.7). Consequently, flight AB plus flight BC is equivalent to flight AC, or **AB** + **BC** = **AC**. Note the shorthand **AB**, etc., which indicates the vector AB, etc. **AC** is called the *resultant* of **AB** and **BC**. A resultant is obtained by adding vectors using the triangle method outlined and not merely by adding the two distances together. By taking a measurement off the triangle, AC = 140 km, which is clearly different from the addition of the two scalars 100 km and 50 km.

It would be possible for the pilot to arrive at C by flying the vector 50 km, 045 first, arriving at D, and secondly to fly along the vector 100 km, 090 from D (Fig. 2.8).

Adding these vectors gives a triangle similar to ABC and the same resultant (Fig. 2.9).

The resultants are the same, since if Figs. 2.7 and 2.9 are superimposed, a parallelogram is obtained (Fig. 2.10).

Thus there are two alternative ways of adding vectors which really amount to the same process.

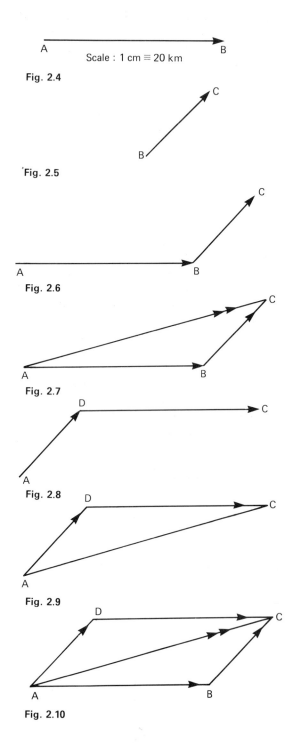

Fig. 2.4

Fig. 2.5

Fig. 2.6

Fig. 2.7

Fig. 2.8

Fig. 2.9

Fig. 2.10

1. The triangle method: Draw the vectors to scale to form two sides of a triangle. The head of the first vector must be joined to the tail of the second. The resultant is the third side of the triangle and its direction is from the tail of the first vector to the head of the last one (e.g. Figs. 2.7 and 2.9).

2. The parallelogram method: Draw the vectors to scale so that they form adjacent sides of a parallelogram. The resultant is the diagonal drawn from the corner with two tails to the corner with two heads – as in Fig. 2.11a not as in Fig. 2.11b.

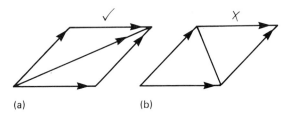

(a)   (b)

Fig. 2.11 Right and wrong application of the parallelogram method of vector addition.

2.10 Comparison of the two methods

It may seem strange to use the parallelogram method since it involves drawing two seemingly unnecessary extra lines in order to obtain a resultant. Yet there are many cases when the parallelogram method is more appropriate as it allows one to understand more easily what is happening. In general, the triangle method may be better when one vector acts first, followed by the second, whereas the parallelogram method may be better when both vectors act during the same period of time, or act at the same point.

How could two displacements occur at once? Imagine a barge moving slowly along a canal (Fig. 2.12). One of the crew starts to walk across the

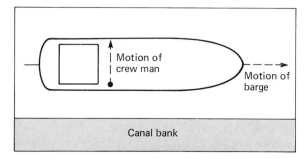

Fig. 2.12 As the barge's displacement along the canal changes so does that of the crewman walking across the barge.

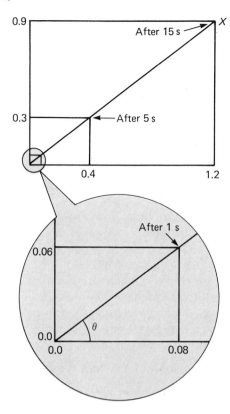

Fig. 2.13 Adding the two displacement vectors.

barge at right angles to the direction of motion. He is moving in two directions at the same time. In 5 s he may walk 0.3 m across the barge and be carried 0.4 m to the right by the barge. Figure 2.13 shows his position relative to the banks of the canal, after 1, 5 and 15 s. He has a speed of 0.06 m s^{-1} across the canal at right angles to the bank and 0.08 m s^{-1} along the canal parallel to the bank. Applying the rules of vector addition, he has a resultant velocity of 0.10 m s^{-1} in a direction which makes angle θ with the bank. Since $\tan \theta$ is 0.9/1.2 or 0.75, θ is 37°.

2.11 Vectors and navigation

An aircraft pilot makes use of the displacement vector when he states his position, and velocity vectors and the rules of vector addition when deciding upon the direction to head the aircraft in order to arrive at a given destination. Suppose the aircraft has a speed of 800 km h^{-1} through the air and that at first there is no wind blowing (Fig. 2.14). If he points the nose of the aircraft due north (or on a heading 000), his track over the ground is also in the direction 000.

If he now finds himself in a jet stream (a high speed westerly wind that occurs at altitudes of

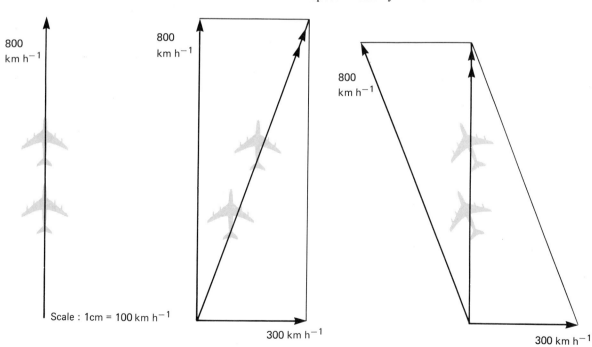

Fig. 2.14   **Fig. 2.15**   **Fig. 2.16**

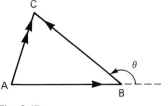

Fig. 2.17

Fig. 2.18   Fig. 2.19

between 7 km and 13 km. At speeds exceeding 300 km h^{-1}) which is blowing with a velocity of 300 km h^{-1} from 270, his resultant motion relative to the ground will be the result of having two velocities at the same time. While the aircraft is moving through the air at 800 km h^{-1}, the air is carrying it eastwards at 300 km h^{-1}.

The resultant can be found by drawing to scale a velocity parallelogram (Fig. 2.15). The resultant velocity is 854 km h^{-1} in a direction 020, and although the aircraft is heading (or pointing) due north, it is moving over the ground in a direction 020 and at a speed of 854 km h^{-1}.

However, the pilot wishes to arrive at a destination which is due north of his starting point. His resultant velocity must be in a direction 000, and a velocity parallelogram can give the direction he must head to achieve this. By heading the aircraft in a direction 338 as shown in Fig. 2.16 he will move over the ground due north and at a speed of 742 km h^{-1}.

2.12 Signs and vectors

Let the two vectors AB and BC be equal in magnitude, and suppose they are added as in Fig. 2.17; the resultant is given by $AB + BC = AC$. Consider what happens when $\theta = 0$. The triangle becomes a line as in Fig. 2.18 and the resultant of the two equal vectors is $2AB$.

$$AB + BC = 2AB$$

Consider next what happens when $\theta = 180°$. The triangle becomes a line, as in Fig. 2.19, and the resultant of the two vectors is zero since A and C coincide. In this case

$$AB + BC = 0 \text{ or } AB = -BC \text{ or } AB = -BA$$

since $BC = BA$ in Fig. 2.19. The significance of a negative sign attached to a vector is that the

direction of the vector is reversed, or turned through 180°.

The application of a sign is frequently used to indicate the direction of motion of an object. For example, suppose an escalator can move upwards or downwards at a speed of 0.6 m s^{-1}, and that we adopt the convention that the upward direction is + or positive, and the downward direction is − or negative. A man, who can walk up a stationary staircase at a speed of 0.4 m s^{-1}, is in a hurry and walks up the ascending escalator. This situation can be described as:

velocity of man
relative to escalator = $+0.4$ m s^{-1},

velocity of escalator = $+0.6$ m s^{-1},

resultant velocity
of man   = $+0.4 + 0.6 = 1.0$ m s^{-1}.

If the escalator is reversed and the man continues to walk upwards the situation is then:

velocity of man
relative to escalator = $+0.4$ m s^{-1},

velocity of escalator = -0.6 m s^{-1},

resultant velocity
of man   = $+0.4 + (-0.6)$ m s^{-1}
              = -0.2 m s^{-1}.

Thus by changing the velocity of the escalator from $+0.6$ m s^{-1} to -0.6 m s^{-1}, i.e. by reversing it, the velocity of the man changes from $+1.0$ m s^{-1} to -0.2 m s^{-1}; the negative sign indicating that he is descending.

2.13 Resolution of vectors

It is often convenient to replace a single vector by two component vectors. For example, we may imagine a truck on rails being pulled by a man

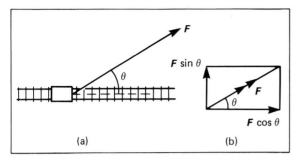

Fig. 2.20   Resolution of a force vector.

using a rope inclined at an angle θ to those rails (see Fig. 2.20a) and exerting a force F (another vector quantity).

We may *resolve* the force F into two component vectors at right angles to each other, F_x along the rails and F_y at right angles to the rails (see Fig. 2.20b). We see at once that, if the law of vector addition is applied,

$$\boxed{F = F_x + F_y.}$$

Moreover we can see from the figure that

$$F_x = F \cos \theta$$

and

$$F_y = F \sin \theta.$$

Of the two component vectors, only F_x is effective in moving the truck since F_y is balanced by the opposing thrust of the rails. F_x is able to accelerate the truck to the constant velocity which develops when it is just equal to the frictional forces opposing it.

These ideas concerning rates of change and vectors will be used often in the work that is to follow.

Summary

Vector quantities possess both magnitude and direction and may be added by the application of the triangle (or the parallelogram) law. Where it is important to the argument such quantities will be represented by the appropriate letter printed in bold type face whilst the magnitude of the quantity will be represented by the same letter printed in ordinary type. For example the magnitude of the force F is represented by F.

UNIT TWO

Motion and Force

Chapter 3

THE INVESTIGATION
OF INTERACTIONS

If we are to develop the particle model for a gas which we introduced in Chapter 1 we shall require an understanding of the ways in which particles behave – either by themselves or as they interact with other particles. This is the proper study of dynamics and this unit will deal with that. The major topics are motion, force, mass, but it starts by considering *momentum*, a word often misused in everyday language.

When a scientist has a complex problem to solve he starts by seeing what factors are involved. He may then attempt to remove or at least reduce the effect of some factors; he may proceed by keeping other factors constant and see if there is a simple law which governs the behaviour of the remaining factors.

What could possibly be simple about anything as wild as a collision or an explosion? At first sight it seems unlikely that there is a simple relationship between the motion of objects before and after a collision.

The collision of two cars is a very complex situation; they are likely to be of different masses and to have different and even changing speeds. A simplification to aid the investigation would be to have two identical cars, each travelling with the same speed, but even this is complicated by the fact that an engine must be working to maintain this constant speed, since friction would otherwise cause a retardation. An idealized colliding system, as far as the scientist is concerned, is one in which no friction exists so that no engine is necessary to maintain constant speeds.

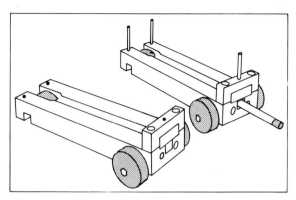

Fig. 3.1 Trolleys on ball-bearing wheels. (Courtesy of Philip Harris Ltd.)

3.1 The reduction of friction in the laboratory

A number of pieces of apparatus have been designed for experimenting with collisions in which the effects of friction have been reduced.

a) Trolleys on ball-bearing wheels

These are designed (see Fig. 3.1) so that two or more trolleys can be stacked and the mass moving varied. Some trolleys are equipped with a spring plunger that can be released by striking a peg. By this means the 'explosion' of two trolleys can be investigated.

b) The linear air track

The length of the track is drilled with small holes from which air, supplied by a vacuum cleaner, is

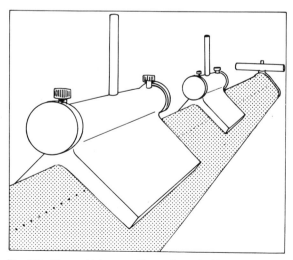

Fig. 3.2 Two vehicles on a linear air-track. (Courtesy of Griffin & George, Ltd.)

blown so supporting the vehicles clear of the actual track (see Fig. 3.2).

c) Pucks on a flat glass plate

These can either be operated by air, supplied by an electrically driven pump carried on the puck itself, or by a small piece of solid carbon dioxide which continuously sublimes to the gaseous state. In both types, the gas pressure underneath the puck supports the puck clear of the glass plate. (See Fig. 3.3.)

In all of these cases, one should question whether friction has been eliminated, or only reduced sufficiently for it to be neglected. No matter how well made are the wheels and bearings

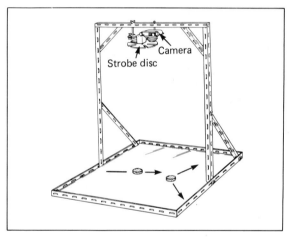

Fig. 3.3 Pucks on a flat glass plate showing one technique for multi-flash photography.

of a trolley, some friction will still remain. Experience tells us that a trolley which has been pushed on a horizontal table will eventually come to rest because friction gradually dissipates the trolley's energy. But what about the air track vehicles and the pucks? They certainly have no wheels yet experiment shows that they gradually slow down and lose energy. Where can the friction be? This time it is caused by viscosity, or gas friction, occurring mostly in the narrow gap between the vehicle and the track or between the bottom of the puck and the glass plate. We will use trolleys, air tracks, and pucks, and accept that in these cases the effect of friction may be neglected.

Problem 3.1   The force of friction tends to retard a moving body. How could you arrange a trolley experiment so that the earth's gravitational pull on a trolley compensated for the frictional force?

Explain whether or not your method could be applied to the case of two trolleys moving in (a) the same direction and (b) opposite directions.

3.2   Methods for measuring speed

In order to determine speed a measurement of a distance and of the time taken to cover that distance is required.

If the speed of a car is being measured, it may be sufficient to use a stopwatch and time it over a distance of a kilometre. A very fast car taking 20 s to cover this distance can be 'hand timed' to an accuracy of ±0.2 s. This represents an error of 1%, which is the worst that can be expected as a slower car will take longer and thus can be measured to a better percentage accuracy. For example, 40 ± 0.2 s represents an error of $\frac{1}{2}$%.

In the laboratory, the distance over which a trolley or a puck is timed is comparatively short; it is not likely to exceed 1 m and in many experiments this distance will be covered in less than 4 s. Consequently, the best that can be attained with a stopwatch is an accuracy of 4 ± 0.2 s or 5%. A trolley going at twice the speed and timed over half the distance will have its speed measured to an accuracy of only 20% if a watch is used. Consequently, other methods are used for measuring the short times involved in laboratory experiments.

a) The ticker-tape timer

This is a familiar piece of apparatus which when driven from a 50 Hz a.c. supply makes a dot every 0.02 s on a piece of paper tape that is attached to a trolley. If the trolley is accelerating, the dots become farther apart. The distance between any two dots is the distance travelled by the trolley in 0.02 s.

b) Photocell and lamp method (see Fig. 3.4)

A phototransistor is used to operate the starting and stopping of an electronic timer which measures time in units of 0.001 s derived from a 1 kHz oscillator. When a light beam shines on the phototransistor the timer does not count. If the beam is interrupted, the timer starts to count

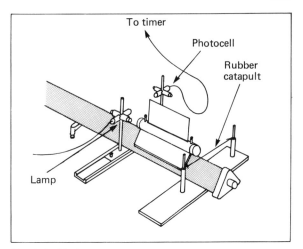

Fig. 3.4   Timing a linear air track vehicle by interrupting a light beam. (Courtesy of Griffin & George Ltd.)

milliseconds until the beam is restored. A card, 0.1 m long, say, mounted on an air track vehicle can be arranged to interrupt the beam as shown in Fig. 3.4. Thus the time taken by the vehicle to travel 0.1 m can be measured. In some electronic timers a microprocessor is used so that the speed is displayed directly. It is also possible to use a microcomputer.

c) Multiflash (stroboscopic) photography

This is a particularly versatile method of analysing the motion of a moving body. As can be seen from Figs. 3.5 and 3.6 a single photograph can record all the motions involved in simple and complex interactions, including motions that take place in two dimensions. Multiflash photography involves the taking of a sequence of pictures, on a single piece of film, at regular intervals. This can be achieved in two ways:

Fig. 3.5   Multi-flash photograph of a collision between a moving air track vehicle (on the left) and a similar stationary one (Experiment B). (Photograph B. T.)

1. The camera shutter is opened in the dark and the event to be recorded is illuminated by a rapidly flashing light which may be obtained from an electronic flash tube. The camera shutter is closed when the event is completed.

2. Alternatively, the trolleys or pucks can be illuminated by a steady light and the film is then exposed at rapid regular intervals by means of a mechanical shutter. This consists of a disc, typically having six equally spaced slots around its edge, which is rotated at a constant rate – typically 5 revolutions per second – in front of the camera lens. When the camera shutter is open, the film obtains a glimpse of the event to be recorded every time a slot passes.

3.3   An investigation of interactions

In order to discover a rule connecting the motion of objects before and after an interaction, some data must be obtained by doing experiments. The problem is first simplified by using two *identical* objects in a variety of interactions in which they are effectively isolated. The objects are either two trolleys or two air track vehicles and any of the timing methods mentioned earlier can be used.

Experiment A: An explosion

Two trolleys are initially at rest and in contact. The spring plunger is released and the trolleys (X and Y) move apart. In this case the card, photocell and clock method was used to give the results quoted in Table 3a.

Table 3a

	Trolley X	Trolley Y
Card length/m	0.10	0.10
Transit time/s	0.27	0.27
Initial velocity/m s^{-1}	0.00	0.00
Final speed/m s^{-1}	0.37	0.37
Final velocity/m s^{-1}	− 0.37	+ 0.37

As mentioned earlier, the sign indicates the direction of motion:

− to the left and + to the right.

Experiment B: A collision with a stationary object

The two identical air track vehicles P and Q are fitted with rubber band buffers. A white drinking straw is attached to each vehicle and these appear as the vertical lines on the multiflash photograph (Fig. 3.5). The space between two lines is proportional to the distance travelled in 1/30 s. Measurements from the photograph give velocities of the images of the straw. (Table 3b.) In this experiment, and in the next, the mass which is moving initially, comes to rest.

Table 3b

	Vehicle P	Vehicle Q
Initial velocity/m s^{-1}	+ 0.066	0
Final velocity/m s^{-1}	0	+ 0.066

Experiment C: An interaction using magnets

The difference between this and the previous experiment is that the interaction takes place *at a distance* by means of the repulsion between the magnets which are mounted firmly on two similar trolleys. The trolleys do not touch. The results are shown in Table 3c.

Table 3c

	Trolley X	Trolley Y
Initial velocity/m s^{-1}	+ 0.55	0.00
Final velocity/m s^{-1}	0.00	+ 0.55

Experiment D: A collision with a stationary object without using rubber or magnetic buffers

The arrangement of the previous experiment is used with the magnets removed so that a 'hard' interaction occurs.

This time, trolley X does not come to rest after the interaction but has a small final velocity. Table 3d shows the results.

Table 3d

	Trolley X	Trolley Y
Initial velocity/m s^{-1}	+ 0.60	0.00
Final velocity/m s^{-1}	+ 0.15	+ 0.45

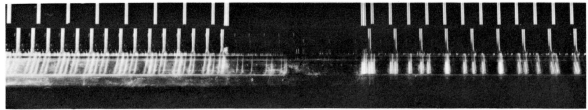

Fig. 3.6   A head-on collision between two similar air track vehicles (Experiment F). (Photograph B. T.)

Experiment E: A collision between two bodies which stick together

In each of the previous experiments, the objects separate after the interaction. Although both are finally moving in Experiment D, trolley Y moves away from X with a relative velocity of 0.30 m s^{-1}. In Experiment E we examine the situation where separation does not occur. This is achieved by having a sharp pin attached to one air track vehicle P that sticks into a small piece of cork fixed to the second vehicle Q, when the collision takes place.

Vehicle P moves from the left and passes the first photocell. P hits Q, which is initially stationary, and both move off together past the second photocell. Table 3e shows the results obtained.

Table 3e

	Vehicle P	Vehicle Q
Card length/m	0.100	–
Transit time of P/s	0.242	–
Transit time of P + Q/s	0.485	0.485
Initial velocity/m s^{-1}	+ 0.413	+ 0.000
Final velocity/m s^{-1}	+ 0.206	+ 0.206

Experiment F: A collision between two moving objects

In the previous experiments, one object was at rest before the second one collided with it. This time, both are in motion moving towards each other for a head-on collision, after which each rebounds and travels back to its starting point. To record this motion, multiflash photography of air track vehicles is utilized. The reverse motion can be photographed without being superimposed on the forward one if a shutter is fitted in front of the track. This takes the form of a hinged flap, about 1 m long and 0.05 m high, which is operated by hand when the interaction occurs. The lower parts of the drinking straws are then revealed and the upper parts concealed from the camera. A typical photograph taken by this method is shown in Fig. 3.6.

In the photograph, the space between two images is proportional to the distance travelled in 1/30 s. Measurements from it give the results shown in Table 3f.

Table 3f

	Vehicle P	Vehicle Q
Initial velocity/m s^{-1}	+ 0.240	− 0.150
Final velocity/m s^{-1}	− 0.120	+ 0.210

3.4   The rule begins to emerge

These experiments provide data about the motion of two similar objects before and after a wide variety of different interactions, including one in which they do not even touch. If the results of each experiment are examined we can see that the following rule applies: (sum of velocities before interaction) = (sum of velocities after interaction). *Note*: Try adding speeds instead by ignoring the signs, and see if the rule works in every case. You should decide that this rule only works if the velocities are treated as vector quantities.

3.5   Extending the rule

We have obtained a satisfactory rule for the simplest possible case, namely two isolated, similar objects, but real life collisions are not always so conveniently arranged. To make our rule more useful we shall change a variable. We shall keep the system isolated and experiment with objects that are not similar. Note that we are only changing one variable. This is the golden rule in scientific investigations. Other variables, such as friction, are not allowed to creep in and confuse the issue. The effect of such forces is treated separately.

So that the interacting objects differ by an obvious amount, the next experiments are performed with stacked trolleys. Thus one trolley can interact with two, two trolleys can interact with three, and so on.

Experiment G: An explosion between a single, and a double trolley

The arrangement is the same as for Experiment A. The symbol Y_2 refers to the double trolley. Table 3g gives the results.

Table 3g

	Trolley X	Trolley Y_2
Card length/m	0.10	0.10
Transit time/s	0.22	0.44
Final velocity/m s^{-1}	-0.46	$+0.23$
Initial velocity/m s^{-1}	0.00	0.00

The sum of the velocities before interaction clearly equals zero. The sum of the velocities after interaction will only be zero if the velocity of the double trolley is doubled.

Experiment H: An explosion between a double, and a triple trolley

The trolleys are referred to as X_2 and Y_3 in Table 3h which summarizes the results obtained using the same technique as in Experiment A.

Table 3h

	Trolley X_2	Trolley Y_3
Card length/m	0.10	0.10
Transit time/s	0.16	0.24
Final velocity/m s^{-1}	-0.63	$+0.42$
Initial velocity/m s^{-1}	0.00	0.00

The sum of the velocities before interaction again equals zero. The sum of the velocities after interaction will only be zero if the velocity of X_2 is doubled and that of Y_3 is multiplied by three.

3.6 The new rule

From the data obtained by these last two experiments a modified rule emerges that applies to explosions between isolated objects of any size. Thus:

$$n_1 u_1 + n_2 u_1 = n_1 v_1 + n_2 v_2 \qquad (3.1)$$

where the symbols have the following meanings:

n_1 and n_2 refer to the number of trolleys in stacks X and Y.

u_1 and u_2 refer to the initial velocities of X and Y.

v_1 and v_2 refer to the final velocities of X and Y.

In words, the sum of the products of the numbers of trolleys and their velocities before the explosion is equal to the sum of the products of the numbers of trolleys and their velocities after the explosion.

Equation 3.1 is a statement of the rule that applies to all of the Experiments A to H, but it requires further testing to see if it is true for *collisions* between objects of any size.

Experiment I: A collision between a triple trolley, X_3 and a stationary single trolley, Y_1

The arrangement is the same as for Experiment D. Table 3i gives the results.

Table 3i

	Trolley X_3	Trolley Y_1
Initial velocity/m s^{-1}	$+0.70$	0.0
Final velocity/m s^{-1}	$+0.45$	$+0.75$

If Eq. 3.1 is tested with these results we shall see it is true for collisions as well as explosions.

3.7 The concept of mass

In Experiment G, H, and I the number of trolleys on each side of an interaction is significant in deciding the velocities after the interaction. In each case, the number of trolleys is a measure of the quantity of matter, or *mass*, involved. The mass of a body governs its behaviour during an interaction. For example, Experiment G shows that after an explosion a body with a large mass moves off with a smaller velocity than one with a small mass. The large mass has more *inertia*. Also,

Experiment I demonstrates that, during a collision, a large mass undergoes a smaller change of velocity than does a small mass. (X_3 underwent a change of velocity of $0.45 - 0.70 = -0.25$ m s^{-1} whereas Y_1 suffered a change of 0.75 m s^{-1}.) Again we can say that the larger mass has more inertia as it tends to maintain its velocity.

■ Mass is a measure of the inertia of a body, as well as being the quantity of matter in the body. Mass is always conserved.

These findings are summarised in

■ *Newton's first law.*

'Every body remains at rest, or moves with constant velocity, unless acted on by an unbalanced force.'

Remember that 'constant velocity' implies that it is moving in a straight line and that its speed is constant.

The meaning of the term 'unbalanced force' is best understood by reference to some examples. Consider a ball at rest on a billiard table. Two forces act upon the ball; the earth exerts a pull on it and the table supports it. These two balance. Then it is struck by a cue. The cue exerts a force on the ball and, as we shall see, the ball exerts an equal and opposite force on the cue. The force exerted by the cue on the ball is unbalanced as far as the ball is concerned. The ball is accelerated and

so acquires a velocity. (At the same time, the cue is decelerated.) Subsequently the ball comes slowly to rest, because friction with the table now provides an unbalanced force.

Consider, too, a mass hanging from a thread. The thread supports it and the earth pulls upon it – there are two balanced forces acting. Now the thread breaks. Under the action of the now unbalanced pull of the earth, the mass falls towards the floor.

Or again, a motor car is accelerated to its maximum speed. The force exerted by the engine through the transmission is unbalanced until, at the maximum speed, it is exactly equal to the sum of all the frictional forces that exist. The forces are in balance as the car moves with constant motion.

3.8 A conservation rule

Returning to Eq. 3.1, the term mass can now be used instead of 'numbers of trolleys.' If m_1 and m_2 mean mass of X and mass of Y respectively, then the equation becomes:

$$m_1u_1 + m_2u_2 = m_1v_1 + m_2v_2 \qquad (3.2)$$

This tells us that the sum of (mass × velocity) before an interaction is equal to the sum of (mass × velocity) after the interaction.

The quantity (mass × velocity), is so important that it is given a special name: *momentum*.

Chapter 4

MOMENTUM AND MASS

As we have seen, momentum is concerned with interactions between bodies and it is in this context that the idea is most fruitful. In everyday speech the word is sometimes used to describe motion, for example, a cyclist free-wheeling down a hill gains 'momentum'. This is, of course, perfectly true; but the speaker probably wishes to say no more than that the velocity of the cyclist is increasing. It would also be correct to say that the cyclist gains kinetic energy. Both momentum and kinetic energy depend on velocity but it would be an error to suggest that these words were nothing more than synonyms for velocity.

Momentum is important because in a collision, whether 'hard' or not, or in an explosion within an isolated system, the total momentum is constant. We say that momentum is conserved.

The *law of conservation of momentum* states:

■ 'The total momentum of an isolated system of bodies, interacting only with each other, remains constant.'

4.1 Momentum as a vector quantity

In arriving at this law from the results of experiments, it was essential to assign directions to the various motions. Likewise in applying the law, it is vital to treat the product (mass × velocity) as a vector quantity. Velocity is a vector quantity and mass is a scalar that is acting as a multiplying factor. Consequently, momentum can also be classed as a vector quantity.

4.2 Isolated systems

An isolated system is composed of a number of masses that only interact among themselves. No external force such as friction 'connects' them to other masses outside of the system. In all of our experiments it is clear which masses form the isolated system: trolley X and trolley Y form one (nearly) isolated system; vehicle P and vehicle Q form a perhaps better isolated system. We can say that these are isolated because, for instance, P only interacts with Q. It does not interact with the air track since there is negligible frictional force between it and the air track. Consequently, the total momentum of P and Q is conserved.

But, suppose Experiment E is performed with the holes under vehicle Q blocked, or covered with thin adhesive tape. There is now considerable friction between Q and the track, and P and Q are no longer the only members of an isolated system. When P interacts with Q it is also interacting with the air track through the medium of friction. Moreover, the track interacts with the bench and the bench . . .

Next, suppose that the air track itself could be arranged to move freely without friction, perhaps by mounting it on special large air pucks. The isolated system now consists of P, Q, and the air track; and when P collides with Q, the initial momentum of P is transferred completely to the final momenta of P, Q, and the now moving air track. The law of conservation of momentum can be applied in this case and indeed to any case in which the members of the isolated system are clearly known.

What about the case of a car colliding with the back of a stationary car? Before impact the brakes would doubtless have been applied and a large frictional force would have acted. Could the law of conservation of momentum be applied here? We believe it could be, so long as all the interacting objects can be specified. In this case there are three involved; the car interacts with the stationary vehicle and also with the road (and therefore the earth). The moving car does not give all of its momentum to the stationary one; while it is braking it is giving some momentum to the earth – momentum the earth lost when the car originally accelerated from rest.

Does the law of conservation of momentum apply to a sprinter as he leaves his starting blocks? He is initially at rest and then suddenly gains some momentum; if the law is applicable an equal but opposite change of momentum must occur. Pre-

sumably the earth undergoes this change and the earth and the sprinter constitute an isolated system.

Some of these points may be illustrated easily with a clockwork or flywheel toy car and a piece of thin wood or card that can roll on double-ended knitting needles or other suitable rods placed parallel to each other. If the car, with its wheels already turning, is placed on the card the car will move forwards at the expense of the card moving backwards. If the car then collides with a small block fixed at one end of the card, the changes of motion will be obvious.

4.3   Making measurements of mass

The conservation of momentum law enables two masses to be compared. Applying Eq. 3.2 to the one-dimensional case, we have

$$m_1 u_1 + m_2 u_2 = m_1 v_1 + m_2 v_2$$

where u_1, etc., are magnitudes only. Rearranging, we get

$$+ m_1 v_1 - m_1 u_1 = -(m_2 v_2 - m_2 u_2) \quad (4.1)$$

thus,

$$\frac{m_1}{m_2} = -\frac{(v_2 - u_2)}{(v_1 - u_1)} = -\frac{\Delta v_2}{\Delta v_1} \quad (4.2)$$

The significance of the negative sign is that, in the case of an explosion, the objects move off in opposite directions, and in the situation where a collision occurs, one body undergoes an increase in velocity while the other suffers a decrease in velocity.

Equation 4.2 relates the masses of two bodies, X and Y, to the change of velocity each undergoes during an interaction. As an example, suppose that two different masses, X and Y, in the form of trolleys are initially at rest, 'explode' apart, and that their velocities become $0.16 \, \mathrm{m \, s^{-1}}$ and $-0.96 \, \mathrm{m \, s^{-1}}$ respectively. Using Eq. 4.2, we obtain

$$\frac{m_1}{m_2} = -\frac{(-0.96)}{0.16} = 6$$

and we see that the mass of X is six times the mass of Y.

As another example, consider an air track vehicle P moving at $0.080 \, \mathrm{m \, s^{-1}}$ which falls to $0.050 \, \mathrm{m \, s^{-1}}$ when P collides with vehicle Q. Since Q was initially at rest it ends up by having a velocity of 0.075 m s^{-1}. Applying Eq. 4.2, we have

$$\frac{m_P}{m_Q} = -\frac{\Delta v_Q}{\Delta v_P} = -\frac{(0.075 - 0)}{(0.050 - 0.080)}$$

$$= -\frac{0.075}{-0.030} = 2.5$$

The mass of P is 2.5 times the mass of Q.

4.4   The unit of mass

In both examples, we have found the ratio of two masses. So that we can speak of the mass of a particular object, it must be compared with some universally accepted standard mass such as the international prototype kilogram. This is kept at the Bureau International des Poids et Mesures which is at Sèvres, a suburb of Paris. This is a truly international establishment and is built on land ceded by the French Government and declared to be international territory. The prototype kilogram is a cylinder, having its height and diameter approximately equal, made of an alloy, 90% platinum and 10% iridium, which has a high density of $21\,570 \text{ kg m}^{-3}$. This was adopted as the standard in 1889 and from it substandards have been made. Copy number 18 is held in Great Britain and from time to time it is returned to Paris for checking. Its mass has remained constant to within 1 part in 10^8. The alloy was chosen because it was thought to be the least liable to be affected by time and environment; experience has since shown this to be true. The objection to the use of platinum–iridium for a standard mass is that because of its high density (i.e. small volume), any slight abrasion will change the mass appreciably, consequently the greatest precautions are taken.

As we have seen, the kilogram is one of the seven base units of SI.

4.5   Inertial mass and gravitational mass

In the earlier example, if trolley Y had been a copy of a kilogram on wheels, then we could have stated that the mass of X is 6 kg. Values of mass obtained in this way, by comparing the inertias of two objects, are called inertial masses.

However, when copy number 18 is taken to Paris for checking it certainly does not undergo any interaction with the international prototype kilogram!

It is common practice to use a pair of scales or some other form of pan balance to compare weights of bodies with the weight of a standard body. In this process the gravitational pull of the earth on the body under test is compared with the gravitational pull of the earth on the standard body. Values of mass found in this way are called gravitational masses.

Now experiments (for example that celebrated experiment which Galileo may or may not have carried out at Pisa) show that a body with a large mass and a body with a small mass which are released together fall to the ground with the same acceleration. It seems that the earth's gravitational pull on the larger mass is larger than that on the smaller but that the material which has to be accelerated is larger in just the same proportion. The resulting acceleration is the same for both.

> 'The pull of the Earth's field is proportional to the amount of matter-to-be-pulled. The mass to be accelerated is proportional to the amount of matter-to-be-accelerated. We have no guarantee – except from the vital experiment which shows the acceleration the same for all – that these two kinds of "amount of matter" are equal, or proportional, for all different materials.'

> *Nuffield Physics (1966)*

The fortunate equivalence of inertial and gravitational mass of a body enables us to compare masses by comparing weights with some form of balance: a vastly more convenient process than that of comparison by inertial effects.

4.6   The cause of momentum change

Consider Eq. 4.1 where

$$m_1 v_1 - m_1 u_1 = -(m_2 v_2 - m_2 u_2)$$

Since momentum is (mass × velocity), this equation in words is:

Change of momentum of X
$$= -(\text{change of momentum of Y})$$
$$m_1 \Delta v_1 = -m_2 \Delta v_2 \qquad (4.3)$$

This equation shows that, during an interaction, each mass undergoes an equal, but opposite, change of momentum. During that interaction each mass exerts a force on the other. It is these forces which produce the changes.

Chapter 5

FORCE

Is there a simple connection between the force which is applied, the time during which it acts and the change in momentum (*mv*)? In dealing with three such variables as these, the technique is to keep any one of the three constant and to examine how the other two are related. We will start by keeping the force itself constant.

Experiment J: To apply a constant force to a single trolley and to examine how the momentum changes with time

A constant force can be applied to a trolley by towing it with a thin length of rubber thread (elastic) kept at a constant stretch. With practice it is possible to keep the end of the elastic that is in the fingers, in line with the front pegs of the trolley even though the velocity of the trolley is increasing. See Fig. 5.1.

The resulting motion of the trolley can be recorded with a ticker-tape timer which makes dots on the paper tape 50 times a second, i.e. at intervals of 0.02 s. If the tape is cut into 10-space lengths which are stuck side by side on to card, the chart shown in Fig. 5.2 is obtained. The length of each strip gives the mean velocity during that particular period of 0.2 s (i.e. 10 × 0.02 s). Examination of this chart shows that each step is of equal size which means that equal changes of velocity occur in equal periods of time, when the force causing the change is constant. It follows that the change in velocity that occurs in, say, 3 s, is three times the velocity change occurring in 1 s. Thus, change of velocity is proportional to the time during which it is changing. In symbolic form $\Delta v \propto \Delta t$ where $\propto$ means 'is proportional to'. Since the mass moving (a single trolley) is constant it is also true that

$$m \, \Delta v \propto \Delta t$$

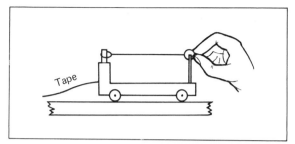

Fig. 5.1 Applying a constant force to a trolley.

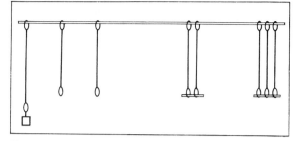

Fig. 5.3

when the force is constant. Consequently the change of momentum is proportional to the length of time for which the force acts.

So far we have investigated momentum and time whilst keeping force constant. We must now apply a different, but nevertheless constant, force to the trolley and see what change of momentum occurs in a given time.

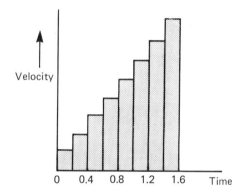

Fig. 5.2 Tape-chart for a trolley moving under the action of a constant force.

Problem 5.1 Before doing the actual experiment with the trolley we must check the elastic threads to see how we may use them to give us three different forces. First we must see whether they are equivalent in their performance. We may do this very simply by supporting them side by side so that they hang vertically and then fastening a small mass to the lower end of one of them. That thread will stretch by some definite amount. Transfer the mass to each of the other threads in turn. What do you conclude if all stretch by the same amount?

Suppose next that you were to link two and then three of them together (Fig. 5.3), in parallel.

What force would you expect to apply to them in order to stretch them by the same amount as in the first of these experiments?

Now we can be sure that, when they are used as in Fig. 5.1 with two and then three attached to the trolley peg in parallel, two and three *units of force* may be applied (taking the case in Experiment J as one *unit of force*).

Experiment K: To apply different, constant forces to a trolley

Tape charts, produced as before, show a constant step size for a particular force. As expected, the largest force produces the greatest step or change in velocity. Some typical results are shown in Table 5a. Allowing for experimental errors, this table shows that the step size is proportional to the force. This means that the force is proportional to the change of velocity and therefore to the change of momentum that results. Thus, $m\,\Delta v \propto F$.

Table 5a

Force	Average step size/mm
1 unit	9.0
2 units	18.2
3 units	26.9

5.1 Newton's second law of motion (Newton II)

Combining this last result with that for Experiment J we have

$$F\,\Delta t \propto m\,\Delta v$$

Or, rearranging

$$F \propto \frac{m\,\Delta v}{\Delta t} \qquad (5.1)$$

This can be interpreted as 'force is proportional to rate of change of momentum' which is incorporated in Newton's second law of motion:

■ 'The rate of change of momentum of a body is proportional to the resultant force causing it and takes place in the direction of that force.'

Equation 5.1 does not lead directly to useful ideas about the sizes of forces. It is rather like saying that the cost c of a tray of eggs is proportional to the number of eggs n in the tray.

$$c \propto n$$

To pay for the eggs we need to know the price p of an egg.

$$c = p n$$

p is the proportionality constant which enables us to calculate the total cost.

Similarly Eq. 5.1 can be written as an equality rather than as a proportionality if we introduce an appropriate proportionality constant k.

$$F = k \frac{m \, \Delta v}{\Delta t}$$

The value of the constant k will decide the number that will describe the size of the force; k can have any value provided all scientists use the same value. For simplicity, the value chosen is $k = 1$.

The equation now becomes

$$F = \frac{m \, \Delta v}{\Delta t} \tag{5.2}$$

This is often written

$$F = \frac{\Delta p}{\Delta t} \tag{5.3}$$

where $p = mv$, the momentum.

It may also be written

$$\boxed{F \Delta t = \Delta p = m \, \Delta v} \tag{5.4}$$

■ The product $F \Delta t$ is known as *impulse*.

If the notation of the calculus is used for the instantaneous rate of change of v with t and of p with t, Eqs. 5.2 and 5.3 are written

$$F = m \frac{dv}{dt} \quad \text{and} \quad F = \frac{dp}{dt}$$

respectively.

5.2 Newton's third law of motion (Newton III)

During a collision between two masses, each mass exerts a force on the other mass. What can be said about the magnitudes of these mutual forces?

The law of conservation of momentum has taught us that in a collision between two bodies of mass m_1 and m_2, moving with velocities v_1 and v_2, each undergoes an equal but opposite momentum change (Eq. 4.3).

$$m_1 \Delta v_1 = - m_2 \Delta v_2$$

The collision involves both bodies and its duration is the same for both. We may therefore write

$$\frac{m_1 \Delta v_1}{\Delta t} = - \frac{m_2 \Delta v_2}{\Delta t}$$

or in calculus notation,

$$m_1 \frac{dv_1}{dt} = - m_2 \frac{dv_2}{dt}$$

The rates of change of momentum of the two interacting bodies are equal and opposite. Since from Eq. 5.1 the rate of change of momentum is proportional to the impressed force, we see that the force on one body is equal and opposite to the force on the other.

This fact is stated in Newton's third law of motion:

■ 'If a force acts on a body, then an equal but opposite force must act upon *another* body.'

In his *Principia* of 1686, Newton gives examples to explain this law in a *scholium* or explanatory note.

'If you press a stone with your finger, the finger is also pressed by the stone. If a horse draws a stone tied to a rope, the horse (if I may so say) will be equally drawn back towards the rope; for the distended rope, by the same endeavour to relax or unbend itself, will draw the horse as much towards the stone as it does the stone towards the horse, and will obstruct the progress of the one as much as it advances that of the other.'

This seems to be a paradox, for it these forces are equal but opposite, how does the horse and stone ever make any progress? The difficulty is resolved by considering the horse and the stone separately and by dealing with *all* the forces involved.

The horse pushes backwards on the ground which pushes forwards on the horse. (This is a pair of Newton III forces.) If the ground's forward push is at least greater than the backward pull of the rope on the horse, then forward motion can occur. (*Note*: Even if the ground's push and the rope's pull were equal, they would *not* constitute a pair of Newton III forces because they both act on the horse and not on different bodies.)

Consider now the stone; it exerts a forward force on the ground and the ground exerts an equal backward force on the stone. (A pair of Newton III forces.) If this backward force is less than the forward pull of the rope, then forward motion occurs. It should be noted that the ground's frictional force and the rope's pull both act on the stone and do not form a pair of forces.

Newton III can also be stated in this form:

■ 'Action and reaction are equal and opposite, and act on different bodies.'

Often, when this form is quoted, the last five words are unfortunately omitted. This can lead to some confusing and fallacious statements when illustrations of Newton's third law are given. For instance, 'If a book is resting on a man's hand, there are two equal and opposite forces acting; the weight of the book acting downwards and the reaction from the hand upwards.' In itself this statement is correct but the forces mentioned are not examples of the 'action and reaction' forces intended in Newton III. They are examples of forces that happen to be in equilibrium, and both are acting on the book (Fig. 5.4a). (Weight is the force acting *on* a mass placed in a gravitational field.) Apart from the fact that both forces act on the *same* body, Newton III does not apply when the hand is suddenly removed. If the upward hand-force becomes zero, the weight certainly does not follow suit! The law has not failed here; it is simply being wrongly used.

To clarify this case, it must be realized that the earth is pulling on the book with a force W and the book pulls on the earth with a force $-W$ (which is equal and opposite). These forces constitute a pair of Newton III forces (see Fig. 5.4b).

When the hand is supporting the book, it pushes upwards on the book with a force $-H$ and

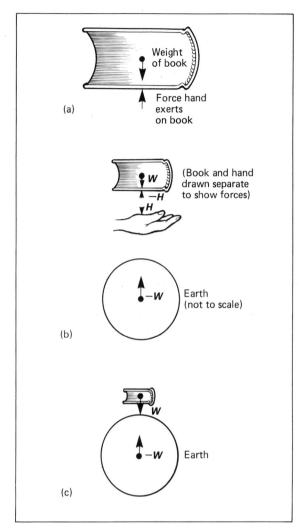

Fig. 5.4

the book pushes downwards on the hand with a force H. These are also a 'pair' of forces and both do become zero when the hand is removed. When this is done the pair of forces, W and $-W$, still remain and cause the book and the earth to fall together or towards one another (Fig. 5.4c). Because of its great mass any movement of the earth is imperceptible.

5.3 Acceleration

In the experiments which led to Eq. 5.2, that is

$$F = \frac{m\,\Delta v}{\Delta t}$$

the bodies involved and their masses remained

unchanged whilst the velocities, the forces and the times were changing. The equation might well be written

$$F = m\frac{\Delta v}{\Delta t} \qquad (5.5)$$

■ $\Delta v/\Delta t$ is the rate of change of velocity or the acceleration produced by the force F acting on the unchanging mass m. Its units will be those of velocity divided by time; and in SI these are $m\ s^{-1}/s$ or $m\ s^{-2}$. This is, of course, a derived unit.

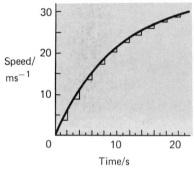

Fig. 5.5 A car starting from rest. Each slope 'triangle' has a base, one second long. The vertical side gives the change in speed in that one second. The acceleration is not uniform; it decreases as the car goes faster.

Acceleration involves velocity, which is a vector quantity, and time, which is a scalar. Consequently acceleration is itself a vector quantity and both the magnitude and the direction must be specified.

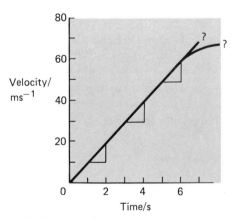

Fig. 5.6 A mass falling freely near the earth's surface. Each slope 'triangle' has the same vertical side and the acceleration is uniform. It is a little less than 10 m s^{-2}.

The acceleration tells us by how much the velocity of a body changes in unit time. Information about acceleration can conveniently be expressed on a velocity–time graph. Some examples are given in Figs. 5.5 and 5.6. It is interesting to note that Experiments J and K show that a mass moving under the action of a constant force does so with uniform acceleration. The implication of the graph (in Fig. 5.6) is that since the falling mass is accelerating uniformly, then the gravitational force on it is constant and is independent of the velocity of the mass. It should be realized that the acceleration will only be uniform while the velocity of the mass is small enough for the resistance of the air to be negligible. As the air resistance becomes greater, the resultant force accelerating the mass becomes smaller until it is zero. The acceleration is then zero and the mass is falling at its maximum, or *terminal velocity*. If the graph had been continued farther, it would have curved over just as the car's graph. The car is also tending to a maximum velocity when the force that the engine exerts through the wheels is exactly balanced by air resistance, etc.

5.4   The unit of force

If we write a for acceleration Eq. 5.5 becomes

$$\boxed{F = ma} \qquad (5.6)$$

and provides us with the means to define a unit of force. If we agree to make the mass $m = 1$ kg and the acceleration $a = 1$ m s^{-2}, then the unit of force is called the newton – a most appropriate name.

■ One newton (N) is the force which will give to a mass of 1 kg an acceleration of 1 m s^{-2}.

This is a further example of a unit derived from the appropriate base units of SI. In terms of those base units the newton is equivalent to kg m s^{-2}. The dimensions of force will be M L T^{-2}.

The equation

$$F = ma$$

can now be used for calculating force in a given set of circumstances. For example, if a mass of 4 kg is found to have an acceleration of 3 m s^{-2}, the force acting on it is 12 N.

Problem 5.2   Look at Fig. 5.5. What is the initial acceleration of the car? If the mass of the car is 800 kg, what force are the tyres exerting on the road during this acceleration? What is your estimate of the acceleration at the end of the tenth second? What is the force used to accelerate the car now? Why is this less than before?

Look at Fig. 5.6. The acceleration near the surface is 10 m s^{-2}. What is the force experienced by a body falling with this acceleration if its mass is 9 kg?

5.5   The effect of a constant force on the linear motion of a body

We have seen that a constant force will produce a constant acceleration when it is applied to a fixed mass.

If a body, already moving in a straight line with velocity v_0 is acted upon by a constant force which gives it a constant acceleration a, the increase in velocity after a time interval t will be at so that the velocity v_t at the end of that time interval will be $(v_0 + at)$.

$$v_t = v_0 + at \qquad (5.7)$$

The mean velocity during that time interval will be

$$\frac{v_0 + v_t}{2}$$

So the distance travelled (s) in that time interval will be

$$s = \frac{v_0 + v_t}{2}\, t \qquad (5.8)$$

Combining Eq. 5.7 and Eq. 5.8 we get

$$s = \frac{(v_0 + v_0 + at)}{2}\, t$$

$$= v_0 t + \tfrac{1}{2}at^2 \qquad (5.9)$$

Note that for a body starting from rest, $s = \tfrac{1}{2}at^2$.

$$\qquad (5.9a)$$

From Eq. 5.7

$$v_t - v_0 = at$$

From Eq. 5.8

$$v_t + v_0 = 2s/t$$

Multiplying these last two equations gives

$$(v_t + v_0)(v_t - v_0) = \frac{2s}{t}\, at$$

Whence $$\boxed{v_t^2 - v_0^2 = 2as} \qquad (5.10)$$

These four equations apply to all cases of motion in a straight line under the action of a constant force, i.e. motion with constant acceleration.

5.6   Reaction from a jet

If you have ever held a hose which is projecting a powerful jet of water you will know that there is a steady force of reaction from the jet. This is the force which is harnessed in some garden sprinklers. A jet engine is a rather more complicated application of this effect. Air enters the front of the engine and leaves as part of the exhaust gases at the rear. The fuel which burns in the engine gives these exhaust gases a high speed with respect to the engine.

Imagine that the engine is attached to an aircraft which is travelling with speed v_0 through stationary air. The air may then be thought of as entering the front of a stationary engine with speed v_0 relative to the engine. At the rear the ejected gases leave with a speed v relative to the engine (Fig. 5.7).

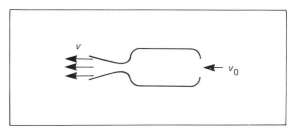

Fig. 5.7

If air passes through the engine at a rate r_{air} (measured in kg s^{-1}) and fuel is burned in the engine at a rate of r_{fuel}, the total rate of change of momentum of the system is

$$r_{fuel}\, v + r_{air}\, v - r_{air}\, v_0$$

The law of conservation of momentum and Eq. 5.3 tell us that this must be equal to the thrust of the engine.

Problem 5.3   The engines of a jet aircraft flying at 250 m s^{-1} take in 800 m^3 of air per second at an operating height at which the density of the air is 0.50 kg m^{-3}. The air is used to burn fuel at a rate of 10 kg s^{-1} and the exhaust gases (including the incoming air) are ejected at 500 m s^{-1} relative to the aircraft. What is the thrust?

The air is passing through the engines at a rate of (800 × 0.50) kg s^{-1} or 400 kg s^{-1}. This is much larger than the rate of burning of the fuel which is 10 kg s^{-1}.

The rate of change of momentum of the exhaust gases is

$$(10 \times 500) + 400(500 - 250) \text{ kg m s}^{-1}$$

or 105 000 kg m s^{-1}.

The thrust is, therefore 105 000 N or 105 kN.

5.7 Rockets

Unlike the jet engine, the rocket engine carries *all* its propellant materials including oxygen. Imagine such an engine in space – so far away that we can ignore any gravitational effects. Then almost the whole of the thrust of the exhaust gases will be available to accelerate the rocket.

Suppose these gases are expelled at speed v_{gas} relative to the rocket. Between time t and time $(t + \Delta t)$, a mass Δm of fuel is burned and expelled from the rocket. As a result the rocket increases its speed relative to an external observer from v to $(v + \Delta v)$ whilst the speed of the ejected gas decreases from v to $(v - v_{gas})$. From the conservation of momentum we have

$$(m + \Delta m)v = \Delta m(v - v_{gas}) + m(v + \Delta v)$$

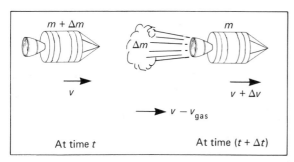

At time t At time $(t + \Delta t)$

Fig. 5.8

So

$$mv + v\,\Delta m = v\,\Delta m - v_{gas}\,\Delta m + mv + m\,\Delta v$$

and

$$m\,\Delta v = v_{gas}\,\Delta m$$

Whence

$$\Delta v = v_{gas}\frac{\Delta m}{m} \tag{5.11}$$

This change takes place in time interval Δt. So the rates of change of velocity and of momentum with time are:

$$\frac{\Delta v}{\Delta t} = \frac{v_{gas}}{m}\frac{\Delta m}{\Delta t}$$

and

$$m\frac{\Delta v}{\Delta t} = v_{gas}\frac{\Delta m}{\Delta t}$$

Since the rate of change of momentum is equal to the thrust (F) on the rocket we have

$$F = v_{gas}\frac{\Delta m}{\Delta t} \tag{5.12}$$

To develop a large thrust, the designer of the rocket will seek to make both v_{gas}, the speed at which the gases are ejected relative to the rocket, and $\Delta m/\Delta t$, the rate at which the fuel is burnt and ejected, as high as possible. In the Saturn V rocket, v_{gas} is about 2500 m s^{-1} and $\Delta m/\Delta t$ about 1.4×10^4 kg s^{-1}. So the thrust can be as high as

$$2500 \times 1.4 \times 10^4 \text{ N or } 35 \times 10^6 \text{ N}$$

Since the mass at lift-off is about 2.8×10^6 kg (2800 tonnes) which corresponds to a weight of 28×10^6 N, the initial upward acceleration of the rocket as it leaves the launch-pad is provided by the difference between the weight and the thrust which is 7×10^6 N. That provides an initial acceleration of about 2.5 m s^{-2}.

As the total mass of the rocket and its fuel decreases, so the acceleration must increase. But not for long! The fuel of the rocket is exhausted in about 150 s.

To meet the need to increase the fraction of the total take-off mass which is payload, multi-stage rockets are used: in these, each complete rocket motor is dumped once the burn is completed and the next motor takes over.

Chapter 6

FORCES AND EQUILIBRIUM

Forces do not always produce accelerations. In, for example, the building industry and in civil engineering, the concern is with forces which are in static equilibrium. In considering such forces the terms *tension* and *compression* are in frequent use. Strings pull – they are in *tension*. And the tension is the same throughout their length. The mechanism lies in the molecular structure of the material from which the string is made; under an applied force, the molecules move a little apart until an equal and opposite force develops.

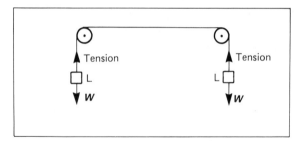

Fig. 6.1

Consider a simple case of a string stretched between two equal loads and passing over two pulleys with frictionless bearings (Fig. 6.1). The two loads are pulled towards the earth by equal forces W – their weights. To keep the system in equilibrium, the supporting string must apply equal forces in the upward direction on each load.

Tensions are not confined to strings and wires; members used in, say, bridge building and cranes may also be under tension. In Fig. 6.2a the member AC is under tension (it could be replaced by a wire); but BC is under compression (deforming slightly until an equal force is established to

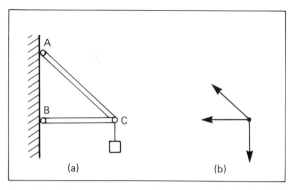

Fig. 6.2   A simple truss supporting a load at C. (a) shows the directions of the three forces acting at C.

resist the force of compression applied). Figure 6.2b shows the three forces which act at the point C.

Problem 6.1   Which of the members AC, BC, etc., of these three systems are in tension and which are in compression? Remember that members in tension could be replaced by strings or wires.

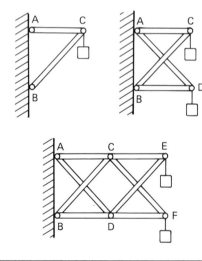

6.1   The conditions for equilibrium

The book (Fig. 6.3), resting on the shelf in the upright position, experiences a force (its weight) towards the centre of the earth; but it remains at rest (in equilibrium) because the shelf deforms slightly bringing an equal and upward force into play. The book is not subject to any net force which could move it bodily. Nor does it rotate.

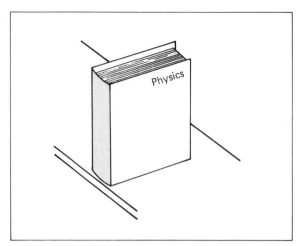

Fig. 6.3

But, push it slightly out of the vertical with an additional force and it will topple on to its back. To get there, the book rotates. And then it is once more in equilibrium.

This simple event illustrates that, to be in static equilibrium, the forces acting on an object must satisfy two conditions:

1. that the object shall not be subject to any net force which would tend to move it bodily (i.e. accelerate it); and
2. that the object shall not be subject to any net force which would tend to rotate or twist it.

The first condition is that for *translational equilibrium* and the second for *rotational equilibrium*.

6.2   Translational equilibrium

A body which has several forces acting upon it but which does not accelerate in any direction is said to be in *translational equilibrium*. Any number of separate forces in the same plane (co-planar forces) may be involved; but the body does not accelerate. In that case, the line of action of all these co-planar forces must meet in a point as in Fig. 6.4a or be parallel to one another as in Fig. 6.4b. Since these bodies can neither accelerate, nor twist nor rotate, we know that the sum of all the co-planar forces must be zero.

If there are two forces (as in the case of the book standing upright) those forces will be equal and opposite.

If there are three or more forces, the vector sum must still be zero. This implies that a vector

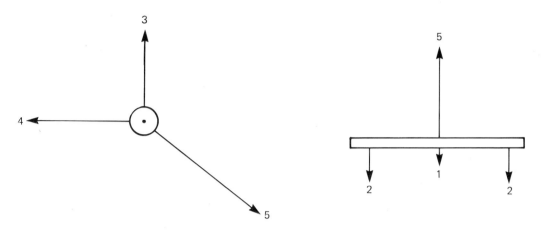

Fig. 6.4 Co-planar forces acting (a) at a point, and (b) parallel to one another. Both systems are in equilibrium.

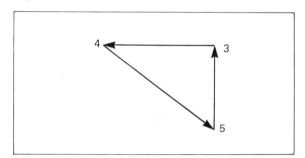

Fig. 6.5 Vector diagram for the three forces shown in Fig. 6.4a.

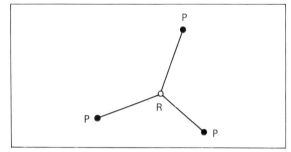

Fig. 6.7

triangle (Fig. 6.5) can be drawn for the case of the three co-planar forces (Fig. 6.4) and a vector polygon for four or more (Fig. 6.6).

To test this, secure three or more rubber bands to a ring (R) (about 2 cm in diameter). Place a sheet of plain paper on a drawing board and push three or more drawing pins (P) into it; their actual positions do not matter. Slip the free end of each rubber band over one of the drawing pins (Fig. 6.7). Mark the position of the centre of the ring and draw lines joining this to the drawing pins. These lines give the directions of the three or more forces on the ring. (Fig. 6.7.)

To find the magnitude of each force, note the extended length of the rubber band responsible and use a dynamometer graduated in newtons to extend the band to that length again. Finally draw the vector polygon.

6.3 Rotational equilibrium

To simplify the argument, we shall assume that the body is in translational equilibrium. That is to

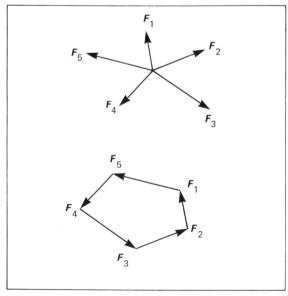

Fig. 6.6 Force diagram with the corresponding vector polygon.

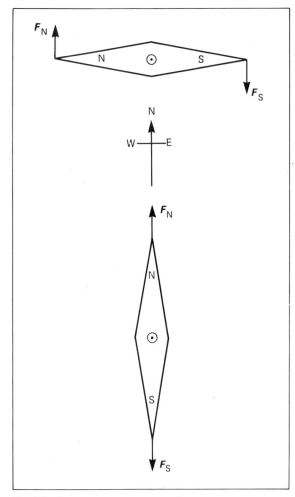

Fig. 6.8 Forces on a compass needle.

say, no net force is acting to provide linear acceleration. A simple compass needle offers a good
example since the earth's magnetic field provides a
pair of equal and opposite forces, one pulling the
north-seeking pole towards the North and other
pulling the south-seeking pole towards the South.
Let us hold the needle so that the axis joining the
two poles lies East–West (Fig. 6.8).

Once the restraining force is released, the
needle rotates under the action of the two equal
opposite forces F_N and F_S towards the N–S line.

It overshoots, oscillates for a short time and
finally settles so that the two equal and opposite
forces are acting in the same straight line (Fig.
6.8b). It is now in rotational equilibrium in the
earth's magnetic field.

6.4 The law of moments

Figure 6.9 offers another example of rotational
equilibrium, this time in the earth's gravitational
field. The bar (assumed to be uniform), which is
supported centrally, is balanced under the action
of two equal and opposite forces applied by equal
loads as shown. The system is symmetrical; the
equal forces balance the bar when applied at equal
distances on either side of the pivot O. Of course,
there are two other forces acting on the bar: its
own weight downwards through O, and an upward
force also through O sufficient to maintain translational equilibrium. What is the magnitude of
this force?

This simple system is precisely that of the
ordinary beam balance.

Experiment shows that rotational equilibrium
can also be achieved when the loads (and therefore
the forces) are unequal. See Fig. 6.10.

The condition is

$$F_1 l_1 = F_2 l_2$$

and this is well known as the law of the lever. The
product of a force and the perpendicular distance
from the line of action of the force to the pivot is
known as the *moment* or *torque* of the force about
an axis through the pivot. In this case force F_1
applies an anti-clockwise moment $F_1 l_1$; F_2 applies
an equal clockwise moment $F_2 l_2$. When the body is
in rotational equilibrium, the algebraic sum of
these two moments is zero.

If we wish to take moments about any point
other than O, we must consider the two forces
through O as well. These are the weight (W) of the
bar itself (downwards) and the reaction of the
pivot to the downward forces ($F_1 + F_2 + W$)
which is upwards. See Fig. 6.11.

If now we take moments about A, we have:

$$\left.\begin{array}{r}\text{clockwise}\\ \text{moment}\end{array}\right\} = F_1 d + W(d + l_1) + F_2(d + l_1 + l_2)$$

$$\left.\begin{array}{r}\text{anti-}\\ \text{clockwise}\\ \text{moment}\end{array}\right\} = (F_1 + F_2 + W)(d + l_1)$$

Since the bar is in rotational equilibrium the two
moments are equal:

$$\begin{aligned} F_1 d + W(d + l_1) &+ F_2(d + l_1 + l_2) \\ &= (F_1 + F_2 + W)(d + l_1) \end{aligned}$$

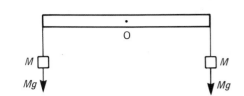

Fig. 6.9

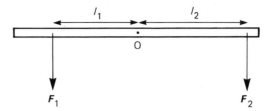

Fig. 6.10

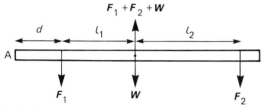

Fig. 6.11

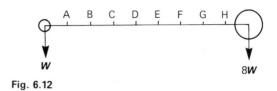

Fig. 6.12

$$F_1 d + W d + W l_1 + F_2 d + F_2 l_1 + F_2 l_2$$
$$= F_1 d + F_1 l_1 + F_2 d + F_2 l_1 + W d + W l_1$$

which reduces to

$$F_2 l_2 = F_1 l_1$$

as before.

This tells us that, if the vector sum of the forces on an object is zero and if the sum of the moments or torques about any one point is zero the sum of the torques about any other point is zero.

■ We may now state the *law of moments* formally: for a body which is in equilibrium the sum of the moments, about any axis, of the external forces acting on the body is zero.

6.5 Centre of Gravity

The example chosen above assumes that the weight of the uniform beam can be thought of as acting downwards through the mid-point. The earth is exerting its pull on each particle of matter in the beam; and since the beam is symmetrical, the sum of the moments of all these forces is zero. Had the beam not been symmetrical, we could still think of the weight as acting through a single point – but this point would not necessarily be central. Figure 6.12 shows two differing masses connected rigidly together by a light (i.e. massless) beam.

Where would you expect to find the *centre of gravity*, that is the point about which the sum of

the moments of the earth's pull on the particles is zero?

■ The centre of gravity of a body is a point through which a single force (the weight) must act to have the same turning effect about any axis as the actual forces on all the particles making up the body.

6.6 Weight – a digression

We have consistently assumed that the earth exerts a force on a load in its gravitational field. And we have called this force the *weight* of that load. The evidence for the existence of this force rests on the simple experience that a spring balance which is supporting such a load stretches and so exerts an upward force. The load is then in translational equilibrium and that tells us that a second force must be acting as well. This second force must be equal in magnitude and opposite in direction to the upward force exerted by the spring. We further assume that this second (downward) force arises because the load experiences a pull towards the earth's centre. This is the force to which we give the name 'weight'.

If now you put a mass (this book perhaps) on your outstretched hand and then lower the hand with acceleration, the book *appears* to 'lose some of its weight'. The upward force exerted by your hand on the mass is reduced, whilst the pull of the earth on the mass (i.e. its weight) remains unaltered.

Fig. 6.13 Weightless conditions in Skylab spacecraft.

If you could accelerate your hand downward at $9.8 \, \mathrm{m \, s^{-2}}$ the upward force exerted by your hand would be zero; but the pull of the earth on the mass remains just the same.

Some would say that, as it falls freely, the mass is 'weightless'. We would prefer to speak of 'apparent weightlessness' as describing the state of the freely-falling mass on which no supporting force is acting. Figure 6.13 shows precisely this situation within Skylab.

6.7 Non-parallel forces and the law of moments

The law of moments is not restricted to forces which are parallel to one another. It is always possible to resolve a force, whatever its direction, into a suitable pair of component forces. Consider, for example, the case of the hand-held load shown in Fig. 6.14.

This shows, diagrammatically, a human arm holding a load stationary with the forearm hori-

zontal. The forearm and its load are in equilibrium under the action of a number of forces: the weight of the load; the weight of the forearm and hand, the tension F_m in the tendon connected to the biceps muscle and the force (F) exerted at the elbow to maintain equilibrium.

We may take the first two (the two weights) together as a single force W acting through the centre of gravity (G) of the combined forearm, hand and load at a distance a from the elbow. The tension (F_m) of the muscle acts at a distance b from the elbow at angle α to the horizontal forearm. The force F which maintains equilibrium is exerted at the elbow in the direction β with the horizontal.

We may think of F_m as made up of two components: $F_m \cos \alpha$ along the forearm in the direction towards the elbow; $F_m \sin \alpha$ at right angles to the forearm (Fig. 6.15).

Since the component $F_m \cos \alpha$ acts through the elbow, the torque it exerts about the elbow is zero. The force which is effective in supporting the arm about the elbow is $F_m \sin \alpha$

Since the forearm is in rotational equilibrium, the net torque is zero. Equating the torques *about*

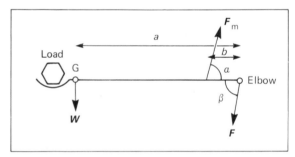

Fig. 6.14

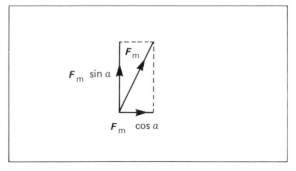

Fig. 6.15

the elbow, we have

$$a W = b F_{\text{m}} \sin \alpha$$

Since a is much larger than b, F_{m} must be much larger than W. The muscle has to exert a large force to support a small load.

The forearm is also in translational equilibrium. In the vertical direction we have

$$W + F \sin \beta = F_{\text{m}} \sin \alpha$$

and in the horizontal

$$F_{\text{m}} \cos \alpha = F \cos \beta$$

Problem 6.2   What will happen to the muscular tension F_{m} if the arm is stretched out so that the angle of the upper arm to the vertical is increased whilst the forearm is kept horizontal? Does your answer accord with your experience?

6.8   The case of the hanging basket

As an example of how these principles may be applied to simple structures, consider a right-angled metal bracket whose purpose is to support a hanging basket full of flowering plants from a vertical wall. Normally two wall bolts would be used to secure the bracket at A and B and, to clarify the argument, we shall assume that these two points are the only points of contact between the bracket and the wall. The cross brace of the bracket, which confers rigidity, is in compression.

The external forces on the bracket are as shown in Fig. 6.16b:

i)   the weight of the load (W) at the perpendicular distance AC ($= l_1$) from the wall: downwards;

ii)   the force F_a pulling the bracket to the wall at A: horizontal;

iii)   the force V_a exerted by the bolt A: upwards;

    (these two forces are the horizontal and vertical components of the single force exerted by bolt A on the bracket);

iv)   the force F_b exerted by the wall on the bracket at B: horizontal;

v)   the force V_b exerted by the bolt B: upwards;

    (these two forces are the horizontal and vertical components of the single force exerted at bolt B);

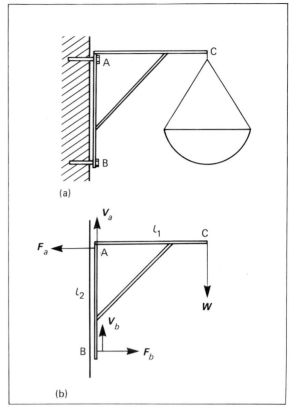

Fig. 6.16   A hanging basket.

vi)   the weight of the bracket itself which we shall ignore since it is very small in comparison with W.

Now take moments about the lower wall bolt B. The moment of each of the forces V_a and V_b is zero.
The moment of F_a is $F_a l_2$ (anti-clockwise), where l_2 is the length AB.
The moment of F_b is zero.
The moment of W is $W l_1$ (clockwise).
Since there is rotational equilibrium,

$$F_a l_2 = W l_1$$

Take moments about the upper wall bolt A. The moment of each of the forces F_a, V_a and V_b is zero.
The moment of F_b is $F_b l_2$ (anti-clockwise).
The moment of W is $W l_1$ (clockwise).
Since there is rotational equilibrium,

$$W l_1 = F_b l_2$$

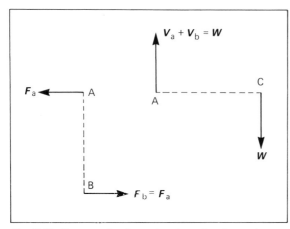

Fig. 6.17   Two couples (i.e. pairs of equal and opposite forces) acting on the hanging basket structure.

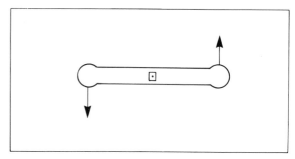

Fig. 6.18   A couple acting on the handle of a tap.

Comparing these two equations we see that F_a and F_b are equal (as they must be, since they are the only horizontal forces involved).

The bracket is also in translational equilibrium. It follows that

$$V_a + V_b = W$$

The analysis shows that there are two pairs of equal but opposite forces acting on the bracket: F_a and F_b in the anti-clockwise direction, W and $(V_a + V_b)$ in the clockwise direction. Such pairs, which are called *couples*, are invariably associated with rotations. Rotational equilibrium requires that there should be pairs of equal, opposing couples acting.

The turning effect of a couple is given by the product of one of the two forces and the perpendicular distance between them.

$$T = Fr \qquad (6.1)$$

In this case the two couples are as shown in Fig. 6.17.

Acting by itself, a single couple will cause the body to which it is applied to rotate. For example, we may apply an anti-clockwise couple to the handle of a tap (seen from above in Fig. 6.18).

The effect of the torque is to turn the tap in the anti-clockwise direction. Similar examples, in which it becomes less and less clear how the separate forces act, include: the screwdriver (we apply a torque and the screwdriver turns); the frictional torque on the bearings of a rotating shaft driven by, say, an electric motor; the torque which induces the armature of the electric motor itself to turn.

Chapter 7

MOTION IN A CIRCLE

7.1   Combining velocities

A dry ice puck sliding across a flat glass plate neither accelerates nor decelerates; neither moves to the left nor to the right. A ball dropped from the hand moves straight downwards, accelerating as it falls. But thrown to one side as it is dropped and the ball follows a path which we recognize as a parabola. (Fig. 7.1.)

That path is the result of the combining of two independent motions: the horizontal velocity, which resulted from the impulse provided by the hand as it threw the ball to the side; and the accelerated motion of free fall.

Let the horizontal velocity be v_x and let us ignore any frictional effects of the motion through the air. After time t, the ball will have travelled a horizontal distance $x = v_x t$.

In the same time t, the ball will have dropped through a vertical distance $y = \frac{1}{2} g t^2$. (See Eq. 5.9a.)

Combining the two equations, we obtain

$$y = \frac{\frac{1}{2} g}{v_x^2} x^2$$

This is the equation of a parabola, having the form $y = k x^2$.

The assumption that the air friction is negligible is only reasonable for dense bodies and short times of flight. Whilst the path of, say, a cricket ball thrown from the hand is nearly parabolic, that of a table tennis ball is not.

Suppose now that the horizontal impulse was increased again and again (still with zero friction), producing, in the balls used, horizontal speeds of v_x, $2 v_x$, $3 v_x$ etc. Figure 7.2 shows the likely result.

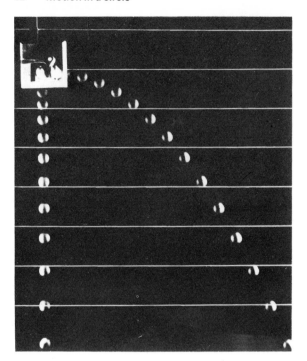

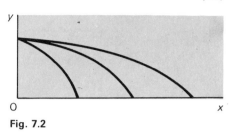

Fig. 7.2

Fig. 7.1 Multiflash photograph of two golf balls released simultaneously from the mechanism shown. One of the balls fell freely. The other was projected horizontally with an initial speed of 2.00 m s^{-1}. The flash rate was 30 Hz. The white lines are 0.15 m apart. (From P.S.S.C. *Physics*, D. C. Heath, 1965 reproduced by permission of Education Development Center, Inc.)

Problem 7.1 How long does it take for each of the projectiles shown in Fig. 7.2 to reach the x axis? How far along the x axis would you expect to find them?

All fall freely in the vertical direction so the time to fall from the point of release to the x axis must be given by $y = \frac{1}{2} g t^2$ and will be the same for all. Since the speeds in the horizontal direction do not change the various values of x will be given by $v_x t$, $2 v_x t$, $3 v_x t$, etc.

Newton imagined just this situation and then asked what would happen if the speed v_x were to be increased more and more. His answer led him to the idea of an earth satellite – the ball going round the earth and returning to its starting point (Fig. 7.3).

The moon, which is such a satellite, moves in an almost perfect circular orbit around the earth. At any instant, its motion can be thought of as a constant speed along the tangent of the orbit and an acceleration towards the centre of the earth. The latter is a result of the gravitational attraction between the two bodies.

7.2 Motion in a circle

The technique of considering a curved trajectory to be the result of two linear motions can be used when discussing the force that must act on a mass if it is to move in a circle with uniform speed.

Consider a simple experiment involving an air puck and a horizontal glass surface. Put a nail into a small block of wood (about $2 \times 2 \times 1$ cm) and tether the puck to it with about 20 cm of thread. The block can be held down on the glass plate with one finger. If the puck is pushed, once, tangentially, it will continue to move in a circular path with a constant speed (see Fig. 7.4). If, when the puck

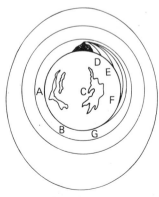

Fig. 7.3 Newton's presentation of the paths of projectiles launched horizontally with various speeds and from various heights above the surface of the earth.

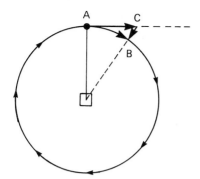

Fig. 7.4   Motion of a tethered puck.

is at A, the finger holding down the block is removed, the puck will travel with the same constant speed but along a straight line AC which is a tangent to the circle.

A flexible thread in tension can only exert a force along its length, and so, when the tether is present it provides a force that acts on the puck towards the centre of the circle, i.e., a *centripetal* force acts *on* the puck. Now, a force acting on a mass gives it an acceleration, and so the circular motion of the puck can be considered as being made up of two *simultaneous* linear motions, (a) a constant tangential speed, along AC, and (b) a constant inward radial acceleration, along CB. Let us calculate the value of this acceleration that changes uniform linear motion into circular motion, i.e., which makes the puck arrive at B on the circle rather than at C. (See Fig. 7.5a.)

Suppose the puck moves around the circle, of radius r, with a constant speed v, and that the time taken to travel from A to B along the arc is t. Let the inward radial acceleration be a.

The journey, A to B, can be replaced by a combination of: uniform motion from A to C where AC = vt, and acceleration from C to B where, applying Eq. 5.9a, CB = $\frac{1}{2} a t^2$. (Fig. 7.5b.)

If B is close to A, then (1) arc AB ≈ line AB ≈ AC = vt and (2) AB̂C ≈ 90° and so the shaded triangles are similar.

Therefore:

$$CB/AC = AD/AO,$$

i.e.

$$\tfrac{1}{2} a t^2 / v t = \tfrac{1}{2} v t / r,$$

and

$$\boxed{a = v^2/r} \qquad (7.1)$$

If the puck has a mass m, then the centripetal force that must act on it if it moves in a circular path is $m v^2/r$.

$$F = \frac{m v^2}{r}$$

This force is provided by the tension in the thread; if the thread breaks, the force on the puck ceases to act and it will move off along the tangent.

The result is a general one, applicable to all cases of motion in a circle. Other examples include the motion of the moon around the earth, the motion of an electron in a uniform magnetic field, the motion of satellites in circular orbits, the motion of planets around the sun. In all such cases the acceleration and the force causing it are directed towards the centre of the circle.

Problem 7.2   A coach is entering a sharp bend to the left and a passenger feels herself sliding on her seat to the right. She tells her neighbour that this is because of centrifugal force. Is she correct? State the forces, specifying their directions, that actually act *on* the passenger.

The passenger sitting on a seat in a coach which is entering a bend to the left will tend to slide along the bench as she obeys Newton's first law and continues to travel in a straight line. It might be better to say that the coach is sliding to the left

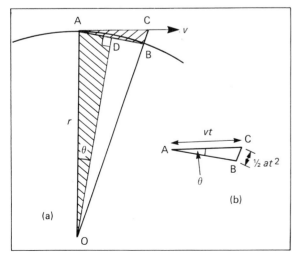

Fig. 7.5

beneath her. Once she reaches the right hand side of the seat she will feel a force constraining her to follow the circle in which the coach is travelling. That force is inward towards the centre of the motion. There is an equal and outward force on the side of the seat.

7.3 Angular velocity

It is often preferable to work with angular velocities rather than with speeds along an orbit. Consider the satellite as a point P moving at a fixed distance r from the centre of orbit O. If the radius OP moves through an angle θ in time t, we say that the angular velocity ω is θ/t. See Fig. 7.6.

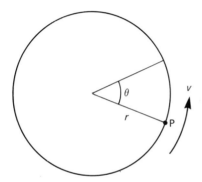

Fig. 7.6

The units of angular velocity are radians per second.

However, the speeds especially of rigid bodies, are often quoted in revolutions per minute or per second or in degrees per second. These must always be converted to radians per second before use in calculation.

$$1° = \frac{2\pi}{360}\,\text{radian} \qquad 1\,\text{rev s}^{-1} = 2\pi\,\text{rad s}^{-1}$$

The speed v of P along the circumference of the circle in Fig. 7.6 is related to the angular velocity by the equation

$$v = \frac{r\theta}{t} = r\omega \tag{7.2}$$

The time T to complete one orbit (or one revolution for a rigid body) is given by

$$T = \frac{2\pi}{\omega} \tag{7.3}$$

and the number of orbits (or revolutions) completed in unit time by

$$n = \frac{1}{T} \tag{7.4}$$

This is known as the *rotational frequency*.

Problem 7.3   What is the angular velocity of a point on the circumference of a 12 inch gramophone disc turning at $33\frac{1}{3}$ revolutions per minute? What is the rotational frequency? What is the angular velocity of a point half way along a radius?

7.4 Angular acceleration

When a body is rotating under the action of a constant torque so that it experiences a constant acceleration from angular velocity ω_0 to angular velocity ω in time t, the angular acceleration α is

$$\frac{\omega - \omega_0}{t}$$

and we see that

$$\omega = \omega_0 + \alpha t$$

The angular displacement (angle turned through) θ is given by

$$\theta = \frac{\omega_0 + \omega}{2}t$$

Substituting for ω this becomes

$$\theta = \frac{\omega_0 + \omega_0 + \alpha t}{2}t$$

$$= \omega_0 t + \tfrac{1}{2}\alpha t^2$$

These equations remind us of those for linear motion under the action of a constant force:

Linear	*Angular*
$v = v_0 + at$	$\omega = \omega_0 + \alpha t$
$s = \dfrac{v_0 + v}{2}t$	$\theta = \dfrac{\omega_0 + \omega}{2}t$
$s = v_0 t + \tfrac{1}{2}a t^2$	$\theta = \omega_0 t + \tfrac{1}{2}\alpha t^2$

7.5   Rotational inertia

Rotating bodies – the earth itself, a grindstone, the wheel of a moving vehicle – all possess inertia. Evidently the mass of the body and its angular velocity play a part in this. But so also does the way in which the mass of the body is distributed about the axis of rotation.

Consider two flywheels with equal mass, but with different diameters and thicknesses. Drive each in turn up to the same angular velocity by means of an electric motor which can also run in reverse as a generator.

When the required speed has been reached, switch off the supply and allow one of the flywheels to drive the motor as a generator and light a lamp.

Repeat the process with the second flywheel in place of the first. How do the times during which the lamp glows compare?

Evidently the flywheel with the larger diameter has more rotational inertia. It is not enough simply to compare masses when dealing with rotating bodies; one must also consider how the mass is distributed about the axis of rotation. This is a matter to which we shall return in Section 8.19.

Energy and Thermal Physics

Chapter 8

ENERGY

In recent years we have grown accustomed to hearing about the energy crisis. Certainly, there is a world-wide problem concerning the supply and distribution of the fuels which human beings need to sustain their civilizations. But fuel crises are not new. Britain experienced a severe one in the late sixteenth century when there was an acute shortage of wood to use as fuel. As commercial prosperity rose more and more timber was needed for house-building, ship-building, salt-making, brewing, iron-smelting and as fuel for use in the home. Trees were being cut down at a far greater rate than they could grow. Fortunately for Britain, there was a solution in the earth itself – coal. So coal-burning superseded wood-burning. In turn other natural resources were utilized – first coal-gas (derived from coal), then natural gas and oil. Unfortunately for mankind, the supply of these 'fossil fuels' is finite – the resource is not renewable. So the present energy crisis is different in kind from earlier ones. Once all the oil has been extracted – no more remains! Once all the coal has been mined – there is no more!

8.1 Some energy arithmetic

Unfortunately, man has been depleting the earth's reserves of these precious fossil fuels at an ever-increasing rate. As the size of the reserves has fallen, the rate of use has risen, thereby hastening the rate at which the remaining reserves are being depleted.

To illustrate the problem let us suppose we have a stock of 1000 items of a product which cannot be replaced because it has gone out of production. If we know that, on average, we sell

100 of these per year, our stock will last for ten years.

But suppose that, for some reason, the selling rate increased from the original 100 by 10% per year. Then our stock and sales figures will look as Table 8a.

Table 8a

Year	Sales	Stock left
1	100	900
2	110	790
3	121	669
4	133	536
5	146	390
6	161	229
7	177	52
8	52	out of stock
9	–	–
10	–	–

Because the rate of sales has gone up, our stock has been depleted in just over seven years.

Problems such as these are examples of *exponential change*. Whilst the sales were increasing exponentially, the stock was decreasing exponentially.

Exponential implies that the change, whether a growth or a decay, occurs at a rate which is a fixed percentage over a chosen time interval. Compound interest is a good example. If I invest £100 in an account offering 10% compound interest over ten years my capital will increase (Table 8b).

Table 8b   £100 at compound interest

Year	Capital/£	Year	Capital/£
Now	100	+ 5	161
+ 1	110	+ 6	177
+ 2	121	+ 7	195
+ 3	133	+ 8	214
+ 4	146	+ 9	236

Table 8c   £100 at simple interest

Year	Capital/£	Year	Capital/£
Now	100	+ 6	160
+ 1	110	+ 7	170
+ 2	120	+ 8	180
+ 3	130	+ 9	190
+ 4	140	+ 10	200
+ 5	150		

My capital will have doubled itself in a little over 7 years.

Had I been foolish enough to invest my £100 at 10% simple interest, the doubling time will be 10 years as Table 8c shows.

To return to the world's fuel reserves. Man has been using these at a rate which has been increasing by about 7% per year. Table 8d shows the effect this will have on the way the resource is used up.

Table 8d

Year	Use of resource	Year	Use of resource
Now	100	+ 6	150
+ 1	107	+ 7	160
+ 2	114	+ 8	172
+ 3	123	+ 9	184
+ 4	131	+ 10	197
+ 5	140		

Table 8d shows that after only ten years the rate of use has doubled. We can say that the *doubling time* is 10 years. If the world used 17×10^9 barrels of oil in this year, in ten years time it will be using 34×10^9 barrels, unless something is done to change the increasing rate of use. The graph (Fig. 8.1) shows the great effect such a rate of growth of use can have over two doubling times. Obviously, the greater the rate of use becomes, the sooner the reserves will run out. And this too is a matter of straightforward arithmetic.

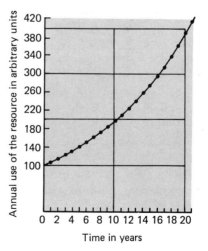

Fig. 8.1   How a 7% growth rate increases the use or consumption of a resource.

The *expiry time* (T_e) of a resource is related to the fractional growth in the rate of use k, the size of the reserve R, and the current rate of consumption r_0 by the equation

$$T_e = \frac{1}{k} \ln \left(\frac{kR}{r_0} + 1 \right)$$

Table 8e shows the result of applying this equation to various estimates. The units used are 10^9 barrels of oil and the initial rate of consumption r_0 was taken to be 16.7×10^9 barrels year^{-1}. A barrel of oil contains approximately 130 kg.

Table 8e How long will our oil reserves last?

Annual growth of use (1%)	Expiry time (T_e) in years assuming reserves R of		
	1691	1881	2451
zero	101	113	147
2	55	59	69
4	41	43	48
7	30	31	35

Perhaps the most interesting thing about this arithmetic is that, whatever reasonable assumption we make about the size of the reserve, for a given annual growth of use, the times left differ but little from one another!

8.2 What we mean by energy

One thing characterizes most of the public discussion about energy: the difficulty of saying precisely what is meant by the word. We may start by noting that, to the public, energy is concerned with fuels and the jobs that can be done with those fuels. When, however, we burn a fuel to, say, lift a load using a petrol engine, we are changing much of the energy content of the fuel into a much less useful form (heat) at the same time as we make some energy available to apply the force on the load. On other occasions we are content with the 'heat' we get directly from burning a fuel – to warm our homes, for example.

These are both 'useful jobs'. Both need fuel: petrol for the engine; natural gas, perhaps, for the central heating boiler. We may wonder whether such fuel-using jobs can always be distinguished from jobs for which no fuel is required. We can support a book by holding it in position with our hands. Fuel (food) is certainly required and tiring muscles soon warn us of this. But we can just as well support that book by placing it on a shelf. No fuel is needed then. Whenever we can replace the man, the horse, the petrol engine, the electric motor by some inanimate prop needing no fuel, we know we are not dealing with a fuel-using job.

It is convenient – and sometimes misleading – to speak of energy as though it existed in a variety of forms all interchangeable one with another. There are the elastic potential energy of a stretched spring, the chemical energy of a petrol – air mixture, the electrical energy of a generator, heat – an important common component – and many others. Then we can show how energy may be followed through a series of changes – the chemical energy of the petrol–air mixture may initiate a chain through 'mechanical energy' of the engine itself to the 'mechanical energy' involved in lifting a load against gravity and terminating in a store of 'gravitational potential energy'. And we observe how, at each stage, waste heat is generated.

As we notice this, we feel a strong urge to measure – but what can we measure? There is one simple case in which we can give an answer of great importance; and that is where a measurable force moves through a measurable distance.

8.3 One way to measure energy – Work

Consider a petrol engine which is driving a truck at a steady speed along a level road. The engine provides a force to the wheel – road system which exactly balances the equal and opposite forces of road and air friction and the truck neither accelerates nor decelerates. Suppose this force is F. If the truck moves through a distance s the engine will use a definite quantity of petrol.

If we were to move the truck through a distance $2s$ we would expect the engine to use twice as much petrol.

If the engine had to exert twice the force to overcome a frictional force which had doubled we would expect the engine to use twice as much petrol. And, certainly, two such engines moving two such trucks would use twice as much fuel as one.

So we shall assume that the fuel used is directly proportional to (i) the distance through which the engine moves, and (ii) the force which the engine exerts.

So petrol used is $\propto$ force F and is $\propto$ the distance s moved in the direction of the force.

Following a argument similar to that already used in Section 5.1 we see that

$$\text{petrol used} = kFs$$

From our point of view the petrol is of no importance except as the provider of energy. The energy content (calorific value) of the petrol used is directly proportional to the mass of the fuel used. We may write:

energy usefully transferred from the petrol $= kFs$

If we then choose units so as to make $k = 1$, the equation becomes:

energy usefully transferred from the petrol $= Fs$

The 'energy usefully transferred' in such a case is usually called *work* so that we have

$$\boxed{\text{work} = Fs} \tag{8.1}$$

If the force is 1 newton and the distance moved is 1 metre, then the work (or energy transferred) is 1 newton metre to which the name 1 joule is given.

■ *One joule* is the work done or the energy transferred when a force of 1 newton moves its point of application through a distance of 1 metre in the direction of the force.

Problem 8.1   A petrol engine is used in a crane to lift a load of bricks weighing 9000 N from the ground to the top of a building 20 m high. How much energy is transferred from the fuel? How much work is done?

Without further information about the efficiency (i.e. the ratio of the useful energy output to the energy input) of the petrol engine we are in no position to answer the first of these questions. But we can say how much work is done in lifting the bricks.

Work = force × distance in the direction of
       the force
    = 9000 N × 20 m
    = 180 000 Nm
    = 180 000 J

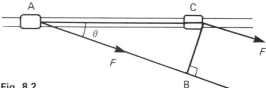

Fig. 8.2

8.4   Which distance?

Whenever distance has been mentioned in this discussion of work and energy, it has always been 'distance along the direction of the force', and there is another way of considering this. Suppose a wagon is on rails, and a man pulls it along using a rope inclined at an angle θ to the rails (see Fig. 8.2). The truck is shown to have moved a distance AC along the rails. However, in the direction of the force it has been displaced a distance AB. Consequently the work done is $F \times$ AB and this is equal to $F \times$ AC $\cos\theta$. This last expression can be rearranged as AC $\times$ $F \cos\theta$. $F \cos\theta$ is the component of F in the direction of motion and so an alternative way of defining the symbols of Eq. 8.1 is:

  W = work done;
  F = component of the force in the direction
       of the movement;
  s = distance moved.

8.5   Power

In all of the discussion so far, no mention has been made of the time needed to perform the work or to transfer the energy. In fact, whether the crane lifts the bricks through 20 m in 10 seconds or 10 hours, it still does the same 180 000 J of work. It is necessary to consider time if we are discussing the *rate* at which the work is done. In the faster case above the power is 18 000 J s^{-1}. The J s^{-1}, the unit of power, is named the watt. Both the joule and the watt are derived units.

■ The *watt* is the rate of working of 1 joule per second.

8.6   Conservation

We have already seen that momentum is conserved so long as we limit ourselves to an examination of completely isolated systems. This was

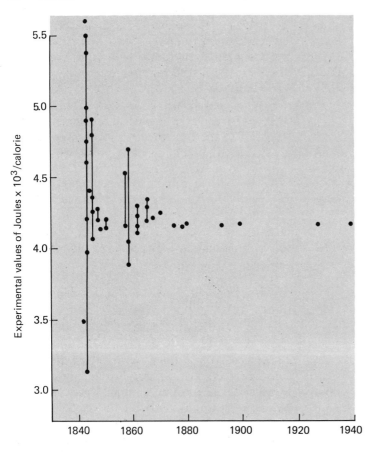

Fig. 8.3 Chart illustrating how improvements in experimental technique in the 19th and 20th centuries led to the acceptance of a fixed rate of conversion for units of energy in the calorimetric and mechanical methods of measurement. Adapted with permission from the *Revised Nuffield Physics Pupils' Text 4*, Longman, 1978.

known and understood by Newton. But the idea of energy and its conservation did not emerge until the nineteenth century with the work of Meyer, Joule, Helmholtz and many others.

As seen at that time, the problem concerned the heat which appeared in all energy changes. Was there some quantitative relationship between the heat developed and the mechanical work performed? Since the measurements of heat and of work were made in two totally different systems of units, the question could be asked in terms of the way in which the two systems were related. If 'heat' is measured in 'calories' and mechanical energy in joules, is there a constant rate of exchange? And what is it?

The technique used was to carry out a long series of conversions from one energy form to another and to make the most careful measurements to trace the fate of all the energy put into the system. The past master at this was James Prescott Joule (pronounced 'Jool' and not 'Jowl') who carried out a long series of experiments between 1839 and 1878 in which many energy conversions were examined. Other experimenters, notably Hirn in France, contributed to the pattern which was building up. This is well illustrated by the chart (Fig. 8.3) which is adapted with permission from the *Revised Nuffield Pupils' Text 4*, page 181.

The chart reveals clearly how the experimenters steadily improved their techniques and how they became convinced that there was a fixed rate of exchange between calories and joules. This implied that heat must be counted among the forms of energy and led directly to the recognition that, in an isolated system, energy too is conserved.

This is the law of conservation of energy which is arguably the most important law in the sciences. It is so firmly believed that whenever an experimental result has arisen in which it seemed that energy may have disappeared, it has led to the recognition of a new form of energy.

Even so, it is not the full story: it says nothing about the availability of the energy at the end of a sequence. That we shall discuss in a later chapter.

8.7 Defining energy

'Of all the physical concepts, that of energy is perhaps the most far-reaching. Everyone, whether a scientist or not, has an awareness of energy and what it means. Energy is what we have to pay for in order to get things done. The word itself may remain in the background, but we recognize that each gallon of gasoline, each therm of heating gas, each kilowatt-hour of electricity, each car battery, each calorie of food value, represents, in one way or another, the wherewithal for doing what we call *work*. We do not think in terms of paying for force, or acceleration, or momentum. *Energy* is the universal currency that exists in apparently countless denominations; and physical processes represent a conversion from one denomination to another. 'The above remarks do not really *define* energy. No matter. It is worth recalling the opinion expressed by the distinguished Dutch physicist H. A. Kramers: "My own pet notion is that in the world of human thought generally and in physical science particularly, the most important and most fruitful concepts are those to which it is impossible to attach a well-defined meaning." The clue to the immense value of energy as a concept lies in its transformation. It is *conserved* – that is the point. Although we may not be able to define energy in general, that does not mean that it is only a vague, qualitative idea. We have set up quantitative measures of various specific *kinds* of energy: gravitational, electrical, magnetic, elastic, kinetic, and so on. And whenever a situation has arisen in which it seemed that energy had disappeared, it has always been possible to recognize and define a new form of energy that permits us to save the conservation law.'

(A. P. French in *Newtonian Mechanics*; Norton, New York, 1971.)

Evidently the search for a definition has its difficulties. The traditional 'energy is the capacity for doing work' tells far too little. And, as we shall see, there are vast stores of energy in the form of low grade heat in the oceans of the world which are quite unavailable to do work. James Clerk Maxwell used a more graphic phrase: 'energy – the "go" of things'. In the quotation given above A. P. French says that 'energy is what we pay for in order to get things done'.

8.8 Storing energy

Once transformed from its fuel form by living beings or engines, energy can be stored in several ways, for example,

a) by changing the position of a mass in a gravitational field,
b) by changing the shape of an elastic body,
c) by changing the motion of a mass.

In considering examples to illustrate these stores of energy we will see that energy is being continually transformed from one form to another. The work which is done in these transformations is a measure of the energy so transformed. There are also other methods of storing energy that involve electric or magnetic fields, or chemical changes, but only the mechanical examples listed above are under consideration in this section.

An early use of mass as a store of energy is to be found in grandfather, or long-case, clocks. Energy from a man is used to raise a large mass inside the clock. The mass – earth system now has more energy than formerly and as the mass falls (taking about eight days to reach the bottom of the case) the chain supporting it causes the gears and the pendulum to operate the hands of the clock.

In a somewhat similar way, mass and the earth's gravitational field are used to store, indirectly, energy derived from electrical generators. The problem is that electricity must be used as it is produced; there is no way of storing large quantities of electrical energy. But there is a need to call on an extra store of energy during the peak periods of use when many factories are operating. At night, on the other hand, there is a surplus of generating capacity. At the Ffestiniog

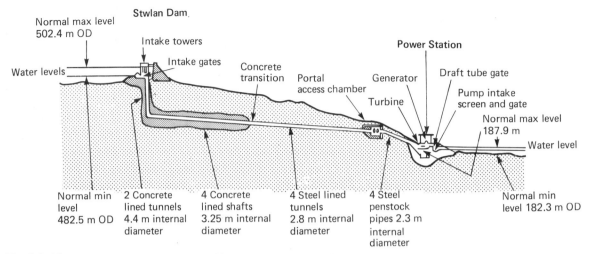

Fig. 8.4   The pumped storage system at the Ffestiniog power station. (Courtesy of the Central Electricity Generating Board.)

power station in Wales, water from one reservoir is pumped into a second reservoir 300 m higher up the mountain during the off-peak periods when the demand for electrical energy is low (Fig. 8.4). During the peak periods of heavy demand for electrical energy the stored water is allowed to flow through four tunnels about 3 m in diameter to the same power station from which it started. But this time the current of water is utilized to drive the four 90 MW generator – motor units. So electrical energy is produced at a rate of 360 MW and transferred to the national transmission line network (Grid).

In this instance the work which is done by the pumps in lifting the mass of the water against the force of gravity can, to a large extent, be recovered. We may with justice, say that energy has been stored in the system.

Spring driven clocks and watches provide further examples of energy storage. In this case the energy is usually stored in a spiral spring whose shape is changed when a force is applied to wind it up. As the spring slowly returns to its original shape, the energy stored within it is used to operate the trains of gears, the escapement and the hands. Of course, electronic watches rely on a tiny battery for their energy supply.

8.9   Potential energy

The water stored in the Ffestiniog reservoir was capable of providing energy almost equal in amount to the energy transformed in putting it there. That energy was capable of coming into action when required. But it was hidden in the system.

The watch spring too possessed energy which had been put there when it was wound up. It could be utilized; and it too was hidden in the system.

These are examples of *potential energy*. In the first case the water in the reservoir possessed potential it did not have at the lower level because of its new position above the pumping station. The force overcome in lifting the water was gravitational; the energy was stored as *gravitational potential energy*.

In the second case, the spring was deformed as the elastic forces were overcome when it was wound up; the energy stored might be called *elastic potential energy*.

The battery used in the electronic watch had a store of *chemical* (potential) *energy*.

8.10   Calculation of energy transfer (work done) when a mass is lifted in a gravitational field

Consider a mass m which is stationary in a gravitational force field such as we find on earth. Let the magnitude of the force on each unit of mass be g. Then the total force acting on the mass is mg.

If now that mass is lifted through a height h in the direction of the lifting force, the energy transfer is the product of the force applied and the distance moved in the direction of the force.

$$\boxed{\text{energy transfer } = mgh} \qquad (8.2)$$

This is the gravitational potential energy transferred to the earth – mass system.

Problem 8.2 The earth's gravitational field exerts a force of nearly $10 \, \text{N kg}^{-1}$ at the surface. How much potential energy is stored in the Ffestiniog pumped storage system when the upper reservoir contains $2 \times 10^9 \, \text{kg}$ of available water? The head of water is 300 m. How long would this operate the four generators at their full output of 90 MW each assuming that the operation is 75% efficient?

The potential energy of the stored water

$$
\begin{aligned}
&= mgh \\
&= 2 \times 10^9 \times 10 \times 300 \, \text{J} \\
&= 6 \times 10^{12} \, \text{J}
\end{aligned}
$$

Rate of working (power)

$$= \frac{\text{total energy output}}{\text{time}}$$

$$\therefore \text{ time } = \frac{\text{total energy output}}{\text{power}}$$

$$= \frac{0.75 \times 6 \times 10^{12} \, \text{W}}{4 \times 90 \times 10^6 \, \text{J}}$$

$$= 1.3 \times 10^4 \, \text{s}$$

(about $3\frac{1}{2}$ hours)

8.11 Graphical treatment

A useful technique for calculating transfers of energy results from plotting a graph of force involved against the distance moved. If the force is constant the graph is a line parallel to the distance axis.

The energy transferred can be calculated by finding the area under the line of the force–distance graph since one side of the rectangle is the force, mg, and the other side is the distance, h.

If the mass is raised a further distance (Δh), that is from a height h to ($h + \Delta h$), then the extra energy transferred is $mg(\Delta h)$. This is shown on the

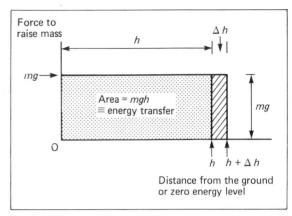

Fig. 8.5

graph (Fig. 8.5) as the area of a narrow strip. This graphical technique is used in the example which follows.

8.12 The calculation of the energy transferred when the shape of an elastic body is changed

The difference between this and the previous case is that the force required to do the work is not constant since it depends upon how much the shape is changed. The simplest case to consider is that of a helical spring for which the force is directly proportional to the extension of the spring beyond its unstretched length (Hooke's law). Figure 8.6 shows such a spring, hooked on to a nail, being stretched along a strip of wood.

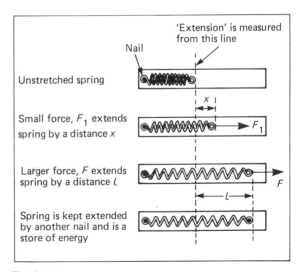

Fig. 8.6

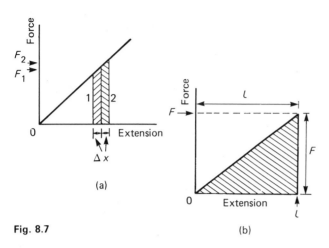

Fig. 8.7

(a)

(b)

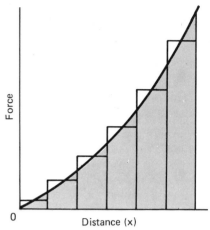

Fig. 8.9

If Hooke's law is obeyed, a graph of force against extension will be as shown in Fig. 8.7a and 8.7b.

The energy transferred to the spring is not Fl as the force is clearly not constant throughout the distance l. To calculate the energy transferred, and stored in the spring, a large number of small steps must be taken; during each one, the change of extension, Δx, is so small that it has negligible effect on the force acting which may then be considered constant. If the force is F_1 while the spring changes length by Δx, then the energy transferred is $F_1\Delta x$, which is the area of strip 1 in

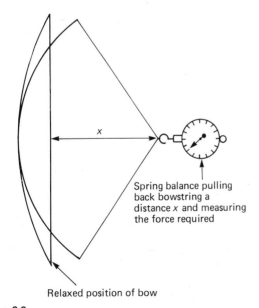

Spring balance pulling
back bowstring a
distance x and measuring
the force required

Relaxed position of bow

Fig. 8.8

Fig. 8.7a. If a further extension of Δx is made during which the force is larger but significantly constant at F_2, then the extra energy transferred is $F_2\Delta x$, which is the area of strip 2. The total energy transferred when the extension changes from zero to l is given by the total of the areas of all the strips involved, which is the same as the area under the line shown shaded in Fig. 8.7b. This area is half that of the rectangle having sides of length F and l. Thus energy stored $= \frac{1}{2}Fl$.

The area method can be extended to the calculation of energy transferred when the force involved does not follow a simple rule like Hooke's law. An example of this occurs with the archer's bow. If the experiment indicated in Fig. 8.8 is performed, a graph can be plotted (Fig. 8.9) of the force required to pull the bowstring back a distance x. This is no longer a linear graph but the energy transferred to the bow can still be determined by dividing the area under the curve into strips and by finding the total area as shown in Fig. 8.9.

8.13   The calculation of the energy transferred when the speed of a mass is changed

When a constant force acts on a mass which is free to move, the mass accelerates. The force is transforming energy into a new form – the motion or *kinetic energy* of the moving mass. The gain of kinetic energy comes from the work performed by the force *while it is acting*. (Fig. 8.10.)

Let the mass be m, the initial speed be v_0, the

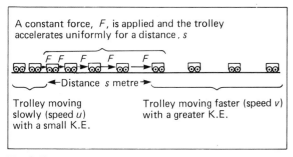

Fig. 8.10

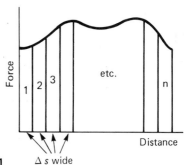

Fig. 8.11 Δs wide

final speed v_t, the acceleration a and the time for which the force acts t.

Gain of energy

$= $ (force) $\times$ (distance in the direction of the force)

$= Fs$

$= mas$ (because $F = ma$)

From Eq. 5.10

$$v_t^2 - v_0^2 = 2as$$

Therefore, the gain of energy

$$= \tfrac{1}{2}m(v_t^2 - v_0^2)$$

As this represents a gain or change of energy, $\tfrac{1}{2}mv_t^2$ is the final kinetic energy, and $\tfrac{1}{2}mv_0^2$ is the initial kinetic energy.

Note: We often say that the kinetic energy E_k of a mass, m, travelling with speed, v, is $\tfrac{1}{2}mv^2$. This is because when the mass is brought to rest from a speed v, it transforms $\tfrac{1}{2}mv^2$ joules of energy from kinetic to other forms.

$$\boxed{E_k = \tfrac{1}{2}mv^2} \qquad (8.3)$$

8.14   The general case

A mass increases its kinetic energy when the force causing this to happen is not constant, as, for example, when a bow transfers its energy to an arrow.

Consider the trolley of the previous case again and let the force that acts vary with distance as the graph, Fig. 8.11, shows. Although the force is not constant, we can take a number of steps chosen to

be so small that the force may be assumed to be constant during each step. Each of the steps or strips shown is Δs wide and for strip (1) the force is F_1, while for strip (2) the force is F_2, etc. While F_1 acts, the speed of the trolley changes from v_0 to v_1, and while F_2 acts, the speed of the trolley changes from v_1 to v_2, etc. While the last strip's force, F_n, acts the speed changes to the final or maximum speed, v_n, from the previous speed, v_{n-1}. Equation 8.3 can now be used to calculate the energy gained during each strip, and by addition, the total energy gained may be found, as shown in Table 8f. If you inspect the last column you will see how, in addition, the terms cancel in pairs.

Table 8f

For strip number:	The work done by the force =	The energy transferred to the mass =
(1)	$F_1\Delta s$	$\tfrac{1}{2}mv_1^2 - \tfrac{1}{2}mv_0^2$
(2)	$F_2\Delta s$	$\tfrac{1}{2}mv_2^2 - \tfrac{1}{2}mv_1^2$
(3)	$F_3\Delta s$	$\tfrac{1}{2}mv_3^2 - \tfrac{1}{2}mv_2^2$
etc.	etc.	etc.
$(n-1)$	$F_{n-1}\Delta s$	$\tfrac{1}{2}mv_{n-1}^2 - \tfrac{1}{2}mv_{n-2}^2$
(n)	$F_n\Delta s$	$\tfrac{1}{2}mv_n^2 - \tfrac{1}{2}mv_{n-1}^2$

The total energy gained by the mass

$= $ total of the final column

$= \tfrac{1}{2}mv_n^2 - \tfrac{1}{2}mv_0^2$

Consequently, whether a mass gains its kinetic energy by means of a constant force, or a varying one, the initial and final speeds and the mass moving are the factors that determine the change in energy.

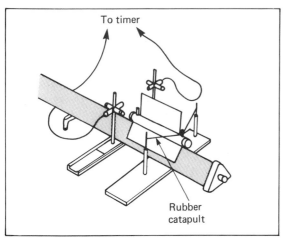

Fig. 8.12

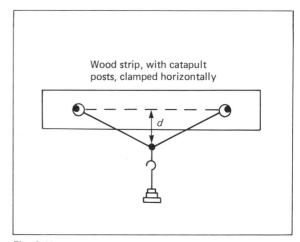

Fig. 8.13

8.15 The quantitative examination of some energy transfers

a) The energy stored in rubber, used as a catapult, can be used to illustrate the equation K.E. $= \frac{1}{2}mv^2$. The arrangement shown in Fig. 8.12 is a linear air track with the usual arrangement of photocell and timer to measure the speed of the vehicle once it has been catapulted along the track. The vehicle carries a card 10 cm long. Trolley elastic threads, stretched between two vertical posts, act as the catapult. Five elastic threads must be selected so that each one, used separately at the same stretch, will give the vehicle the same speed. By using one, or a number of elastic threads, the vehicle can be given 1, 2, 3, 4, or 5, 'unit doses' of energy. Typical results are shown in Table 8g. Inspection of these results, or a graph of energy against v^2 which is a straight line through the origin, show that energy transferred to kinetic energy is proportional to v^2.

Similar experiments performed with a half-length vehicle, and combinations of vehicles, show that kinetic energy is also proportional to the mass in motion.

Table 8h

Force displacing middle of elastic thread/N	Displacement, d/m
0.1	0.01
0.2	0.017
0.3	0.023
0.4	0.029
0.5	0.034
0.6	0.038
0.7	0.042

It is instructive to measure the actual energy stored in the catapult and to see if it is all converted to kinetic energy in the vehicle. To do this a force–distance graph for an elastic thread must be plotted, and to obtain data for this, small masses are hung from its centre, as shown in Fig. 8.13, and for each one, the displacement, d, of the centre is measured. A set of results obtained with one of the elastic threads used above is given in Table 8h.

Problem 8.3 Plot a graph of these or your own results and find the energy stored when $d = 0.033$ m, using the area method described earlier.

This energy should then be compared with the

Table 8g

Number of elastic threads (energy)	Transit time/s	Speed v/m s^{-1}	v^2
1	0.189	0.53	0.28
2	0.133	0.75	0.56
3	0.108	0.93	0.86
4	0.095	1.05	1.10
5	0.085	1.18	1.40

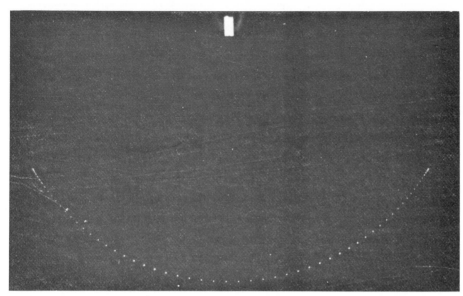

Fig. 8.14 Multiflash photograph of a swinging pendulum bob. The flash rate was 100 Hz. The pendulum was 0.325 m long. The photograph is printed at one fifth full-scale. (Photograph B. T.)

kinetic energy of the vehicle. In this case the vehicle had a mass of 0.048 kg and speed 0.53 m s^{-1} when one elastic thread was used with a displacement of 0.033 m.

b) We can also examine the energy transfers that occur when a pendulum swings. The strobe photograph, Fig. 8.14, is of a pendulum that swings from rest, at A, across to a rest position at C. As can be seen, it has a maximum speed at the lowest point B. When it is at A or C, the mass has more potential energy than when it is at B. In going from A to B, there is a transfer of energy from the potential form to the kinetic form.

Problem 8.4 Take measurements off the photograph (Fig. 8.14), which is printed one fifth full scale to determine the distance h 'fallen' and the maximum speed, and then calculate the energy transferred from potential to kinetic and vice versa. The mass of the pendulum bob was 0.07 kg; the pendulum was 0.325 m long and the flashing rate of the strobe was 100 flashes per second. How do the two energy transfers compare?

Energy is continually transferred from one form to the other without loss apart from that caused by air resistance. That will eventually claim *all* the original potential energy and the pendulum will come to rest at B.

The pendulum bob is a shiny steel ball and the bright points of light recorded in the photograph are the reflections of the xenon strobe lamp in the spherical surface.

c) Again, we can examine the energy transfers occurring when a brick falls. If a brick is allowed to fall through a known distance, its speed can be measured by allowing it to interrupt the light beam falling on to the photocell controlling a timer. The energy transferred from the potential form (mgh) to the kinetic energy form ($\frac{1}{2}mv^2$) can be compared.

In each of these cases, and within the limits of experimental error, the prediction of Eq. 8.3 is confirmed. That equation was a direct consequence of the application of Newton's second law to the case of the application of a constant force to a constant mass. We have interpreted it in terms of the interchange of potential and kinetic energies during the experiments. It is tempting to speculate what happens to the kinetic energy of the falling brick when it reaches the floor? Certainly the kinetic energy disappears. The law of conservation of energy tells us that that energy has been transferred to other forms – a little to sound and most to warning the surroundings.

8.16   Energy 'book-keeping' during collisions

a) Inelastic collisions

In these, there is no relative velocity after the interaction as the colliding masses stick together. Experiment E (Section 3.3) offers an example of this. Another is the case of a lamp of putty that falls on to a table and does not bounce.

Although the total energy is conserved, the kinetic energy obviously is not. What happens to it?

b) Partly elastic collisions

Such collisions are much more common than totally inelastic ones. Experiment D examined such a case. A rubber ball bouncing on the ground offers another good example. The relative velocity between the ball and the ground is reversed in direction and reduced in magnitude during the interaction since the ball does not bounce back to its original height. Kinetic energy is not conserved during the collisions.

c) Elastic collisions

Perfectly elastic collisions, that is, collisions in which the relative velocity after the interaction is reversed in direction and is *equal* in magnitude to the initial relative velocity, are characteristic of the collisions between molecules in a monatomic gas. The kinetic energies are conserved.

In polyatomic gases, the transfer of energy during collisions between the molecules is much more complex but, *on average*, the collisions may be thought of as elastic.

d) Comparing energy and momentum conservation

The laws of conservation of energy and of momentum are universal laws applying to all cases of interaction. In collisions, it is tempting to wonder whether *kinetic energy* is conserved. In most cases it is not. In a very few cases it is; such collisions are the elastic ones.

8.17   Conservative and dissipative forces

A *conservative force* is one that permits the storage of potential energy whereas forces such as those resulting from friction, viscosity, etc., are known as *dissipative forces*.

Conservative forces can be provided by springs and also by such fields as gravitational, electric and magnetic. In these cases, a kinetic energy can be converted to a stored potential energy, e.g., a ball thrown upwards has a kinetic energy that decreases as the potential energy rises. The stored potential energy can be retrieved and converted back directly to its original form – as the ball falls back to the ground.

Work done against a dissipative force, such as friction, causes a temperature rise; this energy is not destroyed or lost for ever. However, it is not directly recoverable as is the energy stored in a spring. To retrieve the energy converted to heat by a dissipative force, it would be necessary to employ some form of engine that changes heat to mechanical energy.

The brakes on a car dissipate unwanted kinetic energy when it is brought to rest from speed. Brakes provide a frictional force against which work is done resulting in a temperature rise. Some kinetic energy is also acquired by the earth itself but this is an infinitesimal part of the whole. Every time a car is stopped from a speed of $30 \, \text{m s}^{-1}$ enough heat is produced in the brakes, often in a few seconds, to boil a kilogram of water. Unfortunately, this energy cannot be used to accelerate the car back to the original speed. The price of stopping the car is paid for in the petrol used when it accelerates again.

In inelastic collisions, the forces are dissipative; in elastic ones, the forces are conservative. In partly elastic collisions, the forces which develop during the interactions are partly dissipative, partly conservative.

8.18   Collisions in two dimensions

Hitherto we have limited our analysis of the fate of the momenta and the energies in collisions to cases in which the motions are linear and the collisions direct. But anyone who has watched a game of snooker or billiards will know that oblique collisions are of great interest too. In those games,

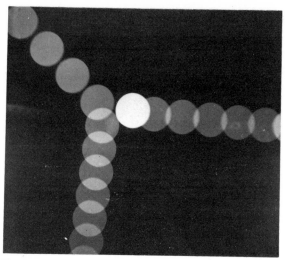

Fig. 8.15 Multiflash photograph of a dry-ice puck colliding elastically with a similar but stationary puck. (Photograph E. J. W.)

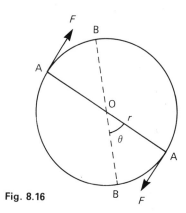

Fig. 8.16

the collisions may be complicated by the spin given to the incident ball. To simplify the situation we shall consider oblique collisions between a moving 'puck' and a stationary one with the same mass. Figure 8.15 shows a 'strobe' photograph of such a collision. The two pucks were floating virtually without friction on dry ice; since they were also powerful magnets there was no touching 'contact' and the collision was elastic. The photograph shows that, after the interaction, the two pucks moved apart on tracks which were at right angles to each other.

The conservation of momentum requires that, vectorially

$$mu = mv_1 + mv_2$$

For the components of the momenta in the x direction

$$mu = mv_1 \cos \alpha + mv_2 \cos \beta$$

For the components in the y direction

$$0 = mv_1 \sin \alpha + mv_2 \sin \beta$$

For an elastic collision between equal masses the conservation of energy requires that

$$\tfrac{1}{2} mv^2 = \tfrac{1}{2} mv_1^2 + \tfrac{1}{2} mv_2^2$$

whence

$$u^2 = v_1^2 + v_2^2$$

By remembering Pythagoras theorem, we can tell straight away from this last equation that the angle between v_1 and v_2 must be a right angle.

Problem 8.5   Given that an incident puck had an initial speed of 240 cm s^{-1}, a final speed of 180 cm s^{-1}; that the other puck acquired a speed of 166 cm s^{-1} as a result of the collision and that the angles α and β were 43° and 47° respectively, show that the conservation of momentum applied to this case.

8.19   Energy and rotation

We have seen that a motor which is driving a shaft develops a torque (Section 6.8). Such a torque can, of course, transform energy from one form to another; it can do work.

Consider first a wheel to which a single couple is applied. This couple is made up of two equal but opposite forces F applied at opposite ends of a diameter. As a result the wheel turns through an angle θ. (See Fig. 8.16.)

The work done by each of the forces
$$= F \times \text{arc AB}$$
$$= F \times r\theta$$
The total work done $= 2Fr\theta$

Now the moment of the couple (or torque) is $F \times 2r$.

So the work done by the couple
= torque × angle of rotation.

Extending this to the general case of a rotating shaft which produces a torque T as it turns at a rate of n revolutions per second, or $2\pi n$ radians per second, we have

$$\text{power} = \text{work done per second}$$
$$= T \times 2\pi n \qquad (8.4)$$

Problem 8.6 An electric drill motor, rated at 350 W, has a maximum efficiency of 35%. What torque will it produce if it is running at 3000 revolutions per minute?

The power output is 35% of the power input, i.e. $0.35 \times 350\,\text{W}$

From Eq. 8.4 the power output $= T \times 2\pi N$
$n = 50$ revolutions per second so power output is $T \times 2\pi \times 50$

$$\text{Torque} = \frac{0.35 \times 350}{2\pi \times 50} = 0.39\,\text{N m}$$

If the motor is fitted with a mechanical gear box which drives the drill at 15 revolutions per second, what is the torque?

The motor is running at the same speed (50 revolutions per second) and the gear box provides a mechanical advantage of 50/15. So the new torque will be

$$0.39 \times \frac{50}{15}\,\text{N m} = 1.3\,\text{N m}$$

To measure the power output of a machine it is necessary to measure the torque provided. The simplest device for this is a band-brake. (See Fig.

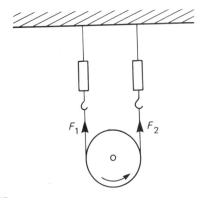

Fig. 8.17

8.17.) Two dynamometers (spring balances) hold a belt tightly against a pulley which is firmly attached to the shaft of the machine. When the shaft is stationary, the dynamometers will register equal forces which are determined by the tension in the belt. But when the shaft rotates the force recorded by one dynamometer will increase whilst that recorded by the other will decrease. (Which is which?) If, when the speed of rotation is stable, the readings are F_1 and F_2 and the pulley radius is r, the net applied torque is $F_1 r - F_2 r$.

We can measure n by counting (at low speeds) or by the use of a flashing stroboscope at high speeds.

8.20 The energy stored in a rotating body

In Section 7.5 we saw that the inertia of rotating bodies depended on the way in which the mass of the body was distributed about the axis of rotation.

Such a body also possesses energy. Consider a flywheel which is rotating about an axis through its centre of mass. Although the wheel is rotating, the centre of mass does not have a speed. In what sense, then, can the equation kinetic energy = $\frac{1}{2}mv^2$ be applied?

The flywheel must be considered, not as a whole, but as being composed of a large number of masses, m_1, m_2, m_3, etc., distributed at various distances, r_1, r_2, r_3, etc., from the axis (Fig. 8.18). Each mass has a speed, v_1, v_2, v_3, etc., which depends upon its distance from the axis. The kinetic energy of each mass can therefore be determined. The total of the energies stored in all the 'elemental masses' that make up the flywheel, gives the energy stored by the flywheel itself.

Let the flywheel rotate at n rev s^{-1}. Since there are 2π radians in 360° or one revolution, then the angular velocity of the flywheel, $\omega = 2\pi n$ radians s^{-1}. Because in general, speed, $v = r\omega$, the speeds of the masses composing the flywheel $= r_1\omega$, $r_2\omega$, $r_3\omega$, etc. The total kinetic energy of the flywheel

$$= \tfrac{1}{2}m_1 v_1^2 + \tfrac{1}{2}m_2 v_2^2 + \tfrac{1}{2}m_3 v_3^2 + \text{etc.}$$
$$= \tfrac{1}{2}m_1 r_1^2 \omega^2 + \tfrac{1}{2}m_2 r_2^2 \omega^2 + \tfrac{1}{2}m_3 r_3^2 \omega^2 + \text{etc.}$$
$$= \Sigma \tfrac{1}{2}mr^2\omega^2 = \tfrac{1}{2}\omega^2 \Sigma\, mr^2.$$

The expression $\Sigma\, mr^2$ is shorthand for the sum of all the terms like mr^2 and is called *moment of*

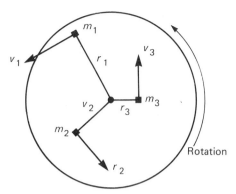

Fig. 8.18

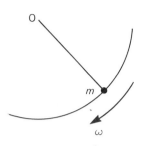

Fig. 8.19

inertia, for which the symbol is I. Thus the kinetic energy of a rotating body is $\frac{1}{2} I\omega^2$.

Since $\Sigma\, mr^2$ depends not only upon mass, but also on its distribution, it follows that a flywheel that has most of its mass concentrated in the rim will possess more kinetic energy than a flat disc having the same mass and radius, and rotating at the same speed. In the case of a flat disc of mass M and radius r, it may be shown that the moment of inertia about an axis through the centre and perpendicular to the plane is given by: $I = \frac{1}{2} Mr^2$. Therefore, the kinetic energy stored in such a disc that is rotating at n rev s^{-1} $= \frac{1}{2} \times \frac{1}{2} Mr^2(2\pi n)^2 = Mr^2\pi^2 n^2$.

The moment of inertia of a rotating body plays the same role in rotational motion as mass plays in linear motion. Where in linear motion we have mass, distance, velocity and force, so in rotational motion we have moment of inertia, angle, angular velocity and torque (the product of force and the arm of the force). This correspondence goes further, however. The rotational counterpart of momentum is *angular momentum*.

■ The angular momentum of a point mass m moving with speed v about an axis through 0 perpendicular to the plane of the motion is the product of the linear momentum and the perpendicular distance from the axis to the line of motion (Fig. 8.19).

$$\text{Angular momentum} = L = (mv)r$$

$$\text{And since } v = \omega r$$

$$\text{angular momentum } L = m\omega r^2$$

$$= (mr^2)\omega$$

Extending the argument to a rigid body as before we see that the total angular momentum of that body is $\Sigma\,(mr^2)\omega$ or $I\omega$.

In rotational motion, angular momentum plays the same part that linear momentum plays in linear motion. For example, Newton's second law takes the form: torque $= \Delta L/\Delta t$. And, like linear momentum, angular momentum is conserved.

Rotation (both revolution as in the case of the moon rotating about the earth and spin as in the case of the earth spinning on its axis) is of fundamental importance in the fields of astro-physics and in both atomic and nuclear physics. For example, within the solar system the sun possesses 99.87% of the mass but only 0.54% of the angular momentum (which means that the planets, with only 0.13% of the mass, have 99.46% of the angular momentum). With the major exception of Venus, the sun and its planets spin on their axes in the same direction; and all the planets rotate about the sun in the same direction with other orbits lying roughly in a single plane. Any theory of the formation of such a system must account for these fundamental facts, of which the extraordinary distribution of the angular momentum is the most important.

To come a little nearer home, try standing on a freely rotating table holding a heavy book in each hand. Get someone to start you spinning. What happens to your angular velocity as you raise and lower your arms? What is happening to your moment of inertia as you do this?

Problem 8.7 A solid cylinder of mass M and radius R rolls from rest down an inclined plane

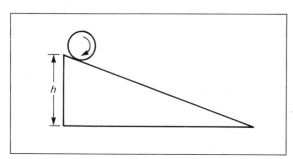

Fig. 8.20

without slipping (Fig. 8.20). Find its speed (linear) when it reaches the bottom.

As it rolls down the plane, potential energy Mgh is converted to kinetic energy both linear and rotational.

$$Mgh = \tfrac{1}{2}I\omega^2 + \tfrac{1}{2}Mv^2$$

where h is the height of the plane, v is the linear speed at the bottom and ω is the angular speed at the bottom.

Now

$$I = \tfrac{1}{2}Mr^2$$

and

$$\omega = \frac{v}{r}$$

So

$$Mgh = \tfrac{1}{2}(\tfrac{1}{2}Mr^2)\frac{v^2}{r^2} + \tfrac{1}{2}Mv^2$$

$$= \tfrac{3}{4}Mv^2$$

and

$$v = \sqrt{\tfrac{4}{3}gh}$$

What would the linear velocity be if this cylinder had slid, without rolling down the slope?

The velocity is given by

$$Mgh = \tfrac{1}{2}Mv^2$$

so that

$$v = \sqrt{2gh}$$

The speed of the sliding cylinder is, therefore, greater than that of the rolling cylinder for the latter has its total energy distributed between rotational and linear kinetic energy.

Chapter 9

HEATING AND WORKING

9.1 Temperature

Matter, including of course gases, has a number of properties which, like pressure, are temperature dependent. Among these are the length of a metal bar (for we know that metals expand on heating), the pressure of a fixed volume of a gas, the volume of a mass of gas whose pressure is kept constant, the resistance of a metallic conductor, the electromotive force generated at a junction of two metals, the colour of a hot furnace. Devices can be constructed using any of these properties and used to indicate the thermal state of a body.

Imagine that such a device is placed in each of two metal cans containing water. One of these cans, A, has been in good thermal contact with the flame of a bunsen burner: the other, B, has been standing on the bench. Now they are brought together and placed in good thermal contact with one another. Experience tells us that, before long, the two similar devices will record that an equilibrium state exists. If, however, the two cans have been separated by an insulating wall, the readings of the devices will continue to vary widely from each other.

Now imagine that we have three such cans, A and B as before but the third, C, has been kept in a refrigerator for some time. The three are placed in an insulated chamber with an insulating wall separating A and C, and a conducting wall between A and B, B and C (Fig. 9.1).

Both A and C are in good thermal contact with B but not with each other. Given the lapse of sufficient time, A, B, and C will come into thermal equilibrium. This state of affairs would not be upset if we were to withdraw the insulating wall from between A and C. This experimental result is often described as the *zeroth law of*

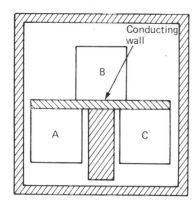

Fig. 9.1

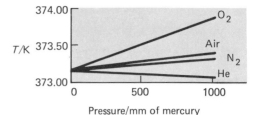

Fig. 9.2 Readings of a constant volume gas thermometer at the temperature of condensing steam, when different gases are used at various pressures. (F. W. Sears and M. W. Zemansky, *University Physics*, Addison-Wesley, 1970.)

thermodynamics and can be stated formally as 'Two systems which are in thermal equilibrium with a third system must be in thermal equilibrium with each other.' Under these conditions we say that all three systems are at the *same temperature*. This gives a very clear meaning to the word *temperature*: it is the property of a system which determines whether or not the system is in thermal equilibrium with other systems.

Problem 9.1 Does a law analogous to the zeroth law of thermodynamics, but not concerned with thermal equilibrium, apply to two pieces of iron and a magnet, and to three men, Smith, Jones, and Robinson?

a) Piece of iron A attracts – or is attracted by – the magnet. So is piece B. Does A attract B? It might or it might not. There is no analogous law.

b) Smith knows Jones. Jones knows Robinson. But does Smith know Robinson? He may do; but then again, he may not. There is no analogous law.

Properties of matter which are temperature dependent are all capable of use in thermometers to measure temperature. But to devise an effective method of providing a temperature scale some fixed reference points must be agreed.

One such temperature-dependent property is the length of a mercury column contained in a glass envelope. This provides the basis of the *'mercury-in-glass' temperature scale*. If we choose as the fixed reference points the 'ice-point' (which we shall call zero degrees Celsius or 0°C) and the 'steam-point' (which we shall call 100°C) of water under strictly specified conditions, we may define

a temperature on this 'mercury-in-glass' scale by the equation

$$\theta = \frac{l_\theta - l_0}{l_{100} - l_0} \times 100$$

where l_θ, l_{100} and l_0 are the lengths of the mercury column at the temperatures θ, 100°C and 0°C.

Similar equations apply to other thermometers using other thermometric properties. For a resistance thermometer a temperature θ will be given by

$$\theta = \frac{R_\theta - R_0}{R_{100} - R_0} \times 100$$

For a constant volume gas thermometer the equation will be

$$\theta = \frac{p_\theta - p_0}{p_{100} - p_0} \times 100$$

and so on.

We are not surprised if two such thermometers using totally different physical properties give different readings when put together into an enclosure at an unknown temperature; unless, of course, that temperature is that of either the 'ice-point' or the 'steam-point'.

What is surprising is that there are thermometers which do approach the same readings. They are all *constant volume gas thermometers* operating at very low pressures. Whether the gas used be oxygen, hydrogen, helium or air, at very low pressures these instruments agree as to temperature (see Fig. 9.2). This agreement sets a standard: that of a constant volume gas thermometer extrapolated to zero pressure. In practice, the best material to use is helium with the advantage that it can be used as low as $-272°C$.

The consistency of behaviour of these low-pressure gas thermometers made it possible to

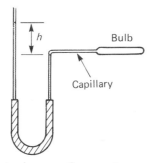

Fig. 9.3   A constant volume gas thermometer.

simplify the definition of a temperature scale.

With any constant-volume gas thermometer working at very low pressures the ratio p_{100}/p_0 always has the same numerical value: 1.366 (to 3 s.f.). The Celsius temperature can therefore be expressed by

$$\theta = \frac{p_\theta - p_0}{1.366\, p_0 - p_0} \times 100$$

$$= \frac{p_\theta - p_0}{0.366\, p_0} \times 100$$

$$= 273\, \frac{p_\theta}{p_0} - 273$$

$$\therefore\ \theta + 273 = 273\, \frac{p_\theta}{p_0} \qquad \dots \text{in }°C$$

This suggested that there would be advantages in defining a new temperature scale with zero at about -273 degrees Celsius and with the *same size of degree interval* as the original Celsius scale. On such a temperature scale the temperature would simply be directly proportional to the pressure in a constant-volume gas thermometer. A further advantage would be that the 'steam-point' would no longer be needed for calibration. It would be necessary only to measure accurately the pressure p_0 at the 'ice-point'. This is the essence of Lord Kelvin's proposal for an *absolute scale of temperature*. For this we have the relationship (to 3 s.f. accuracy):

temperature on the   =   temperature on the   + 273
absolute scale   Celsius scale

In practice, the 'ice-point' is not easily reproducible to the precision required for accurate work. This led to a decision to change that fixed point to the readily reproducible *triple point of water* whilst retaining the *absolute zero of temperature* as the other.

Water can exist as a solid, liquid and gas together in the same vessel at one temperature and one pressure alone. The pressure is 4.58 mm of mercury; the temperature can be assigned arbitrarily. By international agreement, the arbitrary number 273.16 K is assigned to the temperature of the triple point, the K indicating a temperature in *kelvin*.

Having made this arbitrary decision, temperature can be defined in kelvin on the constant-volume gas thermometer scale by the equation

$$T = 273.16\, \frac{p_T}{p_{tr}}$$

where T is the temperature in kelvin, p_T is the pressure of the gas at temperature T and p_{tr} is the pressure of the gas at the triple point.

A constant-volume gas thermometer consists essentially of a bulb containing the chosen gas, which is connected through a capillary tube to a mercury column (Fig. 9.3). The height of the mercury column can be adjusted by raising or lowering a mercury reservoir to adjust the pressure in the bulb. The difference in height of the two mercury columns is measured when the gas is contained in its fixed volume at the temperature of the triple point of water and at the temperature which is to be measured.

When very low values of gas pressure are used, this measure of temperature depends but little on the properties of any particular substance, but on the properties of gases in general. It is called the *ideal gas temperature*.

9.2   Heating and working

As we saw in Section 8.3 *work* has acquired a specific meaning as a measure of energy transformed from one form to another by some mechanical process. For example, the application of an upward force of 50 N to a mass which is moved through a vertical distance of 2 m involves the transfer of 50×2 joules of energy.

Work = force × distance moved in the direction of the force.

In this example, 100 J have been transformed from the fuel which drove the force-producing mechanism to gravitational potential energy, provided that the speed developed was low and the kinetic energy negligible.

The mass which has been lifted through these 2 m may, at some time in the future, return the potential energy stored by falling back to its starting point. In so doing it might drive a paddle wheel contained in a carefully insulated vessel of water as in the experiments of J. P. Joule which have already been referred to in Section 8.6. That long series of experiments revealed that the temperature of the water would rise as a result of this process of energy transformation; and, moreover, that the expenditure of the same amount of work W always produced the same temperature rise in the same mass of water.

The implications of this discovery are far-reaching. Consider for a moment a simple 'thought experiment'. You have two similar insulated systems each containing a vessel of water. Under one of these vessels is a flame; the other contains the paddle wheel and falling mass mechanism. Each also contains a thermometer and, whilst you can see the reading of that thermometer, you cannot see inside the insulated containers which surround the two systems.

Each system is then set in operation. You observe that the temperature of the water rises in each vessel. You have no method of differentiating between the two systems. You know that in one case energy is being transferred to the water from the falling mass and that you can measure this as work. One consequence is that the temperature of the water rises. If you accept that energy is conserved, you must assume that the water is increasing some internal store of energy as this happens and that the temperature rise indicates this. There is no reason why you should believe anything different is happening to the other system. But, commonly, it is said that the flame is transferring 'heat' to that system. Since we take work to be in Maxwell's phrase, 'the transference of energy from one system to another', it would seem sensible to take 'heat', too, as energy in transit. There must be some mechanism which enables the water to store this energy internally just as there must be some mechanism which permits the storage of what is called gravitational potential energy in the lifted mass. That mechanism will be considered later.

The classic series of experiments performed by Joule in the nineteenth century was not limited to the simple case of energy transformations which

occur when gravitational potential energy is transferred to water through a paddle wheel mechanism. Joule also arranged for the falling masses to drive a dynamo which was short-circuited and itself immersed in the water; on another occasion the output current of the dynamo was allowed to pass through a resistor which was immersed in the water. Joule forced water to flow through fine tubes and observed the temperature change which resulted from this fluid friction; he compressed air in a cylinder which was immersed in water and observed the temperature change; he replaced the water by other fluids; he rubbed metal plates together. Always he found that, in a given system, the temperature rise was the same whatever form of energy transformation was responsible. These experiments were extended by other physicists but always with the same result. In consequence, scientists accept that energy is conserved in all such changes. The law of conservation of energy rests firmly on a mass of experience. If we can find some model for matter which can accommodate the internal energy store which we have associated with the temperature rise, then we shall believe even more strongly in the validity of this law.

These experiments revealed that the transfer of a quantity of energy ΔE to a system – whether by mechanical, electrical, or other means – changed the internal energy of that system by an equivalent amount ΔU,

$$\Delta E = \Delta U$$

Since ΔE is normally measured directly or indirectly as a certain quantity of work, ΔW, it follows that

$$\Delta U = \Delta W$$

A process such as this which involves only the performance of work and consequent changes in the internal energy and temperature, is known as an *adiabatic* process. In an adiabatic process, there is no flow of heat into or out of the system.

Now consider the case of the water which was receiving its additional internal energy from the hot flame of the burner. This was an example of a flow of 'heat' without the involvement of work. The only indication we have of change is that the temperature of this water rose. If the rise in temperature in each of the two similar systems in our thought experiment was the same then the

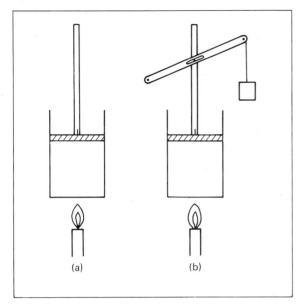

Fig. 9.4

internal energy of the two masses of water increased by the same quantity of energy. We may say that an amount of energy ΔQ entered the system during the heating and that

$$\Delta Q = \Delta U$$

These two processes are very simple ones and are rarely met with in practice. A much more usual situation involves a working substances such as a gas which is contained within a cylinder and enclosed by a piston so that it can expand and contract. Consider such a system and imagine that it can be heated (see Fig. 9.4a).

As energy ΔQ enters the gas, the internal energy rises, the temperature increases and the pressure also increases. The piston therefore moves outwards to allow the gas to expand. In so doing it exerts a force and therefore does work externally. It may, for example, store up some gravitational potential energy (Fig. 9.4b). We can identify three amounts of energy:

a) The energy entering the system from the heat source $= \Delta Q$.
b) The energy leaving the system as work and transferred to gravitational potential energy $= \Delta W$.
c) The energy remaining in the system as additional internal energy $= \Delta U$.

Applying the principle of conservation of energy, we see that

$$\Delta Q - \Delta W = \Delta U$$

Or

$$\boxed{\Delta Q = \Delta U + \Delta W} \qquad (9.1)$$

This statement which applies to all large scale (macro-) systems is known as the *first law of thermodynamics*.

9.3 Heat capacities

Consider two isolated systems A and B (Fig. 9.5). System A has internal energy U_1 and is at temperature T_1. System B has internal energy U_2 and is at temperature T_2.

Whilst remaining isolated from their surroundings they are placed in good thermal contact. Energy is transferred as thermal equilibrium is attained and then system A has internal energy U_3 at temperature T whilst system B has internal energy U_4 at the same temperature T.

From the law of conservation of energy

$$U_1 + U_2 = U_3 + U_4$$

and so

$$U_1 - U_3 = U_4 - U_2$$

which may be written

$$(\Delta U)_A = -(\Delta U)_B \qquad (9.2)$$

Since the change is adiabatic (i.e., no 'heat' energy is transferred to or from the systems from any external system), we may use the first law of thermodynamics to express Eq. 9.2 as

$$-(\Delta Q)_A = (\Delta Q)_B$$

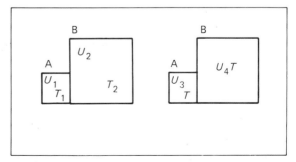

Fig. 9.5

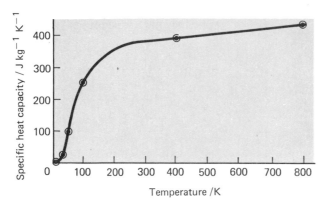

Fig. 9.6 The specific heat capacity of copper as a function of temperature.

That is, the energy lost by A is equal to the energy gained by B in this process.

During the transfer of energy from A to B, the temperature of A falls and that of B rises. Now, experiment shows that, for any small range of temperature, the energy transfer from A is proportional to the temperature fall in A and similarly, the energy transfer to B is proportional to the rise in temperature of B,

$$(\Delta Q)_A \propto (\Delta T)_A$$

The ratio $\Delta Q/\Delta T$ is a constant for this range of temperature. It is called the *heat capacity* of a body (C) and is measured in $J\,K^{-1}$.

Experiment also shows that the heat capacity of a body is proportional to its mass. It is therefore convenient to introduce a *specific heat capacity* (the heat capacity of unit mass). In SI this is the heat capacity of a mass of 1 kg of the substance. Similarly, the *molar heat capacity* of a substance is the heat capacity of 1 mole of the substance.

$$\boxed{\text{Specific heat capacity } (c) = \frac{1}{m}\frac{\Delta Q}{\Delta T}} \quad (9.3)$$

Some typical values are given in Table 9a.

Over small ranges of temperature the values of the specific heat capacity of a substance can be assumed constant. However, experiment shows that the specific heat capacity *is* a function of temperature. This is illustrated in Fig. 9.6, which shows how the specific heat capacity at constant pressure of copper increases to a value of about 390 J kg^{-1}K^{-1} (25 J mol^{-1}K^{-1}) at 400 K and

Table 9a

Substance	Temperature/ K	Specific heat capacity (δ) at constant pressure/ J kg^{-1}K^{-1}	Molar heat capacity (δ) at constant pressure/ J mol^{-1}K^{-1}
Ice	263	2100	37.7
Water	288	4186	75.5
Water vapour	373	2030	36.5
Aluminium	290	878	23.7
Copper	290	381	24.2
Iron	290	438	24.4
Lead	290	126	26.1
Silver	290	235	25.4

changes but little after that as the temperature is increased further. This is typical behaviour for the solid elements (see the discussion of the law of Dulong and Petit in Section 15.6).

Water offers an interesting and exceptional case. In the first place, its specific heat capacity (4190 J kg^{-1}K^{-1}) is exceptionally high – a convenient fact for our central heating systems. In the second place, its behaviour is quite untypical in showing a shallow minimum specific heat capacity at 308 K.

For energy transfers of the type discussed in this section it follows that

$$\Delta Q = cm\Delta T \quad (9.4)$$

This provides a convenient approach to the measurement of the energy transferred in heating and cooling. The energy transfer when a body of mass m and specific heat capacity c has its

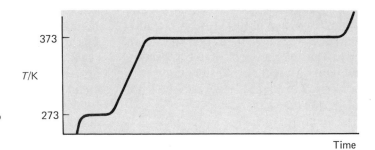

Fig. 9.7 Changing the state of ice to water and then to water vapour.

temperature changed by ΔT is $cm\Delta T$. This is the foundation of the science of calorimetry, a foundation which was laid by Joseph Black in the eighteenth century. Black's investigations into the heat exchanges between hot and cold bodies when mixed together (the method of mixtures), and into the change of phase which occurs when ice melts to water, were based upon a different model of heat from that which we have used. Essentially, heat was then regarded as a fluid (called caloric) which could flow from a hot body into a cold body. This caloric model was satisfactory for the changes to which it was applied. The first substantiated doubts about the model, which eventually led to its replacement by the energy theory, originated in the experiments of Rumford in 1798. Count Rumford (born Benjamin Thompson in Massachusetts in 1753) showed that the amount of heat generated when a cannon was bored out with a blunt drill was proportional to the mechanical energy transferred. As we have seen, this is the style of experiment which was perfected by J. P. Joule later in the nineteenth century and which led to the complete replacement of the caloric model by the energy model.

Problem 9.2   An aluminium kettle weighing 1 kg, containing 1 kg of water at a temperature of 20 °C, is placed on a gas ring and the gas is lit. The water just began to boil after 230 s. How much energy has been transferred to the kettle and its contents?

If the specific heat capacity of water in this temperature range is taken to be 4200 J kg^{-1} K^{-1} and of aluminium 900 J kg^{-1} K^{-1} the energy which

must be supplied to raise the temperature of the water from 20 °C to 100 °C is

$$4200 \times 1 \times (373 - 293) \text{ joules}$$

$$= 336\,000 \text{ joules.}$$

Similarly the energy required to raise the temperature of the kettle itself to 100 °C is

$$900 \times 1 \times (373 - 293) \text{ joules}$$

$$= 72\,000 \text{ joules.}$$

The total energy supply is 408 000 joules.

Problem 9.3   What is the rating in kilowatts of an electric kettle of equivalent performance?

The gas was supplying energy at a rate of 408 000/230 joules per second. This is 1774 J s^{-1} or, approximating, 1.7 kilowatts.

Have a look at the plate on an electric kettle and see if the rating is anything like this.

9.4   Changing the state

To melt ice is to change water from the solid to the liquid state. It is a process which requires an input of energy. Similarly to change water to the gaseous state requires an input of energy.

In an experiment done at normal pressure in which a specimen of ice is warmed from a temperature below 273 K until it has all boiled away as vapour, the temperature – time graph will take the form shown in Fig. 9.7.

If the rate of supply of energy is constant throughout it will be found that it takes much longer to change the liquid to the gas than it takes to change the solid to the liquid. It is usual to refer to the energies supplied during these two transitions as the latent heats of vaporization and fusion respectively. The temperatures of the substance remain constant during these transitions.

When referred to 1 kg of substance, these are known as the *specific latent heats*.

Table 9b

Specific latent heat of fusion of ice	$330\,kJ\,kg^{-1}$ at 273 K
Specific latent heat of vaporization of water	$2300\,kJ\,kg^{-1}$ at 373 K
Energy required to warm 1 kg of liquid water from 273 K to 373 K	420 kJ

Much more energy is required to change water from the liquid state to the gaseous than to change it from the solid to the liquid form.

9.5 The efficiency of engines

Since the time of Newcomen and Watt, steam has been an important vehicle for the transfer of energy from the chemical form to some more immediately useful mechanical form; but the best efficiencies of conversion (i.e., output of work/input of energy) have been around 15% for the typical steam engine and around 40% for the steam turbine. A petrol engine may have an efficiency of about 25% and a diesel engine of about 35%. All four use energy in transit (heat), develop an output of work, and emit hot exhaust gases; all four are heat engines. Is this low efficiency a consequence of incompetence; or is it due to some natural limitation? The first law of thermodynamics placed no such limitation on the possibility of transforming energy from any one form to any other form. Nothing in that law suggests that the total supply of energy cannot be converted to work. It is experience which suggests this.

This theoretical problem about the heat engine was first approached by Carnot, a French engineer, in 1824. His work and that of such men as Clausius and Kelvin led to the realization that there did exist a further natural law and that this law imposed a natural limitation on these energy transformations between heat and work.

We shall not pursue this work on heat engines. However, we shall return to it in Chapter 12 when we approach the problem from the microscopic rather than from the macroscopic angle.

A Model for Matter

Chapter 10

A MODEL FOR A GAS: A SECOND LOOK

Now that we have a working knowledge of dynamics, it becomes possible to quantify the model for a gas proposed in Unit 1 and, in particular, to examine its implications mathematically. It will be recalled that the model was based on a system of randomly moving, widely spaced particles. It was assumed that the collisions between these minute particles were, at least on average, elastic and that interparticle forces were only observable at extremely short ranges.

10.1 A simple mathematical analysis of the model

We have supposed that the model behaves like a gas in a container, exerting pressure by virtue of the impact of the particles on the walls. If we can also show that it does this in a quantitative way as well (for example, by showing that the pressure and volume in the model are related in the same way as they are in a gas), we shall have much more confidence in the model itself.

We must first calculate the pressure these flying particles will exert. To do this rigorously is quite difficult, but some of the difficulties can be avoided by taking a few reasonable short-cuts.

Imagine the 'gas' confined in a rectangular box with sides of length, x, y and z, and that it has been there long enough to be in a steady state (Fig. 10.1). At any instant, the particles are moving in any one of a variety of directions as the arrow on each sample particle is intended to show. Let us concentrate on the impact of the particles on the left-hand end of the box. The particles are approaching this end of the box from all possible directions. All bounce off the end, exerting a

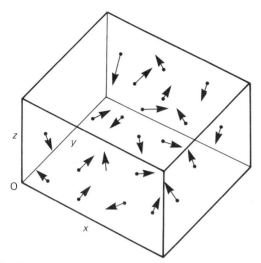

Fig. 10.1

10.3b). Its final kinetic energy is the same as its initial kinetic energy but its momentum is now mc in the opposite direction to its initial momentum. It has suffered a momentum change of

$$mc - (-mc) \quad \text{or} \quad 2mc$$

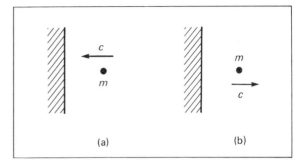

Fig. 10.3

certain force on the face of the box as they do so (Fig. 10.2). How can we set about finding this force?

Consider first of all a single particle of mass m and travelling with velocity c along a line perpendicular to the wall (Fig. 10.3a).

Let us assume that in this case it bounces off elastically in the opposite direction. As the particle approaches the wall, it slows up as the short range intermolecular force of repulsion comes into play, and its momentum is reduced to zero. At this instant, the kinetic energy of the particle has been entirely converted to potential energy. The particle now accelerates away from the wall to velocity c exactly reversed from its approach velocity (Fig.

To bring about this change, the wall must have provided an impulse; a force acting for a short time. We cannot say how big this force is for a single particle hitting the wall. It all depends on the time during which the intermolecular forces between the wall and the particle act. This is a simple application of Newton's second law of motion which states that the force acting on a body is proportional to its rate of change of momentum.

If this particle were the only one in the box, it would eventually collide with the opposite wall and could bounce back again towards the face of the box under consideration (Fig. 10.4).

The next collision on the end of the box to the left will take place time $2x/c$ later. The momentum of the particle is thus changed by $2mc$ at the face of the box on the left in time intervals of $2x/c$ if it

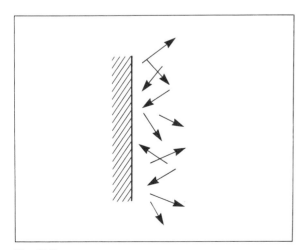

Fig. 10.2

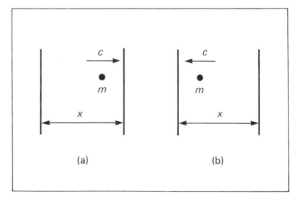

Fig. 10.4

always bounces off in the opposite direction from its approach. The rate of change of the momentum at the face is

$$\frac{2mc}{2x/c} \doteq \frac{mc^2}{x}$$

in momentum units per unit time. By Newton's second law of motion, this is the average force acting on the face if we imagine the forces resulting from this series of impacts to be 'smeared out' to a constant value.

So far we have considered the impact of one particular particle rattling back and forth between two opposite faces of the box. What happens when the box contains N particles all moving with different speeds and in different directions?

The random nature of the motion of the many particles allows us to assume that, on average, the component of the velocity parallel to the wall, $c \sin \theta$ (Fig. 10.5), does not change whilst the component perpendicular to the wall, $c \cos \theta$, is reversed in direction. The only velocity component of interest, as far as impact with the wall is concerned, is the perpendicular one, $c \cos \theta$.

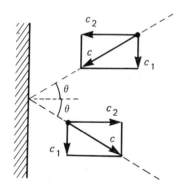

Fig. 10.5

Let us suppose that this direction is the x-direction of a set of rectangular axes. We will call this the x-component of the velocity in the x-direction.

We can resolve the velocity of any particle into three components, c_x, c_y, c_z. These are related to one another and to the velocity c of the particle by an extension of Pythagoras' theorem.

Ox, Oy, and Oz are three mutually perpendicular axes (Fig. 10.6). OF represents a velocity c. FG is a perpendicular from F to the plane Oxy so

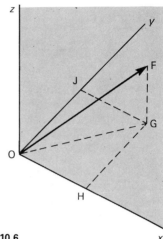

Fig. 10.6

that OGF is a right angle and, by Pythagoras' theorem

$$OF^2 = OG^2 + GF^2$$

GH and GJ are perpendicular to the Oxz and Oyz planes respectively and

$$OG^2 \doteq GJ^2 + GH^2$$

whence

$$OF^2 = GJ^2 + GH^2 + GF^2$$

But GJ, GH, and GF are the components of the velocity c in the x, y, and z directions. So

$$c^2 = c_x^2 + c_y^2 + c_z^2$$

For each of N particles in the box, all of which are moving with a variety of velocities, we may write

$$c_1^{2} = c_{x_1}^2 + c_{y_1}^2 + c_{z_1}^2$$
$$c_2^2 = c_{x_2}^2 + c_{y_2}^2 + c_{z_2}^2$$
$$\cdots\cdots\cdots\cdots\cdots$$
$$c_N^2 = c_{x_N}^2 + c_{y_N}^2 + c_{z_N}^2$$

Add:

$$(c_1^2 + c_2^2 + \cdots + c_N^2) = (c_{x_1}^2 + c_{x_2}^2 + \cdots + c_{x_N}^2)$$
$$+ (c_{y_1}^2 + c_{y_2}^2 + \cdots + c_{y_N}^2)$$
$$+ (c_{z_1}^2 + c_{z_2}^2 + \cdots + c_{z_N}^2)$$

Divide throughout by N:

$$\frac{(c_1^2 + c_2^2 + \cdots + c_N^2)}{N} = \frac{(c_{x_1}^2 + c_{x_2}^2 + \cdots + c_{x_N}^2)}{N}$$
$$+ \frac{(c_{y_1}^2 + c_{y_2}^2 + \cdots + c_{y_N}^2)}{N}$$
$$+ \frac{(c_{z_1}^2 + c_{z_2}^2 + \cdots + c_{z_N}^2)}{N}$$

This may be written:

$$\overline{c^2} = \overline{c_x^2} + \overline{c_y^2} + \overline{c_z^2}$$

The term $\overline{c^2}$ is the average of the square of the speed and is usually called the *mean square speed*. It is a meaningful average to use in this case.

Now collisions among the particles within the box will serve to maintain such an even distribution of the particles that

$$\overline{c_x^2} = \overline{c_y^2} = \overline{c_z^2} \qquad (10.1)$$

(This is our reasonable short-cut. The last few lines form an essential part of our analysis, but the appeal for their correctness is to common-sense. It seems reasonable that a large assemblage of particles should behave in this way – it is mathematically complex to show that they will: we must guard against the specious argument that 'of course' the mean value of all the three components must be the same because we know that in a gas the pressure is the same on all the faces of the box. That is true in a gas. We are trying to show that our model behaves like a gas – we must not assume that it does in the process of setting up a proof.)

From Eq. 10.1, it follows that

$$\overline{c_x^2} = \overline{c_y^2} = \overline{c_z^2} = \tfrac{1}{3}\overline{c^2} \qquad (10.2)$$

For one particle, the force on the left-hand face of the box was $mc_{x_1}^2/x$. For N particles the force is

$$\frac{mc_{x_1}^2}{x} + \frac{mc_{x_2}^2}{x} + \cdots + \frac{mc_{x_N}^2}{x} = \frac{Nm\overline{c_x^2}}{x}$$

from our definition of mean square speed.

Since, from Eq. 10.2,

$$\overline{c_x^2} = \tfrac{1}{3}\overline{c^2}$$

the force is

$$\frac{Nm}{x}\tfrac{1}{3}\overline{c^2} \qquad (10.3)$$

The area of the face is yz. So the pressure, p, on the face is

$$\frac{\text{force}}{\text{area}} = \frac{Nm}{x}\tfrac{1}{3}\overline{c^2}\frac{1}{yz} \qquad (10.4)$$

Now, $xyz = V$, the volume of the box, so

$$pV = \tfrac{1}{3}Nm\overline{c^2} \qquad (10.5)$$

If $\tfrac{1}{3}Nm\overline{c^2}$ is constant at constant temperature, pV is constant and the model will obey Boyle's law. This is encouraging; and it raises an important question about the meaning of the product $\tfrac{1}{3}Nm\overline{c^2}$. Nm is the total mass (M) of the particles in the box. If now we write the right hand side of Eq. 10.5 as $\tfrac{2}{3}(\tfrac{1}{2}M\overline{c^2})$ we are very strongly reminded of the expression for translational kinetic energy. But, when we do this, we must remember that translational kinetic energy is not the only possible form for energy to take. We are excluding the rotational and the vibrational kinetic energies which will also be possessed by molecules of the polyatomic gases.

10.2 An equation of state

Experimental evidence long ago established two further empirical laws descriptive of the behaviour of real gases. The first stems from the work of Charles (1787) and of Gay-Lussac (1802) and relates the volume change of a fixed mass of a gas, maintained at a constant pressure, to the temperature change recorded, say, in K on the mercury-in-glass scale of temperature.

$$\boxed{V = V_0(1 + \alpha\Delta T)} \qquad (10.6)$$

where V and V_0 are the final and the initial volumes of the gas, ΔT the temperature change in K and α is a constant. Interestingly enough, this constant has very nearly the same value for all gases at low pressure and is $0.003\,660\ \text{K}^{-1}$.

The second law concerns the way in which the pressure of a fixed mass of a real gas, maintained at a constant volume, changes as the temperature changes. Using the same temperature scale as above, we find that

$$\boxed{p = p_0(1 + \beta\Delta T)} \qquad (10.7)$$

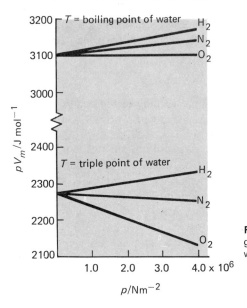

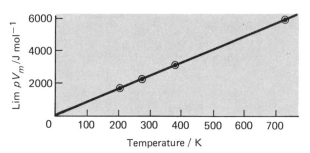

Fig. 10.8   Graph of the limiting value of pV_m against temperature.

Fig. 10.7   Graphs of pV_m against p for different gases at the triple point and at the boiling point of water.

where p and p_0 are the final and initial pressures, ΔT the temperature change in K and β is a constant. This constant β proves to be virtually the same as α and neither differs by much from the fraction 1/273. This is hardly an accident, as we shall see.

Although of considerable historical interest, these two laws lost their fundamental significance once the constant volume gas thermometer was adopted as the standard. The two laws then became a direct consequence of the choice of a definition for temperature. However, they do confirm that the mercury-in-glass thermometer scale of temperature conforms to the gas thermometer scale. It was in Section 9.1 that we observed that gases at low pressures are the best thermometric substances and we there defined temperatures in terms of the constant volume gas thermometer

$$T = 273.16\frac{p_T}{p_{tr}} \dots T \text{ in K}$$

where p_{tr} is the pressure at the triple point of water.

If the product pV_m for a fixed amount of gas – and we shall specify a mole (see Section 2.3) occupying volume V_m (the *molar volume*) – for a number of real gases is plotted against pressure p (Fig. 10.7) a remarkable property emerges. For each temperature, as the pressure approaches

zero, the product pV_m tends to the same value, irrespective of the gas chosen.

Moreover, if values of the limiting values of pV_m are then plotted against temperature, the graph is linear and passes through the origin (Fig. 10.8). The product pV_m is therefore proportional to temperature. And, since these limiting values are independent of the gas used, one equation can be written for all of them. This equation, which brings together Boyle's, Charles' (or Gay-Lussac's) and the constant volume law, is

$$\boxed{pV_m = RT} \qquad (10.8)$$

where R is a constant. Since this is of universal application to gases, R is known as the *universal gas constant*.

$$R = 8.314 \text{ J mol}^{-1}\text{K}^{-1}$$

Equation 10.8 is an equation of state which applies to all gases at low pressures. This suggests that it would be valuable to imagine an 'ideal gas' to which this equation of state would apply under all circumstances. For such a gas we should have

a) $p \propto 1/V$ at constant temperature (Boyle's law),
b) $p \propto T$ at constant volume,
c) $V \propto T$ at constant pressure.

Under ordinary conditions – i.e., for relatively large volumes – the behaviour of real gases differs but little from that of this ideal gas and the simple equation of state of Eq. 10.8 can often be applied.

Problem 10.1 In the graphs (Fig. 10.7 and Fig. 10.8) the units for the product pV_m are given as J mol^{-1}. Explain.

10.3 The ideal gas: model and reality

In Section 2.3, when we discussed the mole, we introduced the hypothesis due to Avogadro that, under conditions of equal pressure and temperature, equal volumes of all gases contained the same number of molecules. The accepted value for this number, which is known as the Avogadro constant or Avogadro's number (N_A), was quoted as 6.022×10^{23} mol^{-1}. Let us now suppose that our model gas contained a mole of particles so that we must write Eq. 10.5 as

$$pV_m = \tfrac{1}{3}N_A m \overline{c^2}$$

Empirical results led to Eq. 10.8 for a mole of an ideal gas

$$pV_m = RT$$

How far can the equation for the model reflect the equation of state?

We have already noted that the expression $\tfrac{1}{3}N_A m\overline{c^2}$ can also be written as $\tfrac{2}{3} \times \tfrac{1}{2}N_A m\overline{c^2}$ in which the product $N_A m$ is the mass of the mole of particles in the box. Therefore, $\tfrac{1}{2}N_A m\overline{c^2}$ is the total kinetic energy of translation of the particles in the box. So, for these particles, we have

$$pV_m = \tfrac{2}{3}\text{ (total kinetic energy of particles).}$$

Further development depends on whether or not we can legitimately compare this result with the known results for an ideal gas. If we do this, do consequences arise which can be tested experimentally?

a. How fast?

Equation 10.5 for the model gives

$$\overline{c^2} = 3pV_m/N_A m$$
$$= 3p/\varrho$$

where ϱ is the density.

$\sqrt{\overline{c^2}}$ is a form of average speed which is known as the *root of the mean square speed* or *r.m.s. speed* and is equal to

$$\sqrt{3p/\varrho}$$

If this equation were to be applied to air, say, with a density of 1.293 kg m^{-3} at s.t.p. (which is 1.013×10^5 N m^{-2} and 273 K) the r.m.s. speed would be

$$\sqrt{\frac{3 \times 1.013 \times 10^5}{1.293}} = 485 \text{ m s}^{-1}$$

How realistic is this figure? We know that the speed of sound in air is in the same order (331 m s^{-1}) which is encouraging. Sound travels through the air as a compression wave and it is very unlikely that the compressions could travel faster than the molecules which are compressed together. We should expect, then, to find that the r.m.s. speed of the molecules is faster, but not much faster, than the speed of sound.

b. Diffusion

If we simply equate RT and $\tfrac{1}{3}N_A m\overline{c^2}$ we could write

$$T = \tfrac{1}{3}\frac{N_A m}{R}\overline{c^2}$$

Taking a mole of two different gases x and y at the same temperature the model requires that

$$\tfrac{1}{3}\frac{N_A m_x}{R}\overline{c_x^2} = \tfrac{1}{3}\frac{N_A m_y}{R}\overline{c_y^2}$$

whence

$$\tfrac{1}{2}m_x\overline{c_x^2} = \tfrac{1}{2}m_y\overline{c_y^2}$$

This suggests that gas molecules of X and Y should have equal average kinetic energies of translation and also that

$$\sqrt{\frac{\overline{c_x^2}}{\overline{c_y^2}}} = \sqrt{\frac{m_y}{m_x}}$$

This would imply that gases with low molecular masses travel faster than those with high ones.

This can be tested experimentally – by *effusion* through small holes into a vacuum and also by *diffusion* through porous walls. This work was done by Graham in the first half of the nineteenth century. He found that the volumes V_x of gas x and V_y of gas y which diffused through a given surface in a given time were, when reduced to the same conditions of temperature and pressure, inversely proportional to their densities.

$$\frac{V_x}{V_y} = \frac{\varrho_y}{\varrho_x} = \frac{m_y}{m_x}$$

This is *Graham's Law of Diffusion* (1830).

The correspondence between Graham's experimentally based equation for the diffusion of gases and the equation suggested by the kinetic model adds further support to the usefulness of the model.

During the 1940s the process of diffusion became of great industrial importance. The analysis above tells us that, if two different gases are placed within a container with porous walls in a vacuum, the lighter of the two will escape at the faster rate. This process was used to separate the fissile isotope uranium-235 from natural uranium. Only 0.7% of natural uranium is U-235. In the gaseous compound uranium hexafluoride the respective molecular masses are 349 for U-235 and 352 for U-238. The relative rates of diffusion are in the ratio 352/349 or 1.0043:1. Small wonder that the process was used in cascade involving some 4000 stages in a plant which covered a vast area.

Experimental evidence of this sort gives the confidence to accept that we can write equations

$$RT = \tfrac{2}{3}N_A \frac{\overline{mc^2}}{2}$$

and

$$T = \tfrac{2}{3}\frac{N_A}{R}\frac{\overline{mc^2}}{2} \qquad (10.9)$$

for a mole of ideal gas. We can now assume that the temperature of such a gas is proportional to the mean kinetic energy of translation of the molecules.

This result is a new interpretation of the meaning of temperature which derives from the application of a model to the reality of gas behaviour. The property of temperature of a gas is directly related to the kinetic energy of translation of the molecules making up the gas. It represents the overall behaviour of the countless tiny particles of which the gas is composed. Just as a pressure gauge comes into equilibrium with the forces exerted by the vast numbers of individual molecules exerting the pressure, so a thermometer

comes into equilibrium with the average kinetic energy of the vast numbers of individual molecules. Both pressure and temperature are statistical statements. To ask what is the temperature of an individual molecule is quite meaningless.

The interpretation in energetic terms is in accord with the observation that temperature can be raised by supplying energy – as heat. The energy supplied is conserved and increases the total kinetic energy of the molecules.

The factor $\tfrac{2}{3}N_A/R$ is a universal constant applying to all gases since it is a combination of N_A, Avogadro's number, and R, the universal gas constant. It is usual to write $\tfrac{2}{3}\frac{1}{k}$ for $\tfrac{2}{3}N_A/R$ where

$$k = \frac{R}{N_A} = \frac{8.31 \text{ J mol}^{-1}\text{K}^{-1}}{6.02 \times 10^{23} \text{ molecules mol}^{-1}}$$

$$= 1.38 \times 10^{-23} \dots \text{ and the unit is } JK^{-1} \text{ per molecule.}$$

The equation for T then becomes

$$T = \frac{2}{3k}\frac{\overline{mc^2}}{2}$$

whence

$$\boxed{\text{r.m.s. speed} = \sqrt{\overline{c^2}} = \sqrt{3k\frac{T}{m}}} \qquad (10.10)$$

Boltzmann's constant, k, is of considerable significance in studies of gas behaviour. Just as R is the universal gas constant per mole, k is the gas constant per molecule. R is used in studies of the macroscopic behaviour of gases; k in studies of the microscopic behaviour.

Boltzmann's constant has even wider significance and will be met with in studies of solids as we shall see in Chapter 12.

There are several other useful forms of these equations.

(i)   The average kinetic energy of a particle

$$\tfrac{1}{2}\overline{mc^2} = \tfrac{3}{2}kT \qquad (10.11)$$

(ii)   The equation of state can also be written

$$pV_m = \tfrac{2}{3}N_A\tfrac{3}{2}kT$$

$$= N_A kT$$

whence

$$p = \frac{N_A}{V_m}kT \qquad (10.12)$$

(iii) N_A/V_m is the number of particles per unit volume or the *particle density, n.*

So $$p = nkT \qquad (10.13)$$

(iv) In this ideal model of a monatomic gas where inter-molecular forces extend over very short distances only, the sole store of internal energy lies in the kinetic energy of translation of the particles. It is $\frac{1}{2}m\overline{c^2}$ per particle at temperature T.

> So the total internal energy
> per mole
> $$= \tfrac{1}{2}N_A m\overline{c^2}$$
> $$= \tfrac{3}{2}N_A kT$$
> $$= \tfrac{3}{2}RT \qquad (10.14)$$

Equation 10.10 gives us a further opportunity to check the arguments with the evidence from experiment. If we could measure the r.m.s. speed of gas molecules we could check the equation directly.

A warning

It is important to remember that these relationships apply only to the model of an ideal gas and to real gases under conditions which make them approximate closely to the ideal. They tell us little about the internal energies in other cases; in particular, we should not assume that they can be applied to any gas at a temperature which is close to 0 K, to vapours in the process of evaporation or to gases under high pressure.

In real gases, the total internal energy will include forms other than translational kinetic energy. There will be potential energies and, for gases other than monatomic ones, such other kinetic energies as rotational and vibrational.

10.4 Molecular speeds

Up to this point, the term *an average velocity* of gas atoms or molecules has been used without careful specification of just what is to be understood by the term. The figure cannot be an arithmetical average for, if the motion is truly random, this must be zero. It is, as indicated above, the square root of the mean of the squares of the speed and it cannot be zero. Evidently in a gas in which molecules are exchanging kinetic energies in a frequent series of collisions, individual molecular speeds must range from zero on the

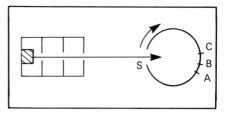

Fig. 10.9 Zartmann and Ko's experiment on the distribution of molecular speeds.

the one hand to very high speeds on the other. Theoretical investigations into the distribution of speed among the molecules were carried out by James Clerk Maxwell in 1860 but experimental techniques for checking Maxwell's description of the distribution were not available until the 1920s when Stern devised a suitable technique which has been improved subsequently. In one method, used by Zartmann in 1931, atoms of bismuth were produced in a vacuum by heating in a furnace. A thin beam of moving atoms was obtained by allowing the main beam to pass through a series of slits. This method of obtaining a rectilinear beam of small dimensions is called *collimation*. The beam of vapour atoms then reaches a drum to which entry can only be gained through a slit S. This drum rotates at some 6000 rev min⁻¹ and is lined with a sensitive film. As the slit S passes through the narrow beam a pulse of atoms enters the drum; and the speeds of the atoms will be distributed amongst a wide range. As the drum rotates the atoms move across its diameter, some taking much longer than others. The fastest atoms will strike the sensitive film at A, slower ones at B and the slowest at C (Fig. 10.9). The density of the

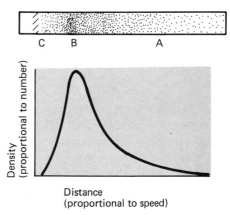

Fig. 10.10

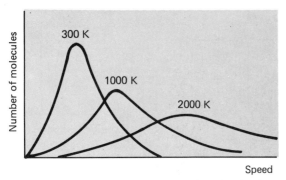

Fig. 10.11 Temperature and the distribution of molecular speeds.

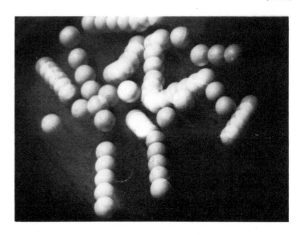

Fig. 10.12 Multiflash photograph of a group of balls on a vibrating table. (Photograph E. J. W.)

film at any point will be a measure of the number of atoms striking that point.

The agreement between the observed results and the results expected from theoretical considerations was found to be remarkably good. This distribution of speeds, as shown in Fig. 10.10 is a familiar one. It is a form of the law of normal distribution which applies to so many things where chance is in control. For example, to the height of the men in a population, to the frequency distribution of a series of measurements and so on. The exact shape of the curve depends on the temperature and the peak shifts as shown in Fig. 10.11 as temperature increases.

10.5   Diffusion and the random walk

We can now assume with confidence that the particles of a gas are moving at random within the space in which the gas is confined; that the velocities of individual particles are distributed about a mean which, even at ordinary temperature, is quite high; a little more than the speed of sound in the gas.

An old objection to this consequence of the model was concerned with the speed of diffusion of a gas into the air. Everyone knows that the smell of, say, natural gas takes time to spread through a room into which it is released. It is a reasonable guess, too, that convection is more effective in conveying the gas from the source to the nose than diffusion is. And yet the model suggests that the molecular velocities are very high.

This is an objection with which the model can deal quite easily and with only a small change in its

construction. Consider the movement of a number of balls confined on a vibrating table (see Fig. 10.12).

This photograph was taken with the camera shutter open for a relatively long exposure time, whilst the balls were illuminated with a flashing strobe light. One can see at once that no ball travels very far in any one direction before

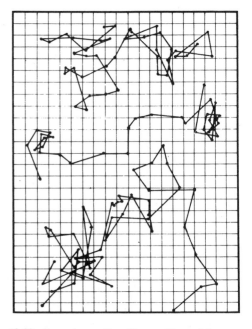

Fig. 10.13 Brownian motion. The positions of three particles, suspended in water, observed at 30 s intervals. (After J. Perrin.)

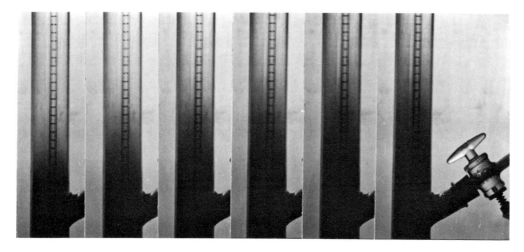

Fig. 10.14 Bromine diffusing into air at atmospheric pressure. The photographs were taken at intervals of 100 s. (Photographs E. J. W.)

colliding with another and suffering a change of velocity. The path of a single ball through the mass of balls must be a series of 'dog-leg' sections.

Because they have a finite size the balls frequently collide and, in consequence, 'diffuse' through the available space quite slowly even though the average velocity is high. We now suppose that the particles of the gas have a small but finite size so that inter-particle collisions occur frequently.

A particle such as this will not travel very far from its starting point for the changes in direction which will occur with each collision are just as likely to move it one way as another. Indeed, it is perhaps surprising that it ever does move away from the starting point. The actual path followed is quite unpredictable but must be something like those pictured on the vibrating table and to the sort of observation which was recorded by Perrin in his study of the Brownian motion (Fig. 10.13).

This motion is highly erratic – a typical random walk. As presented in Fig. 10.13 it is a two-dimensional walk. But in the gas this must be imagined as a three-dimensional random walk. In an experiment on the diffusion of the dense coloured gas, bromine, into air at ordinary temperature and pressure, the series of photographs in Fig. 10.14 reveals the slow spread of the advancing bromine front as it moves into the colourless air. Of course, the advancing front of the bromine reveals a gradual increase in the concentration of

the bromine and the density of the colour, but estimates of a 'half-dark' position suggest that the front is advancing at a rate of about 0.1 m in 500 s. In this time an average bromine molecule can be expected to travel a total distance of $500\,\bar{c}$ where $\bar{c}$ is the magnitude of the *average velocity*, or *average speed*.

If the magnitude of the average velocity $\bar{c}$ is 200 m s^{-1}, this means that on average a path length of 10^5 m has been folded up by the constant collisions into a distance of 0.1 m. The walk is indeed a random one.

Figure 10.15 represents Perrin's drawing of one typical walk.

Consider the first few steps of this walk. Let the molecule start out from O and end up at F.

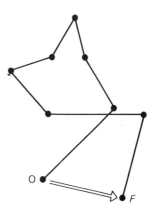

Fig. 10.15

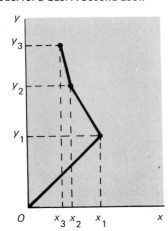

Fig. 10.16

It has travelled a distance OF by making a series of n steps, each of which can be resolved into components in the x and y directions as shown in Fig. 10.16. The final x component of OF is given by

$$x = x_1 + x_2 + x_3 + \cdots + x_n,$$

due allowance being made for signs, of course. The final y component of OF is given by

$$y = y_1 + y_2 + y_3 + \cdots + y_n.$$

The distance OF is given by

$$OF^2 = x^2 + y^2.$$

Now

$$x^2 = (x_1 + x_2 + x_3 + x_4 + \cdots + x_n)^2$$
$$= x_1^2 + (x_1x_2 + x_1x_3 + \cdots + x_1x_n)$$
$$\quad + x_2^2 + (x_2x_1 + x_2x_3 + \cdots + x_2x_n)$$
$$\quad + x_3^2 + (x_3x_1 + x_3x_2 + \cdots + x_3x_n), \text{ etc.}$$

The terms x_1x_2 are just as likely to be negative as positive. So, when very large numbers of steps are considered, the total sum of these terms is likely to be zero. Then

$$x^2 = x_1^2 + x_2^2 + x_3^2 + x_4^2 + \cdots + x_n^2$$

and

$$y^2 = y_1^2 + y_2^2 + y_3^2 + y_4^2 + \cdots + y_n^2$$

whence

$$OF^2 = (x_1^2 + y_1^2) + (x_2^2 + y_2^2) + \cdots + (x_n^2 + y_n^2).$$

If each individual step length is represented by λ_1, λ_2, etc., the theorem of Pythagoras gives

$$\lambda_1^2 = x_1^2 + y_1^2, \text{ etc.}$$

and

$$OF^2 = \lambda_1^2 + \lambda_2^2 + \lambda_3^2 + \cdots + \lambda_n^2$$

Taking λ^2 as the average of the squares of the value of all n individual steps

$$\lambda^2 = \frac{\lambda_1^2 + \lambda_2^2 + \lambda_3^2 + \cdots + \lambda_n^2}{n}$$

$$= \frac{OF^2}{n}$$

So $OF^2 = n\lambda^2$ in which λ is the square root of the average of the squares of the step lengths (a root mean square length). Hence

$$\boxed{OF = \sqrt{n}\lambda} \qquad (10.15)$$

This is a well-known statistical result and it can, in fact, be applied to the three-dimensional case as well as to the two-dimensional case considered above. It can be applied directly to the diffusion of bromine experiment.

The average step-length (*mean free path*) between collisions is also to be found by dividing the total distance $\bar{c}\Delta t$ travelled in time Δt by the number of steps n. So

$$\lambda = \frac{\bar{c}\Delta t}{n}$$

From Eq. 10.15 we have

$$\lambda = \frac{\text{distance made good (OF)}}{\sqrt{n}}$$

In the experiment, then

$$\lambda = \frac{200 \times 500}{n} = \frac{0.1}{\sqrt{n}}$$

whence

$$n = 10^{12}$$

That is to say the bromine molecule which had travelled a total distance up the glass vessel of only 0.1 m had, en route, made 10^{12} collisions. This took 500 seconds and so the number of collisions per second must be 2×10^9. Having determined n to

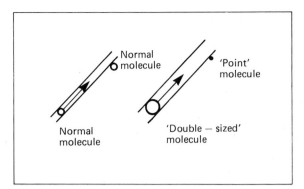

Fig. 10.17

be 10^{12}, it remains to substitute in either of the equations for the mean free path λ to find that

$$\lambda = \frac{200 \times 500}{10^{12}} = 10^{-7}\,\mathrm{m}$$

Problem 10.2   How long would it take an average molecule of a gas with a mean speed of 500 m s^{-1} to travel across a room which was 5 m wide?

Consider now a molecule in motion among its fellows following the series of dog-leg steps we have come to associate with any random walk. When it collides with another molecule the centres of mass of the two are, on average, one 'molecular diameter', d, apart. We can envisage the process more clearly if we imagine the collision to be between one double sized molecule and a point molecule (Fig. 10.17).

The average volume swept out by the 'double-size' molecule before it makes the collision is $\pi d^2 \lambda$.

In this volume, it finds, on average, just one point molecule. In other words, a typical molecule occupies a volume of

$$\pi d^2 \lambda$$

Now liquid nitrogen (which is a safer material than liquid air) has a density of about 800 kg m^{-3} and, when it boils, its volume increases by a factor of 750. It would be sensible to assume that the molecules of the liquid nitrogen now occupy 750 times the space they occupied when in the liquid form. If a molecule of diameter d occupies a cubical box of side d (as though the molecules in the liquid are close packed like apples in a box)

each molecule occupies $750d^3$ when in the gaseous state. So

$$750d^3 = \pi d^2 \lambda$$

and

$$750d = \pi \lambda$$

Taking $\lambda = 10^{-7}\,\mathrm{m}$, d is seen to be about $4 \times 10^{-10}\,\mathrm{m}$ or $4 \times 10^{-1}\,\mathrm{nm}$.

10.6   The atomic model

There is a good deal of evidence to support the suggestion that the basic units of material structure (atoms, ions, or molecules) are only $10^{-10}\,\mathrm{m}$ in diameter. We have noted already that the laws of chemical combination led Dalton to a particulate model for matter. In order to account also for the simple ratios in which the volumes of gases combined, it was necessary to adopt a further hypothesis, due to Avogadro, that equal volumes of gases at the same pressure and temperature contain the same number of particles. This hypothesis is not susceptible to direct proof. Its justification lies in the coherent pictures of atoms, singly and in combination, which developed from it.

On the basis of this hypothesis, and the laws of chemical combination, a detailed model of the particulate state emerges. The ultimate particle (as far as chemical interaction is concerned) is the *atom*. There is a different atom for each of the hundred-odd elements and their masses can be arranged in a table of *atomic masses* (referred originally to the mass of a hydrogen atom as 1, but today to the carbon isotope taken as 12 atomic mass units). In the free, elemental state, these atoms may be grouped into particles consisting of two or more identical atoms. These particles are called *molecules*. In chemical combinations with other atoms, *compound molecules* are formed, consisting of two or more different atoms.

The relative masses combining in a chemical reaction are not the same as either the relative molecular or atomic masses. To two significant figures the relative atomic mass of hydrogen is 1.0, while that of oxygen is 16. Their respective molecular masses are 2.0 and 32. However, in a chemical reaction to form water, each 1.0 g of hydrogen combines with only 8.0 g of oxygen.

This leads to the idea that different atoms have different combining powers. On comparing, either directly or indirectly, the combining powers of each element with hydrogen, we find that each atom of oxygen will combine with two atoms of hydrogen. On the other hand, two atoms of aluminium will combine with three of oxygen, and consequently, aluminium has a combination power of three. This combining power is called the atom's *valency*.

The phenomenon of *electrolysis* shows that many atoms are charged in solution. For the passage of a given quantity of electricity, elements are deposited at anode or cathode in proportion to the ratio: atomic mass/valency. This is *Faraday's second law of electrolysis*.

This strongly suggests that the valency of an atom is related to the degree to which it becomes charged in solution. On this basis, hydrogen atoms carry only one unit of charge, while oxygen carries two, and aluminium, three. No atom has a smaller valency than hydrogen.

Millikan's experiments on the charge carried by oil drops show that the smallest charge which can occur is 1.602×10^{-19} coulomb. This is the charge we eventually associate with an electron. It is tempting to assume that the charge associated with a valency of 1 is also 1.602×10^{-19} coulomb. Let us make this assumption.

Careful electrolysis experiments show that it takes 96 500 coulombs to release 1.008 g of hydrogen. If each hydrogen atom had carried 1.602×10^{-19} coulombs, there must be

$$\frac{96\,500}{1.602 \times 10^{-19}} \text{ atoms in } 1.008 \text{ g of hydrogen}$$

$$= 6.02 \times 10^{23} \text{ atoms in } 1.008 \text{ g of hydrogen}.$$

Similar calculations for other elements lead to the conclusion that there are 6.022×10^{23} atoms in every mole. This is the number to which we gave the name of the Avogadro constant (N_A). This method of arriving at it depends upon two unproved assumptions:

a) Avogadro's hypothesis,
b) a valency of 1 is related to a charge in solution of 1.602×10^{-19} coulomb.

Other methods for finding the Avogadro constant depend on other assumptions. The convergence of the experimentally determined values is part of the consistent picture which has been developed of the atomic state and lends powerful support to the whole theoretical structure.

Problem 10.3   Some particle arithmetic.
a) The Avogadro constant is 6.022×10^{23} particles per mole. Calculate the mass of an atom of a substance one mole of which has a mass of 1.000×10^{-3} kg.

The mass of one atom of this substance is

$$\frac{1.000 \times 10^{-3}}{6.022 \times 10^{23}} \text{ kg} = 1.66 \times 10^{-27} \text{ kg}.$$

b) The mass of one mole of molecular nitrogen is 28.0×10^{-3} kg. What is the mass of a single molecule of nitrogen?

The mass of a single molecule of nitrogen is

$$\frac{28.0 \times 10^{-3}}{6.022 \times 10^{23}} \text{ kg} = 46.5 \times 10^{-27} \text{ kg}.$$

c) What is the average translational kinetic energy of a molecule of an ideal monatomic gas at a temperature of 300 K?

From Eq. 10.11 the average kinetic energy is

$$\tfrac{3}{2}kT = \tfrac{3}{2} \times 1.38 \times 10^{-23} \times 300 \text{ J}$$
$$= 6.20 \times 10^{-21} \text{ J}$$

d) What is the total translational kinetic energy of the particles of 1 mole of such a gas at this temperature?

From Eq. 10.14 the total internal energy is

$$\tfrac{3}{2}RT = \tfrac{3}{2} \times 8.3 \times 300 \text{ J}$$
$$= 3735 \text{ J}$$

e) What is the r.m.s. velocity of a molecule of bromine at 300 K? The mass of the bromine molecule is $160 \times 1.66 \times 10^{-27}$ kg.
From Eq. 10.10

$$\sqrt{\overline{c^2}} = \sqrt{\frac{3kT}{m}}$$

$$= \sqrt{\frac{3 \times 1.38 \times 10^{-23} \times 300}{160 \times 1.66 \times 10^{-27}}} \text{ m s}^{-1}$$

$$= 216 \text{ m s}^{-1}$$

Chapter 11

FLUIDS

Gases share with liquids certain behaviour which provides important models to help us in visualizing other phenomena. Both states have the ability to flow and to take up different shapes. Both are fluid. But there are two important differences. Firstly: whereas gases are easily compressed, liquids, like solids, are very difficult to compress. Secondly: liquids are bound by a surface; gases are not. You can half-fill a can with water; you cannot half-fill it with a gas.

11.1 Pressure in a fluid

We have already seen how the Brownian motion in a gas offers evidence for the kinetic model we developed in the previous chapter. That motion can also be observed in liquids; indeed it was first observed in a liquid. So it is not unreasonable to assume that the pressure within a liquid owes its origin to the bombardment of the wall of the vessel by the particles which make up the liquid.

■ Pressure is defined as the ratio of the perpendicular contact force to the area of contact.

$$p = F/A \qquad (11.1)$$

The unit is derived from those of force (N) and area (m²) and is the newton per square metre or *pascal* (Pa).

Note that the direction in which the pressure acts is always at right angles to the surface upon which it is acting. Pressure is not a vector in the sense that force is.

It is a matter of observation that, within a liquid at rest in a gravitational field, the pressure is

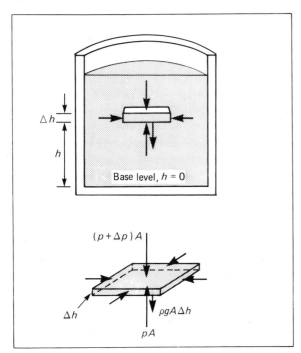

Fig. 11.1 Forces on a slab within a liquid.

the same at all points on the same horizontal line. If this were not so the liquid would perforce flow along that horizontal line until it was!

It is also a matter of observation that the pressure within a fluid increases with depth.

Consider an element of the fluid in the form of a thin slab of thickness Δh and surface area A (Fig. 11.1). If the density of the fluid is ϱ, the mass of that element is $\varrho A \Delta h$ and the weight is $g\varrho A \Delta h$, where g is the gravitational field strength.

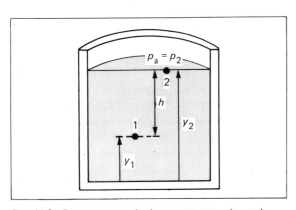

Fig. 11.2 Pressure at a point in an open-topped vessel containing liquid.

The forces exerted are everywhere normal to the surfaces and, by symmetry, the resultant horizontal force on the vertical sides of the slab is zero.

The upward force on the lower face is pA.

The downward force on the upper face is $(p + \Delta p)A$.

The three vertical forces (two due to pressures and one to weight) are in equilibrium, whence

$$pA - (p + \Delta p)A - \varrho g A \Delta h = 0$$

and

$$A\Delta p = -\varrho g A \Delta h$$

$$\boxed{\Delta p = -\varrho g \Delta h} \qquad (11.2)$$

This equation is easy enough to apply to the case of a liquid. But, where the fluid is gaseous (e.g. the atmosphere) we have to note that the density is itself a function of height.

The equation implies that an increase in height above the base level is accompanied by a negative Δp, that is a decrease in pressure. For a liquid, if p_1 and p_2 are pressures at heights y_1 and y_2 above the base level (Fig. 11.2), then

$$p_2 - p_1 = -\varrho g(y_2 - y_1)$$

Figure 11.2 shows a liquid contained in a vessel which is open to the atmosphere. At the point 2 which is in the surface of the liquid, the pressure is that of the atmosphere, p_a.

At point 1 the pressure p_1 will be given by

$$p_a - p_1 = -\varrho g(y_2 - y_1)$$

and so

$$\boxed{p_1 = p_a + \varrho g h} \qquad (11.3)$$

It is important to note that only the depth below the surface of the liquid of density is involved; the shape of the vessel and the area of the liquid water surface are not involved in the pressure at all.

Problem 11.1 Forces against a dam. The water impounded by the dam (Fig. 11.3) exerts a horizontal force on the wall which tends to slide it along its foundation and a torque tending to overturn the wall about the point O.

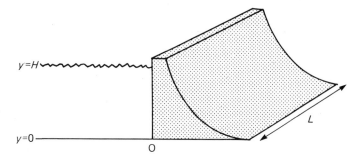

$y=H$

$y=0$

O

Fig. 11.3   Water impounded by a dam.

The atmospheric pressure applies equal and opposite forces on the two faces of the dam.

Since the pressure increases uniformly with depth, the pressure at half-depth (which is the mean pressure) is $\frac{1}{2}\varrho gH$.

Total force on the wall $= p_{\text{av.}} \times$ area
$$= (\tfrac{1}{2}\varrho gH)(LH)$$
$$= \tfrac{1}{2}\varrho gH^2L$$

The total force on the dam depends on the square of the depth. It is not dependent upon the surface area of the water impounded.

Equation 11.3 implies that, if the pressure on a liquid is increased, perhaps by fitting a piston and pressing it down, the pressure at any depth in the liquid will increase by the same amount. This is the basis of *Pascal's law*.

■ Pressure applied to an enclosed fluid is transmitted unchanged to every part of the fluid and to the walls of the containing vessel.

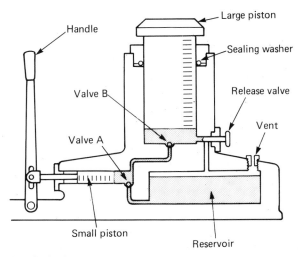

Fig. 11.4   A hydraulic press.

This is the basis of many hydraulic devices. For example, braking systems of cars; in fork-lift trucks; in tractor systems; in car-jacks; in lifts, etc. Essentially a small force applied over a small area of a piston can generate a larger force elsewhere in the system utilizing a larger area. Figure 11.4 illustrates this.

Problem 11.2   An hydraulic press. Figure 11.4 shows some of the features of a hydraulic press. What are the functions of the three valves? If the areas of the two pistons are in the ratio 100:1 what are the mechanical advantage and the velocity ratio of the machine?

Although producing a large increase in force, the machine does not contravene the law of conservation of energy. Explain.

11.2   Upthrust

A balloon filled with hydrogen or helium, or even hot air, is able to float in the atmosphere; a human being can float in water. In each case the body is supported within the fluid and this implies the existence of an *upthrust*.

Figure 11.5 shows a vessel containing a fluid.

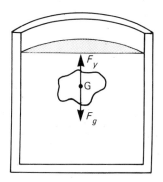

Fig. 11.5

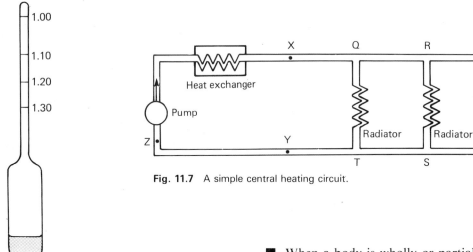

Fig. 11.6 A hydrometer.

Fig. 11.7 A simple central heating circuit.

An irregularly shaped sample of that fluid is shown in outline. The sample is in equilibrium. No net forces, either horizontal or vertical, can be acting.

The weight of the sample F_g acts vertically downwards through the centre of gravity, G. To maintain the observed equilibrium, an equal force F_y (an upthrust) must also act through G.

This upthrust is equal in magnitude to the weight of the fluid in the sample and, as we have seen, it is due to the difference in pressure between the upper and the lower surfaces of the sample.

Now suppose that the sample is replaced by another sample of precisely the same shape but with a different density and therefore a different mass and a different weight. If this weight is less than the upthrust, which is unchanged, the new sample will rise; if more than the upthrust, it will sink.

And, if the line of action of the upthrust does not pass through the centre of gravity of the new sample, that sample will experience a torque and so rotate as well.

For a sample which floats within the fluid like, e.g., a submarine, the upthrust is equal to the weight and moreover the average density of the sample must be equal to that of the fluid in which it is immersed.

These results are summarized in *Archimedes' principle*

■ When a body is wholly or partially immersed in a fluid, the fluid exerts an upthrust on the body equal to the weight of the fluid displaced by it.

Problem 11.3 Hydrometers are used to determine the densities of liquids. Figure 11.6 shows a common type. The glass bulb is loaded so that the hydrometer floats with part of the narrow stem above the liquid surface. The depth to which the device sinks depends on the density.

The narrow stem is usually calibrated directly; and the numbers indicating the density increase downwards. Why downwards?

The scale is not a linear one. Why not?

11.3 Fluids flowing

If a pressure difference develops in a fluid, then the fluid will tend to flow 'down' the pressure gradient from high to low pressure. Obvious examples are the flow of water along a river bed, the flow of hot water in a central heating system, the flow of air around and into an atmospheric low pressure system, convection currents, etc. Less obvious examples are the flow of traffic along a road system, of people around a town centre; of animals in migration.

Consider first a closed system – for example, a central heating system. Figure 11.7 illustrates some of the basic principles.

Water is pumped round the circuit; it is warmed as it passes through the heat exchanger

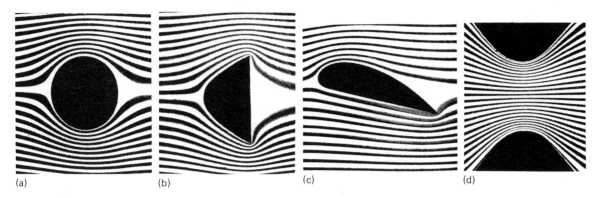

(a)   (b)   (c)   (d)

Fig. 11.8 Stream-line flow (a, b and c) around objects of various shapes and (d) through a constriction in a channel. (From F. W. Sears, M. Zemansky and H. D. Young, *College Physics*, Addison-Wesley, 1980.)

(perhaps a gas, a coal or an oil-fired boiler) before flowing through the 'radiators' which are heat exchangers whose function is to transfer energy from the warm water to the air of the room.

The first point to notice is that the mass of the water in the system is conserved (if it isn't the system has developed a leak!).

The second point to note is that the rate of flow of water (current) measured perhaps in kg s^{-1} is the same at the points marked X, Y and Z in Fig. 11.7.

And the third point is that the water current at Q divides with the sum of the currents on either side of the junction Q being the same. Similarly at R, S, and T.

The system is a transport system; the water transports energy received from the 'boiler' to the rooms in the building. The pump which provides the pressure difference to generate the flow requires an external supply of energy.

The flow of water obeys very simple rules in such a circuit. But the behaviour of the flowing water itself is much more complex. One only has to watch the flow of a river in flood or the motion of the cloud of water droplets rising from a power station cooling tower, or even the smoke rising from a bonfire to realize this. To simplify the analysis, we shall take as our model a fluid which is incompressible and ideal in the sense that there is no internal friction. Liquids approximate quite closely to this ideal.

The path followed by an element of a moving fluid is a *line of flow*. And we shall further assume that these lines of flow are *laminar* or *streamline* – which conforms to our assumption of zero internal friction. Examples of such flows are given in Figure 11.8.

Figure 11.9 shows a 'flow tube' bounded by streamlines. No fluid crosses the side 'walls' of the flow tube. And, since the liquid is incompressible, all the fluid entering the tube at one end must leave it at the other.

In the figure the cross-sectional area changes from A_1 to A_2 as the incompressible fluid flows down the pressure gradient from p_1 to p_2.

In time Δt the mass of fluid entering the flow tube through A_1 is $\varrho A_1 v_1 \Delta t$, where v_1 is the speed of motion of A_1. The mass leaving at the other end in the same time is $\varrho A_2 v_2 \Delta t$.

The volume contained between A_1 and A_2 is constant and, since the flow is steady, the mass leaving the flow tube must equal the mass entering it.

$$\varrho A_1 v_1 \Delta t = \varrho A_2 v_2 \Delta t$$

and so

$$A_1 v_1 = A_2 v_2$$

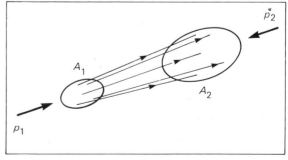

Fig. 11.9 Flow into and out of a flow-tube of varying cross-section.

So, when the cross-section of a flow tube decreases (as in a constriction), the velocity of the moving fluid increases.

For a gas, the density changes and the equation is modified to

$$\varrho_1 A_1 v_1 = \varrho_2 A_2 v_2$$

Problem 11.4

Water is flowing steadily through the tube ABC in which B has a narrower bore than either A or C, which are equal (see Fig. 11.10).

a) How does the speed of the water in B compare with that in A and C?
b) Explain your answer to (a).

Imagine a small cylindrical object being carried, submarine-like, within the water.

c) What will happen to the speed of this object as it moves from A to B and from B to C?
d) What does this imply in terms of the forces on the ends of the cylinder?
e) What does this tell you about the water pressure in A, in B and in C?

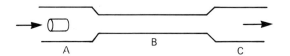

Fig. 11.10

a) It must be faster because (b) the volume of water passing any point in unit time must be the same since water is incompressible.
c) The cylinder must accelerate as it passes from A to B and it must slow down again as it leaves B for C.
d) In moving from A to B, the cylinder experiences a net force from left to right in the figure; and vice versa in moving from B to C (Newton 2).
e) The pressure in A and C must exceed that in B.

This is an example of the application of the principle of Bernoulli, and may be expressed qualitatively: in the flow of a fluid, the faster the flow the lower the pressure.

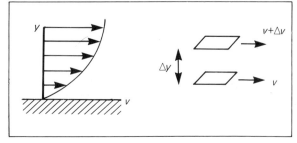

Fig. 11.11

11.4 Viscous effects

The pump which forces the water to circulate in the central heating system described above expends energy as it works against the internal frictional or *viscous* forces which resist the flow. These resistive forces ensure that the layer of the fluid which is nearest to the wall of the tubes in which the fluid is flowing has the lowest speed. In fact, the layer immediately adjacent to the wall will have zero speed. At some point within the following fluid, the speed will be a maximum. Figure 11.11 shows how the velocity vectors change with distance from the boundary layer. We may say that the fluid is being *sheared* – one layer sliding over the layer below it. The forces between such layers are the viscous forces.

To describe a coefficient of viscosity, consider two such layers, the lower with velocity v and the upper with velocity $v + \Delta v$. Each layer exerts a tangential force F_1 and F_2 on the other; the upper tending to accelerate the lower and the lower tending to slow down the upper one. If the system is in equilibrium these two forces must be equal and opposite (Newton 3).

Let A be the area of the layer considered; then F/A is the tangential stress. Let $\Delta v/\Delta y$ be the velocity gradient within the fluid.

The ratio of the tangential stress to the velocity gradient is known as the *coefficient of viscosity*,

$$\eta = \frac{F/A}{\Delta v/\Delta y}$$

The streamline flow of liquids in pipes or tubes is readily examined by experiment using apparatus of the sort shown in Fig. 11.12.

The results show that the rate of flow of volume $\Delta V/\Delta t$ is dependent directly on
(i)   the pressure gradient $\Delta p/l$,

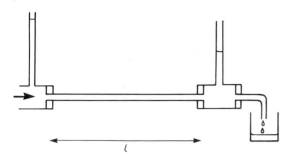

Fig. 11.12

(ii) the fourth power of the radius a of the tube, and inversely on

(iii) the viscosity of the fluid at the temperature of the experiment.

$$\frac{\Delta V}{\Delta t} = k\frac{a^4}{\eta}\frac{\Delta p}{l}$$

It can further be shown that $k = \frac{\pi}{8}$ so that the complete equation is

$$\frac{\Delta V}{\Delta t} = \frac{\pi a^4 \Delta p}{8\eta l} = \frac{a^2}{8\eta}A\frac{\Delta p}{l} \qquad (11.4)$$

where A is the cross-sectional area of the tube.

This equation is known as *Poiseuille's equation* after the French scientist who first derived it in 1842.

For a given tube and for laminar flow the ratio

$$\frac{\text{pressure gradient}}{\text{current}}$$

is a constant which is dependent upon the resistance to the flow in the case considered.

The dependence of the rate of flow on the fourth power of the radius can be of great importance. For example, in blood transfusions, it is better to increase the bore of the needle rather than to increase the height of the bottle. A twofold increase in bore will increase the rate of flow sixteenfold.

11.5 The application of dimensional analysis

Theoretical work on viscosity frequently depends on the technique of dimensional analysis (see Chapter 2). As a further example of the method, let us apply it to the case we have just examined.

The dimensions of viscosity are those of

$N\,s\,m^{-2}$ (see above), or of $(kg\,m\,s^{-2})\,s\,m^{-2}$, or $kg\,m^{-1}\,s^{-1}$. So

$$[\eta] = M\,L^{-1}\,T^{-1}$$

A thought-experiment will suggest that the rate of flow of volume $\Delta V/\Delta t$ depends in some way on the pressure gradient $\Delta p/l$ along the tube, on a the radius of the tube, and on η. Let us write

$$\frac{\Delta V}{\Delta t} = k\left(\frac{\Delta p}{l}\right)^x \eta^y a^z$$

Now

$$\left[\frac{\Delta V}{\Delta t}\right] = L^3 T^{-1}$$

$$\left[\frac{\Delta p}{l}\right] = \frac{[F]}{[A]\,[l]} = \frac{MLT^{-2}}{L^3}$$

$$= ML^{-2}T^{-2}$$

and

$$[a] = L$$

Equating the dimensions of the two sides of the equation gives:

$$L^3T^{-1} = (ML^{-2}T^{-2})^x (ML^{-1}T^{-1})^y (L)^z$$

We obtain, by equating the powers of L,

$$3 = -2x - y + z$$

of M

$$0 = x + y$$

and of T

$$-1 = -2x - y$$

whence

$$x = 1; \; y = -1; \; z = 4$$

So that

$$\frac{\Delta V}{\Delta t} = k\frac{\Delta p}{l}\frac{a^4}{\eta}$$

as before.

11.6 Applying the flow model to other phenomena

In the case of the flow of water (a real fluid) in a narrow tube the current (rate of flow) is proportional to the force due to the pressure gradient and inversely proportional to a resistive term (due to the vicosity of the fluid).

In addition we have seen that, where a complete circuit without branches is involved, the

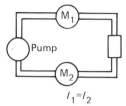

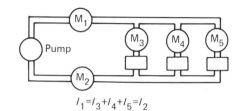

Fig. 11.13   Fluid flow in (a) a series arrangement, and (b) a parallel arrangement of pipes.

$I_1 = I_2$

$I_1 = I_3 + I_4 + I_5 = I_2$

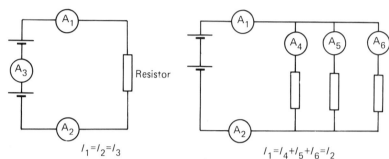

$I_1 = I_2 = I_3$

$I_1 = I_4 + I_5 + I_6 = I_2$

Fig. 11.14   Electric currents in (a) a series, and (b) a parallel circuit.

current of the fluid is the same wherever it is measured (Fig. 11.13a) and that where such a circuit is branching the sums of the currents (I_1, I_2, etc.) on either side of the junction are the same (Fig. 11.13b).

Simple electrical circuits

When similar arrangements of batteries (or cells), ammeters and resistors are wired up, similar results apply (Fig. 11.14).

This, of its own, is sufficient reason for us to talk of electric *currents* in wires in exactly the same sense as we talk of currents in cases of tangible fluids. We shall return to this is Section 23.1.

The conduction of heat

Experiment shows that when heat is conducted along a solid bar (Fig. 11.15), the rate of flow of energy is proportional to the temperature gradient

$$\frac{T_1 - T_2}{l}$$

and to the cross-sectional area (A).

$$\boxed{\frac{\Delta Q}{\Delta t} = \lambda A \frac{T_1 - T_2}{l}} \qquad (11.5)$$

where λ is known as the *thermal conductivity* of the material of the bar. It does depend on the temperature; it is normally quoted in $J\,s^{-1}\,m^{-1}\,K^{-1}$ or in $W\,m^{-1}\,K^{-1}$.

The study of thermal conductivities has acquired a great practical significance as, on the one hand, heating engineers have strived to cut down energy losses from buildings and, on the other hand, have tried to promote efficient energy transfer in heat exchangers.

Current building regulations in the U.K. require that the walls of houses should have a value for λ/l (known as the U value) of less than $1\,W\,m^{-2}\,K^{-1}$. A brick with a thickness (l) of 225 mm will have a U value of about $4.4\,W\,m^{-2}\,K^{-1}$. A cavity wall made of two layers of brick separated by a 50 mm air gap and lined with plaster will meet the regulation.

Table 11a reveals very clearly why pockets of air are so useful when incorporated into material used for lagging and for warm clothing. And styrofoam cups make excellent calorimeters for experimental work in the lab!

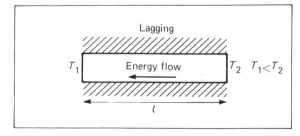

Fig. 11.15   Energy flow along a conductor from high to low temperature.

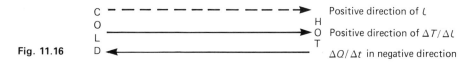

Fig. 11.16

Positive direction of l

Positive direction of $\Delta T/\Delta l$

$\Delta Q/\Delta t$ in negative direction

Table 11a Some thermal conductivities at $T = 273\,\mathrm{K}$

Material	$\lambda/\mathrm{W\,m^-\,K^\pm}$
Copper	400
Aluminium	238
Glass	~1
Brick	~1
Breeze block	~0.4
Styrofoam	0.01
Air	0.24

The conduction of electric charge

Experiment shows that, when electric charge flows through a conductor the rate of flow of charge (or current)

$\propto$ potential gradient $\dfrac{V_1 - V_2}{l}$ applied to the

conductor

$\propto$ area of cross-section (A)

$$\therefore i = \frac{\Delta Q}{\Delta t} = \sigma A \frac{V_1 - V_2}{l} \qquad (11.6)$$

where σ is known as the *electrical conductivity* of the material; it depends on the temperature of

Table 11b Some electrical resistivities at 273 K

Material	$\varrho/10^{-8}\,\Omega\mathrm{m}$
Copper	1.55
Aluminium	2.45
Iron	8.70
Tungsten	4.82
PVC	10^{22}

the conductor. It is more usual to quote the resistivity of the material rather than the conductivity. The *resistivity* of a material is the reciprocal of its conductivity. The unit of resistivity (ϱ) is the ohm metre ($\Omega\mathrm{m}$).

11.7 A comparison of several types of flow phenomena

1. Laminar flow for fluids

$$\frac{\Delta V}{\Delta t} = -\frac{k}{\eta}a^4\frac{\Delta p}{\Delta l} = -\frac{a^2}{8\eta}A\frac{\Delta p}{\Delta l}$$

2. Energy (heat) flow in solids

$$\frac{\Delta Q}{\Delta t} = -\lambda A\frac{\Delta T}{\Delta l}$$

3. The flow of charge in conductors

$$\frac{\Delta Q}{\Delta t} = -\sigma A\frac{\Delta V}{\Delta l}$$

$$= -\frac{1}{\varrho}A\frac{\Delta V}{\Delta l}$$

In each case the rate of flow is proportional to a gradient (of pressure, temperature or potential), to a term involving the cross-sectional area and to some resistive constant which is characteristic of the material through which the flow takes place.

The negative signs in this form of the equations arise because the flow is down the gradient. If, for example, $\Delta T/\Delta l$ is taken as temperature increasing in the direction of increasing l, the direction of heat flow is the direction of decreasing l and conversely (see Fig. 11.15).

Chapter 12

THERMAL EQUILIBRIUM

We first looked at the concept of energy in Chapter 8. That chapter opened by considering the world-wide problem concerning the supply of fuel – by looking, in other words, at our primary energy sources. As we progressed through that chapter we learned to refine our ideas about energy; we learned how to measure it; and we learned that energy is conserved. Why then is there a world fuel crisis? It is not simply the fact that our fuel requirement is rising faster than our ability to provide it. The crisis seems to lie in the *nature* of our energy requirement. Oil, coal, natural gas and nuclear energy are all transformed to other forms, as we make use of them. But it seems that these new forms are not (ultimately) as much use to us as the original fuel was.

In what way does the energy we take from these primary fuels manifest itself, then, if energy is always conserved? Fig. 12.1 shows a flow diagram for a coal-fired power station supplying electrical energy to our homes. It is not hard to see that virtually all the energy originally stored in the coal eventually ends up warming up the surroundings. Almost every other energy flow chart shows the same end-point. So the problem about why there is a fuel crisis turns upon the question 'why are we unable to utilize energy as "heat" with the same effectiveness that we can utilize other forms of energy?'

12.1   One-way processes

The transfer of energy which results simply in things getting warmer is an example of a *one-way process*. We can burn a piece of paper, reducing it to ash and consequently warm up our surroundings slightly. Energy is conserved – all the energy that goes into the surroundings as 'heat' comes from the chemical energy stored in the material of the paper and the oxygen required to burn it – no more and no less. Yet the energy from the sur- roundings never goes back into the ash, to reconstitute the paper. There is nothing in the law of conservation of energy to prevent this, but it never does happen. It is a *one-way* process. Another example of a one-way process is a brick,

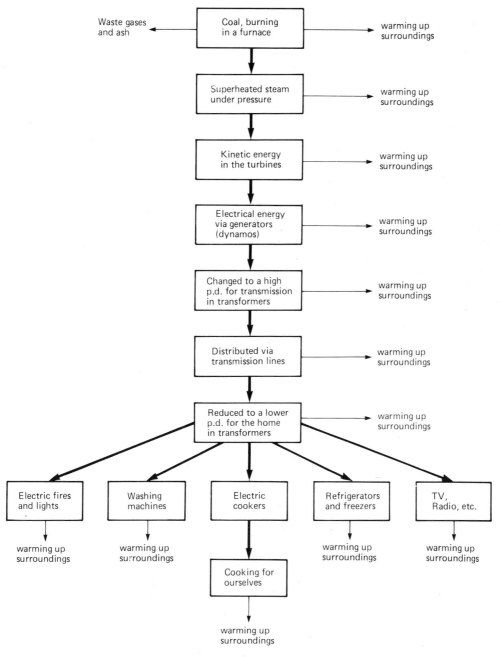

Fig. 12.1 Energy flow from a coal-fired power station to the home.

falling to the ground. The energy it loses is transferred to noise, warmth of the surroundings and distortion of both the brick and the ground. The conservation of energy allows us to say that there is sufficient energy in all these three to raise the brick again from the ground, to end up where it started, in our hands. But that never seems to happen. In fact a good test of a one-way process is to make a film of it and then run the film backwards. The unreality of the backwards running event is a sure test of the process's intrinsic 'one-wayness'.

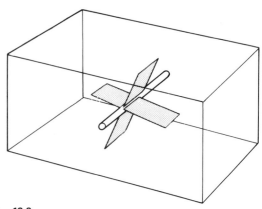

Fig. 12.3

Problem 12.1   Think of three more one-way processes and draw energy flow diagrams for them.

A backward-running film of some events does not, however, appear strange. An example is that of a swinging pendulum, filmed for a few oscillations; another is that of an air-track vehicle moving backwards and forwards along a track. Even some more complex energy transfers do not appear unduly unrealistic. A lead–acid accumulator can drive a motor which in turn speeds up a flywheel. The energy transfer diagram is given in Fig. 12.2. In reverse, the wheel would slow down, turning the 'motor' (which now acts as a dynamo) so 'charging up' the cell – a process that is by no means impossible.

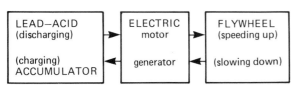

Fig. 12.2

But even these apparently *two-way* processes eventually come to a stop. Why? It is because during each energy transference, a little energy is transferred to the surroundings and this 'little' is never returned.

Problem 12.2   Think of some more (nearly) two-way processes. Is it impossible to make any process *exactly* two way?

12.2   The nature of thermal energy

We have seen in Chapter 10 that the particle model for matter allows us to interpret the nature of thermal energy. In that model it is the kinetic energy of the constituent particles. If energy is transferred to the surroundings it is shared amongst all the particles which go to make up the surroundings. This is the way thermal energy differs from all other forms. The energy of a swinging pendulum remains associated with the pendulum; the spinning flywheel retains all the energy transferred to it. If the energy remains localized it seems possible to get it back. When energy is transferred to the surroundings – or indeed to any object in warming it up, the energy does not remain localized. Instead it becomes spread out and shared between all the constituent particles.

You might say this is equally true of the flywheel; all *its* atoms share in the energy of rotation. The difference is, however, that the atoms all move in the same way, co-operating, as it were, with each other. This is not so when we 'heat up' a gas. Suppose we place a small paddle wheel in a gas and then warm up the gas (Fig. 12.3). The gas particles move more quickly. Can we make the paddle wheel rotate? Certainly not. The particles of gas hit its blades more often and more quickly, but as many come from one direction as come from another and no energy can thus be extracted from the hot gas.

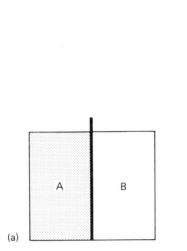 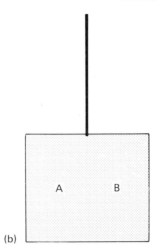

Fig. 12.4 (a) (b)

Energy cannot be extracted from warm surroundings, not just because it is spread out, but also because it is spread out in a *random way*.

Other one-way processes

Transferring energy to warm things up is not the only example of a one-way process. Suppose a little coloured ink is dropped into water. The colour rapidly spreads throughout the water – it *mixes*. The ink *never* spontaneously *un*mixes.

Pour a hundred small white polystyrene spheres into a large beaker and then pour a hundred black ones on top. If the beaker is shaken up, the black and white spheres soon become mixed up. No matter how much the shaking is continued, the spheres never seem to unmix themselves again.

Problem 12.3 Make a list of some more one-way processes involving mixing.

Suppose a film were made of the mixing up of the black and white spheres from their originally ordered state. You would have no difficulty in telling if the film were being run forwards or backwards when you saw it projected. A backwards-running film would confirm that it is quite unrealistic to expect to see order appear out of chaos. If, however, a film were made of just four spheres (two black and two white) being shaken up in a beaker you would have great difficulty in telling whether it was being projected forwards or backwards.

So one-wayness in changes involving mixing involves the jumbling up of large numbers of objects. By comparison the one-wayness of energy transfer involves the sharing of energy amongst large numbers of particles in a random way. As we shall see, these two processes are not unrelated.

12.3 The diffusion of gases

In order to understand why energy transfer to warming things up is a one-way process we shall answer first two rather different questions. These are:

1. Why are events involving the mixing of large numbers of things, one way?
2. Why does thermal energy always flow from a hot body to a colder one – and never, spontaneously, the other way?

We shall start with the first question.

The diffusion of a gas from one container to another is a simple one-way process which is akin to mixing. A large container (Fig. 12.4a) is divided by a gas-tight shutter into two equal halves. Half A contains a gas; half B has been evacuated. If the shutter is raised, gas flows quickly from A to B. What would be the state of affairs a moment later? We would expect to find equal amounts of gas in A and B (Fig. 12.4b).

If we waited, would all the gas end up in B? Or return to A? Experience suggests that this will never happen. This may seem a trivial observation – but its explanation will give us insight into more far-reaching ideas.

12.4 Gas diffusion and the concept of chance

Suppose we have ten counters on a board constructed as in Fig. 12.5.

This is like the gas before the partition is removed. In fact each gas particle, represented by a counter, is moving. It could, at the instant the shutter is removed, be moving in any direction. Will the gas particle be in half A or in half B a moment later? Since it could with equal probability have been moving in any direction, the answer must be that we cannot tell! All we can say is that it has an *equal chance* of being in either half. What then happens, on average, to all the gas particles? We may play a game with dice and the counters on the board which will help us to understand what is happening.

For each counter, throw a die. If a 1, 2, or 3 turns up, a particular counter stays in A: if 4, 5, or 6 it moves to B. After ten throws, how many counters are in A? How many are in B? Now repeat the game for as many times as you have patience, without disturbing the position of the counters at the end of each 'game'. Collect together your results for the number of counters in halves A and B at the end of each game. What distribution of counters came up most often? How often were *all* the counters in one of the halves?

You will probably find that most often there were equal, or almost equal, numbers of counters

in A and B and that you may never, or hardly ever, have got *all* the counters in one half. To understand why this should be so we must consider all the possible arrangements of counters we can encounter. The number of possible arrangements will depend on the number of counters we start with. So Fig. 12.6 shows all the possible arrangements of first *one* counter, then *two, three* and finally *four*.

The number of *possible* arrangements of the counters gets bigger and bigger as their number increases. The number of possible arrangements falls into a pattern which suggests that if we start with N counters, there are 2^N possible arrangements which can be made of them using either side of the board. But in *every* case, only *two* arrangements have all the counters in either A or B. Now the positioning of the counters depends on the fall of a die. Every arrangement has as much likelihood of turning up as every other. The placing of each counter does not depend on where any other counter has been previously placed. So if in 16 games with 4 counters, only 2 arrangements out of the sixteen possible place all the counters on one side or the other we would expect *on average* that 2 games out of sixteen will finish with all the counters on one side of the board.

Problem 12.4 Why do we say 'on average'? Would you be surprised to get *no* games ending with all the counters on one side, out of 16 games played? What if you played 16 000 games?

With 10 counters, we would expect (again on average) that only 2 games in 2^{10} will end with all the counters on one side – that is 2 in 1024.

We believe that the movement of gas atoms follows the same pattern of chance. A box of total capacity 1000 cm³ would contain almost $6 \times 10^{23}/24$ particles at room temperature and pressure (1 mole of any gas occupies about 24 000 cm³ under these conditions). This number is 2.5×10^{22}. Suppose we 'count' the gas atoms in the box once every second. How long would it be on average before we found all the particles in one half of the box? 2.5×10^{22} particles can be distributed between the two sides in $2^{2.5 \times 10^{22}}$ different ways. This is a huge number – we will call it W. Of these ways 2 will have all the particles in one half.

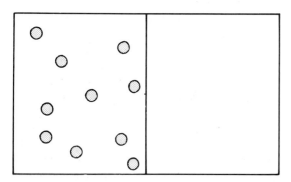

Fig. 12.5

No. of arrangements

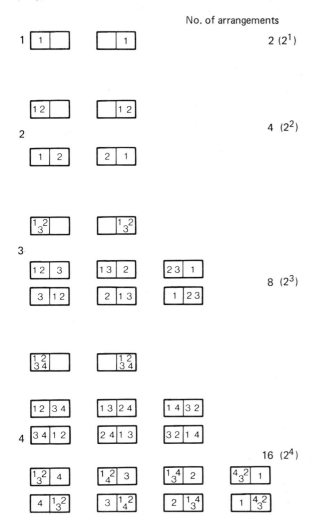

2 (2^1)

4 (2^2)

8 (2^3)

16 (2^4)

Fig. 12.6 Possible arrangements of counters on the two halves of a board.

distribution of the gas particles between the two halves. A little thought will show how in principle we might obtain an answer to this.

The particles do not all appear in one half of the box because there is very little chance of this happenings. All distributions of the particles are equally likely but only *two* distributions out of 2^N give the one required. We do not find this special distribution in practice because it does not happen very often – and it does not happen often because it can only happen in two ways. That which has *few* ways of happening, happens infrequently. The converse of this is that what *is* found in reality must be that which can happen in *most* ways. So gas particles, or counters, are evenly distributed between the two halves because this is the distribution that can be obtained in the most number of different ways. In other words more of the 2^N different ways of arranging the N particles between the two halves give equal, or nearly equal, numbers in both halves than give any other distribution.

We may demonstrate this fact by again playing the game with counters on a board. But this time we will use first 20 counters and then 100 counters. In order to produce more statistically reliable results we will play the game of moving each counter at random 1000 times. It would take a long time to play such a game in real life. To move 100 separate counters each 1000 times would take 100 000 moves. If you could throw a die and make a move in one second, the game would take almost 28 hours continual play. Allowing for the fact that the distribution of counters has to counted after each game and recorded (say another 100 s per game) this makes a total time of 56 hours! So instead of playing it ourselves, we may get a computer to do it for us.

The bar-charts in Fig. 12.7 show the number of counters in the left-hand half of the board plotted against the number of times the 'game' ended in that way. The horizontal scales have been adjusted so that the 100 counter game occupies the same scale length as the 20 counter game. These two bar-charts show that the most likely distribution for both sets of counters is that which achieves equal numbers of counters in both halves and also that the chance of departing from this distribution by more than, say, 10% is much less for the 100 counter game than for the 20 counter

So in $W/2$ seconds we might expect to see this happen once! Of course it might happen on the first observation – or the last – or with equal chance at any time. On average, however, we shall have to wait $\frac{1}{2} \times (W/2)$ seconds – or about $10^{10^{22}}$ seconds. To get some idea of how big such a number is, compare it with the age of the earth which is thought to be 4.5×10^9 years – that is only 10^{17} seconds!

This of course explains why we never do see all the particles in one half of the box. It is not because it cannot happen, but that it is *extremely unlikely* to happen. This argument does not, however, explain why we always find an *even*

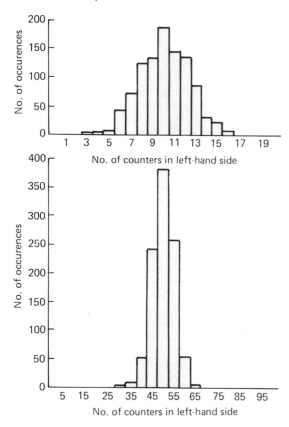

Fig. 12.7 Bar charts showing the distributions for a game with 20 counters (top) and one with 100 counters.

game. A 10% variation in the 20 counter game is a change in number on the left-hand side by ±1 counter. In fact 46.4% of all the games ended with 10 ± 1 counters in the left-hand half. A 10% variation in the 100 counter game is a change in number on the left-hand side by ±5 counters. In this case, 87.6% of all the games ended with 50 ± 5 counters in the left-hand half.

As the number of counters, N, increases, the chances of a departure from a distribution of $N/2$ on each side of the board by any agreed percentage gets smaller and smaller. When N reaches the size of the numbers of gas particles in a box, the chances of a departure from an even distribution by even the smallest fraction of one percent is infinitesimally small.

12.5   The internal energy of a body

Before turning our attention to the flow of thermal energy to see if we can apply the same ideas to that, we must first give some attention to the internal energy of a body. We have seen in Chapter 10 (Eq. 10.11) that a direct consequence of the kinetic model of a gas is that in an *ideal gas* (i.e. one that exactly matches the model) the average kinetic energy of each particles is $\frac{3}{2}kT$. From this it follows that the molar heat capacity of ideal gas is $\frac{3}{2}N_A k$ at constant volume (N_A is the Avogadro constant). In practice, this expression is found to apply only to monatomic gases over a wide range of temperatures, and to some other gases over a narrower temperature range. However, many gases and even most solids have a *molar* heat capacity which can be expressed (to a good approximation) as

$$\frac{n}{2} N_A k \ldots \text{ in J mol}^{-1}\text{K}^{-1}$$

where n is an integer equal to 3 or greater. Table 12a gives some examples. It was James Clerk Maxwell, most famous for his work on electromagnetic waves, who suggested that this might be the result of materials taking up energy in forms different from simple translational movement (that is movement from place to place.) Diatomic molecules can rotate and vibrate; atoms of a solid can only vibrate. In order to explain why different materials had different molar thermal capacities, Maxwell invented a concept of 'degrees of freedom'. The ideas, which modern quantum mechanics has now made largely out of date, need not concern us here except in so far as Maxwell was able to 'explain' the specific heat capacities of many solids by assigning to each atom three independent modes of vibration (in the three dimensions of space) and giving each mode *two* degrees of freedom. Maxwell gave each degree of freedom an energy of $\frac{1}{2}kT$. This made the molar heat capacity of a solid $3 \times 2 \times \frac{1}{2}k \times N_A =$

Table 12a

Gas	C_v/J mol^{-1}K^{-1}	n
Helium	12.47	3.0
Argon	12.47	3.0
Hydrogen	20.42	4.9
Nitrogen	20.76	5.0
Oxygen	21.10	5.1
Carbon monoxide	20.85	5.0
Carbon dioxide	28.46	6.8

$3N_A k$, giving a value of about 24.3 J K^{-1} mol^{-1}. This agrees quite closely with the values for many solids as can be seen from Table 12a.

The importance of this result to us lies in the way it suggests that the internal energy of a solid depends on the vibration of its atoms, and that the average energy associated with each independent oscillation is kT, a result we shall need later.

What happens if we now increase the internal energy of a solid of temperature T by 10%? Does the energy of each atom increase by 10% to 1.1 kT? Again surprisingly, the answer is probably not! Careful experiments described in Unit 11, show that individual atoms cannot increase, or decrease, their energy by any amount as one may add or take water from a jug. Rather, atoms carry energy in much the same way that egg-boxes carry eggs. The quantity of eggs in a box may be increased or decreased by whole numbers of eggs only – half an egg is not possible. We will not go into all the experimental reasons why we believe this to be so. Unit 11 outlines many of the reasons for our belief that energy is transferred between atoms in packets each of which is called an *energy quantum* (or simply a *quantum*). The behaviour of energy is very like that of money, which is transferred in units of currency. When large amounts of money are involved, it matters little that the smallest unit of currency in for example the UK is a half penny. Currency units only become important when small amounts of money are involved – for example, when one needs an exact fare to buy a ticket from an automatic ticket machine.

You will find that paying for energy in energy quanta is rather more complicated than paying for things in currency units. There are a large number of different energy 'ticket machines'. Some accept small units, others will accept only large units (like some automatic petrol dispensers which will accept only £1 notes). Yet others will accept only large units (£5 notes, if you like) at first, but after that will take smaller units. Fortunately for us there is good evidence to show that atoms in vibration take up energy in equal-sized quanta no matter how much energy they already have. Because the energy of an atom can only change in discrete steps, the energy changes possible are often represented as a 'ladder' of energy 'levels'. In this particular case, the 'ladder rungs' (that is

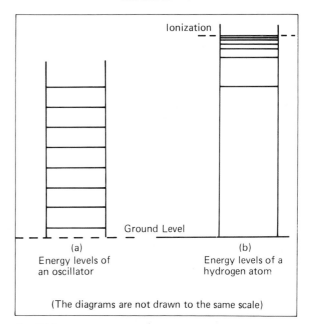

Fig. 12.8

the energy levels) are equally spaced. Figure 12.8 compares the spacing of the energy levels for a vibrating atom in a solid with the more complex energy levels of a hydrogen atom.

12.6 The Einstein model of a solid

In a simple model of a solid we may imagine the atoms attached to each other as though by springs (Fig. 12.9a). Each atom can take up thermal energy by increasing the amplitude or the frequency of its oscillation. In such a model, if one atom started to vibrate more vigorously it would affect all its neighbours, causing their vibrations to change also. The internal energy of a solid would thus be 'smeared out' over all the atoms of the solid. However, we have seen in the previous section that atoms only exchange energy in quanta of well-defined size. An atom can only 'share' energy with its neighbours if it can lose one or more quanta of energy which are subsequently transferred to other atoms.

This sharing is going on all the time as the atoms constantly gain and lose energy to each other. But instead of the energy being smeared out over all the atoms, the energy is more like a swarm of bees with individuals constantly alighting first

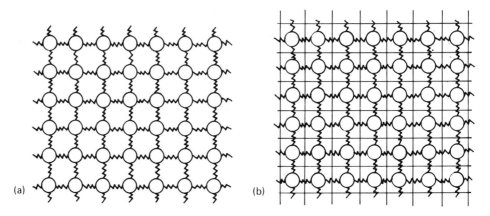

Fig. 12.9 Two dimensions of an Einstein solid.

on one flower and then leaving for another. If, in a particular neighbourhood, six atoms share say two energy quanta then, at best, two will have one quantum each and four none – the 'sharing' cannot be more even than this. To describe this state of affairs, Albert Einstein invented a rather different model of a solid which still bears his name – an Einstein solid. Instead of thinking of the atoms as tightly bound together he thought of them as confined to individual cells with slightly flexible sides (Fig. 12.9b). In this way, atoms could gain and lose energy without necessarily affecting their immediate neighbours (as must be the case if energy is transferred in quanta), while the slight flexibility of the walls enables energy to pass to and from atoms.

12.7 The flow of heat

With the help of this admittedly highly simplified model of a solid, let us see if we can explain why heat energy *always* flows from a hot body to a colder one.

Suppose a particular body consists of 800 atoms, which share 800 energy quanta between them. How will these quanta be arranged amongst the atoms? They *could* all be on one atom, but this seems highly unlikely. More probably you may think there will be just one energy quantum per atom. We will look at the most likely arrangement of the energy quanta later, but for the moment we will consider not the most likely arrangement but the total number of possible arrangements.

To work this out, we need a way of representing the pattern of any particular arrangement of atoms and quanta. To do this, we will represent the solid as a three-dimensional arrangement of atoms as shown in Fig. 12.10.

Starting at the top left-hand corner, we will label the first atom A_1, the next A_2 and so on until the atoms on the front face have been labelled. Then we continue the labelling starting at the top left-hand corner of the next layer. We go on in this way until every atom in the solid has been labelled. Now in a particular arrangement of quanta, each atom will carry a certain number of these energy quanta (which may of course be none). To describe the arrangement, we label each quantum of energy E_1, E_2 and so on and write after the label for each atom, the energy quanta it carries. Thus a particular arrangement of quanta over atoms could be written:

$$A_1 E_1 E_2 A_2 E_3 A_3 E_4 A_4 A_5 E_5 E_6 A_6 \ldots \text{ and so on}$$

(In this arrangement, atom A_1 carries two energy quanta, A_2, carries one, A_3, one, A_4, none, A_5, two, and so on.)

How many different arrangements could we make of the atoms and the quanta? We *must* start with a labelled atom in this arrangement, but having picked one, the remaining 799 atoms and 800 quanta can go in *any* order. Mathematically, making every possible arrangement of 1599 objects is called a permutation, and 1599 objects can be arranged in 1599! different ways. (1599! is called *factorial* 1599 and it represents the number

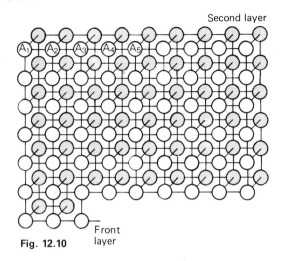

Second layer

Fig. 12.10   Front layer

$1599 \times 1598 \times 1597 \times 1596 \times \ldots \times 1$). However, this procedure will lead to our supposing there are *more* different arrangements than really exist. In the first place the 1599! different arrangements of the letters may include

$$A_1E_1E_2A_3E_4A_2E_3A_4A_5E_5E_6A_6 \ldots$$

as well as

$$A_1E_1E_2A_2E_3A_3E_4A_4A_5E_5E_6A_6 \ldots$$

This is not a different arrangement – this is just a different order of counting up the same atoms (in this case A_3 is counted before A_2). Once the first atom has been chosen, there are 799! ways of arranging the others with the atoms in each ordering carrying the same quanta – and none of these represents a different arrangement. This realization reduces the number of possible arrangements to 1599!/799!.

But this is still more than the true number of different arrangements. To see why we must realize that quanta are not individually distinguishable from each other. We may have labelled the quanta E_1, E_2, E_3, etc., but an arrangement like this:

$$A_1E_1E_2A_2E_3E_4A_4\dot{A}_5E_5E_6A_6 \ldots$$

is not distinguishably different from

$$A_1E_3E_2A_2E_1E_4A_4A_5E_5E_6A_6 \ldots$$

It does not matter whether atom A_1 carries the quanta we have labelled E_3 and E_2 or whether it carries the ones labelled E_1 and E_2 – all that

matters is that A_1 and A_2 each carry two quanta. In any arrangement, the 800 quanta can all be interchanged with one another without affecting the overall distribution – a further 800! ways in all. Thus the total number of *distinguishably* different arrangements of energy quanta over the atoms is

$$\frac{1599!}{800! \times 799!}$$

In future we shall refer to this as the *number of ways* of arranging the quanta and denote it by the letter, W.

Suppose now we add just *one* energy quantum more to the 800 atoms. In how many ways, W^*, can the quanta (now 801) be arranged? Using the same arguments again, it can be seen that

$$W^* = \frac{1600!}{801! \times 799!}$$

So

$$W^*/W = \frac{1600!}{801! \times 799!} \times \frac{800! \times 799!}{1599!}$$

$$= \frac{(1600 \times 1599 \times 1598 \times \cdots \times 1) \times (800 \times 799 \times 798 \times \cdots \times 1)}{(801 \times 800 \times 799 \times \cdots \times 1) \times (1599 \times 1598 \times 1597 \times \cdots \times 1)}$$

$$= \frac{1600}{801} \quad \text{i.e., very nearly 2.}$$

By the same token W will *decrease* by a factor of 2 if just *one* quantum is removed.

We will now repeat the same calculation more generally. If there are N atoms sharing q quanta, then

$$W = \frac{(N + q - 1)!}{q!(N - 1)!}$$

With just one more quantum:

$$W^* = \frac{(N + q)!}{(q + 1)!(N - 1)!}$$

So

$$W^*/W = \frac{(N + q)!}{(q + 1)!(N - 1)!} \times \frac{q!(N - 1)!}{(N + q - 1)!}$$

This simplifies to

$$\boxed{\frac{W^*}{W} = 1 + \frac{N}{q}} \quad \text{if } q \gg 1 \quad (12.1)$$

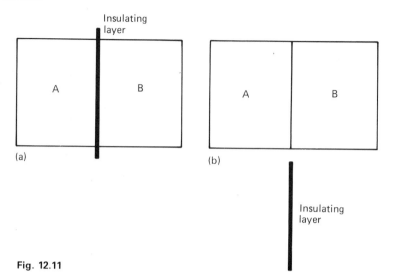

Fig. 12.11

Now let us suppose we have two identical bodies, A and B, each consisting of 800 atoms. (They are of course tiny – but the argument would be no different if we considered each of them to consist of 6×10^{23} atoms – the arithmetic would simply be more difficult!) We will further assume that A shares 800 energy quanta amongst its 800 atoms while B shares only 200 energy quanta amongst *its* 800 atoms. A has on average more energy per atom than B and is thus *hotter* than B. The two bodies are placed close to each other, but are separated for the moment by a thermally-insulating wall (Figure 12.11a). A can arrange its energy quanta in W_A different ways, while B can arrange *its* energy quanta in W_B different ways. Since each arrangement of the quanta in A can occur with each of the arrangements in B, the total number of different arrangements for A and B together is

$$W = W_A \times W_B$$

Now let us remove the insulating layer between A and B (Fig. 12.11b). Suppose just *one* energy quantum is transferred from A to B. The number of ways the remaining quanta can arrange themselves in A is now W_A'. Since W refers to the number of ways of arranging the higher number of energy quanta,

$$W_A / W_A' = 1 + N/q_A$$
$$= 2$$

as we have already seen when $N = q$.

The quanta in B (which have increased by one) can now arrange themselves in W_B' ways, where

$$W_B' / W_B = 1 + N/q_B$$
$$= 1 + 800/200$$
$$= 5$$

So $W_A' = W_A/2$, while $W_B' = W_B \times 5$. The total number of ways the quanta can arrange themselves between A and B is now

$$W_A' \times W_B' = (W_A/2) \times (W_B \times 5)$$
$$= 2.5 W_A W_B$$

By just transferring *one* energy quantum from the hotter body A to the colder body B there are 2.5 times more ways of arranging the quanta amongst the atoms. Transferring another one increases the number of ways by the same factor again, and so on.

Now since the energy quanta are moving around amongst the atoms at random, any one arrangement is as likely to be found as any other (remember the gas diffusion arguments). However, if one particular *distribution* of quanta between the two bodies A and B can be brought about in more *ways of arrangement* of the quanta, than another distribution, this particular distribution will occur more often, on average. (By *distribution of quanta*, we mean the relative proportions of quanta in A and B regardless of which atoms carry particular quanta.) It is clear from what we have shown above that the number of possible arrangements of the quanta between A and B gets greater and greater as energy quanta are transferred from A to B. This is not to say that energy quanta do not move from B to A; indeed movements of this sort must be taking place all the time. But since every possible arrangement comes up with equal probability, the distribution of quanta we shall find in practice is the one associated with the highest number of different arrangements of the quanta.

When does the *net* transfer of quanta from A to B cease? When a transfer of one quantum from A to B does *not* lead to an *overall* increase in the number of ways of their arrangement. This will be when

$$1 + N_A/q_A = 1 + N_B/q_B$$

If there is no net movement of energy from one body to another then the two bodies are in thermal equilibrium with each other. This implies (see Chapter 9) that they are at the *same temperature*. This approach to thermal equilibrium via a consideration of the random movement of energy quanta thus suggests that bodies at the same temperature have the same average energy per atom. We obtained the same result earlier when developing the particle model of an ideal gas.

12.8 The distribution of energy quanta within a body

In the previous section we considered the flow of heat from one body to another and we saw that it was the result of the chance arrangement of energy quanta. The distribution which can happen in the largest number of ways, happens most often. If one particular distribution can be produced in an overwhelmingly large number of ways compared with any other distribution, then this will be the one found in practice.

We have seen that energy quanta move on average from a hot body to a colder one because the number of arrangements of the energy quanta when evenly distributed between the two bodies (N/q the same) is overwhelmingly larger than the number of arrangements associated with any significant departure from this distribution. (A significant departure in everyday energy terms would involve millions of quanta – the energy carried by each quantum is less than 10^{-21} joule.)

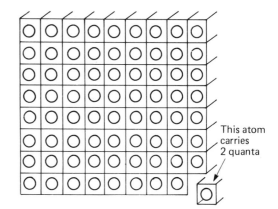

This atom carries 2 quanta

Fig. 12.12

We will now apply the same ideas to the atoms *within* a solid body. We have already calculated the number of ways the energy quanta can be distributed over the atoms. Technically each 'atom' should have been called an *oscillator* since, as we saw in Section 12.5, each atom in a solid can act as three independent oscillators. However, we shall stick to the term 'atom' as some of the results we shall obtain can be applied to liquids and gases as well, and here each energy-carrying unit is indeed an atom, or even a molecule. It is only if we have to *count* the 'atoms' in a particular solid that we must remember that they are really oscillators and thus we must multiply the number of real atoms by three.

Suppose then we have a particular solid of 10 000 'atoms' whose internal energy consists of 10 000 energy quanta. Now let us separate out from this body, *one* atom which happens to carry 2 quanta (Fig. 12.12). The remaining 9999 atoms share the remaining 9998 quanta in W different arrangements. Now let us remove 1 quantum from the particular atom which carries 2 quanta and make it available to the other 9999 atoms. These 9999 atoms now have one *more* energy quantum to share and the number of sharing arrangements increases to W^*, where

$$\begin{aligned} W^*/W &= 1 + N/q \\ &= 1 + 9999/9999 \\ &= 2 \end{aligned}$$

(We could have said $W^*/W = 1 + 10\,000/10\,000$ since the one atom and the quanta it carries have little effect on the total numbers of either. This is of course even more true when realistically large numbers of atoms are being considered.)

Suppose this one quantum of energy is now randomly exchanged by all 10 000 atoms. Since the number of possible arrangements of 9999 quanta amongst 9999 atoms is twice as great as the number of arrangements of 9998 quanta amongst 9999 atoms, we are much more likely at any instant to find this one quantum on one of these 9999 atoms than we are to find it on the particular atom we have singled out for special consideration. In fact we are *twice* as likely to find it amongst the 9999 atoms as on the particular atom chosen. So at any instant, this particular atom is twice as likely to have *one* quantum as it is to have *two*. What is true for one atom must be true for any other atom. At any instant any atom is twice as likely to have one quantum as two. So for all 10 000 atoms taken together, twice as many atoms at any one time will be found carrying one quantum of energy as carry two. (An analogy may help here: if an insurance company says there is a chance of 1 in 10 000 of one of its customers making a claim for fire insurance in one year, it means that if it has 100 000 customers, it expects there will be 10 fire insurance claims in any one year.)

This argument can be repeated for an atom having 2 quanta compared with 3; having 3 quanta compared with 4 and even having 0 quanta compared with 1. In each case, a particular atom may be expected to have the lower number of quanta twice as frequently as it carries the higher number because there are twice the number of ways of arranging all the remaining quanta amongst the remaining atoms when they also share the additional energy quantum. On average, at any one time twice as many atoms will have no quanta as have 1; twice as many will have 1 as have 2 and so on. In general,

$$\frac{\text{number of atoms with } p \text{ quanta } (n_p)}{\text{number of atoms with } p + 1 \text{ quanta } (n_{p+1})} = 2$$

Can this result be justified in practice? To do so, we play another game with counters. In this 'game', a board of thirty-six squares is set up with one counter (representing an energy quantum) on each square (Fig. 12.13). The squares are numbered 1 to 6 horizontally and vertically.

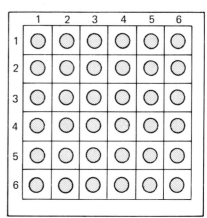

Fig. 12.13

Counters (quanta) are now moved about the board at random. To achieve this, two dice are thrown. On the first throw their numbers designate the row and column of a square on the board. A counter (if there is one) is removed from this square. The dice are thrown again and the counter placed on the new square designated by the dice. The game is played, typically, for 100 moves and the distribution of the counters analysed by counting the number of squares on which there are zero, 1, 2, etc., counters. A graph of number of squares (n) with a particular number of counters (p) is plotted against p. The graph obtained is characteristic of the expected shape with most squares containing no counters, fewer containing one, still fewer containing two, and so on. But with so few squares (atoms) and counters (quanta) a constant ratio between adjacent values of n is unlikely to be found in practice. More convincing results will be found if the game is played on a larger board (say 20 × 20 squares) for a much larger number of moves. Such a game can be played rapidly by a small microcomputer and Fig. 12.14 shows a histogram plotted for such a game. Figure 12.14a shows the result after 1000 moves; Fig. 12.14b shows the result of 10 000 moves. (The second game took the computer over 2 hours to play!) The close similarity between the two figures suggests that both represent an 'equilibrium' distribution – no overall change having taken place over the 9000 moves between the first game and the second. With 400 'atoms' and 400 'quanta', the expected ratio of adjacent values of

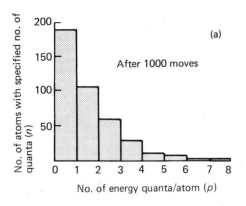
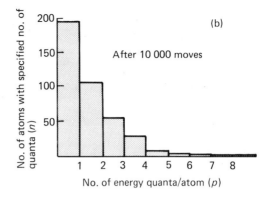

Fig. 12.14   Histograms showing distributions after (a) 1000 moves, and (b) after 10 000 moves.

Table 12b

n	p	n_p/n_{p+1}
196	0	–
106	1	1.8
54	2	2.0
27	3	2.0
6	4	4.5
4	5	1.5
2	6	2.0
1	7	2.0

n is again 2 (i.e. $1 + N/q$). Table 12b gives the values found for the graph in Fig. 12.14b.

Even with a board of this size, there are still considerable departures from the expected ratio due to the relatively small numbers involved.

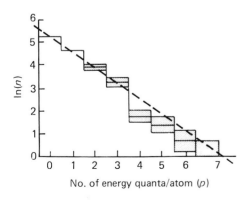

Fig. 12.15   Variation of ln n with p. The shaded areas show the likely error to be associated with each value of ln n.

A distribution in which the successive ratios of n_p/n_{p+1} are constant, is called an exponential distribution (see also Section 26.7). Another test for the expected exponential relationship between n and p is shown in Fig. 12.15, where the natural logarithm of n (ln n) is plotted against p. The straight line again confirms the exponential nature of the relationship and the slope of the line gives the average ratio of n_p/n_{p+1} to be 2.1, which is satisfactorily close to the expected value of 2.

12.9   Bodies of different temperature

So far we have considered only the distribution of energy quanta in a body for the special case in which

$$\frac{\text{no. of ways of arranging } q \text{ quanta among } N \text{ atoms}}{\text{no. of ways of arranging } q + 1 \text{ quanta amongst } N \text{ atoms}} = 2$$

We have shown that this ratio, which we called W^*/W, is the same as the ratio

$$\frac{\text{no. of atoms carrying } p \text{ quanta}}{\text{no. of atoms carrying } p + 1 \text{ quanta}}$$

Thus, in general,

$$n_p/n_{p+1} = W^*/W$$
$$= 1 + N/q.$$

Figure 12.16 shows the results of playing the 'quantum game' with different ratios of q/N. Figure 12.16a repeats the results for $q = N$. Figure

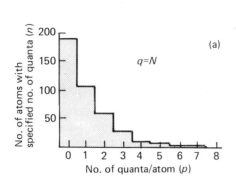

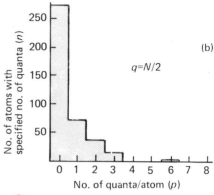

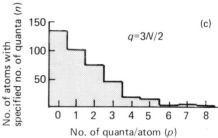

Fig. 12.16 Histograms showing distributions for different values of q/N; (a) $q/N = 1$, (b) a colder body with $q/N = 0.5$, (c) a warmer body with $q/N = 1.5$.

12.16b shows the results for a 'colder' body in which $q/N = 0.5$. As expected the ratio of n_p/n_{p+1} is greater now, in line with the expected value of $1 + N/q = 3$. Figure 12.16c shows the results for a 'hotter' body in which $q/N = 1.5$. The ratio of n_p/n_{p+1} is now smaller. In this case $1 + N/q = 1.7$.

12.10 The Boltzmann factor

Let us call the ratio n_p/n_{p+1}, μ, for the sake of convenience. A consequence of the constancy of this ratio is that

$$\frac{n_0}{n_1} = \frac{n_1}{n_2} = \frac{n_2}{n_3} = \cdots = \frac{n_{p-1}}{n_p} = \mu$$

Thus

$$\frac{n_0}{n_p} = \frac{n_0}{n_1} \times \frac{n_1}{n_2} \times \frac{n_2}{n_3} \times \cdots \times \frac{n_{p-1}}{n_p}$$

$$= \mu \times \mu \times \mu \times \cdots \times \mu$$

$$= \mu^p$$

And so

$$\frac{n_p}{n_0} = \mu^{-p}$$

Taking natural logarithms of both sides:

$$\ln\frac{n_p}{n_0} = \ln\mu^{-p}$$

$$= -p\ln\mu \qquad (12.2)$$

Now $\mu = 1 + N/q$, and at ordinary (say, room) temperatures $q \gg N$. So N/q is a small number, much less than 1. Under these conditions

$$\ln(1 + N/q) \approx N/q \qquad (12.3)$$

This rather surprising result can be confirmed by putting a few values into a calculator. Table 12c shows some results of doing this: Hence $\ln\mu = N/q$ for all practical purposes. Substituting this in Eq. 12.2 gives

$$\ln\frac{n_p}{n_0} = -p\frac{N}{q} \qquad (12.4)$$

Table 12c

N/q	$1 + N/q$	$\ln(1 + N/q)$
0.5	1.5	0.41
0.1	1.1	0.95
0.01	1.01	0.00995
0.001	1.001	0.0009995
0.0001	1.0001	0.00010

12.11   The value of *N/q*

We have already seen from our kinetic model of matter that the average value of the energy carried by an oscillator in a solid is kT. The quanta of energy ϵ are also associated with the oscillations of the atoms of the solid. In order to evaluate N/q for a solid we must now take note of the fact mentioned earlier that, in a solid, each atom acts as three independent oscillators and strictly speaking N is the number of oscillators in the solid. With this interpretation of N, if ϵ is the energy carried by each quantum then the average energy carried by each oscillator is a solid is

$$q\epsilon/N$$

and this as we have already seen is kT. Thus

$$q\epsilon/N = kT$$

$$\boxed{\frac{N}{q} = \frac{\epsilon}{kT}} \tag{12.5}$$

This is an extremely important result, found to be valid for much more complex solids than the Einstein model we have used to derive it. It must be so in view of the way we have chosen to measure temperature. However, today the tables have been turned and this equation has been used to define a new scale of temperature, independent of the nature of any material and called the *thermodynamic scale of temperature*. Its unit is the kelvin (K). To all intents and purposes it is the same as the ideal gas scale of temperature.

Returning to Eq. 12.4 we can now write

$$\ln \frac{n_p}{n_0} = -p\frac{\epsilon}{kT}$$

But $p\epsilon$ is the quantity of energy the n_p atoms carry in excess of that carried by the n_0 atoms. Let us call this energy $p\epsilon$, E and write n_p as n_E, where n_E is the number of atoms with energy E. Then

$$\ln \frac{n_E}{n_0} = -\frac{E}{kT}$$

Hence

$$\frac{n_E}{n_0} = e^{-E/kT}$$

$$\boxed{n_E = n_0 e^{-E/kT}} \tag{12.6}$$

This equation describes the variation in the average number of atoms (or oscillators) in an Einstein solid with the energy each carries and shows the exponential character of this variation. The analysis we have undertaken is greatly simplified and obviously does not apply to materials which *cannot* be represented by this model. The extent to which this equation can be applied to *real* materials is a matter for experiment. There are several behaviours of matter which must depend upon the fact that a small proportion of its atoms have much higher energy than the average. Two examples are the evaporation of liquids and chemical reactions. In both cases, change takes place (liquid to vapour, or reaction) for those atoms or molecules with sufficiently high energy, which is often much greater than the average energy of the particles. The proportion of particles having energy, E, compared with those having no energy is given by

$$n_E/n_0 = e^{-E/kT}$$

If $E > kT$, $e^{-E/kT}$ is a very small number. Quite small changes in T can have a large effect upon n_E in proportional terms, while having little effect on n_0. For example, if $E/kT = 12$, $e^{-E/kT} = 6.1 \times 10^{-6}$. If T is increased by 9% to make $E/kT = 11$, then the new value of $e^{-E/kT} = 17 \times 10^{-6}$ – an *increase* of almost 300%.

Under these conditions we can write

$$\boxed{n_E \propto e^{-E/kT} \text{ for small changes in } T} \tag{12.7}$$

The term $e^{-E/kT}$ is called the *Boltzmann factor*.

When this result is applied to the vapour pressure of liquids, or the rate of many chemical reactions, it is found, surprisingly, correctly to explain the way they both vary with temperature. We say 'surprisingly' because we have derived the result from a model for a solid which has no obvious links with either an evaporating liquid, or chemically-reacting materials. This illustrates another aspect of model-building in physics. Results obtained for a model of limited application can be tested for a wider application and, if found applicable, can lead to new insights: in this case new insights into the process of energy interchange amongst atoms and molecules.

12.12   Entropy

We shall close this chapter by answering finally the question that we started with: 'Why cannot we utilize heat energy in the same way as other forms of energy?' We have seen already that thermal energy flows from a hot body to a colder one because the energy quanta have more ways of arranging themselves with the new distribution than they have with the old. This idea is of great importance. In general, if some particular distribution of energy can be arrived at in more ways than any other distribution, this is the one we shall find in practice. The number of ways of arrangement is described by an important quantity called *entropy*.

If one energy quantum of size ϵ is added to a body already containing q quanta and N atoms, then the number of ways of arrangement of the quanta increases to W^* where

$$W^* = W(1 + N/q) \qquad \text{(See Eq. 12.1)}$$
$$= W\mu \qquad \text{(See Section 12.10)}$$

If *another* quantum is added, the new number of ways of arrangement, W^{**}

$$= W^*\mu$$
$$= W\mu^2$$

If n quanta are added, increasing the internal energy of the body by $n\epsilon$, and provided $n \ll q$,

(no. of new ways of arrangement, W')
$$= (\text{no. of original ways, } W \times \mu^n)$$

So
$$W'/W = \mu^n$$
$$\ln(W'/W) = n \ln \mu$$

Now

$$\ln \mu = \ln \left(1 + \frac{N}{q} \right)$$

$$= N/q, \text{ if } q \gg N \qquad \text{(see Eq. 12.3)}$$

$$= \frac{\epsilon}{kT} \qquad \text{(see Eq. 12.5)}$$

Thus

$$\ln \frac{W'}{W} = n \frac{\epsilon}{kT}$$

$$\therefore \ln W' - \ln W = n\epsilon/kT$$

where $n\epsilon$ is the thermal energy added to the body

of temperature T. We will denote this by ΔQ. So,

$$\ln W' - \ln W = \Delta Q/kT$$
$$k \ln W' - k \ln W = \Delta Q/T \qquad (12.8)$$

Now both ΔQ and T can be measured directly. This equation tells us that the ratio $\Delta Q/T$ is a direct measure of the *change* in the number of ways of arranging the energy quanta as a consequence of adding energy ΔQ to the internal energy of the body. The quantity $k \ln W$ whose change is equal to $\Delta Q/T$ is called the *entropy* of a body. It is denoted by the symbol S. Equation 12.8 above can be re-written as

$$\boxed{\text{change in entropy, } \Delta S = \Delta Q/T} \qquad (12.9)$$

In the particular case discussed above, ΔQ is sufficiently small that its transfer to the body does not measurably affect its temperature. However changes in S can be calculated if T changes as well, although the scope of this application is outside the present book.

We have seen that processes that increase W will happen spontaneously and in one direction. This is so simply because the new distribution is *more likely* to occur than the old one. We now see that another way of saying this is that processes in which the entropy increases will happen one way and in one direction.

When energy changes take place, they do so because the change brings about an overall increase in entropy. Transfers to warming the surroundings always lead to an increase of entropy. Energy cannot *overall* go back again because this would cause an overall *decrease* in entropy – that is an overall decrease in the number of possible arrangements of the quanta. And this, with the numbers of atoms and quanta always involved, is so unlikely as to be impossible.

12.13   The second law of thermodynamics and the efficiency of heat engines

From what has just gone before, we have to accept that *it is impossible for the total entropy of the universe to decrease.* This is one way of stating an important principle called the *Second Law of Thermodynamics.* Let us look at just one consequence of this principle when it is applied to

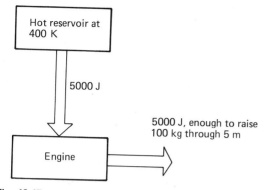

Fig. 12.17　An ideal heat engine with 100% efficiency.

such heat engines as a steam turbine or an internal combustion engine.

Imagine that we were to attempt to construct a heat engine with an efficiency of 100%. Figure 12.17 is a diagram of the energy transfers involved when such an engine raises a load. In this process, the entropy of the hot reservoir will change by an amount

$$\Delta S_H = \Delta Q/T$$
$$= -5000/400$$
$$= -12.5 \text{ J K}^{-1}$$

In a totally isolated system containing only the reservoir, engine and load, there are no other changes of entropy. So the *total* entropy change in the process of raising the load would be -12.5 J K^{-1} – a *decrease* of entropy. And that is *impossible*.

Suppose now we construct our engine with 25% efficiency, transferring some of the heat energy to a cold reservoir (Fig. 12.18). In this new system, the entropy of the hot reservoir still decreases by 12.5 J K^{-1}. However, the entropy of the cold reservoir also changes, by an amount

$$\Delta S_C = +3750/500$$
$$= +12.5 \text{ J K}^{-1}$$

There are no other changes of entropy in the system, so the total entropy change is

$$\Delta S_H + \Delta S_C = -12.5 + 12.5$$
$$= 0 \text{ J K}^{-1}$$

and that is (just) possible.

So the efficiency of a heat engine with a hot reservoir at 400 K and a cold reservoir at 300 K *cannot* exceed 25% if the entropy of the whole system is not to decrease. To raise the efficiency we must raise the temperature difference between the two reservoirs. Let us try making the temperature of the hot reservoir 600 K (Fig. 12.19). The

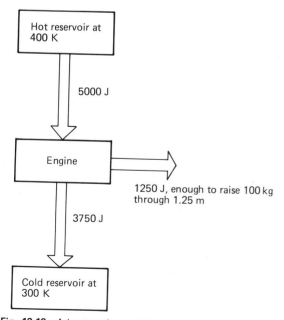

Fig. 12.18　A heat engine working between 400 K and 300 K with an efficiency of 25%.

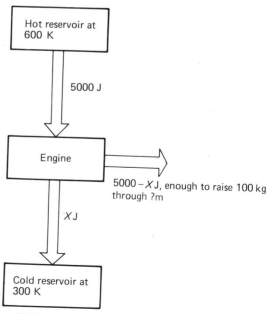

Fig. 12.19　A heat engine working between 600 K and 300 K.

entropy decrease in the hot reservoir is

$$\Delta S_H = -5000/600$$
$$= -8.33 \text{ J K}^{-1}$$

If the overall entropy is not to decrease the entropy of th cold reservoir must increase by *at least* the same amount. Suppose that X J are transfer to the cold reservoir.

$$\Delta S_C = +X/300$$
$$= +8.33$$
$$X = 2500 \text{ J}$$

This leaves $5000 - 2500 = 2500$ J to be transferred to the load.

$$\text{Efficiency} = \frac{\text{useful energy transferred to the load}}{\text{total energy transferred from hot reservoir}} \times 100\%$$

$$= (2500/5000) \times 100$$
$$= 50\%$$

By raising the temperature of the hot reservoir from 400 K to 600 K, the efficiency of the heat engine has been raised from a maximum of 25% to a maximum of 50%.

The question about efficiency which we raised in section 9.5 has its answer. There is a natural and severe limitation on the efficiency of man-made engines, brought about by the second law of thermodynamics.

Problem 12.5 Show that in general if the temperature of the hot reservoir is T_H and that of the cold reservoir is T_C, then the efficiency of such a heat engine cannot exceed

$$\left(1 - \frac{T_C}{T_H}\right) \times 100\%$$

Chapter 13

A MODEL FOR A SOLID

The success of the kinetic model of a gas in which an assemblage of particles has to be endowed with very few properties in order to produce something which behaves much as a real gas behaves is so encouraging that the extension to the liquid and the solid phases is an obvious step to take.

Now we know that all matter can exist in three phases – solid, liquid, and gas. Which phase or state a particular material happens to be in depends on its temperature and also on the external pressure. While gases take up the whole volume of their container and are highly compressible, liquids and solids have a definite volume of their own and, compared with gases, are virtually incompressible. If we are to extend our present model to incorporate liquids and solids, it would appear reasonable to assume that in these two phases, the particles are packed together as closely as possible. Certainly the change in volume of a material, in the liquid phase, is very small as it is cooled down further, compared with the great change in volume which occurs when it is condensed to a liquid from a gas.

Solids possess an important property not shown by air: the property we commonly call strength – the ability to resist not only compressive and extensive forces, but also shearing and twisting forces. It is this strength and rigidity which gives solids their important constructional properties.

None of these features can be explained by the simple model which matched the properties of a gas unless it is assumed that some force acts between the particles, holding them together, whilst some other force acts to keep them apart and prevents further collapse.

13.1 A possible relationship between force, energy and the separation of the particles

Two pieces of smooth metal placed side by side 'in contact' show no sign of attracting each other; even the gravitational force is negligible. If, however, the separation between the two were to be reduced to atomic dimensions (say 0.2 nm), then there would be a powerful attraction between them; so powerful that real metals have great tensile strength.

So the particles of the solid must attract one another with an appreciable force over a range of a few particle diameters. (We assume that the particles are spherical in shape for this is the simplest assumption to make.)

Let us confine ourselves to two particles only, and that these two particles are in equilibrium and separated by distance d as in Fig. 13.1.

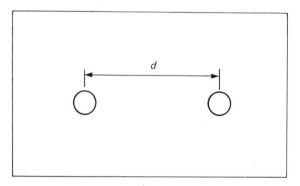

Fig. 13.1

The net force on them is zero: so, in addition to the attractive force there must be a repulsive force. If we apply forces to pull them apart, we have to expend energy to secure their separation. By convention, the potential energy of the two is taken to be zero when they are separated by a great distance (this is an arbitrary choice comparable with the one we make about the potential energy of a very distant mass in the gravitational field of the earth).

The lower part of Fig. 13.2 shows how the attractive force might change during this process and the lower part of Fig. 13.4 shows how the potential energy might change.

On the other hand, if we attempt to force the two particles closer together (to compress them), we find we have to expend still more energy. And,

knowing how difficult it is to produce any appreciable compression in a solid, it appears that the repulsive forces must increase very steeply as the distance between the particles is reduced and also that the range of this force is quite small. The upper part of Fig. 13.2 and the upper part of Fig. 13.4 show how such a force, and how the potential energy, might change in this case.

Adding the two force graphs gives us Fig. 13.3. Adding the two energy graphs gives us Fig. 13.5. Comparing them, we see that equilibrium separation is characterized by zero resultant force and minimum potential energy. Although our argument has rested on a consideration of the forces between two isolated particles, the conclusion also applies to assemblages of particles.

For a large scale model which behaves in a somewhat analogous way, two ceramic ring magnets will illustrate the situation well. Place one magnet in the bottom of a beaker which is a little wider than the magnet. Gently lower a second ring magnet into the beaker so that the magnets repel each other. The downward force on the upper magnet is gravitational; the upward force is due to magnetic repulsion. In equilibrium, with one magnet 'floating' above the other, the two forces are equal and opposite and the potential energy a minimum.

Now let us reintroduce the internal energy. Suppose the minimum potential energy is $-E_0$ and that the internal energy is U_T (written with this suffix to show that it is a function of temperature). If we also assume that E_0 is a function simply of the material and its particles (atoms or molecules) then if $U_T > E_0$, the internal energy of the particles will always be much greater than the potential energy tending to bind the particles together in the collection. So the material will be a gas.

If $E_0 > U_T$ the particles will always be bound together. Extra energy must be put into the material to pull the particles apart. The material will be a solid or a liquid. And in between? Well, if E_0 is only slightly greater than U_T, the particles will be loosely bound together. The material may be thought of as a liquid near to its critical state. Even today, a good model of a liquid has not been obtained. Suffice it to say that the liquid state is something between the two extremes of solid and gas, but nearer to the former than to the latter.

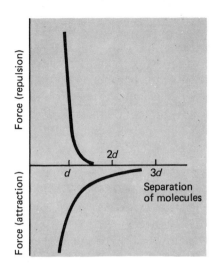

Fig. 13.2 Variations in the forces of attraction and repulsion between two particles with separation.

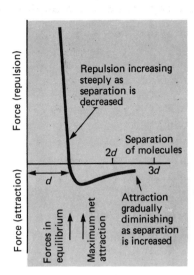

Fig. 13.3 Combining the curves of Fig. 13.2 gives the variation in the resultant force between two particles with separation.

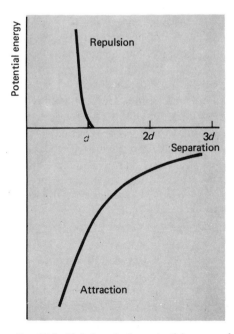

Fig. 13.4 Variations in the potential energy of either particle due to the force fields of the other with separation.

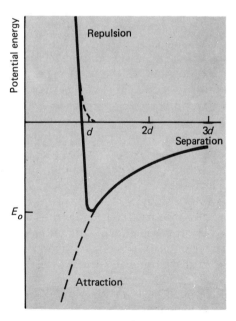

Fig. 13.5 Combining the two curves of Fig. 13.4 gives the variation in the potential energy of either particle with separation.

This we deduce from the observation that the specific latent heat of vaporization is always greater than that of fusion for a substance.

If $U_T \gg E_0$ we may ignore the inter-particle force, and our old model will do (although some of the properties of a gas associated with its departure from the ideal behaviour do depend on E).

Let us now see if this addition of an inter-particle force can help explain the properties of a solid, when $E_0 \gg U_T$.

13.2   The physical properties of solids

The inter-particle force and energy curves have been constructed on the basis of the observed existence of solids and their known ability to resist deformation. It would be unwise now to examine this force curve and show that it apparently predicted the existence of solids and their properties of resisting deformation.

This may seem a rather obvious statement. Nevertheless, you may read accounts of the properties of materials which follow that pattern. Such accounts state as a first principle that the forces between the particles (atoms, molecules, or ions) assumed to make up the material do vary according to Fig. 13.3. This graph is then used to explain the existence of the solid state with its properties of strength and elasticity. It would be possible to justify such an approach to the model, but one would then have to derive the inter-particle force curve from other considerations: say from some model of the structure of the particles. If we digress for a moment to consider such sophisticated models, it would appear that the force tending to bind solids together is not the same in all cases. Consider the three solids: sodium chloride, copper, and diamond. Evidence from the chemical behaviour of these materials and models of their atomic structure suggest that the interparticle forces arise in

a) sodium chloride, from the electrostatic attraction of the oppositely charged ions of sodium (positive) and chlorine (negative).

b) copper, from the interaction of positively charged ions (copper) with a uniformly distributed cloud of electrons,

c) diamond, from the sharing of electrons with neighbouring atoms (covalent bonds).

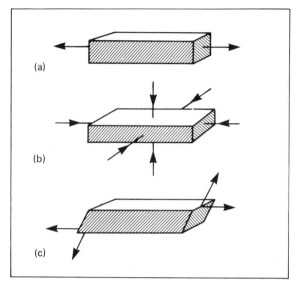

Fig. 13.6 Three ways of deforming a material (a) by stretching in one dimension, (b) by uniform compression, (c) by shearing.

We raise this point now just to point out how careful one must be in seeking to justify a model in terms of its relationships with real properties. One must be clear about which features have been used to set up the model and which features may be predicted *by* the model.

However, it would be naive to assume that all the properties of solids were summed up in their ability to resist deformations. Whereas gases are remarkable for the independence of their characteristic behaviour from their chemical nature, the characteristic behaviours of solids are remarkable for their wide disparity – as the well-known saying 'As different as chalk from cheese' implies. Let us see under what particular headings we normally characterize the behaviour of solids.

a) Stiffness

Solids will deform under the application of any force no matter how small. Figure 13.3 shows that as soon as the particles of the solid are either pulled farther apart or pushed together, a restoring force appears. Solids must deform in order to produce a reaction to the applied force.

There are three basic ways in which a material can be deformed (see Fig. 13.6 a, b, c). It can be stretched or compressed in one dimension (13.6a). It can be uniformly compressed (13.6b). It can be

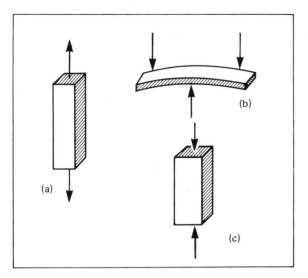

Fig. 13.7

sheared by opposed equal couples (13.6c). Of these, perhaps the most important deformation is that in Fig. 13.6a as it is closely related to the *bending* of materials under applied forces (see Fig. 13.7b). The tendency of a body to resist the deformation characterized by Fig. 13.6a is a measure of what is called its *stiffness*. (See also Section 15.4.)

The extension produced in most solids by a tensile force is proportional to that force, but only for small extensions. This is the most general form of a law attributed to Robert Hooke, who first stated it for springs. The same is true of small compressions.

The extension of any piece of material depends not only on the applied force but also on its length and its cross-sectional area. It is therefore usual to talk of the applied *stress* (equal to the force/unit cross sectional area) and the consequent *strain* (equal to the extension per unit length). The ratio *stress/strain* is a quantity characteristic of the stiffness of the material. It is called the *Young modulus* (E).

This ratio, which is measured in newtons per square metre, tells us how stiff or how flexible a material is. This is quite a different property from the *strength* of the material. Compare, for example, stiff and strong steel with stiff and

weak chalk, or flexible and strong nylon with flexible and weak jelly. Table 13a will enable you to make other comparisons.

b) Strength

By the strength of a material we shall mean explicitly the ability of the material to withstand a force without breaking. The converse of a strong material is a weak material. But there is more to strength than the breaking of a material. We need also to say how the material is going to be broken (see Figs. 13.7a, b, c). The ability of a material to resist breakage in (a) is called its *tensile strength*. Breakage in (b) is closely related to this. Breakage in (c) depends on its *compressive strength*. To say that a material is either weak or strong is not enough. Some materials, brick, for example, are weak in tension but strong under compression.

Table 13a

Material	Young's modulus (E) $N\,m^{-2}$	Tensile strength $N\,m^{-2}$
Aluminium	70×10^9	70×10^6
Mild steel	200×10^9	400×10^6
Cast iron	20×10^9	$30 - 140 \times 10^6$
Nylon	$0.5 - 3 \times 10^9$	70×10^6
Wood	13×10^9	100×10^6 along the grain 4×10^6 across the grain
Glass	70×10^9	$3.5 - 150 \times 10^6$
Carbon fibres	900×10^9	1000×10^6
Concrete	—	4×10^6

c) Stress–strain behaviour

The behaviour of materials under increasing stress up to breaking point is usually characterized by stress–strain curves. Many metals behave typically as in Fig. 13.8. The stress–strain curve is linear up

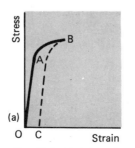

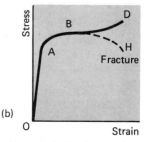

Fig. 13.8 Variation of stress with strain in ductile materials.

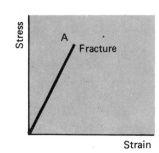

Fig. 13.9 Variation of stress with strain in brittle materials.

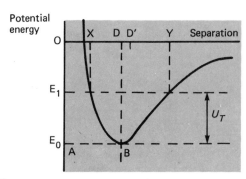

Fig. 13.10

to point A (called the *elastic limit*). Thereafter, the material yields *plastically*, and then on removal of the force, the material is found to have a permanent deformation (OC), (Fig. 13.8a).

Ultimately the material snaps after a plastic deformation many times greater than its initial elastic deformation, which is typically about 1% for engineering materials. Whether the curve follows BD or BH depends whether the average stress along the material is plotted (curve BH), or the real stress at the point of snapping (curve BD) (Fig. 13.8b).

Solids which yield plastically in this way are referred to as *ductile*.

Some other materials such as glass behave as in Fig. 13.9. Fracture occurs at A and there is no plastic yielding. Typically the breaking stress is low compared with that for ductile materials. Materials which behave in this way are referred to as *brittle*.

d) The effect of temperature

Solids expand on heating. We can interpret this expansion in our model using the graph in Fig. 13.10. The equilibrium separation of the particles, centre to centre, is OD.

At temperature T, the particles will have additional energy U_T. Since the particles still have insufficient energy to escape from each other, the form of this energy will be vibrational. As a body vibrates, it constantly re-distributes its total energy between kinetic and potential forms. At X and Y, in Fig. 13.10, when it is momentarily at rest, its energy is entirely potential. In this case, the potential energy $(-E_1)$ must be $-(E - U_T)$. (Remember the total potential energy must remain negative.) The shape of the potential energy curve

will control the amplitude of the vibration. The average position of the vibrating particle will be half-way between these two extremes – at D. OD′ is greater than OD because the potential energy curve is not symmetrical about D.

So the material expands on heating, not solely because the 'thermal wobble' increases but because of the asymmetrical nature of the energy curve. We might expect the curve to be asymmetrical in this way because it is clearly impossible to squash the particles so close that they overlap completely (separation zero), but there is no limit to how far apart they may be pulled. Expansion with rise in temperature is confirmation of this guess.

Apart from expansion effects, temperature also has considerable effects on strength and stiffness. In general, a solid's strength and stiffness decreases with rising temperature.

The brittle characteristics of a solid can be affected by temperature changes. By suitable heating and rapid cooling, a ductile piece of steel can be changed into a brittle piece. Re-heating and slow cooling can reverse the effect.

This is also an appropriate point to note the behaviour of polymeric solids which chemical evidence has suggested are built up from atoms arranged in long chains. At room temperature, rubber shows the typical stress–strain relationship given in Fig. 13.11. However, if the rubber is cooled down a sufficient amount, the stress–strain relationship radically alters to that exhibited in Fig. 13.9, which was typical of the behaviour of glass at room temperature. This change in behaviour and properties takes place at a characteristic temperature called the *glass-transition temperature*. (See Section 13.3).

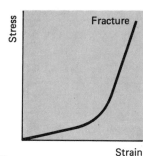

Fig. 13.11 Variation of stress with strain in rubber at room temperature.

e) The effect of time

Time occurs as an effective variable in two important behaviours of materials.

1. Creep. Under the continuing application of a stress, some materials (e.g., timber and many plastics) progressively deform.

In considering earlier stress–strain relationships the observed strain was assumed to take place instantaneously in response to the applied stress. When creep takes place, a progressively increasing extension or other deformation takes place as long as the load is applied. This extension cannot be accounted for in terms of our present model.

2. Fatigue. The repeated application and removal of a stress well below the stress required for fracture can eventually fracture the material. This is a particular feature of metallic behaviour and is the mechanism by which a wire can be broken by bending it backwards and forwards in one's hands.

13.3 Metals and non-metals; glasses and polymers

So far we have considered the general behaviour of solids under stress making no distinction between the various solid forms. But solids do not all behave in the same way. This has led to their classification into groups based on their behaviour under stress. There are, for example, obvious differences between metals and non-metals. Metals are usually (but not always) ductile and plastically distorted by stress; non-metals are usually (but not always – for example many plastics) brittle. A more certain distinction between metals and non-metals is to be found in their electrical and thermal properties under normal conditions. All metals are good conductors of electricity and heat. Non-metals are poor conductors of both electricity and heat.

The distinction between the two classes is never complete, and there have always been materials, such as carbon in its graphite form, which are difficult to classify. In this case, it appears electrically to behave like a metal, but its stress–strain properties are more to be associated with non-metals.

A further group of substances, the semi-conductors, of which germanium is a typical representative, is difficult to classify either as metallic or non-metallic.

There are further distinctions to be made. One group of solid substances will, on heating, pass from the solid to the liquid state without showing sharp transitions. There is no sharp melting point; no involvement of latent heat; no sudden density change. Such substances which are collectively known as *glasses* possess much the same structure in the solid as in the liquid state.

At high temperatures glasses are truly liquid; the atoms are free to move around. As the liquid cools it becomes very viscous and the growth of crystals does not occur. The liquid is super-cooled. As the temperature continues to fall, the random 'liquid-like' structures are 'frozen in' and the glass solidifies without that long-range order which is characteristic of crystals. This occurs at the *glass transition temperature* (T_g) (see Fig. 13.12).

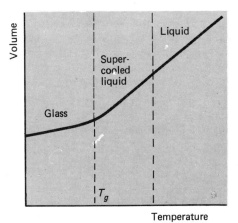

Fig. 13.12 Variation of the volume and state of a glass with temperature.

A further group of materials – many of organic origin (for example, human hair, wool, cellulose, silk), others man-made – have molecules which are formed from very long chains of atoms. The polythene molecule, for example, may have as many as 20 000 carbon atoms (each of which is linked to two hydrogen atoms as well as to the two neighbouring carbon atoms) in a chain. The length of such a chain is in the order of 10^{-3} mm but since it is far from rigid, it will normally be bent, twisted, doubled back and in an altogether tangled state.

Polymers are classed as thermoplastic or thermosetting. The former (which include cellulose acetate, perspex, polystyrene, nylon and polythene) are 'glasses' at low temperature but, at the glass transition temperature, soften, becoming flexible and rubbery and then, with increasing temperature, melt. 'Rubbers' are thermoplastic polymers which are, at normal temperatures, above the glass transition temperature. They are made up of long chain molecules which are held together rather weakly.

Thermosetting polymers are hard and rigid (e.g. bakelite, polyester resin); heating will decompose them but not soften them. In such substances the polymeric chains are cross-linked (unlike the glasses).

13.4 Further investigation of a model for the solid state

While the interparticle force and energy curves can be used to interpret the *inherent* strength and elasticity of solids, they assist but little in the interpretation of the wide qualitative differences in behaviour. See also Section 15.4.

When we considered the gas model, the particles were assumed to be distributed with a random variation in velocity. This provides a reason why all gases behave similarly. There is only one sort of randomness for the particles.

What about the distribution of the particles as the gas is condensed to the liquid and then to the solid phase? We have assumed that they come as close to one another as they can.

If marbles are shaken down into a jar so that they make the most compact arrangement possible, their distribution has an ordered look about it. A completely disordered arrangement takes up more space.

Similarly, we can blow bubbles of a fixed diameter in a soap solution, allowing them to pack close together over the surface. This arrangement is clearly an ordered one. Perhaps the particles in the liquid and solid phases pack themselves into ordered arrangements. And since these particles may not be simply spheres, there may be different arrangements for different materials.

Our model is now becoming more complex. We are having to assume that the constituent particles of matter have certain properties (unnamed) which distinguish them from each other. It is no longer good enough to ignore this fact.

Chapter 14

CRYSTALLINE STRUCTURE

A solid is referred to as crystalline when it exhibits a regularity of form recognized by plane surfaces meeting each other at characteristic angles. By building up ordered structures using large spheres (Fig. 14.1) it is possible to reproduce the macroscopic regularity of form which crystals exhibit. Maybe crystals too have such an underlying

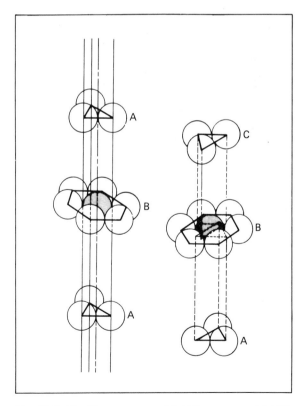

Fig. 14.1 Stacking sequences. (a) ABA sequence. (b) ABC sequence. (From L. H. Van Vlack, *Elements of Materials Science*, Addison-Wesley, 1964.)

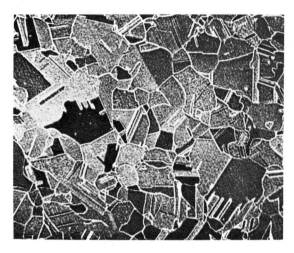

Fig. 14.2 Photomicrograph of hot-worked deoxidized copper. (Courtesy of the Copper Development Association.)

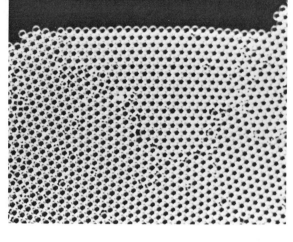

Fig. 14.3 A bubble-raft showing close packing with grain boundaries. (Photograph E. J. W.)

regularity of arrangement of the basic particles (atoms, ions, or molecules).

The fact that many such materials are also possessed of a certain beauty of appearance is not to say that many less notable solids are not also crystalline in their *micro*-structure. Careful examination with a microscope reveals that many substances, whether metallic or non-metallic, are made up of a conglomerate of minute crystals (Fig. 14.2). The outer form of a material may not necessarily reveal an inner regularity of structure.

Before pursuing the question of possible outcomes from a regular or a random arrangement of the particles of a solid, we ought first to see what sort of ordered arrangements we may expect to find. Again, we shall keep our model as simple as possible consistent with the new circumstances. We will assume the simplest shape – a sphere – for the particles and deal only with solids in which a single type of particle is known to be present – namely, elements such as copper and carbon.

In order to help the discussion, we may wish to build large-scale regular structures with large scale spheres (5 cm polystyrene spheres were used in the photographs, Fig. 14.4). These structures are often called 'models', but this is a different use of the term 'model' from that intended when we talk of, say, the 'kinetic model' of a gas. It is much closer to the scale-model builders' use of the word.

14.1   Three-dimensional close packing

A two-dimensional arrangement of soap bubbles, all the same size and packed together as closely as possible, would look like Fig. 14.3. This is the only close-packed arrangement possible. No other arrangement, whether orderly or disorderly, will pack the bubbles so closely together.

A solid is three-dimensional. Close-packing can be achieved by stacking two-dimensional rafts, one on top of the other. To see the sort of thing that can be achieved, we may make structure models from spheres.

Figure 14.1 shows a lower raft, or sheet (A) of close-packed spheres with another one (B) placed above it. Each sphere in the sheet B fits into the spaces between every three spheres in the bottom sheet. If a third sheet, C, is placed upon the sheet B, we find that there are two choices for its position. This will give two different spatial arrangements of the spheres, both close-packed.

If the relative position of the spheres in the first sheet is identified by the letter A, and those in the second sheet by the letter B, then a three-dimensional structure which continuously repeats this form of close packing can be denoted by the letter sequence:

ABABABA ... etc.

Alternatively, each sphere in the third sheet can be placed so that it is directly above neither the spheres in A nor in B. Denoting their relative

Fig. 14.4 Close packed structures made up from polystyrene spheres. Left: hexagonal close packed. Right: face centred cubic.

positions by the letter C, a three-dimensional structure continuously repeating this form of close packing can be denoted by the letter sequence:

ABCABCABCA . . . etc.

Figure 14.4 shows two regular structures built up from 5 cm polystyrene spheres using (on the left) the sequence ABABA . . . and (on the right) the sequence ABCABCABCA . . .). We shall find that these two structures turn out to be rather important. They underlie quite a few crystal structures. Theoretically it is possible to have mixtures like ABCBCBCABCABABABC, etc., but, in practice, they do not occur at all frequently. It will be beyond our present model to explain this. The structure ABA . . . is usually referred to as *hexagonal close packed* (h.c.p.) and the structure ABCA . . . is called *face centred cubic* (f.c.c.).

14.2 Orderly arrangements

Orderly arrangements of any sort can be described in terms of a pattern. The word pattern is commonly used when describing such things as wallpaper or dress fabrics. The essence of order is the constant relationship of any basic unit to its near neighbours. Wallpapers and dress fabrics can differ from one another not only in the design of the basic unit, but also in the relationship of the basic units to each other. The smallest piece of the whole design which displays both the basic unit

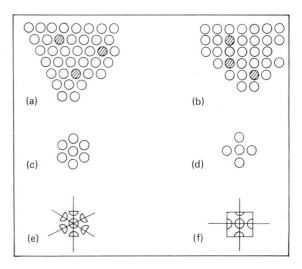

Fig. 14.5

and its inter-relationship with other units is called the pattern.

This idea of a pattern can be carried over to any orderly arrangement. Look at the two-dimensional arrangements of discs in Fig. 14.5a and b. In each case, any one shaded disc bears the same relationship to its neighbours as every other disc within its own arrangement. But the patterns in the two cases are different. Figure 14.5c illustrates the basic pattern in the first case and Fig. 14.5d in the second case. In both cases some discs are shared with neighbouring repetitions of the pattern, so the basic patterns may be drawn as in Fig. 14.5e and f.

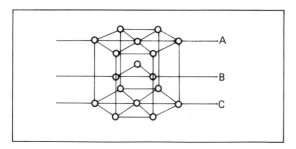

Fig. 14.6 Schematic view of a hexagonal close packed structure showing the location of the atom centres. (From L. H. Van Vlack, *Elements of Materials Science*, Addison-Wesley, 1964.)

In three-dimensional structures, the pattern of the arrangement is the smallest three-dimensional structure which shows the relationship between each atom and its nearest neighbours. The pattern is called the *unit cell* and in the case of the ABA ... structure, is (using dots for the centre of each sphere) shown in Fig. 14.6. Hence the term *hexagonal close-packed*; 'hexagonal' describes the shape of the unit cell.

Figure 14.7 shows the case of the ABCA ... structures. Figure 14.8 shows it as made up solid spheres. (The letters in each diagram refer to the planes identified by these letters in Fig. 14.1.) In this last case, the unit cell is a cube with spheres at each corner and one in each face – hence the term *face centred cubic*.

The shape of the unit cell provides an effective description of the many possible arrangements of the spheres.

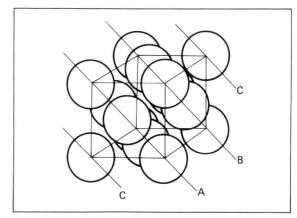

Fig. 14.8 A face centred cubic structure.

14.3 Crystal structure and the properties of solids

We have seen that many of the properties of solids, and, in particular, the wide variations in their properties, cannot be explained in terms of the simple inter-particle force curve. Will the possibility of an ordered arrangement of the particles help?

14.4 Elastic deformation of an ordered structure

In Fig. 14.9, A and B represent sections through two imaginary sheets of atoms in a metal under stress. For simplicity, the stress is shown at right angles to the sheets of the atoms which are arranged in a simple cubic array. The distance between the atoms' centres is x. Under the action of stress the atoms will move apart until the inter-particle force balances the applied force. The distance between the atom centres in the direction of stress increases to $x + \Delta x$.

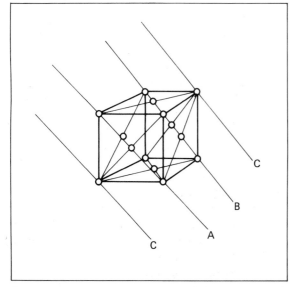

Fig. 14.7 A face centred cubic structure.

If the stress is removed, the atoms will return to their original equilibrium position and the deformation will vanish. This represents the process of elastic deformation which we have already considered.

In such a case the elastic strain is $\Delta x/x$. Since the deformation is elastic, atoms in adjacent planes will be pulled together by a force $nk\Delta x$ where n is the number of atoms in such a layer and k is the force constant. The force per unit area between the two layers will be $(n/A)k\,\Delta x$.

Now the array was a cubical one so that the number of atoms per unit area is $(1/x^2)$. Therefore,

$$\text{force per unit area or stress} = \frac{1}{x^2}k\,\Delta x$$

$$\text{Young modulus } E = \frac{\text{stress}}{\text{strain}}$$

$$= \frac{\dfrac{1}{x^2}k\,\Delta x}{\dfrac{\Delta x}{x}}$$

$$= \frac{k}{x}$$

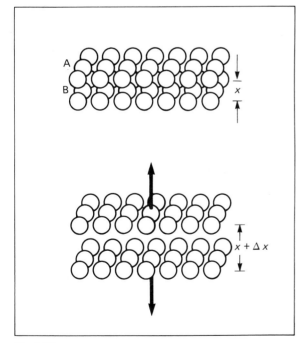

Fig. 14.9

In this case of steel, E is about $2 \times 10^{11}\,\text{N m}^{-2}$ and x is about $3 \times 10^{-10}\,\text{m}$. So the force constant k is about $2 \times 10^{11}\,(\text{N m}^{-2}) \times (3 \times 10^{-10})\,\text{m}$ or $60\,\text{N m}^{-1}$.

The reality of this physical separation of the atoms in a metal under tension has been confirmed by measurements of the inter-atomic spacing in stretched and unstretched specimens, using X-ray diffraction techniques.

Determinations of the speed of pulses along metal bars gives added confirmation of the usefulness of the particle model (see Section 17.5).

14.5 Solids as ordered structures

Evidently the presence of order in solid structures could help to explain some of their important differences. What evidence is there that such order exists? If it does exist, what are its basic patterns?

First, let us attempt to estimate the size of the particles involved in these basic patterns. We will accept the basic assumptions of the atomic model and apply them to a metal, for example, copper. The density of copper is $9000\,\text{kg m}^{-3}$ and its atomic mass 63.6.

Calculate the volume occupied by 63.6 g of copper. Divide this volume by the number of atoms in 63.6 g. This will give you the volume, V, occupied by each atom of copper. If all this space were occupied by the copper, the atoms would be small cubes of side $2.3 \times 10^{-10}\,\text{m}$. If, on the other hand, we assume some close packed structure of copper spheres (it turns out to be face-centred cubic) and so allow for the space not occupied by the copper, the diameter of each atom comes out nearer to $1.3 \times 10^{-10}\,\text{m}$.

Either way, this result supports our assertion as to the small size of these ultimate particles. It is beyond the range of even the most powerful microscopes. We cannot expect to see this ordered arrangement, if it exists, in any direct sense. But we may be able to infer it from some indirect observations which themselves depend on this order. Crystalline structure has already given some hint of underlying order. But the most important materials do not, in general, show any visible external regularity of form.

Instead of trying to see the particle directly, we use a different effect.

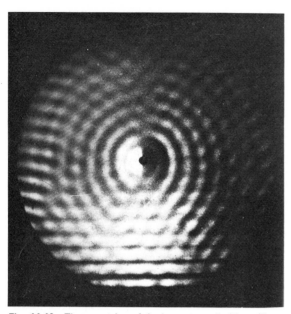

Fig. 14.10 The scattering of ripples at a small object. (From W. Llowarch, *Ripple Tank Studies of Wave Motion*, Oxford University Press, 1961. Copyright Oxford University Press.)

14.6   Interference patterns and crystals

Look at Fig. 14.10. It shows a cylinder placed in the path of plane waves in a ripple-tank. The plane waves are diffracted by the cylinder. The pattern it produces is best described as a plane wave with another circular wave, centred on the cylinder, superimposed upon it.

This pattern is produced only if the cylinder's diameter is of the same order of magnitude as the wavelength of the incident wave. Try it for yourself. What happens if

a) the cylinder is much larger than the wavelength of the plane wave?
b) much smaller?

(See also Section 18.4. To answer, imagine a sea wave passing (a) a large rock and (b) a post.)

In three dimensions, a sphere will set up a spherical secondary diffracted wave centred on the sphere.

An atom will scatter waves in this way if they are of the right order of magnitude of wavelength. We have seen already that atoms are about 10^{-10} m across. We would need radiation of about this wavelength to produce a similar effect to that observed in the ripple tank – X-rays, in fact.

A crystal, we suspect, is a regular array of atoms. Let us see how a regular array of scattering centres would affect radiation whose wavelength is about that of the diameter of a scattering centre.

First consider a simple model in a ripple tank (Fig. 14.11). The scattered radiation is concentrated into specific directions. In this case at right angles to the incident waves. It is no longer evenly distributed as it was from the single cylinder.

In another model (Fig. 14.12) the 2.8 cm radiation is scattered into sharply defined directions only when the model crystal makes particular angles with the incident beam.

This strong reinforcement of the radiation in particular directions is a characteristic behaviour of waves when several waves are superimposed on each other – a behaviour we call *interference*. Each aluminium sphere in the model crystal scatters the electromagnetic radiation in all directions. These scattered wavelets will interfere with each other. We will now see if we can calculate the directions in which we would expect

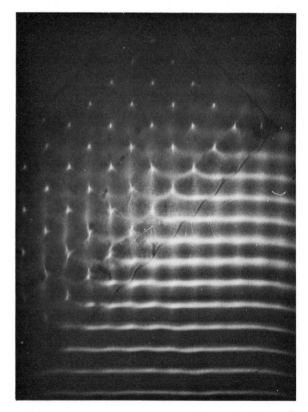

Fig. 14.11 The scattering of plane ripples from a regular array of obstacles in a ripple tank. (Photograph E. J. W.)

Fig. 14.12 Experimental arrangement for the investigation of the scattering of 2.8 cm waves in a model crystal structure. Waves from the transmitter on the left are rendered parallel by the wax lens and then fall on the model crystal which stands on the turntable. The model crystal shown contains 90 aluminium spheres embedded in the polystyrene block. (Courtesy Unilab Ltd.)

strong reinforcement. To simplify matters, consider a section through a regular cubic array of spheres which is placed in the path of an on-coming wave, ONM (Fig. 14.13). AB and CD are planes of atoms and the wave normal NP makes an angle θ with AB.

Let us further suppose that strong reinforcement of the scattered radiation takes place in a direction θ' to AB. If this is so, all the scattered wavelets, from scattering centres in all the planes AB, CD, etc., must be in phase. It may seem a formidable task to write down the conditions for this, but a moment's inspection of the diagram shows that:

a) if PS is in phase with QT, then all waves from scattering centres in plane AB will be in phase with each other. So will all waves from scattering centres in plane CD be in phase with each other.

b) if QT is also in phase with RU, then all waves from scattering centres in AB will be in phase with all waves from scattering centres in CD and so on.

For PS to be in phase with QT,

$$VQ - PW = k\lambda$$

where k is an integer. In triangle PVQ:

$$VQ = a \cos \theta$$

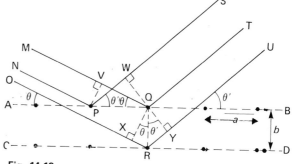

Fig. 14.13

In triangle PWQ:

$$PW = a \cos \theta'$$

So the first condition to be satisfied is

$$a(\cos \theta - \cos \theta') = k\lambda$$

For QT to be in phase with RU,

$$XR + RY = l\lambda$$

where l is an integer. In triangle RXQ:

$$XR = b \sin \theta$$

In triangle RYQ:

$$YR = b \sin \theta'$$

So the second condition to be satisfied is

$$b(\sin \theta + \sin \theta') = l\lambda$$

So for a strong reinforcement of the radiation to take place in a direction θ' to the crystal surface:

$$a(\cos\theta - \cos\theta') = k\lambda \qquad (14.1)$$

$$b(\sin\theta + \sin\theta') = l\lambda \qquad (14.2)$$

These equations are due to von Laue, the discoverer of X-ray diffraction patterns produced by crystals.

The existence of strong reinforcement of radiation falling on a crystal would itself be sufficient to justify our belief in a regular structure. A measurement of angles θ and θ' should enable us to calculate the internal dimensions of the structure. This would be difficult, if it were not for a curious geometrical result which was first worked out in detail by Sir Lawrence Bragg.

He pointed out that there were a large number of ways of dividing up a regular crystal into a number of parallel planes. Let us look again at the simple array of points we have considered.

In Fig. 14.14, the array of points has been divided up in four different ways. You will be able to spot even more. We could describe each of these parallel lines of points in terms of the angle they make with AB.

If α is the slope of lines BU, CT, etc.,

$$\tan\alpha = \frac{b}{a}$$

For rows AG, UH, etc.,

$$\tan\alpha = 0$$

For rows SB, MH, etc.,

$$\tan\alpha = \frac{3b}{a}$$

For rows RD, etc.,

$$\tan\alpha = \frac{4b}{3a}$$

For the vertical rows

$$\tan\alpha = \infty$$

In general, each set of rows can be represented by lines of slope α such that

$$\tan\alpha = \frac{mb}{na}, \text{ where } m \text{ and } n \text{ are integers}$$

Bragg pointed out that in a three-dimensional array one could *always* find a set of planes such that the angle made by the incident radiation with

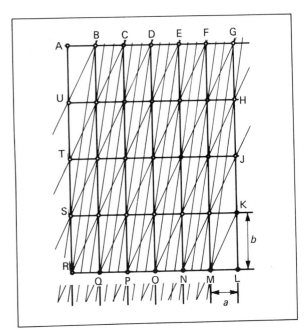

Fig. 14.14

the plane was equal to the angle made by the reinforced radiation with the plane. Let us look again at our regular cubic array.

Suppose $A'B'$, $C'D'$ (Fig. 14.15) is such a set of planes. Then $\theta + \alpha = \theta' - \alpha$. We shall try to prove that these planes include rows of points within the regular array.

Let

$$\phi = \theta + \alpha = \theta' - \alpha$$

Then

$$\theta = \phi - \alpha, \theta' = \phi + \alpha$$

$$\sin\theta = \sin(\phi - \alpha) = \sin\phi\cos\alpha - \cos\phi\sin\alpha$$

$$\sin\theta' = \sin(\phi + \alpha) = \sin\phi\cos\alpha + \cos\phi\sin\alpha$$

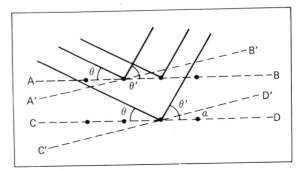

Fig. 14.15

Adding, we obtain

$$\sin \theta + \sin \theta' = 2 \sin \phi \cos \alpha$$

But

$$b(\sin \theta + \sin \theta') = l\lambda \qquad \text{(from Eq. 14.2)},$$

$$\therefore 2b \sin \phi \cos \alpha = l\lambda \qquad (14.3)$$

Similarly we can show that

$$a(\cos \theta - \cos \theta') = 2a \sin \phi \sin \alpha$$

and hence

$$2a \sin \phi \sin \alpha = k\lambda \qquad (14.4)$$

Dividing Eq. 14.4 by Eq. 14.3

$$\frac{2a \sin \phi \sin \alpha}{2b \sin \phi \cos \alpha} = \frac{k\lambda}{l\lambda}$$

$$\therefore \frac{a}{b} \tan \alpha = \frac{k}{l}$$

Hence

$$\tan \alpha = \frac{kb}{al} \qquad (14.5)$$

Since we have already said that k and l must be integers, this is just the condition for the lines to be rows of points within the array.

Let us look again at the condition for constructive interference in terms of a plane for which the incident angle is equal to the angle made by the reinforced beam (Fig. 14.16). $XYZ = n\lambda$ for reinforcement at incident and scattering angles ϕ. In triangles OXY, OYZ,

$$XY = YZ = d \sin \phi$$

Consequently

$$\boxed{2d \sin \phi = n\lambda} \qquad (14.6)$$

This is the Bragg law for crystal diffraction.

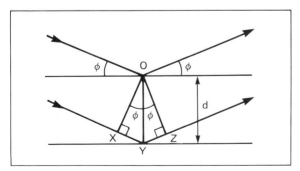

Fig. 14.16

All the strongly reinforced scattered beams can be treated as though they have been reflected from a succession of parallel layers of spacing d. The spacings so worked out will be characteristic of the regular array from which scattering has taken place.

From the set of plane spacings so obtained, it is possible to reconstruct the internal pattern of the array.

14.7 Investigating the structure of real solids

In our discussion above we have limited our attention to an array of known order – a large-scale model. In so doing, the radiation used had a wave of about the same length as the diameter of a particle. Atoms are about 10^{-10} m across. To investigate the structure of solids, we shall need a radiation of similar wavelength. Such radiation exists within the electromagnetic spectrum. It is called *X-radiation*.

The interaction of X-rays with solids has been used both to demonstrate the wave-like nature of X-rays and the regular structure of solids. Here is a case where two models have provided support for each other. This is another interesting example of the development of theories and models in physics.

The techniques of crystal analysis using X-rays are very important. We will content ourselves here by just noting the various methods which can be used.

a) Rotating crystal

The experiment with the 2.8 cm wave and model crystal is an analogue for this method which uses a monochromatic beam of X-rays. These may be detected electronically or, more usually, photographically. The interference maxima form a series of spots which indicate the internal regular structure of the crystal. The basic crystal pattern turns out to be a face-centred cubic. In X-ray diffraction patterns further information about the structure can be gained from the intensity of the interference maxima. The intensities will differ because firstly there may be more than one set of planes in the crystal with the same spacing leading to the same angle θ for the direction of an intense beam, and secondly, scattering off layers which

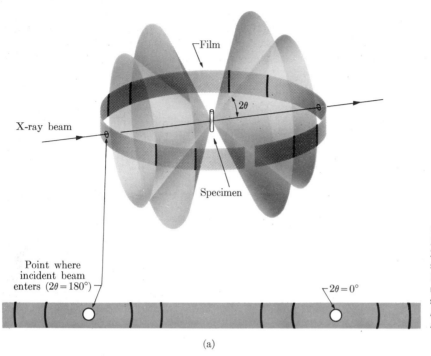

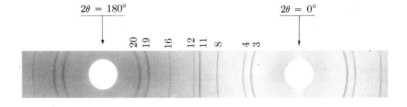

Fig. 14.17 The exposure of X-ray diffraction patterns. Angle 2θ is precisely fixed by the lattice spacing d and the wavelength λ. Every cone of reflection is recorded in two places on the strip of film. (From B. D. Cullity, *Elements of X-ray Diffraction*, Addison-Wesley, 1956.)

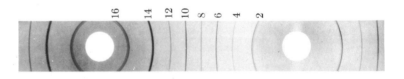

Fig. 14.18 X-ray diffraction for copper. (From B. D. Cullity, *Elements of X-ray Diffraction*, Addison-Wesley, 1956.)

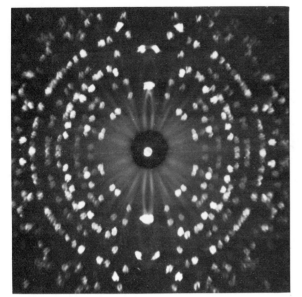 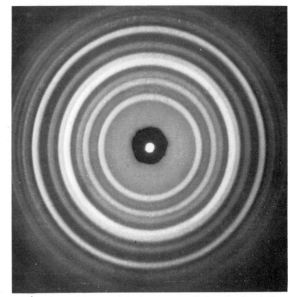

Fig. 14.19 X-ray diffraction patterns for (a) a single potassium alum crystal, and (b) powdered alum crystals. (Photographs Dr H. Judith Milledge, Department of Geology, University College, London.)

are not close packed will in general lead to weaker intensity and thus help to identify the particular planes from which scattering has taken place.

b) Powder method

Instead of rotating the crystal to bring about the necessary conditions for scattering, the crystalline material can be dispersed so that minute crystals are oriented in all possible directions. Some of the planes will always be at the correct angle for their particular spacing. The radiation is scattered into cones as shown in Fig. 14.17. This is a much better method when single crystals are hard to obtain. Figure 14.18 shows a photographic record obtained from copper. The strong interference maxima give evidence of its regular structure: analysis of the photograph shows that copper is a face centred cubic structure. These photographs provide our first real evidence that the basic particles which go to make up metals are packed together in an ordered way just like the spheres of the large-scale model.

c) von Laue's method

Instead of varying θ, it is possible to use X-rays of a wide range of wavelengths, so varying λ. In particular directions there will be a strongly scattered radiation such that

$$n\lambda = 2d \sin \theta.$$

Such pattern are of great historical interest, being the first X-ray diffraction patterns obtained from crystals and thus serving to justify the model of solids and X-rays we have been using.

They were first observed in 1912 by W. Friedrich and P. Knipping following upon a proposal by Max von Laue that, if one assumed that X-rays were electromagnetic radiation and that a regular structure of atoms was a characteristic property of crystals, X-rays penetrating a crystal should behave in a similar way to light striking a diffraction grating. The experiment established both hypotheses. and it led W. H. Bragg to develop the technique of crystal analysis. All three methods are in constant use today.

Figure 14.19 shows X-ray diffraction patterns for alum. The left-hand pattern was made by a single alum crystal and shows the Laue pattern; the right-hand pattern was made by powdered alum crystals. In both cases the X-rays used had a wavelength of 154×10^{-12} m and the patterns are comparable in scale.

Chapter 15

APPLYING THE MODEL FOR MATTER

How does the model we have developed apply to particular cases: to a wire hawser under tension, for example? Or to the special properties of liquid and solid surfaces?

We shall need to bear in mind the force–separation and energy–separation diagrams (Figs. 13.3, 13.5). These indicate that potential energy must be supplied if the strong attractive forces already existing due to be overcome.

15.1 Strain energy in a metal under tension

We will need to examine again the stress–strain graph (Fig. 13.8) for a typical metal. The first part of this graph is linear; the stretching is elastic and Hooke's law

$$\text{stress} \propto \text{strain}$$

applies.

■ Stress has been defined as the force per unit cross-sectional area; strain as the extension per unit length.

$$\text{stress} = \frac{F}{A} \qquad \text{strain} = \frac{x}{l}$$

The graph of force against extension (x) must also be a straight line so long as the extension is elastic (Fig. 15.1). Then $F = -kx$ where F is the force, x is the extension and k is a constant. If, as a result of increasing the force F_1 by a small amount ΔF_1, the extension also increases from x to $x + \Delta x$, the energy transformed may be taken to be $F_1 \Delta x$ since F_1 is, to a close approximation, the mean force acting during this change. It can be

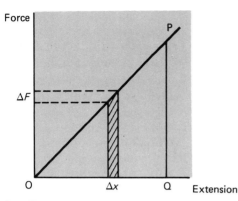

Fig. 15.1

seen that this is equal to the shaded area on the graph in Fig. 15.1.

The total energy involved in the extension of the solid by a length x is the area OPQ on the graph, i.e.,

$$\tfrac{1}{2}OP \times OQ \qquad \text{or} \qquad \tfrac{1}{2}Fx$$

So the energy stored $= \tfrac{1}{2}Fx$

$$= \frac{1}{2}\left(\frac{F}{A}\right)\left(\frac{x}{l}\right)Al \quad (15.1)$$

or

$$\tfrac{1}{2}\,\text{stress} \times \text{strain} \times \text{volume}$$

It is often convenient to express this as the strain energy per unit volume.

$$\frac{\text{strain energy}}{\text{per unit volume}} = \tfrac{1}{2}\text{stress} \times \text{strain}$$

Since $\quad \dfrac{\text{stress}}{\text{strain}} = \text{Young modulus}$

$$\boxed{\frac{\text{strain energy}}{\text{per unit volume}} = \frac{1}{2}\frac{(\text{stress})^2}{\text{Young modulus}}} \quad (15.2)$$

Energy stored as strain energy may be quite considerable. Consider, for example, a steel hawser of cross-sectional area 10^{-3} m^2 and length 10 m which is experiencing a tension of 2×10^6 newtons. If Hooke's law can be applied (and the breaking force for such a hawser might be 3×10^6 newtons so this is not an unreasonable assumption) then

strain energy per unit volume

$$= \frac{1}{2}\left(\frac{2 \times 10^6}{10^{-3}}\right)^2 \times \left(\frac{1}{2 \times 10^{11}}\right) \cdots \text{in J m}^{-3}$$

since the Young modulus for steel is about 2×10^{11} N m^{-2}

$$= \frac{1}{2} \times \frac{4 \times 10^{12}}{10^{-6}} \times \frac{1}{2 \times 10^{11}}$$

$$= 10^7 \dots \text{in J m}^{-3}$$

Total energy stored in the hawser whose volume is $(10^{-3} \times 10)$ m^3 is

$$10^7 \times 10^{-2} \text{J} = 10^5 \text{J}$$

Problem 15.1   What is the speed of a truck (mass 2000 kg) which possesses the same kinetic energy as the energy stored in the hawser?

This is a very considerable amount of energy and, should the hawser break under tension, considerable damage might be done as this strain energy is released and, almost entirely, converted to kinetic energy.

Should this happen, two adjacent layers of atoms must become separated from one another. Two new surfaces are formed. This also involves an energy change.

Each particle in the solid lattice will have a number of near neighbours. Within the body of the solid, these are likely to be regularly distributed in the space around the particle we are interested in. The potential energy of the particle is almost entirely due to the effects of the nearest neighbours, for we have already seen that the energy changes very rapidly with distance. If there are q near neighbours and E_0 is the energy required to separate a particle from its neighbour, the energy which binds it in position is qE_0. To remove such a particle qE_0 units of energy must be supplied. The simplest way of supplying such energy is by heating – and this may result in the sublimation of a solid. In this case the heat of a sublimation is a measure of the energy required to break the interparticle bonds in the solid.

If E_0 is the bond energy and q is the number of near neighbours which each atom possesses, the *energy of sublimation* (L_s) per mole is given by

$$L_s = \tfrac{1}{2}N_A q E_0$$

The factor of $\tfrac{1}{2}$ takes account of the fact that each bond is shared between a pair of atoms. Now

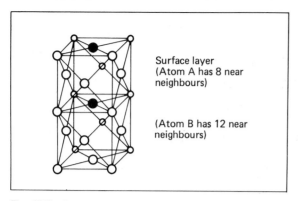

Fig. 15.2   Atoms and their near neighbours in a close-packed structure.

for a close-packed structure q is 12 (see Fig. 15.2), so that

$$L_s = 6N_A E_0 \qquad (15.3)$$

This simple relationship gives a very good guide to the value of the bond energy for certain substances (notably solid neon, argon, krypton, and xenon) but unfortunately not for the metals. This suggests that our model is not without faults and may well require modification later. Another approach must be adopted for metals.

A particle which is near to the surface of a solid has fewer near neighbours than one which is embedded in the interior. We have just seen that, in a close-packed structure, an atom which is well within the crystal lattice has 12 near neighbours. One such atom on the surface of the lattice will have only 8 near neighbours (see Fig. 15.2). Such an atom will have a different potential energy from one in the body of the structure. Now, we have seen that it is convenient to regard the potential energy of an atom as zero when it is at a very large distance from any neighbours; and we have seen how an atom within a lattice is in

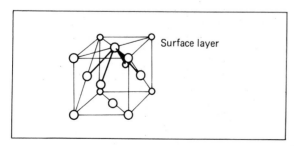

Fig. 15.3

equilibrium about a minimum potential energy E_0. So the atom in the surface has a higher potential energy than the one within, and the surface as a whole must possess surface energy. It follows that the creation of new surfaces by, say, breaking the specimen in two, requires the supply of this energy.

Consider a rectangular bar of metal of cross sectional area A. To break it a total of four bonds must be broken for each atom which exists in the surface (see Fig. 15.3) so that a total energy of $4E_0N_s$ must be supplied. (N_s is the number of atoms in the surface layer.) The area of surface which is created by breaking the bar is 2A so that the surface energy per unit area (γ) is given by

$$\gamma = \frac{4E_0 N_s}{2A} = 2E_0 \frac{N_s}{A}$$

Now we have already seen that the energy needed for the sublimation of N_A atoms is

$$L_s = 6E_0 N_A$$

So the energy needed for the sublimation of N_s/A atoms is $6E_0(N_s/A)$ which is just three times the surface energy of the single layer.

Combining the two equations we have

$$\frac{\gamma}{L_s} = \frac{2E_0 N_s/A}{6E_0 N_A} = \frac{1}{3}\frac{N_s}{A}\frac{1}{N_A} \qquad (15.4)$$

15.2   Liquids and the kinetic model: surface energy and surface tension

Although a liquid is, in many ways, much closer in behaviour to a solid than to a gas we have to remember that the molecules within it can move around quite freely. Within a closely packed solid each molecule has 12 near neighbours, whereas a molecule within the surface has only 8. How far does this apply to a liquid?

The molecules within a liquid are moving in much the same way as we have imagined the molecules to be moving within a gas. The mean free path must be much shorter, but there will still be a distribution of molecular speeds. There is evidence to suggest, however, that at any instant, any one molecule within the largely random arrangement of molecules can be regarded as having 8 to 10 close neighbours. Within the surface there will be fewer. So, as in the case of the

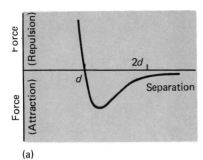

Fig. 15.4   (a)

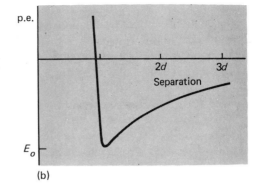

(b)

solid, the energy state of a molecule will depend upon whether it is in the body of the liquid or within the surface layer. Suppose a molecule within the body of a liquid has q_1 near neighbours. The energy required to separate it from those neighbours is on average $q_1 E_0$, where E_0 is the energy required to separate it from one neighbour only. A molecule within the surface has q_2 near neighbours. It is, in consequence, in possession of energy $(q_1 - q_2)E_0$ above its counterpart within the body of a liquid. The surface is said to possess *free surface energy*.

One important consequence of this is that the surface of a liquid tends to occupy the minimum area consistent with the boundaries provided for it. A drop of oil within another liquid of the same density will be spherical in shape so that it has the minimum possible surface area – and the minimum possible free surface energy.

Possession of surface energy implies that the surface is in tension. It may not be immediately obvious that this is so. To analyse the situation, we must first return to the arguments of Section 13.1. There we considered the forces of attraction and repulsion between two molecules in some detail. We saw that the short-range force of repulsion which prevents collapse is in balance with the long-range force of attraction between the two molecules. We saw that, at equilibrium separation d, the two forces were equal. We also saw that the potential energy of the one molecule in the force field of the other was a minimum for equilibrium separation (and zero resultant force). The shape of this potential energy curve is not, however, symmetrical as Fig. 15.4 shows.

We might ask what would happen if a second

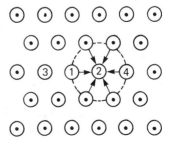

Fig. 15.5   (a) Two molecules in equilibrium. (b) Introduce a second pair of molecules.

pair of molecules is introduced, one on either side of the original pair? See Fig. 15.5.

Molecule 2 is now within the repulsive and attractive force fields of three other molecules. The short range repulsion due to 3 is negligible at this separation. But the longer range attraction of 3 is effective. So molecule 2 moves a little closer to 1. Molecule 1 moves a little closer to molecule 4. The equilibrium separation between the molecules is a little smaller now.

If we extend the argument to the body of the liquid, the same argument will hold. As can be seen in Fig. 15.6 the equilibrium distance between

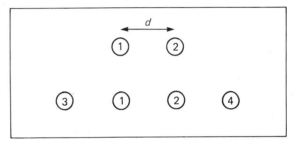

Fig. 15.6   Molecules in the body of a liquid.

the many molecules will be a little less than the separation d of Figs. 15.4a and b. The molecules within the liquid are in compression.

Now neither the molecules in the body of the liquid nor those in the surface layer are static; they are constantly moving – some will be entering the surface layer as others are leaving it. It is easier to leave the surface layer than enter it because the forces on the molecules in the surface are almost entirely due to the molecules within the layer and those below it, there being so few molecules in the vapour above. So, to maintain dynamic equilibrium, there will be fewer molecules in unit volume of the surface than elsewhere. Any one molecule in the surface layer has fewer close neighbours than a molecule within the body of the liquid. These surface neighbours will be a little farther apart than the molecules within the liquid – far enough for the resultant force on each molecule to be attractive. The surface is, therefore, in tension. Moreover, under the action of suitable applied forces, the surface layer can extend in area, with new molecules entering from the body of the liquid. This is a different situation from that existing in a solid under tension and we cannot apply the concept of a modulus of elasticity in this case. Nevertheless the surface tension is a real enough, if small, force.

Surface energy and surface tension

Consider a liquid film (with two surfaces) of width l and length x, bounded by a wire AB which is free to slide (Fig. 15.7). If a force, F, is applied to the wire AB, the two surfaces may each extend by a length Δx so that a new area of surface

Fig. 15.8   A pond skater (*Gerris sp.*) walking on water. (Photograph Lou Gibson/Frank W. Lane.)

appears. This new area will be $2l\Delta x$. The energy provided in this process is $2l\Delta x\gamma$, where γ is the surface energy per unit area. The work done by the applied force is $F\Delta x$. So

$$2l\Delta x\gamma = F\Delta x$$

and

$$\boxed{\gamma = F/2l} \qquad (15.5)$$

The *surface energy per unit area* is equal, numerically, to the force per unit length. This 'line tension' (force per unit length at right angles to a line drawn in the surface of the liquid) provides an alternative measure which is easy to make. Known as the *surface tension* of the liquid, it is quoted in $N\ m^{-1}$. For water, its value is about 7×10^{-2} $N\ m^{-1}$ and the surface energy is about 7×10^{-2} $J\ m^{-2}$.

This surface tension, or if you prefer, the possession of free surface energy, allows a liquid surface to withstand the application of small forces (Fig. 15.8); it is involved in the process of wetting of solid surfaces in contact with it; it controls the shape of liquid drops (Fig. 15.9). Although the effect is a small one, it is of great practical importance. Some of these properties of liquid surfaces are considered below.

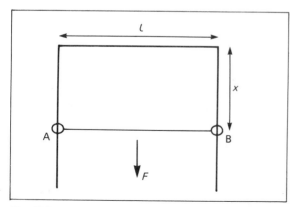

Fig. 15.7

Fig. 15.9 Water drops on a surface which they do not wet.

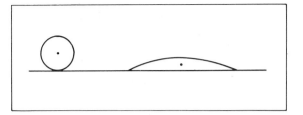

Fig. 15.10 As the spherical drop spreads out across a surface, the centre of gravity falls and gravitational potential energy becomes available.

a) The spreading of liquids and the wetting of surfaces

The shape of a falling drop departs from the spherical, because the drop experiences viscous drag as it falls through the air. Nevertheless, in cloud or fog, the droplets are very nearly spherical. But, under normal circumstances, once they touch the ground they collapse and spread out across the surface (which we shall assume to be horizontal).

This raises a question about the surface energy, which is a minimum for the spherical drop. As the drop spreads, the surface area of the liquid which formed the drop increases and, therefore, its surface energy must have increased. Some of the energy for this will have come from the gravitational potential energy which is made available as the shape changes (see Fig. 15.10).

Now, as we have already noted, surface energy is not a property which is restricted to liquid surfaces. Solid surfaces and vapour surfaces also possess surface energy.

The area of the liquid–vapour surface of the spherical drop increases as the drop spreads out. The solid–vapour surface decreases during this process with a consequent decrease in surface energy. So long as the total potential energy of the

system drop–vapour–solid continues to fall, the liquid will spread across the surface.

The line along which the liquid–vapour surface meets the liquid–solid surface is of particular interest.

Figure 15.11 shows the situation in a glass beaker containing water. Three surface films meet, each of the films being in the order of a few molecules thick. The surface tensions are denoted as γ_{LV} for the liquid–vapour surface, γ_{SL} for the solid–liquid surface and γ_{SV} for the solid–vapour surface. γ_{LV} is usually written γ in tables giving values of surface tensions of liquids. The fourth force A is the adhesive force between the portion of the surface we are considering and the wall.

The shape of the liquid surface near to the solid wall will depend on the difference between the surface energies γ_{SV} and γ_{SL}. In the case shown

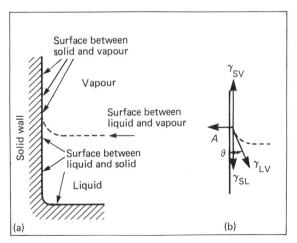

Fig. 15.11 A liquid in a beaker. (a) The three surface boundaries between the solid glass, the liquid (e.g. water) and the vapour. (b) The directions of the forces acting at the boundary.

the *angle of contact* between the liquid surface and the solid wall is θ.

If we consider unit length of an isolated section of the system which is in equilibrium, the vertical components of the forces give

$$\gamma_{SV} = \gamma_{SL} + \gamma_{LV} \cos \theta$$

and the horizontal components give

$$A = \gamma_{LV} \sin \theta$$

The former of these two equations shows that the angle of contact is

$$\cos^{-1}(\gamma_{SV} - \gamma_{SL})/\gamma_{LV}$$

So, if $\gamma_{SV} > \gamma_{SL}$, $\cos \theta$ is positive and θ lies between $0°$ and $90°$. Seen from above, the meniscus is concave and the liquid is said to *wet* the wall surface. If, however, $\gamma_{SL} > \gamma_{SV}$, $\cos \theta$ is negative and the angle of contact is larger than $90°$. Table 15a gives some values of contact angles.

Table 15a

Liquid	Solid wall	Contact angle
Water	Soda–lime glass	0°
Methylene iodide	Soda–lime glass	29°
Water	Paraffin wax	107°
Mercury	Soda–lime glass	140°

The values of the angle of contact are altered greatly by the presence of impurities. Indeed the purpose of a 'detergent' (whether soap or synthetic) is to reduce the angle of contact so that the liquid 'wets' the solid. Conversely the purpose of 'water-proofing agents' is to increase the angle of contact between water and the treated material to 90° or more.

b) Pressure and liquid surfaces

Consider a spherical liquid drop, so situated that it is in equilibrium (Fig. 15.12a).

The spherical shape confers minimum surface area and the pressure within the drop exceeds the pressure outside it. Suppose the drop to have radius r and that the excess pressure is Δp. Now imagine the drop to be cut in half (Fig. 15.12b). The surface tension on the circumference of the cross-section of the right hand half is pulling as shown and the total force along this circumference is $\gamma(2\pi r)$.

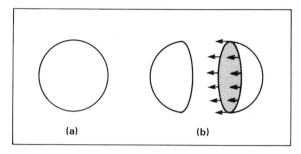

Fig. 15.12

Equilibrium is maintained by the force due to the excess pressure Δp. This is pushing against the internal surface of the right hand hemisphere and everywhere at right angles to that surface. The force component which is effective in maintaining equilibrium at the cut section is the force due to this excess pressure on the projected area of the hemisphere, i.e., on the area shaded in Fig. 15.12b. This area is πr^2.

Equating the two forces we have

$$\gamma(2\pi r) = \Delta p(\pi r^2)$$

Whence

$$\boxed{\Delta p = \frac{2\gamma}{r}} \qquad (15.6)$$

Had we considered a soap bubble (with two surfaces) we would have found that

$$\Delta p = \frac{4\gamma}{r} \qquad (15.7)$$

c) The formation of drops

Table 15b shows the result of applying Eq. 15.6 to a range of drop sizes.

The excess pressure within a cloud droplet is 100 times greater than that within a large raindrop. It is rather difficult to see how the smaller drops can form. In fact, they do *not* form in perfectly pure air even when that air is cooled

Table 15b

	Radius/m	Excess pressure/Pa
An average cloud droplet	2×10^{-5}	7000
A small raindrop	2.5×10^{-4}	560
A large raindrop	2×10^{-3}	70

below the dewpoint (that is the temperature at which condensation normally occurs). But atmospheric air is not pure. Apart from the obvious dust particles, which play no part in the condensation process, there are countless numbers of *condensation nuclei* about which the water vapour may condense. These are hygroscopic in nature; they are often minute particles of sea-salt brought into the air by the evaporation of water from the sea; they are often the products of the combustion of fuels containing traces of sulphur. Such nuclei have radii around 10^{-6} m – that is twenty times smaller than the water droplets which may form around them.

As we shall see, condensation can also occur on ions and this is utilized in the cloud chamber.

Similar arguments apply to the formation of bubbles of vapour within a hot liquid. In the absence of suitable nuclei, bubbles may eventually form explosively at temperatures well above the normal boiling point. Like the drops which form on ions in a cloud chamber, bubbles can also form on ions – and this is the basic mechanism of the bubble chamber.

d) Capillarity

This is the phenomenon of the rise (or fall) of liquids in narrow tubes (and in such materials as soil, bricks, lamp and candle wicks, etc.). It is closely linked with the wetting problem discussed in sub-section (a) above. When the angle of contact is less than 90° (the liquid wetting the tube) the liquid rises in the tube to some equili-

brium height. When the angle of contact exceeds 90°, the level of the liquid is depressed in the tube. In both cases the curved liquid surface is known as the meniscus.

The height to which the liquid rises (or falls) in the tube may be found from a consideration of the pressures involved.

The surface of the liquid in a very narrow tube may be considered as spherical. In Fig. 15.13b, the tube has diameter $2r$, and the liquid surface has a radius R. If the angle of contact between the liquid surface and the wall of the tube is θ we see that

$$R = \frac{r}{\cos \theta}$$

At the point C, which is just outside the liquid surface, the pressure is atmospheric and larger than that just within the surface at B by

$$\Delta p = \frac{2\gamma}{R} = \frac{2\gamma \cos \theta}{r}$$

The pressure at A in Fig. 15.10a is also atmospheric and so the pressure at B, just within the liquid surface, may also be stated as less than atmospheric by the pressure due to the liquid column of height h. If the density of the liquid is ϱ then this pressure difference is $h\varrho g$.

These two statements of the pressure difference across the liquid–vapour surface must be the same, and so we may write

$$h\varrho g = \frac{2\gamma \cos \theta}{r}$$

$$\therefore h = \frac{2\gamma \cos \theta}{\varrho g r} \qquad (15.8)$$

If the liquid–wall angle of contact is zero, then

$$h = \frac{2\gamma}{\varrho g r} \qquad (15.9)$$

This applies, as Table 15a shows, to the case of water in glass tubes.

Problem 15.2 Estimate the height to which water will rise in a glass capillary tube of radius 1 mm. What difference will it make if the tube does not have a uniform bore?

For water ($\varrho = 1000$ kg m^{-3}) in a glass tube the

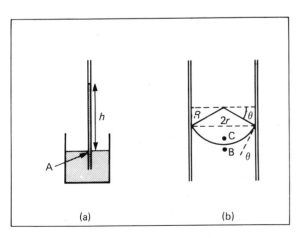

Fig. 15.13 (a) Liquid in a capillary tube which it wets. (b) The meniscus of the liquid.

(a)   (b)

angle of contact is zero, so we may use Eq. 15.9.

$$h = \frac{2 \times 7 \times 10^{-2}}{1000 \times 10 \times .001}$$

$$= 1.4 \times 10^{-2}\,\text{m (14 mm)}$$

The excess pressure depends on the radius of the surface film rather than the bore of the tube. Consequently variations in the tube radius do not affect the height to which the liquid will rise.

Problem 15.3   In an experiment to measure the surface tension of water (for which $\gamma = 7 \times 10^{-2}$ N m^{-1}) a tube of average radius 0.1 mm is placed vertically in the water so that a length of 12 cm is out of the water. What will happen?

This tube has a radius which is one tenth of that in the problem above. The water will be able to rise ten times as far – that is to a height of 140 mm, 20 mm higher than the length of tube provided.

The water will therefore rise to the top of the tube and there form a meniscus of such a shape that the excess pressure will prevent further rise. From Eq. 15.8 it can be seen that this implies that the angle of contact must change. We shall have

$$\cos\theta = \frac{h\varrho gr}{2\gamma}$$

$$= \frac{120 \times 10^{-3} \times 1000 \times 10 \times 0.1 \times 10^{-3}}{2 \times 7 \times 10^{-2}}$$

$$= 0.86$$

and so

$$\theta = 31°$$

---

e) Measuring the surface tension

Capillary rise offers a convenient method. But there are problems; the tube must be thoroughly clean (free of traces of detergent, for example) and the tube radius determined at the level of the meniscus.

15.3 The surface energy of a solid surface

Unlike a liquid, in which the molecules are mobile, the molecules in the surface of a solid are fixed in position. Consequently the surface energy and the surface tension for a solid will be related to each other in a more complex way than for a liquid.

As the area of a surface is increased, the energy transferred per unit area depends on γ, the surface energy and on the way γ changes as the area changes. In the case of a liquid with its mobile molecules the latter change is zero; in the case of a solid it is not. As the surface stretches so the atoms are pulled apart and the surface energy diminishes.

We would expect, then, that the measurement of the surface energy of a solid would be difficult. In the case of copper a reliable estimate has been made by Udin, Shaler, and Wulff (*Metals Transactions*, 1949, p. 186). Fine copper wires support small masses in an evacuated chamber at a high temperature so that the metallic creep is appreciable even under small stresses. If the weight exceeds the surface tension the wire stretches; if not, the wire shrinks. At the critical load, the wire does not creep at all and the surface tension (which, as we have seen, is related to the surface energy) is equal to the weight. The surface energy of copper was found by this method to be about 1.4 J m^{-2}.

In spite of these practical difficulties, a knowledge of the surface energy is well worth having, for, not only does it help us to determine the bond energy but it also enables us to calculate the breaking stress (tensile strength) of a metal and so check the application of the model to a real material.

Consider two adjacent layers of particles separated by a distance x. Then a layer of unit area will have a volume of x. If this material is under stress, the strain energy per unit area of either layer (Eq. 15.2) will be

$$\frac{1}{2}\frac{1}{E}(\text{stress}^2)\,\text{volume} = \frac{1}{2}\frac{1}{E}(\text{stress}^2)\,x$$

If two new surfaces are to be created as the layers are pulled apart, then, this is the amount of energy which must be provided and shared to give surface energy of γ to each. We may write

$$2\gamma = \frac{1}{2}\frac{1}{E}(\text{stress}^2)\,x$$

i.e.,

$$(\text{stress}^2) = \frac{4\gamma E}{x}$$

and   breaking stress $= 2\sqrt{\dfrac{\gamma E}{x}}$ (15.10)

In making this estimate we have assumed that Hooke's law applies right up to the moment of fracture: we know that this is not so. The stress must therefore be too high – but it is unlikely to be more than two or three times too high. If it is twice too high we would expect to find that

$$\text{breaking stress} = \sqrt{\frac{\gamma E}{x}}$$

To test the argument, let us insert the known figures for steel.

Surface energy of steel is about $1\,\text{J}\,\text{m}^{-2}$.
Young modulus for steel is about $2 \times 10^{11}\,\text{N}\,\text{m}^{-2}$.
Atomic spacing in steel is about $2 \times 10^{-10}\,\text{m}$.
So tensile strength (breaking stress)

$$= \sqrt{\frac{1 \times 2 \times 10^{11}}{2 \times 10^{-10}}}\,\text{N}\,\text{m}^{-2}$$

$$= \sqrt{10^{21}}\ \text{N}\,\text{m}^{-2}$$

$$\approx 3 \times 10^{10}\,\text{N}\,\text{m}^{-2}$$

Unfortunately, a typical breaking stress for steel is only about $4 \times 10^{8}\,\text{N}\,\text{m}^{-2}$ and even a specially prepared steel can claim no more than about $3 \times 10^{9}\,\text{N}\,\text{m}^{-2}$.

Moreover, calculations for other metals reveal similar discrepancies. Either the logic of the argument or the model of atoms in layers which we have used must be at fault. Of the two, the argument is the easiest to check, and, even allowing for the guess which we made when we observed that Hooke's law cannot be applied in this case, it does not seem likely that our error could be as much as two orders of magnitude out.

15.4   Plastic deformation and dislocations

The stress–strain curve (Fig. 13.8a) for a typical metal bends somewhat after the linear reversible region (in which Hooke's law applies) is passed. If now the stress is relaxed, the metal returns along a new curve; it has suffered permanent, or plastic, deformation.

This process is often accompanied by a well-known phenomenon called work-hardening. This is easily illustrated. Take about 20 cm of new copper wire of diameter about 3 mm. Bend the end to make a small hook and clamp the other end firmly so that the wire is horizontal. Hook a light-weight carrier to the end and observe what happens as you increase the load. The copper wire will bend rapidly. Now take it between your fingers and bend it between them along its whole length several times. Repeat your test. Now the wire will support a much increased load before yielding. This is known as work-hardening – and its explanation demands yet another modification of the model.

These two phenomena – that of plastic deformation and of work-hardening – can hardly be accounted for in terms of the simplified structure of Chapter 14. When discussing that structure we noted that the model used was simple in the extreme. The layers of atoms which were being examined were assumed to be stressed at right angles to the direction of the applied force. It is far more likely that we should be considering layers of atoms which are inclined to that force.

Layers so stressed are under a shearing stress and, before proceeding, it will be necessary to consider such stresses in general. For simplicity let us take a block of material ABCD and imagine that a pair of forces is applied as shown in Fig. 15.14a. If the block is not merely to turn round, a second pair of forces must be applied as in Fig. 15.14b. The result may be that the material is sheared through the angle θ. This is a measure of the strain.

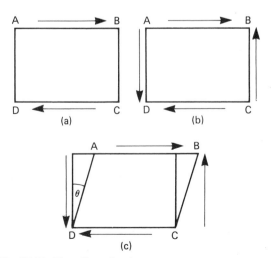

Fig. 15.14   The effect of a shear.

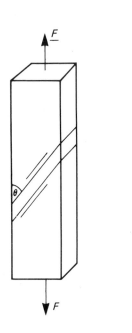

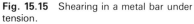

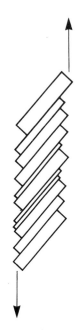

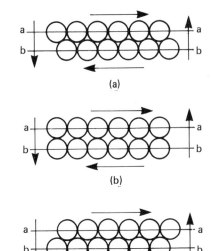

Fig. 15.15 Shearing in a metal bar under tension.

Fig. 15.16

Fig. 15.17

The shearing stress is F/A where A is the area of the surface of the block through which the force F is applied. Experiment shows that

$$\frac{\text{shearing stress}}{\text{strain}}$$

is a constant for small stresses. The name *shear modulus* or *rigidity modulus* (G) is given to this.

Returning now to the case of the metal bar under tension (Fig. 15.15). Suppose that the layer shown represents two such planes of atoms in a bar subject to a tensile stress of F/A where A is the cross-sectional area and F the applied force. On one side of the layer shown there is a force $F\cos\theta$; on the other side $F\cos\theta$ in the opposite direction. This couple and the opposing anticlockwise couple may cause a shear to occur. Should two such layers of atoms move relative to one another under this shearing action, the material may well acquire a permanent elongation (Fig. 15.16).

Let us examine this shearing action in atomic terms, considering a close packed structure (Fig. 15.17a).

If layer a slips relative to layer b by the application of pairs of forces as shown, the resistance to slip will increase to a maximum at some

position intermediate between the initial position of Fig. 15.17a and the position of Fig. 15.17b. In this latter position all resistance to the shear has gone and the layer a is quite likely to slip down to the position of Fig. 15.17c. The angle of shear is now 60°. As a first approximation we can take about one quarter of this as the angle of shear corresponding to the maximum resistance to shear. This is 15° or, in radians, about 0.25. If the material were steel, with a shear modulus of about $0.8 \times 10^{11} \, \text{N m}^{-2}$, we would expect the maximum shearing stress to be

$$\begin{aligned}
&\text{shear modulus} \times \text{shear angle} \\
&\approx 0.8 \times 10^{11} \times 0.25 \, \text{N m}^{-2} \\
&\approx 0.2 \times 10^{11} \, \text{N m}^{-2}
\end{aligned}$$

Unfortunately, experiment reveals that the maximum shearing stress for steel is about one fiftieth of this and so the angle of shear must be much less than 15°. Yet another major discrepancy between a prediction made on the basis of the model and the reality is revealed.

Metals start to deform plastically under stresses which are far smaller than those predicted by theory. It is, in fact, this effect which confers upon them the very important property of ductility. Ductile materials will undergo a considerable amount of irreversible distortion before breaking. In some cases, for example the work-hardening of copper, this may even strengthen the metal. In any case, this ability of ductile materials to accept permanent distortion and therefore to be shaped (whether by cold or by hot working) is a very important property.

It was G. I. Taylor who, in 1934, advanced an explanation of the discrepancy between the calculated and the actual shear strength for metals. He pointed out that the theoretical calculation presupposes a perfect crystal structure. If, however, there are imperfections within the crystal, these imperfections might themselves be able to travel through the material of the crystal and, in so doing, lower the shear strength.

Figure 15.18a shows a linear defect (*edge dislocation*) within a crystal. There is an extra plane of atoms.

If the shearing couple is applied as in Fig.

15.18b, rows of atoms may move relative to one another – the extra row being moved in stages to the edge of the crystal (Fig. 15.18c). The bonds at the end of the extra row will be highly stressed so that the application of quite a small shearing force will cause them to break down and slip to occur. But immediately the bonds are re-established with the next pair of rows and so on until the extra row reaches the edge. Figure 15.19 shows a photograph of some edge dislocations.

Such a defect in the structure can move around quite freely in an otherwise perfect crystal. Sooner or later, however, it will meet another dislocation and this impedes its freedom of movement. A metal with many interlocked dislocations in its crystal structure will be harder than a metal with only a few. It is this which accounts for the work-hardening process.

A simple analogue in two-dimensions is available in the bubble raft of W. L. Bragg (Fig. 14.3 and 15.20).

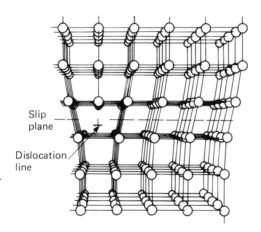

Fig. 15.18a An edge dislocation. (From A. G. Guy, *Elements of Physical Metallurgy*, Addison-Wesley.)

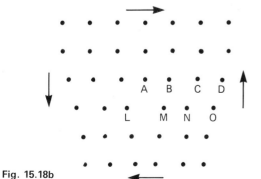

Fig. 15.18b

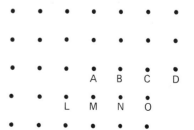

Fig. 15.18c

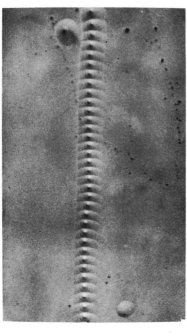

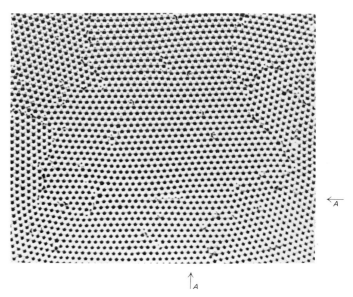

Fig. 15.19 Germanium crystal etched to show the ends of edge dislocations. (From L. H. Van Vlack, *Elements of Materials Science*, Addison-Wesley.)

Fig. 15.20 A bubble raft showing grain boundaries and a dislocation at AA. There are others. (Photograph E. J. W.)

A raft of uniform sized bubbles about a millimetre in diameter is easily formed by blowing gently through a jet held just below the surface of a detergent solution. The depth of the jet below the liquid surface should be kept fixed. These bubbles will adhere to one another, i.e., interbubble attractive forces and also forces of repulsion will exist. The close packed structure of the raft is clearly seen and so is the existence of 'bubble crystal grains'. At a point such as A in Fig. 15.20 a bubble is missing – and a dislocation exists. If, now, shearing forces are applied to such a raft, the motion of the dislocation is very clear to see.

The presence of atoms of a different size from the majority of the atoms which make up the lattice will also play an important part in the interaction of dislocation with dislocation.

15.5 Cracks and brittle fractures

In 1920 A. A. Griffith, who was working with glass fibres, made a remarkable discovery which was to give rise to further modification of the simple model for a solid which we have been studying.

He had been working with thin glass fibres and found these to get stronger as they became thinner. Indeed the strength of a fibre of diameter 2.5×10^{-3} mm was as much as a quarter of the calculated strength, whereas a glass rod of about 1 mm diameter broke at about one eighteenth of the calculated strength.

Griffith suggested that some mechanism must be responsible for concentrating the stress so that, locally in the glass, the stress rose to higher values than expected. This idea of stress concentration had risen seven or eight years before when Professor Inglis was investigating the causes of failure in the hulls of steel ships. Now Griffith extended the idea from the macroscopic to the microscopic scale.

Consider an atomic lattice with its regular array of bonds and particles under tension (Fig. 15.21). Each bond contributes to the whole strength of the material. But suppose that there is a fault or a crack as at C in Fig. 15.22. At the tip of this crack there will be an increase in the

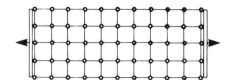

Fig. 15.21

Fig. 15.22

concentration of the stress. The inter-particle bonds are broken along the length of the crack; the whole stress is taken up at the tip. Although the stress which is applied to the material is well below the anticipated breaking stress, just at the tip of the crack the stress will be much higher than we imagine. The bonds between the particles in the vicinity of the crack tip are the ones which are most likely to fail under stress.

Griffith suggested that minute cracks, smaller perhaps than a wavelength of light, would serve to reduce the strength of his glass fibres. It was many years before suitable techniques were developed which enabled men to reveal the crack patterns, but Griffith's analysis had been accepted by many long before this. The particle model could now accommodate brittle fractures of glass.

An excellent example of the effect of a crack in weakening a material is to be found in the breaking of a glass tube or the cutting of a pane of glass. All that the glass worker does is to initiate a crack with a suitable knife and then to apply a suitable stress. The glass breaks cleanly along the line of the crack.

15.6   The thermal and electrical properties of metals

In Section 13.2(d) we observed that the expansion of solids when heated could be explained in terms of the potential energy–particle separation graph of Fig. 13.10. Although the effect is a relatively small one it is of considerable importance.

If a solid bar of length l_0 increases by length Δl as a result of a relatively small rise in temperature ΔT, experiment shows that

$$\Delta l \propto l_0 \Delta T$$

If follows that $\Delta l = \alpha l_0 \Delta T$, where α is a constant for the material of the bar and is known as the *linear expansivity*.

■ The linear expansivity of a substance is the fractional increase in length per unit temperature rise. In SI its unit is K^{-1}.

Typically metals have expansivities between 10^{-6} and $3 \times 10^{-6} K^{-1}$ at ordinary temperatures.

The expansivity of a substance does, however, depend on the temperature. This could be anticipated from the form of the graph shown in Fig. 13.10. As the temperature rises, the potential energy $(E_1 - E_0)$ rises and the asymmetry of the curve within which the particle is confined increases. This means that the average position of the particle from A is increased and this implies a small increase in length.

In addition, the amount of the expansion is roughly proportional to $(E_1 - E_0)$. And, as we have already seen, so too is the specific heat capacity of the material. The graph of expansivity against temperature (Fig. 15.23) and that of specific heat capacity against temperature (See Fig. 9.6) are very similar in shape and this is only to be expected.

As the crystal lattice absorbs energy from a heat source, its internal energy is increased. This energy is available within the vibrations of the atoms as the sum of the potential and kinetic energies of those atoms.

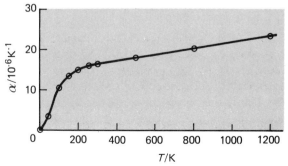

Fig. 15.23 Variation of the expansivity of copper with temperature.

As long ago as 1819, Dulong and Petit observed that most solid elements absorbed nearly the same amount of energy per unit temperature rise if the masses chosen for comparison were proportional to the atomic masses. This has come to be known as Dulong and Petit's law. It is well illustrated in Table 9a in which the molar heat capacities of the metals quoted share a value around 25 J mol^{-1}K^{-1}. See also Section 12.5.

However, as we have seen, experiments show that the value of the molar and the specific heat capacities of substances fall as the solid is cooled. This confirms that, as we have seen in Chapter 12, there must be some restrictions on the way in which the atomic oscillators can accept the energy proffered to them.

If one end of a solid bar is heated energy is conducted away from the heat source mainly by the vibration of the particles which constitute the lattice. One might expect that the thermal conductivities of solids would not differ widely from one another. But this is not so. There are excellent solid thermal insulators and there are excellent solid thermal conductors (especially such metals as copper and silver). The range of thermal conductivities, written λ in Eq. 11.5:

$$\Delta Q/\Delta t = \lambda A \frac{T_1 - T_2}{l}$$

is about 1 to 1000.

Such a wide range of values suggests that the model of thermal conduction based solely on lattice vibrations is inadequate in some way.

This becomes clearer when the electrical properties are considered. On the one hand there are the electrical insulators; on the other there are the electrical conductors (again headed by such metals as copper and silver). The range of electrical conductivities written σ in Eq. 11.6:

$$i = \Delta Q/\Delta t = \sigma A \frac{V_1 - V_2}{l}$$

is about 1 to 10^{10}.

The lattice vibration model is inadequate to account for good thermal conduction. It is difficult to see how it can contribute to electrical conduction in metals and other good electrical conductors. There must be some other mechanism at work.

Problem 15.4 An aluminium kettle containing boiling water is in contact with a gas flame at a temperature of about 550°C. The water is found to be evaporating at a rate of 0.1 kg per min. The base of the kettle is 1 mm thick and its area is 0.015 m^2. Estimate the temperature of the outer surface of the base of the kettle. (Specific latent heat of vaporization of water is 2.27 × 10^6 J kg^{-1} and the thermal conductivity of aluminium at this temperature is 210 J m^{-1}s^{-1}K^{-1}.)

Since the water is vaporizing at the rate of 0.1 kg per min and the specific latent heat of vaporization of water is 2.27 × 10^6 J kg^{-1}, energy is being transmitted through the aluminium base at a rate

$$\frac{\Delta Q}{\Delta t} = \frac{0.1}{60} \times 2.27 \times 10^6 \text{ J s}^{-1}$$

$$= 3780 \text{ J s}^{-1}$$

Substituting in the equation

$$\frac{\Delta Q}{\Delta t} = \frac{\lambda A (T_1 - T_2)}{l}$$

we have

$$3780 \text{ J s}^{-1} = \frac{210 \text{ J m}^{-1}\text{s}^{-1}\text{K}^{-1} \times 0.015 \text{ m}^2 \times (T_1 - T_2)}{10^{-3} \text{ m}}$$

hence

$$T_1 - T_2 = 1.2 \text{ K}$$

Temperature of the outside surface of the base of the kettle is about 1 °C higher than the water temperature within. The metal is 'in contact' with a hot flame at about 500°C. This suggests that the major temperature change must occur in the surface layers of gas.

15.7 Conclusion

The development of theories and models in physics is a very complex process. It is a creative act which goes on alongside the process of observation and experiment. In retrospect, it is always easy to describe a model or theory and then produce an amount of experimental evidence in support of this model.

When a particular model is well established,

this is often the most economical method of approaching it. Such is the case with the model for matter; but further modifications are necessary in order to produce something whose properties correspond closely to the properties of real materials.

The form of the accepted model is now well established in physical theory. An engineer, interested in using materials; a technologist interested in improving their properties; a chemist interested in producing new materials will not concern themselves with the details of the whys and wherefores. They will take, and learn to use, the best model available to achieve the end in view. But these terms *engineer*, *technologist*, and *chemist* are in many ways shades of meaning applied to those involved in the study of physical science. They reflect the particular interest of these people. But all are involved in the processes typified by the work of Boyle. Every so often new observations will be made which do not fit in with accepted models and ideas. The model or theory must be modified. This is the process of creative science. And at these moments it is necessary to ask why science has accepted its present ideas.

There are those scientists today who are engaged in what is often called fundamental research. These are people who continually ask and probe the question 'Why?' There are others – the great majority – who are engaged on the application of present knowledge, extending and using the models already developed. But it is only in recent years that we have seen such an apparent division of 'scientific' study. The history of science suggests that these two activities cannot be successfully divided. Boyle, who is chiefly remembered today for his law, must have been well known in his own day as a theorist and speculator, and not just an investigator and observer. His friend, Robert Hooke, again remembered for his law, was perhaps better known in his time as an architect – a man to whom the *application* of stresses and strains was everyday business.

UNIT FIVE

Waves and Vibrations

Chapter 16

WAVE BEHAVIOUR

After electricity, wave motion is probably the most important topic in modern physics. Some might say that waves appear even more frequently than electricity. Yet if anyone is asked to produce an everyday example of wave motion, almost the only ones that come to mind are the ripples running over the surface of water, or the larger scale effects of waves at sea. It is hard to think of another common example of waves. Yet a proper understanding of waves is essential to anyone investigating communication, whether by speech, radio or television; how we see; how we can make images; what is colour; how musical instruments work; how we navigate; and even why atoms are as they are. How is it that waves are so important, and yet so uncommon in our day-to-day experiences?

It is because wave motion – the behaviour characterized by ripples on the surfaces of a stretch of water – serves as an important *model* when we try to understand so many different aspects of the physical world. Light, radio, sound, even electrons in some of their properties, *behave* in the way waves behave. We cannot see what it is that 'waves' – as we can with the water on the surface of a lake – but then we cannot *see* the tiny particles which make up a gas. We accept models because they provide an explanation of what we *can* observe.

But this is hurrying along too quickly with the story of this Unit. First we shall see what set of properties are characterized by the words 'wave motion'. Then we shall look at a number of other phenomena in physics and see how well they share 'wave-like' properties. Finally we shall try to put these wave models to good use.

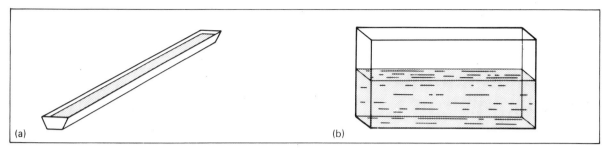

Fig. 16.1 (a) A wave channel. (b) A wave trough.

16.1 Waves on the surface of water

Since ripples on the surface of water are the commonest example of wave motion, we shall use them to start our investigation into wave behaviour.

Water-surface waves can be created in a number of ways. Dropping a stone on to the still surface of a pond can create many of the effects we shall look at, but waves on such a large stretch of water are not easily controlled.

Water waves can be made to run over quite long distances in a laboratory by confining them to a channel, which can be several metres long,

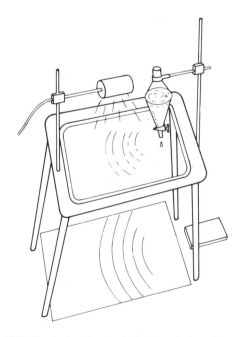

Fig. 16.2 Producing circular ripples in a ripple tank.

made from plastic guttering (Fig. 16.1a). Shorter, transparent tanks are also useful for looking at the up-and-down motion of a water surface as a wave passes across it (Fig. 16.1b). However, neither of these lets us look at a wave spreading out into the space around it – as it does on a pond.

To do this, a wider stretch of water is required. Usually this is arranged in a laboratory in a *ripple tank* (Fig. 16.2). The advantage of such transparent tanks is that a shadow of the ripples can be cast on to white paper beneath the tank and the pattern of the waves more easily seen. Apart from tank and lamp, the only requirement for creating easily-seen ripples is a chemist's separating funnel, which allows drops of water to fall on to the water surface at a controlled rate (about one every two seconds is about right). A number of things can be observed.

The ripples spread out in a series of ever-widening circles. As they cross the water surface a series of 'hills and valleys' (usually called crests and troughs) are formed. A section through the water surface would look something like that shown in Fig. 16.3a.

Such a wave of limited length is called correctly a *wave train* or *wave group*. This distinguishes it from a *continuous wave*, which has neither beginning nor end – and is of course a product of a physicists' imagination! – (Fig. 16.3b), and a *pulse* (Fig. 16.3c) which does not contain any *repeated* up-and-down motion.

The wave train runs across the surface of the water at a steady speed. We shall see what factors control this *wave speed, c,* in a later chapter.

Measurement shows that the distance between successive circles is constant. In fact, the dark

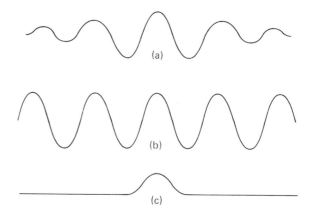

Fig. 16.3 (a) A wave group or wave train; (b) a continuous wave; (c) a pulse.

circles correspond to the 'troughs' in Fig. 16.3. This distance is called the *wavelength*, λ (Fig. 16.4).

The number of crests or troughs passing a particular point on the surface each second is called the *wave frequency, f*. The unit of its measurement is the hertz (Hz) – 1 hertz is the same as one crest or trough passing a particular point each second.

Any one point on a wave of frequency, f, will move forward f complete wavelengths every second – a total distance of $f \times \lambda$. This distance is also equal to the wave speed, c. This leads to the first important result arising from a study of waves:

$$\text{Wave speed, } c = f\lambda \qquad (16.1)$$

Phase

Ripples spreading out from a single point are called *circular* waves. Every point on one particular circle started out at the same time from the source. A particular circle joins up all the points on the wave which started out at the same time

from the source. A line such as this is called a *wave front*. All points on a particular wave front are behaving in the same way at any particular instant of time. They are said to be *in phase*. This word 'phase' will be given a much more precise meaning in a later chapter, but for our present use we need only understand two uses of the word. The first is 'in phase' and the second is *out of phase* (or more precisely 180° out of phase) which will refer to two points on a wave behaving in exactly opposite ways to each other. For example, a crest and a trough are exactly out of phase. Moreover, two points which are *different* distances from the source may be in phase provided the difference in distance of the two points from the source is a *whole number of wavelengths* (Fig. 16.5).

Problem 16.1 Suppose a wave travels out from a point in *three* dimensions. What shape is a wave front? Do the rules about phase given for a two-dimensional wave also apply to one in three dimensions?

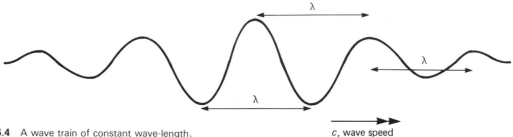

Fig. 16.4 A wave train of constant wave-length.

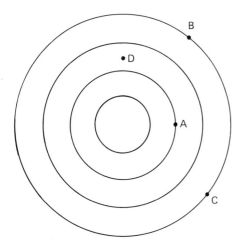

Fig. 16.5 Circular wave fronts emanating from a central point. Points B and C are moving in phase and are on the same wave front; points A and C are also moving in phase but are not on the same wave front. Points B and D are moving exactly out of phase.

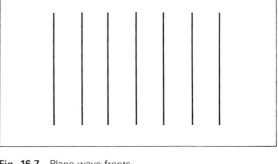

Fig. 16.7 Plane wave fronts.

Plane waves

When a circular ripple is some way from its source, a small section of a wave front is almost a straight line (Fig. 16.6). Because we are often dealing with small sections of a wave front many wavelengths from the source, it is easier to treat the wave front as though it *were* plane (or straight). We draw such wave fronts as in Fig. 16.7. Geometrically it is much easier to handle plane waves than circular waves. Plane wave trains can be created in a ripple tank by placing a 20 cm length of wooden rod in the water and giving it a short sharp push forward (Fig. 16.8).

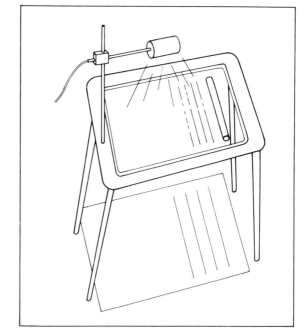

Fig. 16.8 Producing plane wave fronts in a ripple tank.

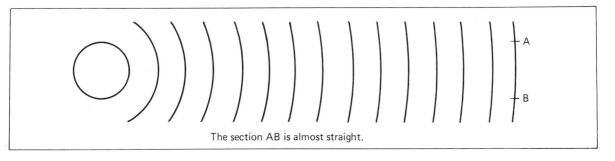

The section AB is almost straight.

Fig. 16.6

Wave energy

As a wave train passes over the surface of water, it does not carry the water with it. This can be seen by placing a few pieces of some light material such as cork on the surface of the water. As the wave passes them, they merely bob up and down. The wave carries *energy*, setting the water in motion – the wave is indeed nothing more than the up-and-down of the water, conveyed to one patch by the movement behind it, nearer to the source, and then passed on to the next patch in front, farther away from the source. *In this way a wave transfers energy without transferring matter*. This property of waves uniquely distinguishes waves from other methods by which energy can be transferred. It is so important that it is worth investigating in more detail. To do this, we will turn our attention to a less common form of wave motion – waves along a stretched spring.

16.2   Waves on a stretched spring

The way a wave travels can be more clearly seen by sending a wave pulse down a long stretched spring. Such springs which can be stretched to many times their original length have been developed from a toy called a 'slinky'. Heavier springs are now also available which, while they cannot be stretched to the same extent, enable quite slow moving pulses to be observed. If pieces of white paper are attached to one or two coils of the spring, the motion of a particular coil can be watched as the pulse passes. It is quite clear that the coils do not move in the direction of the wave but are forced to move up and down by the movement of the coils behind them. Their motion is

then passed on to coils ahead of them and so the pulse is propagated along the spring.

The maximum displacement of the coils from their equilibrium, or rest, position as the pulse passes is called the *amplitude* of the pulse. If a wave passes along the spring, the amplitude of each successive crest and trough may be the same as in Fig. 16.3b in which case the wave is referred to as having a constant amplitude. The short wave train in Fig. 16.9 does not, however, have a constant amplitude.

The energy carried by the wave depends both on its amplitude *and* its frequency. The higher the frequency, the more quickly must the wave medium oscillate between its maximum displacements. In a detailed investigation, which we shall carry out later, we shall see that when a body oscillates up and down, as it does when a wave passes, the energy transferred to it is proportional to the *square of the amplitude*. The energy carried by a wave across unit area per unit time is called the wave *intensity*. The wave intensity is thus proportional to the square of the wave amplitude.

Longitudinal and transverse waves

In the experiment just described, the spring coils oscillate in a direction *perpendicular* to that in which the wave is travelling (Fig. 16.9). Such waves are referred to as *transverse*. Another sort of wave can be sent along a stretched spring in which the coils move backwards and forwards in the *same* direction as that in which the wave is travelling (Fig. 16.10). Such a wave is referred to as *longitudinal*. It is a sort of pressure or shunting wave as one might get if a locomotive reverses quickly into a line of trucks. The wave is transmitted by each part of the wave medium pushing or pulling on the part just ahead of it.

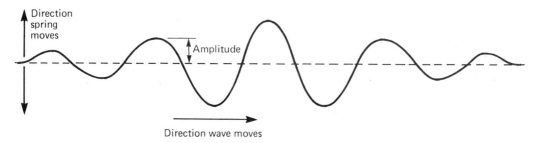

Fig. 16.9   A wave train with varying amplitude.

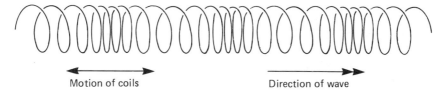

Fig. 16.10   A longitudinal wave on a stretched spring.

Polarization

In a longitudinal wave nothing more can be said about the direction of motion of the wave medium other than that it is in the same direction as the wave itself moves. In a transverse wave, simply saying that the motion of the wave medium is perpendicular to the direction in which the wave moves is not enough. We can ask, for instance, if that motion is up-and-down, or side-to-side (Fig. 16.11). Or it may be that the direction of movement is changing all the time.

Transverse waves, in which the direction of movement of the wave medium is always maintained in the same plane, are said to be *plane polarized*. We shall meet this idea of *polarization* again in a later chapter.

Problem 16.2   Send a short transverse wave train down a slinky spring. Is the wave polarized? Is it possible to send an un-polarized transverse wave down the spring? How could you arrange something that would only allow waves polarized in a particular plane to pass along the spring? (Such a device would be called a *polarizer*.)

Reflection of wave pulses

Before leaving this brief look at pulses on a stretched spring we cannot help noticing that a pulse sent down such a wave medium does not simply disappear at the far end (Fig. 16.25). Indeed the principle of conservation of energy ensures the energy cannot vanish. If the far end is fixed, a pulse will be seen to be *reflected* back along the spring when the incident pulse reaches the fixed end. The pulse is also reflected 'the other way up' – or exactly out of phase with the original, incident pulse. Why this change of phase takes place is something we shall investigate later. For now we return to water waves.

16.3 Huygens' principle

Christiaan Huygens was a Dutch physicist who lived at the same time as Isaac Newton. He was probably the first scientist to make a proper study of wave motion. One of the first things to puzzle him was how a wave in two dimensions moved forward in the way it appeared to. We have just seen that on a spring a part of it is set in motion by the part behind and that this part then sets in

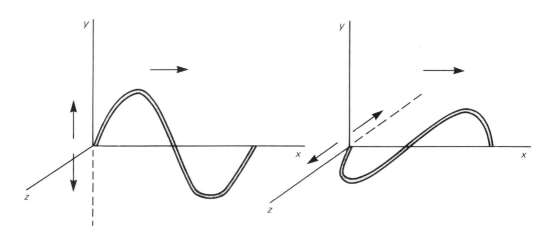

Fig. 16.11   Plane polarized transverse waves.

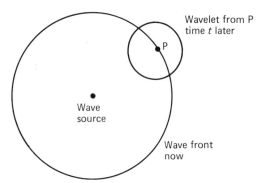

Fig. 16.12   A secondary wavelet from a point on a wave front.

motion other parts ahead of it. Consider now the circular wave front shown in Fig. 16.12. Huygens assumed that a point like P must communicate its movement to *all* points around it. If this is so, there ought, a time *t* later, to be another circular wave front centred on P, of radius *ct*, where *c* is the speed of propagation of the wave motion (Fig. 16.12). But the new wave front does not look like this.

Thinking further, it is clear that all points on the wave front would similarly send out secondary wavelets of the same sort (Fig. 16.13). Each of these wavelets would be a circle of radius *ct* some time *t* after the original wave front arrived at P.

But the new wave front at time *t* later is actually a circle whose centre is the wave source and whose radius is the previous wave front. This is the circle that *touches* all the secondary wavelets

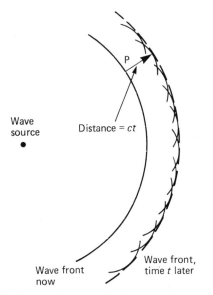

Fig. 16.13   Huygens' construction for a circular wave front.

– a line called the *envelope* of these secondary wavelets. In three dimensions, the secondary wavelets are spherical waves and their envelope is itself a spherical wave centred on the original wave source.

Huygens' principle states that every point on a wave front can be considered to be the source of a secondary wavelet which spreads out uniformly in all directions at the speed of wave propagation. The new complete wave front at some time *t* later is the envelope of all the secondary wavelets.

This is a geometrical construction which is hard to prove from wave theory, but clearly works for a circular wave spreading outwards in all directions. It turns out that the construction can be used to give correctly new wave fronts under much more complex circumstances and we shall find considerable use for the principle in this and succeeding chapters.

16.4   The reflection, diffraction and refraction of water waves

The reflection of water waves can be easily brought about by placing a barrier in the path of the on-coming wave train. Figures 16.14a and b show the reflection of a plane and circular wave train at a straight barrier. In the case of the plane wave, it is always found that the angle between the incident wave and the barrier is equal to the angle between the reflected wave and the barrier. But instead of drawing in the line of the wave fronts it is more usual to draw in the direction in which that wave is travelling (that is the direction of travel of the energy associated with the wave). And instead of measuring angles between the wave front and the reflecting surface, we measure angles between the direction of wave travel and a perpendicular to the reflecting surface (called the *normal* to the surface). Figure 16.15 shows that the angle between a plane wave front and the reflecting surface is the same as that between the direction of wave travel and the normal.

Angle θ is called the *angle of incidence*. The law of wave reflection states that the angle of incidence is equal to the angle of reflection. The reason for this rather complicated way of going about things is that it enables us to apply the same law when we are dealing with *circular* wave fronts and *curved* reflectors.

(a)

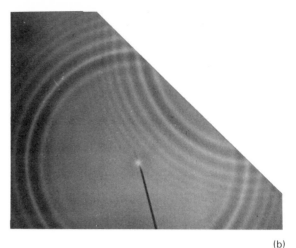

(b)

Fig. 16.14   The reflection of (a) plane parallel and (b) circular ripples at a plane barrier. (From W. Llowarch, *Ripple Tank Studies of Wave Motion*, Oxford University Press, 1961. Copyright Oxford University Press.)

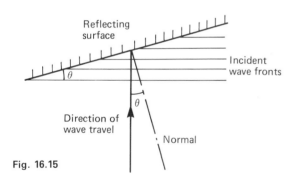

Fig. 16.15

Both the scattered wave and the bending of the wave round an obstacle in its path are examples of *diffraction*. Wave diffraction is the general term used for the effect on a wave front of an obstacle in its path, although scattering is often not regarded as a part of diffraction.

Diffraction

In the previous section the barrier from which waves were reflected was large. To all intents and purposes the wave reached a 'dead end'. What happens if there is only a small obstacle in the path of the wave? Figures 14.10 and 16.16 show what happens when a wave meets an obstacle about the same size as the wave's wavelength. A wave is *scattered* off the object, and the smaller the object, the smaller is the amplitude of the scattered wave. Beyond the object, in the direction of wave travel, the wave front is at first disturbed, but soon it bends around the object and, a little way from the object, the wave front is again re-established.

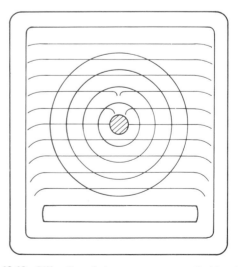

Fig. 16.16   Diffraction of plane waves at a small object (see also Fig. 14.10).

The amplitude of a scattered wave depends on the size of the scattering object compared with the wavelength of the scattered wave.

Diffraction by an aperture in a barrier is shown in the sequence of pictures in Fig. 16.17. These pictures show clearly that when the width of the aperture is small compared with the wavelengths of the waves (wavelength being here the distance between successive bright regions in the photograph of the ripples) the pattern made by the diffracted waves is indistinguishable from that due to a single, weak, point source situated in the centre of the gap. The energy flow associated with the diffracted wave is equally distributed over a 180° sector.

In Fig. 16.17b, where the width of the aperture is some two or three times the wavelength of the waves, we see that the energy flux distribution is not uniform, and the bulk of the energy flux is being propagated in directions similar to the direction of the incident waves. This is the interpretation of the observation that the ripples are strong and distinct only within a narrow sector emerging from the aperture. Also we see, if we look carefully at one particular diffracted ripple, moving our eye along it from left to right, that the amplitude is not simply increasing steadily towards the centre and then decreasing, but that it varies through a series of maxima and minima, although the central maximum is outstandingly the greatest. Figure 16.17c shows the pattern produced when the aperture is many wavelengths wide, and here we see that there is very little energy flux propagated in directions other than the original direction.

Refraction

It is possible to change the speed of a wave in a ripple tank by changing the depth of the water. To obtain a significant change, the wave must move from a deep region to one which is extremely shallow. In Fig. 16.18 a glass plate has been put in a ripple tank and the water depth has been adjusted so that it only just covers the region of the plate. You can see that the plane wave has slowed up as it passes over the shallow region by the way the waves at the *side* of the plate (in the deeper water) have hurried ahead of those passing *over* the plate.

Since just as many waves must pass over the

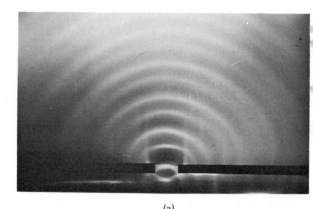

(a)

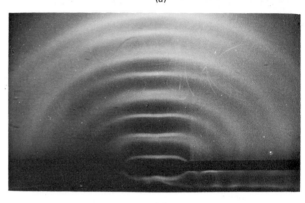

(b)

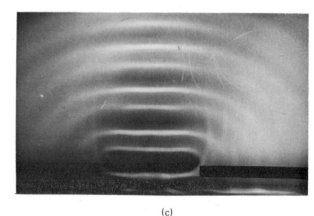

(c)

Fig. 16.17 Diffraction of plane ripples passing through a series of apertures. (Photographs by E. J. W.)

plate each second as come up from behind, in the deeper water, the frequency of the waves is unchanged as the waves cross from the deep region to the shallow one – it is the wavelength which changes. Since $c = f\lambda$, it follows that

$$\text{wavelength} \propto \text{wave speed.}$$

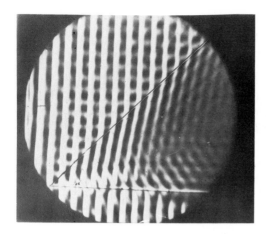

Fig. 16.18 The refraction (and partial reflection) of plane ripples at boundaries separating deeper from shallower water. (From W. Llowarch, *Ripple Tank Studies of Wave Motion*, Oxford University Press, 1961. Copyright Oxford University Press.)

If the boundary of the shallow region is not parallel to the on-coming wave front, the direction of wave travel changes as the wave travels over the shallow region (Fig. 16.19). This change in direction produced by a change of wave speed is called *refraction*. Again angles are referred to the direction of wave travel and the normal to the surface. Angle i (Fig. 16.19) is the angle of incidence, while angle r is the angle of refraction. The relationship between these two angles can be explored experimentally in a ripple tank, but here we will use Huygens' principle to find it.

Applying Huygens' principle to wave refraction

Imagine two points (Fig. 16.20) A and B on an incoming wave front. A just touches the edge of a region in which the wave travels at a speed c_2 which is slower that its original speed c_1. Now let us construct the new wave front at such a time, t, later that a wavelet from point B just touches the boundary edge at C. BC $= c_1 t$. The corresponding wavelet from A will only have reached D, where AD $= c_2 t$. The new wave front (which we have seen in the ripple tank is also plane) must contain D *and* C and so is the straight line joining D to C.

Triangles ABC and ADC are both right-angle triangles with angles of 90° at B and D respectively. The angle of incidence, i, is also the angle between the wave front and the boundary edge just as it was in the case of the reflected wave front. The same is true for the angle of refraction, r.

$$\text{In triangle ABC: } \sin i = \text{BC/AC,}$$
$$\text{and in triangle ADC: } \sin r = \text{AD/AC}$$
$$\sin i / \sin r = \text{BC/AD}$$
$$= c_1 t / c_2 t$$
$$= c_1 / c_2$$

So the ratio $\sin i / \sin r$ is a constant, as measurements made in a ripple tank will confirm. This ratio, which is equal to the ratio of the wave speeds in the two wave media, is called the *relative refractive index* for the two wave media.

Dispersion

If instead of using short wave trains, a continuous wave is set up in a ripple tank by using a mechanical vibrator, the frequency of the incident wave may be varied.

If the experiments on refraction are now repeated, it will be found that the angle of refraction depends not only on the angle of

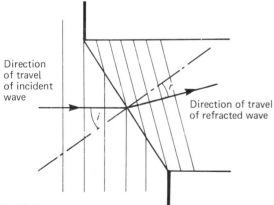

Fig. 16.19

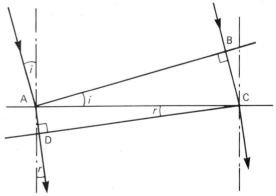

Fig. 16.20 The application of Huygens' principle to refraction.

Fig. 16.21

incidence but also upon the frequency of the incident wave. This comes about because the wave speed in the shallow region depends upon the wave frequency. Wave media for which this is true are said to show *dispersion*. Waves of different frequency are refracted at different angles even though the angle of incidence is the same for them all.

16.5   What happens when two waves cross?

A feature of military displays is the intricate pattern-weaving of 'cross marching'. In this, two lines of soldiers will accurately pass through each other's lines without interrupting either's step (Fig. 16.21). That the effect is always so spectacular is because we know how difficult it is for bodies to pass across each other without colliding and upsetting each other's path.

What happens when two waves cross? This can be investigated using either a slinky spring, a rubber tube, or a long trough of water (Fig. 16.1a). In order to 'label' the waves, one can be

sent as a *single* pulse from one end of the spring or tank, while a *double* pulse is sent simultaneously from the other end (Fig. 16.22). Experiment will quickly show that the two sets of pulses pass right along the spring or tank as though the other did not exist. Both sets of pulses pass straight through each other and are quite unaffected by the process. This is a remarkable property of wave motion. The same effect can be seen, but with greater difficulty because of the smaller space, in a ripple tank.

The principle of superposition

You may have noticed that if two pulses are sent along from each end of a water-filled trough, some care has to be taken that the water does not splash over the trough when the two pulses meet. This occurs, even though the original pulses seem small enough. At the moment of overlap the two pulses seem to coalesce momentarily to make one large pulse before parting to go their separate ways. This effect can be seen even more clearly on a stretched spring or heavy rubber tube.

However, if two pulses of opposite phase are sent from each end of the spring or rubber tube a different effect is observed at the moment of overlap. Figure 16.23 shows a series of pictures of what happens. As the pulses overlap, the spring is momentarily undisplaced. This ability of waves both to add together and to cancel each other is summarized by an important rule called the *principle of superposition*. It states that when two or more waves meet, the displacement produced is the algebraic sum of the displacements each wave would produce if it were acting on its own. The word 'algebraic' simply means that due account must be taken of whether the displacement is positive or negative (e.g. 'up' or 'down').

Reflection of waves on a spring

In order to show one example of this important rule in operation, we shall look again at the

Fig. 16.22

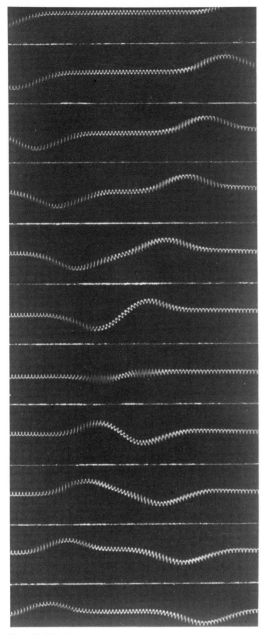

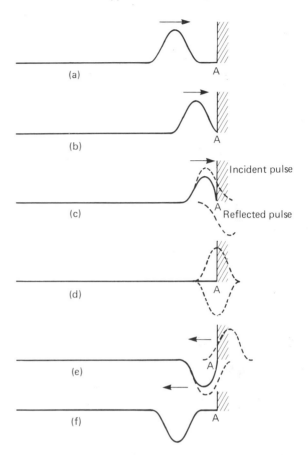

Fig. 16.24 The 'hard' reflection of a pulse.

Fig. 16.23 The superposition of two equal and opposite pulses on a spring. (From P.S.S.C. *Physics*, D. C. Heath 1965, reproduced by permission of Education Development Center, Inc.)

reflection of a wave from the fixed end of a spring. Again, it is simplest to look at a *pulse* rather than a wave train. What is true of a pulse will be equally true of a train of waves.

Figure 16.24a shows such a pulse approaching the fixed end A of a spring along which it is

travelling. As it reaches A, the parts of the spring immediately to the left of A exert a force on A to move it upwards. But point A is fixed. In order to remain at rest an equal and opposite downward force must act on the spring at A which of course will be transferred to neighbouring parts of the spring to the left of A (Fig. 16.24b).

A short time later, the incident pulse has moved farther to the right, while the reaction force at A has sent out a reflected pulse to the left. The two pulses add together according to the principle of superposition to give the resultant pulse shown in Fig. 16.24c. As the incident and reflected pulses overlap more completely, there is a moment when the spring is not displaced at all (Fig. 16.24d). A little while later still, the larger reflected pulse is now largely responsible for the size of the total pulse near A and the original on-coming pulse is seen to have reversed both its direction and its phase (Fig. 16.24e). Thereafter

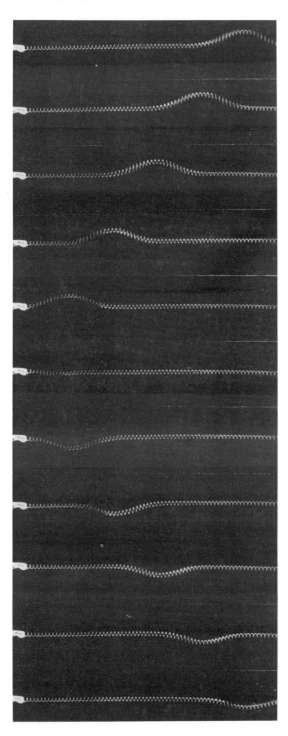

Fig. 16.25 The reflection of a pulse from the fixed end of a spring. (From P.S.S.C. *Physics*, D. C. Heath, 1965, reproduced by permission of Education Development Center, Inc.)

the reflected pulse travels back along the spring (Fig. 16.24f). Figure 16.25 shows a series of photographs of a similar pulse being reflected along a stretched spring.

As we have seen, a wave-pulse is reflected, *with inversion*, at a rigid boundary: this is called a hard reflection. But reflection occurs, in general, whenever there is a discontinuity in the wave medium: this does not have to be a rigid boundary. We can demonstrate a soft reflection by means of a slinky and a long thread, as shown in Fig. 16.26.

The thread keeps the spring in tension but offers negligible resistance to any transverse displacement of the end J. When a pulse is sent along the spring from H to J it is reflected from J, but this time without inversion. If the thread has negligible mass, then it absorbs no energy and the reflected pulse carries the same energy as the incident pulse.

Wave interference

The cancellation and reinforcement of waves in the region in which they overlap is referred to as *wave interference*, or more briefly as *interference*. In some ways this term is something of a misnomer because as we have seen, the effects described as interference come about because waves (quite remarkably) do *not* interfere with each other in the normal sense of the word, but act independently as though the others were not present. However, the phrase is too well entrenched for it ever to be changed, and the effects it produces are some of the most important met in a study of wave motion.

To look in more detail at wave interference, we return to the ripple tank and waves in two dimensions. It is hard to see the effect two waves have on each other if we rely on creating two wave trains in

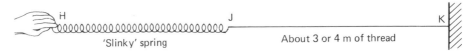

H · J · K

'Slinky' spring   About 3 or 4 m of thread

Fig. 16.26

Fig. 16.27 Interference fringes as seen in a ripple tank. The two vibrating sources are in phase. (From W. Llowarch, *Ripple Tank Studies of Wave Motion*, Oxford University Press, 1961. Copyright Oxford University Press.)

the way we have done previously. For interference effects to be seen clearly it is best to generate two continuous circular waves by means of a vibrator. Having done this, the pattern of waves produced in the region of overlap is best seen using either stroboscopic illumination or stroboscopic viewing at the same frequency as the waves. In this way the wave pattern is 'frozen' and effects like those shown in Fig. 16.27 can be seen. It is readily

observed that the wave disturbance seems to affect only some regions of the water surface; others, extending like 'rays' from the sources appear to be quite still. Such a pattern of disturbance is typical of an 'interference pattern'. In some regions the waves reinforce each other while in others they cancel out.

Figure 16.28a shows two sets of circular waves emerging from sources A and B. The lines represent the crests of the waves while the spaces half-way between are the troughs. At the point R, the disturbances resulting from the two wave fronts are in phase with each other. By the principle of superposition they will add together to give a large, reinforced disturbance. A moment later a trough from A will arrive at R at the same time as a trough from B. So the wave medium at R will continue to undergo large displacements as the two waves cross.

At C, the situation is different. Here a crest from A coincides with a trough from B. If the two amplitudes are the same, cancellation takes place. Waves passing across the surface from A and B will always at C be 180° out of phase and there will be no disturbance of the water. In Fig. 16.28b, the diagram has been redrawn so that there are no wave fronts shown in the regions of cancellation. These spread out like rays from A and B just as they did in Fig. 16.27.

Both the experiment in a ripple tank and drawings such as that in Fig. 16.28 will confirm that if

(i) the two sources A and B are separated further, or

(ii) the wavelength is shortened,

then the successive regions of cancellation and reinforcement get closer together.

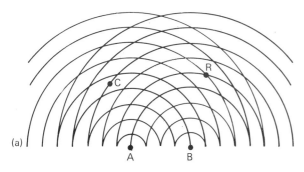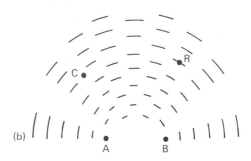

Fig. 16.28 (a) Two sets of ripples from two sources in phase. (b) The interference pattern produced by the two sets of ripples.

Coherence

It was noted above that interference patterns cannot easily be seen in a ripple tank unless two sources producing continuous waves are used. The reason for this is that sources A and B (Fig. 16.28) must remain continuously in step with each other, or if not in step, then maintain a *constant phase relationship* with each other (e.g. always exactly out of step). If they do not the regions of cancellation and reinforcement will continuously shift their position as the phase relationship between A and B changes. This means that for a steady interference pattern to be produced the frequencies of the two waves must also be identical.

Two wave sources which constantly maintain an unchanging phase relationship are said to be *coherent*. If a succession of wave trains is produced from two separating funnels as described in Section 16.1 the waves will be *non-coherent* and a steady interference pattern is not observed. (Try it for yourself – the experiment is well worth doing).

It *is* possible to produce an interference pattern when sources of waves produce a sequence of non-coherent wave trains (as a succession of water drops will do). To do this, a barrier with two gaps in it is placed in front of the source. The wave trains are diffracted at the gaps, spreading out from them as though from two separate sources in the centre of the gaps. An interference pattern like that in Fig. 16.29 is seen on the far side of the barrier. No matter what changes take place in the source of the waves, the two waves will always be in phase with each other at the two gaps.

It is possible in a ripple tank to produce an interference pattern in this way using as a source a series of wave trains from a wooden rod placed behind the gaps in the barrier. Two sets of circular waves are produced in the gaps which are always in phase no matter what irregularities occur in the incident wave. This is an important observation that we shall put into effect later on.

A general rule

The waves from two sources, A and B (Fig. 16.30) produce an interference pattern in the region around them. What is the condition that at point P the waves will interfere constructively, producing reinforcement? If the two waves are in phase with each other when they set out from A and B, they must either have travelled the same distance in reaching P or the *difference* in distance travelled must be a whole number of wavelengths. In general then:

$$|AP - BP| = n\lambda$$

(The vertical lines around AP − BP imply that only the absolute value of the difference has any significance.) This difference in distance travelled by the two waves is called the *path difference*.

For P to be a point at which the waves interfere destructively, producing cancellation (or at least a minimum amplitude), then the difference in distance travelled by the two waves must be an *odd* number of *half*-wavelengths:

$$|AP - BP| = (n + \tfrac{1}{2})\lambda$$

where *n* takes the values 0, 1, 2, etc., as before.

These rules can sometimes be used directly in order to determine a wavelength from measurements made on the interference pattern the waves produce. Suppose that P lies at the centre of a region of wave reinforcement (commonly referred to simply as a 'maximum') and that P_1 lies at the

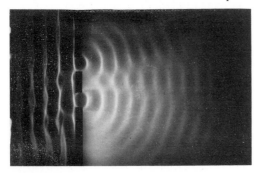

Fig. 16.29 Interference of ripples diffracted at two apertures. (Photograph E. J. W.)

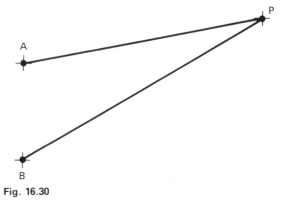

Fig. 16.30

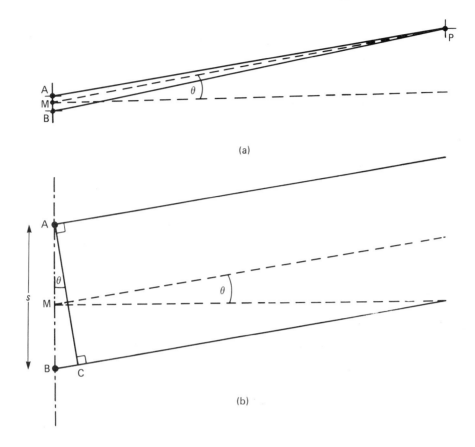

(a)

(b)

Fig. 16.31

centre of the adjacent region of reinforcement (the next 'maximum'). Then

$$|AP - BP| = n\lambda$$

and

$$|AP_1 - BP_1| = (n + 1)\lambda$$

Hence

$$|AP_1 - BP_1| - |AP - BP| = \lambda$$

Often, however, some simplification is possible. It is frequently the case that the two wave sources are very close together compared with the distance the sources are away from the place where interference is being observed (Fig. 16.31a). The two lines AP and BP are almost parallel. If M is the midpoint between A and B, then AP and BP make nearly the same angle as MP does with the perpendicular to AB.

If an enlargement of the diagram is produced (Fig. 16.31b) and a point C found on BP such that AP = CP, then the difference in distance the two

waves travel in reaching P (the path difference) is

$$|AP - BP| = BC$$

If P is a long way from A and B compared with the separation AB of the sources, then the angle ABP will be very small. This means that the angles at the base of the isoceles triangle APC are almost 90°.

In that case, we can take triangle ABC to be right-angled at C. Calling angle BAC $= \theta$

$$BC/AB = \sin\theta$$
$$BC = s\sin\theta$$

If P is in the centre of a region of reinforcement, the path difference BC $= n\lambda$. Hence

$$\boxed{n\lambda = s\sin\theta} \qquad (16.2)$$

This is an important result we shall be using on several occasions. But it can only be used provided that either the point of observation is a long way from the source of the waves compared with the separation of the sources, or, for some reason, the directions of the interfering waves are truly parallel as they leave A and B. This latter situation can be realized by changing the path of the waves after they leave A and B – something which we have seen done already by changing their speed.

16.6   The distribution of energy in an interference pattern

Interference patterns arise owing to the ability of waves to both reinforce and cancel each other. What then happens to the energy carried by the waves? In the region of cancellation, the water surface will not be disturbed at all if the two waves have equal amplitude. There is thus no energy being transferred to the water here.

In regions of reinforcement, the two waves add together to give the water double the amplitude of motion that it would have had if only one wave had been present. But you will recall (see Section 16.2) that the energy given to a body set in oscillation is found to be proportional to the

square of the amplitude. So in the regions of reinforcement the water gains *four times* the energy it would have had, had only one wave acted.

Suppose that one wave transfers W joules of energy per second to the surrounding water. If two similar waves of equal amplitude and frequency travel over the water they should by the law of conservation of energy transfer $2W$ joules of energy per second to the surrounding water. In fact the two waves of equal amplitude and frequency will set up an interference pattern. In the regions of reinforcement, $4W$ joules of energy will be transferred per second, while in the regions of cancellation, no energy will be transferred.

So the average energy transferred to the water per second by two waves is

$$\frac{4W + 0}{2} = 2W \text{ joules}$$

which is just as we would have expected. The wave energy is conserved, but in an interference pattern it is redistributed so that in some areas there is none and in others considerably more than might have been expected.

Chapter 17

MECHANICAL WAVES

In this chapter we are going to look more carefully at the way waves are propagated through a wave medium and derive some formulae which relate wave speed to other properties of the medium through which the wave travels. This will not only help in understanding how waves travel but also gives some results that are important in such widely different fields as the design of bridges and the design of musical instruments.

17.1 What factors influence the speed of mechanical waves?

In the previous chapter, some of the properties of waves were explored using a slinky spring. A transverse pulse passes along the spring because the displacement of one part of the spring causes an unbalanced force to act on the next part, so causing it in turn to be displaced (Fig. 17.1). The displacement of part A of the spring causes a force to act on part B. This in turn is displaced to position B′, causing a force to act on C, which is in turn displaced to C′, and so on. The speed with which a pulse (and thus a wave) is propagated will depend on the size of this unbalanced force. Figure 17.1 suggests that this unbalanced force is proportional to the overall tension, F, in the spring.

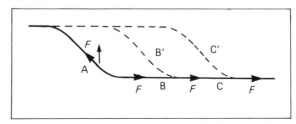

Fig. 17.1 Propagation of a transverse pulse along a spring.

But successive displacements of neighbouring parts of the spring will also depend on how quickly this force F can displace the spring. This is controlled by the *acceleration* given to each part of the spring. Newton's second law related acceleration to force and mass. Hence,

$$a = \frac{\text{unbalanced force}}{\text{mass displaced}}$$

Clearly this mass is not the whole mass of the spring but the amount close to the displacing force. This will depend on the *mass per unit length* of the spring. No other factor would seem to affect the speed with which the spring displacement passes along the spring.

So wave speed, c, is a function of some combination of spring tension, F, and mass per unit length, μ. Writing,

$$c \propto F^a \mu^b,$$

we can use dimensional analysis to see what relationship we may expect between these three quantities.

The dimensions of c are $[LT^{-1}]$ while those of F are the dimensions of force $[MLT^{-2}]$. The dimensions of μ are $[ML^{-1}]$. So,

$$[LT^{-1}] = [MLT^{-2}]^a [ML^{-1}]^b$$

Equating coefficients of
(i) M : $\qquad\qquad 0 = a + b \qquad$ (17.1)
(ii) L : $\qquad\qquad 1 = a - b \qquad$ (17.2)

Adding Eqs. 17.1 and 17.2, gives

$$1 = 2a$$
$$a = \tfrac{1}{2}$$

Substituting this result in Eq. 17.1 gives,

$$0 = \tfrac{1}{2} + b$$
$$b = -\tfrac{1}{2}$$

Consequently the relationship between c, F, and μ should be

$$c \propto F^{1/2} \mu^{-1/2}$$

or

$$c \propto \sqrt{(F/\mu)} \qquad (17.3)$$

We shall now see if we can justify this expression experimentally and then make a more detailed analysis using the equations of dynamics.

17.2 Speed of a transverse pulse along a spring

We expect the speed of a transverse pulse to depend on both the accelerating force of the spring and the mass to be accelerated. A good model of a spring can be made which separates the mass of a spring from its 'springiness' by linking together a line of trolleys (which act as the 'mass') by a series of light springs (which confer the springiness without themselves adding to the mass) (Fig. 17.2).

Transverse pulses (and waves) can be sent along such a trolley-and-springs model and their behaviour is seen to be identical with similar pulses passing along a slinky spring. However, the expression for the wave speed derived above can now be tested by independently changing both the mass per unit length and the tension.

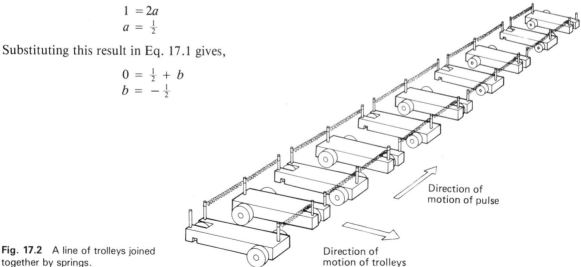

Direction of
motion of pulse

Direction of
motion of trolleys

Fig. 17.2 A line of trolleys joined together by springs.

Changing the mass per unit length

The mass per unit length of the model can be doubled by taping to each trolley a mass equal to that of the trolley. In a particular experiment the mass of each trolley was found to be 0.70 kg. An arrangement is shown in Fig. 17.3 which enables the time a pulse takes to travel along the model to be measured electrically. Using such an arrangement, the following results were obtained:

Time for a pulse to travel along trolley-and-springs model with unloaded trolley (five trials):

Average time = 1.63 s
Average speed = 1.47 m s^{-1}

Time for pulse to travel along trolley-and-springs model with each trolley carrying an additional mass equal to its own (five trials):

Average time = 2.33 s
Average speed = 1.03 m s^{-1}

According to the relationship derived in the previous section,

$$c \propto (\text{mass/unit length})^{-\frac{1}{2}}$$

So

$$c \times (\text{mass/unit length})^{\frac{1}{2}} = \text{a constant},$$

for constant tension.

For unloaded trolleys:

speed $\times$ (mass/unit length)$^{\frac{1}{2}}$ = 2.75 units

For loaded trolleys:

speed $\times$ (mass/unit length)$^{\frac{1}{2}}$ = 2.73 units

This is a difference of only 1%, well within the experimental error of the measurements.

Changing the tension

The tension between the trolleys can be doubled, while maintaining the same separation between them, by doubling the number of springs connecting the trolleys.

Problem 17.1 Why is it essential to maintain the same separation between the trolleys in this experiment?

In this case, the formula predicts that speed $\propto$ (tension)$^{\frac{1}{2}}$. With one set of springs, measurement gave an average speed of 1.47 m s^{-1} for a pulse passing along a line of unloaded trolleys (see above).

Time for a pulse to travel along trolley-and-springs model with unloaded trolleys but with two springs where there was one before (five trials):

Average time = 1.02 s
Average speed = 2.36 m s^{-1}

According to the formula,

$$\frac{\text{old speed}}{\text{new speed}} = \frac{(\text{original tension})^{\frac{1}{2}}}{(\text{new tension})^{\frac{1}{2}}}$$

$$= (1/2)^{\frac{1}{2}}$$
$$= 0.71$$

Measurement gives

$$\frac{\text{old speed}}{\text{new speed}} = 0.62$$

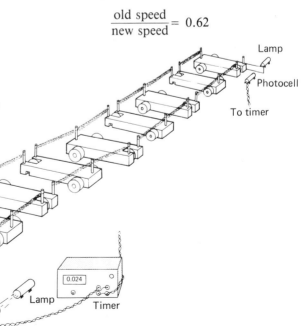

Fig. 17.3 Timing the passage of of a transverse pulse along a line of trolleys.

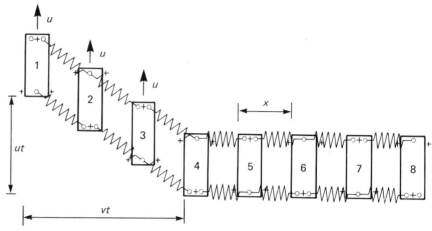

Fig. 17.4 Plan showing how the pulse appears as it travels.

Again the difference is well within the experimental error of the measurements, bearing in mind the variation in spring constant between different springs.

Analysing the pulse movement dynamically

So far we have said nothing about the constant which enables the proportionality sign in $c \propto (F/\mu)^{1/2}$ to be replaced by an equality sign. To do this we must try to find an exact relationship between wave speed and other factors. A full analysis of wave propagation along a spring is difficult, but a simplified treatment gives the same answer and reflects correctly the processes involved.

Turning again to the trolley-and-springs model, suppose that the first trolley is given and maintained at a steady speed, u, in a direction perpendicular to the line of trolleys. Each trolley in turn acquires a speed, u, and thereafter continues to move at that speed. The 'pulse' that passes down the line of trolleys is the change in speed of each successive trolley in the line, marked out as a 'kink' in the line of trolleys. A short time, t, after the first trolley was set in motion, the 'kink' will have passed some way down the line and the trolley-and-springs model will appear something like Fig. 17.4. In this diagram, trolleys 1, 2 and 3 are all moving upwards with speed u. Trolley 4, however, has not yet started to move. Figure 17.5 shows a diagram of the forces acting on this trolley.

In the direction along which the displacement pulse moves the trolleys themselves do not move.

So the forces on the trolley which is about to move are balanced in this direction. Thus

$$F' \cos \theta = F$$

We shall now make an important approximation (which also has to be made in the less approximate approach to this problem). We shall assume that the angle θ is always small, so that $\cos \theta \approx 1$. (The angle θ can be almost as big as 20° before the error in this assumption exceeds 5% – try it for yourself on a calculator.) If this assumption is accepted, then

$$F' \approx F$$

Vertically, the force acting on the trolley is $F \sin \theta$. This force will vary as the angle θ changes. So here again we are going to make a simplification. We shall assume that each trolley changes speed so quickly that we can assume a *constant* force $F \sin \theta$ acts on each in turn, hopping instantaneously from one trolley to the next. This will mean that during the time interval, t, from the instant the first trolley started to move, a constant

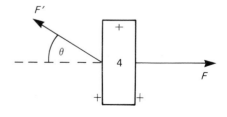

Fig. 17.5 The forces acting on trolley 4.

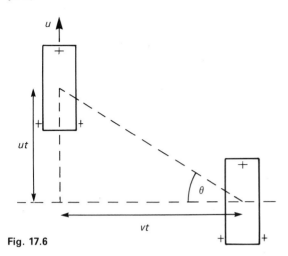

Fig. 17.6

force $F \sin \theta$ has acted, getting first one trolley, then the next, moving – a total impulse of $F \sin \theta\, t$. It is hard to justify this simplification, and all one can say is that a detailed analysis does come up with the same result!

This impulse will have changed the momentum of each of the trolleys in motion from 0 to mu. So,

$Ft \sin \theta = mu \times$ (number of trolleys in motion)

Although in Fig. 17.4, only 3 trolleys had been set in motion, the actual number moving will depend on the speed of the pulse, c, the length of each trolley-and-spring unit and the time elapsed since the pulse started. The number of trolleys set in motion will thus be ct/x, where x is the length of each trolley-and-spring unit. So,

$$Ft \sin \theta = mu \times ct/x \qquad (17.3)$$

It only remains now to find an expression for $\sin \theta$. Figure 17.6 shows that $\tan \theta = ut/ct = u/c$. We have already made an approximation on the assumption that θ is small. For the range of angles for which we could assume $\cos \theta = 1$, we can also assume that $\tan \theta = \sin \theta$. (This time, for an angle $\theta = 20°$ the error introduced is just over 6% – about the same error as is introduced by assuming $\cos \theta = 1$.) So provided θ is small,

$$\sin \theta = u/c$$

Substituting in Eq. 17.3, we have

$$Ft \times u/c = mu \times ct/x$$

Simplifying, gives

$$F/c = mc/x$$

$m/x = \mu$, the mass per unit length of the trolley-and-springs model, and the force, F, is the tension. Hence,

$$\boxed{c = \sqrt{(F/\mu)}} \qquad (17.4)$$

The proportionality constant in the original expression for c was thus 1. This result agrees with that obtained by using fewer approximations and simplifications, but both depend on the assumption that the pulse is small (so that the 'kink' in the spring is small).

17.3 Longitudinal pulses on a trolley-and-springs model

The trolley-and-springs model can be changed to provide a model capable of transmitting longitudinal pulses (Fig. 17.7). The springs must be capable of compression as well as extension and usually rather stiffer springs are used than in the model illustrating the passage of transverse waves.

Before describing experiments which can be performed with this model, we shall find the

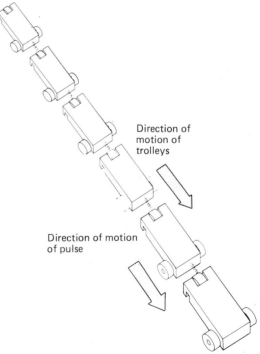

Direction of motion of trolleys

Direction of motion of pulse

Fig. 17.7 Trolleys and springs arranged for the investigation of longitudinal pulses.

expected relationship between the speed of a longitudinal pulse and other characteristics of the model, using the same procedure as that used to analyse the transverse pulses. Again we shall simplify the procedure by working out the speed with which a single impulse passes along the line of trolleys. Since all waves and pulses travel at the same speed, an expression worked out for such a single impulse ought to apply to *all* waves and pulses.

Suppose that the first trolley is given a speed, u, which is maintained in the direction that the impulse travels (that is, along the line of the trolleys). Just as in the case of the transverse pulse, the displacement is passed down the line of spring-connected trolleys and some short time, t, after the instant the first trolley was set in motion, the line of trolleys appears as in Fig. 17.8. In this diagram trolleys 1, 2 and 3 are all moving forward at speed u. Trolley 4, however, has not yet started to move and an unbalanced force acts on it owing to the compression of the spring on the left compared with that on the trolley's right.

As the first trolley moves forward it starts to compress the spring on *its* right. This transfers a force to the second trolley, accelerating it. This in turn compresses the spring on *its* right and starts the third trolley moving. Once the second trolley has reached speed u, the springs on its left and right will be equally compressed and no net force acts on it. Thus it continues to move forward at steady speed u. The maximum extra force applied by one spring depends on the total extra compression it receives. If this extra compression is some distance a, then assuming Hooke's law to apply to

the spring, the extra force applied is ka, where k is the spring constant. To simplify our working we will assume that each trolley applies this maximum extra force to the next one until it reaches speed u at which point this force is transferred instantaneously to the next one. Thus a constant force ka acts for a time t, transferring momentum mu to each trolley in turn. So,

$$ka\,t = mu \times \text{(number of trolleys in motion)}$$

If x is the length of each trolley-and-springs unit, then the number of trolleys in motion after time, t, is again ct/x.

$$ka\,t = mu \times ct/x$$

The distance occupied by the moving trolleys is ut less than the distance they originally occupied. This reduction in distance must be due to the compression of the springs. So the extra compression of each spring

$$= ut \div \text{(number of springs compressed)}$$
$$= ut \div ct/x$$
$$= ux/c$$

and this is the value of a.

Substituting for a:

$$k(ux/c)\,t = mu \times ct/x$$

Simplifying:

$$kx/c = mc/x$$

Hence

$$c^2 = x^2 k/m$$
$$c = x\sqrt{(k/m)} \qquad (17.5)$$

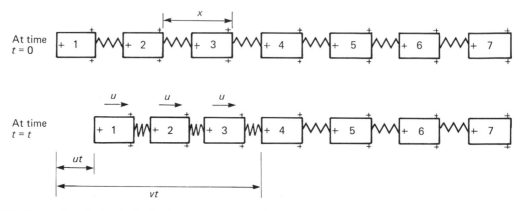

Fig. 17.8 The motion of a longitudinal pulse along a line of trolleys.

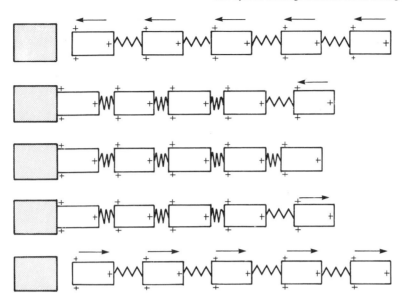

Fig. 17.9 Testing the expression for the speed of longitudinal pulses.

This expression gives the speed of pulses along the trolley-and-springs model. It cannot be so readily reinterpreted in terms of a spring or other medium as could the expression for the speed of transverse pulses. But before seeing how this can be done, we shall first test the result on the trolley-and-springs model.

Testing the expression for the speed of longitudinal pulses

To test this expression, a line of five trolleys and springs are connected together (Fig. 17.9); the front trolley and a massive block are connected separately to an electronic timer. The timer is set to operate when electrical contact is established between the front trolley and the mass. The line of trolleys is set moving at a steady speed towards the metal block. When the front trolley makes contact with the metal block the timer starts to operate. A compression pulse now moves down the line of spring-connected trolleys as the remaining trolleys are brought to rest. As the last trolley comes to rest, the compressed spring between it and the previous trolley accelerates the last trolley in the reverse direction. A 'separation pulse' then moves back along the line, breaking contact between the trolley and the metal block when this pulse reaches the first trolley again. At this point the timer stops, having recorded the time for the pulse to move out and back along *four* trolley-and-spring units (a total distance of $8 \times 0.36 = 2.88$ m in all). (The length of the first trolley is not included in the distance travelled – it acts simply as an 'inertial marker' for the leading edge of the pulse.) From the time and distance data, the pulse speed can be calculated. In a particular experiment the following values were found:

Speed of pulse (mean of five trials) = 2.70 m s^{-1}
Spring constant of springs   = 45.6 N m^{-1}
Mass of each trolley   = 0.81 kg
Length of each trolley and spring
   unit   = 0.36 m
Hence calculated speed   = 2.70 m s^{-1}

This value compares well with the measured speed.

17.4 The speed of longitudinal waves along a metal bar

Hooke's law, which we have assumed in order to derive the speed of a pulse along a line of spring-connected trolleys, applies generally to almost all metals provided the extension is sufficiently small. We have seen in Chapter 14 how this result can be generalized in order to define a coefficient of elasticity, called Young's modulus,

Young's modulus, E = stress/strain

To understand the elastic behaviour of metals at an atomic level we must assume that the atoms interact with each other with forces of attraction and repulsion which for small displacements behave as though the atoms were connected by springs.

Spring-connected particles are a useful model for interpreting the behaviour of many metals under stress. We might thus expect that such a metal bar will transmit longitudinal waves and pulses in the same way as the trolley-and-springs model we have just been examining.

We shall now try to find an expression for the speed of such waves in terms of other measurable properties of a metal.

Relationship between the spring constant and Young's modulus

As we have seen in Section 14.4 the Young modulus (E) and the spring constant (k) are related by the expression

$$E = k/x$$

where x is the separation between the particles. Of course the atoms themselves form the springs as well as the masses, and x, the centre to centre distance between the atoms, is the same as the atomic diameter.

Relationship between m, the atomic mass and the density of the bar

If we imagine the bar with n^2 atoms in cross-section to be l atoms along, the total mass of the bar is $l\,n^2m$.

If the atoms are placed in a regular array so each touches the next atom as shown in Fig. 17.10, then the size of the bar is $nx \times nx \times lx$, where x is the diameter of each atom.

So density, $\varrho = ln^2m/n^2lx^3$
$$= m/x^3$$

Speed of a longitudinal pulse down a metal bar

The speed of a longitudinal pulse passing down spring-connected trolleys is given by

$$c = x\sqrt{(k/m)}$$

Applying this to the metal bar and substituting for k and m

$$c = x\sqrt{\frac{Ex}{\varrho x^3}}$$

So

$$\boxed{c = \sqrt{(E/\varrho)}} \qquad (17.6)$$

This derivation assumes that the atoms of a metal are arranged in a particularly simple way – which in fact they are not. It also ignores any forces of interaction between the *rows* of atoms. However, different arrangements of atoms will only affect x, the spacing between the atoms, and as we have seen x cancels out in the final analysis. Cross-linking of the atoms similarly has no effect on the result as such an arrangement would still *behave* as if the atoms were connected by linear springs of some spring constant, k. It simply means that k is not the actual spring constant which relates the forces between any particular pair of atoms. So it is not surprising to find that this result, obtained under assumptions about a particular arrangement of atoms, is *generally* true.

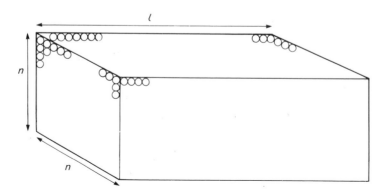

Fig. 17.10

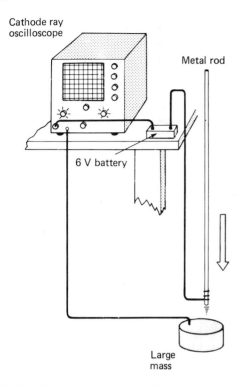

Fig. 17.11 Measuring the speed of a longitudinal pulse in a metal rod.

17.5 Measuring the speed of longitudinal pulses down a metal bar

Essentially the same method can be used as that employed to measure the speed of a longitudinal pulse down a line of spring-connected trolleys. The arrangement is shown in Fig. 17.11.

The cathode ray oscilloscope's time-base is adjusted so that it will only operate when a p.d. of a few volts appears across the input. The bar is dropped on to the heavy block and the time base operates as soon as the bar makes contact. A pulse then travels up to the free end of the bar and back down again before the bar springs off the block. At this point, the signal drops to 0 V. A trace appears on the c.r.o. screen as shown in Fig. 17.12.

By repeating the experiment several times a fairly accurate assessment of the time of contact, t, of the bar on the block can be made and the speed of the pulse thus estimated. This can be compared with the expected value derived from the known values of E and ϱ for the material of the bar.

17.6 Water waves

We cannot end this chapter without giving some attention to water waves, since our investigation into wave motion started with these. However an analysis of wave motion on the surface of water is not easy. In the first place two different forces operate to pull the water back to its equilibrium position: *surface tension* and *gravity*. Secondly careful observation of the surface of water carrying big waves (as distinct from surface ripples) will show that the water does not simply move up and down as a wave passes over it, but round in circles. It is as though it were undergoing transverse and longitudinal wave motion at the same time!

Three types of water waves are usually fairly clearly distinguishable from each other.

a) Deep water waves

These are defined as waves for which the wavelength is much smaller than the depth, but large enough to make surface tension a negligible factor. For such waves, the relationship between the wave speed and other factors is

$$c = \sqrt{\frac{g\lambda}{2\pi}}$$

where g is the acceleration due to gravity and λ is the wavelength. Many ocean waves come into this category. You will see that such waves suffer dispersion – the longer the wavelength, the faster they travel.

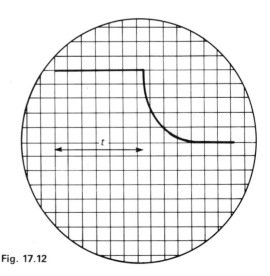

Fig. 17.12

b) Shallow water waves

For these, the wavelength is large compared with the depth, and large enough also to make surface tension a negligible factor. For such waves the relationship between c and other factors is

$$c = \sqrt{gd}$$

where d is the depth of the water. Such waves travel without dispersion.

c) Surface ripples

Surface tension is here the predominant factor and the expression for wave speed is

$$c = \sqrt{\frac{2\pi\gamma}{\varrho\lambda}}$$

where γ is the surface tension and ϱ the density of the water, while λ is the wavelength of the waves. These waves also suffer dispersion, but the longer the wavelength, the slower the wave moves.

Finally, to show how complicated the matter of water waves is, it has to be said that none of these expressions on its own satisfactorily accounts for the motion of waves in a ripple tank!

Chapter 18

WAVES AS MODELS

We have described a water wave as a disturbance of a patch of water brought about by movement of the water behind it and passed on by it to the water in front. A similar description would fit a wave on a stretched spring or a line of spring-connected trolleys. We are aware of the wave motion because of the movement it produces. But this is not the only feature associated with this wave motion. As a consequence of the passage of a wave or a pulse along a spring, energy is also transferred; and this is done without transfer of matter. This constrasts a wave with *any* other mode of energy transfer which always involves transfer of matter.

Waves may not be limited to those in which the displacement can be seen directly. An example of this is an *earthquake*. The *effects* of an earthquake (transfer of energy) can be sensed by an instrument called a seismometer (Fig. 18.1a) many thousands of miles from the source of the earthquake. The trace recorded (Fig. 18.1b) is of an up-and-down movement characteristic of a wave.

There has been a transfer of energy, but not of matter, from the earthquake source. In this case we understand how this may be so. The trolley-and-springs model has shown us how longitudinal and transverse waves may pass through an elastic solid. An application of such a model to earthquake waves has not only served to explain the existence of such waves but has also revealed almost all we know about the structure of the earth's interior.

However, in other circumstances, for example in the transfer of sound and light energy – it is by no means obvious whether the transfer of energy is wave-like (without transfer of matter) or

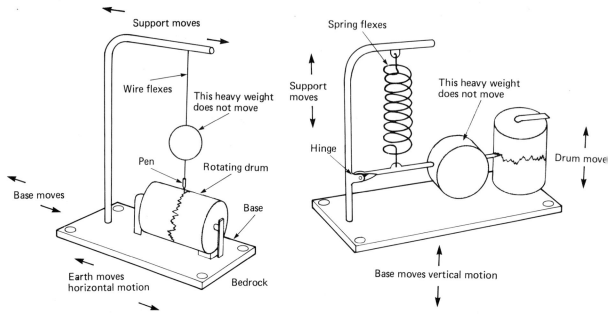

Fig. 18.1(a)   The principle of the seismometer. Horizontal motions of the Earth's crust are detected by the left hand, vertical motions by the right hand arrangement. (From Open University Science Foundation Course. Unit 22, Open University Press, 1971).

particle-like (that is, with transfer of matter, as with water from a hose-pipe). To develop wave models further we need first to compare wave-energy transfer with particle-energy transfer.

18.1   Transfer of energy by waves and particles

We have already seen that the transfer of energy by waves has a number of other properties

associated with it. Waves can be *reflected* and *refracted*. They are *diffracted* by obstacles and apertures, and multiple sources produce *interference* effects. Transverse waves show the further property of *polarization*. To what extent are these properties shared by a stream of particles?

Reflection

Figure 18.2 shows a simple arrangement for exploring the behaviour of moving particles

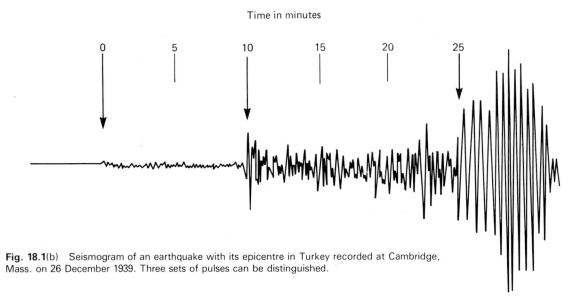

Fig. 18.1(b)   Seismogram of an earthquake with its epicentre in Turkey recorded at Cambridge, Mass. on 26 December 1939. Three sets of pulses can be distinguished.

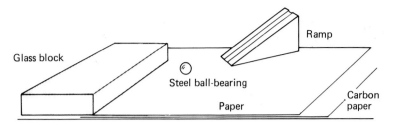

Fig. 18.2 Hard reflection of an elastic particle.

making elastic collisions with some solid object. Figure 18.3 shows a diagram of the particle's path. Conservation of momentum parallel to the reflecting surface gives

$$mv \sin \theta = mv \sin \phi$$

Perpendicular to the surface, it can be shown that if the mass of the reflector is very large compared with m, and if the collision is elastic, then

$$v \cos \theta = -u \cos \phi$$

It follows from this that

$$\theta = \phi$$

and thus particles colliding elastically with a solid reflector obey the same reflection law as do waves.

Refraction

Figure 18.4 shows an arrangement for investigating the behaviour of particles which undergo a change of speed at a boundary. Again the process may be analysed dynamically. Parallel to the boundary (Fig. 18.5)

$$mu \sin \theta = mv \sin \phi$$

since all momentum change takes place in a direction perpendicular to the boundary. So

$$\frac{\sin \theta}{\sin \phi} = \frac{v}{u}$$

This is almost the same rule as that formulated for waves in Section 16.4. We say 'almost' because in fact a careful comparison will show that the ratio of speeds is inverted.

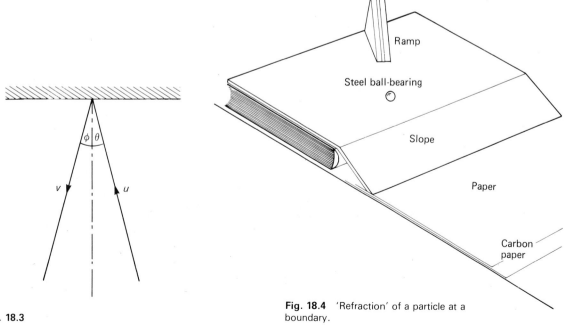

Fig. 18.3

Fig. 18.4 'Refraction' of a particle at a boundary.

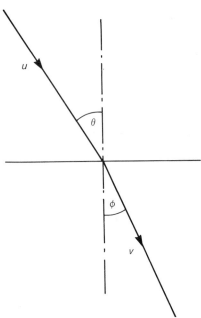

Fig. 18.5

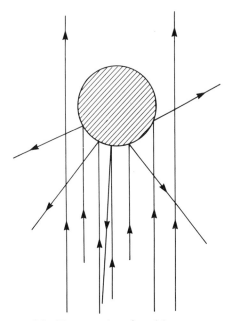

Fig. 18.6 The scattering of particles.

For waves:

$$\frac{\sin \theta}{\sin \phi} = \frac{\left(\begin{array}{c}\text{speed in medium}\\\text{corresponding to angle } \theta\end{array}\right)}{\left(\begin{array}{c}\text{speed in medium}\\\text{corresponding to angle } \phi\end{array}\right)}$$

For particles:

$$\frac{\sin \theta}{\sin \phi} = \frac{\left(\begin{array}{c}\text{speed in medium}\\\text{corresponding to angle } \phi\end{array}\right)}{\left(\begin{array}{c}\text{speed in medium}\\\text{corresponding to angle } \theta\end{array}\right)}$$

Put another way, we can say that for particles the direction of energy flow gets closer to the normal if its speed *increases*. For waves the direction of energy flow gets closer to the normal if its speed *decreases*. This is the first important difference between waves and particles.

Diffraction

A stream of high-speed particles hitting an obstacle will certainly be scattered by it, but they are unlikely to bend round it to fill the space behind it (Fig. 18.6). Similarly, if a stream of particles passes through an aperture those hitting the edge may undergo some scattering, but without the effects shown in Fig. 16.16.

So diffraction of a stream of particles is an unlikely effect.

Interference

Interference effects for waves are a consequence of the principle of superposition. A similar behaviour for particles is impossible. Two particles arriving at the same point will convey the sum of their energies to that point; there is no way in which their effects can be cancelled out at a point, or their energies redistributed over the region of interaction.

The presence of interference phenomena of the sort associated with waves is *unique* to wave motion. *Its existence immediately tells of the transfer of energy by waves.*

We shall now apply these ideas to some other forms of energy transfer where it is not obvious whether a wave or particle stream is the energy carrier.

18.2 The behaviour of sound

It is commonly believed that sound energy is propagated as a pressure wave in air (or in any other medium through which it travels). But what is the evidence for this?

Echoes are a common enough example of the *reflection* of sound, but its *refraction* (which has been found to take place in the atmosphere when particularly loud sounds are transmitted over large distances) is less common.

Being able to hear 'round a corner' is an every-day experience. That sound is easily *diffracted* (that is, bends round obstacles in its path) suggests a wave-like transfer of energy. The crucial test, then, is to set up an *interference pattern* using sound. This can be done in a laboratory using a signal generator and two loudspeakers. The interference pattern produced can be detected not only aurally but also using a microphone and a cathode ray oscilloscope. If the microphone is moved between successive maxima (loud sounds) (Fig. 18.7) the wavelength of the sound can be calculated by a direct application of Eq. 16.2.

Two features of this experiment can sometimes present difficulties. First, sound is also reflected from surrounding walls of the laboratory creating additional maxima and minima by interference. Reducing the volume of the sound does not necessarily help, since the sensitivity of reception has correspondingly to be increased. Often the only successful solution is to conduct the experiment out of doors. Secondly, the two sources of sound are often not placed far enough apart. The important equation $n\lambda = s \sin \theta$ (Eq. 16.2) for the separation of interference maxima also shows the limits of the closeness of the two sources in order to obtain an interference pattern of any sort. In the limit, in which only one inter-ference maximum is observed, $n = 1$ for $\theta = \pi/2$. Under these circumstances,

$$1 \times \lambda = s \times \sin \pi/2$$
$$s = \lambda$$

If the sources are closer than λ no distinguishable maxima and minima will be obtained. For a sound in mid-hearing range, say $f = 1000$ Hz, $\lambda = 34$ cm. This means a considerable source separation if several interference maxima are to be detected. It may well be better to raise the frequency.

Sources of sound, pitch and the range of hearing

The wave nature of sound is further confirmed when we examine sound sources. All of these consist of some sort of vibrator which sets the air in its neighbourhood vibrating in sympathy. Often these vibrations can be clearly seen and felt as in the case of a violin or guitar string, a loudspeaker cone or a tuning fork. When the sound wave arrives at a microphone, its diaphragm is in turn set vibrating. A changing electric potential difference is conveyed to a cathode ray oscillo-scope where its variation can be seen and its frequency measured.

The pattern of these vibrations is often complex (Fig. 18.8a) and a clear frequency is not easily distinguishable. The output of a signal generator (and of a tuning fork) is particularly symmetrical and represents an oscillation of single frequency called a *pure tone*. More complex patterns can often be shown to be made up of a main frequency (the *fundamental*) and a sequence of higher frequencies (called *overtones*). The output from musical sources is of this type (Fig. 18.8b). The fundamental conveys the sense of pitch (see below) while the overtones convey the

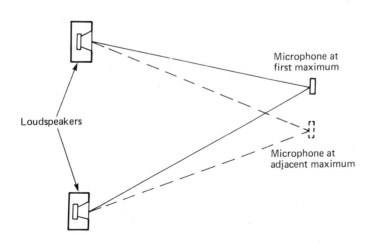

Fig. 18.7 Investigation of the interference pattern caused by two sound sources.

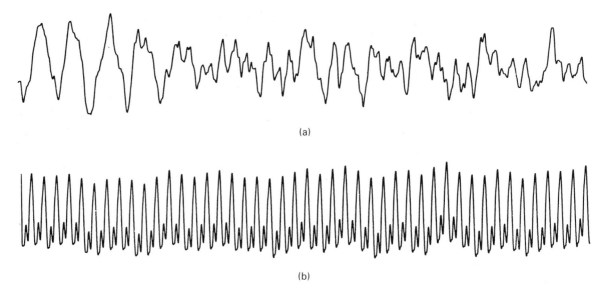

(a)

(b)

Fig. 18.8 (a) Trace of a 0.1 s fragment of the Mendelssohn violin concerto. (b) Trace of the note A on a violin over 0.1 s. (Photographs by Professor C. A. Taylor and reproduced by permission.)

special quality associated with a particular musical sound.

Experiments with a signal generator and a loudspeaker will show that

a) variation in frequency of a sound corresponds with the physiological sensation of a change of *pitch*, and

b) sound waves can be detected aurally over only a limited range.

The bottom of this range is not easily determined, but below 20–30 Hz any vibrations detected do not usually convey a sense of pitch. The upper limit varies from one person to another and also with their age. It is often in the range 16 000 Hz to 20 000 Hz. It should not be assumed, however, that the range of frequencies of sound waves has any particular limit. Those well above the limit of hearing (in the range of 40 kHz up to several hundreds of kHz) are of importance both scientifically and technologically. Sound waves in this region are called *ultrasonic*.

The speed of sound

There are a number of methods, some direct and others indirect, for measuring the speed of sound.

We will consider here one direct method and one indirect method as both are of general application to the measurement of wave speed.

A direct method   In this method, the output pulse of a cathode ray oscilloscope, which occurs as the sweep of the electron beam starts, can be amplified and picked up by a distant microphone (Fig. 18.9a). The output of the microphone can be fed back to the oscilloscope and shows up as a 'blip' on the electron beam trace. The microphone can be moved and the corresponding extra time taken by the sound to move the extra distance can be evaluated from the movement of the 'blip' on the cathode ray oscilloscope screen. From the two measurements of distance and time, the speed of sound can be calculated (Fig. 18.9b).

An indirect method   In this second method, the wavelength of a sound produced by a pair of loudspeakers fed from a signal generator is measured by making measurements of the position of the interference maxima they produce. The frequency of the sound can be measured either by relying on the calibrations of the signal generator or by direct measurement using a microphone and a cathode ray oscilloscope. From these two measurements the speed of sound can be calculated using the formula,

wave speed = frequency × wavelength.

Sound waves in air

A gas will behave elastically if subjected to variation in pressure – a fact that can be readily checked by pressing down on the handle of a cycle pump while keeping the lower end closed. So air has all the properties necessary to transmit a *longitudinal* wave: each part can be pushed and pulled by neighbouring parts and can transmit this push and pull to other regions. There are no elastic forces that would permit a transverse wave to be transmitted through air. Sound waves should thus show no *polarization* effects, and this turns out to be the case.

Longitudinal waves along a trolley-and-springs model gave a relationship between wave speed, spring constant and trolley mass of $c = x\sqrt{(k/m)}$, where x was the length of a trolley-and-springs unit. This changed to $\sqrt{(E/\varrho)}$ for an elastic compression wave travelling along a rod. A similar expression is found to apply to pressure waves in a gas: namely, $c = \sqrt{(K/\varrho)}$, where K is the *bulk modulus* for a gas. (The bulk modulus is defined as pressure/change in volume per unit volume.)

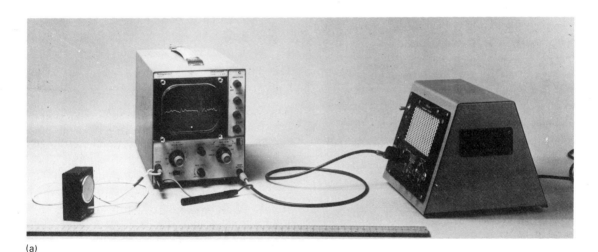

(a)

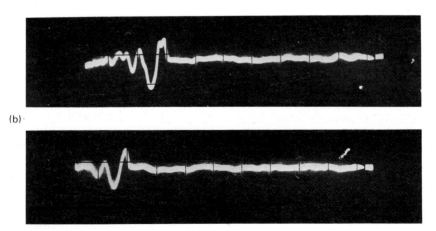

(b)

Fig. 18.9 (a) Direct measurement of the speed of sound. (b) c.r.o. traces obtained as the microphone is moved. (From G. W. Dorling, *Time*, Longman, 1973.)

18.3 A model for light

Controversy over the nature of light is a feature of the history of science from the turn of the eighteenth century until the middle of the nineteenth. Isaac Newton was probably the first scientist to explore systematically the properties of light. As a result of his researches he proposed a model for light consisting of a stream of 'luminiferous particles' emitted by sources of light, scattered by objects in their path and enabling vision by their reception by the eye.

At the same time Christiaan Huygens was proposing a wave-model for light to account for many of the same properties. In order to propagate such a wave, he imagined space filled with 'light particles' that transmitted a longitudinal wave in much the same way that air transmits a sound wave.

Neither model was entirely successful in explaining the then-known properties of light. Huygens was unable to demonstrate any interference effects which would have been crucial in supporting his wave model; Newton was unable to advance any experimental evidence for the speed of light in different substances which could have been critical in supporting his particle model. While neither theory could claim acceptance over the other in terms of the evidence available, such was the standing of Isaac Newton during the next hundred years that his particle model was uncritically accepted by the majority of scientists.

The evidence

We have seen that both waves and a stream of particles making elastic collisions with obstacles will obey the same laws of reflection and refraction.

For light, the relationship between the angle of incidence and the angle of refraction for a beam of light passing from, say, air to water is known as Snell's law, and the constant ratio

$$\frac{\text{sine (angle of incidence)}}{\text{sine (angle of refraction)}}$$

is called the *relative refractive index* from air to water ($_a n_w$).

This leads to the first critical test for the wave- or particle-model. If light is a wave motion,

$$_a n_w = \frac{\text{speed of light in air}}{\text{speed of light in water}} \tag{18.1}$$

If light is a stream of particles,

$$_a n_w = \frac{\text{speed of light in water}}{\text{speed of light in air}} \tag{18.2}$$

Since $_a n_w$ is approximately 1.3, the crucial question is: does light travel faster, or slower, in water than in air?

This is the question Newton was unable to answer. In fact it was not until the middle of the nineteenth century that anyone could answer it. By then a number of experiments were being performed in the laboratory to measure the speed of light and in one of these the equipment was reduced to dimensions small enough for the speed to be measured in water. It was found that light travelled more slowly in water and that the ratio of its speed in air to its speed in water was just that predicted above.

Measuring the speed of light

The speed of light is, today, an important constant. Indeed it represents a limiting speed – the fastest rate at which energy can be transferred as wave or particle. Because of its importance we will consider one modern determination that can be carried out in a school laboratory. The method is one of a family of experiments that depend upon the reflection of a light beam from a mirror rotating at high speed (Fig. 18.10).

Light from a source D passes through a glass plate, or half-silvered mirror to fall on mirror B which, although stationary at present, can be rotated by a motor at a high speed. From there the light is reflected to a distant, curved, mirror, C. The light is reflected back along its same path to return to B. It is partially reflected on to the scale seen from A. The distances are so arranged that the light reflected from C is brought to a focus on the scale and its position identified by a hair-line across the source.

If B is set into rotation at a high enough speed, it may have rotated through only a small angle, θ, during the time the light travels from B to C and back again. This will cause the image of the source to shift a small distance x, say. By measuring this displacement, x, the angle through which the

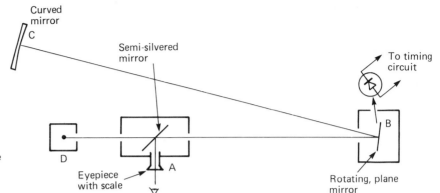

Fig. 18.10 Measurement of the speed of light using a rotating mirror method. (Courtesy Philip Harris Ltd.)

mirror rotated as the light travelled the distance 2BC can be calculated.

A measurement of the mirror's speed of rotation can be made by placing a photosensitive cell in the path of the reflected beam and observing the voltage pulses produced from the cell on a cathode ray oscilloscope. If the mirror makes n rotations per second, the angle θ through which the mirror rotates takes a time $\theta/360n$ seconds. Thus the speed of light is given by the expression,

$$c = \frac{2BC}{\theta/360n}$$

Current measurements of the speed of light give a value for c of $3.00 \times 10^8 \, \text{m s}^{-1}$, correct to three significant figures.

18.4 More evidence – interference and diffraction

The variation of the speed of light in different substances seems to provide strong evidence for its wave-like nature. However, the critical test for this is the production of interference and diffraction patterns. The diffraction of light is recorded by an Italian scientist, Grimaldi, as long ago as the seventeenth century, but the effect was not recognized then for what it was. Interference effects were first produced by an English scientist, Thomas Young, in 1801. Yet, despite the evidence, many scientists still refused to accept the consequences of these observations.

The problems encountered in producing an interference pattern with light are that the wavelength turns out to be very small, and the light from normal sources is neither spatially nor temporally coherent. This means that the phase of the light wave changes abruptly both along what should be a wave front (spatial non-coherence) and randomly with time (temporal non-coherence). In other words the light from a normal source is like the waves produced in a ripple tank by the random dripping of a number of closely spaced water droppers (see Section 16.1). To overcome these problems, we use a narrow line source placed a metre or more from a pair of narrow, closely-spaced slits. By using a distant line source, only a small section of wave front is used, so overcoming some of the problems of spatial coherence. By using two slits this wave front is divided into two, so overcoming the problem of lack of temporal coherence described in Section 16.5. A simple arrangement of light source and slits which works well is shown in Fig. 18.11. Lamp L is a clear glass, straight filament lamp (for example, a 12 V 24 W car headlamp-type bulb). It should be surrounded by a suitable opaque shield to reduce stray light. S is an ordinary microscope slide blackened with colloidal graphite, having a pair of parallel slits scored on it by a fairly sharp pointed instrument such as a fine ballpoint pen. These slits are about 0.5 mm apart, and set parallel to the line of the lamp filament. The distance from the lamp to the screen is not critical; between half a metre and a metre is generally satisfactory. An observing screen, O, is placed about two metres from S. The best kind of screen is a ground- or etched-glass one, or one of translucent plastic or paper. In a well-darkened room, with precautions taken to prevent any appreciable stray light reaching the

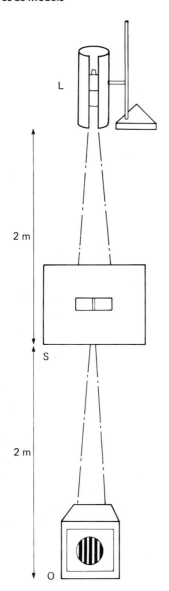

Fig. 18.11 Arrangement of light source and double slits for the observation of Young's fringes.

you should try the effect of using a different colour filter, say green. This will be found to alter the spacing of the bright bands observed on the screen: they are more closely spaced with green light than with red. This we interpret as indicating that green light has a shorter wavelength than red. When a white light source is used the pattern on the screen is the combined effect of sets of bands having different spacings corresponding to the different colours present in white light. This explains the spectral colour tingeing of the bands when white light is used. You should also examine the effect of using different spacings between the two slits on the slide S. On a single slide one can rule three or four pairs of slits, with care, having spacings of about, say, 0.25 mm, 0.5 mm, 0.75 mm, and 1 mm. It is found that the smaller this spacing, the greater the spacing of the bright bands on the screen. These results are consistent with what is observed in a ripple-tank when one varies the relation between wavelength and source-spacing.

We will now see how measurements made in this experiment can be used to find the wave-length of light of a particular colour. Figure 18.12 shows a diagram of the arrangement. We imagine we are looking down on the experiment bench from above. The two slits are at A and B. O is a point midway between them and P is a point on the screen equidistant from the slits. The line OP is a useful axis for reference: we can call it the optical axis of the experiment. Now we imagine that the experiment has been set up in a symmetrical fashion, using monochromatic light, and with the source of light equidistant from A and B. In this way the wave vibrations at A and B can be assumed to be in phase with each other, and since the waves from these two secondary sources have had to travel equal distances to get to P, the vibrations at P must be in phase, so that there is constructive interference and hence a maximum of brightness at P. P therefore marks the central bright band of the pattern on the screen. Q is some other point on the screen. What is observed at Q depends on the distances AQ and BQ. If these distances differ by a whole number of wavelengths, then the wave vibrations arriving at Q will be in phase and there will be constructive interference: maximum brightness again. If the two distances differ by half a wavelength or three

eye, one can see a number of parallel bright bands on the screen, tinged with spectral colours. If a colour filter, (red is a good one to start with) is placed between L and S, the bands become more discrete. When the screen is moved closer to the twin slits, S, the bands become more closely spaced. The bright bands on the screen are regions of constructive interference between the diffracted waves emerging from the slits, and the dark spaces between them are regions of destructive inter-ference. With the screen at a fixed distance from S

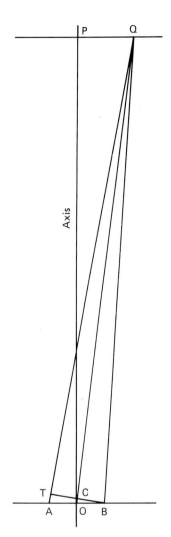

Fig. 18.12

The separation of neighbouring bright lines is

$$PQ/n = \lambda \frac{OP}{s}$$

Measurement of OP is easy enough, while s, the separation of the slits, can be done either by projecting a magnified image on to a distant screen, or by using a travelling microscope. From this, and a measurement of PQ/n, λ can be determined. Its value is found to range from about 650 nm for red light to about 450 nm for violet light at the other end of the spectrum.

Diffraction

Although Grimaldi produced, in the seventeenth century, what are now known to be diffraction effects with light, everyday experience suggests that light travels in straight lines and obstacles in its path cast sharp shadows. However, under the right conditions, diffraction effects are easily seen and many diffraction effects are common enough although not recognised as such. These include the light and dark rings seen round street lights on a misty evening. In this case the diffraction, or bending, of the light is caused by the water droplets in the atmosphere.

It was in fact a diffraction phenomenon which finally told in favour of the wave theory of light. Despite Young's experiments at the turn of the nineteenth century, many scientists still argued vehemently the case for the particle model. Augustin Fresnel, a French physicist argued the case for a wave model in front of a group of distinguished physicists early in the nineteenth century showing how a wave equation could, under appropriate conditions, describe all the effects observed for light. Scorn was poured on his ideas and one physicist, Poisson, pointed out as he saw it, the fallacy of the model, by showing that such a wave model for light predicted that a small round object placed in the path of a light beam should produce a shadow region *with a bright spot in the centre*. Far from assuming that such an effect was impossible, Fresnel set up equipment to see if this was so. We can repeat his experiment today in the laboratory, using a quartz iodine lamp (which gives a concentrated, bright source of light) behind a pinhole. The arrangement is shown in Fig. 18.13. If you look carefully at the shadow of the ball bearing cast on the translucent screen

halves, or any odd number of half-wavelengths, then there will be destructive interference – darkness.

If Q is at the nth bright band from the central band, P, then the path difference, $AT = n\lambda$. We have already seen (Eq. 16.2) that if AB is small compared with OP, then

$$n\lambda = s \sin \theta$$

where $s = AB$ and θ is the angle POQ. Since θ is small, $\sin \theta$ is approximately equal to $\tan \theta = PQ/OP$. Thus,

$$n\lambda = s \frac{PQ}{OP}$$

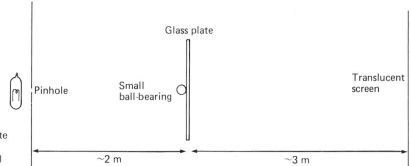

Fig. 18.13 Arrangement to investigate the pattern produced when light is diffracted by an obstacle, here a small ball.

you will indeed see that there is a bright spot at the centre. Fresnel was then able to confirm his theory (and presumably discomfort Poisson) by demonstrating this prediction of his theory.

Thus the behaviour of light in showing both interference and diffraction effects confirms that it is wave-like in nature. That these effects are not in everyday observation arises from the extreme smallness of its wavelength. But a wave of what? The difficulty in imagining a medium so tenuous that it was indistinguishable from a vacuum and yet of such elasticity that light could travel at 300 000 km per second through it was (and indeed still is) a major obstacle to everyone's acceptance of this wave model.

18.5   Transverse or longitudinal?

We shall conclude this search for a model for light by asking finally whether light is a longitudinal or a transverse wave. Transverse waves, as we have seen in Section 16.2, can be polarized so that the wave motion is confined to one plane. In Problem 16.2 we asked how something could be arranged

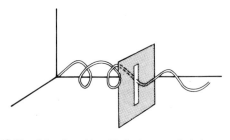

Fig. 18.14 A barrier with a frictionless, vertical slot passes those components of the wave in the string which are polarized in the vertical plane and no others. (From F. W. Sears, M. W. Zemansky and H. D. Young, *College Physics*, Addison-Wesley, 1980.)

so that only transverse waves polarized in a certain plane could be passed along a spring. One way of doing this would be to make a slot in which the spring could slide vertically but not horizontally (Fig. 18.14). Transverse waves passing through this slot will be polarized in a vertical plane. These waves will pass through another vertical slot farther along the spring, but not through a horizontal slot. If we did not know the plane of polarization of the wave passing through the first slot, the orientation of the second slot would tell us. The first slot is called a *polarizer* while the second is usually called an *analyser*. When no wave passes through the analyser after previously passing through the polarizer, the polarizer and analyser are said to be *crossed*. Of course these effects will only be seen if a transverse wave passes down the spring. The polarizer and analyser will pass longitudinal waves no matter what their relative orientations.

Certain crystals are found to have the same effects on light as these slots have on mechanically transverse waves along a spring. The commonest of these are embedded in plastic, and known as Polaroid. Light passing through one sheet of Polaroid can be stopped completely by another sheet. If the second sheet is rotated through 90°, however, the light again passes. This phenomenon can only be explained by assuming that light is a transverse wave.

This is not the only way that light can be polarized, and we will consider briefly two more.

Polarization by reflection

Natural light is unpolarized. This means that the plane of its vibrations is continually changing. Reflected light, however, is often partially

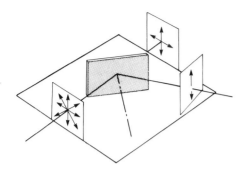

Fig. 18.15 When light is incident on a reflecting surface at the polarizing angle, the light reflected is plane polarised. (From F. W. Sears, M. W. Zemansky and H. D. Young, *College Physics*, Addison-Wesley, 1980.)

While this brief survey of the properties of light serves to confirm its transverse, wave-like nature, it certainly does not do justice to the variety of application of these properties or to the whole topic of image formation. A more detailed study of diffraction will be found in chapter 21.

18.6   Electric waves

The photograph in Fig. 18.17 shows an electrical oscillator, based on a capacitor and an inductor. The frequency of the oscillations produced can be calculated from a knowledge of the value of the capacitance C and inductance L involved from the formula

$$f = \frac{1}{2\pi\sqrt{LC}}$$

(see Section 36.4).

The oscillator in the photograph works at a frequency of 10^9 Hz (1 GHz). If its output is connected to a pair of metal rods, it can be shown that electrical energy can be transmitted into the surrounding space and picked up by the similar pair of rods connected to a galvanometer.

The radiation can be stopped reaching the receiver by interposing a metal plate, which can also be used as a reflector.

Interference effects confirm that this radiation is wave-like. But this time we shall use a rather different arrangement to allow two sets of waves to interact (Fig. 18.18). Since all the radiation emerges from one aerial and the oscillator runs continuously, the radiation is coherent. However, a second source may not have exactly the same frequency, so a stable interference pattern cannot

polarized as can be seen by looking at reflections off any non-metallic surface using a piece of Polaroid. At a particular angle of reflection, it is in fact completely polarized (Fig. 18.15). The polarization of reflected light is the basis upon which Polaroid sun glasses work – cutting out the 'glare' from reflections. Most troublesome reflections come from horizontal surfaces and the polarizing material is set to cut out the light reflected from such surfaces.

Polarization by scattering

The blue light of the sky is produced by the scattering of sun light by air molecules, high up in the atmosphere. This scattered light is partially polarized, polarization being greatest in a direction at right angles to the direction of the sun (Fig. 18.16). Again this can be readily understood if light is a *transverse* wave, since the scattered light can only vibrate in the direction shown.

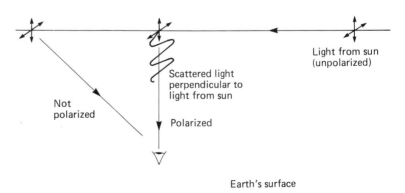

Fig. 18.16 The polarization of light scattered in the atmosphere.

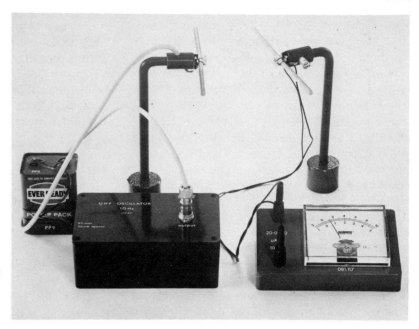

Fig. 18.17 A U.H.F. transmitter and receiver operating at 1 GHz (30 cm). (Courtesy Unilab Ltd.)

be formed by this means. Instead the radiation from the aerial is reflected off a metal plate so that the receiver picks up both the direct and the reflected radiation.

Let us suppose that a strong signal is picked up at B when the reflector is at R, as shown. The path difference of the two waves arriving at B is

$$ARB - AB = 2BR$$

Let the reflector now be moved away from AB until the next strong maximum is picked up at B. The mirror is now at R'. The new path difference is

$$AR'B - AB = 2BR'$$

If we have moved from one maximum to the next, the increase in path difference is one wavelength. So,

$$\lambda = AR'B - ARB = 2RR'$$

Hence this experiment not only suggests a wave-like radiation, but also determines the wavelength.

An interesting side-effect will be noticed if R is moved steadily towards AB. Close to AB, the path difference between the direct and the reflected wave approaches zero; we would then expect a maximum signal at B. Surprisingly it turns out to be almost zero! The explanation for this is that the wave, on reflection at R, changes its phase by 180° – just as the wave on a slinky spring when it

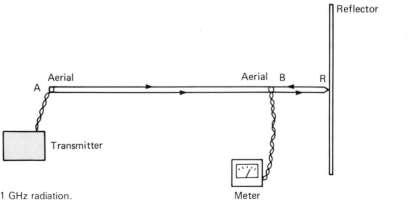

Fig. 18.18 The reflection of 1 GHz radiation.

reaches a fixed end. So if the path difference between the two waves is zero, they arrive at B exactly out-of-step and so cancel each other out.

Knowing the frequency of the oscillator and assuming, not unreasonably, that this is the frequency of the wave, we can calculate the wave speed. In a particular experiment, the reflector was placed so that the signal picked up by the receiving aerial was a minimum. The reflector was then moved back until the received signal had fallen again to a minimum for the fourth time. The reflector was displaced a distance of 61.0 cm. The extra distance travelled by the reflected wave was thus 122 cm – equivalent to four wavelengths. Hence,

$$\lambda = 122/4 \text{ cm}$$
$$\lambda = 0.305 \text{ m}$$

Hence, wave speed $= \lambda f$
$$= 0.305 \times 10^9$$
$$= 3.05 \times 10^8 \text{ m s}^{-1}$$

Within the limits of experimental error this is the same as the measured speed of light.

That this electric wave is *polarized* can be seen readily by rotating either the aerial or the receiver (separately) through 90° about the line which connects them. In neither case is any radiation picked up. The aerial and the receiver can only transmit and pick up electric waves oscillating in a plane parallel to themselves – the waves are thus *transverse* – again the same as light waves.

18.7   Radio waves?

Radio and television transmitters use electrically-oscillating circuits similar to that used in the previous investigation. The wave-like nature of a VHF radio transmission can be shown by an arrangement similar to that given in Fig. 18.19. A large metal plate is placed behind a portable VHF radio so that the radiation from the transmitter is received both directly and by reflection off the plate. Movement of the plate towards and away from the receiver will cause regular variations in the volume of the signal received – an observation which can only be explained by assuming that interference is taking place between two waves. Again, the wavelength of the radiation can be measured in the same experiment. You may find that this experiment works best if performed in the open air.

Figure 18.19 shows a diagram of the receiver receiving waves both directly and from the reflector. The path difference of the two waves arriving at A, the aerial on the receiver, is 2AR. It is easiest to set R so that the signal received is at a minimum in this case. If R is moved to R' so that the signal falls to the next minimum, the increase in path difference is one wavelength.

$$\lambda = 2AR' - 2AR$$
$$= 2RR'$$

Since the frequency of the transmission is determined by the electrical characteristics of the transmitter, we can again use this data to find the speed of the wave. Again, to within the limits of experimental error, this turns out to be the same as the speed of light.

Such waves are also transverse, but it may be harder to show this than it was with the laboratory-based oscillator. Scattering from nearby objects tends to produce waves of a variety of polariza-

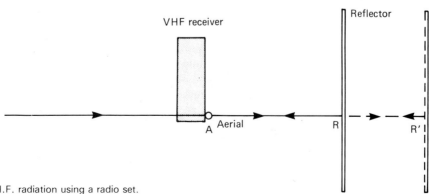

Fig. 18.19   The reflection of V.H.F. radiation using a radio set.

tions. Even so rotation of the aerial in a plane perpendicular to that containing the propagation direction of the wave may well give a noticeable change in signal strength.

If a portable television set is available, the experiment may be repeated using its internal aerial. The radiation can again be shown to be wave-like, by reflecting some of the transmitted signal back on to the aerial. At minima, the signal strength will be so weak that the picture will break up, giving a clear indication of a minimum signal strength.

In a particular experiment, a reflector was moved a distance of 19 cm as the signal received varied from one minimum to the next. This is the distance RR′ in the equation given above for the VHF radio. Thus λ is 38 cm, or 0.38 m. The television set was tuned to Channel 61 and the picture transmission frequency quoted for this channel is 791.25 Hz, a figure which depends on the electrical characteristics of the transmitter.

So, wave speed = λf
$$= 0.38 \times 791.25 \times 10^6$$
$$= 3.01 \times 10^8 \,\mathrm{m\,s^{-1}}$$

Again the speed turns out to be, experimentally, the same as that of light. However, since portable television sets usually use loop aerials, polarization effects may be hard to establish. For this a directional aerial of the indoor type must be used.

Electrical waves have so far been shown to exist over a range of wavelengths of a few metres down to a few tens of centimetres. Radio transmission, however, takes place at frequencies as low as 200 kHz, representing a wavelength of 1500 m. We will finally consider one more experiment which extends this range of electric wavelengths down to a few centimetres.

Figure 18.20 shows apparatus which generates a very high frequency electrical radiation. Interference effects can be established using a reflector as before, but this time we shall, for variety's sake, return to an arrangement similar to that used for light. Two slits can be constructed using three metal plates and an interference pattern picked up by a receiver placed beyond the plates (Fig. 18.21). If the receiver is moved along a line parallel to the slits AB, a series of maxima and minima will be recorded on the meter attached to the receiver. Alternatively, the electric wave can be *modulated* by an oscillation in the range of hearing and the effect registered via an audio amplifier as a variation in sound intensity.

If the receiver, R, is placed on the first

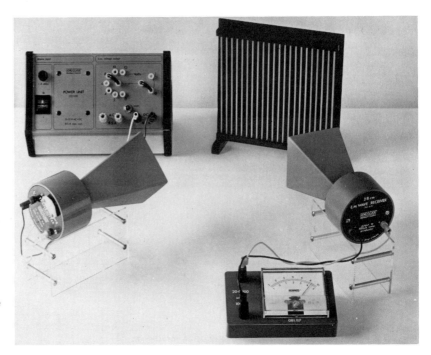

Fig. 18.20 A 10.7 GHz (2.8 cm) transmitter (left) and receiver in use to examine the reflection of the radiation from an array of vertical rods. (Courtesy Unilab Ltd.)

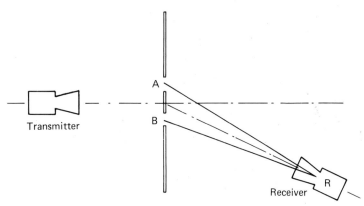

Fig. 18.21   Double-slit interference with 2.8 cm radiation.

maximum after the central one, the path difference of the two waves from A and B is one wavelength:

$$AR - BR = \lambda$$

Thus λ can be measured.

A particularly striking effect may be observed if the detector is set at a point where the signal is at a minimum of intensity. If *one* of the slits is now covered up, the signal strength can be seen (or heard) to *increase,* showing very clearly that with both slits open *destructive* interference was taking place at this point.

The diffraction of the waves can be demonstrated by placing the receiver at an angle of 45° to the direction of the beam from the transmitter, after the beam has passed through a single slit

about 10 cm wide made by two metal plates (Fig. 18.22). Little signal will be received. Narrowing the slit, however, far from reducing the signal strength even further, actually causes the intensity of the radiation received to increase. This can only be caused by diffraction of the radiation as it passes through the slit.

With a wavelength a little less than 3 cm (usually 2.8 cm), the frequency of the radiation cannot be easily determined electrically, so a determination of its speed demands a more direct method. Here we can employ a technique similar to that used for measuring the speed of sound waves. In this case the radiation is *pulsed* so that bursts of radiation are emitted at a frequency of 200 kHz. The frequency of the waves is so high that each pulse will contain many waves. These

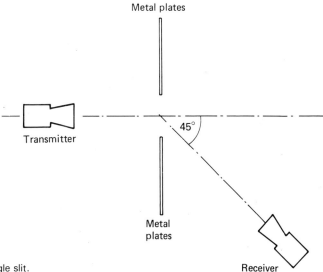

Fig. 18.22   The diffraction of 2.8 cm radiation at a single slit.

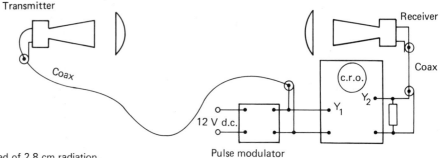

Fig. 18.23   Measuring the speed of 2.8 cm radiation.

waves are too high in frequency to be recorded on a normal cathode ray oscilloscope, but it will record the envelope of the pulse.

One possible experimental arrangement is shown in Fig. 18.23. The transmitted and received pulse are fed to separate beams of a double beam cathode ray oscilloscope. The time base should be capable of a calibrated resolution of at least $0.1\,\mu\text{s cm}^{-1}$. With transmitter and receiver widely separated (several metres) the position of the received pulse is noted on the time base. The transmitter is now moved closer to the receiver, and the new position of the received pulse noted. The distance, d, the transmitter has been moved is recorded. The time taken for the pulse to travel this distance, d, can be calculated from the change in position of the received pulse on the time base.

If the received pulse moved a distance x cm along the time base when set to a calibrated time of $10^{-7}\,\text{s cm}^{-1}$, the time taken for the pulse to travel d is $x \times 10^{-7}\,\text{s}$. Hence the speed of the microwaves is given by

$$c = \frac{d}{x} \times 10^7 \ldots \text{ in m s}^{-1}$$

Since the pulse displacement, x, will be only 2 mm for a movement of the transmitter of 6 m, high accuracy cannot be expected using instrumentation normally available to schools. Nevertheless the experiment is worth doing for it shows again (and by direct measurement) that the speed of these electric waves is experimentally the same as that of light.

The transverse character of these 2.8 cm waves is easily demonstrated by turning the receiver through an angle of 90° while still directing it at the transmitter. In this position, no radiation will be picked up.

18.8   The electromagnetic spectrum

The fact that this wide range of electric waves and light are all transverse waves travelling at the same high speed of $3 \times 10^8\,\text{m s}^{-1}$ suggests that the waves are identical in character, differing from each other only in their frequency. Light is not an obvious electrical oscillation, but even this problem is resolved once the emission of light has been associated with the movement of electrons within an atom (see Section 44.2).

Infrared and ultraviolet radiation

If light from an incandescent source such as a carbon arc is refracted by a prism, it spreads out into a spectrum of colours. We have already seen in the light interference experiment that colour is the physiological sensation associated with frequency. The spectrum is visibly terminated at the red (long wavelength) end and the violet (short wavelength) end. Yet other methods of determining the energy radiated shows that the spectrum does not in fact finish here. If a blackened thermometer is held in the spectrum, the light energy falling on it will produce a rise in temperature. This rise in temperature can be detected *beyond* the red end, showing that there are waves of longer wavelength in the spectrum which are not visible. These are called *infrared* waves. Similarly beyond the violet end, certain fluorescent paints and photocells show the presence of radiation extending to shorter wavelengths than are visible. Waves in this region are referred to as *ultraviolet*.

Electrically-generated waves can now be produced down to wavelengths of a few millimetres ($10^{-3}\,\text{m}$). The near infrared (near to the visible region) has a wavelength of the order of $10^{-6}\,\text{m}$. We must assume that waves exist with wavelengths

Fig. 18.24 An X-ray tube showing the anode construction. (Courtesy Mullard Ltd.)

which fill this gap although they are not commonly observable. This suggests that there is a continuum of wavelengths (a spectrum) extending from wavelengths greater than 10^3 m down to as short as 10^{-7} m (the near ultraviolet). Does this spectrum extend to even shorter wavelengths?

X-rays and γ-rays

X-rays are produced by the rapid deceleration of high-energy electrons (Fig. 18.24). This suggests that these waves – whose behaviour we have already discussed (see Section 14.6) on account of the interference patterns produced when they interact with crystals – may also be electrical in nature and members of the same family of waves. This extends the spectrum to wavelengths of the order of 10^{-10} m.

Even shorter wavelength waves are produced in nuclear interactions. These are referred to as γ-rays and crystal diffraction has confirmed their wave-like nature. These waves have a wavelength of the order of 10^{-14} m. While their production is not obviously electrical, their speed has been measured by generating them from nuclear interactions between artificially accelerated particles and matter. Again it has been shown that they travel at the same speed as light which suggests that they are very short wavelength members of

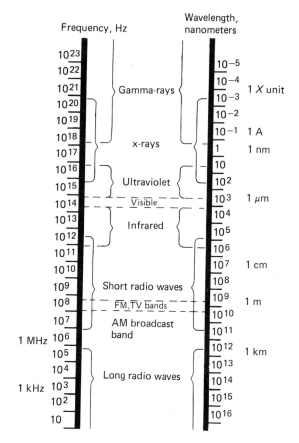

Fig. 18.25 The electromagnetic spectrum. The scales used are logarithmic.

the same family that includes light and radio waves.

The electrical nature of all these waves has led to them being called *electromagnetic* waves. Each is a member of the *electromagnetic spectrum* (Fig. 18.25) which extends from long radio waves with a wavelength of many kilometres (10^3 m) down to γ-rays with a wavelength of only 10^{-14} m. All are transverse waves and all travel at the common speed of 3.00×10^8 m s^{-1}, in a vacuum.

Chapter 19

PERIODIC MOTION

The passage of waves through a medium has constantly drawn attention to repetitive, to-and-fro motion. When a continuous wave passes over the surface of water, any particular patch of water moves up and down in a repetitive fashion, passing on this motion to the next patch. However, patterns of repeated motion of this nature are not confined to those associated with waves. All motion in the physical world can be divided into two broad categories: *periodic* motion and *non-periodic* motion. In periodic motion an object repeats its pattern of motion at regular intervals of time: it is rhythmic motion. The movement of a bird's wing, and a fly's wing, in flight, are examples of periodic motion. One wing-beat of a rook lasts about half a second. The wing-beat of a bluebottle fly lasts about one hundredth of a second. The water level in the Pool of London changes in a periodic way, the time interval between one high tide and the next being just under thirteen hours. Periodic motion is very common in nature although often very complex in form: the precise motion of a rook's wing in flight is a subtle and beautiful thing, and far from easy to describe exactly. In the world of machinery periodic motion is commoner still, and vibrations can be an engineer's nightmare: the violent vibrations of an engine during trial runs may succeed in shattering it if the vibrations cannot be suppressed.

Rotational motion too is a form of periodic motion, but in this chapter we shall be concerned mainly with periodic motion back and forth along a straight line: how to describe this kind of motion, the forces which cause periodic motion, and the analysis of a variety of simple examples of periodic motion.

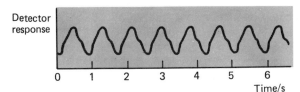

Fig. 19.1 Periodic motion of the human pulse.

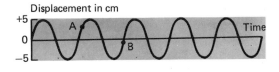

Fig. 19.2 Periodic motion of a mass hanging from a spring.

19.1 Describing periodic motion

The graph in Fig. 19.1 shows the pulse of a healthy human being. It is in fact a record of the movement of the radial artery in the wrist, in response to the pumping action of the heart. It can be seen that the pattern of pressure variation repeats itself regularly. When a doctor feels a patient's pulse he is detecting, with his fingertips, the maxima of blood pressure which distend the artery in the patient's wrist in a periodic way. The unit of the pattern in Fig. 19.1, which is repeated regularly, is termed one *cycle* of the motion. The time duration of the cycle is called the *periodic time*, or more commonly, the *period* of the motion. Thus, a person with a pulse of 72 per minute has a period of blood-pressure variation of $\frac{60}{72} = \frac{5}{6}$ second.

Compare the pattern of Fig. 19.1 with that of Fig. 19.2. In Fig. 19.2 the graph shows a simpler example of periodic motion: the movement of a mass suspended on a spring and oscillating freely up and down, above and below its equilibrium position. Clearly, the pattern of the motion shown in Fig. 19.2 is simpler than that of Fig. 19.1. In fact, as will be discussed later, the motion illustrated in Fig. 19.2 is an example of the simplest kind of periodic motion.

We shall use the motion shown in Fig. 19.2 to define some of the terms used in describing periodic motion. The rhythmic movement of an object back and forth about an equilibrium position, or the rhythmic variation of a measurable physical quantity above and below an equilibrium value, is often termed *oscillation*. The moving object, and the varying physical quantity, can be said to perform oscillations or to execute oscillatory motion. The term *period of oscillation* has already been defined and we may see that, although the oscillatory motions shown in Figs. 19.1 and 19.2 could have the same period, they would still be quite different. We now have to discuss how the precise pattern of an oscillatory motion can be described.

Referring again to Fig. 19.2, we say that when the mass is at its equilibrium position its displacement is zero. The distance of the mass from its equilibrium position, at any given instant, is termed its *displacement*. If the mass is above the equilibrium position its displacement is reckoned as positive; if below equilibrium, it is reckoned as negative. Thus, at the instant represented by point A, the displacement is $+3$ cm. At point B the displacement is -1 cm. The maximum value of the displacement is termed the *amplitude* of the oscillation, and in this example we see that the amplitude is 5 cm.

The graph of Fig. 19.2 does not perfectly represent the free oscillation of a mass suspended by a spring because it does not show the gradual dying-down of the oscillation. We will assume for the present that this dying-down is negligible.

The *frequency* of an oscillatory motion is the number of cycles of oscillation performed per unit time, and the most commonly used time unit is the second. The basic unit used for measuring frequency is known as the *hertz* (Hz). A frequency of one hertz means simply one complete oscillatory cycle per second. Scientists and engineers often have to deal with very high frequencies and use the standard prefixes (kilo, mega, etc.) for these. The vibrations associated with sound waves range from about 20 Hz to 20 kHz; the electrical oscillations associated with radio waves range from about 100 kHz to 100 MHz. When you get the opportunity, look at the tunning scale on a radio receiver, and notice the ranges covered. Not all such scales are marked in frequency units, however; they may be scales of wavelength.

The relationship between the period and the frequency of an oscillatory motion is simple. The symbol generally used for frequency is f, and the symbol for period is T, and one is the reciprocal of the other, that is,

$$T = \frac{1}{f} \tag{19.1}$$

For example, if you swing a small mass tied on a thread about 25 cm long like a pendulum you will find that the period is almost exactly half a second: thus the frequency is almost exactly 2 Hz. To take another example: in a certain radar transmitting antenna the frequency of the alternating current is 200 MHz. This means that the period of oscillation of the electrons is

$$\frac{1}{200 \times 10^6}\text{s} = 5 \times 10^{-9}\text{s}$$

$$= 5\text{ ns (five nanoseconds)}$$

19.2   Isochronous oscillation

A pendulum is a simple form of oscillator. The pendulum has been used for about three centuries as the basis of reliable time-keeping instruments. It is probably not surprising that a pendulum, swinging back and forth with a constantly maintained amplitude, should keep regular time. But it is perhaps rather surprising to observe that, if a pendulum is set swinging freely, as it dies down and the amplitude decreases, the rhythm does not alter appreciably. This is to say that the period of oscillation is very nearly independent of the amplitude of oscillation. Table 19a shows how very nearly constant the period of a simple pendulum is for small amplitudes.

Table 19a

Total angle of swing	Period (s)
10°	1.0005
20°	1.0019
30°	1.0043
40°	1.0076
50°	1.0117
60°	1.0168
70°	1.0227
80°	1.0293
90°	1.0387

This property of a pendulum swinging with small amplitudes, namely, that its period is independent of its amplitude of swing, is called *isochronism*. It was first observed, so the tradition has it, by Galileo Galilei in the cathedral of Pisa. He was attending mass and, during a tedious sermon, timed the swing of a lantern suspended by a long chain from the roof, by counting how many of his own pulse-beats corresponded to each swing of the lantern. He discovered that the time of swing was the same for large and small amplitudes alike.

A pendulum is not a perfectly isochronous oscillator. One might wonder if in fact there is such a thing as a precisely isochronous oscillator. As will be made clear later, an oscillator consisting of a mass attached to elastic springs is very nearly isochronous. The distinction between isochronous and non-isochronous oscillations is very nicely demonstrated by a simple piece of apparatus (Fig. 19.3) consisting of pieces of bent metal curtain-track with a steel ball rolling in each. One is a V-shaped track with a curved apex. Another is in the form of a circular arc. The third is parabolic. A ball is released near the top of one side of the V-shaped track and allowed to roll back and forth. The amplitude of the motion decreases rapidly as a result of energy-loss through friction. The period of oscillation is long at first; then, as the amplitude decreases, it gets shorter until finally the ball is rattling back and forth rapidly in the apex of the track.

By contrast, the motion of the ball in the circular track is quite different. As the amplitude decreases there is a noticeable change in the rhythm of the movement, and the period appears to remain constant right up to the stage at which the motion becomes imperceptible. Thus the ball rolling in a circular arc, like a pendulum bob moving in a circular arc, appears to be a nearly

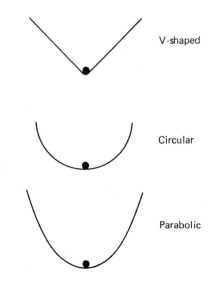

Fig. 19.3

perfect isochronous oscillator. The ball rolling in the parabolic track is found not to be isochronous, although as the amplitude becomes smaller, the period, initially long, becomes less, and seems to approach a limiting constant value.

19.3 The motion of a simple oscillator

Many simple oscillating systems are isochronous. They also share in common the particularly smooth displacement–time graph shown in Fig. 19.2 for the mass on the end of a spring.

It is worth a moment to explain what the term 'simple' means here. For what follows may not be considered by everyone to be 'simple' physics. In that case, 'simple' would mean 'easy'! Physicists, however, often use the term 'simple' to mean 'depending on a few obvious variables' and 'depending on these variables in a straightforward manner'. For instance the 'simple' pendulum is a mass on the end of a string. Its swing is controlled by the pull of gravity – a constant downward force. Air resistance, stretching of the string, etc., are all ignored. A trolley whose horizontal motion is controlled by a pair of springs is also a simple oscillatory system (Fig. 19.4). Friction between the trolley and the surface on which it runs can be ignored, since it affects only the amplitude of the motion. The forces acting on the trolley are 'simply' those of the springs – which obey Hooke's law – a 'simple' (i.e. straightforward) relationship between force and extension.

There are a number of ways of determining the variations of displacement with time of these simple oscillators. The one of universal application is stroboscopic photography. The object under investigation is illuminated by a regularly-flashing lamp and a half-cycle of its motion is recorded on film. Figure 8.14 shows a strobo-scopic photograph of a swinging pendulum.

Several cycles of the displacement–time graph can be drawn for a pendulum by pulling a sheet of

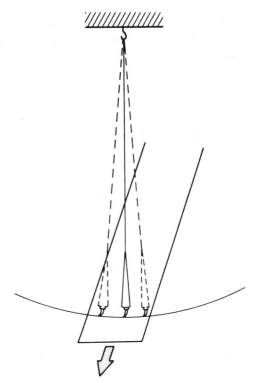

Fig. 19.5 A massive pendulum showing how a displacement/time graph may be obtained.

paper steadily along underneath the heavy pendulum of Fig. 19.5, which has an inked pen attached to it.

The trolley-and-springs oscillator shown in Fig. 19.4 is a particularly useful example to use in analysing the patterns of motion which underlie isochronous periodic motion. A trolley of the kind used in dynamics experiments is tethered by two long helical springs lying along the same axis. At equilibrium both springs are stretched to about twice their original unstretched length so that, when the trolley is pulled to one side and let go, it will oscillate back and forth without the springs ever slackening. The surface upon which the trolley runs needs to be smooth: a sheet of paper, tinplate, formica, or any similar hard, smooth material is suitable. This apparatus may be used for investigating how the period of oscillation depends on the mass of the object oscillating and the stiffness of the springs. But for the present we are concerned simply with the *kind* of motion performed by this oscillator and the forces causing this motion. We will suppose that, for the time being, we can neglect any forces of friction. If

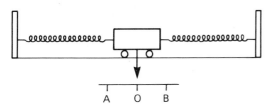

Fig. 19.4 A 'trolley-and-springs' oscillator.

there were no friction, the trolley, once displaced and let go, would continue to oscillate with constant amplitude. In practice the amplitude decays, and a typical trolley will perform 20 or 30 cycles before finally coming to rest.

A half-cycle of the motion of the oscillator can be plotted by using the following procedure. A piece of narrow paper tape is attached to the trolley and passed under the vibrator of a ticker-timer whose operation has been described earlier (Section 3.2). The trolley is pulled to one side and the slack tape is taken up between it and the timer. The timer is started and the trolley is released. A graph of half a cycle of its position can now be plotted from measurements made on the tape. For purposes of later comparisons it will be found convenient to make the ratio (spring constant/ mass of trolley) a whole number. The spring constant, k, is of course the value for *both* springs acting together. It must thus be measured by pulling back the trolley tethered by both springs a known distance, using a dynamometer to measure the force required. The mass of the trolley can then be suitably adjusted by adding mass to it.

19.4 Analysing the motion of the trolley-and-springs oscillator

Returning to Fig. 19.4, when the trolley is in equilibrium, the pointer attached to it is opposite O. Suppose now that it is displaced to B. There is now a resultant force on the trolley to the left, and if the trolley is being acted upon only by the forces of the springs, it must therefore (by Newton's second law of motion) have an acceleration to the left. Suppose that the trolley moves under the action of this force. Its velocity, to the left, will increase. When the trolley is at a point nearer to O than B the force on it is less, therefore the acceleration is less: but this still means that its velocity to the left is increasing although it is increasing at a slower rate than before. As the trolley passes through the equilibrium position, O, the forces on it are balanced so its acceleration is zero. After passing O the force on the trolley is now to the right, and this force decelerates the trolley. The greater the trolley's displacement to the left of O, the greater is this deceleration. At A, which is the same distance from O as B, the velocity of the trolley has been reduced to zero,

and thereafter it accelerates back towards B. So we see that, at every point during the motion, the trolley is acted upon by a resultant force which is directed back towards the equilibrium position. Hence the acceleration (which, by Newton's second law, is directly proportional to the force) is always in the reverse direction to the displacement.

What further conditions have to be satisfied if the motion is to be isochronous? Let us suppose we double the amplitude of the motion by displacing the trolley some further distance from O. If the trolley is to complete its journey back to O in the same time, the average speed with which it moves will have to be doubled. But this doubled speed has to be achieved in the same time as before – so the trolley's acceleration has to be doubled. The condition for isochronous motion of the trolley is then that the trolley's acceleration is proportional to its displacement. Since, as we have just said, force is proportional to acceleration, this also implies that the force tending to restore the trolley to its equilibrium position is proportional to the displacement of the trolley. And this of course is just a statement of Hooke's law for the springs. The trolley oscillates isochronously *because* its acceleration is at all times proportional to its displacement. This is the *general* condition for isochronous motion, and oscillators which obey this simple acceleration pattern are called *simple harmonic oscillators* and are said to undergo *simple harmonic motion*.

In algebraic terms, the force F acting on the trolley when it is displaced a distance x from its equilibrium position is given by

$$F = -kx$$

where k is the spring constant for the pair of springs acting on the trolley. The negative sign is necessary as the force always acts in the opposite direction from the displacement.

From Newton's second law, a mass m acquires an acceleration a under a force F given by

$$a = F/m$$
$$= -(k/m)x$$

We shall now compute the motion of a typical trolley-and-springs oscillator for one quarter-cycle of its motion: from B to O in Fig. 19.4. We will suppose for the purposes of this calculation that the trolley has a mass of 3 kg and that the springs, together, provide a force of 12 N m⁻¹ displacement of the trolley, either side of its equilibrium position.

Suppose that the trolley is displaced a distance x (measured in metres). Then the force on the trolley is $-12x$.

Hence the acceleration, which by Newton's second law is got by dividing the force by the mass, is $-4x$. Here we have, then, an acceleration which is directly proportional to the displacement and in a direction opposite to the displacement.

The motion begins, we shall assume, with the trolley pulled out to a distance 0.200 m and released from rest. We shall consider what happens in successive time intervals of 0.1 s from the start. Our computation will be only an approximation to the truth. How good an approximation it is we shall be able to see afterwards. In this method of computation we build up the displacement–time graph for the oscillator step by step. Figure 19.6 shows this graph completed. Note that the graph is made up of straight line segments. Each segment therefore represents motion with uniform velocity, and there is an abrupt change in velocity between segments. This

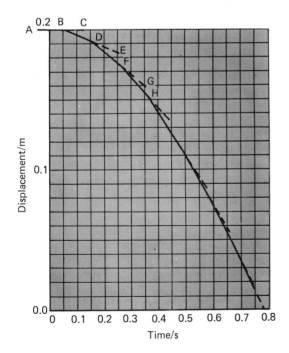

Fig. 19.6 Computing the graph of displacement against time for the 'trolley-and-springs' oscillator.

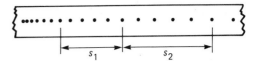

Fig. 19.7

of course is not the truth, only an approximation to it. Let us now study the principles of the computation. It may help to refer to an imaginary ticker-tape record of the motion, during the first quarter-cycle, as shown in Fig. 19.7.

Suppose the distances travelled in successive time-intervals Δt are Δs_1 and Δs_2. Then the average velocities for these two time intervals are $\Delta s_1/\Delta t$ and $\Delta s_2/\Delta t$. The acceleration, a, or rather the average acceleration, is therefore given by

$$a = \frac{\dfrac{\Delta s_1}{\Delta t} - \dfrac{\Delta s_2}{\Delta t}}{\Delta t} = \frac{\Delta s_2 - \Delta s_1}{\Delta t^2}$$

If the object had been moving with uniform velocity it would have travelled equal distances in successive time intervals Δt. Because it is accelerating it has travelled an *extra* distance $\Delta(\Delta s)$ given by

$$\Delta(\Delta s) = \Delta s_2 - \Delta s_1 = a\,\Delta t^2$$

In building up the graph we calculate, and make use up of, the extra distances travelled by the object, during successive intervals of time, as a result of its acceleration.

[Using the notation of calculus $\Delta(\Delta s)$ can be written $\Delta^2 s$, giving

$$\Delta^2 s = a\,\Delta t^2$$

or

$$a = \frac{\Delta^2 s}{\Delta t^2}$$

In the limit as $t \to 0$, we write this

$$a = \frac{d^2 s}{dt^2}$$

a form which is familiar to mathematicians as the second order derivative of s with respect to t.]

The acceleration is not constant, of course: it is proportional to the displacement and in our example is given by $a = -4x$. Our value for Δt is 0.1 s, so that the expression for the extra distance travelled during each time interval becomes

$$\Delta(\Delta s) = a(0.1)^2 = \frac{-4x}{100}$$

We can now build up the displacement–time graph (Fig. 19.6) for our oscillator. During the first 0.1 s time interval the oscillator's velocity has increased from zero to some value in the negative x-direction: and we know that in fact this increase has taken place smoothly, but we assume, for our approximation, that there has been an abrupt increase in velocity halfway through the time interval. We know that the gradient of the displacement–time graph represents the velocity: this means that the segment AB (gradient = 0 and hence velocity = 0) represents the velocity at time $t = 0$. Segment BD represents the velocity at time $t = 0.1$ s. If the velocity of the oscillator had remained zero after the start, then it would be at point C on the graph after the second time interval: but its velocity has in fact changed and instead of travelling zero distance it has travelled an extra distance which we calculate using our formula derived above

$$\Delta(\Delta s) = \frac{-4x}{100}$$

Here $x = 0.200$ m, so $\Delta(\Delta s) = -0.008$ m. Therefore the distance CD on the graph must be 0.008 m: this enables us to plot point D and then draw the line BD. Going on from point D, if the oscillator travelled from there with constant velocity it would arrive at point E: but since it is in fact accelerating it will arrive at point F and to locate this point on the graph we must compute the extra distance travelled in the 0.1 s time-interval around time 0.2 s. This extra distance is $-4/100 \times 0.192 = 0.008$ m (correct to the third decimal place). Thus the extra distance, EF on the graph, is 0.008 m so that point F can be plotted and the line DF drawn in.

You should continue the process and produce the graph until it cuts the time-axis. If this is done carefully it will be found that the line crosses the time axis very close to time 0.78 s. If the process were continued further it would be found that the line, having gone below the time-axis and curved upwards, cuts the time-axis again at 3×0.78, then again at 5×0.78, 7×0.78 s, and so forth. The period of oscillation is thus $4 \times 0.78 = 3.12$ s very nearly. The curve we have obtained by this method looks very much like the first part of a cosine graph. We can test this numerically: then we shall deal with the theory.

The equation to our graph of Fig. 19.6 can be written, we will assume, thus: $x = 0.200 \cos \theta$ since the initial value of x is 0.200 m. Now $\cos \theta = 0$ when $\theta = \pi/2$, $3(\pi/2)$, $5(\pi/2)$, etc. The first zero of our curve occurs when $t = 0.78$ which is very nearly $\pi/4$. So, in order for our graph to be $\cos \theta$ plotted against t, θ must equal $2t$, so that when $t = \pi/4$, $\theta = \pi/2$ as we require. Let us try how good a fit the function

$$x = 0.200 \cos 2t$$

is to our curve. Take a trial value, say, time = 0.40 s. When $t = 0.40$, $\cos 2t = \cos 0.8$ (that is, 0.8 radians which equals 45.8°). Therefore $x = 0.200 \cos 45.8° = 0.200 \times 0.698 = 0.140$. This agrees very closely with the graph.

19.5   Sinusoidal motion

We shall now try to justify this link between the displacement–time graph for a simple harmonic oscillator and a cosine curve. To do this we shall first show a link between simple harmonic motion and *circular motion*. This correspondence can be demonstrated with the apparatus illustrated in Fig. 19.8. We set up a slow-running turntable close to a vertical screen. On the turntable we place a thin vertical rod mounted upon a heavy base and supporting a small ball at the top end. A

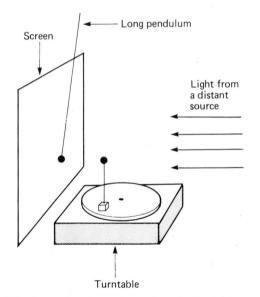

Fig. 19.8 Comparing the motion of the bob of a simple pendulum with that of a sphere moving in a horizontal circle.

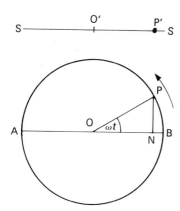

Fig. 19.9

small, intense light source is placed several metres away, on the same horizontal level as the ball, so that it casts a shadow of the ball on the screen. The light from the distant source is very nearly parallel and so the shadow of the ball should be sharp and negligibly larger than the ball itself. When the turntable is set moving at a steady speed the shadow is seen to move back and forth in a straight line on the screen. Is the motion sinusoidal? Well, it is easy, half-closing one's eyes, to imagine that the shadow is not that of a ball going in a circle, but of the bob of a long pendulum swinging to and fro. Later in Section 19.8, it is proved that the motion of a simple pendulum approximates very closely to sinusoidal motion provided that the angular amplitude of swing is small. We can mount a long pendulum – a small heavy ball suspended by a thread – to swing near the screen in a plane parallel to it, and by adjusting its length arrange that the shadow of the pendulum bob moves with the same period as the period of revolution of the ball on the turntable. This needs some patience to achieve exactly, but makes an impressive demonstration. The two shadows will be seen to move with almost exactly the same motion.

We can explain the correspondence demonstrated above by a simple geometrical argument. In Fig. 19.9 imagine that the small ball is travelling anti-clockwise around the circumference of the circle. A clock is started when it is at B. At time t the ball is at point P. The ball is travelling around the circle with constant speed. If its angular velocity is ω then the angle PON is ωt. A shadow of the ball is being cast on a screen SS, the line SS being parallel to AB, by a parallel beam of light in a direction perpendicular to the screen.

Thus the shadow of the ball is at P'. As the ball travels around the circle the shadow P' travels back and forth in a straight line on the screen. Clearly, the displacement of the shadow-ball, measured from the central point (O') of this motion, is equal to the distance ON at every distant. It is easier now for us to fix our attention upon the motion of N rather than on P'.

What we have established is that if a point P moves round a circle at a *uniform* speed, then its *projection*, N, on to a diameter of the circle moves along the diameter with simple harmonic motion.

This argument can be put into reverse using the ticker tape record of the trolley-and-springs oscillator. The distance from one extremity of the motion to the other is measured and a semicircle of diameter equal to this distance is drawn, centred on the bisector of the line of motion (Fig. 19.10).

Perpendicular lines are now drawn from each recorded dot on the tape to meet the semicircle. Lines are then drawn from the points of intersection to the centre, O, of the circle. It will be seen that the angle between each line and its neighbour is constant. The dots are spaced apart equally in time and thus as the oscillator executed one half of its cycle, the corresponding point on the semicircle was moving around it at a steady speed.

Returning now to Fig. 19.9, we can use this diagram to derive an expression relating the displacement of N to time. The amplitude, a_0, of the motion of N is equal to the radius of the circle, OA. Thus the displacement, x, of N, measured from O, is given by

$$x = a_0 \cos \omega t \qquad (19.2)$$

where ω is the angular velocity, in radians per second, with which P was moving around the circle. Thus we have demonstrated that the

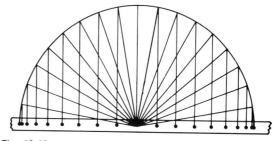

Fig. 19.10

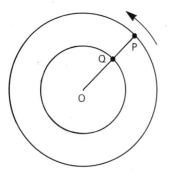

Fig. 19.11 Two circular motions in phase.

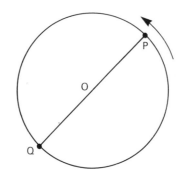

Fig. 19.12 Two circular motions in antiphase (the phase difference is π).

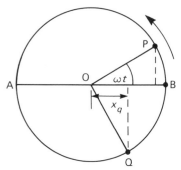

Fig. 19.13 Two circular motions in quadrature (the phase difference is $\pi/2$).

displacement–time graph for a simple harmonic oscillator does fit a cosine curve as we expected it might.

19.6 Phase

We have used the term 'phase' in earlier chapters to express the idea that waves may be 'in step' or 'out-of-step' with each other. Using the apparatus of Fig. 19.8, we can now give a precise meaning to the word 'phase'.

On the turntable we place another ball on a stick as shown in the plan view of Fig. 19.11 at Q. With the turntable rotating the two shadows move back and forth. For convenience we will use the symbol P′ for the shadow of P, and Q′ for the shadow of Q. When P′ is at its maximum leftwards displacement so also is Q′. Likewise, P′ and Q′ both reach their maximum displacement to the right at the same instant, and have zero displacement at the same instant. The amplitudes are different: that of Q′ is smaller than that of P′. But the two motions are *in phase*, and therefore we could represent them by the equations

$$x_P = a_P \cos \omega t$$

$$x_Q = a_Q \cos \omega t$$

Now arrange P and Q as shown in Fig. 19.12. As they go around the circle (this time the amplitudes are equal) the displacement of P′ will be at every instant equal in magnitude but opposite in sign to that of Q′: the two motions are *in antiphase*, or of *opposite phase*.

A case of special interest is when the motions of P′ and Q′ are *in quadrature*. This is achieved

by arranging P and Q as in Fig. 19.13. For this demonstration it is best to have the balls P and Q on different horizontal levels so that their shadows are easily identifiable on the screen. As P and Q go anticlockwise around the circle the shadow P′ leads the shadow Q′: and Q′ gives the impression of lagging behind like a lazy dog on an elastic lead. Observing carefully, one notices that when P′ has maximum displacement Q′ has zero displacement, and vice versa. Now, if the graph of Fig. 19.14b represents the motion of P′, which of the other

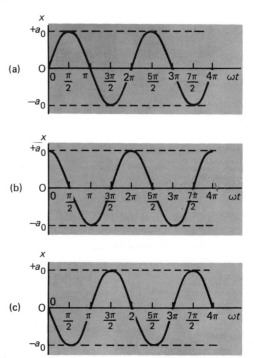

Fig. 19.14 The variation of displacement of a sinusoidal oscillator with time.

two graphs in Fig. 19.14 represents the motion of Q'? It is Fig. 19.14a: you should check this carefully. Thus the motions of P' and Q' in this situation can be described by the equations

$$x_P = a_0 \cos \omega t$$

$$x_Q = a_0 \sin \omega t$$

(We write the amplitude, a_0, here because we arranged the amplitudes to be equal in the demonstration.) Readers having a good knowledge of trigonometrical functions will know that $\sin \theta = \cos(\theta - (\pi/2))$, and so the two equations could be written

$$x_P = a_0 \cos \omega t$$

$$x_Q = a_0 \cos \left(\omega t - \frac{\pi}{2} \right)$$

But this can easily be seen by studying Fig. 19.13 in which the displacement x_Q equals $a_0 \cos Q\hat{O}B$, but $Q\hat{O}B$ is of course equal to $\omega t - (\pi/2)$. In the two equations above, the term $-(\pi/2)$ is the *phase difference* between the two motions. We can say that Q' has a *phase lag* of $\pi/2$ radians behind P', or that P' *leads* Q' by $\pi/2$ radians.

If we wanted Q' to lead P' with a phase-difference of $\pi/2$ radians, then, using the same expression as before to represent the motion of P', we would write the expression for Q' thus:

$$x_Q = a_0 \cos \left(\omega t + \frac{\pi}{2} \right)$$

In general, therefore, we can write the expressions

$$x_P = a_0 \cos \omega t$$

$$x_Q = a_0 \cos (\omega t + \phi)$$

where ϕ is the phase difference between the two motions, and can be expressed in radians (or in degrees).

The concept of phase is vital in describing the relative motions of two or more sinusoidal oscillators. But it also has relevance when describing the motion of a single oscillator, as is shown by the example which follows.

Problem 19.1 A particle is performing sinusoidal oscillation about a fixed point with amplitude 5 cm and frequency 8 Hz. A clock is started when the displacement of the particle is $+2.5$ cm. Find an expression for its displacement as a function of the time, t, registered by the clock.

We can straight away write the expression in the form

$$x = a_0 \cos (\omega t + \phi)$$

and we have to find the appropriate values for a_0, ω, and ϕ. Now $a_0 = 0.05$ m, and $\omega = 2\pi f = 2\pi \times 8 = 16\pi$ radians per second. To find ϕ we reason as follows.

We require that when $t = 0$ s, $x = 0.025$ m. Therefore

$$0.025 = 0.05 \cos \phi$$

hence

$$\cos \phi = +0.5$$

and so

$$\phi = 60° \text{ or } 300°$$

But how does it come about that ϕ can have two possible values? You should be able to reason this out for yourself, perhaps with the help of sketch graphs, and bearing in mind that the displacement of the particle has the value $+2.5$ cm twice during every cycle.

The most general expression for sinusoidal motion is

$$\boxed{x = a_0 \cos (\omega t + \phi)} \qquad (19.3)$$

or, using the sine function, $x = a_0 \sin (\omega t + \phi)$. You will meet both the cosine and sine forms, in different contexts, and should be ready to handle both. From this point onwards in this chapter we shall keep to the cosine form, and, for simplicity's sake, usually assume when we are dealing with a single sinusoidal oscillator that our clock (usually an imaginary clock) has been started when the oscillator had its maximum positive displacement, so that the expression we use is $x = a_0 \cos \omega t$.

19.7 Period

The period of a simple harmonically oscillating body is an important constant of its motion. We shall now try to relate this to other constants of the motion. Referring back to the diagram in Fig. 19.9, we can see that the point N will complete one cycle of its oscillation in the time P takes to go once round the circle. If P moves at ω radians per second,

$$T = 2\pi/\omega$$

Since $T = 1/f$ the frequency of this oscillation is given by $f = \omega/2\pi$. It now remains to relate ω to other constants.

The acceleration of P is $\omega^2 a_0$ (which is equivalent to v^2/a_0, see Sections 7.2 and 7.3) and is directed towards the centre of the circle. The acceleration of N is thus the component of $\omega^2 a_0$ parallel to AB (Fig. 19.9). This is $\omega^2 a_0 \cos \omega t$. But we have already proved that $a_0 \cos \omega t = x$, the displacement of N from O.

So,

$$\boxed{\begin{aligned}\text{acceleration of N} &= -\omega^2 a_0 \cos \omega t \\ &= -\omega^2 x\end{aligned}} \quad (19.4)$$

(bearing in mind that x is in the opposite direction to the acceleration).

For the trolley-and-springs oscillator,

$$a = -(k/m)x$$
$$\omega^2 = k/m$$

Hence

$$T = 2\pi\sqrt{\frac{m}{k}}$$

We can justify this expression, in part, by the following argument. What could we do to the trolley-and-springs oscillator to double the period, T, of its oscillation? If T is to be doubled, any motion from one extremity to the equilibrium position (B to O in Fig. 19.4) will take twice as long – so the average speed is halved. Since this halved speed is attained in *twice* the time the acceleration must reduce by a factor of 4. Bearing in mind Newton's second law of motion, this could be achieved *either* by increasing the mass of the trolley by a factor of 4, *or* by reducing the spring constant by a factor of 4 – just as is predicted by the equation above for the period.

Generalizing the expression for T found above, we can say that for any simple harmonic mechanical oscillator

$$T = 2\pi\sqrt{\frac{\text{mass}}{\text{restoring force/unit displacement}}}$$

$$(19.5)$$

We shall now examine some other simple oscillatory systems which fulfil the conditions required for the performance of simple harmonic motion; and we shall thereby be able to find by theoretical analysis the factors which determine their periods of oscillation. It should be emphasized that very few oscillatory systems are *truly* simple harmonic, thus allowing easy analysis. However, many systems approximate well to being simple harmonic oscillators provided that the amplitude of oscillation is kept small. The reason for this is that, although it is rarely the case that the restoring force is directly proportional to the displacement (which is the condition for simple harmonic motion), the restoring force may be approximately proportional to the displacement over a limited range of displacements. A good example of an oscillator which is not truly simple harmonic is the simple pendulum.

19.8   The simple pendulum

A small, massive object suspended by a thread of negligible mass from a rigid support-point, allowed to swing in a vertical plane, is termed a *simple* pendulum. This is to distinguish it from the commoner, *compound* pendulum in which the mass has considerable spatial extension and cannot be considered as all being concentrated at one point as in the simple pendulum.

If you have not already done so you should investigate experimentally two important properties of a simple pendulum. The first of these is the very nearly perfect isochronism for small amplitudes of swing. It is most impressively demonstrated by setting up side by side two pendulums consisting of lead balls, one or two centimetres in diameter, suspended by threads about two metres long. Having made sure the threads are of equal length, set one pendulum swinging with a very small amplitude, say, about 5 cm, and the other with an amplitude of about ten times as much. It should be found that they do not become noticeably out of step over a time of ten cycles or so. The other important property is that the period of swing does not depend on the mass of the pendulum bob, but only on the length of the suspension. To demonstrate this, replace one of the pendulum bobs by a much more massive one, ten, fifty, or even a hundred times the mass of the

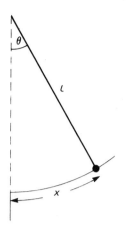

Fig. 19.15

first. Again, if the two pendulums are set swinging in phase with each other they will be found to keep in phase for several cycles. The analysis which follows will show how these two properties arise.

Figure 19.15 shows a simple pendulum displaced from its equilibrium position. The displacement can be measured either by the angle θ made by the thread with the vertical, or by the distance, x, through which the centre of mass of the pendulum bob has moved along the arc of its path. The length of the suspension, measured from the point of support to the centre of mass of the bob, is l. The mass of the bob is m, and the gravitational field strength (i.e. the gravitational force per unit mass) is g.

First of all we want an expression for the force acting on the bob in its direction of motion. The tension in the suspending thread is always acting in a direction perpendicular to the motion of the bob so that it has no effect on the bob's motion. The weight or force of gravity on the bob is mg, and the component of this parallel to the direction of the bob's motion is $mg \sin \theta$. Thus:

restoring force at displacement $x = -mg \sin \theta$

Now θ (measured in radians) $= x/l$. If θ is small we can use the approximation $\sin \theta = \theta$. How good an approximation this is, we shall consider later. Using the approximation we can write

restoring force at displacement $x = -mg \dfrac{x}{l}$

and so we see that the restoring force is directly proportional to the displacement. Also, it is clear that the force is always in the reverse direction to

the displacement. The conditions for simple harmonic motion are therefore fulfilled, and so we should expect the oscillations to be isochronous, so long as the angle θ never becomes so large that the approximation $\sin \theta = \theta$ is invalid.

It was established earlier (Section 19.7) that, for a simple harmonic oscillator,

$$\text{period } T = 2\pi \sqrt{\frac{\text{mass}}{\text{force per unit displacement}}}$$

hence, for our simple pendulum,

$$\text{period } T = 2\pi \sqrt{\frac{m}{mg/l}}$$

hence

$$T = 2\pi \sqrt{\frac{l}{g}} \qquad (19.6)$$

and we see that the mass, m, does not appear in this expression, which predicts that the period is independent of the mass of the bob.

19.9 Some more applications

Problem 19.2 Oscillation of a vehicle on its spring suspension.

In a certain automobile, when five passengers get in, their total mass being 360 kg, the springs of the suspension are compressed a distance 4 cm. The total mass including passengers supported by the suspension is 900 kg. Assuming that the displacement of the springs is directly proportional to the compressive force in them, calculate the period of oscillation of the loaded car on its suspension, neglecting any effects of damping (which of course is very heavy in a well-designed suspension system!). (Take $g = 10 \text{ N kg}^{-1}$.)

First we find the force per unit displacement of the suspension springs. The weight of the passengers is $360 \times 10 \text{ N} = 3600 \text{ N}$, and since this causes a compression of 4 cm, which equals 0.04 m, the force per unit displacement is $3600/0.04 \text{ N m}^{-1} = 90\,000 \text{ N m}^{-1}$. Using the relation

$$\text{period} = 2\pi \sqrt{\frac{\text{mass}}{\text{force per unit displacement}}}$$

we get

$$T = 2\pi \sqrt{\frac{900}{90\,000}} = 2\pi \times \frac{1}{10} = 0.628 \text{ s}$$

Problem 19.3 Oscillation of a climber dangling on a rope.

A typical mountaineer's rope, about 35 m long, will 'give' a distance of about 1.6 m under the weight of a climber hanging freely on the end. Assuming the climber's mass is 80 kg, estimate the period of vertical oscillation of the climber when dangling freely on the end of the rope. How does this compare with the period of swing if the climber 'pendules' (that is, swings on the rope like a pendulum-bob)?

The solution goes on the same lines as the previous problem. The weight of the climber is $80 \times 10 = 800$ N, and so the force per unit displacement provided by the rope is $800/1.6$ N m$^{-1} = 500$ N m^{-1}. Hence the period is given by

$$T = 2\pi \sqrt{\frac{80}{500}} = 2\pi\sqrt{0.16} = 2\pi \times 0.4$$

$$= 0.8\pi = 2.5 \text{ s}$$

The period of a simple pendulum is given by

$$T = 2\pi \sqrt{\frac{l}{g}}$$

Substituting our numerical values we get

$$T = 2\pi\sqrt{\tfrac{35}{10}} = 2\pi \times 1.87 = 12 \text{ s}$$

very nearly.

Problem 19.4 Bobbing float in water. Any object floating in water, partly immersed, will, if displaced and let go, perform oscillations which quickly die away because of damping. What factors determine the period of such oscillation?

One can get an appreciation of what they are by analysing a simplified situation. Figure 19.16 shows a uniform straight stick with a counter-weight at its lower end to make it float upright. It

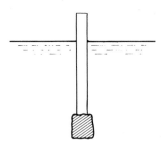

Fig. 19.16

has a total mass m and the area of cross-section of the stick is A. It is floating in a liquid of density ϱ in a place where the gravitational field strength is g.

To find the period of small vertical oscillations we use the relationship

$$\text{period} = 2\pi \sqrt{\frac{\text{mass}}{\text{force per unit displacement}}}$$

To get an expression for the force per unit displacement in this case we use Archimedes' principle: a body totally or partially immersed in a fluid experiences an upthrust equal to the weight of fluid displaced. When our stick is in equilibrium the pull of gravity on it is balanced by the upthrust of the liquid. If we displace it a distance x vertically downwards there is an increase in upthrust equal to the weight of extra liquid displaced. This increase in upthrust is the net upwards restoring force on the stick.

a) Consider the stick displaced vertically downwards a distance x. Write an expression for the net upwards force on it. Is this force directly proportional to x?

b) Consider the stick displaced vertically upwards a distance x, and write an expression for the net downwards force on the stick. Is *this* force directly proportional to x?

c) Write an expression for the force per unit displacement and hence derive an expression for the period of small vertical oscillation.

d) If the period in pure water is 1 second exactly, what is it in thin oil whose density is 0.81 g cm^{-3}?

a) When the stick is displaced downwards a distance x it displaces an extra volume of liquid Ax. The weight of this extra displaced liquid is $Ax\varrho g$. A, ϱ, and g are all constant. By Archimedes' principle, this expression gives the net upwards force on the stick and clearly it is directly proportional to the displacement x (Fig. 19.17).

b) When displaced a distance x upwards, the upthrust on the stick is reduced by an amount $Ax\varrho g$, following the same lines of argument as in (a). Since, before the stick was displaced from equilibrium, the upthrust balanced the stick's weight, this means that we have a net downwards force, tending to restore the stick to equilibrium,

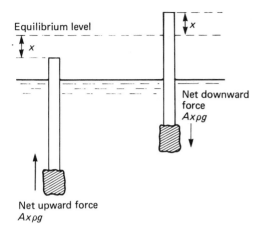

Equilibrium level

Net downward force $Ax\rho g$

Net upward force $Ax\rho g$

Fig. 19.17

of amount $Ax\varrho g$. Thus, once again, we have a restoring force proportional to the displacement.

c) Displaced either above or below equilibrium, the stick experiences a restoring force per unit displacement $A\varrho g$. Thus the period of oscillation is given by

$$T = 2\pi \sqrt{\frac{m}{A\varrho g}}$$

d) The expression for the period shows that $T \propto (1/\sqrt{\varrho})$. If T_1 is the period of oscillation in liquid whose density is ϱ_1, and T_2 the period in liquid of density ϱ_2, then

$$\frac{T_2}{T_1} = \sqrt{\frac{\varrho_1}{\varrho_2}}$$

If T_2 is the period in the oil, then substitution of the numerical data yields $T_2 = 1.1$ second.

19.10   The energy of a simple harmonic oscillator

The important idea that will be developed in this section is that a simple harmonic oscillator's energy is in two forms, kinetic and potential, and that the energy is changing periodically from the one form into the other.

Consider the trolley-and-springs oscillator, as illustrated in Fig. 19.4. As was said earlier, if there is no energy loss as a result of friction or any other resistances to motion, the oscillator once started will continue to move back and forth with constant amplitude. This is an ideal situation

never achieved in practice: but for the purposes of our argument we will suppose that we have such an ideal oscillator. At the ends of the motion, at maximum displacement, the velocity of the trolley is zero and therefore the kinetic energy (E_k) is zero: the energy is entirely potential in the stretched springs. We define the amount of potential energy (E_p) as the work done in displacing the trolley from equilibrium to its extreme position. At the mid-point of the oscillation the potential energy is zero (by the above definition), but the velocity is a maximum and so the kinetic energy is a maximum, and the total energy of the oscillator is in kinetic form at this point. Thus we see that during one complete cycle of oscillation the energy of the oscillator changes thus:

$$E_p \rightarrow E_k \rightarrow E_p \rightarrow E_k \rightarrow E_p$$

At points intermediate between the equilibrium position and the extremities of the motion, the energy is shared between potential and kinetic forms. We shall now derive an expression to show how the potential energy varies with displacement.

The potential energy at a displacement, x, is given by the work done in displacing the object through a distance x from its equilibrium position. The force required to do this increases uniformly from zero at the equilibrium position to some maximum value, F. Using Hooke's law,

$$F = kx$$

So the potential energy gained

= work done
= *average* force × displacement
= $\frac{1}{2}(0 + kx)x$

$$\boxed{E_p = \tfrac{1}{2} kx^2} \tag{19.7}$$

A graph of the variation of potential energy with x is plotted in Fig. 19.18.

The variation of kinetic energy with x can now be obtained. Since energy is conserved,

kinetic energy = total energy − potential energy

When the object reaches the extremity of its displacement from the equilibrium position, all its energy is in the potential form. The displacement now is a_0, the amplitude of the motion.

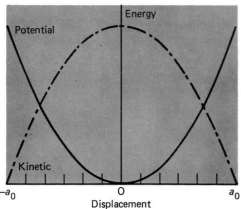

Fig. 19.18   Variation of E_k and E_p with displacement in s.h.m.

So, total energy $= \frac{1}{2}ka_0^2$
Thus, kinetic energy $= \frac{1}{2}ka_0^2 - \frac{1}{2}kx^2$

$$E_k = \frac{1}{2}k(a_0^2 - x^2) \qquad (19.8)$$

The variation of kinetic energy with displacement is also shown in Fig. 19.18.

Another expression for the total energy of an object undergoing s.h.m. can be obtained by writing down an expression for its kinetic energy as it passes through the equilibrium position. At this point, the speed of the object is the same as the speed for the equivalent point moving in a circle, whose projection is the position of the object at any time (see Fig. 19.19). The angular velocity of this point is ω, so its speed at any time is $a_0\omega$. Hence the kinetic energy of the object as it passes through its equilibrium position is given by:

$$\text{kinetic energy} = \frac{1}{2}m(a_0\omega)^2$$
$$= \frac{1}{2}ma_0^2\omega^2$$

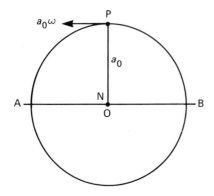

Fig. 19.19

This result ought to equal that obtained for the maximum potential energy. We can check this by making the substitution, $\omega^2 = k/m$ in the expression above.

$$\text{total energy} = \frac{1}{2}ma_0^2\omega^2$$
$$= \frac{1}{2}ma_0^2(k/m)$$
$$= \frac{1}{2}ka_0^2$$

This is of course the expression obtained earlier for the maximum potential energy.

19.11   Waves and simple harmonic motion

We have seen already that when a continuous wave passes through a medium, the parts of the medium undergo a rhythmic to-and-fro motion. However, the pattern of this motion is often complex. The source of a wave is itself an oscillating object and the complex patterns of oscillation in the medium reflect the complexity of the oscillations of the wave source.

We have noted one case, however, in which the wave form is particularly symmetrical – that is the one associated with a pure tone. The shape of the wave form is very similar to the displacement–time graph of a simple harmonic motion, so it is not surprising to find that waves of this sort are associated with sources which are oscillating simple harmonically.

While simple harmonic waves are uncommon, their study is specially important because it turns out that *all* waves, no matter how complex their pattern, can be broken down into a series of simple harmonic waves (often referred to as 'sine waves'). The proof of this was first illustrated by a French mathematician, Baron Jean Baptiste Fourier at the turn of the eighteenth to nineteenth century. The sequence of simple harmonic terms that go to make up any particular oscillation is known as its *Fourier series*. The process of splitting up an oscillation into simple harmonic terms is called *Fourier analysis*.

We shall now put together what we know about waves and about simple harmonic motion to derive an equation for a simple harmonic wave. Figure 19.20 is a 'snapshot' of the medium through which a (transverse) wave is passing. The wave is moving in the z-direction. Its displacement at any time is x.

At the source of the wave ($z = 0$) the medium

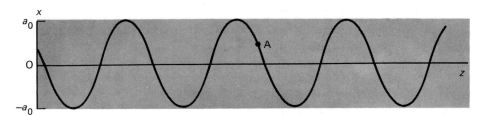

Fig. 19.20

is oscillating in step with the oscillation of the source. The variation of x with t is thus given by,

$$x = a_0 \cos \omega t$$

What can we say about the oscillation of a point A in the medium, a distance z from the source? The wave disturbance will have taken a time z/c to reach A from the source (c is the wave speed). This means that A is doing *now* what the source was doing a time z/c seconds ago. It thus *lags* behind the source in phase. To write down an expression for the variation of x with t at the point A, we need to find some way of writing down the phase difference between *its* oscillation and that of the source. Referring back to Fig. 19.20 we can see that for every increase in z of λ (the wavelength of the wave) the phase of the oscillation changes by 2π (one complete cycle).

If the point A were a distance λ from the source, then the phase difference between the oscillation at A and the oscillation of the source at $z = 0$ would be 2π.

If, on the other hand, A is some distance z from the source, then the phase difference between the oscillation at A and the oscillation of the source at $z = 0$ is

$$2\pi \frac{z}{\lambda}$$

and this is the phase difference we require.

Thus we can now write down an expression for the oscillation expected at *any* point, A, a distance z from the source, as

$$x = a_0 \cos\left(\omega t - 2\pi \frac{z}{\lambda}\right)$$

This then is the wave equation for a simple harmonic wave.

There are a number of different ways of writing this result. One useful way makes the substitution

$$\omega = 2\pi f$$

(If the wave frequency is f, then the angular frequency of the oscillation it produces will be $2\pi f$.)

$$x = a_0 \cos(2\pi ft - 2\pi z/\lambda)$$
$$= a_0 \cos \frac{2\pi}{\lambda}(\lambda ft - z)$$

Substituting $c = \lambda f$:

$$\boxed{x = a_0 \cos \frac{2\pi}{\lambda}(ct - z)} \qquad (19.9)$$

Chapter 20

FORCED OSCILLATIONS AND STANDING WAVES

20.1 Forced oscillations

A very simple experiment serves to introduce the idea of a *forced* oscillation. Set up a simple pendulum (any small heavy object on a thread will do) with your hand as the support. You will find that you can excite it into oscillation by very small rhythmic movements of your hand from side to side. If the movements of your hand are in rhythm with the natural, free vibrations of the pendulum, you can build up an amplitude which is very much greater than the amplitude of your hand's motion. If now you keep your hand perfectly still the amplitude of oscillation will of course decay as a result of damping until all motion ceases: but you can maintain oscillation at a constant amplitude by keeping your hand moving very slightly, all the time in rhythm with the natural swinging of the pendulum. The movement of your hand may be almost imperceptibly small. In a pendulum clock the amplitude of swing is kept constant by the escapement mechanism which delivers small impulses to the pendulum, in rhythm with its natural swing. The loss of energy due to damping is made good by energy fed from the wound-up clock spring (or the raised weights, as in a grandfather clock), *via* the gear trains and the escapement, to the pendulum.

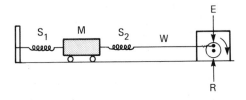

Fig. 20.1 A demonstration of a forced oscillation using the 'trolley-and-springs' oscillator.

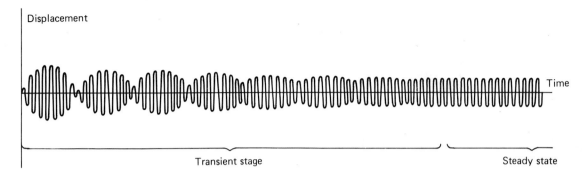

Fig. 20.2   Variation of displacement with time at that start of a forced oscillation.

Figure 20.1 shows a way of demonstrating forced oscillations in the laboratory, using the trolley-and-springs oscillator. The trolley, M, is attached by spring, S_1, to a rigid support. The other spring, S_2, is joined to a wire W whose other end is bent to form a loose-fitting ring which is slipped over a rod E mounted eccentrically on a small wheel R. The wheel is rotated by a motor which can be set to run at any desired speed. When the system is started from rest the motor quickly reaches a steady speed after being switched on. But the movement of M, which should be studied closely, is puzzling: it takes a fair length of time for the amplitude of oscillation to settle to a steady value. The graph of Fig. 20.2 shows how the displacement varies with time at the start of the process. The initial stage of the motion during which the amplitude is not constant is known as the *transient* stage: after that we have the steady-state oscillation.

For the time being we shall concentrate attention on the steady-state oscillation. When we experiment with the effect of changing the speed, setting it at different speeds and then observing the steady-state oscillation which is attained at each motor speed, we find that the steady-state amplitude depends on the motor speed in a sensitive way. The amplitude is greatest when the frequency of the motion of the rod E (which is the 'driver' in this situation) is equal to the frequency of free oscillation of the trolley and springs arrangement (which we can call the 'responder'). At any other driving frequency the amplitude of the responder is less, and the greater the difference between the driving frequency and the natural frequency of the responder, the smaller is the

amplitude. Figure 20.3 shows a typical result from this experiment.

The terms *driver* and *responder* are engineering language and very useful in this context. 'Driver' refers to whatever agency is providing a periodic force which is acting upon the responder: the 'responder' being the complete oscillatory system which has its own characteristic frequency (or frequencies) of oscillation when allowed to oscillate freely (i.e. undriven).

The effect of damping on forced oscillations is of prime importance. In all the oscillatory systems so far analysed it has been assumed that there is no energy loss due to friction or any other resistances to motion: thus the amplitude of oscillation has remained constant. In any real-life mechanical oscillatory system performing free oscillations, there *are* resistances to motion which cause a loss

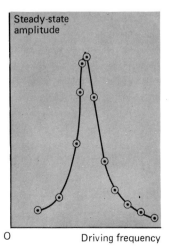

Fig. 20.3   Variation of the amplitude of the responder with the driving frequency.

Displacement

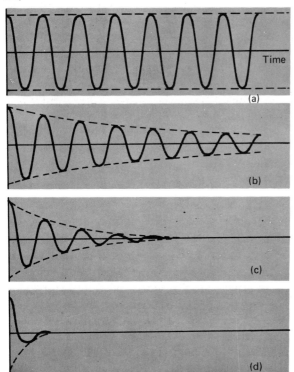

Fig. 20.4 Variations of displacement with time for (a) an undamped, (b) a lightly damped, (c) a heavily damped and (d) a very heavily damped oscillator.

of energy and a decay of the amplitude of oscillation. Oscillations decaying in this way are known as *damped* oscillations, and the term *degree of damping* is generally used to signify how rapidly the amplitude decays. A heavily damped oscillation decays rapidly in amplitude, and a lightly damped oscillation decays slowly. Figure 20.4 shows the displacement–time graph for (a) an undamped, (b) a lightly damped, (c) a heavily damped, and (d) a very heavily damped oscillator.

An experiment in which one can vary the amount of damping applied to a forced oscillator is based on the arrangement shown in Fig. 20.5. The responder is the mass attached to the spring, and controllable damping is provided by the metal plate moving in glycerol. The driver is a small wheel with an eccentrically mounted rod E, the linkage from it to the responder being the cord which runs over the two pulleys. An experimental run can be done with light damping, using a metal plate of small area in the glycerol, and measuring the steady-state amplitude of oscillation for different settings of motor speed. The driver

frequency is measured by having a revolution counter attached to the wheel R and counting the number of revolutions made over an interval of one minute or so. The experiment can be repeated with heavy damping provided by a plate of large area. Provided the plates are thin and have very small mass compared with the oscillating mass permanently attached to the spring, the extra mass of the larger plate will have negligible effect on the frequency of free oscillation of the system. Figure 20.6 shows how the steady-state amplitude depends on driver frequency for light and heavy damping.

When the steady-state amplitude has its maximum value, that is, when the driving frequency is equal to the natural frequency of free oscillation of the system, we have the condition known as *resonance*. The system can be said to *resonate* to the driving agency. With light damping a sharp resonance is obtained: the variation of amplitude with change in frequency is very marked when the driving frequency is close to the natural frequency of the responder. With heavy damping a broader resonance is obtained. In fact the resonant frequency itself is slightly altered when the degree of damping is varied, but the difference is scarcely perceptible in this practical example, and here we shall ignore it.

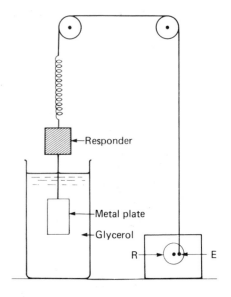

Fig. 20.5 Apparatus for investigating the effect of damping in a forced oscillation.

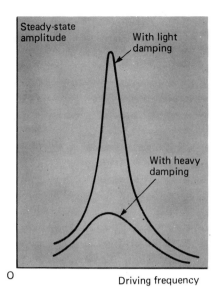

Fig. 20.6 Variation of the steady-state amplitude with driving frequency.

The degree of damping also affects the way in which the phase-difference between driver and responder varies with the driving frequency. The graph in Fig. 20.7 shows this. These phase relationships can be observed, using the apparatus of Fig. 20.5, by attaching an eye-catching horizontal pointer (e.g. a thin strip of brightly-coloured paper) to the load on the end of the spring, and a similar pointer to the part of the string between the top of the spring and the pulley

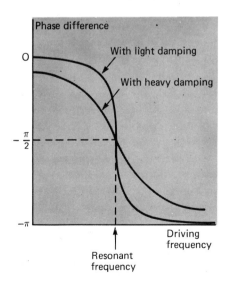

Fig. 20.7 Variation of phase difference between driver and responder with frequency.

above it. The relative motion of these two pointers shows clearly that driver and responder move very nearly in phase with each other when the driving frequency is much less than the resonant frequency, and with almost exactly opposite phase when the driving frequency is much higher than the resonant frequency. At resonance there is one quarter-cycle phase difference between driver and responder. Figure 20.7 shows the phase-difference between driver and responder as a function of driving frequency, for lightly and heavily damped forced oscillations.

Barton's pendulums

A very striking example of forced oscillations and resonance can be obtained using the arrangement of pendulums shown in Fig. 20.8. In this demonstration, the frequency of the *driver* is fixed, and a sequence of *responders* of differing natural frequencies are forced into oscillation. The responders consist of a series of pendulums which vary in length from about one metre to a few centimetres. The driver is a pendulum about half a metre long. In order that the responders do not take up too much of the energy of the driver, the pendulum bobs of the former consist of small paper cones, while that on the latter is a small lead sphere. Coupling of the driver to the responders takes place along the string upon which all the pendulums are suspended.

When the driver is set swinging, all the responders are forced into oscillation. They settle down rapidly to a steady-state oscillation. The amplitude of those whose natural frequency is close to the driven frequency is seen to be large, while those whose natural frequency is remote from the driven frequency oscillate with only

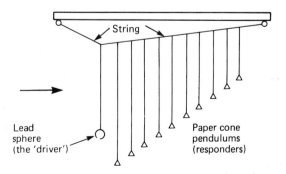

Fig. 20.8 Arrangement of Barton's pendulums.

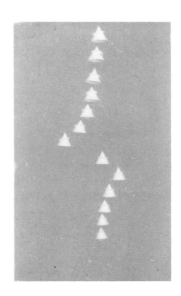

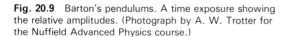

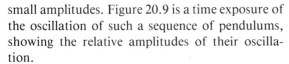

Fig. 20.9   Barton's pendulums. A time exposure showing the relative amplitudes. (Photograph by A. W. Trotter for the Nuffield Advanced Physics course.)

Fig. 20.10   Barton's pendulums. An instantaneous photographs showing the phase relationships. (Photograph by A. W. Trotter for the Nuffield Advanced Physics course.)

small amplitudes. Figure 20.9 is a time exposure of the oscillation of such a sequence of pendulums, showing the relative amplitudes of their oscillation.

The variation in phase between driver and responders is also very clear as the photograph in Fig. 20.10 shows. Pendulums whose natural frequency is close to the driven frequency oscillate one quarter of a cycle ($\pi/2$) out of phase with the driver. Those driven at a frequency much lower than their natural frequency are in phase while those driven at a frequency much higher than their natural frequency are half a cycle (π) out of phase with the driver.

The damping of the responders may be *decreased* by increasing the mass of their bobs. This can be achieved by placing a small split plastic curtain ring on each paper cone. Much sharper resonance will then be observed.

20.2   Resonance

The phenomenon of *resonance* is one of far-reaching importance both in theoretical physical science and in the world of engineering, and in this section we shall discuss a few examples.

The oscillatory system shown in Fig. 20.5 has one and only one natural frequency of oscillation.

It resonates at that frequency and only that frequency, when it is subjected to a periodic forcing agency of variable frequency. This implies that it is a very simple mechanical oscillatory system, and engineers very rarely deal with systems as elementary as this! Even a simple structure, such as a metal table with four legs and diagonal bracing struts, for supporting a piece of machinery, is capable of oscillating in different modes at many different frequencies. These frequencies can be calculated if one knows the dimensions, masses, and elastic properties of the various parts of the structure; but the calculation cannot be done without the use of a computer and even then the task of writing the computer program is long and intricate. In practice an engineer is more likely to make vibration tests on the structure itself, or on a scale model of the structure, by observing how it responds to being driven by a vibrator of variable frequency. Resonance in such a structure is generally highly undesirable, because the large amplitudes of oscillation of the structure at resonance may be sufficient to shake its fixings loose or even to damage the machinery supported by the table. In actual use the structure may be excited into resonant oscillations by the action of the machinery supported by the structure.

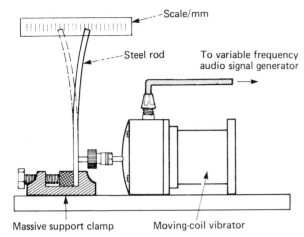

Fig. 20.11 Investigating the application of a forced vibration to a steel rod clamped firmly at one end.

A very simple experiment illustrates well the existence of more than one resonant frequency in a mechanical structure, although the system in this example is so simple that it scarcely deserves the name *structure*, being no more than a uniform, springy, steel rod clamped firmly at one end with the other end free to vibrate. This rod is forced into vibration by a moving-coil vibrator which is driven by an audio-frequency signal generator. The arrangement is illustrated in Fig. 20.11. To start with, the frequency is set at its lowest value (say about 10 Hz). The rod is observed to vibrate, and after a while it attains a steady-state oscillation and the amplitude of vibration of the end of the rod can be measured on the scale of millimetres. Rough measurements of steady-state amplitude are made at different frequencies, and it is observed that there is a distinct resonant

frequency at which the rod vibrates with very large amplitude in the mode (a) of Fig. 20.12. As the frequency is increased further, the amplitude decreases, but at a certain higher frequency the rod begins to oscillate in a different mode, (b) in Fig. 20.12. A third mode (c) can also be observed at a higher frequency still. If a rough graph of amplitude and driving frequency is made, it will look like the graph of Fig. 20.13.

In the experiment which yielded the results displayed in Fig. 20.13 the first resonance (fundamental mode) occurred at a frequency of about 14 Hz, and the second one at 90 Hz. The third resonance was of much smaller amplitude than the others, but nevertheless a distinct maximum of amplitude, at a frequency of about 220 Hz. Resonances at higher frequencies were detectable by ear as distinct peaks in the volume of sound emitted by the vibrating rod at certain frequencies, but the amplitude of vibration was too small to be observable with the naked eye. It is of interest to note here that the modes of vibration of a rod rigidly clamped at one end (what engineers call a *cantilever*) are the same modes as those of a tuning-fork's prongs. When a tuning-fork is struck the fundamental mode is the most prominent and long-lasting and is responsible for the characteristic note emitted by the fork; but the higher modes of oscillation are excited when the fork is struck and these are responsible for the 'clang' which dies away rapidly, much more rapidly than the fundamental vibration.

A rod clamped rigidly at both ends has characteristic modes of vibration, each with its own natural frequency: these are different from the modes of vibration of a rod which is clamped only at one end. A uniform flat plate (*a lamina*) of elastic material – any reasonably springy metal

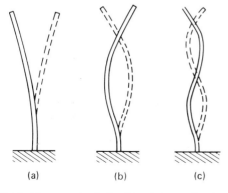

Fig. 20.12 Three mods of vibration of the steel rod.

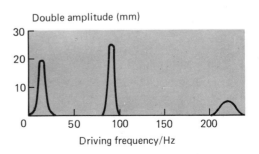

Fig. 20.13 Variation of amplitude with driving frequency.

will do – has characteristic modes of vibration which have a two-dimensional pattern. Fascinating experiments can be performed with Chladni's plate (Fig. 20.14). The plate can be made to resonate in its many modes at different frequencies. It can be regarded as the two-dimensional counterpart of the vibrating rod of Fig. 20.12. Although the amplitude of vibration is generally too small to be observed, the pattern of the vibration is revealed beautifully by sprinkling a thin layer of fine dry sand evenly on the plate. When the plate is vibrating in one of its normal modes the sand is shaken away from the regions of maximum amplitude and collects in regions of minimum amplitude. Intriguing symmetrical patterns are revealed at different frequencies.

Damping and selectivity

We return now to the simple forced oscillatory system with viscous damping of Fig. 20.5 to consider the shape of the resonance curves (see Fig. 20.6) which are characteristic of that system. We have observed that, the smaller the degree of damping, the sharper is the curve. It will now be shown that a quantitative measure of the width of a resonance curve is a useful concept. We define the width in terms of the graph of (amplitude)^2 against frequency, as in Fig. 20.6. The square of the amplitude is chosen, rather than the amplitude itself, because the energy of the oscillator is proportional to (amplitude)^2, and energy is generally the important parameter in such a

system. The width of the resonance peak is defined as the difference in frequency between the points at which the squared amplitude is half the value of the squared amplitude at resonance, as can be seen in Fig. 20.15. If the damping is not too heavy, the shape of the peak between points P and Q is very nearly symmetrical about the resonant frequency, so that the frequency at point P can be written $\omega_0 - \Delta\omega$, and at point Q, $\omega_0 + \Delta\omega$, where ω_0 is the resonant (angular) frequency. Thus the width of the peak is $2\Delta\omega$. It must be emphasized that the concept of width of a resonance peak is applicable only to situations in which the degree of damping is small enough for the shape of the resonance curve to be very nearly symmetrical about the resonance frequency.

Resonance in atoms and molecules

In certain conditions individual atoms and molecules behave like oscillatory systems having a number of natural modes of oscillation. The phenomenon of resonance is well known to the scientist who uses the technique known as *infrared spectroscopy*. When a beam of high-frequency electromagnetic radiation – infrared radiation – passes through a sample of a substance which is in the molecular state it is found that the sample absorbs energy from the radiation very strongly at certain precise frequencies. By measuring the frequencies at which the strong absorptions occur the spectroscopist can identify the types of chemical bond between atoms in the molecule. In

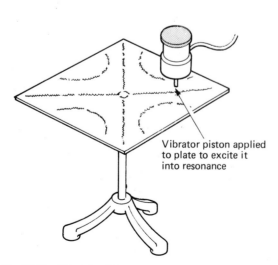

Fig. 20.14 Chladni's plate.

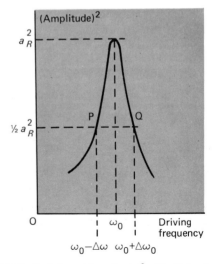

Fig. 20.15 Variation of (amplitude)^2 with driving frequency.

(a) Bond stretching

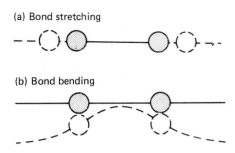

(b) Bond bending

Fig. 20.16

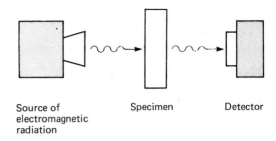

Source of   Specimen   Detector
electromagnetic
radiation

Fig. 20.17   The principle of the absorption spectrometer.

this way infrared spectroscopy is a powerful tool for use in determining the structures of substances in the molecular state.

A molecule is built of atoms bonded together by electrical forces which behave to some extent like springy elastic links between the atoms. A pair of atoms can vibrate in such a way that the bond between them alternately stretches and shortens, as shown schematically in Fig. 20.16a. (See also Section 13.1 for potential energy/distance graphs.) The bond between a pair of atoms can bend back and forth, as shown in Fig. 20.16b: the situation here is analogous to a pair of lead balls fixed to either end of a spingy steel rod. With each of the two types of distortion described there is a vibratory motion and the characteristic frequency of the vibration depends upon the masses of the atoms at either end of the bond and upon the stiffness of the bond. The two types of motion described are not the only kinds with which a precise, characteristic frequency is associated but those two must serve to illustrate the principle in this extremely brief account.

The exact manner in which the electromagnetic radiation interacts with the molecules can only be satisfactorily described with the help of quantum theory. The basic principle of an absorption spectrometer (not merely an infrared spectro-meter, which works in the infrared region of the electromagnetic spectrum) is shown in Fig. 20.17. We have already discussed how the rate at which energy is absorbed by an oscillator depends upon the driving frequency. The rate of energy absorption is a maximum at resonance. This means that, in the system shown in Fig. 20.17, the energy of the electromagnetic wave reaching the detector will vary with the frequency of the source, and at resonance it will be a minimum

because the energy flux reaching the detector is a minimum. A graph of detector response plotted against the frequency of the radiation will look like the curve shown in Fig. 20.18: this would be a very simple example of an *absorption spectrum.*

It is not possible in this book to discuss in detail any further examples of resonance. In concluding this section we briefly mention some other examples of importance in the study of the structure of matter. Subatomic particles which have electric charge (protons and electrons for instance) also have spin and thus they behave like small magnetic dipoles. Now a magnetic dipole which is free to move, when placed in a steady magnetic field, is an oscillatory system (compare a pivoted compass needle, how it flutters, when placed between the poles of a magnet), and it has a characteristic frequency of free oscillation. The particle can be forced into oscillation by means of

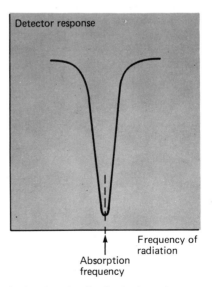

Detector response

Frequency of
radiation

Absorption
frequency

Fig. 20.18   An example of a simple absorption spectrum.

a periodically varying magnetic field: for example, the magnetic field of an electromagnetic wave. If the frequency of the wave is the same as the natural frequency of the particle in the steady field, then the particle will be excited into resonance. If a very large number of particles are thus excited at the same frequency (there will be several billions of such particles in a milligram of matter), they may absorb energy at an appreciable rate from the electromagnetic radiation which falls upon them, and an absorption spectrum may be observed. *Electron spin resonance* is one example of this type of resonance, and *nuclear magnetic resonance* is another. Study of these phenomena can yield useful information about the subatomic structure of substances.

20.3   Stationary waves

The experiment shown in Fig. 20.19 demonstrates another form of forced oscillation. The vibrator is connected to a signal generator whose frequency can be varied. Starting at a very low frequency of oscillation, the frequency of the vibrator is slowly increased. At particular frequencies it will be found that the stretched string is set into violent oscillation. The string vibrates with a large amplitude and forms a number of fuzzy loops. At the lowest frequency at which large amplitude oscillations are set up along the string, a single loop is formed as shown in Fig. 20.20a. Increasing the frequency slightly causes the amplitude of oscillation of the string to become almost imperceptible. Further increase of frequency will bring about a second large amplitude of oscillation whose pattern is shown in Fig. 20.20b. Figures 20.20c and 20.20d shows the pattern of subsequent large amplitude oscillations. These

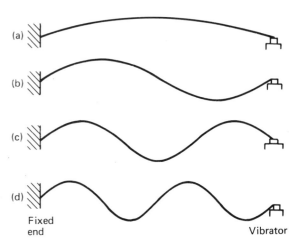

(a)

(b)

(c)

(d)

Fixed end   Vibrator

Fig. 20.20   Four modes of vibration of a stretched string.

amplitude resonances clearly have something in common with the situation shown in Fig. 18.8.

To follow the motion of the string in detail we can use a stroboscopic lamp, adjusting the flashing rate so that the string appears to move very slowly. We then observe that the string performs a periodic sequence of movements as illustrated in Fig. 20.21. These string resonances are a special kind of *wave motion* called a *stationary*, or *standing* wave. The crests do not move along the

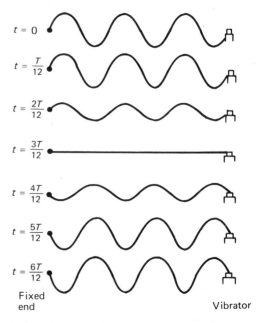

$t = 0$

$t = \dfrac{T}{12}$

$t = \dfrac{2T}{12}$

$t = \dfrac{3T}{12}$

$t = \dfrac{4T}{12}$

$t = \dfrac{5T}{12}$

$t = \dfrac{6T}{12}$

Fixed end   Vibrator

Fig. 20.21   'Glimpses' of a stationary transverse sinusoidal wave at time intervals of $T/12$. where T is the period of the vibration.

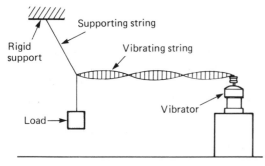

Rigid support

Supporting string

Vibrating string

Vibrator

Load

Fig. 20.19   A standing wave in a stretched string (Melde's experiment).

string, as they do in a travelling wave. Instead they shrink, turn into troughs, and back into crests again. Within any one loop of the pattern all the particles of the string are oscillating in phase with each other, and exactly out of phase with all particles in the two neighbouring loops. The amplitude of oscillation of any chosen particle is constant, and at certain places on the string this amplitude is a maximum. Such places are called *antinodes* of the stationary wave pattern. At other places the amplitude is nearly zero. Points of zero amplitude are called *nodes*. You will notice that the vibrator itself is at a node for the large amplitude oscillations. The amplitude of the vibrator is very low compared with that at an antinode.

Before attempting to analyse the process by which a stationary wave is generated, it is worth returning to the slinky spring, and attempting to produce a stationary wave pattern with it. With one end of the slinky spring fixed, and shaking the other end rhythmically from side to side, with gradually increasing frequency, you will find that at certain frequencies the wave-pattern appears to cease to travel along the spring: instead, a side-to-side wiggle of large amplitude begins and you have a stationary wave pattern. By adjusting the frequency, it is possible to form single, double, treble and even higher loop patterns.

We have seen already (Section 16.5) that any wave disturbance, excited at one end of, and travelling along, a stretched string (or spring) is reflected from the anchored end. This suggests that the stationary wave arises from the combined effect of an outgoing and a reflected wave. It is, in other words, a form of *wave interference*. Let us try to predict what will be the result of superposing a continuous sinusoidal wave of constant amplitude, travelling from right to left along a string, and the wave which results from this outgoing wave being reflected at the fixed end of the string. Fig. 20.22 shows the profiles (broken lines) of the two waves, at successive instants in time. The solid line shows the result of their superposition. So here is our stationary wave, predicted by this geometrical argument. You will notice that all points within any loop oscillate in phase with one another and that particles in neighbouring loops are 180° out of phase with each other, as we saw for the stationary wave on a string. This

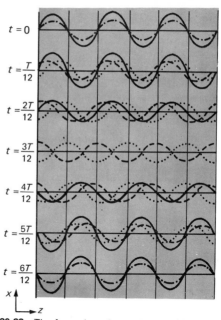

Fig. 20.22 The formation of a stationary wave. The dashed line represents a travelling sinusoidal wave going towards the left and the dotted line represents a wave going to the right. The full line is the stationary wave pattern which results from the superposition of the two travelling waves (compare Fig. 20.21).

superposition of two travelling waves also provides the additional information that successive nodes (and successive antinodes) are separated by half a wavelength.

Now we shall carry out the same process algebraically. Let us suppose that the waves in Fig. 20.22 are travelling along the horizontal z-axis and that the incident wave is reflected at $z = 0$. The wave displacement is along the vertical x-axis. For the diagram at $t = 0$; $x = 0$ when $z = 0$ for both incident and reflected wave.

In Section 19.11 we showed that a travelling wave could be represented by the equation:

$$x = a_0 \cos \frac{2\pi}{\lambda}(ct - z)$$

However, this represented a wave for which $x = a_0$ when $t = 0$ and $z = 0$. It also referred to a wave travelling in the z-direction from left to right. This equation has to be modified if it is to represent the incident and reflected waves in Fig. 20.22. In fact the following equation correctly represents the incident wave:

$$x_i = a_0 \sin \frac{2\pi}{\lambda}(ct + z)$$

(Check that this correctly matches the diagrams in Fig. 20.22 by finding x_i when (a) $t = 0$ and $z = \lambda/4$, and (b) $t = T/4$ and $z = 0$.) Since the reflected wave is travelling in the opposite direction,

$$x_r = a_0 \sin \frac{2\pi}{\lambda}(-ct + z)$$

(You can check this by making the same substitutions.)

The resultant displacement, x, for any value of z and t, is given by

$$x = x_i + x_r$$

which yields, after sine terms have been expanded,

$$x = 2a_0 \cos \frac{2\pi ct}{\lambda} \sin \frac{2\pi z}{\lambda} \qquad (20.1)$$

This expression represents a simple harmonic oscillation whose amplitude is a function of z. Since it is the amplitude variation which is of interest to us, we will write

$$x = A \cos \frac{2\pi ct}{\lambda} \qquad (20.2)$$

where,

$$A = 2a_0 \sin \frac{2\pi z}{\lambda} \qquad (20.3)$$

The maximum amplitude is $2a_0$ and occurs at the places where the value of z is such that $\sin 2\pi z/\lambda = +1$ or -1. For this to be true,

$$2\pi z/\lambda = n\pi + \pi/2$$

where $n = 0, 1, 2, 3$, etc. Hence maxima will occur at distances $\lambda/4$, $3\lambda/4$, $5\lambda/4$, etc., from the reflecting end.

Similarly $A = 0$ at places where the value of z is such that

$$\sin \frac{2\pi z}{\lambda} = 0$$

For this to be true.

$$2\pi z/\lambda = n\pi \qquad (20.4)$$

where $n = 0, 1, 2, 3$, etc. Hence nodes occur at distances, 0, $\lambda/2$, λ, $3\lambda/2$, etc., from the reflecting end. You will notice that the reflecting end is a node (as it must be if the analysis is correct)

and that neighbouring nodes and antinodes are separated by half a wavelength.

While this analysis does produce a stationary wave with all the features observed for the wave on the stretched string, there are apparently no restrictions on the possible frequencies of such a wave. In order to explain why the stationary waves set up on the stretched string existed at only particular frequencies we must take account of the conditions that must be met at *both* ends of the string. So far the only 'boundary condition' we have imposed is that the two waves must form a node at the anchored end. But we have seen that the end to which the vibrator is attached is also a node. This restricts the stationary waves that can be set up to those for which a whole number of loops can be fitted into the length of the string. Thus if the string is of length, l, only stationary waves for which $l = n(\lambda/2)$, where $n = 1, 2, 3$, etc., can be set up. So the possible wavelengths are restricted to those obeying the relationship

$$\boxed{\lambda = 2l/n} \qquad (20.5)$$

This analysis of stationary waves has arisen by considering the forced oscillations of a stretched string. However, it leads to no new ideas about resonance itself. The string *resonates* when the frequency of oscillation is equal to the natural frequency of the string. The importance of this experiment is that it shows us that a stretched string (and indeed many other systems) has a number of discrete natural frequencies of vibration. The analysis of these natural frequencies in terms of stationary waves shows a way in which we may evaluate the wavelength of the waves and thus the frequency of these natural vibrations.

With each different mode of vibration of the string, characterized by the number of loops in the pattern, one particular frequency is associated, often called the *eigen frequency* of that mode. The mode of vibration associated with the lowest eigen frequency is often called the *fundamental* mode, and its frequency the *fundamental frequency*.

The set of eigen frequencies associated with the vibration of a particular string or air column is called its *harmonics* – the first harmonic being the fundamental frequency of the vibration. The use of the term 'harmonic' is particularly associated with the modes of vibration of musical instruments.

The second and higher harmonics are also called *overtones*.

We will now apply these ideas of a stretched string of length, l, fixed at both ends. If the string is plucked it will vibrate. The natural frequencies of its vibration are those given by the associated stationary waves of wavelength,

$$\lambda = 2l/n$$

To find the eigen frequencies, we must use the two relationships:

$$c = f\lambda$$

and, for a stretched string, from Eq. 17.4

$$c = \sqrt{(F/\mu)}$$

Making substitutions, gives

$$f = \frac{n}{2l}\sqrt{\frac{F}{\mu}} \qquad (20.6)$$

All of these possible modes will be excited at the same time, resulting in a very complex mode of vibration. The relative amplitudes of each mode depend on the position and way the string is set into vibration. The fundamental mode will have the largest amplitude, however, and will determine the *pitch* of the note produced from the string. The higher eigen frequencies (harmonics) give the characteristic quality or timbre to the note.

The application of the foregoing principles to stringed musical instruments is of great importance, but there is no room to treat them in detail here.

In the case of a stretched string the eigen frequencies of the normal modes of vibration are all whole number multiples of the fundamental frequency. This is by no means always the situation. The prong of a tuning fork, for instance, has overtones whose frequencies are non-integral multiples of the fundamental. One prong vibrates in the modes which were shown earlier in the discussion of mechanical resonance: you should look back at Fig. 20.12 and 20.13 where you will see that the three lowest modes of vibration have eigen frequencies which are related in a fashion very different from that for a stretched string. A metal bell is a good example of a system in which the eigen frequencies are not generally integral multiples of the fundamental. Although the bell-makers of centuries ago could not perform a mathematical analysis of the oscillatory behaviour of a bell, they did know that certain overtones could have frequencies which caused a very discordant sound when heard in conjunction with the fundamental frequency. In a sweet-sounding bell the most discordant overtones are suppressed, by the skill of the bell-maker. A drumskin has many normal modes of vibration, and the overtones again have eigen frequencies which are not integral multiples of the fundamental – similarly, the wooden bars of a xylophone, or the metal bars of a glockenspiel, or metal chimes. Also, a prime example is shown in the delightful experiments with Chladni's plate, described in Section 20.1.

Stationary waves can often be seen in water. A cup of tea, when jiggled around, can show this, and stationary waves can be excited in a small tank, for instance your bath. Do you think that the eigen frequencies of deep water in a narrow rectangular tank, for instance, would be whole number multiples of a fundamental? The formulae for water wave speeds given in Section 17.6 can help you answer this.

Chapter 21

INFORMATION AND IMAGES

One of the most important facts about waves is that we clearly depend upon them for almost all the information we receive about the world in which we live. Our *two* major senses of sight and hearing depend respectively on the reception of light and sound, both of which turn out to be wave-like. In recent years, increasing amounts of information have reached us via radio and television, both of which depend upon electromagnetic waves of rather longer wavelengths than light. Our knowledge of the universe in which we live depends almost entirely on the reception of electromagnetic waves, and that knowledge increases as we increase the range of frequencies over which the information can be received. Even our understanding of the structure of the earth depends very much on the reception and analysis of another kind of wave – the seismic wave. Navigation, which at one time depended solely on sight, now depends largely on radar, which is the scattering and reception of electromagnetic waves in the microwave (10^{-3} m) region of the spectrum.

How then do we receive and interpret the information carried by all this wave-like energy? In this chapter we shall survey some of the important features.

The waves we receive may come from one of two sources: they are either a *direct transmission* from the original energy source; or they have been *scattered* from some object in their path. It is useful for a moment to look back at the various wave transmissions that are of value to us and ask which source of the waves is important in each case. *Sound* waves arrive at our ears largely by direct transmission. Scattering certainly affects the nature of the sound, but for certain reasons related to its wavelength, little information is received in this way. For some animals such as bats the situation is different and they depend almost entirely on *scattered* sound. The majority of the information we receive from *light* comes from scattered waves; direct viewing of a light energy source is usually distracting and even harmful. Navigation by radar depends entirely on the reception of scattered waves, but television and radio depend on reception of waves from the original source, although these may have undergone reflection and diffraction before reaching us. Reception of scattered waves here can seriously degrade the information received.

A wave, whether scattered or direct, can convey information about its source in only a limited number of ways – surprisingly limited in view of the wealth of information we receive. One wave may differ from a companion wave in the following ways:

in location	in phase
in amplitude	in polarization
in frequency	

Of course the last-named can only apply to transverse waves. The ability to derive information from the state of polarization of electromagnetic waves is of importance to radio-astronomers and only recently has it been shown that some insects depend upon this feature for navigation. However, we shall not consider this further in this brief account. We shall now consider first the world of sound in relation to the first four items in the list.

21.1 The reception of sound

The information we receive from sound depends almost entirely on variations with time in the *amplitude* and the *frequency* of the waves received.

Amplitude

Variations in amplitude are linked directly to variations in the intensity of a sound. This in turn affects its *loudness*, but not in any very simple way. The sensitivity of our ears varies with frequency, and very low pitched and very high pitched sounds are not as loud as mid-range sound of the same intensity.

Increasing steps of loudness are not linearly related to increasing steps in intensity even if the frequency remains constant. Suppose a particular sound source of fixed frequency emits energy at a rate of P_0 watts. Doubling this power to $2P_0$ watts will bring about a certain increase in apparent loudness. To increase *further* the loudness by an *equal* amount, it is found that the power of the source has to be doubled again to $4P_0$ (*not* increased by an equal step to $3P_0$). A further increase in loudness by a similar step requires the power to be *doubled* again – to $8P_0$.

Thus the steps in loudness increase are proportional to the ratio of the power increases, the same *ratio* of powers giving the same *step* in loudness increase. For this reason, a scale of loudness is based on the logarithm of the power of the sound. A logarithmic scale of the powers of sound transmitted will go up in steps which are proportional to the apparent change in loudness. So if a sound source increases in transmitted power from P_0 to $2P_0$,

$$\begin{aligned}
\text{Changes in loudness} &= \log_{10} 2P_0 - \log_{10} P_0 \\
&= \log_{10}(2P_0/P_0) \\
&= \log_{10} 2 \\
&= 0.3010
\end{aligned}$$

The unit of changes in loudness is the *bel*; 1 bel increase in loudness represents a change in 1 of the difference between the logarithms to base 10 of the transmitted power. Thus the change in loudness above is 0.3 bel. In fact the bel is rather a large unit for normal measurements of loudness, and commonly a unit one-tenth of a bel – the decibel (dB) – is used. The change in loudness above is one of about 3 dB. By fixing an arbitrary zero at the threshold of hearing, it is possible to construct a scale of loudness such as that shown in Table 21a.

Frequency

The range of frequencies over which normal hearing takes place ranges, as we have seen earlier, from about 25 Hz up to 16–20 kHz. This is the widest range of frequencies in terms of the ratio of highest to lowest of any of the waves we use for communication. A doubling in frequency represents a pitch change of one *octave*. The word 'octave' is often used in any discussion of frequency to express a doubling of the frequency. The range 25 Hz to 20 kHz is a change of about

Table 21a

Sound	Intensity/Wm^{-2}	Loudness/dB	Description
Threshold of audibility	10^{-12}	0	Silence
Breathing	10^{-11}	10	Quiet
Whisper	10^{-10}	20	
Quiet conversation	10^{-9}	30	
Inside quiet car	10^{-8}	40	
Normal conversation	10^{-7}	50	Average
Average radio	10^{-6}	60	
Loud radio	10^{-5}	70	
Heavy traffic	10^{-4}	80	Loud
Noisy factory	10^{-3}	90	
Underground train	10^{-2}	100	
Thunder	10^{-1}	110	Very loud
Threshold of feeling	1.0	120	

nine and a half octaves. (You should check this calculation for yourself.)

Our ability to differentiate changes with time in the amplitude and frequency of a sound is what makes speech possible. An investigation of the structure of the ear shows how it is we are capable of receiving such a wide range of frequencies and how we are able to make fine distinctions between the pitch of two sounds of different frequencies.

Location

We cannot, with any degree of accuracy, locate sources of sound. We can do so to some extent, as is shown by the popularity of the stereo reception of radio transmissions and the play-back of recorded music. The ability roughly to locate sound sources in space seems to add to the quality of music. However, precisely how we are able to make this location of sound is the subject of continuing research and some doubt has been cast on the extent to which many listeners *can* distinguish between stereo and mono reception.

Neither are we able to make much use of sound scattered by other objects. Our ability to do this or to locate accurately sources of sound is a consequence of the wavelength of sound waves compared with the dimensions of common objects around us and the size of our ears. The problem of location we will leave until we tackle the more important topic of images, but the lack of scattered sound refers back to the ripple-tank experiments described in Section 16.4. Here we saw that all objects would scatter a wave falling on them, but the amplitude of the scattered wave was very small compared with the amplitude of the incident wave if the object was about the same size, or smaller than, the wavelength of the wave falling on it.

Most sounds we hear are in the frequency range of 100 Hz to 2000 Hz. Using the formula, $c = f\lambda$, we can calculate the wavelength range:

$$\text{For } f = 100 \text{ Hz}, \lambda = 340/100$$
$$= 3.4 \text{ m}$$
$$\text{For } f = 2000 \text{ Hz}, \lambda = 340/2000$$
$$= 0.17 \text{ m}$$

Such a wavelength range suggests that we may get substantial scattering by an object the size of a house, but very little from something the size of a car.

If an animal, such as a bat, is to navigate by sound it will need to utilize a wave of much smaller wavelength. This it does by producing its own sounds with a frequency of about 50 kHz. Again using $c = f\lambda$,

$$\lambda = 340/50\,000$$
$$= 6.8 \times 10^{-3} \text{ m}$$
$$= 6.8 \text{ mm}$$

Such a wave will be strongly scattered by most objects that it will be important for a bat to avoid, or to find, as prey.

While the small wavelength will help a bat in locating the *direction* of incidence of a scattered wave, this does not in itself help in pinpointing the distance of the scattering object. Here, experiment has shown that many bats use a technique now employed in radar location. Instead of emitting a continuous sound wave, the sound is emitted in

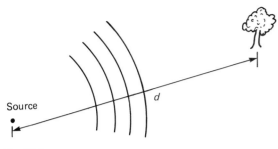

Fig. 21.1

pulses. There will then be a *delay* between the sending of the pulse and the reception of the scattered pulse. This delay will depend upon the distance to the scattering object (Fig. 21.1). By sensing the length of the delay the bat, and radar, can assess the distance to the scattering object.

This technique, originally known as echo-location, has also been employed for many years for detecting underwater objects. The process, called *sonar*, uses a high-frequency sound wave which is strongly scattered by objects of appropriate size. In this way ships at sea are able to detect and locate moving underwater objects, assessing both their location and their speed. In war-time this has been used for the detection of submarines (an early device was known as ASDIC which were the initials of the anti-submarine detection investigation committee). Sonar is also used for the location of shoals of fish.

Phase

Because of the continuous variation of amplitude and frequency of most transmitted sound, differences in phase are rarely maintained unchanged for long enough for the transmission of information in this way. However, there is one consequence of phase differences that contributes both to the quality of received sound and is utilized for the comparison of frequencies. The phenomenon is referred to as *beats* and arises

when sounds of two different, but close, frequencies are produced simultaneously. The result is a note, the intensity of which varies in a periodic manner. Beats can be produced easily using the apparatus shown in Fig. 21.2. As the frequency of one signal generator approaches the other, the combined sound emitted from the two loudspeakers will be found to rise and fall in intensity. The period of this rise and fall will get longer as the two notes approach each other in frequency.

We shall now find a relationship between the frequency of the rise and fall in intensity – called the *beat frequency* – and the frequencies, f_1 and f_2 of the two notes 'beating' together. Let us suppose that the two sound waves are of equal amplitude, that $f_1 > f_2$ and that at the instant $t = 0$ crests from the two waves coincide at the point of reception, thus producing a maximum intensity in the sound received.

A moment later the intensity will be less as the two waves are no longer in phase (Fig. 21.3). The intensity continues to fall, reaching zero when the two waves are 180° out of phase at the point of reception. Thereafter the intensity continues to rise, reaching a maximum at time $t = T$, when the two waves are again in phase at the point of reception. During the periodic time, T, of one beat there will arrive at the point of reception *one more* cycle of the higher frequency wave of period T_1 than of the lower frequency wave of period T_2. Since T/T_1 complete cycles of the higher frequency wave and T/T_2 complete cycles of the lower frequency wave will arrive in this time,

$$T/T_1 - T/T_2 = 1$$

Hence

$$1/T_1 - 1/T_2 = 1/T$$

Since wave frequency = 1/wave period, it follows that the beat frequency, f_b, is given by

$$f_1 - f_2 = f_b$$

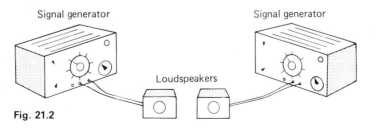

Fig. 21.2

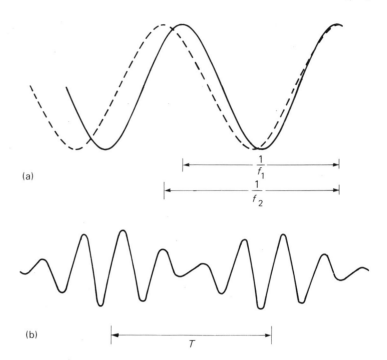

Fig. 21.3 (a) Two waves with equal amplitudes but slightly different frequencies which, when added together, give the variation in intensity shown in (b).

A measurement of the beat frequency thus gives the difference between the two frequencies of the notes beating together. Practically, the two notes can be brought into unison by adjusting one until the beat frequency is as low as can be practically detected. This technique is also used for 'tuning' some shortwave band radio receivers.

21.2 The reception of light

The use we make of information conveyed by light is generally quite different from that of sound. Variation in amplitude and frequency with time are only of limited value in an everyday context. The ability of our eyes to detect differences in either is very poor when compared with our ear's ability to detect changes in the amplitude and frequency of sound. The ability to utilize phase changes in conveying information is a very recent technological development (usually called *holography*) and not a matter of everyday experience. This leaves only *location* – and here our eyesight is unparalleled. The power of sight lies in the ability to locate in space the source of each scattered wave received by the eye: the ability, in other

words, to form *images*. This amazing property far outweighs the limitations of our eyes to detect variation in amplitude, frequency and phase. We shall give a great deal of attention to this power of image formation since in understanding it we learn how we may utilize the process over a range of the electromagnetic spectrum far outside the range of sight. And in understanding the limitations of the natural image-forming process, we can learn how to overcome them and thus derive even more detail about the world in which we live. But before that we will give some attention to the matters of amplitude, frequency, and phase.

Amplitude

While we can clearly detect differences of amplitude in space – 'light and shade' – our eyes are unreliable indicators of the variation of amplitude of light waves with time. This is due to their self-compensating ability to limit or increase the amount of light received (depending on the brightness of the source) by variation in the size of the eye pupil. If it is important to make comparisons of amplitude, this must be done by other means – either photographically or by using light-sensitive electronic components. These technological aids also have the advantage of being able to reach beyond the range of intensities that the human eye can detect.

We shall say little more about variations in amplitude, but this should not imply that the subject is unimportant. The relationship between the amplitudes of scattered light waves and the amplitude of the incident wave is a complicated one; the relationship between the amplitude of all such waves and their consequent effect on photographic film is a science in itself. Opto-electronic components which depend on the amplitude of the light wave falling on them, and often on its frequency are becoming increasingly important. However, we shall concentrate in this chapter on matters of even more general importance.

Frequency

Vision operates over a band of wavelengths from about 6.5×10^{-7} m to 4.5×10^{-7} m. This represents a frequency range of 4.6×10^{14} Hz to 6.7×10^{14} Hz – considerably less than an octave. We have already seen (Section 18.8) that differences in frequency are sensed as differences in colour. In general, our ability to sense differences between frequencies of two sources of light is limited to not much more than the differences between the colours of the spectrum: red, orange, yellow, green, blue, and violet. Our eyes are in fact only equipped to differentiate two broad frequency ranges: high and low. It is by the relative response of these two different types of cells that we seem able to differentiate colour. Theories of colour *vision* usually describe red, green and blue as *primary colours* while the sensations produced by mixing any pair – namely yellow, cyan (peacock blue) and magenta – are termed *secondary colours*. Such theories are of importance in fields such as *colour photography* but it must be emphasized that they relate to colour *vision* rather than the actual frequencies of light waves.

The frequencies associated with light sources (and indeed other sources of electromagnetic waves) are intimately related to the nature of the emitting source. An analysis of these frequencies, called *spectral analysis*, is of great importance in deriving information about the nature of sources. Indeed a very large part of astronomy depends upon such a process. Again it is clear that eyesight cannot be relied upon to perform this analysis. So important is it, however, that we will spend a little time showing how this analysis *can* be performed.

The diffraction grating

When white light is passed through a double-slit a series of dark and light bands is formed on a distant screen (Section 16.5). Careful examination of the bands shows that their edges are coloured. This has been ascribed to the fact that white light can be thought of as composed of a range of wavelengths, all forming interference maxima at slightly different angles. If light of just two colours, say red and blue, falls on a double slit, the two sets of interference bands can be observed. A graph of the variation of intensity with the sine of the observation–direction angle would appear as in Fig. 21.4. (In fact the intensity of each maximum falls away either side of the maximum. We shall understand the reason for this later, but for the moment it has been disregarded.) The angles of the interference maxima are given, as we have already seen (Eq. 16.2), by $\sin \theta = n\lambda/s$. A diagram expressing this fact is shown in Fig. 21.5a.

If now the number of slits through which the light passes is increased notable changes take place in the observed interference pattern. If the separation between neighbouring slits is unchanged, then

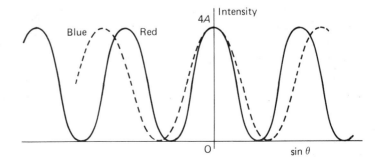

Fig. 21.4 Spatial variation of intensity for beams of red and blue light interfering after passing through a pair of slits (assuming uniform wave amplitudes).

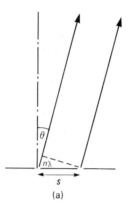
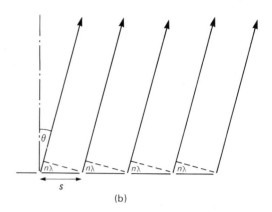

Fig. 21.5 Light emerging from (a) a pair of slits and (b) a number of slits in the direction θ and producing an interference maximum (assuming uniform wave amplitudes).

(a)   (b)

there will be no change in the position of the interference maxima, as Fig. 21.5b shows. However, the distribution of intensity with angle now appears as in Fig. 21.6. The vertical scale has been reduced and the progressive fall off in the amplitude of the maxima either side of the central maximum has again been ignored.

It is a helpful experiment to look at the pattern produced by two, three, four, etc., slits of the same width and separation and see the progressive brightening and sharpening of the maxima.

A device consisting of many closely spaced slits is called a *diffraction grating*.

There are two main classes of diffraction gratings – reflection gratings and transmission gratings. The action of a transmission grating, which is simply an array of equally spaced similar apertures in a barrier, is demonstrated in a ripple-tank. Figure 21.7 shows the result of placing a barrier containing a number of equally spaced slits in the path of a train of continuous straight ripples. The secondary waves emerging from the

slits interfere in such a way as to produce not only a train of very nearly straight ripples travelling in the same direction as the original ones, but also trains of very nearly straight ripples travelling in other directions, on either side of the central train. Thus the energy flux associated with the waves is concentrated into certain specific directions including the original direction. If the number of slits is increased the emerging ripples become more nearly exactly straight, and thus the specific directions of energy flux become more sharply defined.

For a simple demonstration of the action of an optical diffraction grating one can use an inexpensive plastic replica grating, which is manufactured in much the same way as a gramophone record. In Fig. 21.8 the lamp L, a clear glass straight filament lamp like the one recommended earlier for the Young's slits experiment, is placed at the focus of a converging lens, C. F is a colour filter (red is good for a start). This arrangement produces a parallel beam of nearly monochromatic light. The

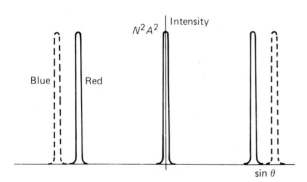

Fig. 21.6 Spatial variation of intensity for beams of red and blue light interfering after passing through a number of slits.

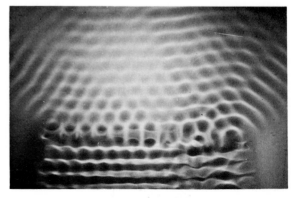

Fig. 21.7 Interference of ripples diffracted from an array of regularly spaced apertures. (Photograph E. J. W.)

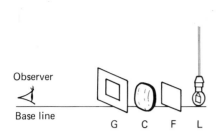

Fig. 21.8 Arrangement of lamp, filter, lens and grating to produce a parallel beam of light of one colour.

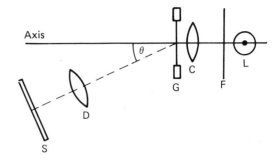

Fig. 21.9 Arrangement of Fig. 21.8 modified to produce a sharply focused first or second order spectrum.

beam passes through the grating at G. If you position your eye on the same horizontal level as the grating you will observe not only plenty of light coming straight through the grating, but also light emerging obliquely, on both sides of an axis perpendicular to the grating. On removing the colour filter F and allowing the full white light through the grating you see that light of different colours is channelled into different directions: thus spectra are produced by the grating. A good way to observe these spectra is to use a second converging lens, D, as shown in Fig. 21.9 to converge the diffracted beam and form an image on a small screen, S, placed in the focal plane of this lens. It is generally possible to see more than two spectra either side of the axis. The first pair of spectra appearing either side of the axis are known as the *first order spectra*, the next pair as the *second order,* and so on.

By placing the light source at the focal point of the lens C the light reaching the grating consists of *plane* waves. The light emerging from each slit is

thus in phase with the light emerging from every other slit. By placing the screen at the focal point of lens D, the interfering waves travel parallel to each other after emerging from the grating (Fig. 21.10). So the formula

$$n\lambda = s \sin \theta \qquad (21.1)$$

is *exactly* true for a grating set up in this way. With an accurately-made grating, for which the value of the grating spacing, s is known to a high order of accuracy, and a spectrometer to measure θ with high precision, one can make very exact determinations of the wavelength of light from a source.

We can use an argument based on the conservation of energy to explain why the diffraction grating maxima are so sharp. We have already seen (Section 16.2) that the energy associated with a wave of amplitude A is proportional to A^2. Thus, in a two-slit interference pattern, the energy at the maximum is increased by a factor of four over that which would have resulted from one slit on its own. If the light passes through N slits, the amplitude at the maxima must be NA and so the intensity is increased by a factor of N^2 over that due to one slit alone. Yet the total transmitted energy has only increased by a factor of N. This implies a substantial redistribution of energy with angle. Suppose (Fig. 21.11), none of the energy were redistributed. Then the whole area in front of the slits would be uniformly illuminated with light of intensity NI, where I is the intensity due to one slit alone.

In Fig. 21.12 the interference maxima are approximated to isosceles triangles. All light within the angle range $\pm \sin^{-1}(\lambda/s)$ of a particular

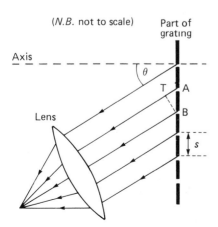

(N.B. not to scale)   Part of grating

Fig. 21.10

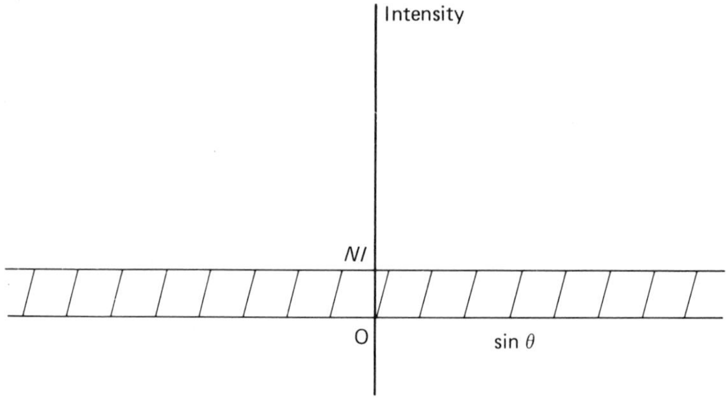

Intensity

NI

O $\sin \theta$

Fig. 21.11

maximum is channelled into the triangle of half-angle δ (δ is so small that $\sin \delta \approx \delta$, if δ is measured in radians.), So, as energy is conserved, the area of the rectangle ($= NI \times (\lambda/s)$) is equal to the area of the triangle ($= N^2I \times \delta$). Hence

$$\delta = \lambda/Ns$$
$$= \frac{1}{N} \times \text{(separation of maxima)}$$

So the bigger is N, the smaller is δ. It is important to note that N is the number of slits through which light passes *to form the observed maxima*. This is not necessarily the same thing as the total number of slits on the grating.

In Section 21.5 you will find a further discussion of the width of the interference maxima produced by a diffraction grating in which the same result is derived by applying the principle of superposition to the wavelets which arise from each slit in the grating.

We have said that, in general, you can observe more than one pair of spectra when using a grating. What determines how many spectra you see? We can answer this by analysing a typical

practical situation. A certain type of plastic replica grating in common use has 6000 rulings to the centimetre specified by the manufacturers. This means that the size of the grating element is given by

$$s = \frac{1}{6000} \text{ cm} = 1650 \text{ nm}$$

Let us now find out what happens to monochromatic light of three particular colours when it goes through this grating: red light for which $\lambda = 650$ nm, yellow light for which $\lambda = 580$ nm, and violet light for which $\lambda = 450$ nm. (Red and violet light with the wavelengths quoted mark approximately the visible limits of the spectrum of white light obtained with a grating like the one described.) To find the values of the angle θ at which the light will appear in the different orders of spectra we use equation 21.1

$$n\lambda = s \sin \theta$$

putting $n = 1$ for the first order, $n = 2$ for the second order spectrum, and so on. Table 21b shows the result of doing this. The calculations predict that we should be able to see the complete

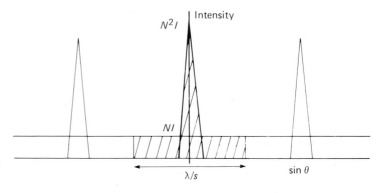

Fig. 21.12   The re-distribution of energy by a grating with N slits.

Table 21b

	Violet		Yellow		Red	
	$\sin\theta$	θ	$\sin\theta$	θ	$\sin\theta$	θ
$n=1$	0.290	17°	0.374	22°	0.419	25°
$n=2$	0.580	35°	0.748	48°	0.839	57°
$n=3$	0.870	60°	1.12	–	1.26	–
$n=4$	1.10	–				

first order and second order spectra, and part of the third order one. But it is not possible to see, for instance, the yellow in the third order spectrum. Why not? The mathematically minded will say that this is because $\sin\theta$ cannot be greater than unity. But it is more convincing, perhaps, to see how this works out geometrically. Refer back to Fig. 21.10. The optical path difference between adjacent paths is AT. As the angle θ is increased, TBA remaining all the time a right angled triangle. AT becomes greater. When θ is very nearly 90° the length of the side AT is very nearly equal to AB which equals s, the grating spacing. Thus s is the upper theoretical limit for the optical path difference, and if this is less than three whole wavelengths for yellow light, then no yellow light will be observable in the third order spectrum.

For the most accurate spectroscopic work reflection gratings are generally used. These are made by very high-precision machines which rule minute parallel grooves in a highly polished metal mirror surface. The grooves are effectively non-reflecting, and the narrow strips of untouched metal between reflect the incident light, acting as sources of secondary waves, just as the adjacent apertures in a transmission grating act as sources of secondary waves.

The equally spaced layers of ions in a perfect crystal can act as the elements of a kind of reflection grating for very short wavelength electromagnetic waves, namely X-rays. This is the basis of X-ray crystallography and is described further in Chapter 14.

21.3 Location

It is perhaps misleading to list 'location' as a feature of a wave alongside amplitude, frequency, phase and polarization, since it is not itself a property of a wave. Waves do not carry with them any indication of the direction in which the energy flux travels. In a transverse wave, indeed, the disturbance associated with the wave is in a direction perpendicular to that in which the wave is travelling. The only possible 'information' that reaches the receiver is that of amplitude, frequency and phase of the disturbance produced by the wave. How then are we able to locate the source of a wave in space – in other words, to form images? To answer this question we must look in detail at the processes involved in image formation.

Scattered light

The observation of self-luminous bodies plays little part in vision. In order to 'see' non-self-luminous bodies we depend upon them *scattering* light which arises from other sources. We have already seen in the ripple-tank experiments that all objects will scatter waves falling on them, producing a circular (or in three dimensions, spherical) wave that travels out uniformly in all directions. This scattering is only substantial for objects several times bigger than the wavelength of the waves falling on them. The average wavelength of light is 5×10^{-7} m, so all objects within our own scale will produce substantial light scattering. Indeed for the purpose of image formation, we must think of every *point* on such objects as scattering light falling on them into a spherical wave centred on that point (Fig. 21.13). It is this scattered light which will eventually form the image.

Recombination

The scattered light will not form an image on its own. If a piece of card is held so that scattered light from the tree (Fig. 21.13) falls on it, the card

Fig. 21.13 Light scattered from three typical points on an object.

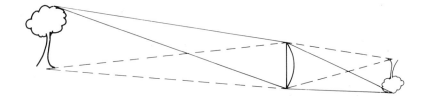

Fig. 21.14 The use of a lens to form an image.

is uniformly illuminated. No point-for-point image of the tree falls on the screen. In order to make an image of the tree on the card we should have to use a *lens*. Figure 21.14 shows a conventional diagram of a lens 'bending' the light so that light from every point on the tree falls on a corresponding point on the card, so forming an image.

This is a process of recombination. The lens recombines the waves so that all those from one point on the object seem to converge to another point on the screen. The procedure of forming a sharp image is often callled 'focusing'. We will now try to understand this procedure from a 'waves' point of view.

The process of scattering and recombination

It is important to realize that these two procedures are quite independent. The lens has no effect on the scattering process – how could it? So all the information which goes to make up the image *must already be present in the scattered waves* when they arrive at the lens. This is an important statement and its recognition has led to new ways of recording and producing images called *holography* (see Section 21.8). A good simple way of understanding the distinction between scattering and recombination is to think of how a slide

projector works. Light from a lamp passes through a slide and then the light, after being scattered by the image on the slide, passes through a lens to form an image on a distant screen (Fig. 21.15). If the lens is removed, the screen is simply uniformly illuminated. And yet all the information which went to make up the image must still be there. The light still passes through the slide and is scattered by it. Why then is there no image?

A moment's thought will show that without the lens, light from every point on the slide is arriving at *every point* on the screen (Fig. 21.16). As a consequence every point on the screen is receiving light from *every point* on the object (the slide). This can be easily tested by taking a long-focal length lens and holding it near the screen so that a small sharp image falls on the screen (Fig. 21.17). Although the lens is picking up only a small part of the light which has passed through the slide, nevertheless an image of *all* the slide is formed on the screen. This image can be formed anywhere on the screen by picking up just the light travelling towards a small part of it.

The pin-hole camera

One way then to form an image would be to arrange matters so that only light from a particular point on the object fell on only one parti-

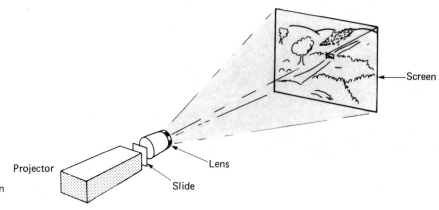

Fig. 21.15 The formation of an image of a transparency.

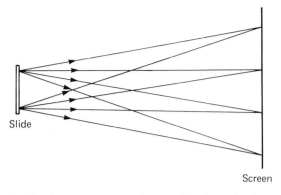

Fig. 21.16 In the absence of a lens, light from all points on the transparency reaches all points on the screen.

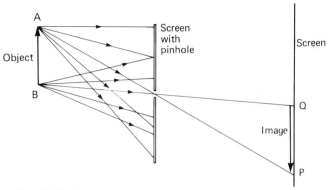

Fig. 21.18 A pin-hole camera.

cular area of the screen. This could be done by placing a small hole in the path of the light (Fig. 21.18). Light from a point A on the object forms a patch of light at P on the screen. Light from B cannot reach P. Instead light from B forms a patch at Q. So light scattered from the object at A falls only on the screen at P; light from B falls only at Q and an image is formed. This is the basis of the pin-hole camera.

Reducing the aperture of the pin-hole will reduce the amount of light reaching the screen, but it may also reduce the size of the light-patch at P and Q so giving a sharper image. But let us return for a moment to the ripple-tank photo-

graphs (Fig. 16.17). If the size of the aperture approaches the wavelength of the waves, the light passing through the hole spreads out uniformly in all directions. Under these circumstances light from A will again uniformly cover the screen – and so will light from every other point on the object. Thus reducing the size of the hole will not necessarily sharpen the image. It may do so at first but ultimately, as it gets fainter and fainter, the image will get more and more 'blurred' as more and more spreading of the light takes place as it passes through the pin-hole.

For some radiations, e.g. X-rays, no lens exists to recombine the waves. Consequently this pin-hole technique is the only way in which images can be formed.

The action of a lens

If light from all points of the object, such as an illuminated slide, falls on all parts of the screen, what information is carried by the waves that enables an image to be formed by a lens? To answer this, let us consider light scattered by two points A and B on an object and arriving at some point P on a screen (Fig. 21.19). Although two

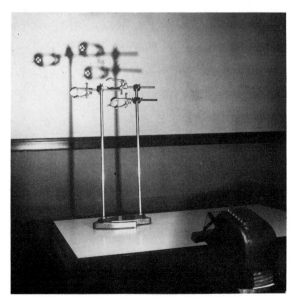

Fig. 21.17 Wherever a lens is placed in the beam, a complete image of the transparency is produced. (From C. A. Taylor, *Images*, Taylor and Francis, 1978)

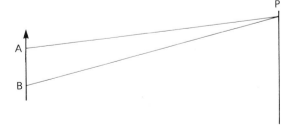

Fig. 21.19 Light scattered from the points A and B reaches the screen at P.

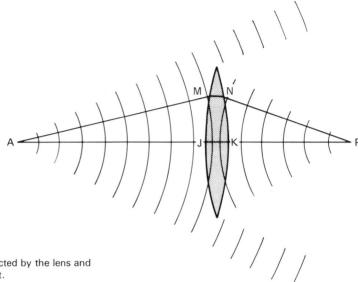

Fig. 21.20 Wave fronts from point A are refracted by the lens and arrive at P in phase to form a sharp image point.

waves arrive at P, only *one* disturbance is produced. The nature of that wave-produced disturbance depends on the principle of superposition. In other words, the two waves will produce an *interference pattern* and the nature of the disturbance at P depends on the path difference BP–AP. So this is the additional information conveyed by the waves which arises from the different *locations* of their sources, they will *differ in phase* at almost all points on the screen. So at P

waves arrive from every point on the object, and their original location is coded by their phase relationships to each other. It is this code that a lens seems able to unravel.

Figure 21.20 shows the wave fronts from a point A passing through a lens and arriving at P where A is sharply focused. What is the path difference (in terms of wavelengths) for the part of a wave travelling along AMNP and AJKP? It turns out that the answer, when due allowance has been made for the thicker glass at JK compared with MN, is zero! And the same goes for every other path. A lens gives a sharp image of A at P because P is the one point for which waves along all paths from A arrive in phase. At every other point the waves interfere destructively. We cannot prove this point here, but that is what detailed analysis shows to be the case. Point B forms a sharp image at Q (Fig. 21.21) because this is the point for which all wave paths from B have zero path difference. Hence the lens produces a point-for-point image of the object. And it will only do it at a particular distance from the lens – this distance depending on the position of the object.

The diagram drawn in Fig. 21.20 can of course be justified empirically by placing a lens-shaped piece of glass or plastic in a ripple tank and allowing waves from a point source to cross it. The wave fronts change direction as they cross the

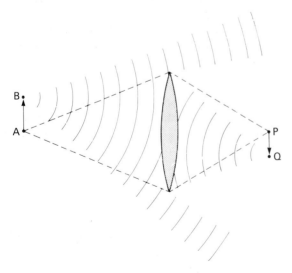

Fig. 21.21 Wave fronts from point B arrive at Q in phase to form a sharp image point.

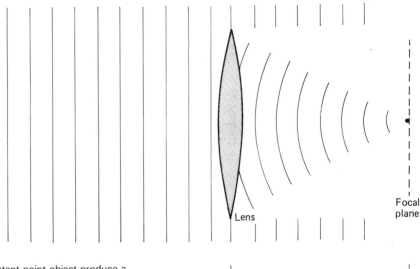

Fig. 21.22 Wave fronts from a distant point object produce a sharp image point in the focal plane of the lens.

shallow region and form new circular wave fronts which converge on an image point, just as is shown in Fig. 21.20.

If the object is very distant from the lens, the waves arriving at the lens are effectively plane. These converge to produce a sharp image at a distance from the lens called the *focal length* (Fig. 21.22). The image is then formed in the focal plane of the lens. The converse of this is that any image formed in the focal plane of a lens is produced by plane waves falling on the lens. This was the basis upon which the lens was used to form the interference maxima produced by the diffraction grating (Fig. 21.10).

Defects in the image

a) Spherical aberration In practice it is not possible to make a lens for which the path difference of waves travelling from, say, A along all routes to P (Fig. 21.20) is *exactly* zero when the lens has a large aperture. This results in P not being a sharp image of A; a defect referred to as *spherical aberration*. Minimization of this defect is one of the arts of lens manufacture.

b) Chromatic aberration Light waves of differing frequencies do not produce image points in exactly the same position from identical object points (Fig. 21.23). This defect can produce coloured fringes around the image point. Again careful manufacture can minimize and even eliminate this defect.

There are a number of 'higher order' aberrations all of which are related to the geometrical difficulty of achieving zero path difference for different routes from object points to corresponding image points for all positions of the object. For their detailed description the reader would have to refer to more advanced and specialized texts.

c) Diffraction Waves from any object point are scattered in all directions; only part of each wave front passes through the lens. We have seen already (Section 16.4) that under such conditions

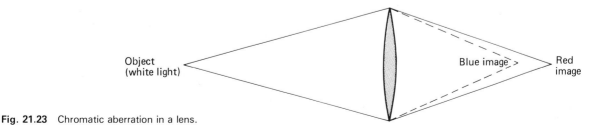

Fig. 21.23 Chromatic aberration in a lens.

the waves are diffracted and spread out into the space beyond the aperture. The effect of this is to *impose* the diffraction pattern produced by an aperture on the waves passing through a lens. This results in a 'blur patch' at P rather than a sharp point, and this limits the ultimate sharpness of the image. Because diffraction ultimately determines the quality of all images, the matter is of great importance and we shall now give some attention to the diffraction pattern produced by an aperture when a light wave passes through it.

21.4   Diffraction by a slit

The diffraction of light when it passes through a slit is easily seen by making a slit from the halves of a broken razor blade and looking at a lamp with a line filament through it (Fig. 21.24). A bright central band will be seen, flanked on either side by one or two less bright bands which will have coloured edges. If the slit is narrowed, by moving the blades closer together, the bright central band becomes wider and the whole pattern spreads out. This is consistent with the diffraction patterns produced by waves in a ripple-tank.

The same effects can be produced objectively on a screen using a laser. Light from a laser passes through a single slit and falls on a distant screen. The same pattern will be observed on the screen, but many more subsidiary maxima will be seen flanking the central maximum.

We shall now use what we know about waves to try to explain this diffraction phenomenon. For simplicity's sake we shall assume that the wave incident on the slit is plane and that the diffraction pattern is formed so far from the slit that we can also treat the waves arriving at the screen as plane. Both these conditions, which produce what is known as a Fraunhofer diffraction pattern, are commonly met in practice. Within the aperture the

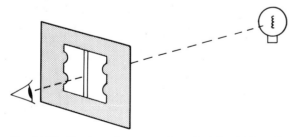

Fig. 21.24 Viewing a line source through a slit of adjustable width.

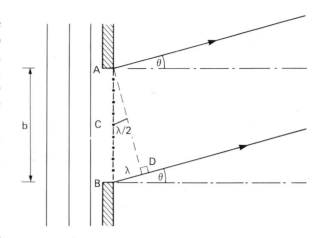

Fig. 21.25 Application of Huygens' principle to the wave fronts emanating from a single slit.

plane wave extends from one edge A to the other at B (Fig. 21.25). We shall now use Huygens' principle to construct the new wave front. To do this, we imagine each point on the incident wave front to give rise to a secondary, circular wave front. Clearly a plane wave front will move forward in the direction the original incident wave was travelling. However, subsidiary wave fronts are found to be produced in other directions, which are not there if the plane wave continues indefinitely beyond A and B. This is because the limited number of secondary sources can produce constructive and destructive interference in other directions. First let us consider the direction making an angle θ with the incident direction such that the path difference between a wavelet from A and one from B is one wavelength, λ (Fig. 21.25).

BD is the path difference of the two wavelets since we are assuming the screen receiving these wavelets to be sufficiently distant compared with the aperture, b, for the waves arriving there to be plane. Imagine now a secondary source at C, such that AC = CB. A wavelet from this secondary source is 180° out of phase with one from A, and can interact with it to give cancellation of the wave. Similarly for every secondary wavelet between A and C there is a corresponding wavelet between C and B with which it can interfere destructively. It follows then that all the light from half AC of the slit can 'cancel' all the light from half CB. So no light energy flows in the direction θ, for which $\sin \theta = \lambda / b$.

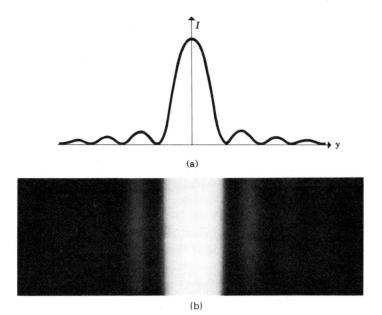

Fig. 21.26 The Fraunhofer diffraction pattern of a single slit: (a) intensity distribution (b) the pattern. (From F. W. Sears, M. W. Zemansky, H. D. Young, *College Physics*, Addison-Wesley, 1980.)

Similarly no light energy flows in the direction θ' for which BD $= 2\lambda$, or in direction θ'' for which BD $= 3\lambda$ and so on. At points in between these angles, not all light is cancelled and subsidiary maxima appear. From the angle $\theta = 0$ to $\theta = \sin^{-1}(\lambda/b)$ progressively more and more of the secondary wavelets can be paired with others 180° out of phase and the intensity of the light falls as θ increases. A graph of intensity against $\sin\theta$ thus appears as in Fig. 21.26a.

One of the consequences of this analysis is that the angular width of the central maximum is *twice* that of the subsidiary maxima. Experiment will quickly confirm that this is indeed the case.

Since the angular width of the entire pattern is often small, it is possible in these cases to write $\sin\theta = \theta$ in radians, and hence the angles in radians at which minima of intensity are found are $\pm\lambda/b$; $\pm 2/\lambda b$; $\pm 3/\lambda b$, etc., from the centre of the pattern.

Both the angular width of the central diffraction maximum *and* the angular separation of successive subsidiary maxima depend on the wavelength of the light – this is the reason the subsidiary maxima produced by the diffraction of white light are coloured. Both the dependence of the width of the central maximum and the separation of the maxima on the wavelength of light can be investigated by placing colour filters across the slit forming the diffraction pattern.

For $b = \lambda$, $\sin\theta = 1$ (it is not appropriate to make the approximation here that $\sin\theta = \theta$). Under these conditions the central maximum occupies the entire space beyond the slit; for $b < \lambda$, we approach the condition in which light spreads uniformly over all the space available to it, as we have previously seen with waves in a ripple tank.

Of course, light waves are rarely limited by a slit in this way, and the only reason for adopting it here is the fact that the diffraction can be treated one-dimensionally. The analysis for a circular aperture is harder and we will content ourselves by quoting the result for the positions of the diffraction minima:

$$\sin\theta = 1.22\lambda/b \qquad (21.2)$$

This means that a point of light giving rise to plane waves at a circular aperture of diameter, b, will produce a *fuzzy patch* on a distance screen of angular diameter $2 \times 1.22\lambda/b$.

Problem 21.1 A 35 mm camera of normal design has its image plane 50 mm behind the lens when photographing a distant object. Set at its smallest aperture, b is about 3 mm. Estimate the diameter of the blur patch produced by a distant point source.

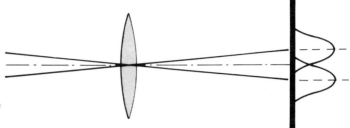

Fig. 21.27 Two point sources are just resolved if the maximum of the diffraction pattern of one source just coincides with the first minimum of the other, (Rayleigh's criterion).

Angular diameter, ϕ, of the blur patch

$$= \frac{1.22 \times 5 \times 10^{-7}}{3 \times 10^{-3}}$$

$$= 4.07 \times 10^{-4}$$

If the diameter of the blur patch $= d$, then $d/50 \times 10^{-3} = \phi$ to a good approximation.
    Hence $d = 50 \times 10^{-3} \times \phi$
$$= 2 \times 10^{-5}\,\text{m}$$
$$= 0.02\,\text{mm}.$$

21.5 Resolving power

If the directions from a lens to two points on an object make an angle less than ϕ in the example above, then they will produce blur patches that in fact overlap (Fig. 21.27). If the patches just touch (centres of the object points separated by an angle ϕ) then clearly the two patches will be distinguishable. If the angular separation is less than ϕ, they may not be distinguishable. At what point do they become indistinguishable? There is no absolute answer to this question. It depends on so many other factors such as contrast of the subject, overall illumination, etc. But clearly there *is* some limit to the extent to which two objects can

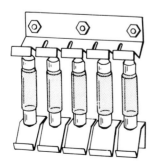

Fig. 21.28 A set of closely-spaced line sources.

approach each other and yet remain distinguishably separate. This limit is said, quite arbitrarily, to be $\phi/2$: in other words the angle between the centre of the diffraction disc and first minimum for the diffraction pattern of a point source. This is a good approximation to the truth for 'average' viewing conditions and was first proposed by Lord Rayleigh, since when it has been known as the *Rayleigh criterion*. Fig. 21.27 shows the distribution of intensity with angle for the central diffraction discs of two sources which are just on the limits of resolution according to the Rayleigh criterion.

Two object points can just be resolved if their angular separation is $1.22\lambda/b$. The smaller the angle, the higher the *resolving power* of the optical instrument forming the image. The way the resolving power depends on both the aperture b and the wavelength λ can be investigated experimentally by viewing a series of closely-spaced line sources through a narrow slit (Fig. 21.28). With a red filter across the lamp, the slit can be narrowed until the separate sources cannot quite be distinguished. If the red filter is replaced by a blue filter it will be found (owing to the shorter wavelength) that the sources can now be distinguished. Figure 21.29 shows a series of photographs illustrating the variation in resolution achieved by using different apertures.

The concept of resolving power and the Rayleigh criterion can be applied to circumstances other than the formation of an image point by a lens. Consider for example the diffraction grating mentioned earlier. We have seen that the angular separation between the centre of an interference maximum and the point at which the intensity falls to zero is λ/Ns, where s is the grating spacing and N the number of slits in use. We shall now derive this result another way.

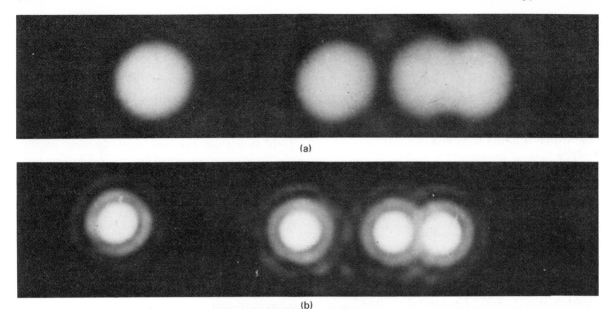

(a)

(b)

Fig. 21.29 Diffraction patterns of four
'point' sources, with a circular opening
in front of the lens. In (a), the opening
is so small that the patterns to the right
are just resolved, by Rayleigh's criterion.
Increasing the aperture size decreases
the size of the diffraction patterns, as in
(b) and (c). (From F. W. Sears, M. W.
Zemansky, H. D. Young, *College
Physics*, Addison-Wesley, 1980.)

(c)

At angle θ (Fig. 21.30) the path difference
between light from the first slit and light from the
Nth slit is $N\lambda$ if θ is the direction to the first order
maximum of wavelength, λ. Suppose the angle is
now increased by a small amount, δ, so that the
path difference between light from the 1st slit and
the Nth slit is $N\lambda + \lambda$. The path difference
between light from the first slit and the $N/2$th slit
is thus

$$\frac{1}{2}(N\lambda + \lambda) = \frac{N}{2}\lambda + \frac{\lambda}{2}$$

Light from these two slits is 180° out of phase and
cancellation occurs. Similarly, light from every
other slit in the first half of the grating can be
paired with light from a corresponding slit in the
second half so that both are 180° out of phase. So
in the direction $\theta + \delta$ cancellation occurs and this
is the direction in which the intensity of the first
order maximum falls to zero. From Fig. 21.30,
$\delta \approx DC/AB$ and DC is λ.

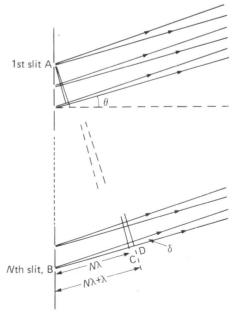

Fig. 21.30

Hence, $\delta = \lambda/Ns$,

as proved before.

If it is possible just to resolve two wavelengths λ and $\lambda + \Delta\lambda$ in the first order, then, using the Rayleigh criterion, light of wavelength $\lambda + \Delta\lambda$ must form its first order maximum at the position where the maximum of wavelength λ just falls to zero – that is at the angle $\theta + \delta$.

For light of wavelength $\lambda + \Delta\lambda$,

and, as we see from Fig. 21.30, $BD = N\lambda + \lambda$

$$BD = N(\lambda + \Delta\lambda)$$
$$N(\lambda + \Delta\lambda) = N\lambda + \lambda$$
$$N\lambda + N\Delta\lambda = N\lambda + \lambda$$
$$N\delta\lambda = \lambda$$

The resolving power for the grating is usually expressed as $\lambda/\Delta\lambda$. So

$$\lambda/\Delta\lambda = N. \tag{21.3}$$

It is not difficult to show that for the nth order of interference,

$$\lambda/\Delta\lambda = nN$$

and the resolving power increases with order of interference.

Again it must be emphasized that the resolving power realized in practice is measured by the number of slits, N, through which the light actually passes. This may be far less than the *theoretical* resolving power of the grating. For this reason, diffraction gratings should be mounted in spectrometers in order to make full use of all the slits available.

21.6 Diffraction and the double slit pattern

Before leaving the matter of diffraction and resolving power it is worth noting that the introductory work on the diffraction grating and the double slit assumed that the light is uniformly diffracted in all directions as it passes through each slit. In fact this is not so.

It is spread out into a diffraction pattern whose extent is governed by the width b of each slit. Fig. 21.31 shows the way the intensity of light from a double slit of *separation*, s and *slit width, b* varies with angle. The solid line shows that the maxima (and the minima) are separated by angles $\theta = \lambda/s$ (making the approximation $\sin\theta = \theta$ for small values of θ). The dotted envelope of these maxima is the diffraction pattern produced by the diffraction of light through each slit. Minima occur in this pattern at angle $\pm n\lambda/b$ either side of the central maximum.

You will notice that as a consequence of the diffraction pattern of each slit, the interference maxima tend to fade out, only to reappear again in the subsidiary maxima of the diffraction pattern. It is not easy to see this when looking at a white light source through a slit, but the effect is very readily seen if light from a laser is passed through a double slit system.

21.7 The limits of vision

The limited resolution of a lens and thus of most optical instruments is brought about because only a part of each scattered wave is received by the

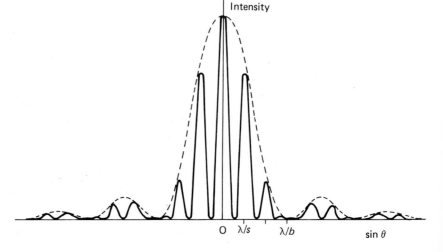

Fig. 21.31 Variation of intensity from a double slit with $\sin\theta$ showing the combined effects of interference and diffraction at a single slit.

lens. In a high quality microscope, however, the lens is so close to the object that the lens is so close to the object that virtually all the scattered wave passes through the lens. There are, however, still limits to the fineness of detail that can be observed. It will be recalled that a lens is able to decode the original location of a wave by the phase difference produced by two scattered waves arriving at the lens. But suppose, in Fig. 21.19 that the points A and B were only one wavelength apart. We have already seen that under these circumstances no maxima and minima can be formed within the region of scattering. In other words, there is too little distinction in phase for the lens to be able to differentiate the location of A and B even if it receives all the scattered waves. Hence the finest detail observable is of the same order as the wavelength of the radiation used. This is about 5×10^{-7} m for light. To observe finer detail, shorter wavelength radiation must be used. This is the reason X-rays are used to analyse the structure of crystals (see Chapter 14).

21.8 The formation of images using coherent light

In seeking to explain the action of a lens in forming an image, we have pointed out that scattered waves from two object points separated in space will interfere with each other producing sets of interference fringes, whose spacing depends on the original separation of the sources of the scattered waves (Fig. 21.19). In the absence of a lens we might thus expect a screen receiving light scattered from some extended object, such as the tree in Fig. 21.13 to be covered in a complex pattern of fringes produced by the interference of all the wavelets scattered from each point on the tree. In fact we have described such a screen as uniformly illuminated. Even the most detailed investigation fails to show any interference fringes. The reason for this is that the light falling on the tree (from the sun, for example) is *non-coherent*. We have already seen that non-coherent wave radiation does not produce an observable interference pattern (Section 16.5). Scattered wavelets from two object points retain a fixed phase relationship for only a fraction of a second (something like 10^{-9} s) and then the phase relationship makes a discontinuous jump as the phase

of one or both of the scattered waves changes.

However, this non-coherence does not invalidate the description of the operation of a lens which collects together light from the scattered waves and provides just *one* set of paths for the light scattered from one object point, all of which have (in wavelengths) the same path length. Any changes in phase for the wave fronts following these paths is the same for all of them and the resulting image point remains unchanged.

Using non-coherent light the lens is the only method of forming and thus recording (photographically) an image. However, we have today increasingly powerful sources of *coherent* radiation, extending over a wide band of the electromagnetic spectrum. The devices producing such radiation in the visible region are called *lasers* (*l*ight *a*mplification by the *s*timulated *e*mission of *r*adiation). Their operation is described in Section 44.6.

If an object is illuminated by coherent radiation, the phase relationship between the scattered wavelets should remain fixed. Such scattered light ought to form interference patterns on a screen on which it falls and such proves to be the case. This phenomenon has led to a method of image recording and reproduction called *holography*. An object (see Fig. 21.32) is illuminated by light from a laser, some of which is also reflected at the partial reflector, M, to form a reference beam. Scattered light from the object falls on a photographic plate at P, in combination with the reflected coherent radiation from the mirror, M. Interference fringes are formed on the plate, the position and spacing of these fringes depending on the *phase* of the scattered wavelets relative to the reference beam. In the absence of the plate, the object could be seen at E via the light that it scatters in that direction. A different view of it could be obtained at E′ via light scattered in *that* direction.

If the photographic plate is now developed and a positive transparency made (so that there are transparent patches on the plate where the light was originally most intense) an image of the original object can be reconstructed by illuminating the plate with light from a second laser. The light from the second laser must fall on the plate in the same direction as the original reference beam. A little thought will show that the photographic

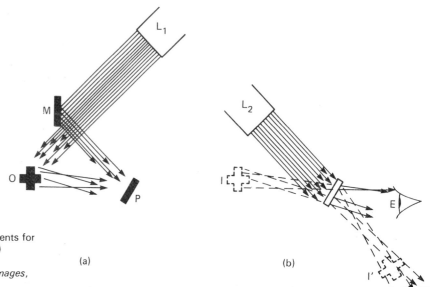

Fig. 21.32 Schematic arrangements for (a) producing a hologram and (b) reconstructing an image from a hologram. (From C. A. Taylor, *Images*, Taylor and Francis, 1978.)

(a)   (b)

plate will modify the beam so that the light travelling in the direction to E is the same as in the original set-up. As far as an eye at E is concerned, it *is* the same and so it sees an image of the original object as I. One of the great advantages of this method of image formation is that the image is three-dimensional. Light now travelling in a direction towards E′ will be the same in its distribution of amplitudes as light originally travelling in that direction. Consequently it will give a view of the original object that would have been obtained in *that* direction.

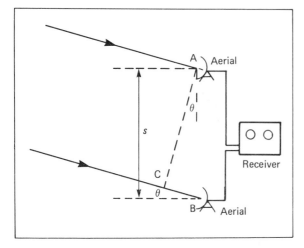

Fig. 21.33

21.9   Location of astronomical radio sources

We complete this section on location by considering one more technique using phase differences to locate sources of radiation. This technique is now much used in radio-astronomy. Figure 21.33 shows a diagram of the set-up. Radiation from a distant radio source is picked up by *two* widely spaced aerials, A and B, and their signals are combined at the receiver. Radiation from the distant source will consist of plane waves at the earth's surface and there may be a path difference between the two signals. In the diagram, this path difference is CB

$$CB = s \sin \theta$$

If CB is a whole number of wavelengths a maximum signal will be picked up at the receiver, while if CB is an odd number of half-wavelengths, no signal will be received.

The two aerials, being mounted on the earth, are rotating with it, so the angle θ is continually changing. Thus the signal picked up passes through a series of maxima and minima which are usually recorded on a chart recorder. Analysis of the interference pattern enables an accurate value of the direction of the source of radiation to be made. The bigger d is the more closely spaced the fringes and the more accurately can the direction of the source be fixed. Base lines (i.e., values of *d*) of over 100 km have been used although there are great technical difficulties in doing this.

21.10 Image transmission by scanning

Images can be recorded for subsequent viewing photographically, but their transmission over long distances cannot yet be done 'in one piece'. Both photography and electric sensors respond only to the intensity of a wave at a given point; they do not record the phase. Transmission of images takes place by recording the intensity of each point on an image of the object and transmitting this as an electromagnetic wave whose amplitude (or frequency, in frequency modulation) is a measure of the intensity recorded. The image is *scanned* point by point and built up again in the receiver by a reversal of the process. Thus an image which consists of a *spatial* variation in intensity is converted into a signal which is a *time* variation in intensity. For moving pictures, the scanning has to be done at high speed, so that each picture is built up in something like 1/25th of a second. Repetitions of pictures at such a rate will be seen by the eye as a continuous and changing scene. On such principles does television depend.

If only 'still' pictures are required, the information may be transmitted over some considerable period of time. This is the technique used in the Voyager space mission where photographs are recorded electronically on board the spacecraft and then transmitted piece by piece back to earth under favourable transmission conditions. A similar but not quite so slow build up of pictures is achieved in the transmission of weather pictures back from meteorological earth satellites.

UNIT SIX

Electric Currents and Charges

Chapter 22

INTRODUCTION

An understanding of basic electrical theory is necessary to understand the structure of matter. Moreover, if we wish to understand the uses of electricity: heating, lighting, motive power, radio, television, computers, micro-processors and a host of others, we must first understand the behaviour of electric currents in conductors and circuits.

In this chapter we shall meet the principles which are necessary for understanding how electricity is made to serve useful ends. As an introduction we shall raise some questions which are relevant to the present-day and to the future technological world by considering two particular areas of development in electrical technology which have radically changed the pattern of life in the twentieth century. We shall discuss very briefly the growth of electrical power in Britain, and the miniaturization of electronic circuits.

Michael Faraday discovered in 1831 that an electric current could be generated by the relative motion of a magnet and a conductor. In the years that followed, this effect, *electromagnetic induction*, was explored by many people with a view to designing a machine which could generate electrical energy in useful quantities. In 1886 John Hopkinson, engineer physicist, wrote the first treatise on dynamo design. By about 1900 electrical generators were in common use: multipolar generators driven by reciprocating steam engines, many producing several thousand kilowatts of electrical power. Electrical energy was all generated, at this time, by private companies, mostly for the sole purpose of supplying street lighting. In 1906 the total capacity of all electrical generating plants in the United Kingdom was about 1000 MW, and in that year the number of kilowatt-hour units sold was 530 million, but the average cost of generation and distribution was high. In 1910, S. Z. de Ferranti, addressing the Institution of Electrical Engineers, expounded the idea of making an abundant and cheap supply of electrical energy available all over the country, mainly for the purposes of conservation of coal resources, to encourage production of homegrown food, and for better utilization of labour.

The demands imposed upon industry by the First World War spurred on the government to consider ways of interconnecting generating stations to economize on plant, coal, and other costs. In 1926 the Central Electricity Board was constituted. The Board set about getting the private electrical generating companies to co-operate in the establishing of a national network of transmission mains, originally called the 'Gridiron', but now known as the National Grid. At that time the total capacity in the country was about 3000 MW, with about 5000 million units (5 GWh) being sold per year, at an average unit price of about £0.01. The Grid was essential if generation and distribution of electrical energy was to grow in an economically viable fashion. This is for two main reasons: firstly, it is not practicable to store appreciable quantities of electrical energy – it must, nearly all, be generated as and when needed; secondly, it is an expensive and slow business to start up and shut down a generating plant in response to a rapidly fluctuating demand.

In the 1950s, the maximum capacity of a single generator was 60 MW: today it is 660 MW. Figure 22.1 shows the growth of electrical energy output from all generating plants in the United Kingdom.. From this it can be seen that the consumption in 1980 was more than 240 GWh, over forty times what it was when the National Grid was proposed.

The chart shown in Fig. 22.1 is clearly not one of exponential growth (see Chapter 8). Although the number of units available for use has grown substantially over the last twenty years, the rate of growth is tending to diminish. Students of recent British history might like to speculate on the reasons for this and for the peaks which appeared in 1973 and in 1979. The figures are taken, with permission, from the Annual Abstracts of Statistics for the period.

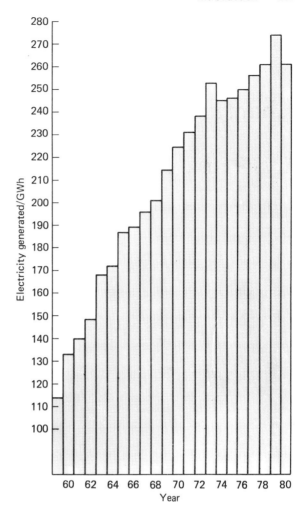

Fig. 22.1 The growth of the output of electrical energy from 1959 to 1980.

The present output of the electricity industry is summarized in Table 22a, and the figures in it should be compared with the ones quoted earlier.

We may now ask some questions about these statistics.

a) Why is it not practicable to store electrical energy other than in pumped storage systems (see Section 8.8)? If it were, then a national supply could be a much simpler system than it is now.

b) Why are the efficiencies of the processes of transforming the energy in steam to the electrical form so low? (See Section 12.12.)

c) What fraction of the electrical energy generated is wasted in transmission from generator to consumer?

Table 22a   The generation of energy and the capacities of power stations in the United Kingdom, 1979–1980

	Energy generated in 1979/ GWh	Percentage of total	Output capacity/ MW
Nuclear plant	34 604	12	5527
Other steam plant	239 317	86	57 307
Gas turbines and oil engines	758 ⎞		3704
Pumped storage plant	1175 ⎬ 2		1060
Other hydroelectric plant	3628 ⎠		1284
Total	279 482		68 882

Note: The average thermal efficiency of conventional steam stations rose from 20.9% to 31.7% between 1948 and 1980. The average thermal efficiency of the twenty most efficient stations was 33.7% in 1980.

d) What is the future of the supply (at present about 12%) from the nuclear power plants (see Chapter 49)?.

The answers to these questions largely determine the economics of the electrical power industry. We cannot answer them now, but what is presented in this book should enable you to see how to answer them, although you may need to go to other sources of information for facts and figures.

At the other extreme from heavy current electrical engineering is the electronics industry which, in its brief life, has already been through several revolutionary stages of development.

'Micro-miniaturization' is a term which briefly sums up the revolutionary change that is happening in electronics.

The period of rapid progress in making electronic circuits smaller and more rugged, began in 1939, at the outbreak of the Second World War. Military electronic equipment, especially airborne radio and radar, had to be compact and highly reliable. Electronic vacuum tubes (valves) were made as small as possible, some of the smallest being known as *acorn valves*, which gives an idea of their size. After that war the change to transistors and other semiconductor devices marked a real turning point in the electronic revolution. The first transistors were manufactured in the early 1950s, and large sums of money were invested in developing the transistor and other semiconductor devices. The advantages of transistors over vacuum tubes can be summed up briefly as follows:

a) They do not deteriorate with time, whereas vacuum tubes do.
b) They waste much less electrical power than vacuum tubes.
c) There is no warm-up period after switching on.
d) They are physically much more robust than tubes.
e) They can be made very much smaller than vacuum tubes which perform a similar function.

Without the advent of miniature semiconductor devices such as transistors and, latterly, integrated circuits, it would have been impossible to develop computers beyond a rudimentary stage. In the early 1950s the primitive electronic computer in Britain known as ENIAC contained 18 000 vacuum tubes and the failure rate of these was such that it spent more time out of commission than it spent working. It was, by modern standards, a very limited machine. It filled a whole room: its modern counterpart can be made small enough to fit in your pocket. The effect of the computer revolution upon our lives need not be dwelt upon here.

There is no doubt that electronic circuits will continue to get smaller and smaller, without losing reliability. Some present students of physics will have one day to use, or even design, such circuits. It is for this reason that the treatment of electric circuit theory in this unit is planned in such a way as to lead the student to be concerned more with the function of an electric circuit element rather than its internal structure, and to think of ways of combining circuit elements to perform a desired function which cannot be performed by a single element alone.

In this unit we introduce the concepts of electric current, potential difference (voltage), resistance, power, charge, energy, capacitance, and the ideas needed to understand electromagnetism and alternating current theory. The questions at the end of the book not only provide exercises in using the concepts, but describe important modern applications and thus aim at widening the range of your knowledge, not simply at testing and reinforcing your understanding.

Chapter 23

STEADY DIRECT CURRENTS

23.1 The simple circuit

It is not possible to tell whether a metal wire is carrying an electric current merely by looking at it. There are, however, two detectable effects which we ascribe as being due to an electric current. Firstly, a heating effect. One of the main practical uses of electricity is to heat a wire. A nichrome alloy wire, heated by an electric current, forms the basis of all electric room heaters, immersion heaters and other heating elements. A tungsten wire, kept white hot by an electric current, in an atmosphere of inert gas, is the filament of an electric lamp. The other effect, the magnetic effect, is less readily detected. Oersted announced in 1820 his discovery of the effect. He used a battery made of copper and zinc electrodes in sulphuric acid, a wire, and a pivoted magnetic compass needle.

Both effects have been employed in instruments for measuring the strength of an electric current. The hot-wire ammeter, in which the expansion of a wire when heated by the current was made to move a pointer over a scale, is now obsolete. The moving-coil and moving-iron meters both make use of the magnetic effect for their operation. The construction of the former instrument is discussed in detail in Section 34.4.

To investigate the properties of steady direct currents, one can do some experiments with simple circuits and ammeters. Suppose we have a circuit arranged as shown in Fig. 23.1, consisting of 3 torch lamps connected in series to 3 dry cells. A current, we say, flows when the switch S is closed, and we have a complete circuit. It ceases to flow, it seems (because the lamps go out), at the

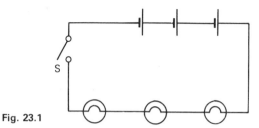

Fig. 23.1

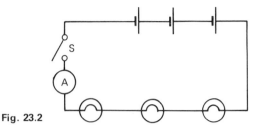

Fig. 23.2

instant we open the switch S. So much will probably seem obvious to you. In our investigation of the behaviour of electric current, we ask two questions.

a) When the circuit is made, what is the current strength at different points in the circuit?

b) Does the current rise from zero to a steady value *instantaneously* when switch S is closed, and fall to zero instantaneously when the switch is opened? If not, how does the current change with time during the switch-on and switch-off processes?

The second question we shall leave until Section 26.7, when we have a fuller understanding of the nature of the electric current. The first can be tackled straight away.

When we switch on the circuit we observe that the three lamps are lit to an equal, steady brightness. If now we break the circuit and insert an ammeter into that break (Fig. 23.2) we see no detectable change in the brightness of the lamps. We can assume therefore, that the presence of the ammeter has had negligible effect. If now we insert that ammeter into any other point in the circuit – between any two of the lamps, between any two of the cells – we find that it gives the same reading as in the first case.

This reminds us of the behaviour of liquid flow (see Section 11.4). Evidently we can use a flow model for the electrical behaviour in the

circuit even though we cannot see anything flowing in the wires.

It is clear, too, how we may compare one ammeter with another. We establish a circuit such as Fig. 23.3 in which A_s is a standard ammeter. The ammeter A is to be calibrated and the variable resistor or rheostat R can be used to control the current. It is legitimate to enquire how the standard ammeter itself was calibrated. This is a matter to which we shall return in Section 34.2.

Experience with such series circuits as these will quickly reveal the need to observe the appropriate polarity when connecting an ammeter into a circuit. The terminals of the battery will be marked as positive or negative; the ammeter (unless of the centre-zero type) will be similarly marked. For correct use, the positive terminal of the ammeter must be nearest (in circuit terms) the positive terminal of the battery.

Unlike a liquid circuit (in which the direction of flow is evident) there is no obvious direction of flow in an electric circuit. In consequence, an arbitrary direction has been assigned and we agree to speak of current as though it flowed out from the positive pole or terminal of the battery, through the circuit elements and so back to the negative pole of the battery. This is the conventional current direction indicated by arrowheads in circuit diagrams.

To check our hypothesis that a flow model is an appropriate one we can set up a parallel circuit as in Fig. 23.4. Here a single cell is providing the current to light the three similar lamps which are connected 'in parallel'. The readings I_0, I_1, I_2 and I_3 of the four ammeters A_0, A_1, A_2 and A_3 will confirm that

$$I_0 = I_1 + I_2 + I_3$$

This behaviour is typical of flow in a branching circuit as we have already seen in Section 11.6.

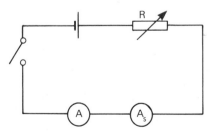

Fig. 23.3 Comparing an ammeter with a standard ammeter.

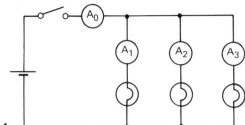

Fig. 23.4

It is described by *Kirchhoff's first law:*

■ The total current into a junction point in a circuit is equal to the total current flowing out of that junction point.

The evidence gathered so far suggests that, in electrical circuits, there is a flow of what we might call electricity but which is better named *electric charge*. That flow conforms to the laws we noted when we examined a flow model earlier. The function of the ammeter is to measure the quantity of electric charge which flows through it in time t. If Q is the quantity of electric charge passing through the ammeter in time t, the current I is given by

$$I = \frac{Q}{t}$$

whence

$$Q = It \qquad (23.1)$$

We have used the ampere (A) as the unit of current so it follows that the unit of electric charge must be the ampere-second. This is usually known as the *coulomb* (C).

■ The coulomb is the quantity of electric charge passing a given point on an electrical conductor in one second when a steady current of 1 ampere is flowing in the conductor.

For the definition of the ampere itself see Section 34.2.

23.2 Potential difference

Armed with the concepts of electric current and electric charge, we can proceed to the important discussion of the energy transformations in electrical circuits. As an introduction, consider the simple circuit shown in Fig. 23.5. Electric charge circulates at a steady rate around the circuit and the lamp is steadily lit. In the cell chemical energy is being transformed; in the lamp energy is being given out in the forms of heat and light. It is reasonable, therefore, to think of the electric charge, as it circulates around the circuit, as being the agency which carries the energy from the cell to the lamp. Since the lamp is giving out energy at a steady rate and charge is circulating at a steady rate, it makes sense to assume that each coulomb of charge is carrying a certain quantity of energy with it: it collects this energy from the cell and delivers it to the lamp. Immediately this raises the question of *how much* energy is being carried by each coulomb of electricity. Consider what would be the effect of altering the strength of the current in the circuit of Fig. 23.5. If the current is increased the lamp glows more brightly: obviously energy is being delivered to the lamp at a faster rate. Thus, clearly, the rate at which energy is delivered to the lamp depends upon the strength of the current. But the current is not the only factor which determines the rate at which the energy is delivered. An excellent demonstration of this fact is an experiment in which two lamps of very different power are connected in series to the mains. In Fig. 23.6 L_1 is a 240 V 60 W lamp. L_2 is an ordinary torch lamp designed to work at a

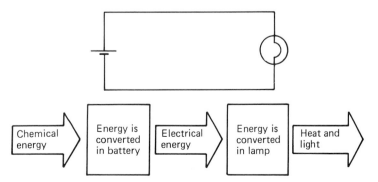

Fig. 23.5 Energy conversions in a simple circuit.

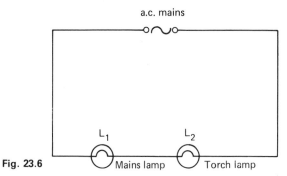

Fig. 23.6

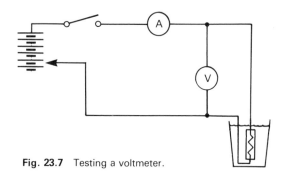

Fig. 23.7   Testing a voltmeter.

voltage of about three volts and to take a current of about a quarter of an ampere. What do you think will happen when the supply is switched on? Both lamps will light to approximately normal brightness, although the layman would probably be willing to bet on the smaller lamp being burned out almost instantly! There is no possible doubt that energy is being converted at a much faster rate in the large lamp than in the small one: but we know, because the two lamps are connected in series, that the same current must be flowing through each. The same current is being accompanied by very different rates of energy conversion: therefore the strength of the current flowing through a device is not the only factor determining the rate of energy conversion in the device. (*You are advised not to try this experiment at home.*)

We now have to find what other factor, or factors, determine the rate at which energy is delivered to electrical devices. The most suitable device is an electric immersion heater since we can be sure that it will convert electrical energy into heat – and heat can be measured by calorimetry. Unfortunately calorimetry experiments are not easy to perform with accuracy for there is always considerable loss of the energy by conduction, convection, radiation and even evaporation. There are, of course, techniques of allowing for these losses. Nevertheless we shall consider a very simple experiment since the principle involved is of such importance. As an experiment of principle it shows how voltmeters are calibrated in the first place.

Figure 23.7 shows a suitable circuit in which the voltmeter under test is connected in parallel with a small immersion heater. A known current (I, typically not more than 5 A) is allowed to flow

through the heater which is immersed in some water (mass m, typically 0.2 kg) contained in a polystyrene cup. The rise in temperature which results after time t (usually about 100 s) is noted. This requires that the water should be carefully stirred and that the initial and final temperatures should be read as carefully as possible. Since the specific heat capacity of water is (to two significant figures) 4200 J kg^{-1} K^{-1}, the energy (in joules) received by the water as it warms up by ΔT is

$$(4200 \text{ J kg}^{-1}\text{ K}^{-1})\, m\Delta T$$

The electric charge which delivers this energy to the water is

$$\text{current} \times \text{time}$$

It follows that each coulomb of charge delivers

$$\text{energy } (4200 \text{ J kg}^{-1}\text{ K}^{-1}) \times \frac{m\Delta T}{It}$$

to the water.

The experiment is repeated several times (always of course with cool water) and with different currents obtained by choosing different numbers of cells in the battery. If the results are compared with the readings of the voltmeter under test, it will be seen that there is reasonably close agreement between them. The voltmeter is giving a direct reading of the number of joules transformed in the heater for each coulomb of electric charge which passes. Table 23a shows some typical results.

Since, as we have said, calorimetric experiments, although of great fundamental importance, are neither easy to perform with accuracy nor convenient, an instrument which can measure electrical energy directly would provide an

Table 23a   Mass (m) of water used in each case: 0.2 kg. Time for which charge passed: 100 s

Cells used	$\Delta T / K$	Energy transferred to water/J	Current/A	Charge/C	Energy transferred charge/J C^{-1}	Voltmeter reading/V
6	7.0	5900	5.0	500	11.8	12.1
5	4.8	4030	4.2	420	9.6	10.0
4	3.1	2600	3.3	330	7.9	8.1
3	1.6	1340	2.5	250	5.4	6.0

opportunity for easier experimentation in a wider context. Such instruments are known as joule-meters (Fig. 23.8a); the industrial equivalent is the domestic meter which one finds in every private house and which, from time to time, a representative of the regional Electricity Board comes to read. The domestic meter is designed to work on a.c. at the mains voltage and to measure energy in units called kilowatt hours (Fig. 23.8b). One kilowatt hour equals 3 600 000 joules. The question of the units in which electrical energy can be measured will be dealt with in detail later: for the time being it is sufficient to appreciate that, when an electrical supply is connected to a load, e.g. a heater, via a joulemeter, the meter will register, in joules, the amount of electrical energy transferred from the supply to the load.

Before using a joulemeter it is wise to make an approximate check of its calibration. Precisely the same technique is used as when calibrating a voltmeter.

An electric immersion heater is connected to the output (or load) terminals of the meter, and a suitable low-voltage supply is connected to the input terminals. The energy output of the immersion heater during a given time is measured by measuring the temperature rise in a measured mass of water, knowing that the specific heat capacity of water is 4200 J kg^{-1} K^{-1}. The energy output, measured in this way, can then be compared with the reading of the joulemeter.

An instructive experiment is to take a selection of low-voltage lamps, say, a 12 V, 6 V and a 2.5 V lamp, and to connect each in turn into the circuit shown in Fig. 23.9. By adjusting the rheostat R it is possible to arrange that the same current flows in each lamp when it is connected. In an experiment of this kind the results displayed in Table 23b were obtained, the supply being switched on for one minute in each case.

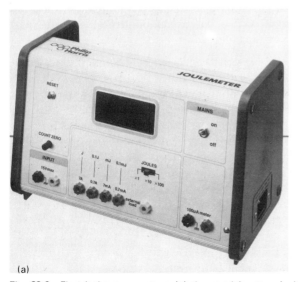

(a)

(b)

Fig. 23.8   Electrical energy meters: (a) shows a laboratory joulemeter (courtesy Philip Harris Ltd.); (b) shows a domestic single phase kWh meter. (Courtesy Ferranti Measurements Ltd.).

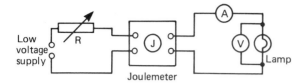

Fig. 23.9

The current through each lamp was 0.20 A and it flowed for a time of 60 s. Thus we know that the charge which has passed through each lamp is 0.20 × 60 = 12 C. If the supply had been switched on for *two* minutes in each case, then twice the amount of energy would have been delivered to each lamp. This suggests that the lamps differ in this important respect: the amount of energy converted per coulomb of charge is different for each lamp. These amounts have been calculated and entered in the right-hand column of Table 23b. It should not be surprising, now, that these amounts correspond roughly to the *voltage* markings on the lamps. We are now in a position to elucidate the precise meaning of the term *voltage*. The word itself means, literally, 'the thing which is measured in volts'. The 'thing', or rather, 'the physical quantity' which is measured in volts, is the *number of joules per coulomb of charge*. Thus, in the smallest of the three lamps used in the experiment described above, for every coulomb of charge which has passed through the lamp fila-ment, approximately 2.5 joules of energy have been transformed into heat and light: and that lamp was marked 2.5 V (meaning 2.5 volts). The actual measurement made with the joulemeter gave 2.7 joules per coulomb. This is explained (assuming the joulemeter has been accurately calibrated) by the fact that the filaments of mass-produced lamps vary one from another in their characteristics, and it should not be surprising that a bulb marked 2.5 V 0.2 A may use a little more or

Table 23b Current in each lamp: 0.20A. Time switched on: 60 s

Lamp type	Energy utilized/J	Charge passed/C	Joules per coulomb
2.5 V 0.2 A	32	12	2.7
8.0 V 0.2 A	97	12	8.1
12.0 V 0.2 A	154	12	12.8

a little less than precisely 2.5 joules per coulomb, even when the current through the filament is maintained at exactly 0.2 A. In the experiment we could have measured this quantity by a much more direct method, using a voltmeter (which has probably occurred to you already!) and it is useful to repeat the experiment of Fig. 23.9, this time connecting a voltmeter in parallel with each of the three lamps being tested. The voltages indicated by the voltmeter should then be found to agree well with the values calculated in the right-hand column of Table 23b.

The quantity which is measured in volts is properly termed *potential difference* (p.d.). A voltmeter is always connected *in parallel with*, or, as it is often expressed, *across* an electrical device, in order to measure the p.d. between the terminals of the device. One often refers to 'the p.d. across' an electrical device when signifying the p.d. thus measured.

The exact definitions of the term p.d. and the unit in which it is measured are as follows:

■ The p.d. across an electrical device is the amount of energy converted per unit charge passing through the device.

■ The volt is the unit of p.d. such that one volt = one joule per coulomb.

It should now be possible for you to work out how, in the circuit of Fig. 23.9, one could find the amount of energy converted in a lamp in a given time without using the joulemeter. These facts give the clues:

a) The voltmeter measures the number of joules of energy converted in the lamp per coulomb of charge passed through the lamp. For this quantity we use the symbol V.

b) The ammeter measures the number of cou-lombs, passing through the lamp per second. For this we use the symbol I.

c) The clock measures the time, t, during which the supply is switched on.

Using the symbol W for the amount of energy, we can write the solution thus:

(number of joules)
  = (number of joules per coulomb)
    × (number of coulombs per second)
    × (number of seconds),

or, in symbols:

$$W = VIt$$ (23.2)

Problem 23.1 An electric room heater takes a current of 6 A from the 250 V mains supply. Calculate (a) how much energy it transforms when switched on for a period of ten hours, and (b) the cost of having the heater switched on for ten hours when the cost per *unit* is 6 p. (1 unit = one kilo-watt hour = 3 600 000 J.)

To compute the answer by the method outlined above we must first express the time in seconds.

  a) Ten hours = $10 \times 60 \times 60 = 36\,000$ s

     energy = W = VIt

             = $250 \times 6 \times 36\,000$ J

             = $54\,000\,000$ J

             = $\dfrac{54\,000\,000}{3\,600\,000}$ kWh

             = 15 kWh

  b) Cost = 15×6 p.

         = 80 p.

It is important for every householder, indeed for every consumer of electrical energy, to appreciate that what he pays for is neither the potential difference (the volts) nor the current (the amperes) separately, but the quantity of electrical energy utilized, i.e., the quantity measured by product VIt. This can always be calculated by the method of Problem 23.1 if the loads are resistive in nature.

23.3 Resistance

One might wonder, referring back to the experiment of Fig. 23.9, how it comes about that the same current flows through each of the three tungsten filament lamps although the p.d. across each lamp is different. The difference can only be due to different characteristics of the tungsten filaments. The property of the filament which determines the relation between the p.d. across it and the current flowing through it is its *resistance*. We can begin investigating this property by means of a simple experiment with a tungsten filament lamp. In the circuit of Fig. 23.10 the current

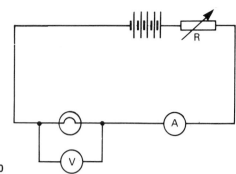

Fig. 23.10

flowing through a lamp can be varied by the rheostat R. There are suitable meters in the circuit to measure the current and the p.d. A typical set of measurements (Table 23c) shows more clearly the relationship between the p.d. and the current. In trying to analyse these results, however, one must bear in mind that in this situation we have not just *two* varying physical properties (p.d. and current) but *three*. The third property is the temperature of the filament, which varies during the course of the experiment from around room temperature to a temperature of around 2000°C. Although the temperature of the filament has not been measured in the experiment, it is reasonable to suppose that the very marked variation in temperature has contributed in some way to the pattern of results obtained. To carry out a meaningful investigation of the relationship between p.d. and current in a conductor (e.g., a metal wire) we must begin by controlling any variable factors other than p.d. and current. It is not practicable to keep the temperature of a lamp filament constant while performing this kind of experiment with it, but a similar experiment can be done with a length of wire of a suitable type immersed in oil to keep its temperature constant. In such an experiment it is found that, regardless

Table 23c

p.d./V	Current/A
0.4	0.60
0.9	0.90
2.0	1.25
3.9	1.70
6.5	2.30
10.0	2.85
12.0	3.10

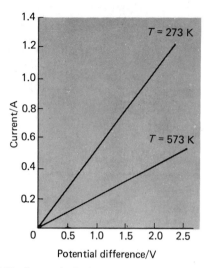

Fig. 23.11 Current/p.d. characteristics for 5 m of tungsten wire of diameter 0.4 mm at temperatures of 273 K and 573 K.

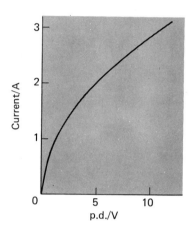

Fig. 23.12 Current/p.d. characteristics for the filament of a lamp.

of the material of which the wire is made, the current is directly proportional to the p.d. over the whole range of measurement. If the experiment is then repeated with the wire maintained at a steady, higher temperature (by heating the wire in a sandbath, for example) then again it is found that the current is proportional to the p.d., but for any given value of p.d., the current is less than what it was in the cool wire. The measurements of p.d. and current, when plotted on a graph, then appear as in Fig. 23.11 which refers to 5 m of tungsten wire of diameter 0.4 mm. This kind of change in the current–p.d. characteristic as a result of change in temperature is a property of *all* wires made of pure metals within a normal working range of temperature. The reader should now be able to appreciate why the current–p.d. graph for the lamp filament, in Fig. 23.12, is not a straight line but a curve of decreasing gradient, remembering that the greater the current, the higher the filament temperature.

We have referred previously (Section 11.6) to the analogy between the flow of a fluid in tubes and the flow of electricity in a circuit. We found that both fluid flow through a tube and the flow of electric charge in a conductor experienced a resistive effect. In the case of a fluid Eq. 11.4 could be written

$$\frac{\Delta p}{\Delta V/\Delta t} = -\frac{8\eta l}{\pi a^4}$$

So, for a given tube and a given fluid, the ratio pressure difference/flow rate gave a measure of the resistance to flow.

In the electrical case, Eq. 11.6 could be written

$$\frac{\Delta V}{\Delta Q/\Delta t} = -\frac{\varrho l}{A}$$

For a given conductor, the ratio potential difference/current gave a similar measure of *electrical resistance*.

In symbols

$$R = V/I \qquad (23.3)$$

and this is the defining equation for resistance.

The unit of electrical resistance is then the volt ampere^{-1} or *ohm*. The international symbol for the ohm is Ω.

In the experiment illustrated in Fig. 23.10 values of p.d. and current were measured for a tungsten filament lamp. These values are tabulated again in Table 23d and the resistance of the filament has been calculated for each pair of values. The resistance is observed to increase as the current through the filament increases. But look at the results of a similar experiment made with a metal wire kept at approximately constant temperature, in Table 23e. It is clear in this case that resistance of the wire is constant, regardless

Table 23d   Tungsten filament lamp

P.d./V	Current/A	Resistance/Ω
0.4	0.60	0.7
0.9	0.90	1.0
2.0	1.25	1.6
3.9	1.70	2.3
6.5	2.30	2.8
10.0	2.85	3.5
12.0	3.10	3.9

of the strength of the current flowing in the wire. This implies that, if one knows the resistance of a wire, one can easily predict what the current in the wire will be for any given value of p.d. applied to it, or what the p.d. will be for any given current, provided that there is no change in temperature. A length of metal wire at constant temperature is one example of a *linear* electrical device: a device in which the current is always directly proportional to the applied p.d. The particular practical uses of linear and non-linear electrical components will be discussed in detail later in this chapter.

The relationship of direct proportionality between the current and the p.d. in metal wires held at constant temperature was first convincingly demonstrated by Georg Simon Ohm, a German mathematician and physicist. The results of his investigations appeared in a paper published in 1826. The proportionality apparently held good for a wire of any dimensions made of any metal, over the whole range of currents which he was able to obtain, provided that the temperature of the wire was not allowed to vary.

■ *Ohm's law*: a steady current flowing through a metallic conductor is directly proportional to the p.d. between the ends of the conductor, provided that the temperature and other physical conditions are kept constant.

Table 23e   Length of resistance wire

P.d./V	Current/A	Resistance/Ω
1.2	0.50	2.4
2.9	1.20	2.4
4.1	1.65	2.5
5.2	2.10	2.5
6.3	2.50	2.5
7.6	3.05	2.5

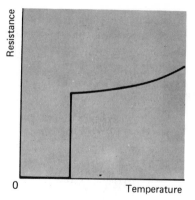

Fig. 23.13   The variation of resistance of a super-conductor with temperature at temperatures near to the absolute zero.

The law is found to be obeyed very closely by all metals, and by certain non-metallic conductors and electrical devices. These can therefore conveniently be referred to as *ohmic conductors* and *ohmic devices*: alternatively (as mentioned earlier) they can be called *linear* because of the linearity of their characteristic current–p.d. relationship.

We must note that at very low temperatures most metals depart radically from the simple behaviour described by Ohm's law. Indeed their resistance falls to zero. In this state they are superconductors. Figure 23.13 shows graphically how the resistance of a superconductor changes at this transition.

The phenomenon, which was discovered by Kamerlingh Onnes in 1911, has the interesting implication that once a current is started in a superconducting ring, it may continue indefinitely. But, of far more significance is the application of the phenomenon to the design of super-conducting magnets. Such magnets can be made to be far more powerful and yet cheaper to run than conventional powerful electromagnets. This suggests important consequences for electric power generation and for electric motors if the upper limit of superconductivity can be raised. At present this stands at 21 K. A further rise to, say, 30 K could revolutionize electrical technology.

23.4   The resistance of wires

G. S. Ohm was also the first person to investigate how the resistance of a wire in a circuit depended upon the *length* of the wire. To investigate the

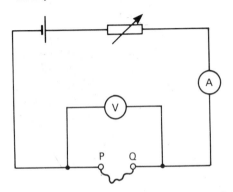

Fig. 23.14   Circuit for the investigation of variation of the resistance of a wire with length. The wire under test is connected between P and Q.

of the wire in such a way that if a wire has twice the thickness of a second wire, having the same length, then the resistance of the first wire is one quarter of the resistance of the second. The resistance is inversely proportional to the area of cross-section. These facts are summed up in a formula for the resistance, R, of a wire in terms of its length, l, and its area of cross-section, A:

$$R = \varrho \frac{l}{A}$$ (23.4)

relationship between resistance and length for a wire nowadays one could use the circuit of Fig. 23.14. It is found that the resistance is directly proportional to the length, provided that the wire has uniform cross-section. To investigate how the resistance depends upon the thickness of the wire one can use samples of wire made of one particular material, all samples having the same length but different thickness. It is found that the resistance depends upon the area of cross-section

The symbol ϱ represents the *resistivity* of the material of the wire, and is constant for a given material at a specified temperature. If R is measured in ohms, l in metres, and A in square metres, then the unit of ϱ must be: ohm metre (abbreviated: Ωm). The resistivities of some common materials used in electrical circuits are shown in bar-chart form in Fig. 23.15. The enormous range of resistivity from the metals ($10^{-8}\,\Omega$ m) to insulators ($10^{14}\,\Omega$ m) makes it necessary to use a logarithmic form for the chart. The scale of resistivity is given in powers of ten. In this

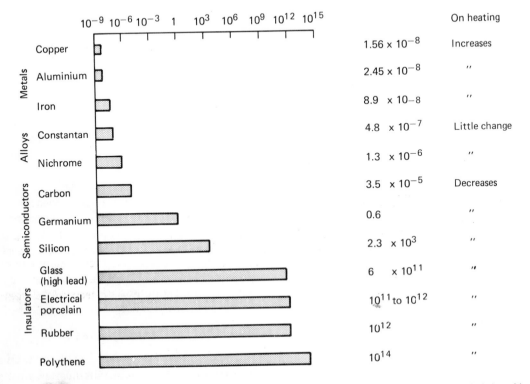

Fig. 23.15   Bar diagram of the resistivities of a number of substances in Ω m at room temperature. The scale is logarithmic.

way, the 22 orders of magnitude (from 10^{-8} to 10^{14}) can be fitted into the space available. Actual values are quoted to the right of the chart together with an indication of how the values change with temperature.

The formula $R = \varrho(l/A)$ can be used to calculate the resistance of any wire or cable of uniform cross-section.

In some applications it is more convenient to use the *conductivity* σ of the material. This is the reciprocal of the resistivity and, as has been pointed out in Chapter 15, varies very widely when considered over the whole range of insulators and conductors.

Problem 23.2   The heating element of an electric toaster is made of nichrome tape, thickness 0.05 mm, width 1 mm. The length of the tape is 4 m. Calculate the resistance of the element at $0°$ C.

Cross-section area of tape

$$= 10^{-3} \times 0.05 \times 10^{-3}\,\text{m}^2$$
$$= 5 \times 10^{-8}\,\text{m}^2$$

Resistivity of nichrome $= 130 \times 10^{-8}\,\Omega\,\text{m}$.
   Using the formula:

$$R = 130 \times 10^{-8} \times \frac{4}{5 \times 10^{-8}}\,\Omega$$

$$R = \frac{520}{5}\,\Omega$$

$$\therefore R = 104\,\Omega$$

If one wanted to find the resistance of a cable consisting of several strands of wire of one particular material, then one would simply substitute the total cross-section area of all the strands for A in the formula. If the strands are made of different materials, as for instance in the steel and aluminium conductors used in the overhead power cables of the National Grid in Great Britain, then one cannot use the simple formula alone: another formula, which gives the resistance of a number of resistors connected together in parallel, must be used as well.

Problem 23.3   The resistance of overhead cables of the National Grid. Estimate the resistance per

kilometre of one type of aluminium-and-steel conductor used on the 132 kV grid system, given the following data and guide-lines.

The cable consists of a core of 7 steel strands, each 0.28 cm diameter: the total cross-sectional area of these steel conductors is therefore about 0.4 cm². Around this core are 30 aluminium strands of the same diameter: the total cross-sectional area of aluminium is thus about 1.8 cm².

Take the resistivity of aluminium as 3×10^{-8} Ω m, and that of steel as $12 \times 10^{-8}\,\Omega$ m.

a) Using the appropriate formula, calculate the approximate resistance of 1 km of the aluminium conductors.
b) Repeat for the steel conductors.
c) Compare the resistance of the two types of conductor: you will notice that one is considerably greater than the other. Most of the current will flow in the conductor of lower resistance, and in our rough estimate we can neglect the presence of the high-resistance conductor: which conductor is this?
d) Finally, state the resistance per kilometre to one significant figure.

The results of these calculations will be used in Problem 24.4 where the power loss in the conductors is considered.

From Eq. 23.4

$$R = \varrho \frac{l}{A}$$

a) For 1 km of aluminium:

$$R = 3 \times 10^{-8} \times \frac{10^3}{1.8 \times 10^{-4}}\,\Omega$$

$$= 0.17\,\Omega.$$

b) For 1 km of steel:

$$R = 12 \times 10^{-8} \times \frac{10^3}{0.4 \times 10^{-4}}\,\Omega$$

$$= 3\,\Omega.$$

c) The resistance of the steel is thus about twenty times that of the aluminium, so most of the current flows through the aluminium.
d) The overall resistance of the cable will be slightly less than 0.17 Ω (because of the presence of the steel). Call it 0.2 Ω per km, correct to 1 significant figure.

Problem 23.4 In the article on microelectronics in the *Electronic Engineers' Reference Book* it is stated that, in a typical thin-film circuit, in which resistors are made by the condensation of vaporized nichrome alloy on to borosilicate glass, a 10 kΩ resistor would be approximately 1 cm long and 0.25 mm wide. The resistor is a thin, flat strip of the metal. From this information, and using the value for the resistivity of nichrome given in Fig. 23.15, estimate to one significant figure the thickness of the strip of metal, and express the answer in nanometres (1 nm = 10^{-9} m). Then, given that the diameter of single atoms in a metal like nichrome is very roughly 0.2 nm, express the thickness of the resistor strip in terms of atomic diameters.

Let the thickness of the resistor strip be x.

Using the formula $R = \varrho(l/A)$ and substituting, we get

$$10^4 = 1.3 \times 10^{-6} \times \frac{10^{-2}}{0.25 \times 10^{-3}x}$$

$$= \frac{130 \times 4 \times 10^{-7}}{x}$$

hence

$$x = \frac{520 \times 10^{-7}}{10^4} \text{ m}$$

$$= 5.2 \times 10^{-9}\text{ m} = 5.2 \text{ nm}$$

or, to one significant figure, 5 nm.

If one atomic diameter is approximately 0.2 nm, then the strip is about 5/0.2 = about 25 atoms thick!

Layers as thin as this are achieved by condensing the vapour of the metal, in a near vacuum, on to the surface of the substrate material (typically, borosilicate glass). It is possible to control the minute thickness very accurately since the material is being deposited, literally, atom by atom.

The resistivity of any pure metal increases with its temperature. Figure 23.16 shows the effect of temperature on the resistivities of some metals. Over a limited range of temperature the graphs are approximately linear, so that a *temperature coefficient of resistivity* can be specified for the metal

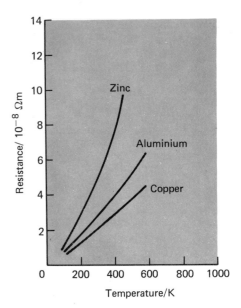

Fig. 23.16 The variation of resistivity with temperature for zinc, aluminium and copper.

concerned. This coefficient is defined as the

$$\frac{\text{increase in resistivity per }°\text{C or K}}{\text{resistivity at } 0\,°\text{C}}$$

We may write

$$\frac{R_\theta - R_0}{R_0} = \alpha\theta \qquad (23.5)$$

where R_θ and R_0 are the resistances at temperatures θ (in °C) and 0 °C respectively and α is the temperature coefficient of resistivity.

Certain alloys have a high resistivity and a very small temperature coefficient of resistivity in comparison with pure metals. Examples of such alloys are constantan and manganin which are the metals most often used for making resistors of the wire-wound variety: and in this application it is very convenient that variation of resistance as a result of change in temperature is generally negligibly small. Table 23f shows the temperature coefficients of resistivity of some metals, and here it should be noted that the coefficients all have the same order of magnitude for pure metals. The coefficients for resistance alloys are in the order of one hundredth of that for a pure metal. So far as is known, it is impossible to find a 'recipe' for an alloy which shows no change in resistivity whatsoever when its temperature changes.

Table 23f

Material	Temperature coefficient of resistivity/K^{-1}
Pure metals	
Aluminium	4.5×10^{-3}
Copper	6.6×10^{-3}
Silver	4.1×10^{-3}
Zinc	4.2×10^{-3}
Alloys	
Constantan	$\pm 0.02 \times 10^{-3}$
Nichrome	0.1×10^{-3}

The elements of electric heaters are generally made of an alloy called nichrome which (like constantan and manganin) has a high resistivity but a small temperature coefficient. Its main virtue is that it does not deteriorate mechanically or chemically when maintained at a high temperature (e.g., red heat) for prolonged periods of time. Often a heating element is made of a flat strip of the metal rather than of round wire (see Problem 23.3 above), as in an electric toaster. The flat strip form of conductor combines a high resistance with a large surface area for rapid dissipation of heat.

Chapter 24

POWER, ENERGY AND ELECTROMOTIVE FORCE

24.1 Power

When you switch on the headlamps of a motor car, radiant energy streams out forwards and lights the way ahead. The energy flow process was discussed in Section 23.2 in which the term *potential difference* was defined. The rate at which energy is supplied by the battery is the power supplied; likewise the rate at which energy is utilized by the lamps is the power utilized. The word *power* means *rate of conversion of energy.*

The unit of power is the *watt* (W). If we say that a lamp has a power of one watt, we mean that electrical energy is being converted into heat and light at the rate of one joule per second. (See Section 8.5). We can write:

$$1\ \text{W} = 1\ \text{J s}^{-1}$$

Let us see now how the power consumed by a passive electrical device such as a lamp is related to the current flowing through it and the potential difference across it. The current is the *quantity of electric charge flowing through the filament per unit time,* and the p.d. is the *amount of energy delivered per unit charge*: thus, if we multiply together the current and the p.d. we get the *amount of energy delivered per unit time,* which is the *power* delivered to the lamp. Expressed in terms of the units for these quantities, the product is written:

$$\text{J C}^{-1} \times \text{C s}^{-1}$$

which is simplified to J s^{-1} or, by definition, W. The conventional symbol for power is P, and we write the formula

$$\boxed{P = VI} \qquad (24.1)$$

Table 24a Power ratings of some domestic electrical
appliances in watts

Tape recorder	30
Reading lamp	60
Colour T/V receiver	150
Electric blanket	200
Vacuum cleaner	750
Iron*	625–1000
Kettle	2000
Room heater*	2750–3250
Immersion heater	3000
(150 litre tank)	

*Maximum and minimum power ratings are generally
marked on such appliances as these which are fitted
with means of controlling the energy supplied.

We can express the power in terms of the resistance, R, of the device if we use the relationship
$R = V/I$, which defines resistance. Substituting
$V = IR$ into Eq. 24.1 we get $P = I^2R$; alternatively, substituting $I = V/R$ we get $P = V^2/R$.

Table 24a shows the power ratings of some
electrical devices in common use. The three
quantities will, of course, be observed to satisfy
the relationship $P = VI$ in each case.

Multiples and sub-multiples of the basic unit,
the watt, are in common use (see Appendix 1), the
former in the literature of electrical power generation and use, the latter in the world of electronics
and telecommunication.

The power of electrical devices, both domestic
and industrial, is nearly always marked on them.
On a domestic lamp one may find, for example,
the correct operating voltage and the power
indicated thus: 240 V 60 W.

The power output of a generating station is
likely to be specified in megawatts. A typical large
station may have a capacity of 2000 MW.

24.2 Electromotive force

The role of the battery in an electric circuit has
been compared with the role of the pump in a fluid
circuit. It must, however, be much more than this,
for not only is it necessary for the maintenance of
the flow of electric charge but it is the major
source of the energy which is transformed to heat,
light, etc., in the circuit components.

Consider a simple circuit made up of a variable
resistor R, an ammeter, a voltmeter of high

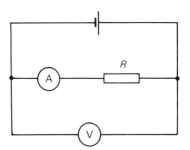

Fig. 24.1 Circuit used to investigate the role of a battery.

sensitivity and resistance and a new $1\frac{1}{2}$ V cell (Fig.
24.1). As R is varied, corresponding readings of
the voltmeter are recorded; see Table 24b.

The rise in the potential difference across the
resistor R is noteworthy; it immediately raises the
question of the fate of the volts 'lost' when the
resistor R has low values.

Table 24b

R/Ω	V/V	I/A
5	1.42	0.28
10	1.48	0.15
20	1.52	0.076
40	1.535	0.038
60	1.54	0.026
80	1.545	0.019
100	1.55	0.016
200	1.56	0.008
1000	1.56	0.0007
5000	1.56	0.0003

The graph of the values of the potential difference across the resistor R and the current
flowing (Fig. 24.2) is a straight line, provided that
the external resistance is not allowed to fall too
low.

The difference between the maximum potential
difference (in this case 1.56 V) and the potential
difference across the resistor is proportional to the
current drawn from the cell. Evidently there is a
further resistor to be considered. We know that
the copper connecting wires themselves can contribute very little to the total resistance of the circuit
so we conclude that the cell itself has some internal
resistance.

Problem 24.1 What is the internal resistance (r)
of the $1\frac{1}{2}$ V cell used in this experiment?

Since the graph (Fig. 24.2) is a straight line, the potential difference recorded by the voltmeter is proportional to the current drawn from the cell. The difference between the highest and the lowest values of the p.d. recorded during the experiment (the 'lost' volts) is (1.56 − 1.42) volts, i.e., 0.14 V.

This p.d. of 0.14 V is responsible for the current flowing through the cell and so

$$\text{internal resistance } r = \frac{0.14\text{ V}}{0.28\text{ A}} = 0.5\ \Omega$$

This result suggests that a fresh cell could give a short-circuit current of 3 A. This value will fall away very quickly as the cell polarizes.

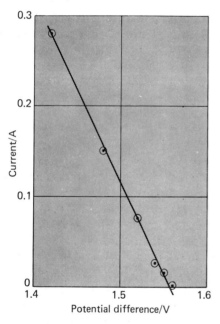

Fig. 24.2 Variation of the current in circuit 24.1 with potential difference for a range of values of resistance R.

Problem 24.2 It was claimed that the resistance of the copper connecting leads contribute very little to the total resistance of the circuit. What is the order of magnitude of the resistance of the leads in an experiment such as this one? The resistivity of copper is $1.6 \times 10^{-8}\ \Omega$ m

In such an experiment perhaps 2 m of connecting wire were used. Typically, such wire is made of stranded copper with a total cross-sectional area of 0.75 mm² (which is 0.75×10^{-6} m²). Applying Eq. 23.4 we have

$$\text{Resistance of the leads} = \frac{1.6 \times 10^{-8} \times 2}{0.75 \times 10^{-6}}\ \Omega$$

$$= 4 \times 10^{-2}\ \Omega$$

Order of magnitude of resistance of the leads is $10^{-2}\ \Omega$.

_____

There seems to be no way in which the voltmeter used in this experiment can be used to measure how energy is utilized or dissipated within the cell itself. We would need to devise another experiment in which the heat produced in the cell could be measured to discover this and, in principle, this could be done. We would then be able to state, exactly, the maximum total energy per unit charge which the complete circuit (cell included) can utilize or dissipate. This maximum voltage we may call the *electromotive force* (e.m.f.) of the cell. We could, in principle, determine it by immersing the complete circuit in a calorimeter and measuring the current and the rate of production of heat.

Now, such a cell can provide energy to a simple circuit for a long period of time without change. The energetic conditions are steady. The amount of energy per unit charge generated by the source is equal to the total amount of energy per unit charge which is utilized in the whole circuit. So we see that the electromotive force of the cell is equal, also, to the energy per unit charge which is generated within it.

Let us now further develop these energetic aspects of the electric circuit in which a steady direct current is flowing. Consider a battery (or, indeed, any suitable electrical source) with e.m.f. E and internal resistance r connected to an external resistance R. Let the current be I.

The e.m.f. E is equal to the external energy transformed per coulomb plus the energy transformed in the battery itself per coulomb. We can measure the first of these.

During an interval of time t, It units of charge pass and the energy transformed in the external circuit is I^2Rt whilst the energy dissipated in the battery is I^2rt.

From the definition of electromotive force as the energy per unit charge utilized in the entire current

$$E = \frac{\text{total utilized energy}}{\text{charge passing}}$$

$$= \frac{I^2Rt + I^2rt}{It} \qquad (24.2)$$

$$= I(R + r)$$

$$\therefore I = \frac{E}{R + r} \qquad (24.3)$$

Multiplying both sides by R

$$IR = E\frac{R}{R + r}$$

This is the voltage V which we can measure with a voltmeter. So

$$V = E\frac{R}{R + r}$$

Dividing both numerator and denominator of the fraction by R we have

$$V = E\frac{1}{1 + \dfrac{r}{R}}$$

As R tends to an infinite resistance, so the p.d. V tends towards E. This is the justification for the direct measurement of e.m.f. by the use of a high resistance voltmeter or a potentiometer.

Equation 24.2 may be written

$$EIt = I^2Rt + I^2rt$$

or

$$EI = I^2R + I^2r$$

which may be written as

$$EI - I^2R - I^2r = 0$$

In this form we are using a sign convention: we are allotting a positive sign to terms representing electrical power applied to the circuit, and a negative sign to terms representing electrical power utilized in the circuit and in the battery. The left-hand side of the equation rewritten in this form then represents the net gain of electrical power in the circuit and battery, which is zero. The importance of using this sign convention becomes apparent in Section 24.6, where one has to be careful to distinguish regions of a circuit in which power is supplied from regions where power is utilized.

24.3 Energy changes

The dry cell used in our experiment will eventually cease to provide a useful e.m.f. and, at this point, is discarded. It is not rechargeable. The constituents of this cell (zinc case, ammonium chloride paste, a mixture of powdered carbon and manganese dioxide, carbon rod) are involved in a chemical reaction which makes it possible to transfer some energy to the circuit. We would hardly expect all the available energy to be transferred usefully: some causes the temperature of the cell to rise when in use. The e.m.f. (which is 1.5 V) describes the maximum transfer of energy to useful forms. When supplying a current to an external circuit the ratio of the energy transferred usefully to the total energy available falls and this is revealed as a fall in the external p.d.

Other dry cells use different constituents; these include the alkaline–manganese cells and the mercury cells. The latter type is particularly useful in applications which require a long life and a high ratio of energy to volume (e.g. in watches, calculators, pace-makers).

Certain cells are reversible and chargeable. The most useful of these include the lead accumulator (so commonly used in car batteries), the nickel–iron or Nife cell and the nickel–cadmium or Ni–cad cell. In such cells, when in use, the e.m.f. is in the same direction as the current and energy is being converted to what we may call an electrical form. But, when on charge, the e.m.f. is in opposition to the current, and electrical energy is being transformed to another form which we may think of as chemical. In this case, too, such other energy sources as the exchange of energy with the surroundings resulting from temperature gradients are involved.

Problem 24.3 A 3 V torch battery has an internal resistance of 0.8 ohm. It is connected to a lamp which draws a current of 0.25 A.

a) Calculate the p.d. across the lamp.

The battery is then used to supply a number of lamps connected in parallel, and they take a total current of 1.25 A from the battery.

b) Calculate the p.d. across the lamps in this case.

a) From Eq. 24.3

p.d. across load $= IR = E - Ir$

$= 3 - 0.25 \times 0.8$ V

$= 3 - 0.2$ V

$= 2.8$ V

b) Similarly,

p.d. across load $= 3 - 1.25 \times 0.8$ V

$= 3 - 1.0$ V

$= 2.0$ V

In Problem 24.3 case (a), only 0.2 V is utilized within the battery; in case (b) 1.0 V. The greater the current taken from the battery, the greater is the utilization of energy within the battery. In practice this is not likely to be a serious problem because one would not normally take such a large current as in case (b) from such a small battery because the useful lifetime of the battery would be inconveniently short in this situation: for, in general, the smaller the chemical cell the larger is its internal resistance. It is partly a consequence of the fact that batteries have internal resistance that torch bulbs rated at 2.5 V are used with dry batteries whose e.m.f., when new, is 3.0 V. The small battery of a pocket torch may have an internal resistance of as much as 5 Ω, and if such a battery is supplying a current of 0.1 A, the 'lost volts' will be 0.5 V, which in this example would be just right. However, internal resistance is not the only consideration in this practical situation: there is also the fact that, during use, the e.m.f. of a battery decreases.

We now have an important theoretical point to clarify. Looking at the circuit of Fig. 24.3 we see that we have two principal elements in the circuit, a device which converts energy from some other form into electrical form (the generator or cell), and a device which converts the energy from electrical into some other form (the load). In each case the rate of energy conversion – the power – can be

calculated from the p.d. across the device and the current flowing through it. But can these measurements of p.d. and current reveal whether power is being supplied by the device or utilized by it? They can, provided that one takes into account the polarity of the p.d. across the device and the direction of the current. In the circuit of Fig. 24.3 positive electric charge flows in the direction of the arrows. This charge, when it moves through the generator, goes from a region of lower potential to a region of higher potential and thus gains energy. In the generator, then, the direction of current flow and the direction of increasing potential are the same. It is convenient to think of the potential rise across the generator, going from its negative to its positive terminal. Thus, the rule is that if the current and potential rise are in the same direction for an electrical device, then the device is supplying electrical energy to the circuit. Consider now the load. It is a passive device, and in it the electric charge loses potential energy as it flows. The current flow is in the opposite direction to the potential rise, and here electrical power is utilized. These ideas are important when considering the circuit of Fig. 24.4. Here we have a storage battery (accumulator, or re-chargeable battery) being charged from a source of direct current. Here, the direction of current flow in relation to the polarity of the battery looks wrong: but the explanation is that, in the battery here, we are converting electrical energy back into chemical energy. Notice that the potential rise across the battery is in the opposite direction to the current: hence (by our rule) the battery is utilizing electrical energy. The internal resistance of the battery, r, is also utilizing electrical energy but, of course, not converting it into chemical form! We can write the equation for conservation of energy in the circuit thus:

$$VI - EI - I^2r = 0$$

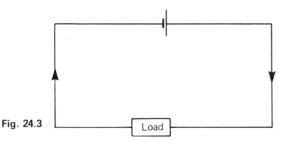

Fig. 24.3

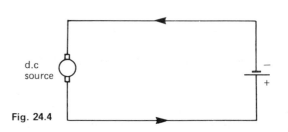

Fig. 24.4

allotting positive signs to the terms representing electrical power *supplied* and negative signs to power *utilized*, using the sign convention explained above and the symbols of Section 24.2.

Problem 24.4 Power loss in the overhead conductors of the National Grid.

The solution of Problem 23.3 gives an estimate of the resistance per kilometre of one type of aluminium-and-steel conductor used on the 132 kV transmission system of the National Grid: to 1 significant figure it was 0.2 Ω km^{-1}. A calculation accurate to 2 significant figures would have given 0.16 Ω km^{-1}.

Suppose that the loading is such that the effective value of the current (that is, the equivalent value of steady direct current) flowing in one of these conductors is 400 A.

a) Calculate the voltage drop along a 20 km run of cable.
b) Roughly what percentage of the working voltage of the system is this?
c) Assuming that the voltage at the supply end is actually 132 kV, and the effective value of the current is still 400 A, what is the power input to the system, to the nearest megawatt?
d) From the answer to (a), and knowing the current, calculate the power dissipated as heat in one conductor, to the nearest tenth of a megawatt.
e) What percentage of the power input is lost as heat in one conductor?

You will see from these calculations, if you have done them correctly, that the power loss is small in comparison with the input power. It is small because a very high voltage (132 kV) is being used. Suppose that the voltage were one tenth of this value: we shall now consider the consequences of this.

f) If the voltage is to be 13.2 kV, what must the current be, in order to have the same power input to the system? (Remember, power = voltage × current.)
g) What is the voltage drop along a 20 km run of cable now?
h) Compare the answer to (g) with the new supply voltage: they are almost the same. This means we have lost nearly all the volts along

the line! This reasoning need be pursued no further – you should now see the full implications of this argument.

a) The resistance of a 20 km run of cable is 20 × 0.16 = 3.2 Ω. The voltage drop is given by $V = IR$, whence $V = 400 × 3.2$ V = 1280 V.
b) The voltage drop is 1280/132 000 × 100 V which is roughly 1% of the working voltage.
c) The power input is 132 000 × 400 = 52 800 000 W = 53 MW to the nearest megawatt.
d) In one conductor the power dissipated = 1280 × 400 W = 0.5 MW to the nearest tenth of a megawatt.
e) The percentage power loss is 0.5/53 × 100, that is, roughly 1% per conductor.
f) If the voltage is reduced to one tenth, then the current must be increased ten times (that is, to 4000 A) if the power input is to be the same, because power = the product of voltage and current.
g) The voltage drop, given by $V = IR$, will be ten times what it was before because the current I has been increased ten times (R, we assume, remains unchanged): thus the voltage drop is now 12 800 V.
h) The new supply voltage is 13.2 kV, that is, 13 200 V: so the voltage drop is only slightly less than this. Presumably therefore the power loss in the cable is also only very slightly less than the power input, and the system would be useless in this form. We see from this that a change in the supply voltage by a factor of *ten*, subject to the constraint of keeping the power input unchanged, means that the power loss is altered by a factor of about *a hundred* (from approximately 1% to 100%).

24.4 Matching source and load

The presence of internal resistance in an electrical generator sets a limit on the amount of electrical power which can be delivered by the generator into a load. We shall now see how this limitation arises, for it has important consequences not only in electrical power engineering but in electronics.

When an electric generator is connected to a load, some of the generated power is dissipated in the internal resistance, and hence wasted, and the

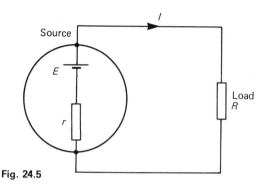

Fig. 24.5

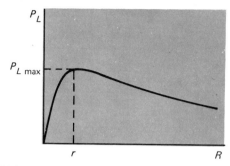

Fig. 24.6 Variation of the power dissipated in a load with the resistance of the load.

rest is delivered to the load. The amount of wasted power depends on the current which is being drawn from the generator, and this current depends on the resistance of the load. Alteration of the load resistance will thus cause a change in the amount of power wasted by dissipation in the internal resistance of the generator. In many practical situations it is both possible and desirable to adjust the value of load resistance so that the greatest possible power is transferred from the source to the load. We shall show here what conditions are required for maximum transfer of power from source to load.

In the circuit of Fig. 24.5 the current is given by

$$I = \frac{E}{R + r}$$

and the power (P_L) dissipated in the load, by

$$P_L = I^2 R = \left(\frac{E}{R + r} \right)^2 R \qquad (24.4)$$

Assume now that the e.m.f. and internal resistance of the source cannot be altered, but that the load resistance is adjustable. What value of R makes the power dissipated in the load a maximum? Figure 24.6 shows how the power, P_L, given by Eq. 24.4, varies with R. It is seen that the power is greatest when the load resistance is equal to the internal resistance of the source. Under these conditions the power wasted (by dissipation in the internal resistance of the source) is equal to the power delivered to the load, and each is equal to $E^2/4r$.

If we wish to find the value of R which makes

P_L a maximum, we must apply a calculus technique and differentiate the expression

$$P_L = \left(\frac{E}{R + r} \right)^2 R$$

$$\frac{dP_L}{dR} = E^2 \frac{d}{dR} \left[\frac{R}{(R + r)^2} \right]$$

$$= E^2 \left[\frac{(R + r)^2 - 2R(R + r)}{(R + r)^4} \right]$$

Equating this to zero gives

$$r^2 - R^2 = 0$$

Hence

$$R = \pm r \qquad (24.5)$$

Since only the result $R = +r$ has physical meaning in this context, we conclude that the condition for P_L to be a maximum is that $R = r$ as indicated in the previous paragraph.

This case is analogous to the transfer of energy when two masses collide elastically. In that case, the maximum transfer occurred when the masses were equal to one another. (See Section 43.2.)

When a source of electrical energy is connected to a load in such a way that the maximum power is transferred from source to load, we say that *the source is matched to the load*. Matching a source to a load is essential in certain branches of electronics. For example, the aerial of a television receiver behaves like a source of electrical energy having a certain internal resistance, and the aerial input socket on the television receiver functions as if it had a certain resistance between its terminals and it can be regarded as the load when the aerial is connected to it. A usual value for this is 75 Ω.

The power of the 'signal' entering the aerial is so very weak that it is important to get as much of this power as possible to enter the receiver circuit, otherwise the quality of the reception may be reduced. Thus, in a properly designed system, an aerial which functions like a source whose internal resistance is 75 Ω is connected to the receiver.

Another situation in which the matching of source to load is desirable is where very high electrical power is involved. In a large radio transmitting station, for instance, where radio waves are sent out carrying a power of hundreds of kilowatts, any loss of power means appreciable loss of money.

In Chapter 41, on amplification, you will meet several examples in which the power transferred from a source to a load is calculated.

24.5 Kirchhoff's second law

It is necessary now to find out what happens when there are two or more electrical devices in a circuit. Look at the circuit of Fig. 24.7. This circuit has been used earlier in elucidating the behaviour of electric current: now we shall use it to throw more light on the concept of p.d. To measure the p.d. across the lamp L_1 we would connect a voltmeter to points A and B: we can say alternatively that we are measuring the p.d. between points A and B. If we then measure the p.d. between points A and C we find that it equals the sum of the p.d. between A and B and the p.d. between B and C. This result can be expressed in symbols thus:

$$V_{AC} = V_{AB} + V_{BC}$$

Similarly it is found by experiment that

$$V_{BD} = V_{BC} + V_{CD}$$

and, as one would expect (and as must follow logically from the previous two results),

$$V_{AD} = V_{AB} + V_{BC} + V_{CD}$$

Another way of expressing these results would be to say that the voltages add up as one goes around a circuit. The same rule is found to be obeyed by the cells when the voltmeter is connected across each cell separately, then across two of the cells, then across all three. One could then express the result thus: the cell voltages when added up equal the lamp voltages added up. But in fact the situation is a little more subtle than this. To clarify the subtlety we need to make voltage measurements on the circuit of Fig. 24.7, using a centre-zero voltmeter. With such a voltmeter, if a cell whose e.m.f. is 1.5 V is connected to the meter with the correct polarity the needle swings to the right of zero, registering + 1.5 V. If the cell is connected with reverse polarity (i.e., with the connections to the meter interchanged) then the needle swings to the left of zero, registering − 1.5 V. Suppose now that we go around the circuit of Fig. 24.7 with our centre-zero voltmeter, starting with the meter's negative terminal connected to point A and its positive terminal to B. The results of a typical set of measurements are shown in Table 24c.

As expected, the negative voltages in the table of results (which represent the p.d.s across the three cells), when added together, equal the sum of the positive voltages (the p.d.s across the three bulbs). We can say, in a sense, that 'the voltage goes up' as we go from A to B around the circuit, and then 'goes down' from D, via E and F, back to A. But our crude language, indicated by inverted commas, is not really precise enough,

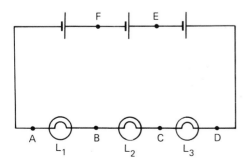

Fig. 24.7

Table 24c

Negative terminal of meter connected to point	Positive terminal of meter connected to point	Voltmeter reading
A	B	+ 1.2
B	C	+ 1.5
C	D	+ 1.4
D	E	− 1.3
E	F	− 1.3
F	A	− 1.5

because the ideas of 'voltage going up' and 'voltage going down' are not exact. The word 'voltage' so far has been an alternative word for 'potential difference'. The word 'difference' is important to appreciate here: there is a certain potential at point A, and another potential at point B, in the circuit of Fig. 24.7, and the difference in potential between A and B is 1.2 V. What is more, the potential at B is 1.2 V higher than the potential at A. Alternatively one can say that the potential at A is 1.2 V lower than the potential at B (not surprisingly, perhaps). Thus, in going from A to D, we find that the potential increases by a certain amount; and in going from D via the cells back to A we find that the potential decreases by the same amount. We can, if we choose, call the potential at point A zero. If we do this, then we can tabulate the potentials at different points in the circuit as in Table 24d. From one point of view the results may seem obvious: one does not expect to arrive back at point A, having gone around the circuit, to find a potential there different from what it was when one started the journey at point A!

Table 24d

Point	Potential/V
A	0
B	+ 1.2
C	+ 2.7
D	+ 4.1
E	+ 2.8
F	+ 1.5
A	0

The results of this experiment are an illustration of *Kirchhoff's second law*, which can be expressed as follows:

■ 'The sum of potential rises (taken as positive) and potential drops (taken as negative) along any closed path is zero.'

The rises occur in suitably connected active components; the drops in passive ones, if one goes round the circuit with the current.

We have considered only a simple series circuit, constituting a single closed path: but Kirchhoff's second law can be applied to any circuit, however complex. Figure 24.8 shows a bridge circuit (this has various applications, as we

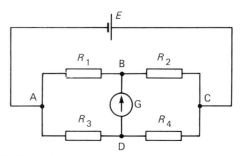

Fig. 24.8 A bridge circuit. There are seven distinct closed paths in the circuit.

shall see). There are several closed paths which can be followed round in this circuit:

a) $A - R_1 - B - G - D - R_3 - A$
b) $B - R_2 - C - R_4 - D - G - B$
c) $A - R_1 - B - R_2 - C - R_4 - D - R_3 - A$
d) $A - E - C - R_2 - B - R_1 - A$
e) $A - E - C - R_4 - D - R_3 - A$

Each of these paths has been followed around clockwise, but it could just as well have been followed around anticlockwise. See whether you can trace two more, perhaps less obvious, closed paths in this circuit. If we were to go around any one of these closed paths with a centre-zero voltmeter in the manner already described, we would find that Kirchhoff's second law is obeyed.

Kirchhoff's second law often enables us to write equations relating the potential differences across the several devices in a complex electrical circuit, and together with Kirchhoff's first law (the current law) it provides a powerful means of analysing complex circuits.

Before we proceed to seeing how simple circuit problems can be solved, we must consider one further fact about circuits in general. In the simple series circuit of Fig. 24.7 the current is flowing clockwise around the circuit. In the lamp L_1 the current flow is therefore from right to left, from B to A, and we note also that the potential at B is higher than at A. Similarly for the other two lamps: in 'going across' a lamp in the same direction as the current flow, we always experience a drop in potential. And this is a general fact; that the current flow through any passive electrical device (that is to say, any device in which electrical energy is converted into some other form) is from the high potential side of the device to the low potential side. But in going across a cell (and neglecting for the moment any internal resistance

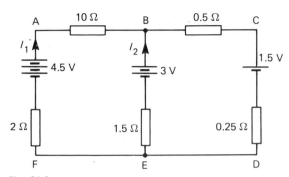

Fig. 24.9

it may have) from its negative terminal to its positive terminal, there is a potential rise; another way of expressing this fact is to say that the positive terminal of the cell is always at a higher potential than the negative terminal. Such a potential rise, associated as it is with an active driving source, is the e.m.f. of the source.

24.6   Solving circuit problems

We shall now apply the laws and the ideas discussed above to the solving of a not-too-complex circuit problem.

The problem we shall set ourselves is to determine the current in each part of the circuit of Fig. 24.9. The small resistors just below the cells represent the internal resistance of the cells. In the diagram we first indicate the assumed directions of the current. In practice, it does not matter whether these directions are the correct ones or not. If we guess wrongly, the current will emerge with a negative sign. We shall start by drawing an arrowhead above the 4.5 V cell and labelling it I_1; which is one of the currents to be calculated. Then, above the 3 V cell, we indicate the current I_2. Now we apply Kirchhoff's first law to the junction point B: the current flowing away from B, towards C, must be equal to the sum of the currents flowing into B. Thus the current flowing from B to C must be $(I_1 + I_2)$. Now we apply Kirchhoff's second law, firstly to the closed path A-B-E-F, which will give us one equation containing the unknown quantities I_1 and I_2; and secondly to the path B-C-D-E, which will give us a second equation containing I_1 and I_2. Thus we shall obtain a pair of simultaneous equations which can then be solved to yield the values of I_1 and I_2.

Consider first the path A-B-E-F. We shall write potential rises as positive quantities and potential drops as negative. Going from A to B we are travelling in the same direction as the current, so we have a potential drop which can be written $-10I_1$. From B to E we have firstly a potential drop as we go across the cell from its positive to its negative terminal: thus we write this potential drop as -3 V. In going down across the internal resistance of the middle cell we are going in a direction *opposite* to that of the current, and so we have a potential rise: this will be written $+1.5I_2$. From E to F we have a direct connection with (we assume) no resistance, so there is no change in potential. From F towards A, across the internal resistance of the left-hand cell, we are going in the same direction as the current, so here is a potential drop, written $-2I_1$. Finally, in returning to A across the cell, we have a potential rise of $+4.5$ V. Now, Kirchhoff's second law says that the sum of potential rises and potential drops along any closed path is zero: so we can write, for the path A-B-E-F:

$$-10I_1 - 3 + 1.5I_2 - 2I_1 + 4.5 = 0$$

And for the path B-C-D-E (you should work through this equation term by term):

$$-0.5(I_1 + I_2) - 1.5 - 0.25(I_1 + I_2) \\ - 1.5I_2 + 3 = 0$$

These two equations can be simplified and rewritten as follows:

$$-12I_1 + 1.5I_2 + 1.5 = 0$$
$$-0.75I_1 - 2.25I_2 + 1.5 = 0$$

If we multiply the upper equation by 1.5, then the equations can be added together and the terms in I_2 will disappear:

$$-18I_1 + 2.25I_2 + 2.25 = 0$$
$$-0.75I_1 - 2.25I_2 + 1.5 = 0$$

Adding the equations together, we obtain

$$-18.75I_1 + 3.75 = 0$$

therefore

$$I_1 = \frac{3.75}{18.75} \text{ A}$$

$$= 0.2 \text{ A}$$

Substituting this value for I_1 in one of the

equations above containing I_1 and I_2 enables us to find I_2, and the result is

$$I_2 = 0.6 \text{ A}$$

The current $(I_1 + I_2)$ flowing from C to D, via the 1.5 V cell, is thus 0.8 A. You will note that we have here a rather curious situation: the 1.5 V cell, if it were operating normally in a circuit, would cause current to flow in a direction opposite to what we have here. In this particular circuit it is, as it were, being overpowered by the effect of the other two cells. This is a situation which can happen in practice. What really happens is that whereas the 4.5 V and 3 V cells are being dis-charged (or more precisely, chemical energy is being converted into electrical energy) in the 1.5 V cell the reverse process is happening, with electrical energy being converted back into chemical energy. This is called *charging*. If the 1.5 V cell is a rechargeable cell, e.g., a lead–acid accumulator or a Nife cell, then indeed it *will* be charged by this process: but an ordinary dry (Leclanché-type) cell is not ordinarily rechargeable since the reaction is not fully reversible.

Having computed the currents in the circuit we can now determine the potential difference between any two points we choose. For example, the p.d. between B and C is simply the product of the current in the resistor and the resistance: $0.8 \times 0.5 \text{ V} = 0.4 \text{ V}$. The potential at B is higher than at C because the current is flowing from B to C. Suppose we wish to find the p.d. between B and E, and to determine which point is at the higher potential. In going from B downwards across the internal resistance of the cell, against the direction of the current, we find a rise in potential of an amount $0.6 \times 1.5 \text{ V} = 0.9 \text{ V}$. So, *down* by 3 V and then *up* by 0.9 V means a net drop of 2.1 V. Thus the point B is at a potential of 2.1 V higher than point E.

Now you should check your understanding of the ideas, by trying Problem 24.5, and carrying out similar calculations for other pairs of points in the circuit of Fig. 24.9. It is a good idea, when solving the problem, to copy the circuit diagram, choose one point in the circuit and make it the zero of potential (say point F), and then to label the other points in their circuit with their potentials when they are calculated.

Problem 24.5

a) Find the p.d. between points A and F. (See Fig. 24.9.) Which of the two points is at the higher potential?
b) Find the p.d. between C and D. Which of these two points is at the higher potential?
c) Calling the potential of F, E, and D zero, label the points A, B, and C with their appropriate potentials. Hence calculate the current in AB and in BC: these results should check with the values of current already calculated.

Problem 24.6

a) Suppose you have a 12 V d.c. supply, and from this you wish to light two lamps, one rated at 6 V, 0.05 A and the other at 2.5 V, 0.2 A. Their voltage and current ratings are different. Because their current ratings are different they cannot be connected simply, in series. The circuit shown in Fig. 24.10 could be used. Using Kirchhoff's two laws, find the values of the two resistors needed, R_1 and R_2.
b) If the 6 V lamp blows, then it becomes an open circuit and takes no current. Try to predict, by calculation, what happens to the 2.5 V lamp. You may assume that the resist-ance of the 2.5 V lamp does not alter: then, afterwards, you may need to take into account that in fact the resistance will increase if the lamp glows more brightly.
c) Consider what would happen if the 2.5 V lamp were to blow: would the 6 V lamp's behaviour alter? Again, first assume that the resistance of the 6 V lamp does not alter.

d) Let us label some of the points in the circuit, as shown in Fig. 24.11. Now we know that the p.d. between C and D must be 6 V: hence also the p.d. between B and E. Thus the p.d.

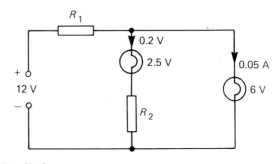

Fig. 24.10

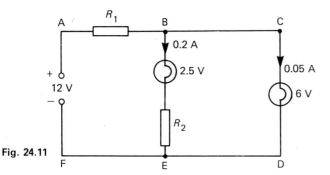

Fig. 24.11

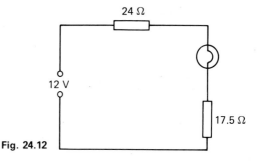

Fig. 24.12

across R_1 must be $(12 - 6)$ V = 6 V. The current flowing from A to B divides at junction B, hence this current is $(0.2 + 0.05)$ A = 0.25 A. Thus we can find R_1, since

$$R_1 = \frac{6}{0.25} \frac{V}{A} = 24 \, \Omega$$

Now to find R_2. The p.d. between B and E is 6 V. But we must have 2.5 V across the lamp, so the p.d. across R_2 must be $(6 - 2.5)$ V = 3.5 V. The current flowing through it is of course 0.2 A. Hence

$$R_2 = \frac{3.5}{0.2} \frac{V}{A} = 17.5 \, \Omega$$

b) If the 6 V lamp blows, then the circuit reduces to the one shown in Fig. 24.12. Assuming that

the resistance of the 2.5 V lamp (which is $2.5/0.2$ = 12.5 Ω) does not alter, then the total circuit resistance is now $(24 + 12.5 + 17.5)$ Ω = 54 Ω, hence the current is $12/54$ = 0.22 A. This implies a slight increase in the brightness of the lamp: 10% over-running, but since its resistance will increase as a result of the higher temperature, the current will not be quite so large as the value calculated. The precise value is indeterminate.

c) If the 2.5 V lamp blows, then we have simply the 6 V lamp, whose resistance in normal operation is $6/0.05$ = 120 Ω, in series with R_1 across the supply. The current we thus calculate as $21/(24 + 120)$ = 0.083 A. Therefore, we might well expect this lamp to blow also.

Chapter 25

SOME USEFUL CIRCUITS

25.1 Combining resistances

When several resistors are connected together in series (see Fig. 25.1) the resistance of the combination is equal simply to the sum of the individual resistances. This can easily be proved as follows. Consider these resistors, R_1, R_2, R_3, connected together in series, and connected to some source of e.m.f. so that a current, I, flows through them. We already know that the current in each resistor must be the same. If the p.d. between the ends of the combination is V, then (from energy considerations)

$$V = V_1 + V_2 + V_3 \qquad (25.1)$$

where V_1, V_2, V_3, are the p.d.s across the individual resistors R_1, R_2, and R_3. But, by the definition of resistance

$$R_1 = \frac{V_1}{I}, \quad R_2 = \frac{V_2}{I}, \quad R_3 = \frac{V_3}{I}$$

Dividing Eq. 25.1 by I throughout we get

$$\frac{V}{I} = \frac{V_1}{I} + \frac{V_2}{I} + \frac{V_3}{I}$$

but V/I is the resistance of the whole combination, which we can call simply R. Thus

$$\boxed{R = R_1 + R_2 + R_3} \qquad (25.2)$$

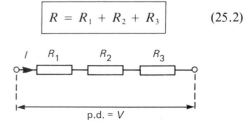

Fig. 25.1 Resistors in series.

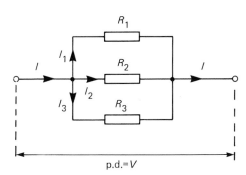

Fig. 25.2   Resistors in parallel.

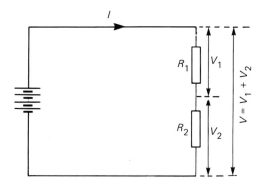

Fig. 25.3   A simple potential divider.

If three resistors are connected in parallel, as in Fig. 25.2, then the resistance of the combination is given by a formula different from the one above. We derive it as follows. Calling the currents in each of the three resistors I_1, I_2, and I_3, we know that the current entering the combination and leaving it, at the junction points, is given by

$$I = I_1 + I_2 + I_3 \qquad (25.3)$$

But, as follows from the definition of resistance,

$$I_1 = \frac{V}{R_1}, \quad I_2 = \frac{V}{R_2}, \quad I_3 = \frac{V}{R_3}$$

And if we call the resistance of the combination simply R, then $I = V/R$. Hence, substituting in Eq. 25.3 we get

$$\frac{V}{R} = \frac{V}{R_1} + \frac{V}{R_2} + \frac{V}{R_3}$$

and dividing this throughout by V we obtain

$$\boxed{\frac{1}{R} = \frac{1}{R_1} + \frac{1}{R_2} + \frac{1}{R_3}} \qquad (25.4)$$

which we can call a reciprocal addition law, for resistances in parallel.

A knowledge of these two laws, for resistances in series and for resistances in parallel, is useful to anyone building electronic circuits. Suppose you require a 1 kΩ resistor for some particular purpose, but you have not got one. However, you find that you have a 330 Ω and a 680 Ω resistor, each one marked with a silver ring indicating that it has a tolerance of 10%. Then, connecting the two resistors in series, you know that you have now a resistance of (330 + 680) Ω ± 10%, that is, 1010 ± 101 Ω. If a 10% tolerance resistor is good

enough for your purpose, then this combination is good enough, because the total resistance could lie anywhere in the range 909 Ω to 1111 Ω. Similarly, if you wanted a resistance of 500 kΩ, but had not got this particular value, or anything near it, in your stock, you could make do with a pair of 1 MΩ resistors connected in parallel.

25.2   The potential divider

The purpose of a 'voltage dropper' is to split the supply voltage into two parts: one part is the voltage across some electrical device – the device's correct operating voltage – and the other part is the voltage drop across a resistor. It is often desirable to divide a supply voltage into two or more (not necessarily equal) parts by means of two or more resistors connected in series across the supply. The simplest example of this potential divider is shown in Fig. 25.3. Here we have two resistors connected in series to a battery. We know that the sum of the p.d.s across R_1 and R_2 must equal the p.d. across the battery terminals, V.

Suppose the current in the circuit is I. Then since

$$V = V_1 + V_2$$
$$V = IR_1 + IR_2$$

and so

$$I = \frac{V}{R_1 + R_2} \qquad (25.5)$$

The p.d. across R_1 is V_1 which is equal to IR_1. Substituting from Eq. 25.5

$$V_1 = \frac{V}{R_1 + R_2} R_1 = \frac{R_1}{R_1 + R_2} V \quad (25.6)$$

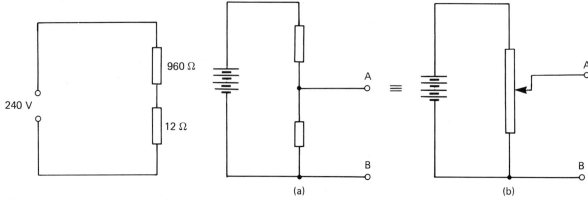

Fig. 25.4   **Fig. 25.5** The principle of the potentiometer

Similarly the p.d. across R_2 is V_2 and

$$V_2 = \frac{R_2}{R_1 + R_2} V$$

Therefore the ratio

$$\frac{\text{p.d. across } R_1}{\text{p.d. across } R_2} = \frac{\dfrac{R_1}{R_1 + R_2} V}{\dfrac{R_2}{R_1 + R_2} V}$$

$$= \frac{R_1}{R_2} \qquad (25.7)$$

Thus the p.d. provided by the battery has been divided in the ratio $R_1:R_2$ across the two resistors. By choosing suitable values of R_1 and R_2 we can obtain any voltage we wish, between zero and V, across one of the resistors.

This analysis assumes that *no* current is drawn by the circuits which may be connected across R_1 and R_2. Problem 25.2 deals with a case in which currents are drawn from the system.

Problem 25.1 In Fig. 23.6 two lamps, one marked 240 V, 60 W and the other 3 V, 0.75 W were connected in series across the 240 V mains supply. Both lamps lit and appeared to be normal. We can now see that they constituted a potential divider. What was the potential difference across each of the two lamps?

When operating normally each lamp takes a current of 0.25 A. It follows that the resistance of the mains lamp was 240/0.25 = 960 ohm and the

resistance of the other lamp was 3/0.25 = 12 ohm. Figure 25.4 shows the system.

Applying Eq. 25.6, the p.d. across the main lamp was

$$\frac{960}{960 + 12} \times 240 \text{ V}$$

or 237 V; the p.d. across the low voltage lamp was (12/972) × 240 V or 2.96 V. No wonder they appeared to be lit normally!

If a continuously variable voltage source is required, the pair of fixed resistors (Fig. 25.5a) can be replaced by a *potentiometer*: this is basically a resistor to which a sliding contact (a *wiper*) has been added. If the potentiometer is made from a length of resistance wire (often coiled around a former), then we have a linear potentiometer, so called because the resistance between the wiper and one end of the resistance wire is directly proportional to the distance between the wiper and that end. Hence, in Fig. 25.5b the p.d. between the points A and B is directly proportional to the distance of the wiper from the lower end of the resistance track. You may wonder what determines the value of resistance which a potentiometer should have. This depends on the particular application for which the potentiometer is being used. If the resistance is small, a large current is drawn from the battery, and this may be undesirable (one should bear in mind that the current flowing in the resistor heats it, and this energy, which must be dissipated, is wasted). If, on the other hand, the resistance of the potentiometer is large, then although only a small current is taken from the battery, it is impossible to take much current from the terminals A and B. However, in many applications of the potential divider, the current taken from the terminals A and B is much smaller than the current flowing through the resistance track: under these conditions no difficulty arises in using it to provide any desired voltage within the range from zero to the full battery voltage.

Potentiometers are available in types ranging from the familiar laboratory bench type, handling currents of a few amperes, to the small wirewound types so familiar as volume controls in radio sets. They may be straight, they may be helical. They may be wire wound; they may have carbon tracks. They may be fitted with turn-counting dials which enables the potentiometer to be calibrated.

Problem 25.2   If the e.m.f. of the battery in Fig. 25.5 is 12 V and the potentiometer has a total resistance of 15 Ω with a maximum current rating of 3 A, how can it be used to light a 6 V, 3 W lamp fully?

The problem is to determine the resistances above and below the potentiometer slider as shown in Fig. 25.5b.

The lamp requires 0.5 A and, when fully lit, has a resistance of 12 Ω. We may redraw the circuit as in Fig. 25.6.

The equivalent resistance of the lamp and R_2, which are in parallel, is given by

$$\frac{1}{R_e} = \frac{1}{12} + \frac{1}{R_2}$$

whence

$$R_e = \frac{12R_2}{R_2 + 12}$$

Since we require 6 V across R_e (and therefore across the lamp), 6 V must also appear across R_1.

The current is the same in R_1 and in R_e, so the two resistances must also be the same.

$$R_e = R_1 = \frac{12R_2}{R_2 + 12}$$

And we know that

$$R_1 + R_2 = 15$$

Combining the two equations we find that

$$15 - R_2 = \frac{12R_2}{R_2 + 12}$$

This reduces to the equation

$$R_2^2 + 9R_2 - 180 = 0$$

The roots of this quadratic equation are $+9.65$ and -18.65. The latter can have no meaning in this situation. So the slider must be placed so that the resistance R_1 is $(15 - 9.65)$ ohm and that the resistance R_2 is 9.65 ohm.

The current flowing in the section R_1 will be $6/5.35 = 1.12$ A; in the section R_2 it will be $6/9.65 = 0.62$ A; and in the lamp the current will be $1.12 - 0.62 = 0.5$ A as required.

It would be simpler to use a voltmeter and to adjust the potentiometer by trial and error!

One important use of a two-resistor potential divider is to supply the correct bias voltage to the base of a transistor, e.g. the one used in the amplifier module discussed in Section 41.1. Figure 25.7 shows how this may be done. Typical values

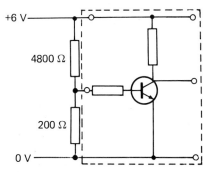

Fig. 25.7   The use of a potential divider to provide a bias voltage to the base of a transistor.

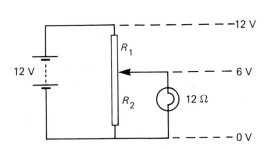

Fig. 25.6

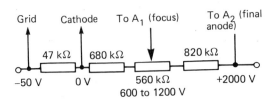

Fig. 25.8 A potential divider which might be used to supply the appropriate potentials to a cathode ray tube (see Fig. 33.8).

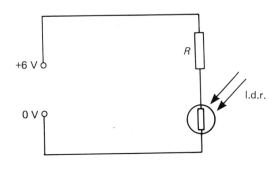

Fig. 25.9

transistor is here maintained at a positive potential with respect to the negative rail whose potential is zero. The p.d. between the base and the negative rail is

$$\frac{200}{4800 + 200} \times 6\,V = 0.24\,V$$

This voltage has been calculated assuming that no current flows between the base of the transistor and the junction of the two resistors. In practice a small current will flow, from the base into the junction of the two resistors, and the effect of this is to reduce the p.d. across the lower resistor of the pair. But again, in practice, this current is generally much smaller than the current flowing in the two resistors already, and so the calculation above is only slightly in error.

A potential divider sometimes consists of a chain of several resistors. Figure 25.8 shows an arrangement used to maintain some of the electrodes of a cathode ray tube at the required operating potentials with respect to the cathode. In this instance the currents taken by the electrodes are very small, and so large values of resistance can be used, drawing very little current from the high voltage supply.

Any point in such a chain may be taken as 0 V and all the other potentials measured with respect to this point. In the example shown the cathode is chosen to be 0 V; the final anode A_2 is then at 2000 V. Equally well, the final anode A_2 could be made 0 V by connecting it to 'earth' and then the cathode is at -2000 V. Whichever is chosen, the potential *difference* between the cathode and the final anode remains the same, i.e.,

$$(2000 - 0) = 0 - (-2000) = +2000\,V.$$

The potential divider principle can be used to convert changes in resistance into changes in voltage. For example, a certain type of photosensitive cell, known as a light-dependent resistor (which we shall abbreviate l.d.r.), has a resistance which depends upon how strongly it is illuminated. Variations in the intensity of the light falling upon the l.d.r. cause variations in its resistance. If the l.d.r. is connected in series with a fixed resistor across a battery (or other voltage source) then variations in illumination give rise to changes in p.d. across the l.d.r. These changes in p.d. can be used to trigger an electronic switching circuit and thus to switch on (or off) some piece of apparatus. Problem 25.3 concerns an l.d.r. and gives you an appreciation of the magnitudes of the variations involved.

Problem 25.3 A typical cadmium sulphide light-dependent resistor has a resistance which is about $100\,\Omega$ when it is strongly illuminated and about $1\,M\Omega$ when in almost total darkness. By using the l.d.r. as one member of a two-resistor potential divider, as shown in Fig. 25.9, it is possible to convert the changes in resistance into changes in voltage for the purpose of operating a voltage-controlled switching circuit. By choosing a suitable value of R one can arrange that the voltage across the l.d.r. is nearly zero when it is illuminated, and very nearly the full supply voltage, 6 V, when unilluminated. Suggest a suitable value for R and explain why it is a suitable value.

For the voltage across the l.d.r. to be nearly zero when the l.d.r. is illuminated we require that its resistance – call it x_{bright} – should be much smaller

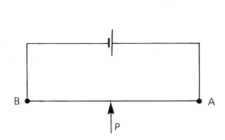

Fig. 25.10   The slide-wire potentiometer.

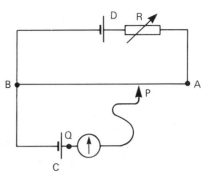

Fig. 25.11   Applying the slide-wire potentiometer to the comparison of e.m.fs.

than R. We also require that x_{dark} should be much greater than R. In terms of ratios, we require:

$$\frac{x_{bright}}{R} \ll 1 \text{ and } \frac{x_{dark}}{R} \gg 1$$

or,

$$\frac{R}{x_{dark}} \ll 1$$

Suppose we simply equate the two ratios which are to be much less than unity, we get

$$\frac{x_{bright}}{R} = \frac{R}{x_{dark}}$$

hence

$$R^2 = x_{bright} \times x_{dark} = 100 \times 10^6$$

hence

$$R = 10^4\,\Omega = 10\,k\Omega$$

The two ratios thus become $1/100$, so that when the l.d.r. is illuminated the voltage across it is only about 1% of the supply voltage, and 99% of the supply voltage when it is dark.

25.3   The slide wire potentiometer

For certain applications, a potential divider is made very easily, using a length of uniform resistance wire AB, connected directly across a battery, as shown in Fig. 25.10. In a simple laboratory version of this, the wire AB is 1 m in length and is mounted directly above a metre scale divided into centimetres and millimetres. A movable contact, P, can be set at any position on the wire. In this way, a potential can be obtained between P and B,

having any desired value in the range from zero to the full p.d. provided by the battery. The usefulness of this type of potentiometer lies in the fact that the p.d. between P and B is directly proportional to the length of the section of wire between P and B so long as no current is taken from P. This is a consequence of the wire's uniformity of thickness which means that the resistance of any section of the wire is directly proportional to that section's length.

The slide-wire potentiometer is used chiefly for comparing voltages, that is to say, finding the ratio of two voltages. It cannot be used for making absolute measurements of voltage. Let us see how the instrument can be used for comparing the e.m.f.s. of two chemical cells. If one of the cells is a standard cell, for example, a Weston Cadmium Cell which has an e.m.f. of 1.0159 V at 20 °C, then the e.m.f. of the other cell can be calculated accurately when the ratio of e.m.f.s has been found by means of the potentiometer. Figure 25.11 shows the arrangement for these measurements. To understand its operation, we shall consider typical numerical values in an experimental situation. D is the battery which is used simply to maintain a steady current flowing in the line AB, and thus a constant p.d. between A and B. For D one can use a lead–acid accumulator, giving a p.d. of about 2 V. The exact value of this p.d. is unimportant: what is essential is that it should remain constant throughout the experiment. We find that when the movable contact, P, is touched to the wire near to B, the needle of the centre-zero galvanometer is deflected strongly one way: let us say this deflection is to the left of zero. When P is close to A, a strong deflection to the right is obtained. As one would expect, there is one point on the wire at which, when P is touched on to that point, the galvanometer shows no deflection. We can interpret these results as follows:

Galvanometer deflection	Interpretation
To the left	Potential at P less than potential at Q, so current flows Q → P.
To the right	Potential at P greater than potential at Q, so current flows P → Q.
Zero	Potential at P the same as at Q, so no current flows between P and Q.

In the third of the situations above, P is at what is often called the balance point. In this condition, since there is no p.d. between P and Q, the p.d. across the section of wire PB must be equal to the p.d. across the cell C_1. Also, in this condition of balance, no current is being drawn from the cell C_1, and the p.d. across it is its open-circuit p.d., that is, its e.m.f.

Let us suppose that the length PB which was obtained with cell C_1 was 60.3 cm, and that C_1 was a small dry cell. We are intending to find its e.m.f. accurately. We now replace the dry cell C_1 with a standard cell C_2, and find the balance-point by the same method as before. The result of doing this, together with the previous result, is tabulated. Because the p.d. across any section of the potentiometer wire is directly proportional to the length of the section and no current is drawn from P, the ratio of the e.m.f.s of the two cells must be equal to the ratio of corresponding lengths, that is,

$$\frac{E_1}{E_2} = \frac{l_1}{l_2} \qquad (25.8)$$

Using the numerical results of Table 25a, we find

$$\frac{E_1}{1.0159} = \frac{60.3}{43.1}$$

whence $E_1 = 1.48$ V. Thus we have found that the e.m.f. of our dry cell is 1.48 V.

With a potentiometer wire one metre long, measurements of length can easily be made to the

Table 25a

Length PB for balance/cm	e.m.f. of cell/V
$l_1 = 60.3$	$E_1 = ?$
$l_2 = 43.1$	$E_2 = 1.0159$

nearest millimetre, so the length-measurements are reliable to 3 significant figures. Since the e.m.f. of a standard cell is reliable to at least 4 significant figures, our calculated e.m.f. is reliable to 3 significant figures. This degree of accuracy is greater than can be obtained with even a very expensive moving-coil voltmeter.

Note that a problem will arise if the e.m.f. of the cell under test is not very different from that of the driver cell D. It is best to arrange matters so that the balance point lies between about 20 cm and 80 cm from the end of the potentiometer wire. This can be done by adjusting rheostat R.

25.4 How good is a voltmeter?

Potential dividers are often constructed so that they reduce the applied voltage to some given fraction. Figure 25.12 shows a simple 'black box' for which the output voltage will always be one third of the input voltage. Voltage reducers of this sort are often called *attenuators*. An attenuator is often used in electronic circuits; for example, connected to the input of an amplifier it can prevent the over-loading of the amplifier by too large an input signal.

By means of external voltage measurements alone we could discover the voltage-reducing property of this particular box. However, the

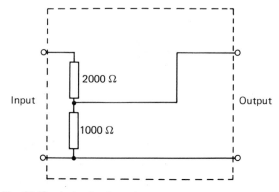

Fig. 25.12 A simple attenuator.

same voltage–reduction ratio could be achieved with any pair of resistors inside the box provided that their resistances were in the ratio 2:1. Clearly to determine the actual value of the resistances we should need to make resistance measurements across each pair of terminals separately. Question 6.9 at the end of the book involves deducing the contents of a similar 'black box' from the results of external measurements.

We now have to tackle a practical difficulty which can arise when making voltage measurements on circuits of the type we are discussing. This difficulty is best illustrated by the results of a simple experiment which present an apparent paradox: by solving the paradox we shall be able to understand why voltage measurements on certain circuits can be misleading.

The situation is this. We have three voltmeters: the best, and most expensive (a), is a large multi-range meter; the second (b) is a cheap laboratory voltmeter; and the third (c) is a small voltmeter salvaged from some old radio equipment. Each of the three reads to 5 volts full scale (the multi-range meter has its selector switch set on the 5 V range). First of all we check the three meters by connecting them in turn to a 4.5 V dry battery (comprising three 1.5 V dry cells). We find that they all read 4.5 V, and so we are satisfied with their accuracy. Then we decide to make some voltage measurements with the 'black box' of Fig. 25.12. We connect a 12 V battery to the input terminals of the box, and measure the output voltage with each of the voltmeters in turn. The results we obtain are as follows:

a) multimeter	4.0 V
b) lab. voltmeter	2.4 V
c) salvaged voltmeter	3.5 V

Here we have discrepancies which are certainly serious! You may have guessed a reason for the disagreement: but consider the results of further measurements. The multimeter seems in order but the two other meters seem to be reading low. Suppose now we connect meters (a) and (b) simultaneously (i.e. in parallel) to the output terminals of the box: we find that they both read 2.4 V. Likewise, when we connect (a) and (c) simultaneously they both read 3.5 V. Then, when (b) and (c) are connected simultaneously we find that they agree with each other, but this time the

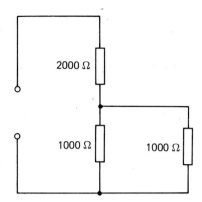

Fig. 25.13

reading is 2.2 V, slightly less than the reading obtained with (a) and (b) together. The truth of the matter is that, although none of the meters is inaccurate, the very act of connecting a voltmeter to the terminals of the box alters the voltage between those terminals. This happens because the voltmeter itself has a finite resistance and when we connect it across the terminals we are thereby adding a resistance in parallel with the output terminals of the box. Let us see exactly how this happens in one particular instance. Figure 25.13 represents the situation when voltmeter (b), which incidentally has a resistance of $1000\,\Omega$, is connected to the output terminals. We shall predict the p.d. which should appear across the terminals. The combined resistance of the lower resistor of the voltage divider and the voltmeter is given by

$$\frac{1}{R} = \frac{1}{1000} + \frac{1}{1000}$$

whence

$$R = 500\,\Omega$$

Hence the p.d. across this parallel pair is given by

$$V = \frac{500}{2000 + 500} \times 12 = \frac{12}{5} = 2.4\ \text{V}$$

We can safely say, therefore, that this particular voltmeter is simply not suitable for measuring voltages in this particular situation. The voltmeter (c), which (we can now reveal) has a resistance of $5000\,\Omega$, does not cause so serious a disturbance of the p.d.s in the circuit, but it is nevertheless unsuitable. The multirange meter (a), however, has a resistance of 10^5 ohm: connecting it in parallel with the $1000\,\Omega$ resistor of the 'black box' produces a combined resistance of very slightly less than 1000 Ω, and the change in voltage caused by this is negligibly small. The meter is nothing like sensitive enough to detect such a change. An ideal voltmeter would be one which possessed infinite resistance. A good approximation to this ideal is the *electronic voltmeter* which is a volt-meter incorporating a d.c. amplifier so that its resistance becomes effectively thousands of megohms, or even greater. In the absence of such a very high resistance voltmeter, what determines whether or not a particular meter is suitable for a particular situation? The foregoing discussion should make the answer clear. If the connection of a voltmeter to a circuit is not to alter appreciably the p.d.s in the circuit, the voltmeter's resistance should be in the order of about a thousand times the resistance of the part (or parts) of the circuit across which the p.d. is to be measured. Thus, a voltmeter like (b), having a resistance of only 1000 Ω, is really only suitable for measuring the p.d. across something which has a resistance of not more than a few ohms at most. It is satis-factory for measuring the p.d. across the terminals of a battery because the internal resistance of a battery whose e.m.f. is 5 V or less is not likely to be greater than an ohm or so. But suppose you want to check the voltages supplied to the electrodes of a cathode ray tube in an oscilloscope, where the voltages are derived from a voltage divider in which the resistors are in the order of one megohm. An electronic voltmeter would be the only suitable instrument in this case.

Having learned to exercise caution in the choice of a voltmeter when making measurements on a live circuit, we may wonder if similar care is needed with ammeters. The problem of choosing a suitable ammeter is generally much simpler to solve. When we use an ammeter to measure the current in a part of a circuit we have first to break the circuit and insert the ammeter into the break. We want the current which flows when the ammeter is present to be no different from the current when the ammeter was absent. If the addition of the ammeter appreciably increases the resistance of the circuit in which it has been inserted, then it will, in general, reduce the current appreciably. So we require simply that the resist-ance of the ammeter should be much less than the resistance of the part of the circuit in which it is to

be inserted. But, typically, the resistance of a cheap ammeter reading up to 5 A will be as little as 0.01 Ω, and this is considerably lower than the resistance of *most* circuits in which one is likely to want to measure current with a cheap ammeter. Problem 25.4 concerns a situation where the measurement of current in a circuit would be difficult because of the finite resistance of the ammeter.

Problem 25.4   Suppose you are going to make measurements of current and voltage on some electrical devices. You have a milliammeter whose resistance is 5 Ω (reading up to 15 mA) and a voltmeter whose resistance is 1 kΩ (reading up to 15 V). Let us say that the device X has a resistance of about 5 kΩ, and that you have connected up the circuit as shown. Calculate the readings of the meters in the situation shown in Fig. 25.14a when the supply voltage is 12 V.

Consider the alternative arrangement shown in Fig. 25.14b. In what way (if any) is it better than the previous circuit? Explain carefully.

For what kind of electrical device would the first arrangement of ammeter and voltmeter be suitable?

Problem 25.5   A typical moving-coil galvano-meter has a resistance of 5 Ω and needs a current of 100 μA to produce full scale deflection. It can be used, in certain limited applications anyway, as a voltmeter for measuring voltages, because the current in the meter, and hence the scale-reading, will be proportional to the voltage applied to its terminals, because the meter itself is an ohmic conductor.

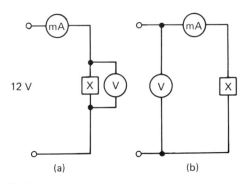

(a)   (b)

Fig. 25.14

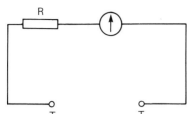

Fig. 25.15

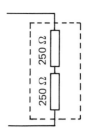 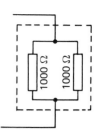

Fig. 25.17

a) Calculate the voltage which, applied to the meter terminals, produces full scale deflection, expressing the answer in millivolts. To convert the meter into a voltmeter capable of reading higher voltages we connect a resistance in series with it, as in the diagram (Fig. 25.15).

b) Calculate the value of R needed to make the meter give full scale deflection when the voltage applied to the terminals is 5 V.

c) In a multi-range voltmeter a chain of resistors is used, as shown in Fig. 25.16. By means of the rotary switch S different resistances can be connected in series with the meter. Suppose we want voltage ranges of 5, 25, 100, 250, and 500 V. Calculate the values of the resistors needed, as shown in the diagram.

a) Using $V = IR$, we get $V = (100 \times 10^{-6}) \times 5 = 0.5 \times 10^{-3} = 0.5$ mV.

b) $5 = (100 \times 10^{-6}) \times (R + 5)$, hence $R + 5 = 5 \times 10^4 = 50\,000\,\Omega$, and so $R = 49\,995\,\Omega$.

c) R_1 is clearly the resistor which gives the 5 V scale, and so it is $49\,995\,\Omega$, as already calculated.

For the 25 V scale the second position of

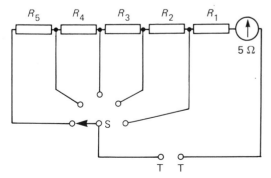

Fig. 25.16   The circuit of a multi-range voltmeter.

the switch is used (going anticlockwise), bringing R_1 and R_2 into the circuit. Hence

$$R_1 + R_2 + 5 = \frac{25}{100 \times 10^{-6}} = 250\,000\,\Omega$$

and since $R_1 + 5 = 50\,000\,\Omega$, this means that R_2 must be $200\,000\,\Omega$ (200 kΩ). To find R_3, we know that $R_1 + R_2 + 5 = 250\,000\,\Omega$, so

$$R_3 + 250\,000 = \frac{100}{100 \times 10^{-6}} = 1\,000\,000\,\Omega$$

Hence $R_3 = 750\,000\,\Omega$ (750 kΩ). In a similar way we find that $R_4 = 1.5$ MΩ and $R_5 = 2.5$ MΩ.

25.5 'Black boxes' as circuit elements

So far we have concerned ourselves with circuit elements and their relationship between potential difference and current. And that is all we need to know in order to use them in a circuit. Generally speaking, we do not need to know how the element produces the V/I relationship. This has led to the idea of the 'black box'; that is, some device with particular input and output characteristics about whose internal workings we need to know nothing. This approach is particularly helpful when extended to modern small-scale integrated circuits. So far as the designer of a circuit incorporating such a device is concerned, only the input and output characteristics are required to be known. Two devices with the same set of characteristics can do the same job but how they achieve this is irrelevant.

Consider this simple example. Figure 25.17 shows two resistive 'black boxes' each offering a resistance of 500 ohm. Internally they are quite different.

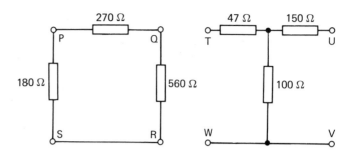

Fig. 25.18

Problem 25.6 Show that the two 'black boxes' (a) and (b) in Fig. 25.18 have the same input and output characteristics.

If we apply a p.d. of V to the terminals P and S, the p.d. which appears across QR is

$$\frac{560}{270 + 560} V \text{ or } 0.67 V$$

If we apply the same potential difference V to terminals T and W the p.d. which appears across UV is

$$\frac{100}{47 + 100} V \text{ or } 0.68 V$$

We might have applied the potential difference to QR and then to UV. In that case the p.d. across PS will be

$$\frac{180}{270 + 180} V \text{ or } 0.4 V$$

and across TW will be

$$\frac{100}{150 + 100} V \text{ or } 0.4 V$$

again.

Looking at the effective resistances across QR and UV we see that these are given by

$$\frac{1}{R_1} = \frac{1}{560} + \frac{1}{450}$$

and by $R_2 = 100 + 150$. So that $R_1 = R_2 = 250$ ohm.

The three tests we have applied show that the two 'black boxes' do have the same input and output characteristics.

25.5 Measuring small differences in resistance: bridge circuits

An ammeter and a voltmeter used together provide an obvious method for the measurement of resistance. But the method has limitations owing to the current drawn by the voltmeter, the p.d. across the ammeter and to the accuracy of the instruments themselves. Normal bench indicators cannot be relied on to an accuracy better than 2% – and then only near to the full-scale deflection. In many modern applications it is necessary to measure small differences in resistance which are in the order of 0.1%; and the ammeter–voltmeter technique is not adequate. However, round about 1843, Charles Wheatstone devised a method which is capable of such measurement.

A network of four resistors is used, together with a suitable galvanometer and a simple power supply (often a dry battery) as in Fig. 25.19a. The

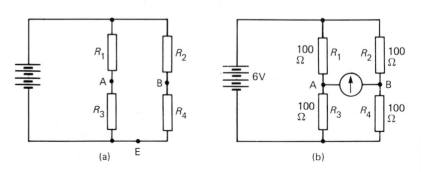

Fig. 25.19 The principle of the Wheatstone bridge.

(a)

(b)

two pairs of resistors constitute two potential dividers in parallel. If all four resistors have precisely equal resistances (100 Ω, say), the potential difference between A and E must be exactly the same as that between B and E. A and B will be at the same potential and a sensitive galvanometer connected between them will show a zero reading (Fig. 25.19b).

Suppose now that the resistance of resistor R_3 increases from 100 to 100.5 Ω (a change of 0.5%).

The potential difference across R_3 becomes

$$\frac{100.5}{(100 + 100.5)} \times 6 = 3.007 \text{ V}$$

So now there is a p.d. between A and B of 7 mV. This will cause an appreciable current (7 μA) to flow through a typical laboratory galvanometer with a resistance of 1000 Ω and graduated 50-0-50 μA. With a more sensitive galvanometer of the light beam type very much smaller changes can be measured.

This arrangement of four resistors in a 'bridge' can solve our problem of detecting and measuring small changes in resistance. Let us therefore consider the general case (see Fig. 25.19a).

We require the potential difference between A and B to be zero. This implies that:

p.d. across R_1 = p.d. across R_3

and that

p.d. across R_2 = p.d. across R_4

So

$$\frac{\text{p.d. across } R_1}{\text{p.d. across } R_2} = \frac{\text{p.d. across } R_3}{\text{p.d. across } R_4}$$

If, when balanced, I_a is the current flowing in R_1 and R_2 and I_b is the current flowing in R_3 and R_4, then

$$\frac{I_a R_1}{I_a R_2} = \frac{I_b R_3}{I_b R_4}$$

therefore

$$\boxed{\frac{R_1}{R_2} = \frac{R_3}{R_4}} \qquad (25.9)$$

We do not need to work with equal resistances: it is the ratios of the resistances which matter, for not only is

$$R_1/R_2 = R_3/R_4$$

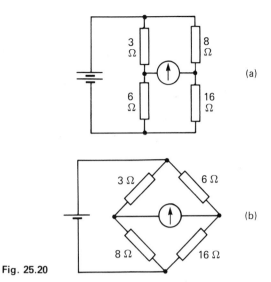

Fig. 25.20

but $R_1/R_3 = R_2/R_4$

Figure 25.20 shows some balancing arrangements.

The Wheatstone bridge is of considerable practical importance, not only for the straightforward comparison of resistances but for the control of servomechanisms. We will examine an application to the measurement of strain in structures with 'strain gauges'.

When a wire is stressed, it lengthens and simultaneously gets thinner. Its electrical resistance therefore increases. There is a further effect which causes the resistance to increase by increasing the metal's resistivity. This is known as the piezo-resistive effect.

A strain gauge is often made of a thin foil of metal which is cut away by etching to leave a long zigzag of the form shown in Fig. 25.21. This is laid on to thin paper or plastic (or insulator) and

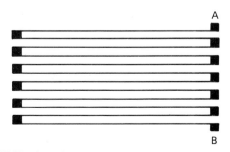

Fig. 25.21   A strain gauge.

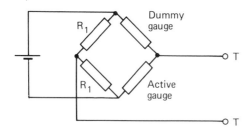

Fig. 25.22 Using a pair of strain gauges in the circuit of a Wheatstone bridge.

electrical connections can be made to the points A and B.

In use, a pair of these gauges is glued to the member (perhaps a load-bearing strut of a bridge). One, the active gauge, is secured so that its length lies along the direction of stress in the strut; the other, the dummy, is secured very close to the active gauge in such a way that it is not affected by the stress. Keeping the two gauges close together ensures that the changes in resistance due to temperature variations will be the same in each gauge.

The pair of gauges is connected into two of the arms of a Wheatstone bridge when measurements are required (see Fig. 25.22).

When the active gauge is unstressed, the voltage developed across the points TT will be zero. Once the active gauge is stressed, its resistance changes, the bridge becomes unbalanced and a potential difference appears across TT. This potential difference is a measure of the strain experienced by the active gauge and by the member to which it is secured.

25.6   Some more circuit elements

Resistors are examples of *circuit elements*: they are self-contained devices which are commonly

used in electrical circuits. So far we have been concerned principally with *what* they do rather than *how* they do it. We have not considered, for instance, the question of what actually happens, on the atomic scale of size, within a resistor when it conducts an electric current.

The conducting behaviour of a circuit element which has only two terminals can often be concisely described by its *voltage–current characteristic* (often referred to simply as *characteristic*). Figure 25.23 shows the voltage–current characteristics of three devices. The characteristics of nearly all circuit elements are dependent upon temperature, as Fig. 25.23c makes clear. This is often of great practical importance with semiconductor devices, as will become apparent later.

The voltage–current characteristic tells a circuit designer what current will pass through the element when a given p.d. is applied to it, or, put another way, what p.d. appears across it when a given current passes through it. The skill of the circuit designer lies in combining different circuit elements so that they produce the required overall function, and often much ingenuity is exercised in counteracting shortcomings in the behaviour of one circuit element by combining its function with another element which annuls, partly if not totally, those shortcomings. For instance, the often very undesirable sensitivity of transistors to changes of temperature can be counteracted by pairing suitable transistors together so that the temperature effects cancel each other.

We shall now discuss a few simple circuit elements and try to show how their functions are useful in practice, and to explain the principles of the designing of circuits.

The first is the *rectifier diode*. The name *rectifier* is given to any device which converts an alternating current into a direct current. 'Diode' is

Fig. 25.23 Current/potential difference characteristics for (a) a thermionic diode, (b) a Germanium diode and (c) a thermistor.

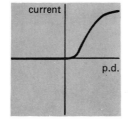

(a) Thermionic diode

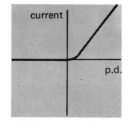

(b) Germanium diode

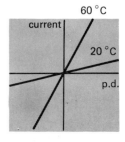

(c) Thermistor

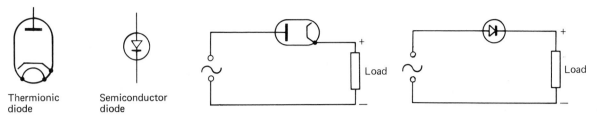

Thermionic   Semiconductor
diode   diode

Fig. 25.24   The symbols for a thermionic and a semiconductor diode.

Fig. 25.25   Half-wave rectification circuits.

a modern word constructed from Greek roots to signify 'two electrodes'. The two main categories of rectifier diode are *thermionic* diodes and *semi-conductor* diodes. In the thermionic diode one electrode, the cathode, has to be heated in order for the diode to function. Its action is described in Section 33.2 where the thermionic emission of electrons is discussed. For low-power rectification of a.c. thermionic diodes are nowadays much less used than semi-conductor diodes. Figure 25.23a, b shows the voltage–current characteristics of typical diodes of each type. Figure 25.24 shows the conventional circuit symbols for the two kinds of diode. In Fig. 25.25 each type is shown incorporated in the simplest 'half-wave' rectifying circuit.

In such a circuit, since the diode conducts only in one half of each cycle of the a.c., the d.c. output across the load is 'half wave' (Fig. 25.26a). To provide 'full-wave' rectification, four similar diodes are arranged in the form of a bridge (Fig. 25.26b).

In the half-cycle in which the input terminal T_1 is positive with respect to T_2, current can flow through diode 2 so that the output terminal T_4 is positive. The external circuit is completed through terminal T_3 and the diode 3.

In the half-cycle in which the input terminal T_2 is positive with respect to T_1, current can flow through diode 4 so that terminal T_4 is again positive. The external circuit is completed through terminal T_3 and diode 1.

An 'ideal' diode would conduct perfectly a current in one direction, that is, it would have zero resistance to current in one direction; and it would not conduct at all a current in the reverse direction. The characteristic of such a diode would therefore be a straight line lying exactly along the voltage axis, the whole length of the negative voltage region, becoming at the origin of the graph a straight line lying precisely along the positive current axis. Although the graph of Fig. 25.23a does not show it, the thermionic diode's characteristic has a slight 'tail' stretching into the negative voltage region, because it still conducts a slight forward current even when small reverse voltages are across it. Any semiconductor diode conducts *slightly* in the *reverse* direction, when reverse voltages are applied, the reverse current being greater for larger reverse voltages. Typically, a small semi-conductor diode which is used to conduct forward currents in the order of one ampere will have a reverse current in the order of several microamperes.

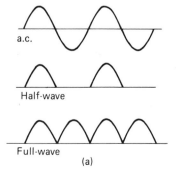

a.c.

Half-wave

Full-wave

(a)

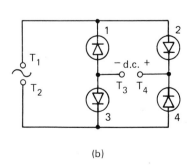

(b)

Fig. 25.26   (a) Wave-forms for a.c., half-wave rectification and full-wave rectification. (b) Circuit to provide full-wave rectification.

The *thermistor* is a semiconductor device whose resistance depends upon temperature in a sensitive way. The characteristic of a typical thermistor is shown in Fig. 25.23c. Unlike a metal conductor, its resistance decreases with rise in temperature, falling by about 75% as the temperature rises from 25 °C to 60 °C. Contrast the temperature sensitivity of a pure metal, such as copper; the resistance of a pure copper wire would increase by approximately 15% for the same rise in temperature.

A thermistor, unlike a diode, has no rectifying property: it conducts equally well a current in either direction through it. Certain types of thermistor are used in electrical temperature-measuring or controlling devices; other types are used to limit surge-currents in circuits.

Another useful semiconductor device is the *light-dependent resistor* (l.d.r.), sometimes called a *photosensitive resistor*, generally made of cadmium sulphide. It behaves as an ordinary resistor, but the resistance depends upon the amount of illumination incident upon it: the greater the illumination, the smaller the resistance. A number of ingenious photo-electronic devices can be designed around an l.d.r., ranging from burglar alarms to exposure meters. Problem 25.3 concerned such a resistor.

Problem 25.6 Errors can arise, when using a moving-coil meter, as a result of changes in resistance of the coil (generally made of copper wire) owing to changes in temperature. Firstly we shall see how great these error can be, and then we shall see how they can be overcome in one particular instance.

a) Consider a moving coil meter whose coil has a resistance of 10.00 Ω at a temperature of 20 °C. The temperature coefficient of resistivity of standard annealed copper is $43 \times 10^{-4} \, K^{-1}$: this is also the *fractional* change in resistance, of any particular copper wire, per degree change in temperature. Use this data to find the increase in resistance when the coil warms from 20 °C up to 40 °C, and hence find the resistance (to four significant figures) at 40 °C. We have to remember, however, that the equation defining the temperature coefficient of resistivity (Eq. 23.5) requires a

knowledge of R_0. This has first to be calculated.

b) Suppose our meter gives full scale deflection for a current of 100 μA. What must the voltage, in microvolts, across the coil be, at 20 °C, to produce full scale deflection?

c) Suppose the same voltage as calculated in (b) is applied to the coil when the temperature is 40 °C: what will the current be now? Express the difference between this reading and the reading at 20 °C as a percentage of the reading at 20 °C. This error is excessive if the meter is a good-quality one in which the true current is guaranteed to be within, say, ±2% of the scale reading.

d) The effect of this temperature-dependent change in resistance can be reduced by connecting in series with the coil a resistor which changes very little with temperature, one made of constantan, for example.

Suppose we have a constantan resistor of 90.00 Ω connected in series with the coil, and that we can neglect any changes in resistance of this resistor due to changes in temperature. The total resistance at 20 °C is thus 100.00 Ω. Calculate the voltage which must be applied to the combination coil + resistor to produce full-scale deflection at 20 °C now.

e) Suppose this same voltage is applied to the meter when the temperature is 40 °C. What current flows now? Express the difference between this current and the current at 20 °C as a percentage of the current at 20 °C, and comment on the result.

The constantan series resistor is generally known as a 'swamping' resistor, and such a resistor is generally used in a meter designed to have a low-resistance shunt connected to it, to convert it into an ammeter, as in Fig. 25.27.

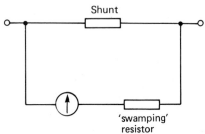

Shunt

'swamping' resistor

Fig. 25.27

a) From Eq. 23.5

$$\frac{R_\theta - R_0}{R_0} = \alpha\theta$$

Therefore

$$R_0 = \frac{R_\theta}{(1 + \alpha\theta)}$$

Putting in numerical values, $R_0 = 9.208\,\Omega$. Repeating the calculation to find R_{40} we have $R_{40} = 10.79\,\Omega$.

b) Using $V = IR$. Voltage needed to produce full-scale deflection $= (100 \times 10^{-6}) \times 10.00 = 10^{-3}\,\text{V}$, or 1 mV.

c) At 40 °C the current is found by using $V = IR$, hence $I = V/R$, thus:

$$I = \frac{10^{-3}}{10.79} = 92.7 \ldots \text{ in } \mu\text{A}$$

The discrepancy is thus $100 - 92.7 = 7.3\,\mu\text{A}$, and expressing this as a percentage we find it to be

$$\frac{7.3}{100} \times 100 = 7\%$$

to nearest percent.

d) With a total meter resistance of $100.00\,\Omega$ we need a voltage of $(100 \times 10^{-6}) \times 100 = 10^{-2}$ V, or 10 mV, to produce full scale deflection.

e) If this voltage is applied to the meter at 40 °C, when the resistance will be $100.79\,\Omega$, the current will be given by

$$I = \frac{10^{-2}}{100.79} = 99.22 \ldots \text{ in } \mu\text{A}$$

Thus the error is now $0.8\,\mu\text{A}$, and consequently a percentage error of 0.8%, or 1% to the nearest percent, and this is generally within the acceptable limits.

Problem 25.7 Figure 25.28 shows the simplest form of meter for measuring resistance directly. The resistance to be measured is connected to the terminals T T, the scale of the milliammeter being calibrated to read the resistance directly. Suppose we have a meter which reads to 15 mA full-scale, an angular deflection of the needle of 90°; the coil resistance of the meter being $5\,\Omega$, and the battery voltage 1.5 V. We shall now see, working step-by-step, how to construct the ohm-meter.

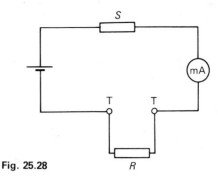

Fig. 25.28

a) Clearly, the larger the resistance R connected to the terminals T T, the less the current. Hence if the current scale is calibrated, as is usual practice, from left to right, the resistance scale will have its zero on the right. We want the zero of the resistance scale to correspond to full-scale deflection. Thus when the angular deflection of the needle is 90°, $R = 0$. The current flowing in the circuit here is, of course, 15 mA. Calculate the value of S needed.

b) What is the value of current in the circuit when $\theta = 75°$? Knowing this, calculate the value of R which will produce this deflection of the needle. Enter the answers in a table like the one below.

c) Repeat the calculation for the other values of θ given in the table. Make a rough sketch of the new meter scale, marked off in intervals of $50\,\Omega$, and comment upon the form of this scale.

Pointer deflection, θ	0°	15°	30°	45°	60°	75°	90°
Current/mA							
R/Ω							

When you next see a multimeter (with an ohms scale), look at it critically.

a) Using $V = IR$. The total circuit resistance (assuming the battery has negligible internal resistance) is $(S + 5\,\Omega)$: thus $1.5 = (15 \times 10^{-3}) \times (S + 5)$, hence $S + 5 = 100$, therefore $S = 95\,\Omega$.

b) When $\theta = 75°$ the current is $\frac{75}{90} \times 15 = 12.5$ mA. To find the value of R:

total resistance $= S + 5 + R = \dfrac{1.5}{12.5 \times 10^{-3}}$

but $S = 95\,\Omega$, therefore

$$100 + R = 120$$

hence

$$R = 20\,\Omega$$

c) We can save time in the long run by making a formula for R in terms of θ. The current needed to produce a deflection θ is $\theta/90 \times 15\,\text{mA}$, that is $\theta/6\,\text{mA}$, or $\theta/6 \times 10^{-3}\,\text{A}$.

Proceeding as in (b), we can write

$$100 + R = \dfrac{1.5}{\dfrac{\theta}{6} \times 10^{-3}}$$

which reduces to

$$R = \dfrac{9000}{\theta} - 100 \dots \text{in } \Omega$$

It is a quick business now to substitute values of θ into this and find the corresponding values of R.

Chapter 26

ELECTRIC CHARGE

We must now return to the question of the electric current itself. We have assumed this to be a flow of something we called electric charge. Our evidence for this application of the 'flow' model was the observed behaviour of the 'current' in series and parallel circuits. In all those cases, the circuits were complete.

The question now arises as to the behaviour of the current in circuits which are not complete. Close the switch in the circuit of Fig. 26.1 and the galvanometer will fail to indicate a current. But attach two large metal plates to the points A and B (Fig. 26.2) and an effect will be observed. Two moderately sized plates (say 20 cm square) with a small air gap between them and, of course, insulated from one another will be adequate. At the instant of switching on, there will be a momentary deflection of the galvanometer needle.

If the switch S is opened and the point P connected to the negative terminal of the battery, a momentary deflection of the same magnitude, but in the opposite direction, is observed. This suggests that electric charge flows on to the plate

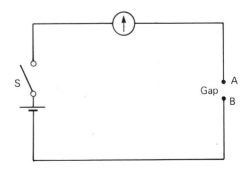

Fig. 26.1

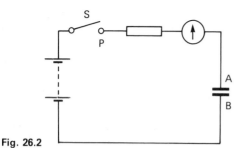

Fig. 26.2

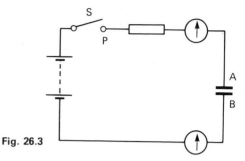

Fig. 26.3

A when S is closed and then flows back when P is connected to plate B (or, possibly, that the electric charge flows away from A when S is closed, then back on to A when P is joined to B).

The experiments in Fig. 26.2 show that having the plates A and B closer to each other results in larger deflections on the galvanometer when the switch S is closed. The investigation can be extended by inserting a second galvanometer into the circuit, as in Fig. 26.3. When S is closed (the plates having first been discharged by joining P to the negative terminal of the battery), the two galvanometers show deflections, usually of about equal magnitude, and in such a direction that if electric charge has flowed on to A it must have flowed away from B, alternatively if electric charge has flowed away from A then it must simultaneously have flowed on to B. We say that plate A has become *positively charged*, and B *negatively charged*, as a result of closing S. When S is opened and P is joined by a wire to the negative terminal of the battery, by-passing the battery, then the two galvanometers show equal deflections in the opposite direction to what they showed when S was closed. We interpret this result as the discharge of the plates A and B.

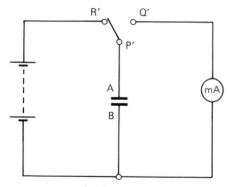

Fig. 26.4   Circuit for investigating the behaviour of a parallel-plate capacitor.

26.1   Capacitors

We proceed now to investigate the action of this parallel-plate arrangement as a holder of electric charge. The investigations may conveniently be carried out using the apparatus of Fig. 26.4.

S is a reed-switch. This is a single pole, two-way switch contained in a glass capsule and capable of a fast rate of switching. Figure 26.5 shows a section through the switch.

PP' is a flexible magnetic strip which is sprung so that the end P is normally in contact with the end R of the fixed non-magnetic strip RR'. QQ' is a fixed magnetic strip. If a magnet is brought near to this capsule, QQ' and PP' will be magnetized and P moves into contact with Q so completing any circuit connected to the terminals P' and Q'. In this usage, the switch is operated by the proximity of a magnet. It is known as a proximity switch.

If, however, the capsule containing the switch is placed within a solenoid connected to, say, a 50 Hz supply, the reed PP' will oscillate between R and Q. It is usual to feed the a.c. into the solenoid through a rectifier diode. Then the switch moves from R to Q 50 times each second.

At a switching rate of 50 Hz, the charges from the plates will cause the needle of the milliammeter to give a constant reading. This reading depends on the charge stored on the plates each time the reed is to the left (Fig. 26.4).

The two plates are separated and insulated from each other by flat spacers of equal thickness and area about 0.5 cm^2, made of plastic insulating material: celluloid, polythene, PVC, are all

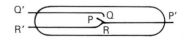

Fig. 26.5   The principle of a reed-switch.

suitable if clean and dry. The plates can be separated by three or four spacers evenly spaced from each other. The separation of the plates can be varied by using more spacers, stacked on top of each other.

The area of overlap of the two plates can be varied by sliding the top one horizontally away from the lower one. The nature of the medium between the plates also affects the amount of charge which can be stored at a given voltage: this can be verified by inserting a sheet of the insulating material which was used to make the small spacers in such a way that the distance between the two plates is not altered.

If the bottom plate is removed altogether (leaving the negative terminal of the battery still earthed, say, to a water-pipe, gas-pipe, metal mains cable sheath, or mains power-point earth socket and the upper plate insulated) it is found that the top plate alone can still store charge although considerably less than it did when the lower plate was present close to it. With the top plate alone it can also be shown that the presence of any other conducting object, close to the top plate but insulated from it, increases the charge-storing ability of the plate.

We can now investigate how the charge-storing ability of the parallel-plate arrangement depends on:

a) The voltage used to charge the plates.
b) The distance between the plates.
c) The area of overlap of the plates.
d) The insulating or dielectric material between the plates.

But before proceeding to accurate measurements, bring your hand close to the upper plate while the reed-switch is running. Note the reading of the milliammeter. Then move your hand away, watching the milliammeter as you do so. That will show clearly the effect of nearby earthed conductors on the ability of the plates to store charge.

Suppose the charge released by the plates when they are discharged through the milliammeter is Q. The reed-switch causes n discharges per unit time. Thus the average current through the milliammeter is nQ and we will assume that this is the steady reading indicated by the milliammeter.

Using a low-voltage a.c. supply from the 50 Hz a.c. mains ensures that n is constant and so it is possible to calculate Q from the current reading (I) of the milliammeter since $nQ = I$.

The first accurate measurement involves setting the plates at a fixed separation and using different values of battery voltage V. It is found that the charge on the plates Q is directly proportional to V. Then, using a constant charging voltage, and setting the plates at different distances d apart, one finds that, provided d is not too large, that $Q \propto 1/d$. Then, keeping constant d and voltage, but altering the area of overlap, A, one finds that $Q \propto A$. Finally, keeping the geometry of the plates constant and using a constant voltage, but putting a solid insulator instead of air between the plates, one discovers that Q is increased. The amount by which it is increased depends on the *relative permittivity* of the material, a term which will be defined in Section 26.3.

The pair of parallel plates which we have used for these experiments is one example of a *capacitor*. A capacitor is any arrangement of a pair of electric conductors, insulated from each other, and designed as a device for storing electric charge. Its ability to store charge is specified by its *capacitance*.

■ Capacitance is defined as the amount of charge stored per unit potential difference between the plates. (One generally refers to the conductors as 'plates' even if they are in fact rolls of metal foil.)

If the charge stored is Q when the p.d. is V then the capacitance is given by

$$C = \frac{Q}{V} \qquad (26.1)$$

and, as the experiments have shown, this is a constant for a pair of conductors having a given geometry. The unit of capacitance is the farad (F).

■ One farad is one coulomb per volt.

The farad is not a conveniently sized practical unit. The parallel-plate arrangement used in the experiments described has a capacitance of only a few millionths of a millionth of a farad. Submultiples in common use include the microfarad (μF) and the picofarad (pF). (See Appendix 1.)

A parallel-plate capacitor of this sort is of little practical use: its importance lies in what it can reveal in experimental use about the nature of the electric field. Electric field theory is dealt with in Unit 7. The uses of capacitors of various types in electric circuits are what concern us now.

26.2   Dielectrics and capacitors

The capacitance of a pair of parallel metal plates, area 400 to 900 cm^2 and spaced a few millimetres apart, is in the order of a few hundred picofarads. This is far too small a capacitance for most applications of capacitors in electronics. A quick look at the lists of capacitors in the catalogue of a manufacturer of electronic components gives a good idea of the ranges of values most frequently used. If one attempted to make a one microfarad capacitor in the form of parallel plates with an air-gap in between, then the area of each plate would have to be hundreds of square metres. But because the capacitance can be increased by reducing the separation of the plates and substituting a solid insulating material for the air between the plates, a capacitor can be made compact by having two layers of thin metal foil separated from each other by a very thin layer of insulating material. Waxed paper has been used for many years, and is still used, in the type of capacitor known simply as a paper capacitor.

Instead of waxed paper, mica and various ceramic materials (for example, titanium oxide) are used commonly as the dielectric material in capacitors. Mica is a naturally occurring mineral which has a crystalline structure such that it can split into very thin sheets of almost perfectly uniform thickness. It can withstand high voltages, and high temperatures, and it is suitable for use in capacitors in circuits where there are ultra-high frequency (that is, in the order of hundreds of megahertz) alternating currents. The ceramic materials specially developed for use in capacitors can also withstand high temperatures and voltages: they also have very high relative permittivities. The meaning of this term is discussed below. Paper, mica, and ceramic capacitors are not generally made with capacitances greater than about one microfarad. For appreciably larger capacitance one generally uses an electrolytic capacitor, in which the thickness of the dielectric

layer between the layers of aluminium foil is very much less than can be achieved using paper, mica, or ceramic. In an electrolytic capacitor the layer of dielectric, which can be less than a thousandth of a millimetre thick, is produced by electrolysis. Paper impregnated with a solution of aluminium borate is used to separate the layers of aluminium foil. This solution is a conductor; but when a p.d. is applied to the two plates of the capacitor, the positive plate becomes *anodized*, that is, a very thin layer of aluminium oxide is deposited by electrolytic action on the surface of the metal. It is this layer of oxide which is the insulator between the two plates, and it is of course sandwiched between the positive aluminium foil and the layer of impregnated paper which makes direct electrical contact with the negative foil. This deposit of aluminium oxide has remarkably good insulating properties and, although it is so thin that one might suppose that it would break down when only a small p.d. is applied to the plates, in fact an electrolytic capacitor of this type can be made to work at voltages of several hundreds of volts. Electrolytic capacitors cover a range of capacitance from about 1 μF to several thousands of microfarads. It is essential to connect such capacitors into circuits the right way round. (Why is this so?)

The different insulating materials used in capacitors have different *relative permittivities*. To understand what the term *relative permittivity* (or *dielectric constant*) means, consider what was done in the experiment described in Section 26.2. The factors determining the capacitance of a parallel-plate capacitor were investigated, and it was shown that inserting a slab of solid insulating material between the plates, without altering their spacing, increased the capacitance. The dielectric, air, was replaced by a solid dielectric. If the plates were in a vacuum we would find that the capacitance was not noticeably different from that when air was present. In fact the capacitance with air is very slightly greater than with a vacuum.

■ The relative permittivity ϵ_r of a material is defined thus:

$$\epsilon_r = \frac{\left(\begin{array}{c}\text{capacitance of an ideal parallel-plate}\\\text{capacitor with material between plates}\end{array}\right)}{\left(\begin{array}{c}\text{capacitance of same capacitor with}\\\text{plates in a vacuum}\end{array}\right)}$$

The relative permittivities of some commonly used insulating materials are given in Table 26a. These are simply numerical constants.

Table 26a

Material	Relative permittivity
Polythene	2.3
Perspex	2.6
Paper (waxed)	2.7
Paraffin	2.7
Paper (oil impregnated)	3.6
PVC	4
Mica	7
Barium titanate	1200
Water	80

The name relative permittivity arises in the following way. The capacitance of an ideal parallel-plate capacitor can be expressed in terms of the geometry of the plates thus:

$$C = \epsilon \frac{A}{d} \qquad (26.2)$$

where A is the area of each plate, d is the distance between the plates, and ϵ is a constant for the insulating material between the plates, called the *permittivity* of the material. If the plates are in a vacuum the constant is written ϵ_0 and is called *permittivity of free space*. The significance of this constant, for free space, and how it measured, are discussed in Unit 7. If the medium between the plates is anything other than a vacuum the value of the constant is greater than ϵ_0. If we use the symbol ϵ_x for the permittivity of a medium X, then the relative permittivity ϵ_r of medium X is defined as the ratio ϵ_x/ϵ_0. Now suppose we have a parallel plate capacitor in a vacuum, and nothing between the plates, then we can write, for its capacitance

$$C_0 = \epsilon_0 \frac{A}{d} \qquad (26.3)$$

Then if we insert a material whose permittivity is ϵ_x between the plates, the capacitance is given by

$$C_x = \epsilon_x \frac{A}{d}$$

Then the relative permittivity

$$\epsilon_r = \frac{C_x}{C_0} \qquad (26.4)$$

From Eq. 26.2 we see that

$$\epsilon_0 = \frac{d}{A} C_0$$

and so ϵ_0 will have units metre farad/metre², or farad per metre ($F\, m^{-1}$). As we shall see in Chapter 30, the permittivity of free space ϵ_0 is 8.854×10^{-12} or $1/36\pi\ F\, m^{-1}$. Since the relative permittivity of air differs but little from unity, we may use ϵ_0 when the charges are separated by air rather than a vacuum.

So far nothing has been said about the fundamental physical reason why the insertion of a slab of insulating material between the plates of a capacitor has the effect of increasing its capacitance. The fact is that the electric field between the plates is modified by the presence of the dielectric. The atoms and molecules of all materials contain positively- and negatively-charged particles. In an insulating material these particles are not free to move away from their mean positions, but they can move a little, so that the distribution of positive and negative charge in the material can be altered slightly. In the presence of an externally produced electric field the distribution of charge in the material is changed. This process is known as *polarization*.

The uses of capacitors can be roughly classified as follows:

a) To store electric charge which, after charging, leaks through some resistance in a circuit at a rate determined by the capacitance and the resistance combined. This process can be employed in electric circuits for determining a time-interval. This category includes the smoothing action of a capacitor, used, with a rectifier, in the conversion of a.c. to d.c.

b) To store a small quantity of electrical energy. A large electrolytic capacitor is used in an electronic flash-gun, for use in photography. The capacitor is charged and then discharges very rapidly through a gas discharge tube (usually containing xenon gas). A small quantity of electrical energy is thus converted into light and some heat.

c) To filter out a.c. from d.c. In many applications a capacitor is used in such a way that it has, in effect, a very low resistance to an a.c. but a very high resistance to a d.c.

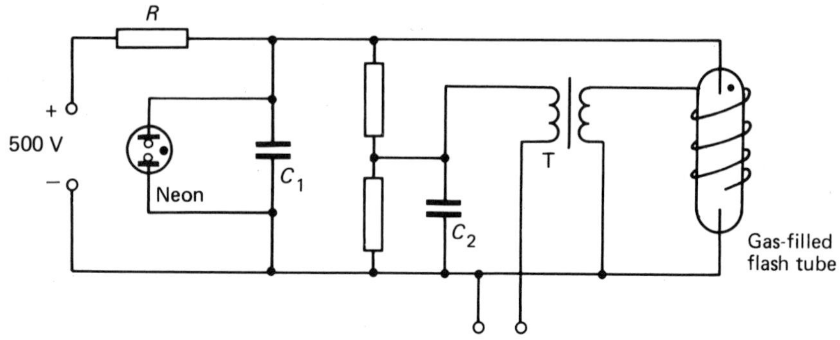

Fig. 26.6   Circuit for an electronic flash-gun.

Camera shutter contacts

26.3   The electronic flash gun

An electric discharges takes place in a suitable gas (often xenon) with the emission of an intense flash of light of short duration (about 1 ms). If such a device is to be operated from, say, a 6 V dry battery, the circuit must first generate a high voltage (300 to 500 V). This is achieved by using a transistorized oscillator which converts d.c. to a.c. which can then be stepped up to a high voltage in a transformer before being rectified to d.c. once more. This high voltage d.c. is then fed to the capacitor C_1 (Fig. 26.6) through a resistor R_1 which limits the current. After several seconds the capacitor will have stored sufficient energy to operate the flash – if enough ions are already present in the flash tube.

The ions are obtained by passing a brief pulse at about 10 kV across the gas tube. This is obtained by using some part of the energy stored in C_1 to charge the capacitor C_2 to about 150 V. When the camera shutter contacts close, this energy provides a current pulse through the primary of the step-up transformer T. The output from T is applied to a transparent conducting coating on the wall of the flash tube and a spark passes. This spark provides the ions within the gas which allow the flash discharge to follow.

26.4   Combining capacitors

Capacitors may be used in circuits in both series and parallel arrangements. *Connected in parallel* (Fig. 26.7) it is self-evident that the effect is to add together the areas of the plates and we expect that the combined capacitance will be $C_1 + C_2 + C_3$.

In a formal proof we note that the application of a potential difference V will cause a total charge Q to accumulate in the capacitors and the total capacitance C will be given by

$$C = \frac{Q}{V}$$

The total charge Q is the sum of the separate charges Q_1, Q_2 and Q_3.

$$\begin{aligned} Q &= Q_1 + Q_2 + Q_3 \\ &= C_1V + C_2V + C_3V \\ &= V(C_1 + C_2 + C_3) \end{aligned} \quad (26.5)$$

Comparing Eq. 26.1 with Eq. 26.5, we have

$$\boxed{C = C_1 + C_2 + C_3} \quad (26.6)$$

Connected in series, as in Fig. 26.8, each of the capacitors will carry the same charge as shown. The potential difference across the three capacitors will be given by

$$V = \frac{Q}{C} \quad (26.7)$$

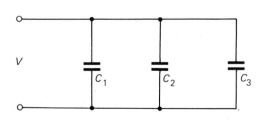

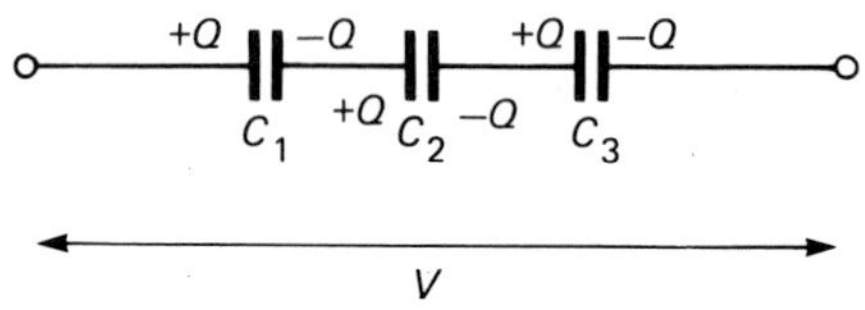

Fig. 26.7   Capacitors in parallel.

Fig. 26.8   Capacitors in series.

But $V = V_1 + V_2 + V_3$

$$= \frac{Q}{C_1} + \frac{Q}{C_2} + \frac{Q}{C_3}$$

$$= Q\left(\frac{1}{C_1} + \frac{1}{C_2} + \frac{1}{C_3}\right) \qquad (26.8)$$

Combining Eq. 26.7 and Eq. 26.8, we see

$$\boxed{\frac{1}{C} = \frac{1}{C_1} + \frac{1}{C_2} + \frac{1}{C_3}} \qquad (26.9)$$

26.5 Storing energy in a capacitor

If a capacitor is connected to a battery through a resistor, charge will flow from the battery on to the plates and a graph relating the potential difference developing across the plates to the charge on the plates will take the form shown in Fig. 26.9. It is a straight line since Q is proportional to V.

Initially no energy is expended by the battery as it transfers charge ΔQ to the uncharged plates. But, as Q increases, so does the p.d. between the plates. When, for example, that p.d. has reached V_1, the energy required to add a further small charge ΔQ to the plates is $V_1\Delta Q$. This is represented in Fig. 26.9 by the unshaded area.

Once fully charged, the energy supplied must have been the sum of all such areas, which is $\frac{1}{2}QV$. This is also $\frac{1}{2}CV^2$ and $\frac{1}{2}Q^2/C$.

This energy is stored in the electric field between the plates (see Chapter 30).

Problem 26.1 The capacitor with the largest capacitance in a particular manufacturer's catalogue is described as '25 V working' with a capacitance of $22\,000\,\mu F$. How much energy can this capacitor store?

Energy stored $= \frac{1}{2}CV^2$
$\qquad\qquad = \frac{1}{2} \times 22\,000 \times 10^{-6} \times 25^2$
$\qquad\qquad = 6.9\,J$

If the capacitor is discharged through a small motor arranged so as to lift a small mass, to what height *might* a 0.5 kg mass be raised?

Energy available $= 6.9\,J = mgh$
$\qquad\qquad\qquad = 0.5 \times 10 \times h$

Whence

$\qquad\qquad h = 1.4\,m$

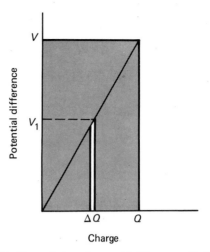

Fig. 26.9 The variation of potential difference with the charge stored on the plates of a capacitor.

26.6 Discharge of a capacitor

The discharge of a capacitor is never an instantaneous process, although we assumed, when doing the experiments with the apparatus of Fig. 26.4 that the time taken for the capacitor to discharge itself was negligibly small. And we assumed this in spite of the fact that the reed-switch was discharging the capacitor fifty or even one hundred times per second. We shall now make a careful investigation of the discharge process. This is conveniently done using a capacitance of about $500\,\mu F$. Figure 26.10 shows the circuit and suitable component values.

The capacitor is charged by closing the switch. With a 10 V supply the microammeter reads $100\,\mu A$ since the circuit resistance is $100\,k\Omega$. When the switch is opened, the ammeter continues to show a reading as the charged capacitor slowly discharges through the resistor. The discharge is rapid at first, continuously becoming less rapid.

If a stop-clock is started as the switch is opened, values of the discharge current I can be taken every ten seconds. The graph of these results (Fig. 26.11) appears to be a curve of the exponential decay type (see Section 46.4), but rather than take this for granted we shall make a theoretical analysis of the process.

At any instant during the discharge, the p.d., V, across the capacitor and the charge, Q, are related by $Q = CV$, where C is the capacitance.

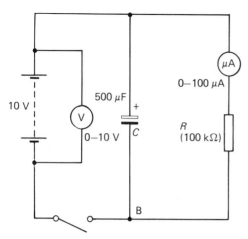

Fig. 26.10   Circuit for the investigation of the variation of the charge on a capacitor with time.

The current, I, flowing in the resistor R is given by $V = IR$. But the current is the rate of loss of charge from the capacitor, and thus we may write

$$I = -\frac{\Delta Q}{\Delta t}$$

the negative sign indicating that as time, t, increases, the charge, Q, decreases.

From the equation for Q we see that $V = Q/C$.
From the equation for V we see that $I = V/R = Q/RC$.
And so

$$I = -\frac{\Delta Q}{\Delta t} = \frac{Q}{RC}$$

Or

$$\boxed{\Delta Q = -\frac{Q}{RC}\Delta t} \qquad (26.10)$$

[Using the notation of calculus this expression is written

$$\frac{dQ}{dt} = -\frac{Q}{RC}\,]$$

This is a differential equation of the first order for which we shall carry out a step-by-step approximate numerical solution (Compare Section 19.4). There are two good reasons for doing this: firstly, it is a technique which does not require a knowledge of the calculus; secondly, it is a method easily programmed on a digital computer. Having obtained numerical values of Q one can plot a

graph and study the properties of the curve.
The steps of the computation are as follows.

a) Calculate the initial charge on the capacitor at time $t = 0$, using $Q = CV$. Plot this point on a charge–time graph.
b) Use the initial value of the current to calculate the charge which leaves the capacitor in the first five seconds using $\Delta Q/\Delta t = I$.
c) Calculate the charge remaining on the capacitor at the end of this 5 s period and plot this on your graph. What fraction of the initial charge is this?
d) Calculate the voltage on the capacitor now – and thence the new discharge current.
e) Calculate the charge remaining on the capacitor at the end of this second 5 s period and plot this. What fraction of the charge remaining at the end of the first 5 s period is this?
f) Continue the computation for several further 5 s intervals. It should be possible to use a calculator for this if you have spotted the way in which the charge remaining decreases with time.
g) Plot the graph of charge against time for, say, 50 s, joining your points with straight lines. Table 26b may be used to check the steps of calculation.
h) Compare your computed and your experimental curves. Why does the former run down more steeply?

In drawing the curve, the decrease in the charge Q was $0.1Q$ on each occasion. The charge at the commencement of each 5 s interval was ($0.9 \times$ the charge at the start of the time interval).

Table 26b

Time/s	Voltage/V	Charge/C	Current $(\Delta Q/\Delta t)/\mu A$
0	10.00	5×10^{-3}	100
5	9.00	4.5×10^{-3}	90
10	8.10	4.05×10^{-3}	81
15	7.29	3.65×10^{-3}	72.9
20	6.56	3.29×10^{-3}	65.6
25	5.90	2.96×10^{-3}	59.0
30	5.31	2.66×10^{-3}	53.1
35	4.78	2.39×10^{-3}	47.8
40	4.30	2.15×10^{-3}	43.0
45	3.87	1.94×10^{-3}	38.7
50	3.48	1.75×10^{-3}	34.8

The charge is reduced by the same fraction of its existing value in successive time intervals. This is characteristic of an exponential decay in which a quantity x is reduced by the same fraction in successive equal intervals. If the quantity is halved from x to $x/2$ in time $t_{1/2}$, then it will fall from $x/2$ to $x/4$ in the next time interval $t_{1/2}$, and from $x/4$ to $x/8$ in the next interval $t_{1/2}$ and so on. In this case $t_{1/2}$, the time taken for the quantity to halve its value is known as the *half-life* of the decaying quantity. If the graph for the capacitor discharge is plotted far enough it will be discovered that the charge is halved from 5 mC to 2.5 mC, from 4 mC to 2 mC, from 3 mC to 1.5 mC in very nearly equal intervals of time, which are, in this example, 33 s to the nearest second.

The exponential law is of great importance in the sciences. Reference has already been made to its use in Chapter 8 in connection with the world's energy resources and usage. In Chapter 12 it was found to describe the distribution of energy quanta among the atoms of an Einstein solid. We shall meet it yet again when dealing with the growth and decay of the radioactive elements.

The exact solution to Eq. 26.10 is

$$Q = Q_0 e^{-t/RC} \qquad (26.11)$$

and the half-life $t_{1/2}$ is found by setting $Q = Q_0/2$ and solving for $t_{1/2}$, thus:

$$\frac{Q_0}{2} = Q_0 e^{-t_{1/2}/RC}$$

$$\therefore e^{t_{1/2}/RC} = 2$$

$$\therefore t_{1/2} = RC \ln 2 = 0.693\,RC \qquad (26.12)$$

The product RC is known as the *time-constant* of the circuit, and in this example it is

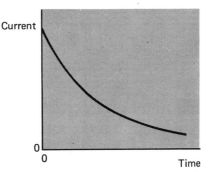

Fig. 26.11   Variation of discharge current with time.

$$10^5 \times (500 \times 10^{-6})\,\text{s} = 50\,\text{s}$$

Hence

$$t_{1/2} = 0.693 \times 50\,\text{s} = 34.7\,\text{s}$$

The result obtained from our approximate numerical solving of the equation was less than this, the discrepancy being about 5%. We could make a better approximation to the truth by using a shorter time interval in the computation, using, for instance, $\Delta t = 2\,\text{s}$ instead of $\Delta t = 5\,\text{s}$.

The time-constant of a circuit containing resistance and capacitance is an important parameter in electronics. At first it may seem surprising that multiplying together these two physical quantities, the one measured in ohms and the other in farads, produces units of time. It is easily shown that this must be so by writing each unit in terms of units from which it is derived, thus:

$$\Omega = V\,A^{-1} = V\,(C\,s^{-1})^{-1} = V\,C^{-1}\,s$$

and

$$F = C\,V^{-1}$$

therefore

$$\Omega \times F = (V\,C^{-1}\,s) \times C\,V^{-1} = s$$

Fields

Chapter 27

FIELDS OF FORCE

27.1 Introduction

Physicists use the word field to mean 'a region of influence' and a familiar example is the magnetic field around a magnet which is taken to be the region where some magnetic effect can be detected. This is not very different from everyday usage. When we talk of a field of grass, we refer to a region in space where grass is to be found. Similarly, when we talk of an electric field, we refer to a region in space in which an electric force can be found. (You will notice the scientist's shorthand here – we ought really to say 'electric force field' and not just 'electric field'. The word 'force' is invariably omitted and must be understood as being meant.)

27.2 Action at a distance

When, in Unit 2, we looked at the way forces set things into motion, or failed to move them, or changed the direction in which they moved, we did not consider how these forces were produced. They were just 'pushes' or 'pulls' – the result usually of a direct contact between something that moved and whatever was trying to produce the movement.

By looking at the effect these forces have when they act, we learn to define a force as something 'trying to change motion'. This is an important generalization of many happenings in the physical world. Once this generalization has been made, we are confident we can always recognize when a force acts. So when we bring a magnet up towards an iron nail, and we see the nail move even when the magnet is some distance away, we say that the magnet exerts a *force* on the nail. We let go of a plate we are carrying and it crashes to the floor. The motion of the plate has been changed. We say that the earth exerts a *force* on the plate.

We do not always have to see a change in motion to assert confidently that forces are acting. We demonstrate our knowledge of physics by rubbing a blown-up balloon on our clothes and sticking the balloon to the ceiling. We think, ordinarily, it ought to fall – because of the pull of the earth on it. So we say another force must pull the balloon towards the ceiling. We learn eventually to call this an electric force.

All these forces, however, seem rather mysterious. They are not like the pushes and pulls we use to generalize our idea of a force. They are not forces of contact. The thing which causes the force is not in direct contact with the thing which moves. The force acts – that is undeniable – but nothing seems to be pushing or pulling. This mystified scientists for a long time; they felt that something ought to be visibly pushing or pulling and went to great lengths to prove that invisible mechanisms were involved.

J. C. Maxwell was one of the first scientists to point out that far from being mysterious, these sorts of forces were probably the only ones which exist. As an example, think of the following: from a footballer's point of view, a foot kicking a ball involves a force of contact. From the ball's point of view, the force is also one of direct contact – direct contact between ball and foot. But from the point of view of an atom on the surface of the football, it is not direct contact at all. The atom's outer electrons experience a repulsive electric force from the outer electrons of an atom on the surface of the boot as this approaches. This electric force pushes the ball-atom away. Other forces from neighbouring atoms help keep all the atoms in their same relative position within the ball. What has become of the force of contact?

Today, rather than try to explain how these forces act, as soon as we meet them, we accept their existence and explore their properties. Newton worked in this way when he made his important discoveries about gravitational forces. He was not the first to give serious attention to forces such as these, but his approach was much

the most successful. You must agree with this, if success is measured by the ends to which we can put his laws, in trips to the moon and launching earth satellites. Yet science is no nearer *explaining* the force of gravity in terms of other fundamental properties. Before turning to an exploration of the properties of forces which act at a distance let us recall what Maxwell said, about two hundred years after Newton's work:

'But if we leave out of account for the present the development of the ideas of science, and confine our attention to the extension of its boundaries, we shall see that it was most essential that Newton's method should be extended to every branch of science to which it was applicable – that we should investigate the forces with which bodies act on each other in the first place, before attempting to explain *how* that force is transmitted. No men could be better fitted to apply themselves exclusively to the first part of the problem, than those who considered the second part quite unnecessary.'

[*Reprinted from Proc. Royal Inst.* 7, 1873.]

27.3 Measurement of gravitational field strength

If we are to explore the properties of fields, then measurements must be taken; this is done by measuring the force acting on a suitable 'detector' placed in the field. There is nothing sophisticated about this since for gravitational fields a suitable detector is simply an object whose mass is known.

To measure a gravitational field strength, suspend the mass from a spring balance which has been calibrated in newtons by accelerating a mass of 1 kg with it. If the force indicated by the balance is F and the mass is m, then the gravitational field strength, g, is defined thus:

$$g = \frac{F}{m} \qquad (27.1)$$

The units of g are N kg^{-1}, and on earth $g = 9.8$ N kg^{-1}, whereas on the moon $g = 1.6$ N kg^{-1}.

■ The gravitational field strength at a point is the force per unit mass acting at that point.

Problem 27.1 On earth, with what acceleration will a 4 kg mass fall?

Force acting on mass = mass × field strength
$$= 4\,\text{kg} \times 9.8\,\text{N kg}^{-1}$$

From Eq. 5.6

$$\text{acceleration} = \frac{\text{force}}{\text{mass}} = \frac{(4 \times 9.8)\,\text{N}}{4\,\text{kg}}$$

$$= 9.8\,\text{m s}^{-1}$$

Repeat the above calculation with a different mass and you will find that the acceleration due to gravity is independent of the mass falling. This fact is well demonstrated by the 'guinea and feather' experiment in which a coin of Newton's day called a guinea and a feather fall at the same rate, side by side, in a vacuum. The usual slow fall of a feather is caused by air resistance.

27.4 Electric field strength or electric intensity

The principle used for measuring the strength of an electric field is the same as for gravity, only the detector used must be a known charge in coulombs rather than a mass. This test charge must be infinitesimally small so that its presence does not affect the field being measured.

■ The electric field strength at a point is the force per unit charge acting at that point.

If the charge is Q, and the force acting on it is F, then the electric field strength, E, is given by:

$$\boxed{E = \frac{F}{Q}} \qquad (27.2)$$

The units of E are N C^{-1}, and later in Section 30.2 this will be shown to be the same as V m^{-1}.

It is often important to know the electric field between parallel plates; for instance, the cathode ray tube in an oscilloscope has two pairs of plates to provide electric fields for deflecting the electron beam. Figure 27.1 illustrates how such an electric field could be measured using the definition given above. The glass spring acts as an insulator and also supports, through a small hole in the top plate, the test sphere that is made of expanded polystyrene coated with conducting paint. If the

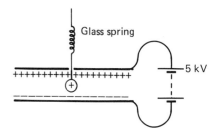

Fig. 27.1

test sphere is given a positive charge it will be repelled by the top plate and attracted to the lower plate; this tiny force will stretch the spring slightly. The force constant of the spring and the charge on the test sphere will need measuring before calculating the electric field strength. These measurements are not easy to perform accurately and Section 30.2 describes an alternative, but related, method.

27.5 Magnetic fields

The basic detector for magnetic field measurement is the *current element* which is a length of wire carrying a current. Such a wire placed at right angles to a magnetic field will experience a force in a direction at 90° to both wire and field. See Fig. 27.2 which illustrates Fleming's left-hand rule, a useful 'memory aid'.

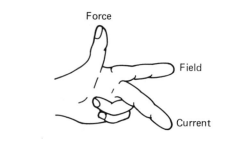

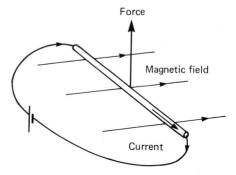

Fig. 27.2 Fleming's left-hand rule.

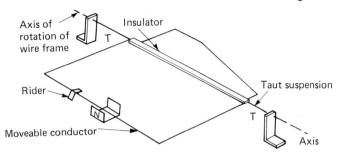

Fig. 27.3 A current balance.
(Courtesy Griffin and George
Ltd.)

To investigate how this force depends upon the current and the length of wire in a magnetic field, the apparatus shown in Fig. 27.3 may be used. This is a *current balance*; it consists of a horizontal, rectangular frame that can pivot about one of its long sides by means of a pair of taut suspension strips. These strips also carry the current to and from the straight conductor which forms the other long side of the frame. Thus the conductor can move up and down in the field produced by one, two or three magnets. Assuming the magnets to be identical, the number used varies the length of the current-carrying conductor being acted upon by the magnetic field. The directions of the field and current are arranged so that the wire between the poles is pushed upwards; a length of paper tape can be bent and used as a rider on the wire to return it to its rest position. This provides a simple way of determining force. With a constant current, 1, 2, and 3 magnets require 1, 2, and 3 units of force, and so $F \propto l$, where F is the force, and l represents the length of wire in the magnetic field. Likewise, keeping the length constant, and varying the current, it is found that the force on the wire is proportional to the current strength, I. Thus $F \propto I$, and combining the two results we may write $F \propto Il$.

Obviously if we had used stronger magnets we would have obtained greater forces. Thus, for a given value of I and a given value of l, the force on a wire lying at right-angles to the direction of the magnetic field can be used as a measure of the strength of the field. (Note that we have said *wire lying at right-angles*. As will be shown, the angle between the wire and the direction of the magnetic field affects the size of the force.) A convenient way of measuring the strength of a magnetic field, which we suppose for the time being to be a *uniform* field, is by defining it as the force per unit length per unit current upon a straight wire lying at right-angles to the field. If the actual current in

the wires is I and the length affected by the field is l, then the strength of the field is given by

$$B = \frac{F}{Il} \qquad (27.3)$$

The SI unit of this quantity is the newton per ampere metre, or tesla.

As we shall see in Unit 9, the term *magnetic field strength* is more commonly called *magnetic flux density*.

The product Il is called a *current element* and so

■ the magnetic field strength at a point can be defined as the force acting on a unit current element placed at that point (force, current and magnetic field being mutually perpendicular).

It is worth noting that for each of the fields, gravitational, electric and magnetic, the definitions of field strength all the take the same form.

What happens when a wire carrying a current lies in a magnetic field but not at right-angles to it can be investigated using the form of current balance shown in Fig. 27.4. A magnet and yoke are placed on the pan of a top pan balance capable of reading to 0.01 g. In this arrangement it is the magnet that moves and the wire loop, with the horizontal current element at the bottom, that is fixed. It is a simple matter to rotate the magnet and so explore the effect of having the current element at an angle θ to the field instead of 90°.

The horizontal part of the loop is short: this is so that it is affected only by the limited, nearly uniform, region of field between the magnet poles. The vertical parts of the loop only produce horizontal forces which cannot affect the readings. Thus the only measured force is that on the current element.

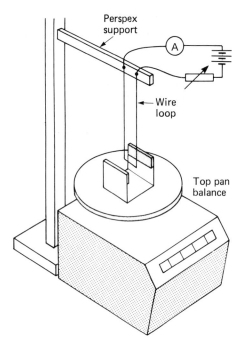

Fig. 27.4 Using a top-pan balance to measure the force on a current element in a magnetic field.

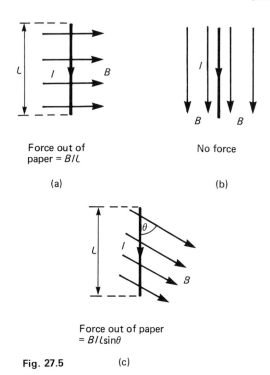

Fig. 27.5   (c)

With a constant current the force is greatest when the wire makes a right-angle with the field. It is zero when the wire is parallel to the field. Careful measurement will reveal that the force is proportional to the sine of the angle, θ, between wire and field (see Fig. 27.5). Thus the complete expression for the force on a straight conductor of length, l, carrying a current, I, in a field of strength, B, is:

$$ F = BIl \sin \theta \qquad (27.4) $$

The direction of this force is always at right-angles to the plane containing the direction of I and the direction of B. The general case (c) in Fig. 27.5 can be arrived at by treating B as a vector quantity as

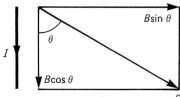

Fig. 27.6

shown in Fig. 27.6. B can be resolved into a component $B \sin \theta$ at right-angles to the current element, and $B \cos \theta$ parallel to it. The $B \sin \theta$ component at right angles to the current produces the force, while the $B \cos \theta$ component produces no effect.

27.6 A circular experiment?

A knowledgeable student, on reading Section 27.5, might object that it is pointless investigating how force depends on current, as measured by a moving coil ammeter. He or she reasons that the meter is using exactly the same effect as the effect we are trying to investigate and that the argument is therefore circular. Is the student correct?

The student is correct in saying that both parts of the circuit are using the same effect (that of the force on a current in a magnetic field) but this does not necessarily invalidate the results. It is still possible to demonstrate that the force is proportional to the current flowing even without a calibrated ammeter. Four identical uncalibrated moving coil instruments are needed. First connect them all in series with a suitable supply and pass a

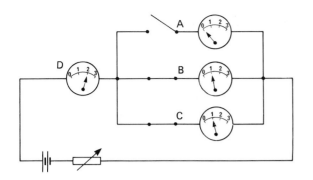

Fig. 27.7

current through them so that a deflection of about $\frac{1}{3}$ full scale is produced; mark this deflection on each meter. Then connect them as shown in Fig. 27.7. Close switches B and C and adjust the current until B and C point to the mark just made. Meter D will now register two units of current and this can be marked on its scale. Repeating the process with all three switches closed will provide a third mark on D for three units of current. Meter D can now be used in the current balance experiment; although its markings are not amperes, they are proportional to the current flowing.

Chapter 28

THE GRAVITATIONAL FIELD

28.1 Introduction

We have already referred to Newton's theory of gravitation in Unit 1. It suggests that the force between two point masses M and m separated by a distance, r, is given by:

$$F \propto \frac{Mm}{r^2}$$

and that

$$F = G\frac{Mm}{r^2} \qquad (28.1)$$

where G is called the gravitational constant and has a value $6.67 \times 10^{-11} \, \mathrm{Nm^2\,kg^{-2}}$.

Newton had no means of determining this value experimentally but made a close estimate of it by using a remarkably accurate guess of the mean density of the earth in order to estimate its mass. His great triumph, however, was the establishment of the inverse square law embodied in Eq. 28.1.

The story of this law's development began with the work of Tycho Brahé (1546–1601), a practical astronomer who spent his lifetime making accurate observations of the planets and stars. These records were used for navigation but, more important, when Brahé died they passed into the keeping of his assistant, Johannes Kepler (1571–1630), who after years of study formulated three laws concerning the orbits of planets.

28.2 Kepler's laws

■ The laws may be summarized thus.

    1. The orbit of each planet is an ellipse with the sun at one focus.

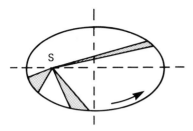

Fig. 28.1 Elliptical path of a planet round the sun, S. The eccentricity is greatly exaggerated.

2. The arm drawn from the sun to a planet sweeps out equal areas in equal periods of time. See Fig. 28.1.

3. The squares of the periods of revolution are proportional to the cubes of the average distances of the planets from the sun.

Modern observations on the solar system, on the moons of the planet Jupiter and on earth satellites confirm the success of the three laws as summaries of the data. See, for example, Table 28a.

Table 28a Solar system data and a test of Kepler's third law

Planet	Mean radius of orbit (r)/m	Period of orbital motion (T)/s	$\dfrac{r^3}{T^2}$
Mercury	5.79×10^{10}	7.60×10^6	3.36×10^{18}
Venus	1.08×10^{11}	1.94×10^7	3.35×10^{18}
Earth	1.49×10^{11}	3.16×10^7	3.30×10^{18}
Mars	2.28×10^{11}	5.94×10^7	3.36×10^{18}
Jupiter	7.78×10^{11}	3.74×10^8	3.36×10^{18}
Saturn	1.43×10^{12}	9.30×10^8	3.37×10^{18}

Kepler's three laws can only be accounted for if the law of force which applies to the sun and the planets is an inverse square law. The vital case is, in fact, that described in Law 1. Only an inverse square law of force can lead to planetary orbits which are elliptical with the sun at one focus. The second law is valid for any centrally directed force.

Kepler did not live to see his laws used by Newton to establish the inverse square law of gravitation. It is the third law that leads most directly to this $1/r^2$ relationship. For example, let

us start by supposing that gravitation follows an 'inverse n' law, i.e., the force is proportional to $1/r^n$ where n is an integer whose value we want to find. If a planet of mass m moves in a circular orbit of radius r around the sun of mass M, the central attraction must provide the centripetal force necessary to retain that planet in its orbit. If T is the period of an orbit, then:

$$\frac{GMm}{r^n} = \frac{mv^2}{r} = \frac{m}{r}\left(\frac{2\pi r}{T}\right)^2$$

or,

$$\frac{GMm}{r^n} = \frac{m4\pi^2 r}{T^2}$$

Rearranging,

$$\frac{GM}{4\pi^2} = \frac{r^{(n+1)}}{T^2}$$

Now, $GM/4\pi^2$ is a constant for the Solar system and so $r^{(n+1)}/T^2$ is also constant for the system. Kepler's third law then gives $n + 1 = 3$, or $n = 2$. Thus the central attraction is proportional to $1/r^2$, and Law 3 is consistent with the inverse square law of gravitation.

28.3 Newton's test of the inverse square law

In this test Newton considered the circular motion of the moon around the earth and reasoned that the centripetal acceleration it must have was because of the gravitational attraction of the earth. For the moon, the centripetal acceleration towards earth

$$= \frac{v^2}{r} = \frac{1}{r}\left(\frac{2\pi r}{T}\right)^2 = \frac{4\pi^2 r}{T^2}$$

Now, the radius of the moon's orbit is 3.844×10^8 m and the period is 27.3 days (2.36×10^6 s) (27.3 days is the mean sidereal month – meaning 'with the stars' – i.e., the time taken by the moon to complete 360° around the earth. The lunar month of 29.5 days is the time between successive full moons, and includes the time to complete an extra 29° because of the motion of the earth around the sun.) Using these values the acceleration of a mass at the moon's orbit is 2.725×10^{-3} m s^{-2}.

Now, the radius of the earth is 1/60 of the distance from earth to moon, so, if the inverse square law applies, the force and therefore the acceleration of a mass close to the earth's surface should be 60^2 or 3600 times greater than if it were at the distance of the moon. Thus the estimated acceleration on earth $= 2.725 \times 10^{-3} \times 3600 = 9.81 \text{ m s}^{-2}$.

The measured acceleration of falling masses, including apples, has just this value on earth and the inverse square law is verified.

This is also an experimental justification for assuming that the earth behaves as a *point* mass. The mass of the earth is distributed over its whole volume but it can be looked upon as a collection of many small masses, each contributing to the total attraction. Newton was able to confirm theoretically that any spherically symmetrical distribution of mass behaves at points outside itself as if all its mass were concentrated at its centre.

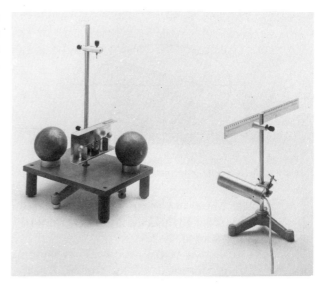

Fig. 28.2   Apparatus for the determination of G, the gravitational constant. (Courtesy Griffin and George. Ltd.)

28.4   Measurement of the gravitational constant (G)

The gravitational constant G cannot be determined from the force law between sun and planets. To do this, we should need to know their masses. The application of this very law is the only means of determining that. And for this we need to know G.

The measurement of the gravitational force between two known terrestrial masses was first achieved by Cavendish in 1798. Subsequently, his apparatus was improved by Boys, who was able to use the combined strength and low torsion of quartz fibres to reduce the scale of the apparatus.

Figure 28.2 shows a photograph of a modern, simple version.

Two small lead masses are attached to a bar suspended from a thin wire with a very low torsional restoring force when twisted. The displacement of these masses by gravitational attraction to the large masses seen outside the instrument can be measured.

Figure 28.3 shows, in plan view, the bar (length $2k$) deflected through an angle θ from its rest position with the centres of either of the pairs of masses separated by a distance r. Once equilibrium has been reached in one direction the large spheres are moved across to the positions (shown

dashed) and a deflection is obtained in the reverse direction.

In equilibrium,

Couple due to attraction of the adjacent masses $=$ opposing torque of suspension wire

If the torsional constant c is the couple exerted per degree of twist of the suspension, then when the twist is θ the couple or torque is $c\theta$.

Thus:

$$\frac{GMm}{r^2} 2k = c\theta \qquad (28.2)$$

The torsional constant, c, is too small to measure directly and so it is determined by timing the

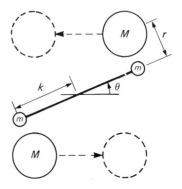

Fig. 28.3

period, T, of horizontal torsional oscillations of the bar and its masses. This motion is simple harmonic (see Section 19.4) and the period is given by:

$$T = 2\pi\sqrt{\frac{I}{c}} \qquad (28.3)$$

where I is the moment of inertia of the oscillating system. It can be calculated from

$$I = 2mk^2 \qquad (28.4)$$

Combining the three previous equations, we have

$$G = \frac{4\pi^2 r^2 k\theta}{T^2 M} \qquad (28.5)$$

All the quantities on the right-hand side of Eq. 28.5 can be measured and a value of G obtained. The accepted value is $6.670 \times 10^{-11}\,\text{N m}^2\,\text{kg}^{-2}$.

28.5 Orbital motion of planets and satellites

The value of G is of importance because once it is known, together with other available information like orbital radii and periods, then it is possible to determine the mass of the sun, of the earth, and indeed the mass of any planet so long as it has an observable satellite.

Let the masses of the planet and of the satellite be M and m, the radius of the orbit r and the period of the orbit T. Let the orbital speed of the satellite be v.

For circular motion,

gravitational attraction
= mass × centripetal acceleration

$$\frac{GMm}{r^2} = \frac{mv^2}{r}$$

Or:

$$\frac{GM}{r} = v^2 \qquad (28.6)$$

Also $v = \dfrac{\text{circumference of orbit}}{\text{period of orbit}} \qquad (28.7)$

$$= \frac{2\pi r}{T}$$

The following problem illustrates the use of these equations:

Problem 28.1 Find the mass of the sun, given that its mean distance from earth is 1.5×10^{11} m.

Eliminating v from Eqs. 28.6 and 28.7 we get:

$$\frac{GM}{r} = \frac{4\pi^2 r^2}{T^2} \Rightarrow M = \frac{4\pi^2 r^3}{GT^2} \qquad (28.8)$$

Using known quantities in Eq. 28.8, gives $M = 2 \times 10^{30}$ kg.

In this problem the earth is the satellite of the sun. The same process can give the mass of the earth because it has its own satellite, the moon, which is 3.844×10^8 m away.

28.6 The gravitational field strength of a planet

Suppose you are planning to land a spacecraft on the planet Mars. It would be necessary to know the gravitational field strength on the Martian surface as it governs weight there. Fortunately this can be easily calculated from observations made on earth before setting off.

We shall assume, as Newton showed (Section 28.3), that the planet is a sphere and that its mass M can be considered to be concentrated at the centre. If r is the planet's radius then a mass m placed on the surface is at a distance r from the mass M. By Newton's law of gravitation, the force acting on m is:

$$F = \frac{GMm}{r^2}$$

Gravitational field strength,

$$g = \frac{\text{force}}{\text{mass}} = \frac{F}{m}$$

Therefore,

$$\boxed{g = \frac{GM}{r^2}}$$

$$(28.9)$$

It has been assumed that the planet is not rotating. The effect of rotation on g is discussed in Section 28.7.

It is possible to find an actual value for g on the Martian surface because the planet has two satellites, Phobos and Deimos. See Question 7.14 at the end of the book.

28.7 The variation of the earth's gravitational field strength, *g*, with latitude

If the gravitational field strength is measured at the poles (g_p) its value will be found to be 9.83 N kg^{-1}, whereas at the equator, g_e is 9.78 N kg^{-1}. This difference occurs for two reasons:

1. The earth is an oblate (flattened) spheroid and not a sphere; the equatorial radius (r_e) is greater than the polar radius (r_p). Consequently a mass on the equator is farther from the centre than a mass at one of the poles, and so gravitational attraction at the equator is less.
2. Because of the earth's rotation a mass on the equator is moving in a circle of radius (r_e) at 464 m s^{-1} (just over 1600 km per hour), whereas a mass at the poles does not have this circular motion. Part of the gravitational attraction provides the necessary centripetal acceleration.

Consider a mass, *m*, hanging from a spring balance first at a pole, and then at a point on the equator. At a pole, the forces acting on the mass are:

a) $GMm/(r_p)^2$ towards the centre of the earth due to gravitation.
b) mg_p away from the centre of the earth due to the spring balance.

Note that mg_p is the reading of the spring balance that we call the *weight* of the mass.

As these two forces are in equilibrium, their resultant is zero

$$\frac{GMm}{(r_p)^2} - mg_p = 0$$

or

$$g_p = \frac{GM}{(r_p)^2} \qquad (28.10)$$

However, at a point on the equator, the two corresponding forces acting on the mass are $GMm/(r_e)^2$ towards the centre, and mg_e away from the centre and this time their resultant is not zero but must provide the centripetal force mv^2/r_e

Thus

$$\frac{GMm}{(r_e)^2} - mg_e = \frac{mv^2}{r_e}$$

so,

$$g_e = \frac{GM}{(r_e)^2} - \frac{v^2}{r_e} \qquad (28.11)$$

Inspection of Eqs. 28.10 and 28.11 will reveal why g_e is less than g_p.

Problem 28.2 Assuming the earth to be a sphere, of radius 6.36 × 10^6 m, show that the difference, $(g_p - g_e)$ is 0.034 N kg^{-1}.

28.8 The variation of gravitational field strength, *g*, with depth

On the earth's surface,

$$g = \frac{GM}{r^2}$$

By similar reasoning, at a point above the surface at a distance r_a from the centre of the earth,

$$g_a = \frac{GM}{(r_a)^2}$$

Since *r* is smaller than r_a it follows that the gravitational field strength is larger on the surface of the earth than it is farther out in space.

What happens if you go down a very deep mine and so get closer to the centre of the earth? This is a different situation and we cannot assume that *g* continues to increase. First consider the gravitational field *inside* a hollow spherical shell, at a point such as P in Fig. 28.4.

Imagine cones spreading out from P having the same axis and equal angles. Where the cones meet the shell they subtend sections of the shell,

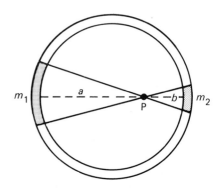

Fig. 28.4 The gravitational field inside a hollow spherical shell.

i.e., masses m_1 and m_2 at distances a and b respectively. These masses are proportional to the areas subtended which in turn are proportional to (distance)^2 from P. (See Section 30.12.)

Therefore

$$\frac{m_1}{m_2} = \frac{a^2}{b^2} \quad \text{or} \quad \frac{m_1}{a^2} = \frac{m_2}{b^2} \quad (28.12)$$

If the angle of the cones is small, m_1 and m_2 can be considered to be point masses and Newton's law of gravitation applied. Place a mass m_3 at P and find the resultant force on it.

F_1 = Force towards left due to m_1

$$= \frac{Gm_1m_3}{a^2}$$

F_2 = Force towards right due to m_2

$$= \frac{Gm_2m_3}{b^2}$$

Now, using Eq. 28.12,

$$F_2 = \frac{Gm_1m_3}{a^2}$$

Therefore the resultant force on $m_3 = F_2 - F_1 = 0$. Of course, this only applies to two small sections of the shell at opposite ends of a diameter, but the process can be repeated for the rest of the shell by drawing other pairs of cones through P, and although the geometry becomes more difficult the result will be the same, namely, the total gravitational field strength *inside* a shell is zero. (Reasoning like this can also illustrate why there is no electric field *inside* a hollow charged conductor.)

Now we can see that g decreases below the surface of a sphere of uniform density even though one is approaching the centre (see Fig. 28.5). The point B can be thought of as being on the surface of a solid core, shown shaded, but is on the inside of the shell that surrounds the core. The shell provides no force on a mass placed at B. The weight of this mass is due only to the core's gravitational attraction which depends on the inverse square law and is proportional to the mass of that core. If B goes deeper it will weigh less because the core beneath it is of smaller radius and therefore smaller mass. It follows that at the centre of such a sphere the mass would appear to be weightless.

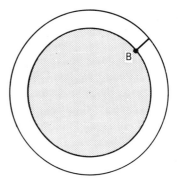

Fig. 28.5 B is a point below the surface of a sphere of uniform density.

The earth itself is not a sphere of uniform density. Its core is much denser than the surface rocks, which average about 3000 kg m^{-3} increasing to 6000 kg m^{-3} at the base of the mantle and, at a depth of nearly 3000 km, suddenly changing to 9400 kg m^{-3}. Consequently a mass lowered into a deep mine experiences an increase in weight as it gets a little nearer to the dense core material.

Problem 28.3 Show that, below the surface of a sphere of uniform density, the gravitational field strength is directly proportional to the distance from the centre.

28.9 Spaceflight and weightlessness

One remarkable consequence of the gravitation force law is the fact that for any mass, m, orbiting about the other attracting mass M at a distance r from it, the period of rotation is independent of the mass m. See Eq. 28.8. It follows that all masses moving around M in an orbit of radius r will have the same speed v (see Eq. 28.6) and therefore all have the same acceleration towards M.

This leads to the phenomenon referred to as *weightlessness*. An astronaut in a space capsule in orbit around the earth experiences no reaction to his weight in the capsule, since the pull of gravity on him is just sufficient to maintain him at the same orbiting speed as the capsule, at the capsule's distance from the earth.

Let us consider three 'thought-experiments':

Experiment 1

If two large masses are allowed to fall from rest they will move with the same acceleration. If they start at the same instant the two may be expected to fall together even though their masses may be different, as the 'guinea and feather experiment' shows. Suppose that, when they are released they are just touching, then they will continue to fall in contact.

Experiment 2

The two masses can together be thrown horizontally. This means that as well as having the same vertical acceleration they also have the same velocity in a horizontal direction. As Fig. 7.1 illustrates, such a combination of linear motions results in a parabolic trajectory and so the two masses will follow two such identical paths. If they start in contact they will remain in contact.

The term 'in contact' has been used in the same sense that they are just touching but not exerting a force on each other, for, if they did, they would move apart.

Experiment 3

If the masses are given a higher common velocity they will again travel in contact but will strike the ground farther away. If they are given a sufficiently high horizontal velocity they will go into earth orbit. (The earth's gravitational attraction provides the centripetal force necessary for circular motion.) This is Newton's 'thought experiment' with a cannon on a mountain already discussed in Section 7.1.

Suppose that one of the masses is now put inside a metal box and the three previous experiments are repeated, i.e., the box is dropped, thrown, put into orbit. As before, the box and the ball are each in the same gravitational field and have the same motion; consequently they are travelling together and are exerting no forces on one another.

This is the situation an astronaut finds himself in when he is in a space capsule which is in free flight (i.e., the rocket engines are not burning). Since the capsule is exerting no force on him, he describes himself as *weightless*. This is perhaps not the best word to use because he is in the gravitational field of a planet (or planets) and must there-

fore have weight. It might be better to describe an astronaut's condition as *reactionless* rather than weightless.

On the other hand, like all earth-bound creatures, the astronaut is accustomed to a gravitational field acting throughout the volume of his body while the surroundings exert a supporting force on a small area of his surface. This supporting force, or reaction, is recognized as being the result of his weight. When he is in orbit no supporting force exists and so he describes his state as weightless. This approach is especially excusable if he has tried to find his weight by connecting himself to his surroundings, i.e., the capsule, with a spring balance, for this will read zero!

28.10   The tides

A triumph of Newton's work on gravitation was that he was able to explain that the tides resulted from the attractions of the moon and sun. The moon is closer and provides the major effect. The moon attracts the water on the earth and raises the water nearest to it into a hump. This hump is fixed relative to the moon, and as the earth rotates beneath it every 24 hours so each part in turn experiences the change in depth. If you are by the sea you will be aware that during the day there are two high tides and two low tides. This implies that there is a second hump of water diametrically opposite on the side of the earth facing away from the moon. This is quite a puzzle because on this far side the water is being pulled by the earth and the moon in the same direction and so it might seem that the combined pull there ought to cause a hollow in the sea rather than a hump.

To explain this double hump (see Fig. 28.6) we

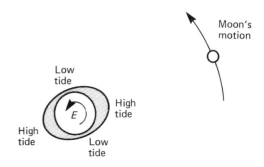

Fig. 28.6   The double tidal hump.

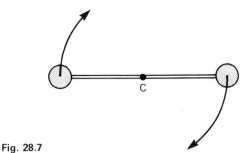

Fig. 28.7

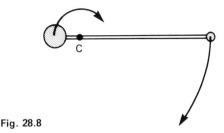

Fig. 28.8

must think of the combined motion of the moon and earth rather than just think of the moon orbiting an earth with a stationary centre.

Consider the following experiment: take a stick, about 30 cm long, and to each end fix a ball of plasticine or clay about the size of an apple. Throw this into the air giving it a flick as it goes. As it spins in the air, the balls will appear to rotate about a point C on the stick roughly midway between them (see Fig. 28.7).

Now replace one of the balls with a much smaller one and again spin the stick into the air (see Fig. 28.8). This time it will rotate about a point C near to the heavier ball. This point is the centre of mass of the spinning system. Another simple way of demonstrating this effect is to spin a hammer into the air and it will rotate about a point on the handle close to the head of the hammer.

The earth–moon system behaves in a similar way and rotates about the common centre of mass, so rather than think of the moon orbiting around the earth, we should think of them as moving around each other as shown in Fig. 28.8. Of course, while this is happening the earth is spinning about its own axis once in 24 hours, but as this motion does not contribute to the raising of

tidal humps it will not be brought into the discussion. It does, however, help to delay their arrival at a point on the earth's surface by about 6 hours. All we are aiming to explain is why there is a tidal hump of water on the far side of the earth.

Both the moon and the earth rotate eccentrically about point H (see Fig. 28.9) once in 27.3 days. If the earth's radius is r, then by imagining the earth and the moon to be joined by a rod and taking moments, it will be found that $OH = \frac{3}{4}r$, approximately. Thus $HA = \frac{1}{4}r$ and $HB = 1\frac{3}{4}r$. (Data for checking this is given in Problem 28.4 at the end of the section.)

Because of the rotation about H, point A moves in a circle of radius HA and B moves in a circle of radius HB; and so masses, either solid or liquid at these places must have the appropriate acceleration towards the centre of rotation H. Now, $HB = 7HA$ and so a mass at B moves 7 times faster than a mass at A. Consequently the necessary centripetal acceleration, and force, at B must be 7 times greater than at A. Using mv^2/r (or $mr\omega^2$) it may be shown that at A, a 1 kg mass must have a force 1.1×10^{-5} N acting towards H, and at B the corresponding force is 7.7×10^{-5} N towards H.

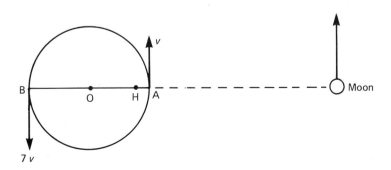

Fig. 28.9

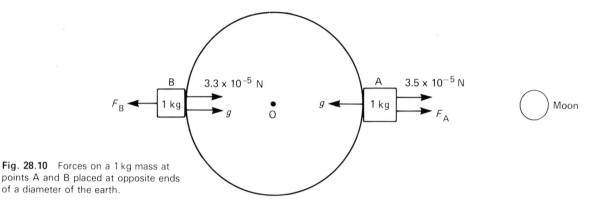

Fig. 28.10 Forces on a 1 kg mass at points A and B placed at opposite ends of a diameter of the earth.

Each of these forces on the 1 kg mass is provided by a combination of three forces:

a) The gravitational attraction of the moon which at the distance of the earth is 3.5×10^{-5} N at A, and 3.3×10^{-5} N at B. Both of these forces act towards the moon, i.e., towards the right in Fig. 28.9.

b) The gravitational attraction of the earth per unit mass, which is g, acting towards O, the centre of the earth.

c) The supporting force the earth exerts on the mass. At A let this force be F_A acting away from O, and at B let the force be F_B also acting away from O.

All of these forces are shown in Fig. 28.11.

The resultant force in newtons towards O, at A

is $g - F_A - 3.5 \times 10^{-5}$

and this provides the 1.1×10^{-5} N necessary for circular motion. Thus,

$$g - F_A - 3.5 \times 10^{-5} = 1.1 \times 10^{-5}$$

so

$$F_A = g - 4.6 \times 10^{-5} \ldots \text{ in N} \quad (28.13)$$

The resultant force in newtons towards O, at B is

$$g + 3.3 \times 10^{-5} - F_B$$

and this provides the 7.7×10^{-5} N required for circular motion. Thus,

$$g + 3.3 \times 10^{-5} - F_B = 7.7 \times 10^{-5}$$

so

$$F_B = g - 4.4 \times 10^{-5} \ldots \text{ in N} \quad (28.14)$$

Equations 28.13 and 28.14 show that the presence of the moon has reduced the effective gravitational field strength at both A *and at B*.

If the 1 kg masses were hanging from spring balances, the presence of the moon would cause the springs to be stretched less and the masses move away from O at both A and B. Water behaves in the same way and so the tide rises at A *and at B*.

The calculation above shows that it is the change in the gravitational field strength of the moon across the diameter of the earth which is responsible for the tides. Although the field of the sun at the earth is 180 times greater than that of the moon, the change across an earth diameter is very small. So a solar tide is much smaller than a lunar one. When the two tides are in phase the combined tide is a maximum – a *spring tide*.

Problem 28.4 Use the following data to check the various quantities quoted in Section 28.10.

Mass of earth = 5.98×10^{24} kg
Mass of moon = 7.35×10^{22} kg
Radius of earth = 6.37×10^6 m
Radius of moon = 1.74×10^6 m
Distance between centres of earth and moon
    = 3.84×10^8 m
$G = 6.67 \times 10^{-11}$ N m² kg^{-2}

1. Take moments to show that HB $\approx$ 7·HA.
2. Calculate the centripetal acceleration at A, and at B.
3. Calculate the gravitational attraction of the moon at A, and at B.

Chapter 29

POTENTIAL IN THE GRAVITATIONAL FIELD

29.1 Potential energy and potential

The concept of *field strength* was introduced so that we could describe a measure of the force that a field might exert on a mass, a charge or a current element placed at any point within it. This description was independent of any particular mass or charge or current element. It was a description of a general property of the field.

Any object situated in a force-field, and therefore acted upon by a force, possesses *potential energy*.

For example, in discussing the case of the gravitational potential energy of a mass at rest at some height h above an arbitrary zero (for example, the floor) we have seen that this energy is expressed by the product mgh. This assumes that h is small compared with the earth's radius so that g can be considered to be constant.

Just as we defined gravitational field strength at a point as the force experienced per *unit mass* at the point in question, we may now define a *gravitational potential* at a point as the potential energy per *unit mass* possessed by a mass placed at that point. Taking the surface of the earth as an arbitrary zero (Fig. 29.1a), the gravitational potential (V) at a height h is gh. This assumes that h is small compared with the earth's radius so that g can be considered to be constant.

In this example, points with the same gravitational potential can be connected by a series of lines parallel to the earth's surface. These are called *equipotential lines*. In three dimensions, points with the same potential all lie in surfaces, called *equipotential surfaces*.

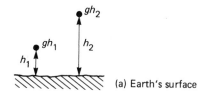 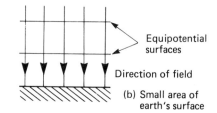

Fig. 29.1 (a) The gravitational potential and (b) equipotential surfaces near to the surface of the earth.

(a) Earth's surface

(b) Small area of earth's surface

The shape of a gravitational equipotential surface close to the earth's surface is spherical if we assume the earth is a spherically symmetrical mass. Within a small region the shape of the equipotential surface is indistinguishable from a plane as in Fig. 29.1b.

29.2 Gravitational potentials in the field around the earth

The calculations of potential in a uniform field are easy enough. But practical problems may not be so simple when the field is non-uniform.

Consider a spacecraft of mass m. How much energy must be given to it in order to get it clear of the pull of the earth? If the earth's gravitational field strength remained constant at all distances from the earth's centre, the question would have no answer. But the field is not uniform; it diminishes according to an inverse square law. This is shown in Fig. 29.2 in which the gravitational field strength g is plotted against distance from the earth.

When the spacecraft is 9 earth radii from the earth's surface (about 6×10^7 m – a sixth of the way to the moon) the gravitational force has fallen to about 1% of its value at the surface. The pull of the earth has fallen to about 0.1 N for each kilogram of spacecraft. Compared with the amount of energy already utilized in getting this far, very little more is required to increase the distance further.

At a distance of 12×10^7 m the pull of the earth is about 0.25% of its value at the surface, and is only 0.025 N per kilogram.

Ultimately, the spacecraft can get so far away that the gravitational pull due to the earth is smaller than any stated amount. We may then say that the spacecraft is so remote from the earth that no further energy has to be provided to free it from the gravitational pull of the earth.

The question is, how much energy has to be supplied to get the spacecraft into such a position?

Consider the spacecraft at a distance r from the earth's centre (Fig. 29.3). At this distance the field intensity g is given by

$$g = G \frac{M}{r^2}$$

In order to move the craft a further distance Δr along the radius, additional energy has to be transformed to gravitational potential energy. The energy gained *per unit mass*, $\Delta E = g\Delta r$.

We can calculate this with the help of the graph (Fig. 29.2) of field strength against distance.

At the distance of 1×10^7 m, $g \approx 4$ N kg^{-1}. If we take $\Delta r = 0.2 \times 10^7$ m, then ΔE is given by the area of the shaded strip under the graph between

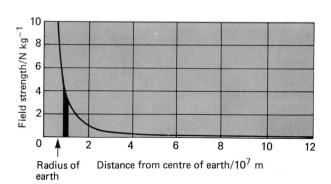

Fig. 29.2 Variation of the gravitational field strength in the vicinity of the earth.

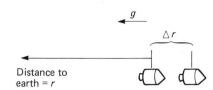

Fig. 29.3

distance values of 0.9×10^7 m and 1.1×10^7 m. Thus $\Delta E \approx 4\,\text{N kg}^{-1} \times 0.2 \times 10^7\,\text{m} \approx 8 \times 10^6$ J kg^{-1}. Therefore, at this distance of 1×10^7 m, every kilogram of the spacecraft requires 8×10^6 J of energy to move it 0.2×10^7 m. As you can imagine from the graph, strips, or elements of similar width, but farther from the earth will have a smaller area indicating the need for less energy; the farther away we go the easier it becomes, thanks to the inverse square law of gravitation.

To find the total energy per kilogram required to take the spacecraft from earth to a distance of 12×10^7 m, over 50 such computations would have to be performed and added together. A quicker but less accurate result will be obtained by taking a larger value for Δr. We shall take $\Delta r = 1 \times 10^7$ m, a large increment indeed.

Table 29a shows the first three summations for $g\Delta r$. You should complete it as far as 12×10^7 m and plot a graph of gravitational potential of the spacecraft against r, the distance from the centre of the earth.

Your graph should approximate to that shown in Fig. 29.4 which is an accurate one produced by using very much smaller increments than those suggested for Table 29a.

Figure 29.4 tells us that to get the spacecraft 12×10^7 m away from earth we have to provide a

total of 5.9×10^7 J for each kilogram of its mass. For a load of, say, 5000 kg (about the mass of an Apollo command module) the total energy to be supplied must be 2.95×10^{11} J.

The spacecraft requires still more energy to get completely free of the earth's pull. The potential energy–distance graph does not seem to be reaching a limit yet. In fact it approaches the limiting energy very slowly. If we go on with our calculations, it turns out that the limit is 6.23×10^7 J but we should have to make calculations to one hundred earth radii from the earth to get within 1% of this value.

This graph of potential energy per unit mass against distance (Fig. 29.4) is also a representation of the way in which the potential at points in the earth's field changes with distance. The potential is measured relative to the earth's surface taken as an arbitrary zero. It would be more rational to transfer the zero to some point far out in space where we could agree that no additional energy is required to move the mass farther away from the earth. We shall, in future, refer gravitational field potentials to this position (see Section 29.3).

To change from the former system to this new one, we need to subtract 6.23×10^7 J from all the

Table 29a

Average value of r/m	Value of g at r/ N kg^{-1}	Δr/m	$g\Delta r$/ J kg^{-1}	Cumulative energy total/ J kg^{-1}
1×10^7	4	1×10^7	4×10^7	4×10^7
2×10^7	1	1×10^7	1×10^7	5×10^7
3×10^7	0.44	1×10^7	0.44×10^7	5.44×10^7

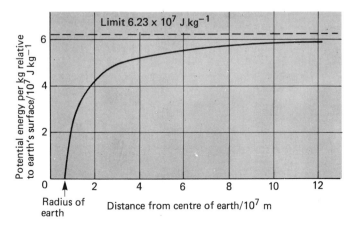

Fig. 29.4 Variation of the gravitational potential energy of unit mass with distance from the centre of the earth.

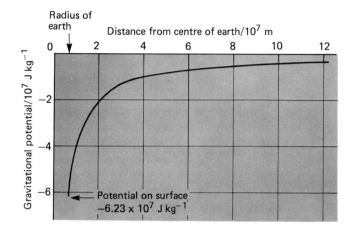

Fig. 29.5 Variation of the gravitational potential with distance from the centre of the earth. Zero potential is at an infinite distance.

numbers calculated, and replot the graph (see Fig. 29.5).

The potential of each point (or the potential energy of unit mass) is, of course now negative; all that has been done is to shift the graph axis. However, the shape of the curve is the same, and potential *differences* are the same, e.g.: find the change of potential when a spacecraft goes from 1×10^7 m to 4×10^7 m away from the earth. Remembering that:

change = (final state) − (initial state)

from Fig. 29.4, ΔE = $(5.2 \times 10^7) − (2.2 \times 10^7)$
= 3×10^7 J kg^{-1}
from Fig. 29.5, ΔE = $(-1 \times 10^7) − (-4 \times 10^7)$
= 3×10^7 J kg^{-1}

29.3   Choice of sign convention and datum for potential

The use of negative potentials is illustrated by the following 'thought-experiment'.

Three students are working in a tall building and are each measuring the potential energy change when a small mass of 0.1 kg is carried up, and down, stairs.

Student A has her laboratory on the ground floor and so she takes this as her zero energy datum. Every time she goes up a floor the mass gains 3 J of energy, and A records her results as in Fig. 29.6a.

Student B decides that his laboratory, half way up the building, will be his zero energy datum. When he goes up a floor the mass gains 3 J, but on going down, it *loses* 3 J per floor. To distinguish gains from losses B makes the latter negative and

his results are shown in Fig. 29.6b.

Student C works on the top floor, as far from the earth's centre as she can get, and this is her zero datum. She can only go down, and to be consistent, all her results are negative (see Fig. 29.6c).

C is really doing the same as we did when drawing the graph of Fig. 29.5 – taking the zero energy point as far from the earth as possible; but in the case of the spacecraft, the distance is infinite.

Asked, 'What is the energy change when the mass falls from the top floor to the ground?', all three students will give the same reply.

change = (final state) − (initial state)

For A, ΔE = $(0) − (15$ J$)$ = -15 J
For B, ΔE = $(-6$ J$) − (9$ J$)$ = -15 J
For C, ΔE = $(-15$ J$) − (0$ J$)$ = -15 J

With our convention, the negative sign signifies a *loss* of energy by the mass.

29.4   Potential changes and potential gradients

A body which is moving in a gravitational field of force may lose or gain potential energy. Whenever the body is displaced in the direction of the force which is exerted on it in the field, it will lose potential energy, changing this energy to some other form.

If the body, assumed to have unit mass, moves through a distance Δr in the direction of the force g due to the field, then the resultant change in the potential energy is given by

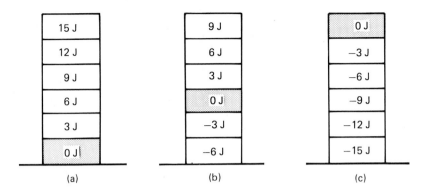

Fig. 29.6   (a)   (b)   (c)

$$\Delta V = -g\,\Delta r \qquad (29.1)$$

and this defines the change in potential, written ΔV.

The change in potential is negative (representing a loss) when the displacement Δr is in the same direction as the force. To get the correct sign for the change in potential when a body moves in a field of force we must pay due regard to the convention which gives rise to this formula.

(a) Sign convention

The positive direction is taken as the direction of increasing distance from a fixed point. Thus, in the case of the earth, any displacement away from the centre is positive, whereas the field, g, which is towards the centre is negative (Fig. 29.7).

As an example let us consider a unit mass which is moved 30 m away from the earth at a point where the gravitational field is $6\,\mathrm{N\,kg^{-1}}$. We apply the sign convention,

$$\Delta r = +30\,\mathrm{m} \quad \text{and} \quad g = -6\,\mathrm{N\,kg^{-1}}$$

Therefore,

$$\Delta V = -(-6\,\mathrm{N\,kg^{-1}})(+30\,\mathrm{m})$$
$$= +180\,\mathrm{Nm\,kg^{-1}} = +180\,\mathrm{J\,kg^{-1}}$$

This is positive and represents an increase in potential of the mass. Or, looked at from the point of view of a zero which is placed at an infinite distance, the potential has become less negative. See Fig. 29.5. If, instead, the mass moves towards the earth, the displacement is in the direction of decreasing distance and is negative, i.e., $\Delta r = -30\,\mathrm{m}$. Applying Eq. 29.1, we obtain

$$\Delta V = -(-6\,\mathrm{N\,kg^{-1}})(-30\,\mathrm{m})$$
$$= -180\,\mathrm{Nm\,kg^{-1}} = -180\,\mathrm{J\,kg^{-1}}$$

This is negative and represents a decrease in potential of the mass. In other words, the potential has become more negative.

Examples

a) Suppose a field intensity of magnitude g is in the direction PO in Fig. 29.8, where the direction OP is regarded as the positive direction. How does the potential change as we move away from O?

$$\Delta V = -(-g)(+\Delta r)$$
$$= +g\Delta r$$

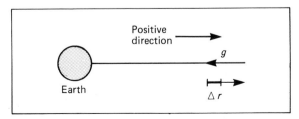

Fig. 29.7

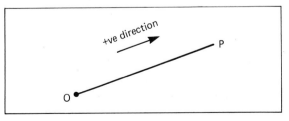

Fig. 29.8

So the potential increases.

b) Suppose now that the field intensity is in the direction OP in Fig. 29.8. How does the potential change as we move from P towards O?

$$\Delta V = -(+g)(-\Delta r)$$
$$= +g\Delta r$$

So the potential increases.

c) Energy is required to move a body from P towards O. What is the direction of the field?

In this case the change in potential is to be positive whilst Δr is negative. So the product $-(g)(-\Delta r)$ must be positive. It follows that g must be positive, and so is in the direction OP.

It should be noted that when a mass is moved in the direction of the gravitational force, its gravitational potential energy decreases whereas when it moves against the direction of the force, the mass gains potential energy.

b) Potential gradient

We have, in the general case of a gravitational field,

$$\Delta V = -g\Delta r$$

So

$$g = -\frac{\Delta V}{\Delta r}$$

In the limit as

$$\Delta r \to \text{zero}$$

$$\boxed{g = -\frac{dV}{dr}} \qquad (29.2)$$

dV/dr is known as the *potential gradient* at the point chosen. We see that the field intensity at a point is equal to the negative potential gradient at that point. Thus if in Fig. 29.4 (or Fig. 29.5) you measure the gradient of the curve corresponding to a distance of 2×10^7 m, you should find it to be about $1 \text{ J kg}^{-1} \text{m}^{-1}$, i.e., 1 N kg^{-1} (the tangent there has a slope of $45°$) which checks with the value of g which can be read off from Fig. 29.2. You could also check that at 4×10^7 m, $g = 0.25 \text{ N kg}^{-1}$ (slope of the tangent is $14°$).

The minus sign that appears in Eq. 29.2 of course indicates that the field, g, acts towards the centre of the earth and not away from it.

29.5   Potential at a point in the earth's gravitational field: the general case

As a unit mass moves through a distance Δr away from the earth the change in potential is given by

$$\Delta V = -g\Delta r$$

At a distance r

$$g = -\frac{GM}{r^2}$$

So

$$\Delta V = \frac{GM}{r^2}\Delta r$$

In the limit

as $\Delta r \to 0$

$$dV = \frac{GM}{r^2}dr$$

The total energy change in moving from $r = r_1$ to $r = r_\infty$ is

$$\int_V^0 dV = \int_{r_1}^\infty \frac{GM}{r^2}dr$$

$$[V]_V^0 = GM\left[-\frac{1}{r}\right]_{r_1}^\infty$$

$$= -GM\left[\frac{1}{\infty} - \frac{1}{r_1}\right]$$

$$0 - V = -\left(0 - \frac{GM}{r_1}\right)$$

$$\therefore V = -\frac{GM}{r_1}$$

If the mass moves from a point distant r from the centre of the earth to infinity

$$\boxed{V = -\frac{GM}{r}} \qquad (29.3)$$

For the earth, $GM = 4 \times 10^{14} \text{ J m kg}^{-1}$ and so Eq. 29.3 becomes:

$$V = -\frac{4 \times 10^{14}\,(\text{J m kg}^{-1})}{r(\text{m})}$$

$$= -\frac{4 \times 10^{14}}{r} \ldots \text{in J kg}^{-1} \qquad (29.4)$$

Equation 29.4 gives the potential at any point

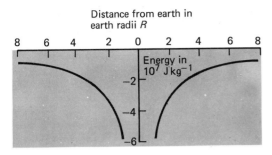

Distance from earth in
earth radii R

Fig. 29.9   The gravitational potential well of the earth.

distance r away from the centre of the earth. All values calculated will be negative because zero potential is taken to be at infinity.

Problem 29.1   Use Eq. 29.4 to check the points used for plotting Fig. 29.5. Also, plot a large version of this graph, and determine g at various distances by finding the gradients at a number of points; see Eq. 29.2. Check your answers using Eq. 28.9, which for the earth becomes

$$g = \frac{4 \times 10^{14}}{r^2}$$

29.6   A potential well

A spacecraft, almost stationary a long way from earth, is attracted towards it. As it falls towards earth, its kinetic energy increases as its potential energy decreases, just as a pail of water will fall down a well, if the rope to which it is attached breaks. This analogy has led to the variation of potential in an attractive force field being referred to as a *potential well*. Figure 29.9 is a graph of the earth's potential well.

A mass in a potential well is acted on by an attractive force towards the centre of the well. We

have so far only considered what will happen to a spacecraft held stationary a long way from the earth. Under these conditions, when the restraining force is removed, the spacecraft falls towards the earth with increasing speed. Is this the only possible motion for a mass in a gravitational potential well?

Here is an experiment you can do, to find out. Set up a large funnel (at least 30 cm in diameter at the mouth) so that its axis is vertical as shown in Fig. 29.10. The funnel forms a potential well. Hold a ball-bearing at the edge and then release it. It runs down the funnel (and out through the central hole, if it is small enough). This is something like the spacecraft hurtling in towards the earth. Now hold the ball-bearing some way inside the lip and give it a small impulse in a direction making an angle with the line to the centre (Fig. 29.10b).

The ball no longer drops down to the centre but moves on an ellipse. Sometimes the ball is close to the centre; sometimes it is far away.

When the ball was given its initial impulse, it had a total energy E.

$$E = E_p + \tfrac{1}{2}mv^2$$

E_p is its potential energy. This is negative if we reckon potential energies at the lip to be zero.

E must remain constant throughout the motion. (Why?)

Thus the lower the ball the more negative is E_p and so the larger is $\tfrac{1}{2}mv^2$. Do your observations confirm this?

29.7   The special case of circular motion

In one particular case, this ellipse becomes a circle. Can you set the ball-bearing into a circular orbit in the funnel? The forces on the ball have a resultant towards the centre of the funnel, in the

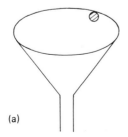

(a)

Fig. 29.10

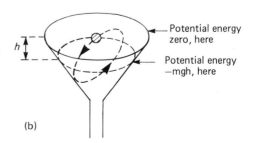

(b)

Potential energy zero, here

Potential energy $-mgh$, here

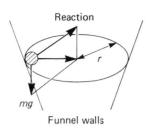

Reaction

r

mg

Funnel walls

Fig. 29.11

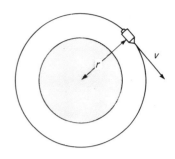

v

r

Fig. 29.12

plane of the orbit, as shown in Fig. 29.11. Ideally the ball maintains a constant distance from the attracting centre. What in fact happens to the ball? Why does it behave like this?

In the special case of motion in a circle, we can work out the value for the total energy E for a spacecraft or satellite in the earth's gravitational field. If the craft is moving with velocity v in a circle radius r as shown in Fig. 29.12,

$$E = E_p + \tfrac{1}{2}mv^2$$

$$= -G\frac{Mm}{r} + \frac{1}{2}mv^2 \qquad (29.5)$$

From Newton's laws of motion applied to motion in a circle (see Section 7.2)

$$\frac{mv^2}{r} = G\frac{Mm}{r^2}$$

Hence

$$mv^2 = G\frac{Mm}{r}$$

Substituting this value for mv^2 in the expression for E, we get:

$$E = -G\frac{Mm}{r} + \tfrac{1}{2}G\frac{Mm}{r} \qquad (29.6)$$

$$= -G\frac{Mm}{2r} \qquad (29.7)$$

29.8   A paradox in space

While a satellite is orbiting, friction with the outer layers of the earth's atmosphere causes a gradual reduction of its total energy. The radius of the orbit decreases and the satellite actually goes faster! This behaviour is not what is normally expected from a reduction of total energy.

This paradox can be understood by considering Eqs. 29.5 and 29.6 which show that the total energy is part potential

$$\left(-\frac{GMm}{r}\right)$$

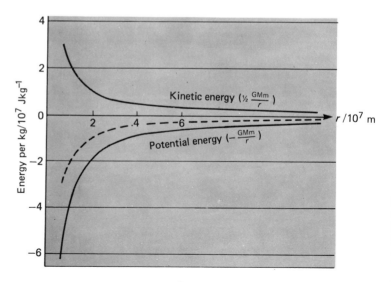

Fig. 29.13   Variation in E_p and E_k for a satelite in orbit with distance from the centre. The dashed curve represents the total energy.

and part kinetic

$$(\tfrac{1}{2}mv^2 \text{ or } \tfrac{1}{2}\frac{GMm}{r})$$

These two parts are plotted on the graph of Fig. 29.13, the dotted curve being the sum of the two, i.e., the total energy

$$(-\frac{GMm}{2r})$$

which is given by equation 29.7.

If the total energy is reduced, that is, made more negative, the result is a decrease in the radius of the orbit. If this happens the graph shows that the kinetic energy increases and so the satellite must go faster.

The reason that the satellite can lose total energy yet gain kinetic energy is that the gravitational potential energy decreases at twice the rate that the kinetic energy increases; thus there is an overall loss of energy. Figure 29.13 illustrates this.

Chapter 30

THE ELECTRIC FIELD

30.1   Exploration of electric fields

In Section 27.4 the principle for measuring the strength of an electric field was established by placing a small charge in the field and measuring the force per unit charge acting on it. In practice it is not easy to accomplish an actual measurement by this method directly, but the principle does give us a simple way of exploring electric fields and seeing how they vary, or if they are uniform.

A very thin flexible foil, usually of metal-coated plastic, about 15 mm long by 3 mm wide is stuck to the end of a long insulating rod (see Fig. 30.1). If this probe is placed between two vertical charged plates that are mounted on insulators, as in Fig. 30.2, no deflection of the foil should be seen. If now the foil is touched against one of the plates it will take charge from that plate and be deflected away to the opposite plate. The force exerted on the charge by the field and the opposing force developed by the flexing of the foil will result in the foil bending through a definite angle. This provides a rough indication of the field strength in the space between the plates.

If the indicator is now moved around in the space between the plates the foil will remain deflected by the same amount except when it is near the edges of the plates. There the deflection will be less. The deflection soon falls to zero as the

Fig. 30.1   A foil indicator for use in an electric field.

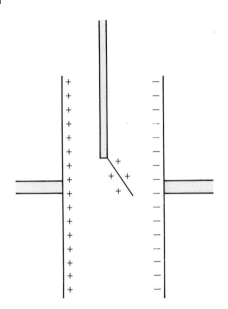

Fig. 30.2 A foil indicator in a uniform electric field.

probe is taken well away from the space enclosed by the edges of the plates. However, in the middle region of the space, the deflection is constant whether the foil is near one plate, in the centre, or near the other plate.

This elementary experiment shows that, apart from edge effects, there is a uniform electric field between charged parallel plates. Establishing that it is uniform is important because calculating the strength of the field then becomes simple as will be seen in Section 30.2.

What factors affect the strength of this field? Perhaps the area of the plates, or their separation,

or the potential difference between them which governs the charge they carry. The following experiment examines these possibilities. Consider a pair of parallel conducting plates that can be maintained at a known potential difference of up to a few kilovolts by means of a continuously variable high voltage supply; a foil indicator is placed between the plates (see Fig. 30.3).

Initially the plates may have, for example, a separation of 0.030 m and a p.d. of 2000 V. The foil is charged and its deflection is recorded by projecting its shadow on to a screen. If now the separation of the plates is changed, it will also be found necessary to change the potential difference in order to produce the same foil deflection, and therefore the same strength of field. In a typical experiment the values given in Table 30a were obtained.

Evidently the quotient V/d is constant within the limits of experimental error for a particular value of electric field strength. We shall see in the next section that this quotient gives a numerical value for the field strength.

Table 30a

Separation of plates (d)/m	Potential difference V/V	Ratio V/d/V m^{-1}
0.030	2000	67 000
0.040	2700	68 000
0.051	3500	69 000
0.063	4200	67 000
0.073	5000	68 000

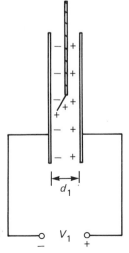
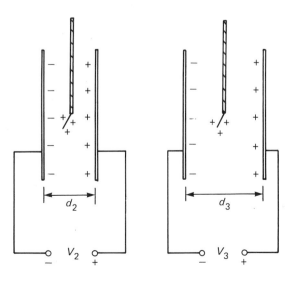

Fig. 30.3 A foil indicator in the field of a pair of charged, parallel plates.

To examine the effect of area on the field it is only necessary to replace the plates by a pair of, say, double the area. If these are given the same separations and potential differences as in Table 30a then the same constant deflection of the foil will result. (This assumes that the foil carries the same charge as before.) We can therefore conclude that area does not affect the electric field strength between parallel plates.

30.2 Calculation of the field between parallel plates

Suppose a positive test charge Q is in the field between the parallel plates A and B and that the force acting on it is F as shown in Fig. 30.4. Our initial foil experiment has established that F is constant right across from plate A to plate B provided Q is not too near the edges. Consequently if the charge is moved from A to B through the distance d between the plates then the energy gained by Q is Fd. Therefore the energy per unit charge $= Fd/Q$.

Now, this is the energy transfer per unit charge passing between the plates or, in other words, the potential difference between the plates. Thus

$$\frac{Fd}{Q} = V \quad \text{hence} \quad \frac{F}{Q} = \frac{V}{d}$$

But, $F/Q = E$, the electric field strength (see Section 27.4). Therefore,

$$\boxed{\frac{V}{d} = E} \tag{30.1}$$

Note that the unit of E can be either $N\,C^{-1}$ or $V\,m^{-1}$.

So, instead of trying the difficult task of determining a field strength as in Section 27.4 by measuring the force on a test charge directly, it is much more convenient to make the measurements with a voltmeter and ruler, and to calculate V/d.

30.3 Potential gradient

Equation 30.1 is that for a *potential gradient*. This concept was met before in Section 29.4 where a change of gravitational potential was given by $\Delta V = -g\Delta r$ (Eq. 29.1); the minus sign resulting from the sign convention adopted.

A similar relationship for electric fields is $\Delta V = -E\Delta r$, where ΔV represents a change in electrical potential, E is the electric field strength, and Δr is the displacement. One difference that the sign convention must accommodate is that whereas masses can only attract, charges can both attract or repel.

The sign convention used is:

a) the positive direction is the direction of increasing distance from a fixed point or datum,
b) the field direction is given by the force acting on a positive charge.

To illustrate this convention, consider a unit positive charge Q that is moved through a small distance Δr near a large charged sphere S as in Fig. 30.5. The positive direction is taken as the direction of increasing distance from S.

If the movement is 0.03 m away from S, then, applying the sign convention, $\Delta r = +0.03$ m. Let the average field strength for this displacement be 200 V m^{-1}. Since this acts away from S we can write $E = +200$ V m^{-1}. Therefore,

$$\Delta V = -(+200 \text{ V m}^{-1})(+0.03 \text{ m}) = -6 \text{ V}$$

This is negative and represents a loss of energy by the charge.

Next consider the case when S has a negative charge and again Q is moved away. Now the field acts towards S and so $E = -200$ V m^{-1}. As

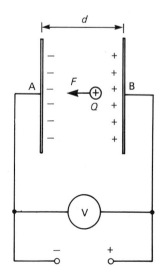

Fig. 30.4

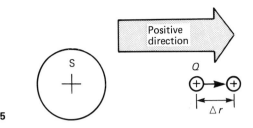

Fig. 30.5

before, $\Delta r = +0.03$ m. Therefore

$$\Delta V = -(-200 \text{ V m}^{-1})(+0.03 \text{ m}) = +6 \text{ V}$$

This is positive and represents a gain in energy by the charge.

Note that when a charge is moved in the direction of the force acting on it, then its electrical potential energy decreases; when it moves against the direction of the force on it, then its electrical potential energy increases.

We have in the general case of an electric field

$$\Delta V = -E\Delta r \qquad \text{or} \qquad E = -\frac{\Delta V}{\Delta r}$$

In the limit as $r \to$ zero,

$$\boxed{E = -\frac{\mathrm{d}V}{\mathrm{d}r}} \qquad (30.2)$$

$\mathrm{d}V/\mathrm{d}r$ is the potential gradient and the electric field strength at a point is equal to the negative potential gradient at that point.

30.4 The flame probe

In Section 29.4 on gravitational fields we saw that if a graph of potentials at various distances was available then the field strength at any point could be obtained by finding the gradient there. The same can be done for electric fields. But how can electric potentials be measured, for instance, in the space between charged plates? Some means of making an electrical 'connection' with a point in empty space has to be devised and this is achieved with a flame probe. A gold leaf electroscope is then used to measure the actual potential.

One form of flame probe is shown in Fig. 30.6. It consists of a 1 m length of about 12 mm diameter rigid plastic tube (PVC water pipe is suitable) one end of which is equipped with a bung and pipe for connection to a gas supply. The other end is fitted with a bung having a hole from which a hypodermic needle protrudes. A thin wire making connection with the needle is taped to the side of the tube and is connected to the electroscope. In use the tube is held by clamping it at the gas inlet end so that the needle is as far away as possible from benches, etc., because the presence of earthed objects will affect the potentials being measured. The gas supply is turned on and once the air has been driven from the tube it will be possible to light a flame at the end of the needle. The gas flow should be adjusted until the flame is about 3 mm long.

The electroscope is first calibrated as a voltmeter by connecting its cap and case to the positive and negative respectively of a power supply capable of being continuously varied up to 5 kV. The best method of providing a scale is to cast a shadow of the leaf on to the electroscope's frosted glass window. Marks can be made on this as the p.d. between cap and case is raised in steps of 200 V up to about 1400 V. The electroscope is now a voltmeter that does not require a continuous flow of current.

To help understand the action of a flame probe an introductory experiment (shown in Fig. 30.7) should be performed. A length of bare copper wire is made into a coil of 5 or 6 turns and about 1 cm across. This is taped to a plastic strip support. This is a simple way of providing a small hollow conductor and it is arranged so that the tip

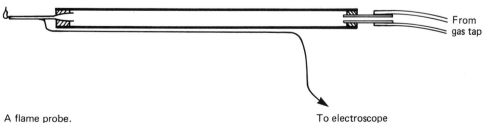

Fig. 30.6 A flame probe.

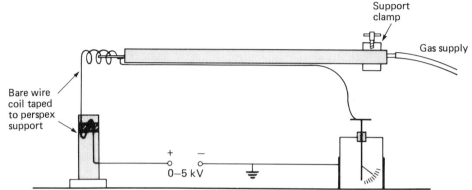

Fig. 30.7 Testing the action of a flame probe.

of the flame probe needle is just inside. The coil can be kept at a known potential V with respect to earth by a variable 0–5 kV supply. The calibrated electroscope measures the potential of the flame probe needle, also with respect to earth.

Initially allow the needle to touch the inside of the coil, and, with no flame burning, raise and lower the potential V and note that the electroscope and voltmeter readings agree.

Return the potential to zero and move the coil slightly so that the needle tip no longer touches its inside. Now when the potential is raised the electroscope reading remains at zero or nearly so.

Finally, light the tiny flame within the coil and, although there is no physical contact there, the electroscope reading will once again agree with the voltmeter reading as the supply is varied. The flame ionizes the air at the tip of the needle and then the ions will allow charge (of appropriate sign) to flow. Charge will only flow when a potential difference exists and so the flow continues until the tip is at the same potential as its surroundings.

The flame probe is thus able to measure the potential of the space within the coil. We shall use it to measure potentials at points in the space between charged plates and near charged spheres.

30.5 The potential gradient between parallel plates

The flame probe can now be used to measure potentials at points *between* the parallel plates of Fig. 30.4. The electroscope case is earthed and connected to the negative plate which is the zero datum for potential and distance measurements.

A suitable value for the p.d. between the plates is 1200 V.

Typical graphs for two different values of the plate separation d are shown in Fig. 30.8. The closer the plates are, the greater the potential difference and the larger the field strength. Within the limits of experimental error the relationship between V and r is linear, indicating a constant value for dV/dr and so a uniform field strength, a fact established earlier with a foil detector and now verified by alternative reasoning.

30.6 Field strength and potential near a charged sphere

A conducting sphere is suspended from the ceiling by an insulating nylon line and is prevented from swinging by the lead that connects its surface to a 1.5 kV power supply. The negative socket of the supply is 'earthed' so that all potentials measured are relative to the potential of the earth which is chosen as the zero datum. Any apparatus and

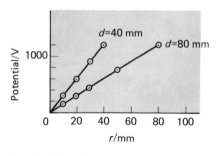

Fig. 30.8 Variation of potential with distance between pairs of parallel plates at two different plate separations.

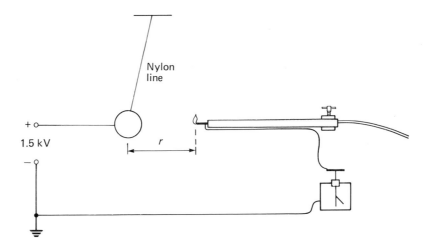

Fig. 30.9 Investigating the potential near a charged sphere with a flame probe.

furniture is at least 1 m away from the sphere so that its field is not affected. If a foil detector (Section 30.1) is charged and held at various distances from the sphere it will be clear that the field is non-uniform.

Figure 30.9 shows how a flame probe may be used to measure the potential at a distance r from the centre of a charged sphere. Some typical results are shown in Fig. 30.10.

Readings taken from this graph will show that the product Vr is a constant and so

$$V \propto \frac{1}{r} \qquad (30.3)$$

The same relationship holds for gravitational potentials and the graphs of Fig. 30.10 and of Fig. 29.5 have the same form; the obvious difference between them is the result of the positive charge on the sphere causing repulsion, whereas gravitation always gives an attraction. If a negative charge had been used instead, then all the potentials measured would have been negative ones and when plotted would give a graph exactly like Fig. 29.5.

We must now examine how potential depends on the charge carried by the sphere. For this a permanent connection to the power supply is not required but the sphere must be well insulated, as before, by suspending it with a nylon line. It is charged by touching it momentarily with the lead from the power supply, set at, for example, 1000 V. This charge is then measured by transferring it to a 0.01 μF capacitor, which is such a large value compared with the capacitance of the sphere that very nearly all of the charge will be transferred. The problem is now reduced to one of measuring the small potential difference across the capacitor. This p.d. is approximately 1 V but we cannot use a moving-coil meter as it will have a resistance of, perhaps, 10^4 ohm and would rapidly discharge the capacitor (time constant $RC = 10^{-4}$ s). Instead we can use an *electrometer*. This is

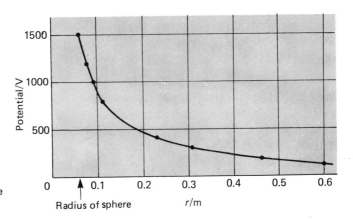

Fig. 30.10 Variation of potential with distance from the centre of a charged sphere.

an electronic device which is effectively a 0–1 V voltmeter with a very high input impedance of about 10^{14} ohm. Now the time constant $RC = 10^6$ s, and it would take about 11 days to discharge the capacitor to one-third of its initial charge! Therefore the electrometer will give a steady reading of the p.d. across a capacitor instead of a rapidly falling one and if such a p.d. is 0.65 V then the charge stored in a 0.01 μF capacitor is given by

$$Q = CV = 0.01 \times 10^{-6} \times 0.65 \text{ C}$$
$$= 6.5 \times 10^{-9} \text{C}$$

This was the charge stored on our sphere when its potential was 1000 V and by taking readings with different initial potentials it can be shown experimentally that the potential of the surface of the sphere is proportional to the charge on it and, in general, $V \propto Q$. Combining this with Eq. 30.3 we can write $V \propto Q/r$, or, introducing a constant,

$$V = k\frac{Q}{r} \qquad (30.4)$$

The value of k can be obtained by inserting known values of V, Q, and r into Eq. 30.4. For instance, on the surface of the sphere $r = 0.06$ m, $V = 1000$ V, Q was found to be 6.5×10^{-9} C. This gives $k = 9.2 \times 10^9$ V m C^{-1}.

In Section 30.15 we shall see that

$$k = \frac{1}{4\pi\epsilon_0}$$

where ϵ_0 is the permittivity of free space (or electric field constant). Since this is 8.85×10^{-12} F m^{-1}, $k = 8.99 \times 10^9$ V m C^{-1}. The equation for the potential at a distance r from the centre of a sphere carrying a positive charge Q is

$$V = \frac{1}{4\pi\epsilon_0}\frac{Q}{r} \qquad (30.5)$$

Problem 30.1 Using values taken from Fig. 30.10 and Eq. 30.4, estimate the charge on the sphere. Use this value of charge to calculate the potential at (i) 0.12 m, and (ii) 1 m, from the centre of the sphere.

(Answers: 10^{-8} C, 750 V, 90 V.)

30.7 The inverse square law for electric field strength

As with gravitational fields (see Section 29.4b) the strength of an electric field at a point can be found if a graph of potential against distance is available and $E = -\,dV/dr$ is applied. Figure 30.10 is such a graph and it (or a large scale version of it drawn using $V = 90/r$) can be used to find the potential gradient at $r = 0.15$ m and 0.30 m. The corresponding electric field strengths are $E = 4000$ V m^{-1} and 1000 V m^{-1}. Thus doubling the distance results in reducing the field strength to a quarter: the characteristic of an inverse square law. If this law applies, what will be the field strength at $r = 0.60$ m? Check your answer by finding the gradient at that point.

Another way of finding a field strength is to differentiate the expression for potential in order to get its rate of change with respect to distance, i.e., to find $E = -\,dV/dr$, where

$$V = k\frac{Q}{r} = kQr^{-1}$$

Therefore,

$$E = -kQ(-1\,r^{-2})$$

hence,

$$E = k\frac{Q}{r^2} \qquad (30.6)$$

The field strength E is the force acting per unit charge. For a charge Q_2 placed at a distance r from a charge Q_1, the force acting is:

$$F = Eq = k\frac{Q_1 Q_2}{r^2} \qquad (30.7)$$

This is a statement of the inverse square law of force between point charges. It is also known as Coulomb's law after the French scientist who investigated the force between electrically charged spheres in 1784.

30.8 Experimental test of Coulomb's law

To investigate the way in which the force between two charges varies with those charges and with their separation we shall describe two experiments. In the first, the light metallized sphere

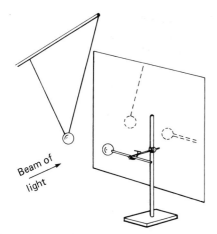

Fig. 30.11 A test of Coulomb's law.

(about 1 cm in diameter) is fixed to an insulated stem and given a charge which is to remain unaltered throughout the experiment. This may be done by using either a power supply providing several kilovolts, or a Van der Graaff generator. A second similar sphere, which is suspended from a well-insulated nylon trapeze (see Fig. 30.11) is also charged. The first sphere, held rigidly as shown, is brought up to the suspended sphere. Repulsion occurs if the charges are of the same sign. To measure the separation of the two spheres without bringing a rule near to them it is convenient to throw a shadow of them on to a screen using light from a distant source. The angle of deflection (θ) of the nylon threads is measured. The separation of the spheres (r) is noted and the charge (Q_1) on

the suspended ball is discharged to the capacitor and electrometer mentioned in Section 30.6, and so measured. The experiment is repeated with different charges on the suspended sphere but with the same separation of the spheres and with the same charge on the fixed sphere.

If F_E is the force between the two spheres, F_T the tension in the thread and m the mass of the sphere, x the displacement shown in Fig. 30.12 and l the length of the thread we have, at equilibrium

$$F_E = F_T \sin \theta$$

and

$$mg = F_T \cos \theta$$

Dividing

$$F_E/mg = \tan \theta$$

so

$$F_E = mg \tan \theta$$

$$\approx mg\, x/l$$

provided that θ is small.

The deflection x of the nylon suspension is therefore directly proportional to the force of repulsion between the two charges.

Figure 30.13 shows the relationship between the charge on the sphere and the deflection of the suspension, which, as we have seen, is a measure of the force. We conclude that

$$\text{force} \propto \text{charge}$$

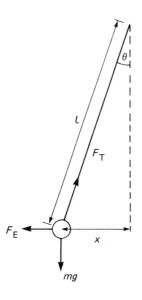

Fig. 30.12

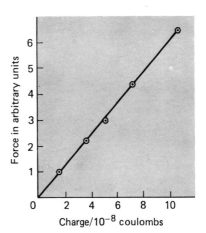

Fig. 30.13

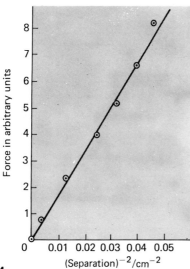

Fig. 30.14

In the second experiment, the charges on the two spheres are kept the same and the separation r is changed. Figure 30.14 shows the graph which is obtained under these circumstances when the deflection (recorded as the force) is plotted against

$$\frac{1}{r^2}$$

We conclude that

$$\text{force} \propto \frac{1}{r^2}$$

More accurate experiments confirm that the force of one charged sphere on the other varies inversely as the square of their separation. We already know that the force is proportional to Q_1, the charge on the suspended sphere. Further experiment can show that it is also proportional to the charge, Q_2, on the sphere fixed to the insulated stem.

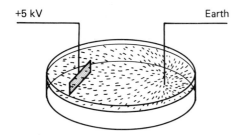

Fig. 30.15 Displaying electric fields.

So we find

$$F_E \propto \frac{Q_1 Q_2}{r^2}$$

and therefore

$$F_E = k \frac{Q_1 Q_2}{r^2}$$

where k is a constant of proportionality.

30.9 Electric field patterns

The plotting of magnetic fields around a magnet with iron filings is familiar enough. Similarly, the visualization of the shape of electric fields near charged conductors can be carried out using electrodes in a shallow dish as in Fig. 30.15. The dish contains a layer of clear oil upon which has been sprinkled a layer of grass seed, semolina, or short hairs about 1 mm long chopped from a brush. The electrodes are connected to a power supply capable of an output of 5 kV, and the semolina particles or hairs will align themselves with the direction of the electrostatic force at each point in the field. Various electrode shapes can be constructed and Fig. 30.16 shows some typical field patterns. Figure 30.16a shows a uniform field characterized by straight parallel lines in the central region. The field becomes curved and non-uniform at the edges of the 'plates'. Figure 30.16b represents a radial field which, in three dimensions, results in an inverse square law of force. Figure 30.16c is also a radial field between concentric electrodes, but note how there is no field within the central electrode. This relates to the zero gravitational field inside a spherical shell discussed in Section 28.8.

In all cases field lines are perpendicular to the electrode at each end of the line. This point should be remembered when sketching electric fields. The reason for this is discussed in the next section.

30.10 Equipotential lines and surfaces

A contour line on a map is an example of an equipotential line. Any point on a particular line is at the same potential, or height above sea level. Likewise, equipotential lines (or in three dimensions, equipotential surfaces) can be drawn in electric fields and you can imagine the path traced

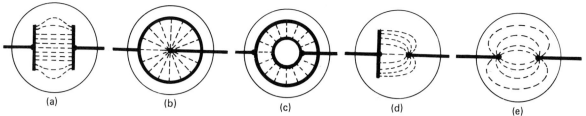

Fig. 30.16 Examples of electric field patterns.

out by a flame probe that is moved so that the indication of potential on the gold leaf electroscope is constant at some chosen value. If this is done with the sphere of Fig. 30.9 the paths traced out will be a series of concentric spheres. Figure 30.17 illustrates equipotentials around a sphere and also between parallel plates. An important point to note is that when equipotential lines and field lines cross, they do so at right angles. This is because the potential energy of a charge does not change if it is moved at right angles to a force, i.e., to a field line.

Because an electrode is a conductor, if there is no current flowing then all points on its surface must be at the same potential and so the surface is an equipotential. This is why field lines enter and leave conductors at 90° to the surface.

30.11 A further look at field potentials

In Section 30.2 and 30.3 we have chosen to consider only energy changes for displacements along the direction of the field. What will happen

if movement takes place along any path? To discuss this we shall find it convenient to use the ideas of the equipotential line and surface (see Section 30.10). Figure 30.17 shows equipotential lines in an inverse square law field and in a uniform field.

Consider the paths shown in Fig. 30.18. The field between the two plates is uniform.

How much energy will be gained if a test charge is moved from A to B? We can break the path down into any number of zigzag steps (Fig. 30.18b) which either go along the field direction or along an equipotential surface. Along the latter, no changes in energy take place. So the net change in energy will be exactly the same as though the test charge had moved in the direction of the field from one equipotential line through A to an equipotential line through B.

From this, it follows that if the test charge returns from B to A by *any* path the energy lost will be identical with the energy gained in moving from A to B. Thus the potential at any point in an electric or a gravitational field has a unique value

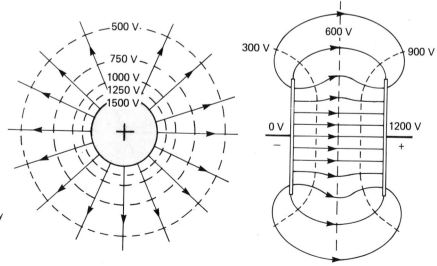

Fig. 30.17 Examples of equipotential surfaces (shown by dashed lines) and the corresponding field lines (arrowed).

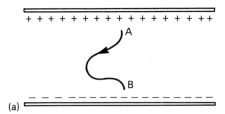
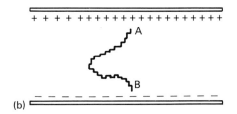
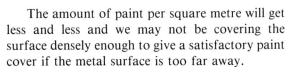

Fig. 30.18   (a)   (b)

relative to any chosen zero. It does not depend on the path a charge takes to reach it. Fields which behave like this are called *conservative* (see Section 8.16).

30.12   Flow, and the inverse square law

We have seen that radial fields, whether electric or gravitational, follow an inverse square law. This is a law that is prominent in physics and it applies to cases other than force fields. For example, the intensity of gamma radiation falls off inversely with the square of the distance from the source, and the same applies to the energy radiated by the sun. Neither of these are field strengths. We will study the inverse square law by considering the flow of paint particles from an aerosol paint spray. This will lead to the flux model which is a useful concept when considering gravitational and electric fields.

Let us suppose we want to spray a patch of colour on a piece of metal held perpendicular to the path of the spray, which is fixed in one position (Fig. 30.19). We apply these restrictions because we are trying to make the problem as simple as possible. Clearly we shall cover a much bigger patch if we place the metal at BB′ than if we place it at AA′. We could cover an even bigger patch if we moved the piece of metal farther away still. Can we paint bigger and bigger areas satisfactorily by just moving the surface to be sprayed farther away from the can?

The amount of paint per square metre will get less and less and we may not be covering the surface densely enough to give a satisfactory paint cover if the metal surface is too far away.

Why is this? Because in all cases, the same total amount of paint falls on the surface in unit time. It is just spread out more thinly as we move the painted surface away from the spray. How much more thinly? Consider Fig. 30.20. The area covered by the spray at AA′ = πs^2, where s is the radius of the cone of spray at distance r. The area covered by the spray at BB′ = πS^2, where S is the radius of the cone of spray at distance R. If an amount of paint P leaves the spray in unit time, the density of paint

$$\text{at AA}' = P/\pi s^2$$
$$\text{at BB}' = P/\pi S^2$$

But

$$s:S = r:R$$

Hence

$$\frac{\text{density of paint at A}}{\text{density of paint at B}} = \frac{S^2}{s^2} = \frac{R^2}{r^2}$$

The density of paint sprayed to any distance r from the source is proportional to $1/r^2$.

So the density obeys an *inverse square law*. This law is a consequence of the geometry of space.

A law of the same form applies to light

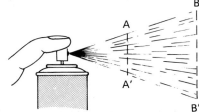

Fig. 30.19 The aerosol paint law.

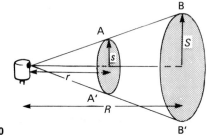

Fig. 30.20

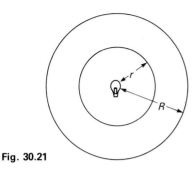

Fig. 30.21

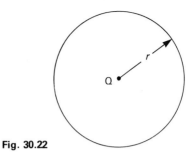

Fig. 30.22

intensities. The amount of light energy I provided by a point source per second, will under normal circumstances, be the amount passing through the surface of the sphere of radius r and also through the sphere of radius R (Fig. 30.21).

If the light is emitted uniformly in all directions the intensity of illumination (or illuminance) at a distance r is

$$\frac{I}{4\pi r^2}$$

and at distance R is

$$\frac{I}{4\pi R^2}$$

It thus follows that the light intensity at any distance r from the source is proportional to $1/r^2$.

30.13 Flux and flux density

Flux is merely another word for flow and indeed in workshops, *flux* is the name of a material used in soldering and brazing processes to help the molten metal to flow freely and to spread evenly. The inverse square law is a law of flow. When any quantity flows outwards from a point source uniformly in all directions at an unchanging rate then the amount passing through parallel surfaces of the same area at different distances from the source will obey an inverse square law. The amount flowing out is known as a *flux* – in the examples considered, a flux of paint and an energy flux in the case of light.

Under these conditions, the flux per unit area diminishes inversely as the square of the distance from the point source. The flux per unit area is known as the *flux density*. In the case of the paint spray, the flux density measures the degree to

which the paint covers the surface to be painted; in the case of light, the flux density is a measure of the illumination of the surface on which the light falls.

30.14 The flux model for the electric field

Our experiments confirm that the electric field intensity obeys an inverse square law and, as we have seen, this represents the typical behaviour of a flux density. We will suppose then, that E, the field intensity, does behave like a flux density.

We have no need to enquire into the reality of this analogy between the two phenomena but we can use it. It provides us with a useful model for discussion. It must be stressed that nothing actually flows in the case of the electric field, and yet the idea of flow is very helpful.

What would be the total flux from a charge Q in this case? This total electric flux, ψ, must be the flux density, E, at the sphere (Fig. 30.22) multiplied by the area of the surface of the sphere.

$$\psi = E \times 4\pi r^2$$

$$= \frac{kQ}{r^2} \times 4\pi r^2 \qquad \text{(using Eq. 30.6)}$$

$$\psi = 4\pi k Q \qquad (30.8)$$

We were able to calculate ψ very easily because we chose a particularly simple and symmetrical case. But the shape of the surface surrounding Q should not affect the electric flux associated with Q, if our model is a good one. To return to the flux law of light, the placing of a book relative to a lamp will certainly control how much light falls on its pages, but it does not affect how much light leaves the lamp. Similarly if we spray a surface with paint by holding the paint spray close to it we shall

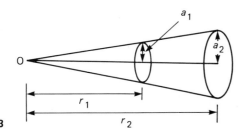

Fig. 30.23

get a dense layer of paint. If we hold it a long way off we shall get a thin layer of paint – but the amount of paint coming from the spray per second is the same in both cases.

So we can generalize our model and assume that any charge Q is a source of electric flux, $4\pi kQ$. The electric field at any point is then the density of the electric flux at that point. This is a simplified statement of Gauss's law. We will look at some consequences and applications of this model shortly. But before that, let us again consider the steps we have taken in setting up this model.

We have used the fact that flux densities, like paint films and illumination, obey inverse square laws in certain symmetrical cases. Why they do so is simply a property of the geometry of space. For the cone shown in Fig. 30.23,

$$a_1/a_2 = r_1/r_2$$

$$\text{area } A_1 = \pi a_1^2 \quad \text{area } A_2 = \pi a_2^2$$

Thus

$$A_1/A_2 = a_1^2/a_2^2$$
$$= r_1^2/r_2^2$$

The areas are in proportion to the square of their distances from O.

The validity of our model of the electric field depends upon the fact that experimentally the electric field E obeys the inverse square law whenever a flux density model suggests that it will do so.

Thus ultimately our model depends upon the trust we can place in the power of 2 in the expression

$$E \propto Q/r^2$$

We shall see in Chapter 47 that this relationship is applied successfully down to the distances involved in collisions between sub-atomic particles.

30.15 The relationship between k and ϵ_0

Equation 30.8 gives the total flux arising from a charge Q. Now consider the uniform field between parallel plates of area A and separation d, having a p.d. of V between them. Since there is a charge of magnitude Q on each plate, if the plates are close together so that we may assume there is negligible field beyond the area of the plates, then the flux between the plates = $4\pi kQ$.

$$\text{Flux density} = \frac{\text{flux}}{\text{area}} = \text{field strength} = \frac{V}{d}$$

Therefore

$$\frac{4\pi kQ}{A} = \frac{V}{d} \quad \text{and so} \quad k = \frac{AV}{4\pi Qd}$$

In Section 26.3 we saw that for a parallel plate capacitor

capacitance,

$$C = \frac{Q}{V} = \frac{\epsilon_0 A}{d}$$

and so,

$$\frac{AV}{Qd} = \frac{1}{\epsilon_0}$$

Therefore,

$$\boxed{k = \frac{1}{4\pi\epsilon_0}} \quad (30.9)$$

The permittivity of free space, ϵ_0, = $8.85 \times 10^{-12}\,\text{F m}^{-1}$ and so $k = 9 \times 10^9\,\text{m F}^{-1}$.

The constant ϵ_0 is the *permittivity of free space*. For media other than a vacuum, the constant is written $\epsilon_r\epsilon_0$, where ϵ_r is the relative permittivity of the medium. Since the relative permittivity of air differs little from unity, we may use ϵ_0 when the charges are separated by air rather than a vacuum.

The value of the permittivity constant ϵ_0 is

$$8.854 \times 10^{-12} \quad \text{or} \quad \frac{1}{36\pi} \times 10^{-9}$$

and its units are often quoted as $\text{J}^{-1}\,\text{C}^2\,\text{m}^{-1}$ or as F m^{-1}.

You should check these units against those which you would expect from the equation

$$E = \frac{1}{4\pi\epsilon_0}\frac{Q}{r^2}$$

30.16   The capacitance of a sphere

The potential, V, of a sphere of radius r that is carrying a charge Q is given by:

$$V = \frac{1}{4\pi\epsilon_0}\frac{Q}{r}$$

If C is the capacitance of the sphere then

$$C = \frac{Q}{V}$$

and so, for an isolated sphere,

$$\boxed{C = 4\pi\epsilon_0 r} \qquad (30.10)$$

30.17   Summary of relationships

This unit 7 has introduced many relationships and this summary illustrates the similarities between the gravitational and electric cases. The textual equation numbers are given in brackets in Table 30b.

Table 30b

	Gravitation		Electric	
Potential gradient or field strength	$-\dfrac{dV}{dr}$	(29.2)	$-\dfrac{dV}{dr}$	(30.2)
Potential at distance r	$-G\dfrac{M}{r}$	(29.3)	$\dfrac{1}{4\pi\epsilon_0}\dfrac{Q}{r}$	(30.4)
Potential energy at distance r	$-G\dfrac{Mm}{r}$		$\dfrac{1}{4\pi\epsilon_0}\dfrac{Qq}{r}$	
Field strength at distance r	$G\dfrac{M}{r^2}$	(28.9)	$\dfrac{1}{4\pi\epsilon_0}\dfrac{Q}{r^2}$	(30.6)
Force at distance r	$G\dfrac{Mm}{r^2}$	(28.1)	$\dfrac{1}{4\pi\epsilon_0}\dfrac{Qq}{r^2}$	(30.7)

These expressions are not difficult to remember if you note the patterns. Potential follows a $1/r$ rule, whereas field and force have a $1/r^2$ rule. Gravitation formulae contain the gravitational constant G, whereas the corresponding constant in the electric cases is $1/4\pi\epsilon_0$.

Potential and field are both 'per unit mass' or 'per unit charge', whereas the expressions for potential energy and force include the mass (m) or charge (Q) on which the appropriate field acts.

It should be stressed that, strictly speaking, these expressions apply to *point* masses or *point* charges.

UNIT EIGHT

Electricity and Matter

Chapter 31

ELECTRIC CURRENTS AND THE PARTICLE NATURE OF CHARGE

31.1 Introduction

When the two terminals of an electric battery are connected by wire to electrodes separated by a liquid, such as a dilute acid or a solution of a salt, changes occur. The circuit is now surrounded by a magnetic field; the components warm up; chemical action takes place in the conducting liquid and in the battery itself. These effects are all related to the electric current.

The word 'current' suggests a flow – in this case, a flow of charges. As we have seen in Chapter 23 measurements of the current do suggest a behaviour not unlike that which we associate with other examples of a flow. In the simple series circuit for example, the current is the same wherever it is measured. At junctions in circuits (Fig. 31.1) the current divides or combines so that as much current enters the junction as leaves it.

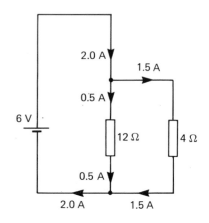

Fig. 31.1

It is possible to build a simple model of such a current of electric charge. And when we add to such a model the concept of the volt as a measure of the energy dissipated per coulomb of charge passing a section of a circuit, it becomes very tempting to elaborate this model of electrical 'stuff' circulating round a circuit, transferring energy in chemical reaction, in heating, and in causing motion through the interaction of magnetic forces. It is not unlike a constant procession of bottles which start full of milk and gradually lose it as they travel from the dairy to the home and back again round the circuit; or, maybe, a fluid which flows around carrying energy as hot water in a heating system carries energy. Imagery of this sort is useful because it provides a framework within which to think about physical problems.

The electrical phenomena of the simple circuit force us to ask questions about electricity. Among them: what is it that flows in a circuit? If charge, is it positive or negative or both? How fast does it flow? What is electric charge – a form of matter or a property of matter? Is it particulate as is matter?

It is the purpose of this Unit to examine possible answers to such questions.

One of the difficulties we experience in building a convincing model of the electric current is that we can have no direct experience of what is going on. All our evidence is indirect. But some can come from experiments in which the passage of electricity is associated with the movement of matter.

31.2 The shuttling ball

This experiment demonstrates that what we observe on a galvanometer, and so associate with an electric current, can be attributed to the continuous transfer of definite quantities of electric charge. See Fig. 31.2

Two metal plates about 150 mm in diameter and mounted on insulating supports are separated by 100 mm. A table tennis ball that has been coated with colloidal graphite to give it a conducting surface is suspended midway between the plates by a long thin nylon thread (a very good insulator). The plates are connected in series with a 5 kV supply and a light beam galvanometer that is capable of detecting currents in the order of a microampere.

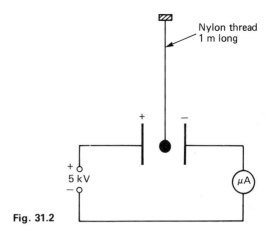

Fig. 31.2

If the ball is uncharged, nothing will be observed even though it is in the strong electric field between the plates. However, when the ball is moved (using the nylon thread to avoid shocks) so that it touches the negative plate, it gains a negative charge and now experiences a force that causes the ball to swing towards the positive plate. The ball touches the positive plate, gains a positive charge and swings back towards the negative plate where it collects a negative charge once again. The ball oscillates continuously back and forth carrying a negative charge when it moves towards the left and a positive charge when it moves towards the right.

While this is happening the galvanometer shows that a current is flowing in the circuit because it gives a reading that is reasonably steady so long as the oscillations of the ball are frequent enough.

Problem 31.1

1. How can we check experimentally that the galvanometer reading is due to both positive and negative charges moving (i) in the space between the plates, and (ii) in the connecting wires?

2. Formulate a model in which only one type of charge can move that can explain the flow of current and also how an insulated conductor can be either positively or negatively charged.

1. (i) The charge on the ball can be tested with a gold leaf electroscope, after contact with each plate, to confirm experimentally that the ball alternately carries positive and negative

charges. (ii) Although both types of charge can be seen to move in one part of a series circuit it cannot be deduced that the same process necessarily happens in a different part, especially in a wire where the process cannot actually be seen.

2. A moving positive charge hypothesis: suppose all uncharged conductors consist of equal numbers of positive and negative charges, and that the positive ones are free to move while the negative charges are fixed. This enables the following to be explained:

a) Current flow: if the terminals of a cell are joined, its positive pole repels the positive charges in the wire and these move towards the negative pole, i.e., in the conventional direction of current flow. A continuous supply of positive charges comes from the chemical action in the cell.

b) An uncharged insulated conductor: this has equal numbers of mutually cancelling positive and negative charges.

c) A positively charged insulated conductor: this has extra positive charges added to it so that the number of positive charges exceeds the number of negative charges.

d) A negatively charged insulated conductor: this has had some of its positive charges removed which leaves the conductor with an excess of negative charge.

Thus all of the usual effects can be explained on the basis that it is only the positive charges that can move through a conductor. Of course, an equally satisfactory hypothesis can be formulated based on fixed positive charges and moving negative charges.

The shuttling ball experiment shows that what is observed on a galvanometer as a flow of current in a circuit is consistent with the passage of many separate or discrete charges. The experiment gives no indication concerning the sign or the size of these charges, or indeed if the discrete charges are all of the same size.

31.3 The basic charge: the electron

The hypothesis of discrete charges originally arose from work on electrolysis and was crystallized by G. Johnstone Stoney in an address to the British Association in 1874. Stoney spoke of a natural unit of electric charge associated with such ions as those of hydrogen, chlorine, etc. In 1891 he went further, and was able to estimate the size of this unit at about one tenth of the value accepted today; he named this unit of electric charge the *electron*, and assumed that all electrical charges were whole multiples of this basic charge. Subsequent to Stoney's invention of the term it became clear that the electron was not just a charge but that it had a mass associated with it. The evidence for this arose as a result of experiments carried out on the conduction of electricity through gases at low pressure (Section 33.1) which culminated in the work of J. J. Thomson who measured the specific charge e/m (the ratio of charge to mass) of the electron (Section 33.3). Thus it became clear that what we now call an electron is a particular negatively charged speck of matter.

31.4 Measuring the charge

Many successful attempts were made to measure directly the small charges associated with atoms of matter. Nevertheless, it was R. A. Millikan who produced definitive confirmation of the existence of a basic unit of electric charge in 1909, even though most physicists at that time would have said that the case was already proven.

The essence of Millikan's experiment was to measure the individual charges on many tiny oil droplets that had been produced by a spray. Friction during the spraying process gave several of the droplets various amounts of charge. By analysing his results, Millikan deduced that the charges measured were all multiples of a fundamental unit of charge.

As an analogy, let us suppose that we are presented with the masses of a number of packages. We are told that, as a working hypothesis, we may assume that each contains a different number of otherwise identical marbles. Table 31a gives a list of such masses (due allowance has been made for the packaging). If the hypothesis is correct then each mass should be an integral multiple of the mass of a single marble. By inspecting the table of masses we may be able to select the smallest multiple. Our task is aided if one of the packages contains only one marble.

Table 31a

Package	Mass/g
A	25.3
B	17.5
C	6.0
D	13.0
E	12.7
F	31.0
G	105
H	42.2
I	18.6
J	12.9

Look at the table and try to select a lowest multiple. Then work out how many multiples of this basic unit are contained in each package.

When you have decided how many basic units (marbles) there are in each package, divide the mass of each package by the number of marbles you think it contains to find the average mass of each marble in the package.

The success of your hypothesis must be judged against the degree of consistency in your answers and the expected experimental error in the measurements.

How many marbles did you find in package G? There are actually 17. You probably found it very difficult to decide whether there were 16, 17, or 18. Would it have made any significant difference to the average mass of each marble in this particular package, whichever figures you had chosen? With large numbers· of marbles in a package it is very difficult to decide how many marbles are contained within it.

The success of Millikan's method was due to the very small quantities of charge he was measuring which represented only small multiples (mostly less than 10) of the basic unit.

31.5 Millikan's apparatus

Figure 31.3 is an example of the type of apparatus that Millikan used. Two metal plates, about 50 mm in diameter, are separated 15 mm apart by an insulating ring equipped with a small window for viewing the oil drops, and another window (not shown) for illuminating the interior from one side. The upper metal plate has a hole 0.5 mm in diameter through which a few oil drops can fall from the spray box above. This has a lid so that a small quantity of oil can be sprayed in with a simple spray. The plates are connected to a switch and a variable power supply. The electric field between the plates can be upwards (lower plate positive), downwards (top plate positive), or zero when the switch connects the plates together.

When a fine spray of oil droplets is produced, frictional charging can happen, perhaps as the oil leaves the jet of the spray. The process is completely random; some drops will be positive, others negative, the size of the charge will vary from drop to drop, and some drops will be uncharged. Also, the weights of the drops will not all be the same.

Droplets that have fallen through the hole can be observed through the microscope as tiny points of light like distant stars. If the plates are shorted together the droplets will all be falling in the same direction. When the power supply is connected some droplets will continue to fall at their original speed, some will fall faster and others will be seen to rise. The ones that rise have charges that are of the same polarity as the lower plate. The power

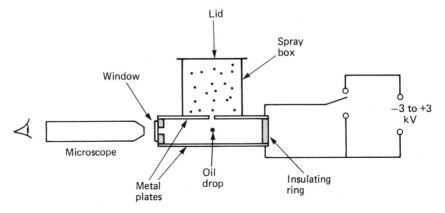

Fig. 31.3 Millikan's apparatus.

supply can be adjusted until a selected drop is stationary and after a few minutes all the other ones will have moved from the field of view. Such a drop may have a charge Q and mass m and if a uniform field E gives an upward force on the charge which just balances it against the pull of gravity, then

$$EQ = mg$$

The field E is governed by the potential difference V between the plates and by the plate separation, d.

$$E = \frac{V}{d}$$

so

$$\frac{V}{d}Q = mg$$

hence

$$Q = \frac{mgd}{V} \qquad (31.1)$$

Thus Eq. 31.1 allows the tiny charge on an oil drop to be calculated from measurable quantities. The next step is to alter the charge on the same drop many times, each time adjusting V until the drop is held stationary and recording its value, so that the corresponding charge can be calculated. Once about ten determinations of the various charges on the drop have been made it should be possible to see that one never finds a charge of less than 1.6×10^{-19} C and that all the values are whole number multiples of this basic unit of charge.

31.6 Some experimental details

We have mentioned changing the charge on the drop and also measuring its mass, but how can this be done with something so minute as a single droplet from the mist from an oil spray? Changing the charge is simple and it involves ionizing the space between the plates by exposing it briefly to the radiation from a small radioactive source. (Millikan also used X-rays for the same purpose.) Initially the ions produced will be rapidly collected up by the positive and negative charges on the plates and nothing else will happen. However, if the potential difference between the plates is removed by connecting them together briefly then the ions will drift slowly around and soon one or two will touch the drop and transfer their charges

to it. Whether the drop's charge increases or decreases depends on the sign of the charge on the ion that it touches. Of course, while the plates are connected together, the drop is falling under gravity but it only does so slowly and it can easily be re-positioned and re-balanced as soon as the potential difference across the plates is switched on again.

To illustrate the problem of finding the weight of a droplet, let us first estimate its mass assuming the diameter to be about 10^{-6} m and the density 1000 kg m^{-3}. Treated as a cube the drop has a volume of 10^{-18} m^3 and therefore a mass of the order of 10^{-15} kg. Use of a chemical balance is clearly out of the question since most can only manage 1 mg (10^{-6} kg).

The method used involves a measurement of the terminal speed with which the oil drop falls through the air. A model of the situation is provided by a ball from a ball-point pen falling through water. This quickly attains a maximum constant or terminal speed, as the opposing forces (the weight downwards and the buoyancy upthrust and viscous drag upwards) reach equilibrium. Although air has a viscosity which is only about 1/50th of that of water, the oil droplet also reaches a terminal speed, taking several seconds to fall 1 mm. At this rate the droplets can easily be timed with a stopwatch to give their terminal speeds.

The problem now is to relate the speed of the falling droplet to its radius. We can use the method of dimensions for this.

It is reasonable to assume that the viscous force on the falling drop will depend in some way on the viscosity η of the air. We already know that this has dimensions $M L^{-1} T^{-1}$ (Section 11.6). It will also depend in some way on the terminal speed v and on the radius a. We write

$$F = k(\eta)^x (v)^y (a)^z$$

where x, y and z are the powers to which η, v and a are to be raised and k is a dimensionless constant. Equating dimensions:

$$M L T^{-2} = (M L^{-1} T^{-1})^x (L T^{-1})^y (L)^z$$

$$= M^x L^{-x + y + z} T^{-x - y}$$

Equating the powers of M, L and T respectively we find that x, y and z are all unity. So the equation for the viscous drag on a drop has the form:

$$F = k\eta va$$

Stokes, after whom this equation is named, showed that the constant k was, in fact, 6π. So we have Stokes' equation

$$\boxed{F = 6\pi\eta va} \qquad (31.2)$$

available for application to Millikan's oil drops.

Now let us consider some data taken from one of Millikan's experiments and published in 1911.

An oil drop fell through 1.010 cm in 23.2 s.

Its terminal speed was 0.01010 m/23.2 s $= 4.354 \times 10^{-4}$ m s^{-1}

The viscosity of air (η) is 1.836×10^{-5} N s m^{-2} at 25 °C.

The density of the oil used (ϱ) was 896 kg m^{-3} at 25 °C.

The separation of the plates (d) was 0.016 m.

The p.d. between the plates (V) was 7730 V.

The gravitational field strength (g) was 9.8 N kg^{-1}.

The mass (m) of the drop = volume $\times$ density
$$= \tfrac{4}{3}\pi a^3 \varrho$$

The weight of the drop = mg
$$= \tfrac{4}{3}\pi a^3 \varrho g \qquad (31.3)$$

Assuming that, at the terminal speed, the viscous force (determined by Stokes' law) balances the weight, we have $\tfrac{4}{3}\pi a^3 \varrho g = 6\pi\varrho av$

Multiplying both sides of this equation by $3/2\pi a$ gives:

$$2a^2\varrho g = 9\eta v \quad \text{whence} \quad a = \sqrt{\frac{9\eta v}{2\varrho g}} \qquad (31.4)$$

Problem 31.2

(a) Use the data above to calculate the radius of the oil-drop used. (You should find that the radius is in the same order of magnitude as the spacing between the lines on a 6000 line per cm diffraction grating. There is no doubt about the smallness of these drops!)

(b) Use the radius you have just calculated and Eq. 31.3 to find the weight of the drop and then use this with Eq. 31.1 to determine the charge carried.

Table 31b

Drop	$Q/10^{-19}$ C	n	$\dfrac{Q}{n}/10^{-19}$ C
A	6.31	4	1.58
B	9.51	6	1.59
C	7.90	5	1.58
D	14.26	9	1.58

The charge on this drop was 6.31×10^{-19} C. Millikan subsequently changed the charge on this drop by the ionization technique and obtained a series of values for the charge carried. This is listed in Table 31b.

Examination of these values reveals that the charges increase or decrease by integral multiples of a basic charge and that the drop carried whole numbers of these charges indicated by the n column. The basic unit of charge, the charge on the electron, is accepted today to be 1.602×10^{-19} C.

Problem 31.3   As an exercise in algebra, use Eqs. 31.1, 31.3, and 31.4, to show that the charge on an oil drop is given by:

$$Q = \frac{18\pi d}{V}\sqrt{\frac{\eta^3 v^3}{2\varrho g}} \qquad (31.5)$$

Use this equation to check your answer to the previous problem.

Precautions to ensure accuracy

a) A single oil drop may be in use for over an hour and an oil that does not evaporate readily should be chosen so that the drop's weight remains constant. The oils used in vacuum pumps have a low vapour pressure and consequently are suitable.

b) The droplet displaces some air and allowance must be made for the resulting upthrust for very accurate results.

c) The viscosity of air varies with temperature and *increases* by approximately 4% for a 10 K rise. The value used must correspond to the temperature of the experiment.

d) Stokes' law is not well obeyed when the radius of the sphere is comparable to the mean free path of the air molecules, and a correction factor must be applied to the viscosity of air to take this into consideration.

Chapter 32

ELECTRIC CURRENTS IN SOLIDS AND ELECTROLYTES

32.1   Introduction

In the previous chapter it was established that an electric current is consistent with the flow of particles of charge and that the basic unit of charge, the charge carried by the electron, had a constant value that could be measured. The aim of this chapter is to consider some of the questions that remain to be answered: Are the charges positive or negative? How fast do they travel? How many charge carriers are there per unit volume of conductor?

The answers come from an effect discovered in 1879 by E. H. Hall, and named after him.

32.3   The Hall effect

Hall showed that when a conductor was placed in a magnetic field and constrained from moving, a p.d. appeared across the dimension of the conductor perpendicular to both field and current. Thus in Fig. 32.1, where the conductor is shown as

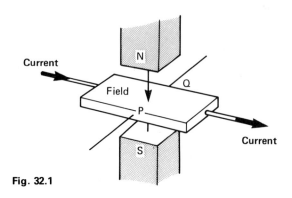

Fig. 32.1

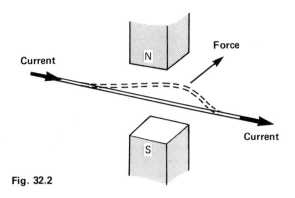

Fig. 32.2

part of a flat ribbon, the p.d. appears across PQ.

The explanation of the Hall effect comes from considering the force acting on a horizontal wire that is carrying a current at right angles to a vertical magnetic field. If the wire is thin and free to bend, it would move in a horizontal plane to the position shown dotted in Fig. 32.2, because the moving charges it carries experience forces that are perpendicular to both the field and the current. The same forces act on the moving charge carriers in the flat ribbon and as this is not free to move, the charges themselves move across to the far side Q. There will now be a deficit of these charges at side P which will have an opposite polarity from Q. Thus a p.d. is set up across the conductor, and this is called the *Hall voltage*. Its size is small, being only a few microvolts in the case of metals even when large fields and currents are used.

32.3 The sign of the charges

The sign of the Hall voltage can provide a direct test of the sign of the charge carrier involved in the electric current within a conductor or semiconductor. The direction of the magnetic deflecting force is independent of whether the current consists of positive charges moving in the direction of the conventional current, or negative charges moving in the opposite direction. The Hall effect immediately tells us the sign of the charges involved, for if Q is positive and P negative, then the charges shifted across must be positive ones. In the case of many metals, e.g., copper, aluminium, silver, gold, platinum and the alkali metals, it is found that Q is negative and P positive; it can therefore be deduced that the charge carriers in these metals are negative. Furthermore, these negative charges can be associated with the electron, since in the Millikan experiment the initial charge on the oil drop is obtained by a frictional transfer of charge between the oil and the spray nozzle. Thus Millikan gives the size of the basic charge while the Hall effect provides the sign.

Although it is now established that an electric current in a wire is a flow of negative electrons going from the negative of a cell around a circuit to the positive terminal, the opposite or conventional direction is still retained. It is not unlike a country deciding that its traffic should drive on the left, for once that has been decided some order, and some laws, can be established. It was decided arbitrarily in the absence of other evidence that the convention should be a flow of positive charges around a circuit from positive to negative, and on this basis, the corkscrew rule, Fleming's left-hand rule, etc., were formulated. Although attempts have been made to change such rules and adopt 'electron flow' they have not generally found favour, and conventional current flow is widely used – after all, in practice, it is not often that one needs to know what mechanism is involved in the flow of electricity; what it actually does is usually of more immediate importance and for this, conventional flow serves us well.

32.4 The speed of electron flow in a conductor

Consider the arrangement of Fig. 32.3 in which a ribbon, or slice of a conductor, of breadth b is placed in a magnetic field of strength B. In their journey through the slice at speed v, electrons, each with charge e, migrate across to side Q and build up a Hall voltage V. This establishes an internal electric field of strength V/b and a force on each electron of Ve/b acting in the direction Q to P. This is in opposition to the magnetic deflecting force which acts from P to Q. As we shall see in Section 34.5, this force has magnitude Bev. In equilibrium,

$$Bev = \frac{Ve}{b}$$

hence

$$v = \frac{V}{Bb} \qquad (32.1)$$

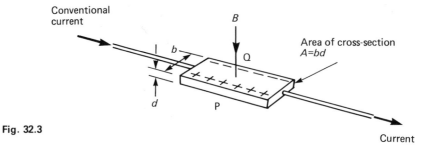

Fig. 32.3

Experimentally it is possible to measure B, V and b and a value for the *mean drift speed* of the electrons is obtained. The reason for the word 'drift' is partly because the electron motion is likely to be haphazard as it travels along and also because of its numerical value, as will become apparent in the problem following.

Problem 32.1 In an experiment with a piece of thin aluminium foil of width 1.1 cm placed between the poles of an electromagnet, a magnetic field of 1.7 tesla produced a Hall voltage of 14 μV. Show that the mean speed of the electrons is 0.75 mm s^{-1}.

There are two remarkable things to note about this result:

(i) the slow mean speed of the electrons, and
(ii) the fact that this speed has been obtained without needing to know the size of the current flowing!

If the current flowing is measured together with the thickness of the conductor the Hall effect yields further information about the material.

32.5 Number of electrons per unit volume

Consider a wire of cross-sectional area A made of a material that has n free electrons per unit volume. Let the charge on each electron be e and their mean velocity v. In t seconds, all the

electrons in a length vt will pass point X (Fig. 32.4).

$$\text{volume of section XY} = Avt$$
$$\text{number of electrons in XY} = nAvt$$
$$\text{total charge in section XY} = nAvte$$

Thus in t seconds, a charge ($nAvte$) passes any point in the wire.

$$\text{current} = \text{rate of flow of charge}$$
$$= \frac{\text{charge passing}}{\text{time}}$$
$$= \frac{nAvte}{t}$$

$$\boxed{I = nAve} \qquad (32.2)$$

We will now apply this general result to the situation shown in Fig. 32.3, noting that the thickness of the foil is d and so the area of cross-section $A = bd$.

Combining Eqs. 32.1 and 32.2 gives:

$$I = nAve$$
$$I = n(bd)\left(\frac{V}{Bb}\right)e$$
$$I = \frac{ndVe}{B}$$

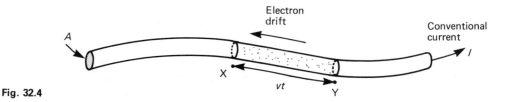

Fig. 32.4

Rearranging, we obtain

$$n = \frac{IB}{Ved} \qquad (32.3)$$

Problem 32.2

(a) Calculate the number of electrons per unit volume in aluminium given the data of Problem 32.1. The current flowing is 5.0 A and the thickness of the foil is 0.02 mm. Take $e = 1.6 \times 10^{-19}\,$C. (Answer: $1.9 \times 10^{29}\,$m^{-3})

(b) If the nucleon number (mass number) of aluminium is 27 and its density is $2700\,$kg m^{-3}, calculate the number of aluminium atoms per unit volume and hence estimate the number of free electrons per atom. The Avogadro constant is $6 \times 10^{23}\,$mol^{-1}.

In 27 g of aluminium there are 6×10^{23} atoms.
But 2700 kg of aluminium occupy 1 m^3.
Therefore 27 g of aluminium occupy $10^{-5}\,$m^3.
Therefore in $10^{-5}\,$m^3 there are 6×10^{23} atoms, and so in 1 m^3 of aluminium there are 6×10^{28} atoms.

From (a) there are 1.9×10^{29} free electrons in 1 m^3 of aluminium.

Therefore, number of free electrons per atom

$$= \frac{1.9 \times 10^{29}}{6 \times 10^{28}} = \frac{19}{6} \approx 3$$

32.6   Factors affecting the size of the Hall voltage

Rearranging Eq. 32.3 gives:

$$V = \frac{IB}{ned} \qquad (32.4)$$

Thus to make V large, the current and magnetic field must be made as large as possible, and the denominator as small as possible. It is easy to select thin slices of the material used, so making d small. Equation 32.4 shows that the width has no effect on V, as b does not appear. The fact that n must be small implies that the best Hall effect is obtained with poor conductors, i.e., those with relatively few free electrons per unit volume.

Particularly suitable materials are semi-conductors like silicon and germanium that have been lightly 'doped' with impurities to make them conduct slightly. It is not possible to state precisely the number of charge carriers per unit volume for such semi-conductors since the number depends on the impurity concentration and also upon temperature; a rise in temperature increases the number of carriers and therefore produces a marked fall in resistivity. The best that can be said is that for semi-conductors the number of charge carriers per unit volume lies within the range 10^{17} to $10^{23}\,$m^{-3}, and therefore the Hall voltage is very much greater than in the case of metals for which n is about $10^{29}\,$m^{-3}. Because of this, semi-conductor Hall devices are used for detecting and measuring very small magnetic fields.

32.7   p-type and n-type semi-conductors

The Hall effect provides some insight into the conduction processes occurring in semi-conductors, and in particular it confirms experimentally the bonding theories involved when minute quantities of an impurity are added (e.g. about 1 impurity atom per 10^6 or 10^7 host atoms.) If phosphorus is the impurity added to silicon then it should provide extra electrons which are available for conduction. Since a flow of negative charges is involved, this particular 'doped' semi-conductor is called *n-type*. However, if boron is the impurity then on theoretical grounds there should be fewer electrons available for conduction than in normal silicon. Vacant sites lacking electrons now exist and these are called *holes*. A positive hole can be filled by an electron in a neighbouring atom dropping into it, but this then leaves the neighbour with the vacancy or hole. The process continues and the hole moves from one end of the semi-conductor to the other in the opposite direction to the electrons, each of which only moves 'next door'. The overall effect is that of a positive charge moving in the conventional direction and so such a material is called a *p-type* semi-conductor. It must be emphasized that no positively charged particles are actually moving.

This can be confirmed by placing slices of each type in a magnetic field, passing a current through them in series, and measuring the Hall voltage. The results are shown in Fig. 32.5. In the case of

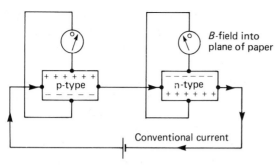

Fig. 32.5 p- and n-type semiconductors in a magnetic field.

the p-type material it is positive charges, or holes, that are moving from left to right and these are deflected across to the far side to make that positive. In the n-type material, electrons are moving from right to left and it is these negative charges that are deflected across to the far side to make it negative. It should be noted that some metals – cobalt, zinc, cadmium, iron – exhibit a positive Hall effect and that the theory of mobile charged particles in a crystalline metal can account for this.

32.8   Hall effect measurement

Figures 32.1 and 32.6a show in outline the basic arrangement for Hall effect experiments and whether the sample is a metal foil or a slice of semi-conductor, a practical difficulty has not yet been mentioned. This concerns making contact with the sides of the sample at points P and Q which are exactly opposite each other, so that when there is no magnetic field, but a current is flowing in the sample, then zero p.d. exists between P and Q. The slice is equivalent to the circuit shown in Fig. 32.6b, so far as any potential difference across the specimen between P and Q is con-

cerned. In the absence of a magnetic field through the specimen, the potential difference V between the points P and Q will be zero if

$$\frac{R_1}{R_2} = \frac{R_3}{R_4}$$

(See Section 25.5.)
The values of these ratios depend on the exact points of contact at P and Q and are most unlikely to be equal to each other. To compensate, a small potentiometer (1 to 2 ohm) is placed across the specimen as shown in Fig. 32.6c. By adjusting this potentiometer, the potential difference V can be made equal to zero in the absence of the field.

32.9   Ions in an electrolyte

The speed of ions

We have seen in Section 32.4 that electrons travel very slowly when an electric current flows in a metal. This section describes an experiment which shows a similar slow motion of ions in an electrolyte; it must be said that the result is not conclusive and the demonstration should be interpreted with caution. There is no doubt that what is observed travels slowly but there is no direct evidence that it is the motion of the charged ions themselves.

A strip of filter paper which has been moistened with ammonia solution is supported on a glass slide and connected to a d.c. power supply by means of spring clips (e.g., crocodile clips) (Fig. 32.7). A few small crystals of copper sulphate and of potassium permanganate are scattered over the surface of the paper and about 100 volts applied to the electrodes. The coloured stains which spread out from the crystals lack symmetry, the blue stains from the copper

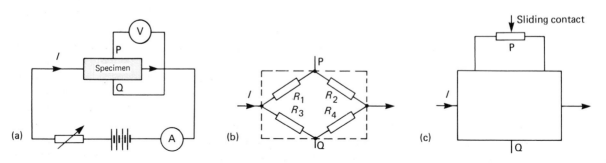

Fig. 32.6   Circuit used to investigate the Hall effect.

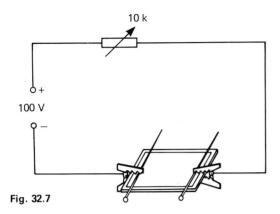

Fig. 32.7

sulphate spreading towards the negative electrode and the purple stain from the potassium permangante spreading towards the positive electrode. Switch the supply off, and the stains spread symmetrically. What do you imagine will happen if you reverse the polarity of the connections and again switch on?

The blue and the purple stains are related chemically to the atoms making up the crystals. Their spread towards the positive and negative terminals of the battery suggests that the passage of the electric current is associated with a movement of atoms from the crystals within the dampened filter paper.

It seems likely that the carriage of electric charge across the paper is, in part at least, associated with the atoms or groups of atoms which cause the coloured stains. Such electrically charged atoms or groups of atoms are called *ions*. This conclusion does not, of course, exclude other processes and it may well be that some other mechanisms are also responsible.

32.10 Faraday's laws of electrolysis and the ionic model

The chemical reactions resulting from the passage of electric charge through conducting liquids were extensively investigated by Michael Faraday, whose name is now remembered in the two laws of electrolysis. He introduced technical terms into his accounts of his researches which are still in use. Liquids, whether molten solids or solutions, which conduct electric currents are called *electrolytes*; chemical reactions accompany the process which is called *electrolysis*. The conductors (usually solid metal or carbon) which connect the electrolyte to

the rest of the circuit are *electrodes*, that which is connected to the positive pole of the supply being the *anode* and that which is connected to the charged particles within the solution. Those carrying a positive charge or charges (e.g., H^+, Cu^{2+}, Na^+) which move towards the cathode are often called *cations*. Those with a negative charge (e.g., O^{2-}, Cl^-, SO_4^{2-}) which move towards the anode are called *anions*.

As an example of an electrolytic reaction, consider the passage of an electric current through a solution of copper sulphate. A 12 V d.c. supply is connected in series with an ammeter, a rheostat and two copper electrodes supported in a beaker containing a strong solution of copper sulphate.

The copper cathode is carefully cleaned with an abrasive cloth and weighed before connection into the circuit. A current of 1 A is passed through the circuit for, say, 10 minutes. The cathode is removed, dried and re-weighed. This procedure is repeated for a range of values of both current and time.

The experimental results show that the mass (m) of copper deposited is proportional to the current (I) for constant times and to the time (t) for constant currents. So

$$m \propto I, \text{ with } t \text{ constant}$$
$$m \propto t, \text{ with } I \text{ constant}$$
$$\therefore m \propto It$$

Now the product It is a measure of the quantity of charge passing through the electrolyte.

In general it is found that the mass of material liberated at the anode or the cathode in electrolysis is proportional to the quantity of charge passing. This statement expresses *Faraday's first law of electrolysis*.

It is difficult to avoid making the hypothesis that the electric charge passing through the electrolytic cell is carried by particles of matter: Faraday's ions.

The case chosen above is a simple one, for only copper appears as an end product. The experiment can readily be adapted to investigate any changes which may occur at the anode. You may wish to investigate this yourself. You could try weighing the anode before and after a charge is passed.

A more complex example involves the electrolysis of water. Now pure water is a very poor conductor of electricity whereas water containing a small quantity of an electrolyte such as sulphuric acid is a very good conductor. It seems that the acid forms ions in the water and these provide the charge carriers. But the complex reactions which occur at the anode and cathode result in the liberation of oxygen and hydrogen, the constituents of water.

Experiment shows that the ratio of the mass of hydrogen liberated to the mass of oxygen is 1.05:8. This is almost identical to the ratio of the masses of hydrogen and oxygen which combine in chemical reaction and is just one half of the ratio of atomic masses of hydrogen and oxygen (1.08:16).

Faraday and many other workers in this field made many such comparisons of the masses of atoms involved in electrolysis and found that the quantities of electricity required to liberate one mole of different elements bear a simple relationship to one another. This is a statement of *Faraday's second law of electrolysis*.

Accurate experiments have given a value of 96 500 C as the charge needed to liberate 1 mole of hydrogen and 193 000 C as the charge needed to liberate 1 mole of oxygen.

Since 96 500 C are needed to liberate 1 mole (1.08 g) of hydrogen, it follows that each hydrogen ion in the acidulated water carries on the average $(96\,500/N_A)$ C, where N_A is the Avogadro constant. The evidence will not allow us to say that each ion carries exactly this charge; but there is no good reason why different hydrogen ions should carry different charges so we shall assume that the charges are identical and are each $96\,500/N_A$ C.

Building an assumption such as this into the model is an acceptable device so long as we remember that the assumption was made. In Chapter 10 when we discussed the speed of each molecule in considering a kinetic model for gases, we were careful to point out that the measured speed was only an average of a wide range of possible speeds. In that instance, that was the most reasonable assumption to make for it was implicit that the molecular velocities were changing continuously as a consequence of the many collisions which were taking place; and even had they all been the same at one moment, they would soon have become randomized.

Since the Avogadro constant is 6.02 ×

10^{23} mol^{-1} the charge carried by the hydrogen ion (which proves to be the smallest quantity associated with any single ion) is

$$+\frac{96\,500}{6.02 \times 10^{23}}\,C$$

This is $+1.6 \times 10^{-19}$ C. The number is already familiar to us as the numerical value of the charge on the electron (e). The hydrogen ion carries a positive charge equal to that of the electron. The oxygen ion carries charge $-2e$.

The number 96 500 is known as the Faraday constant (F) and has units: C mol^{-1}.

32.11   Applying the current model

We have seen indirect evidence that charge carriers in an electric current move slowly, and yet our experience is that when a lamp is switched on it lights up very quickly. What model of a current can reconcile these observations? Fig. 32.8 shows such a circuit. The wires contain a very large number of free electrons (about 10^{29} m^{-3}) and before the switch is closed, the source of e.m.f. has repelled electrons to one side of the switch making that negative, and removed electrons from the other side which is positive. No flow occurs and the free electrons throughout the wires move in random directions.

Once the switch is closed and the circuit completed, all the electrons start to drift in the same direction, including those in the thin tungsten wire which forms the lamp filament. Thus one does not have to wait for electrons to leave the battery and go all the way slowly to the lamp before it lights up. The electrons are already in the wire and ready to go; as soon as the battery pushes the first one, they all start to move, just like a locomotive shunting a long line of trucks.

In their journey through the ionic lattice of the

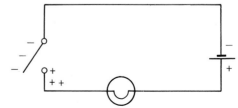

Fig. 32.8

metal wire the electrons will collide with irregularities such as impurities and discontinuities in the lattice. In such collisions electrons transfer energy to the lattice itself; the wire gains internal energy and becomes warm.

A similar process occurs in the filament which becomes very hot because the electrons are travelling much faster and so transfer more kinetic energy. The higher speed is because the filament wire is considerably thinner than the connecting wires, and since they are in series they must be conveying the same current. Considering Eq. 32.2 $I = nAve$, if n is taken to be the same for both parts of the circuit, and I and e are the same, then the product Av must be a constant. Therefore for the thin filament wire, the cross-sectional area A is small and so the electron drift speed v must be large.

Problem 32.3   The diameter of the connecting wires is 10 times that of a lamp's filament. Show that the kinetic energy of the electrons in the filament is 10^4 times greater than elsewhere in the circuit.

Chapter 33

CHARGE CARRIERS IN SPACE

33.1 The conduction of electricity through gasses

Gases at atmosphere pressure are non-conductors of electricity unless the electric field to which they are subject is very high indeed. In air at atmospheric pressure the field must exceed $3 \times 10^6 \text{V m}^{-1}$ before a current, in the form of a violent and noisy spark, passes between the two electrodes providing the field.

The lightning flash is the best example of such a discharge. Experiments using photography suggest that the structure of the flash is complex. A leading downstroke, which moves quite jerkily and slowly, ionizes the air leaving a conducting path in its wake. This path then carries the main discharge upwards at very high speed (about $40\,000 \text{ km s}^{-1}$) in a current of from 10^4 to 10^5A. Since the duration of this intense flash is about 10^{-4}s the quantity of electricity involved is in the order of 10C. The potential difference between cloud and cloud or between cloud and ground may be as high as 10^8V, so that 10^9J may be involved in a typical discharge.

If low pressures are maintained in a discharge tube, a quieter steadier passage of electricity can take place between the two electrodes. In this case also the current is accompanied by the emission of light as in a neon sign.

33.2 Thermionic emission and the electron gun

Figure 33.1 shows a section through a highly evacuated glass envelope in which the pressure is about 10^{-4}Pa (equivalent to about 10^{-6}mm of

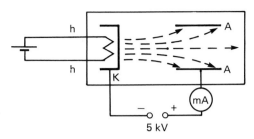

Fig. 33.1 A hot cathode and a cylindrical anode in an evacuated glass envelope.

mercury). At one end there is a metal electrode K which can be raised almost to white heat by passing a current through the heater coil 'hh' which is contained within it.

When a metal is very hot, some of the free electrons within its lattice will have enough energy to leave the surface of the metal. This is not unlike the evaporation of molecules from a liquid surface. These electrons do not wander far but gather in a cloud around the electrode from which they have been emitted. This is because they have taken their negative charge with them and have left the electrode with a positive charge. It is this positive charge which attracts the freed electrons and keeps them within the cloud.

The electron cloud is not static; electrons closest to the electrode are constantly being recaptured by it as others are being emitted. A state of dynamic equilibrium will be reached when the number of electrons returning in unit time equals the number leaving in that time. The outermost electrons in the cloud are prevented from getting nearer to the electrode by the repulsion from the 'space charge' of the electrons closer in. If the heater is switched off and the electrode cools, the emission of electrons ceases; the electron cloud returns to the electrode, which becomes electrically neutral once more.

This process of producing electrons in space by heating is known as *thermionic emission*; it was discovered by Elster and Geitel in the 1880s. Cathode ray and television tubes are entirely dependent upon this process.

To return to Fig. 33.1, the second electrode, A, is cylindrical in form. If made positive with respect to the electrode K by connection to a 5 kV supply, it will attract electrons from the electron cloud and a current will be indicated by the milli-

ammeter. Evidently the positively charged electrode (or anode) is collecting electrons from the space charge and consequently from the heated negatively charged electrode (or cathode). An electron stream flows across the gap between the electrodes from K to A.

Problem 33.1 What will happen if (a) the heater is switched off but with the 5 kV supply still connected, and (b) if the heater is switched on again after the connections to the 5 kV supply have been reversed so that K is positive with respect to A?

The emission of the charge carriers is associated with the heater and therefore with the electrode K. Current flows only when electrode A is positive with respect to K. It seems then that only negative charge carriers are involved.

Problem 33.2 If air is allowed into the envelope of Fig. 33.1 and the heater is on, it is found that a current will flow when the anode is positive *and* when it is negative. Explain why.

(*Hint*: You may have seen an experiment in which two insulated metal plates are placed 5 cm apart and are in series with a 5 kV supply and a galvanometer. If a candle flame or α-source is placed between the plates, a current will flow.)

Not all of the electrons are collected up by the anode; those that are initially close to the axis of the anode will miss its far end and continue diverging to the end of the envelope (see Fig. 33.1). There, this diverging beam of cathode rays or electrons will cause fluorescence especially if the glass is coated with a material such as zinc sulphide or calcium tungstate.

In the tube shown in Fig. 33.2 a metallic object in the shape of a Maltese cross placed in a divergent beam forms a shadow on the fluorescent screen, confirming rectilinear propagation of cathode rays. The beam can be deflected by a magnet and shown to be consistent with a stream of negative charge. The white-hot cathode also emits light and casts a shadow but one that is undisturbed by the magnetic field because light is an electromagnetic wave and not a stream of charged particles.

Fig. 33.2 A hot cathode 'Maltes Cross' tube showing the shadow cast by the electron beam. (Courtesy Teltron Ltd.)

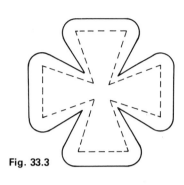

Fig. 33.3

Problem 33.3 The modern design of a Maltese Cross tube allows a connection to be made to the cross itself. If an additional power supply is used to make the cross negative with respect to the anode then the sharp-cornered shadow shown broken in Fig. 33.3 becomes larger and with rounded corners rather like a four-leaf clover. Explain this effect.

Originally Perrin showed directly that cathode rays consisted of negatively charged particles by actually collecting them and testing their sign. A modern version of his tube is shown in Fig. 33.4. In this tube the cathode rays are formed into a narrow beam by passing through a small hole at the end of the anode which acts as a crude form of *electron gun*. A small can-like electrode is offset from the centre line and is connected to a gold-leaf electroscope. When a magnetic field is applied to deflect the beam into the collecting can the electroscope leaf deflects showing that a charge has been collected. The electroscope is next isolated from the collecting can and a test on the charge carried by the leaf shows it to be negative.

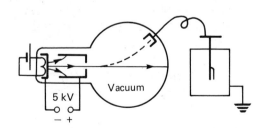

Fig. 33.4 A hot cathode 'Perrin' tube. (Courtesy Teltron Ltd.)

Problem 33.4 State carefully the direction of the magnetic field if the electron beam is to be deflected as shown by the broken line in Fig. 33.4. Explain your reasoning.

A stream of negative charges moving to the right is equivalent to a conventional current moving towards the left. The magnetic field moves this current up the page. Use the left-hand rule to find the field direction.

33.3   The nature of cathode rays

The experiments referred to so far suggest that there is a basic unit of charge of magnitude 1.6×10^{-19} C which is the charge on the electron. Charge separation can be achieved by friction and by ionization; and the Hall effect offers evidence for believing that the charge carriers in at least some metals have negative charges. Experiments with thermionic emission demonstrate that heating can cause negative charges to leave a metal. It is a small step to associate these thermionically-produced negative charges with the electrons produced by friction and ionization in Millikan's experiment where their charge was measured.

Figure 33.5 shows a thermionic tube (a fine beam tube) which gives us even more information about the nature of electrons. Instead of a near perfect vacuum, this tube contains a little hydrogen or helium at very low pressure (about 1 Pa). When the filament is heated and a potential difference of about 100 V is applied between cathode and anode, a fine beam of cathode rays

Fig. 33.5 A 'fine-beam' tube. The electron beam, made visible by the re-combination of the ions it has produced, follows a circular path in the magnetic field produced by the current in the coils. (Courtesy Teltron Ltd.)

emerges from the anode, its path being made visible by the ionization of the gas atoms it hits. The evidence of the other experiments suggests that the beam is made up of a stream of electrons. Let us assume that each particle has mass m and a negative charge e. Their motion results from the application of an electric field within the gun so that the kinetic energy acquired in the field is equal to the potential energy provided. If the potential difference applied is V, the potential energy provided is Ve and the kinetic energy acquired is $\frac{1}{2}mv^2$ where v is the velocity with which the particles leave the gun. So

$$Ve = \tfrac{1}{2}mv^2 \qquad (33.1)$$

The resultant beam of moving charges is deflected by a magnetic field B produced by the pair of coils alongside the tube. Since the field B is at right angles to the direction of motion of the particles, the beam is deflected into a circle of radius r. The deflecting force (F) bends the path of particles of mass m and speed v into a circle of radius r.

$$F = \frac{mv^2}{r} \qquad (33.2)$$

This force is provided by the interaction of the magnetic field and the moving charges and is given by:

$$F = Bev \qquad (33.3)$$

(see Section 34.5).
Combining Eqs. 33.2 and 33.3 we get

$$Bev = \frac{mv^2}{r}$$

$$Be = \frac{mv}{r} \qquad (33.4)$$

It follows from Eq. 33.1 that

$$v = \sqrt{\frac{2Ve}{m}} \qquad (33.5)$$

Substituting in Eq. 33.4 we get

$$Be = \frac{m}{r}\sqrt{\frac{2Ve}{m}}$$

and simplifying, we obtain

$$\frac{e}{m} = \frac{2V}{B^2r^2} \qquad (33.6)$$

Since V, B, and r can all be measured, we can calculate a value for e/m. Since the term *specific* means 'per unit mass', the ratio e/m is called the *specific charge of the electron*.

Problem 33.5 In a typical experiment, B was found to be 6×10^{-4} tesla, and V was 100 V. The consequent value of r was 5.7 cm. Calculate the value of e/m.

In his original experiment for measuring the specific charge (e/m) J. J. Thomson (1897) used a 'null-deflection' method to measure the speed of

Fig. 33.6 A modern version of the 'Thomson' tube. The electron beam is seen undergoing deflection in the electric field between the two horizontal plates. (Courtesy Teltron Ltd.)

the electrons. Figure 33.6 shows a modern version of his tube, designed for use in demonstrations of the principle.

The beam is moving from left to right and its path can be seen on a fluorescent screen marked out with a grid so that its radius of curvature may be found. At the top and bottom of the screen are horizontal plates between which an electric field can be established to deflect the beam. If the field strength is E then the force acting on an electron is Ee and this can be arranged to act downwards as in Fig. 33.6. A pair of circular coils provides a horizontal magnetic field which can deflect the beam independently upwards, the force acting on an electron being Bev. Since the two deflecting forces are arranged to act in opposition they can be adjusted until they counterbalance and the beam goes straight on undeflected.

Under the conditions of balance:

$$Ee = Bev$$

where E the electric field between the parallel plates may be calculated from the known separation of the plates and the measured potential difference between them. Therefore

$$v = \frac{E}{B} \qquad (33.7)$$

Substituting this value for speed in Eq. 33.4 gives the specific charge, e/m, directly, without making any assumptions about the energy gained by the electrons while they are being accelerated in the electron gun.

Problem 33.6   Look at Fig. 33.6. (i) What is the polarity of the top plate if the deflection is due to the electric field alone? (ii) If the deflection is due to the B-field alone, is this acting (a) up the page, (b) down the page, (c) into the plane of the page, or (d) out of the plane of the page?

(Answers: (i) negative; (ii) c.)

J. J. Thomson experimented extensively with such electron beams and showed that the value of e/m was the same (within the limits of a large experimental error) whether the electrons were produced by cold-cathode discharge, the thermionic effect or by the irradiation of the cathode with ultraviolet light (another discovery by Elster

and Geitel). Nor did the ratio depend on the nature of the gas in the tube. Also, later measurements using beta particles from a radioactive source gave a similar value for e/m once corrections had been applied. This was necessary because of relativistic effects that become significant with the high speeds of beta particles.

33.4   Specific charge and the mass of the electron

The accepted value of the specific charge of the electron is $1.76 \times 10^{11}\,\text{C kg}^{-1}$. Once this, and the value of e is known, the mass of the electron follows. Thus:

$$m = \frac{e}{e/m} = \frac{1.602 \times 10^{-19}\,\text{C}}{1.76 \times 10^{11}\,\text{C kg}^{-1}} = 9.1 \times 10^{-31}\,\text{kg}$$

33.5   Comparison of electrons with hydrogen ions

Experiments on the electrolysis of water lead readily to a value for the specific charge of the hydrogen ion. This proves to be $9.57 \times 10^{7}\,\text{C kg}^{-1}$. This is nearly 2000 times smaller than the specific charge of the electron. Either the electron has a much larger charge than the hydrogen ion or a much smaller mass. Other evidence shows that both have the same sized charge, but with opposite signs. So we are forced to conclude that the electron is much less massive than the hydrogen ion; $9.1 \times 10^{-31}\,\text{kg}$ compared with $1.67 \times 10^{-27}\,\text{kg}$. The ratio m_p/m_e is approximately 1840 to 1.

33.6   Calculating the speed of electrons

Knowledge of the specific charge of the electron enables us to calculate the speed of electrons that have been accelerated through a known potential difference, since Eq. 33.5 only requires us to know the value of e/m, and the p.d. V.

Problem 33.7 Calculate the speed of the electrons that have been accelerated through
(i)   100 V as in the experiment of Problem 33.5
(ii)   5 kV as in the Maltese Cross experiment
(iii) 25 kV as in the tube of a colour television set.

(Answers: (i) $5.9 \times 10^{6}\,\text{m s}^{-1}$; (ii) $4.2 \times 10^{7}\,\text{m s}^{-1}$; (iii) $9.4 \times 10^{7}\,\text{m s}^{-1}$.)

33.7 A pause for thought

In physics we meet numbers of all sizes ranging from 6.6×10^{-34} J s (Planck's constant) and the diameter of an atom, 2.4×10^{-10} m, through atmospheric pressure, 10^5 Pa, right up to the mass of the earth, 6×10^{24} kg, and of the sun, which is 2×10^{30} kg. It is difficult to grasp the significance of such numbers whether they be very small or exceedingly large, and it is sometimes helpful to try to relate them to more familiar situations.

It is all too easy to be unimpressed by the size of the answers to Problem 33.7, and to pass on; but stop and think. If you travelled at the speed of the *slowest* electron in the list you could go right around the earth in under 7 seconds and could reach the moon in just over a minute! All three of these answers are enormous, as speeds go. In the colour television tube these tiny pieces of matter have started from rest yet they arrive at the screen travelling at about $\frac{1}{3}$ of the speed of light.

Two questions emerge:

a) How is it possible for masses to be accelerated to such high speeds, and in the short length of the electron gun (e.g. 5 cm)?

b) If only 25 kV can cause them to go so fast why is it not possible to make electrons travel at the speed of light and beyond? After all, about 256 kV should do the trick and such a p.d. can be produced by a small Van der Graaff generator – even a school one.

The large acceleration is a result of the high specific charge of the electron. Consider an electron between two plates separated by 5 cm and having a potential difference between them of 100 V.

Electric field between the plates

$$= \frac{100 \text{ V}}{0.05 \text{ m}} = 2000 \text{ V m}^{-1} \text{ (or N C}^{-1})$$

Force acting on an electron

$$= 2000 \text{ NC}^{-1} \times 1.6 \times 10^{-19} \text{ C}$$
$$= 3.2 \times 10^{-16} \text{ N}$$

This is a rather small force, but when it is acting on a mass of only 9.1×10^{-31} kg, the resulting acceleration is:

$$a = \frac{F}{m} = \frac{3.2 \times 10^{-16} \text{ N}}{9.1 \times 10^{-31} \text{ kg}} = 3.5 \times 10^{14} \text{ m s}^{-2}$$

This is incomparably greater than a good racing car can manage (e.g. 10 m s^{-2}) or a tennis ball which, while being hit hard, accelerates at about 2×10^3 m s^{-2}.

If this electron starts from rest and maintains this acceleration over a distance of 5 cm, then the speed reached can be calculated using $v^2 = u^2 + 2as$. Thus

$$v^2 = 2 \times 3.5 \times 10^{14} \times 0.05 = 35 \times 10^{12} \text{ m}^2 \text{ s}^{-2}$$

hence

$$v = 5.9 \times 10^6 \text{ m s}^{-1}$$

which is the answer (i), obtained before, using energy conservation.

To understand why electrons, or any matter, cannot in practice be made to go faster than 3×10^8 m s^{-1}, the speed of light, one has to know something of *relativity* (see Chapter 39). One of the conclusions that Einstein reached was that energy has mass. As a body with rest mass m_0 is accelerated to speed v, its kinetic energy and therefore its mass increase. The new mass is given by:

$$m = \frac{m_0}{\sqrt{\left(1 - \frac{v^2}{c^2}\right)}} \qquad (33.8)$$

where $c = 3 \times 10^8$ m s^{-1}.

This equation has been used to calculate the mass increase of various particles travelling at high speed and the results are shown in Table 33a.

Table 33a

Particle	Speed	Mass
Spacecraft	10 times speed of sound $= 3 \times 10^3$ m s^{-1}	$1.00000000005\, m_0$
Electron	0.1 c $= 3 \times 10^7$ m s^{-1}	$1.005\, m_0$
Faster electron	$\frac{1}{3} c = 10^8$ m s^{-1}	$1.06\, m_0$
Fast beta particle	0.99 c	$7.1\, m_0$

The table illustrates that the effect of the mass-increase only becomes significant at speeds approaching the speed of light so little wonder that we do not normally notice it! However, the important result from this is that as the speed of a mass grows nearer and nearer to the speed of light, the accelerating force encounters more and more mass, and so further acceleration becomes steadily more difficult. Electrons given ever-increasing amounts of energy, approach the speed of light, but cannot reach it.

Equation 33.5 was developed in the context of a mass being given energy to make it accelerate, as in a laboratory dynamics experiment; and then it was used in cases where very high speeds were involved. In conditions well beyond normal experience, Eq. 33.5 does not give the right answers, for if it did, the speed of light should be attainable by an electron that has accelerated through a potential difference of 256 kV. Extending a theory from an initial situation to a very different one is rather like extending a graph well beyond the plotted experimental data and expecting it to work. Such extrapolation is allowable with caution, but too often, unforeseen factors produce invalid results if the extension is carried too far. An example is the familiar graph of extension of a wire against load which is, initially, a straight line but soon departs from it. This illustrates the danger of always assuming that what happens in our normal experience (e.g. when mass is constant) will continue to occur in situations that are beyond our experience. Once the limitations of Eq. 33.5 are known it is a very useful relationship that yields answers close to the truth; electrons going as fast as $10^8\,\mathrm{m\,s^{-1}}$ only increase their mass by 6%.

33.8   The electron volt

This is a non-SI unit of energy that is particularly convenient when describing the energy of a charged particle.

■ The electron volt is the energy gained by a charge of $1.6 \times 10^{-19}\,\mathrm{C}$ (an electron) that has passed through a potential difference of 1 volt. Thus $1\,\mathrm{eV} = 1.6 \times 10^{-19}\,\mathrm{J}$.

As an example, if an electron, or a proton, moves through a potential difference of 200 V then it can be described as having gained 200 eV of energy, which is equivalent to $3.2 \times 10^{-17}\,\mathrm{J}$. Commonly used multiples are

$$\mathrm{keV} = 10^3\,\mathrm{eV} = 1.6 \times 10^{-16}\,\mathrm{J},$$
$$\mathrm{MeV} = 10^6\,\mathrm{eV} = 1.6 \times 10^{-13}\,\mathrm{J}$$
$$\mathrm{GeV} = 10^9\,\mathrm{eV} = 1.6 \times 10^{-10}\,\mathrm{J}.$$

If an alpha particle has been accelerated through $2 \times 10^6\,\mathrm{V}$ (2 MV) then it gains 4 MeV of energy since it carries *two* proton charges; this is equivalent to $6.4 \times 10^{-13}\,\mathrm{J}$.

However, if an alpha particle is described as having an energy of 2 MeV then this equals $3.2 \times 10^{-13}\,\mathrm{J}$ and could have been gained by the particle accelerating through a p.d. of $10^6\,\mathrm{V}$ (or 1 MV).

Problem 33.8   Assuming that the energy of a particle is all in the form of kinetic energy, calculate the speed of:

(i)   a 2 keV electron (mass = $9.1 \times 10^{-31}\,\mathrm{kg}$);
(ii) a 2 keV alpha particle (mass = $4 \times 1.67 \times 10^{-27}\,\mathrm{kg}$);
(iii) an alpha particle that has been accelerated through a p.d. of 2 kV.

(Answers: (i) $2.65 \times 10^7\,\mathrm{m\,s^{-1}}$; (ii) $3.1 \times 10^5\,\mathrm{m\,s^{-1}}$; (iii) $4.38 \times 10^5\,\mathrm{m\,s^{-1}}$.)

33.9   Focusing an electron beam

In the tubes shown in Figs. 33.4 and 33.6 a crude form of electron gun was used in which a narrow parallel beam was produced from a divergent beam by passing it through a narrow hole, just as light is passed through a slit in a raybox. This is not a very efficient method since so many of the electrons are stopped by the anode and relatively few form the actual beam. Figure 33.7 shows how cylindrical electrodes may be arranged to form an 'electron lens' which causes the whole divergent beam to become one that converges to a spot on the distant fluorescent screen. A_2 is called the accelerating or final anode, while A_1 is the focusing anode. Relative to the cathode, A_1 is at about 800 V while A_2 is kept at 2000 V for example. Consequently, electric fields are established between the electrodes as shown by the broken lines and electrons *tend* to follow the directions of these. Electrons would only follow the lines exactly if they had zero mass. Initial acceleration is caused by the diverging field between the cathode and A_1 but the electrons soon

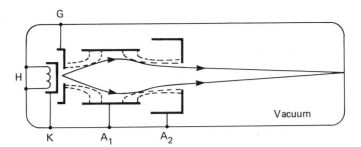

Fig. 33.7 An 'electron lens'.

meet the converging field between A_2 and A_2
which provides further acceleration and also
makes the beam converge instead. By adjusting
the potential of A_1 the shape of the electric fields
can be altered until the beam comes to a point on
the screen; thus A_1 is called the focusing anode.

Figure 33.7 also shows an electrode G, or grid,
that partially surrounds the cathode. If this is
made negative with respect to the cathode,
electrons are repelled back to the cathode and so
fewer are available to form the final beam. Thus
the grid is able to control the number of electrons
striking the screen per second and therefore
control the brightness of the spot observed.

Figure 33.7 is typical of the electron guns
found in the tubes of television sets and cathode
ray oscilloscopes.

33.10   The cathode ray tube

The discovery of thermionic emission led, several
decades later, to the development of a major
industry – the electronics industry. And although
many of the thermionic vacuum tubes which were
produced have now been replaced by solid state

devices and integrated circuits there remains a
significant group of tubes in everyday use. This
group includes the television picture tube in the
home and the cathode ray tube in the workshop
and laboratory.

Freeing the electron provided a charged
particle of minute mass (and therefore low inertia)
which could respond very rapidly indeed to
changes in electric and magnetic fields. Moreover,
these particles could be given sufficient energy to
cause fluorescence to appear when they struck a
suitable screen. It was this behaviour which is put
to use in the *cathode ray tube* (Fig. 33.8.) See
Section 25.2 for a resistor chain which could
provide the operating potentials for such a tube.

The beam from the electron gun (see Section
33.9) passes between two pairs of flat parallel
plates which are located between the anode and
the screen. A potential difference between the first
pair (*Y* plates) will cause the beam to deflected in
the vertical plane; a potential difference between
the second pair (*X* plates) will cause deflection in
the horizontal plane. Using the two pairs together,
it is possible to direct the beam to fall on any point
on the screen.

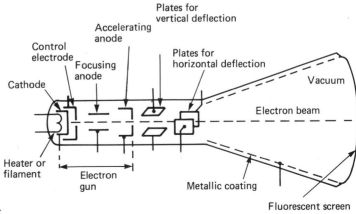

Fig. 33.8   The electrodes in a cathode ray tube.

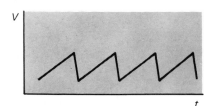

Fig. 33.9 A saw-tooth voltage for a time base.

It is usual to apply a potential difference which varies with time (in the way shown in Fig. 33.9) to the X-plates. Thus provides the time base. Then the spot moves across the screen in a horizontal direction at a steady speed taking 1 ms per cm. When the spot reaches the end of the screen, it is brought back to the starting point very quickly and repeats its travel. If now an alternating potential difference is applied to the Y plates, the spot will trace out a path which displays the waveform of the alternations. If the time base is set at 1 ms per cm, say, and the frequency of the applied alternating potential difference is 5 kHz (i.e., 1 cycle in 0.2 ms) 5 cycles will appear for each cm of trace.

The uses of such a tube which, with the appropriate circuitry, is the heart of the *cathode ray oscilloscope*, include the display of waveforms, the measurement of voltages (both peak-to-peak and direct), direction and range-finding (radar), navigational aids, echo-sounding; indeed almost any phenomenon can be investigated provided that it can be expressed in terms of a change in electrical potential. But, of course, it is the television picture tube which affords the commonest example of the application of thermionic electrons.

The television picture tube differs from that shown in Fig. 33.8 in that it does not have X and Y plates; deflection of the beam is achieved magnetically using two pairs of coils placed outside the neck of the tube. Also, a colour tube has three electron guns, one for each of the colours: red, green and blue. Mixtures of these primaries can produce any required colour.

33.11 The X-ray tube and the discovery of X-rays

This penetrating radiation was discovered by accident when W. C. Röntgen in Würzberg was doing research on the nature of the cathode rays.

He had enclosed his discharge tube completely with black card to absorb the visible light of the discharge. A translation of his paper was published in *Nature* on January 23rd, 1896. It began:

'A discharge from a large induction coil is passed through a Hittorf's vacuum tube, or through a well-exhausted Crookes' or Lenard's tube. The tube is surrounded by a fairly close-fitting shield of black paper; it is then possible to see, in a completely darkened room, that paper covered on one side with barium platinocyanide lights up with brilliant fluorescence when brought into the neighbourhood of the tube, whether the painted side or the other be turned towards the tube. The fluorescence is still visible at two metres distance. It is easy to show that the origin of the fluorescence lies within the vacuum tube. It is seen, therefore, that some agent is capable of penetrating black cardboard which is quite opaque to ultra-violet light, sunlight, or arc-light. It is therefore of interest to investigate how far other bodies can be penetrated by the same agent. It is readily shown that all bodies possess this same transparency, but in very varying degrees. For example, paper is very transparent; the fluorescent screen will light up when placed behind a book of a thousand pages; printer's ink offers no marked resistance. Similarly the fluorescence shows behind two packs of cards; a single card does not visibly diminish the brilliancy of the light. So, again, a single thickness of tinfoil hardly casts a shadow on the screen; several have to be superposed to produce a marked effect. Thick blocks of wood are still transparent. Boards of pine two or three centimetres thick absorb only very little. A piece of sheet aluminium, 15 mm thick, still allowed the X-rays (as I will call the rays, for the sake of brevity) to pass, but greatly reduced the fluorescence. Glass plates of similar thickness behave similarly; lead glass is, however, much more opaque than glass free from lead. Ebonite several centimetres thick is transparent. If the hand be held before the fluorescent screen, the shadow shows the bones darkly, with only faint outlines of the surrounding tissues.'

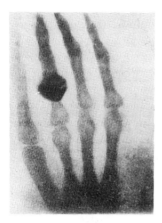

Fig. 33.10 Röntgen's photograph of the bones of a human hand. (From *Nature*, January 23, 1896. Reproduced by permission of the publisher.)

Röntgen, seeing the significance of his original observation, abandoned his other research and, in a matter of seven weeks, discovered almost all the properties of X-rays known to us today. His work was quickly translated and published in scientific journals all over the world. The fascination of the new discovery and especially of the ability it provided to produce shadow photographs of opaque objects captured the imagination of the scientific community. This is hardly surprising for later in his paper Röntgen wrote:

'The justification of the term "rays", applied to the phenomena, lies partly in the regular shadow pictures produced by the interposition of a more or less permeable body between the source and a photographic plate or fluorescent screen.

I have observed and photographed many such shadow pictures. Thus, I have an outline of part of a door covered with lead paint; the image was produced by placing the discharge-tube on one side of the door, and the sensitive plate on the other. I have also a shadow of the bones of the hand (Fig. 33.10), of a wire wound upon a bobbin, of a set of weights in a box, of a compass card and needle completely enclosed in a metal case, of a piece of metal where the X-rays show the want of homogeneity, and of other things.

For the rectilinear propagation of the rays, I have a pin-hole photograph of the discharge apparatus covered with black paper. It is faint but unmistakable.' (Röntgen, 1896.)

X-rays are produced whenever high speed electrons are suddenly slowed down or brought to rest as for example when they hit the anode of a discharge tube, or a cathode ray tube screen. The tube of a television set is a potential source of X-rays but they are not very penetrating, and the glass front absorbs them. Figure 33.11 shows a vacuum X-ray tube in which the source of electrons is a heated cathode and the anode is a disc of tungsten embedded in solid copper that extends outside the tube to cooling fins. Electrons are accelerated to high energies by a p.d. of between 25 to 100 kV and on hitting the tungsten anode only about 0.5% of their kinetic energy is converted into X-radiation, the remainder increasing the internal energy of the anode. Hence the need for a high melting point metal at the place of impact and a good conductor leading to the cooling fins to dissipate the thermal energy.

Problem 33.9

(i) Calculate the energy in joule gained by a single electron that has been accelerated through 50 kV in an X-ray tube.

(ii) How many electrons are hitting the anode per second if the tube current is 10 mA.

(iii) Estimate the thermal power that must be dissipated from the tube.

(Answers: (i) 8×10^{-5} J per electron; (ii) 6.25×10^{16} electrons per second; (iii) 500 W.)

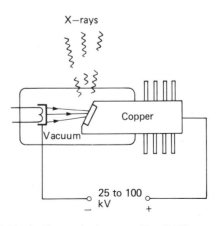

Fig. 33.11 An X-ray tube (compare Fig. 18.24)

The penetrating power of X-rays depends on the energy of the electrons producing them and consequently on the accelerating p.d. The *intensity* of the radiation from a tube depends, however, on the number of electrons leaving the cathode per second and this in turn is governed by the cathode temperature.

33.12   The properties of X-rays

1. X-rays are not deflected by electric or magnetic fields and so cannot be charged particles.
2. They can be diffracted by a crystal lattice (see Section 14.6) and therefore have a wave nature. Their wavelength is very short, ranging from 10^{-9} m to 10^{-14} m. Like other electromagnetic waves they travel at 3×10^8 m s^{-1} in a vacuum.
3. X-rays affect photographic emulsions and cause fluorescence in a number of materials including glass.
4. They cause ionization to occur in gases.
5. X-rays can produce damage in living tissue.

The uses of X-rays are not confined to the photographic examination of the human body but extend to the examination of metal castings, the inspection of canned products, the examination of baggage at airports, and as we have already seen in Chapter 14 to the examination of crystal structures.

Even the very real disadvantages of the last property listed has been put to good use in the treatment of cancer. For this, very high energy X-radiation is produced by electrons that have gained up to 4 MeV of energy in a linear accelerator. These X-rays can be carefully directed and used to destroy cancer cells within the body. Cancer cells are of a simpler structure from normal body tissue and are unable to regenerate themselves after receiving radiation therapy.

Electromagnetism and Alternating Currents

Chapter 34

MAGNETIC FORCES ON CURRENTS AND MOVING CHARGES

34.1   The force between two parallel currents

In this section we explain how the electromagnetic force between a pair of conductors, with electric currents flowing in them, can be investigated; we also explain the definition of the unit of current, the ampere.

In 1820 the Danish physicist Hans Christian Oersted announced his discovery that an electric current in a wire influenced a magnetic compass-needle. Figure 34.1 illustrates his findings. If a current is flowing along the wire from north to south, the north-seeking pole of a compass-needle placed *above* the wire is deflected to the west: it is deflected to the east when placed *below* the wire. This, and further experiments, showed that the pattern of the magnetic field produced by a current flowing in a straight conductor is in the form of concentric circles around the wire.

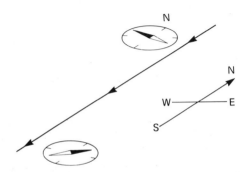

Fig. 34.1 Oersted's discovery: with the compass needle above the wire, the deflection of its north-seeking pole is westwards and with the needle below the wire, the deflection is eastwards.

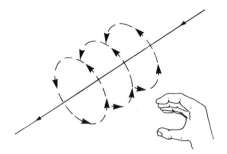

Fig. 34.2 The magnetic field due to a current in a long straight wire can be represented by concentric circles in planes perpendicular to the wire.

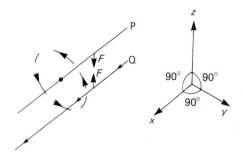

Fig. 34.3 The forces between parallel currents P and Q. One field line of the magnetic field of P is shown; where this line meets Q the field is in the positive y-direction.

The direction of a magnetic field was originally, and can still be, defined by the direction of the force on a north-seeking magnet pole placed in the field. It can be deduced from Fig. 34.2 that the direction of the field around a straight wire obeys a 'right-hand grasping rule': imagine grasping the wire with your right hand, the thumb pointing in the direction of the current; your fingers, curled around the wire, give the direction of the magnetic field. Also, this is the direction in which an ordinary screw with a right-hand thread would rotate when advancing in the direction of the current (the 'corkscrew rule').

In Chapter 27 we studied how the strength of a magnetic field can be measured by measuring the force on a suitable conductor, with current flowing in it, placed in the field. This technique can be used to measure the magnetic force which one conductor exerts upon another when a current flows in each of them.

At this point we should mention that it was André Marie Ampère, whose name was eventually chosen for the unit for electric current, who first investigated the magnetic force between wires carrying electric currents. In a memoir published in 1820, he discusses the differences between the effects of *static* electricity and *current* electricity, and wrote:

'But the differences which I have recalled are not the only ones which distinguish these two states of electricity. I have discovered some more remarkable ones still by arranging in parallel directions two straight parts of two conducting wires joining the ends of two voltaic piles; the one was fixed and the other,

suspended on points and made very sensitive to motion by a counterweight, could approach the first or move from it while keeping parallel with it. I then observed that when I passed a current of electricity in both of these wires at once they attracted each other when the two currents were in the same sense and repelled each other when they were in opposite senses.'

The apparatus which we use to investigate these effects today, soon to be described, is very little different from Ampère's original apparatus which used 'voltaic piles'. These were batteries consisting of stacks of metal discs separated by pieces of cardboard soaked with electrolyte. They were invented by Alessandro Volta in 1800.

Before attempting to measure the force between two conductors we shall predict the direction of this force, using the ideas developed in Chapter 27 and the information given above about the magnetic field around a straight conductor. Figure 34.3 shows a pair of horizontal straight wires, in the same vertical plane, with a current flowing along each one of them. Just one of the magnetic field lines of the field around P is shown; you should verify that this obeys the right-hand grasping rule. The set of axes: x, y, z, all at right angles to each other, in Fig. 34.3, is useful for defining directions in three-dimensional space. These axes are the conventional Cartesian axes as used in three-dimensional geometry. The two currents are in the positive x-direction. The magnetic field of wire P, where it meets with Q, is in the positive y-direction. Now, to predict the direction of the force on wire Q, use the Fleming left-hand rule (Section 27.5). You should find that

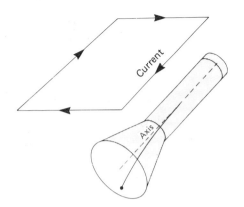

Fig. 34.4 To show the effect of a magnetic force on an electron beam. The beam is repelled by the current in the wires above it.

this force is in the positive z-direction (upwards).

Of course we can just as well consider the magnetic field of wire Q and the force which this field produces upon wire P. By this reasoning we find that the force on P is in the negative z-direction. This is in instance of Newton's third law: the force which P exerts upon Q is equal and opposite to the force which Q exerts upon P. Our predictions imply that two wires, carrying currents flowing in the same direction, should attract each other.

If the current in one of the wires is reversed when the direction of its magnetic field will also be reversed, and thus the direction of the force on the other wire will be reversed. So we can infer that two wires, with opposite-going currents flowing in them, will repel each other.

The simplest apparatus for demonstrating the existence of these forces comprises a pair of long strips of thin aluminium foil hanging loosely side by side, with means for passing a current of several amperes through each of them, and a means of reversing the current in one or both of the strips. The force is weak, but sufficient to cause a visible change in the spacing of the strips.

These forces exist even when there is no conductor to carry the current. A stream of charged particles travelling through empty space is surrounded by a magnetic field, and consequently charged particles moving in a magnetic field experience a force (provided that they are not travelling parallel to the field). Figure 34.4 shows a demonstration using a cathode ray tube. A large current is needed in the conductor lying parallel to the axis of the tube, and so a bundle of wires is used. About 20 thicknesses of wire with a current of about 2 A in the wire is equivalent to a single current of about 40 A, and this is enough to cause a noticeable deflection of the electron beam, either towards or away from the wires. The demonstration can be used to confirm that the electron beam is equivalent to a conventional current going in the opposite direction to the beam, hence that the beam must consist of negative charges.

34.2 An Ampere balance and the definition of the unit of current

This instrument, an example of which is shown in Fig. 34.5, is one type of *current balance* (see also

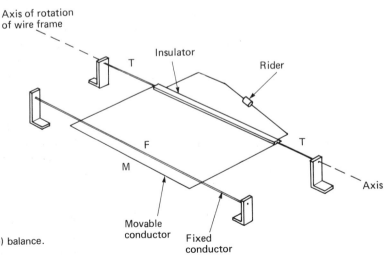

Fig. 34.5 A simplified current (or Ampère) balance. (Courtesy of Griffin and George Ltd.)

Section 27.5). The instrument is similar to the apparatus used by Ampère himself and described in a memoir he presented to the French Academy of Sciences in 1820: for this reason we call it an *Ampère balance*. We use the Ampère balance to measure the force between a pair of parallel currents, and to make absolute measurements of currents in amperes.

In the Ampère balance one conductor is fixed (F) and the other movable (M). The movable conductor is one side of a rectangular wire frame which is pivoted by means of two springy metal tapes (T,T) which are in a state of tension. The frame can be balanced in an horizontal plane by means of a small movable counterweight (a 'rider'). Current flows in and out of the wire frame by way of the two metal tapes.

The Ampère balance can be used to find how the force between a pair of parallel straight conductors depends upon the distance between them, when the current in each conductor is kept constant. A circuit is made comprising a steady d.c. supply, a variable resistor, an ammeter, and the fixed and movable conductors of the Ampère balance, all connected in series. The purpose of the ammeter is simply to ensure that a constant current is maintained. Table 34a shows the results of an experiment in which the ammeter reading was 2 A. These results are consistent with the supposition that the force is inversely proportional to the distance between the conductors, because, for each setting, the product of the force and the distance is very nearly constant. Experiments with apparatus capable of greater precision have shown that this is exactly the case, provided that the thickness of the conductors is negligible compared with their spacing, and that 'end-effects' can be neglected. By 'end-effects' we mean deviations of the magnetic field pattern near the ends of the

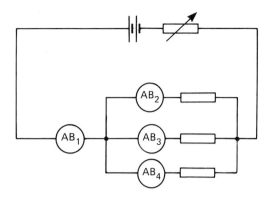

Fig. 34.6 Testing Kirchhoff's first law using Ampère balances.

straight conductors, where they change direction. These end-effects are negligible if the spacing of the conductors is very much smaller than their lengths.

Suppose now that we set and keep a constant spacing between the two conductors of the Ampère balance, and keep a constant current flowing in one of the conductors (the fixed one, let us suppose). We can then use the force between the conductors as an indication of the strength of the current in the movable conductor. Suppose now that we had four of these Ampère balances, all identically arranged so that they produce the same force when the same current flows through the movable conductor of each of them. This would be verified, in practice, by connecting the four movable conductors in series with a suitable supply and resistance. These four Ampère balances are then arranged in a circuit as shown in Fig. 34.6, where we use the symbol AB to represent the *movable conductor only* of each balance. The same constant current flows through the *fixed* conductor of all four balances. This is provided by a separate supply.

Imagine now that we measure the force between the conductors in each of the four Ampère balances, keeping the spacing of the conductors the same for all of them. We would find that the relationship between the four forces is

$$F_1 = F_2 + F_3 + F_4$$

This demonstrates Kirchhoff's first law (see Section 23.1), and also indicates that the force between the conductors in each balance is directly

Table 34a

Distance (d) between centres of conductors/ mm	Force (F) on movable conductor/μN	Product $F \times d$
12	13	156
10	16	160
8	20	160
6	27	162
4	40	160

proportional to the current in the movable conductor. The experiment could be performed just as well with the symbol AB representing the fixed conductor in each balance, and having the four movable conductors connected in series to a separate supply.

It follows that the force between a pair of conductors must be proportional to the current in each of them. And because (assuming end-effects are negligible) this force is uniformly distributed along the length of wire upon which it acts, we can infer that the force is directly proportional to this length. For instance, if we halved the length of the shorter, movable conductor, we should expect the force to be halved, provided that we do not alter the current or the spacing of the conductors.

We can now write an expression for the force between a pair of parallel straight conductors:

$$F \propto \frac{I_1 I_2 l}{a}$$

where I_1 and I_2 are the two currents, l is the length of the pair over which the force acts, and a is the distance between them. The expression can also be written:

$$F = \text{(a constant)} \frac{I_1 I_2 l}{a} \qquad (34.1)$$

The value of the constant depends on how the unit for current is defined. The unit is appropriately named the *ampere*, as you already know. It is one of the *base units* of SI:

■ The ampere is that steady current which, flowing in two infinitely long straight parallel conductors of negligible circular cross-sectional area placed one metre apart in a vacuum, causes each conductor to exert a force of 2×10^{-7} newton on each metre of the other conductor.

In Eq. 34.1, if we substitute $I_1 = I_2 = 1$ A, $l = 1$ m, and $a = 1$ m, then, by the definition of the ampere, the force must be 2×10^{-7} N. Thus the value of the constant in the expression must be 2×10^{-7} N A^{-2}. This means that we can use the Ampère balance to make a measurement of a current by measuring a force, a length, and a distance. Such a measurement is termed an *absolute measurement* because the method of measurement is directly related to the definition of

the ampere: we are not using an instrument which first of all has to be *calibrated* to read current in amperes.

Problem 34.1   A parallel-wire Ampère balance was connected, with both its conductors in series, in a circuit with an ammeter believed to be wrongly calibrated, a battery, and a variable resistor. These measurements were made:

Length of movable conductor (the shorter of the two): 200 mm

Spacing between centres of conductors: 8 mm
Force between conductors: 28 μN
Ammeter reading: 2.5 A

Use Eq. 34.1, and the numerical value for the constant given above, to calculate the current. Comment on the correctness or otherwise of the calibration of the ammeter.

$$F = 2 \times 10^{-7} \text{N A}^{-2} \frac{I_1 I_2 l}{a}$$

Since the two conductors are connected in series we can write: $I_1 = I_2 = I$. Therefore

$$I^2 = \frac{28 \times 10^{-6} \text{N} \times 8 \times 10^{-3} \text{m}}{2 \times 10^{-7} \text{N A}^{-2} \times 200 \times 10^{-3} \text{m}}$$

hence

$$I = 2.4 \text{ A (to two significant figures)}$$

But this does not mean that we can say that the ammeter is wrongly calibrated. Consider the measurement of the spacing of the conductors: this has been given to the nearest millimetre only. A variation of ± 0.5 mm in this spacing would make a considerable difference to the result. If, in the calculation above, we substitute 7.5 mm instead of 8 mm the calculated current becomes 2.2 A. If we put 8.5 mm instead, we get 2.5 A.

This illustrates the point that the simple parallel-wire Ampère balance of Fig. 34.5 is not capable of enough precision to calibrate ammeters, although it does show the principle of how such calibration can be made, and also demonstrates the law of force between parallel straight conductors.

34.3   Forces on a coil in a magnetic field

Up to now our discussion of electromagnetic forces has been limited to the forces on straight

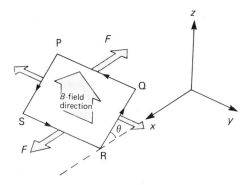

Fig. 34.7 The forces on a plane rectangular current-loop in a uniform magnetic field. The forces labelled F produce a torque. The unlabelled forces are in equilibrium and do not produce a torque.

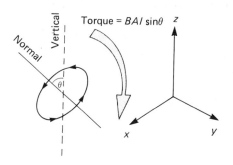

Fig. 34.8 A plane circular current-loop in a uniform magnetic field acting in the Z-direction. Its orientation is defined by the direction of the normal. The area enclosed by the loop is A.

conductors in magnetic fields. In order to understand the principles of certain types of electric meter, of dynamos and motors of various kinds, we need first to know about the forces which act upon a *coil* in the presence of a magnetic field.

Figure 34.7 shows a single rectangular loop of wire PQRS. A set of Cartesian axes, xyz, is shown in the diagram so that we can define directions in three-dimensional space. The sides PQ and RS of the rectangle are parallel to the y-axis. We shall suppose that the xy plane is the horizontal plane. The plane of the rectangle PQRS is tilted in relation to the xy plane. Let us suppose also that there is a uniform magnetic field, strength B, parallel to the positive z-direction. We imagine that there is a current I flowing around the rectangle, although, for simplicity, we have not shown any connections to an electrical supply.

Consider side PQ first. The current I is flowing in it from Q to P, in the negative y-direction. The magnetic field is at right angles to this direction. The Fleming left-hand rule can be used to determine that the force on PQ is in the negative x-direction. Similar reasoning for side RS shows that the force on this is in the positive x-direction. If l is the length of each of these sides, then there are two equal and opposite horizontal forces acting and, as shown in Chapter 27, the magnitude of each force is given by

$$F = BIl$$

The perpendicular distance between the lines of action of these forces is $w \sin \theta$, where w is the length of side PS and side QR. Thus the torque

produced by this pair of forces is given by

$$T = BIl \times w \sin \theta = BIlw \sin \theta \quad (34.2)$$

Consider now the forces acting upon the sides PS and QR. The force on PS is the negative y-direction, and the force on QR is in the positive y-direction. The two forces have a common line of action midway between sides PQ and RS: thus they counterbalance each other and produce no torque.

The torque on the loop, due to the forces on sides PQ and RS, will be greatest when $\theta = 90°$, and zero when $\theta = 0°$. If the coil were free to move it would rotate and eventually come to rest with its plane parallel to the xy-plane (or, to express this another way, with its plane perpendicular to the magnetic field).

Note that in Eq. 34.2 the product lw is the area enclosed by the wire rectangle PQRS. Thus, if we call this area A, the equation can be written

$$T = BIA \sin \theta \quad (34.3)$$

Although we shall not do so here, it can be proved theoretically (and verified experimentally) that Eq. (34.3) is correct for a plane loop of any shape, not only rectangular.

The angle θ in Eq. 34.3 is the angle shown in the diagram on Fig. 34.7; in this diagram it is indicated as the angle between the side QR of the rectangle and the x-axis. Now suppose we were dealing with a circular loop instead of a rectangular one, again with its plane tilted with respect to the horizontal xy-plane. How would we indicate the angle in that case? Figure 34.8 shows this situation. The normal to the plane of the loop ('normal' here meaning a line perpendicular to the plane of the loop) makes an angle θ with the magnetic field which, once again, is parallel to the z-axis. In three-dimensional geometry the orientation of a plane, such as PQRS, in three-dimensional space is usually described by specifying the direction of a normal to the plane. In the case of a circular coil of wire we often refer to the *axis* of the coil, and the direction of this axis will always be very nearly the same as the direction of the normals to the planes of all the loops which go to make up the whole coil.

34.4 The moving coil galvanometer

The term 'galvanometer', derived from the name of the Italian physicist Galvani, has survived as the name used for an instrument which measures small electric currents. The name d'Arsonval – often associated with the moving coil galvanometer – was a French medical physicist who developed the instrument for use in industry, education, and medical research. It still forms the basis of many current-measuring instruments, and voltmeters. For certain applications it is rapidly being superseded by fully electronic instruments with digital displays. By contrast with a digital instrument, a moving coil instrument is often termed an 'analogue' measuring instrument, because the angular position of the coil is an analogue of the electrical quantity being measured.

A moving coil galvanometer makes use of the torque on a coil wrapped around a light rectangular frame and mounted on pivots in the field of a strong permanent magnet. Figure 34.9 shows the moving part of a typical instrument, with the pivot bearings and their supports. Figure 34.10 shows the fixed magnet assembly of the same instrument. The central, fixed cylinder is a solid piece of soft iron and forms a part of the magnet-assembly. This design of magnet-assembly ensures that, whatever position the coil settles in, the direction of the field in the annular air-gap is parallel to the plane of the coil: in other words, the field is in a radial direction. If this is the case, then the torque acting upon each loop of the coil is simply: BIA. To see the reason for this, look again at Fig. 34.8. If the loop is set with its plane parallel to the field,

then θ is 90° and so the torque is $BIA \sin 90° = BIA$. If the coil has N turns then the torque on its is $BIAN$.

The coil is not generally allowed to swing through a total angle of more than 60° at most. In this way the fact that the field becomes appreciably non-radial near the gaps between the poles is of no consequence.

In a typical milliammeter such as is used in school laboratories the magnetic flux density in the air-gap is likely to be in the order of 10^{-1} tesla. The sides of the coil, assumed square, might be about 10 mm. The number of turns in the coil could be, say, 50. Suppose that the current producing full-scale deflection is 100 μA. We can estimate the torque on the coil as follows.

Area enclosed by coil:

$$10 \text{ mm} \times 10 \text{ mm} = 10^{-4} \text{ m}^2$$

Torque on one turn:

$$BIA = 10^{-1} \text{T} \times 10^{-4} \text{A} \times 10^{-4} \text{m}^2$$
$$= 10^{-9} \text{N m}$$

Torque on whole coil:

$$50 \times 10^{-9} \text{N m} = 5 \times 10^{-8} \text{N m}$$

This is about the amount of torque which would

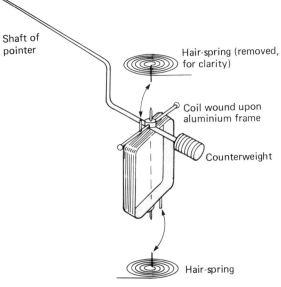

Shaft of pointer

Hair-spring (removed, for clarity)

Coil wound upon aluminium frame

Counterweight

Hair-spring

Fig. 34.9 The coil assembly of a moving coil meter. The two hair-springs provide the electrical connections to the coil as well as the counter-balancing torque.

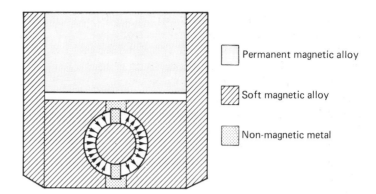

Fig. 34.10 The magnet assembly of a moving coil meter provides a radial field in the cylindrical air-gap.

result from a small ant sitting on one side of the coil! In practice, the torque on the coil due to the current is counterbalanced by the torque produced by two spiral hair-springs: these also serve as electrical connections to the coil. They are wound in opposite directions to counteract effects of thermal expansion. As a rule, the outer end of one of these hair-springs can be moved by means of an adjusting screw in order to set the pointer to the zero of the scale when no current flows.

When a current flows in the coil, the coil turns until the torque due to the springs (often called the *restoring torque*) counterbalances the torque due to the current. The torque due to the springs is proportional to the angle through which the coil has rotated from its zero position. Calling this angle α, we can then call the torque provided by the springs $c\alpha$, where c is torque per unit angular rotation, or *torsional constant* of the pair of springs. Assuming that any friction is negligible, when the coil reaches equilibrium:

torque due to current = torque due to springs
$$NBIA = c\alpha$$

From this equation we obtain an expression for the current sensitivity of the meter. The current sensitivity is defined as the *angular deflection per unit current* and it is given by:

$$\frac{\alpha}{I} = \frac{NBA}{c} \qquad (34.4)$$

This expression can be regarded as a 'design formula' for a moving-coil galvanometer. Question 9.4 at the end of this book considers this 'design formula' further.

The voltage sensitivity of a galvanometer is defined as the *angular deflection per unit p.d.* If the resistance of the coil is R and a p.d. V is applied to it then the current in the coil is given by:

$$I = \frac{V}{R}$$

The voltage sensitivity is given by:

$$\frac{\alpha}{V} = \frac{\alpha}{IR} = \frac{NBA}{cR} \qquad (34.5)$$

34.5 Forces on moving charged particles

If you have the opportunity to study the inside of a television receiver (when it is switched off and unplugged from the mains), look for the *deflection coils* which are situated close against the outside of the picture tube, as in Fig. 34.11. One

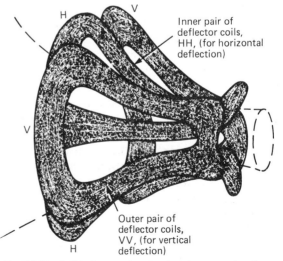

Fig. 34.11 A TV picture tube showing the two pairs of deflection coils.

pair (V, V) produces a magnetic field in a vertical direction and this deflects the electron beam horizontally. The other pair (H, H) produces a horizontal magnetic field for deflecting the beam vertically.

In the realm of atomic physics, huge electromagnets are used to keep electrons, protons, or other charged particles travelling in circular paths in the particle accelerating machines called *synchrotrons*. *Mass spectrometers* employ the same principle, and these will be met later in Chapter 48.

We now proceed to study the forces which act upon moving charged particles in a magnetic field.

In Chapter 32 we developed the model of an electric current as a flow of charged particles. We use this model also in deriving an equation for the force on a single charged particle moving in a magnetic field. In Section 32.5 we obtained the equation

$$I = nAve$$

for the current in a conductor, where n is the number of electrons per unit volume, A is the area of cross-section of the conductor, v is the drift velocity of the electrons each of which carries a charge e. Consider a length l of the conductor, and suppose the conductor is straight and that it lies at an angle θ to a magnetic field of uniform strength B. The number of electrons in this length of conductor is nAl. The force on the length l is given by

$$F = BIl \sin \theta$$

and so the mean force acting upon each particle is

$$\frac{B(nAve)l \sin \theta}{nAl} = Bev \sin \theta$$

We can now suppose that the magnetic force acting upon the conductor as a whole is the combined effect of the magnetic forces acting upon all the individual particles. The direction of the force on a particle is therefore presumably in a direction perpendicular to the plane which contains the direction of the magnetic field and the particle's velocity.

We shall now discuss what shape of path is followed by a charged particle in a uniform magnetic field, when no other forces are acting upon it.

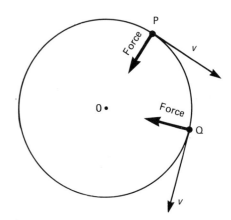

Fig. 34.12 The motion of a positively charged particle in a uniform magnetic field which is perpendicular to, and out of the plane of the diagram.

The simplest case to consider is when a particle is moving in a plane perpendicular to the direction of the field. This is represented in Fig. 34.12 where we have to imagine that there is a uniform magnetic field in a direction perpendicular to the plane of the diagram, and directed up out of the plane of the diagram. P and Q are two successive positions of the particle, assumed to be carrying positive charge e. At P we suppose that the direction of the particle's velocity v is in the plane of the diagram, so that the force on it, which has to be perpendicular both to the field and to the direction of v, is also in the plane of the diagram, and (by Fleming's left-hand rule) directed towards O. Because this force is perpendicular to the particle's motion at every instant, the velocity of the particle will keep constant magnitude, but will continuously alter direction. Because the magnitude of the velocity is constant and the field is constant, the force will have constant magnitude, and therefore the particle will travel in a circular path. The magnetic force provides constant centripetal acceleration and so, if the radius of the circle is r, we can write:

$$Bev = \frac{mv^2}{r}$$

$$r = \frac{mv}{Be} \qquad (34.6)$$

This equation shows that the radius of the particle's path is directly proportional to the momentum of the particle, and inversely propor-

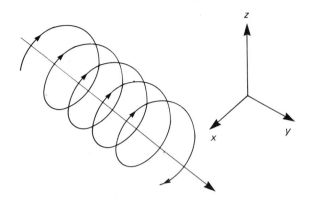

Fig. 34.13 Helical path of charged particles in a uniform magnetic field (in the positive *y*-direction).

tional to its charge. This implies that, if we have a stream of particles all of one kind, each with the same mass and charge, entering a magnetic field in a plane perpendicular to the field, each will travel in a circular path whose radius is directly proportional to the particle's speed. This prediction can be tested in an impressive way using a 'fine beam tube', in which the track of a thin beam of electrons is made visible by the presence in the tube of hydrogen gas at a very low pressure. Such a tube is shown in Fig. 33.5 in connection with its use in the determination of the specific charge of the electron. An electron gun projects a fine beam of electrons vertically upwards. A magnetic field is provided by a pair of coils, one in front of the tube and one behind, with their axes perpendicular to the plane of the figure. This arrangement of coils produces a field which is approximately uniform in the region inside the tube where the electrons travel, and the direction of the field is perpendicular to, and directed down into, the plane of the figure. The two coils are connected in series in such a way that their fields reinforce each other, and the strength of the field can be altered by varying the current in the coils. The speed of the electrons is controlled by the p.d. between cathode and anode in the electron gun.

A simple demonstration with the fine beam tube proceeds as follows.

a) Dependence of path radius *r* upon electron speed *v*. Eq. (34.6) predicts that $r \propto v$, if *B*, *m*, and *e* are constant. The electron 'gun

voltage' (that is, the p.d. between cathode and anode) is increased, and it is observed that the circular beam increases in size. This observation is consistent with the prediction (although no quantitative measurements have been made here).

b) Dependence of path radius *r* upon field strength *B*. Eq. (34.6) predicts that $r \propto 1/B$, if *m, v* and *e* remain constant. Without altering the gun voltage, the field strength is increased (by increasing the current in the coils), and we observe that the path radius decreases: and this is consistent with the prediction.

We have already seen how the fine beam tube is used to determine the specific charge (*e/m*) of the electron. (Section 33.3.)

One further simple experiment with the fine beam tube should be tried: the effect of 'injecting' the electrons into the magnetic field in a direction which is not in a plane perpendicular to the field. This is easily done by twisting the tube a little in its holder so that the electron gun's axis is at a small angle to the vertical. Then, especially if you look down on the tube from above, you will see that the path of the beam is helical in shape. Figure 34.13 shows the path of a beam of charged particles, all having the same mass, charge, and speed, in a uniform magnetic field when they are not travelling in a plane perpendicular to the field.

We can predict that the path will have this helical shape, as follows. At every instant, the particle's velocity can be regarded as made up of two velocity components, one component parallel to the field and the other component perpendicular to the field; and the effect of the field on these two components can be considered separately. We know already that a charged particle travelling in a direction parallel to a magnetic field experiences no magnetic force: we can therefore suppose that there will be no contribution to the force on the particle, in Fig. 34.13, arising from its velocity component parallel to the field. However, its velocity component perpendicular to the field will give rise to a magnetic force in a plane perpendicular to the field and this force will cause circular motion in this plane. But since there is no force on a particle in a direction parallel to the field, then the particle's velocity component parallel to the field (by Newton's first law) will remain unaltered. Hence the particle's motion is a combination of uniform motion in a circle in a plane perpendicular to the field, and constant velocity parallel to the field. Thus a particle travels as if it were going along the groove in a screw-thread of uniform pitch and diameter. The name for this shape of path is *helix* and the adjective is *helical*. The shape is not a *spiral*: a spiral is a curve of continuously changing radius in one plane.

A magnetic field can 'capture' moving charged particles, by forcing them to travel in circles or helical paths around the direction of the field. This phenomenon is widespread in the universe. For example, the earth's own magnetic field 'captures' some of the ions and other charged particles which are continually, although irregularly, being emitted by the sun. These particles are most highly concentrated, and lose their energy most rapidly, near the earth's poles, and they cause the faint, coloured glowing streamers in the night sky known as the *Aurora borealis*, or *Northern Lights*, in the northern hemisphere. The *Van Allen belts* are regions, girdling the earth, in which charged particles are held 'trapped' by the earth's magnetic field. The characteristic shape of an eruptive *solar prominence* reveals the pattern of the magnetic field which tends to keep the charged particles contained.

Chapter 35

ELECTROMAGNETIC INDUCTION

35.1 Basic experiments and theory

One of the best introductions to electromagnetic induction is to dismantle a bicycle dynamo: the sort which is driven by a small wheel pressing against the side of the tyre.

Figure 35.1 shows the two main parts. One part is the *rotor*, which is simply a permanent magnet mounted on a spindle. It may not look like other magnets you have seen. To find where its poles are, put it in a thin plastic bag and dip it into iron filings; or better still, explore the magnetic field around it with a small magnetic compass. The other part is the *stator*, the stationary part, which is clearly like an electromagnet, comprising a coil of wire wrapped around a core made of iron. Note that its pole-faces are so shaped that the air-gap between them and the rotor is very small. The cylindrical magnet of the rotor has its poles at opposite ends of a diameter, and when it spins an alternating e.m.f. is generated in the

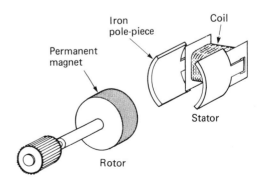

Fig. 35.1 A dismantled bicycle dynamo.

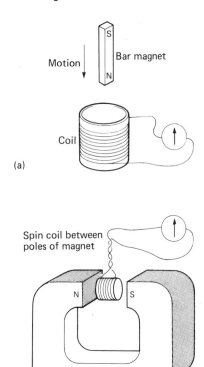

(a)

(b)

Fig. 35.2 Simple demonstrations of electromagnetic induction.

stator coil. What exactly is happening? We shall answer this question in Section 35.3. First of all we must study electromagnetic induction using some specially-made apparatus.

We know that a current-carrying conductor in a magnetic field generally experiences a force, and Chapter 34 was concerned with the practical applications of this force. When a current flows through a wire a magnetic field is set up around the wire, and the interaction of this field with some other magnetic field usually gives rise to a force on the wire. The converse effect, in which a magnetic field causes a current to flow in a wire, is known as *electromagnetic induction*, and was first fully investigated by Michael Faraday. If you have not done so before you should try the simple demonstrations of electromagnetic induction illustrated in Fig. 35.2. In each experiment it is found that the galvanometer registers a small current when the coil and the magnet are in relative motion, and that this current ceases as soon as the relative motion stops. The direction of the current is seen to be related to the direction of

the relative motion of coil and magnet, and the strength of the current is related to the speed of the relative motion: the greater the speed, the greater the current. Also, if you increase the number of turns of wire in the coil in each case, you will find that this increases the current generated.

The simple demonstrations just described do not enable us to investigate precisely what factors determine the strength of the induced current, and a carefully designed experiment is necessary. Figure 35.3 shows an arrangement for moving a wire (or rather, a bundle of wires) at a steady, measurable speed between the poles of an electromagnet. It is found that a current is generated while the wires are moving, and only while they are moving. In this experiment we can alter:

a) the strength of the field,
b) the speed of the wire,
c) the number of turns in the coil,
d) the resistance of the circuit.

We can alter each parameter singly, and we find the following.

a) If the strength of the field is increased, by stepping up the current in the electromagnet, the induced current is greater, with the coil moving at the same speed. If one uses a current balance to measure the flux density (strength) of the magnetic field (see Section 27.5) then one can show that the strength of the induced current is directly proportional to the flux density of the magnetic field.

b) Increasing the speed of the wire increases the

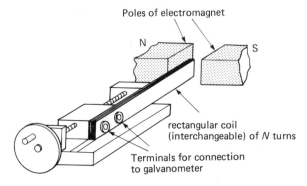

Fig. 35.3 A 'flux-cutter'. The interchangeable rectangular coil of N turns can be moved into and out of the magnetic field by turning the drive screw. (Courtesy Griffin and George.)

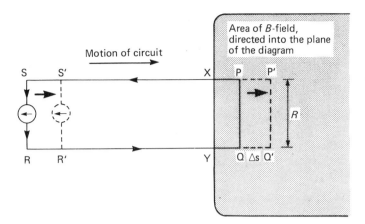

Fig. 35.4 The circuit PQRS moves through distance $\triangle s$ in time $\triangle t$.

induced current, and it is easily shown that the current is directly proportional to the speed.

c) If different numbers of turns in the coil are connected into the circuit with the galvanometer, then it is found that, for motion with the same constant speed, the current is directly proportional to the number of turns. Hence, presumably, the total length of wire *interacting with the field* is the important factor.

d) Varying the total circuit resistance, while keeping parameters (a), (b), and (c) constant, shows that the greater the resistance the smaller the current. Therefore it is the e.m.f., not the current, which is determined by (a), (b) and (c). Thus the induced e.m.f., E, is given by

$$E \propto Blv \qquad (35.1)$$

where B is the field strength, l is the total length of wire interacting with the field, and v is the speed.

We shall now see how the expression for the induced e.m.f. can be derived theoretically.

Consider the situation shown in Fig. 35.4 in which the conductor PQ carries free, positively charged particles. The charge on each particle is e. PQ is part of a circuit PQRS which moves from left to right at a speed v into a uniform magnetic field of strength B. The direction of this field is perpendicular to the plane of the diagram and down into this plane. XY is the boundary of the field region.

As each particle moves along with speed v it is acted upon by a magnetic force (as discussed in Section 34.6) which acts at right angles to the direction of B and to the direction of v, and by

Fleming's left-hand rule we find that the force is directed from Q to P. The magnitude of the force is given by

$$F = Bev$$

The positive charges will therefore be pushed up from Q to P and will then travel around the circuit. The amount of energy converted when one charge goes from Q to P is the product of the force and the distance it goes, and so is given by

$$Fl = Bevl$$

This means that an e.m.f. is being generated in the circuit, because (Section 24.2) e.m.f. is defined as the *total energy per unit charge* that the complete circuit can utilize, or dissipate. Hence, considering a single particle as it goes around the complete circuit:

$$\frac{\text{energy converted}}{\text{charge}} = \frac{Bevl}{e}$$

and so the e.m.f. is given by:

$$E = Bvl \qquad (35.2)$$

So we see that the constant of proportionality in Eq. 35.1 is in fact unity.

If we imagine the free charges in the moving conductor PQ, in Fig. 35.4, to be negative instead of positive, then the magnetic force will push them downwards towards Q, and they will travel around the circuit QRSP in a clockwise direction. But the direction of conventional current (positive charge flow) will be the same as before, and the amount of energy gained by the charges will be the same as before, and so the same final expression for the induced e.m.f. will be obtained. Thus the magnitude of the induced e.m.f. does not depend upon whether the charge-carriers are positive, or negative, or even a mixture of both.

As soon as there is an induced current flowing around the circuit PSRQ, in Fig. 35.4, there will be a force on the wire PQ, just as there is a force on any conductor which is at right angles to a magnetic field, when a current flows in the conductor. We have already reasoned that the direction of (conventional) current is anti-clockwise around the circuit: so along PQ the current direction is from Q towards P. Applying the Fleming left-hand rule, we find that the force on PQ must be towards the left, opposite to the direction of motion of the circuit. The implication of this is that, in order to keep the circuit moving, a force equal and opposite to this leftwards force must act upon the circuit, and that mechanical work must be done in moving the circuit along.

This must be so, in order to satisfy the principle of conservation of energy. When a steady current is flowing in the circuit, electrical energy is being converted into heat, because the circuit has electrical resistance. This energy is being provided by whatever agency is pushing the circuit along. If the direction of motion were reversed, the induced current would likewise be reversed in direction, and so would the direction of the magnetic force on the wire. This force would now be directed towards the right; but once again it would *oppose* the circuit's motion. We have here an instance of Lenz's law:

■ The direction of the induced current is always such as to oppose the change which causes that induced current.

35.2 The concept of magnetic flux

Let us refer once again to Fig. 35.4 and suppose that the circuit PQRS has moved forwards a distance Δs, and that its leading side is going from PQ to $P'Q'$, in a time Δt. Then its velocity is given by:

$$v = \frac{\Delta s}{\Delta t}$$

consequently

$$Bvl = B\frac{\Delta s}{\Delta t}l$$

Inspection of the diagram shows that

$$B\Delta s\, l = B\,(\text{area } P'Q'YX - \text{area } PQYX) \quad (35.3)$$

and this is the change in the product:

(field strength) × (area enclosed by circuit)

This product is the amount of magnetic flux 'passing through' the area. In this illustrative example the direction of the field is *normal* (that is, perpendicular) to the plane of PQYX. In situations where the field direction is not normal to the plane of the circuit the magnetic flux 'passing through' the circuit is the product:

$$\left(\begin{array}{c}\text{component of field strength} \times \text{area enclosed}\\ \text{normal to plane of circuit} \qquad \text{by circuit}\end{array}\right)$$

The unit for magnetic flux could be expressed as the product of the unit for field strength, the tesla; and the unit for area, square metre. But because the concept of magnetic flux is regarded as being fundamental in the theory of electro-magnetism it has a unit of its own: the *weber* (Wb). Thus the unit *tesla* can be expressed in terms of the weber in this way:

$$1\,\text{T} = 1\,\text{Wb m}^{-2}$$

Furthermore, the term *magnetic flux density* is commonly used instead of *magnetic field strength*, in electromagnetism. Here, the term 'density' is used in the special sense of meaning 'per unit area' (unlike the way the term is used when saying: 'the density of pure water is $1000\,\text{kg m}^{-3}$'). So, from here onwards, we shall use the term *magnetic flux density*, although we shall continue to use the *tesla* unit where appropriate.

An amount of magnetic flux can be pictured, in the mind, as a number of *lines of flux*, and in Fig. 35.4 we can think of the wire PQ slicing through these lines of flux as it goes along, like a reaping machine cutting stalks in a field of corn. The amount of flux cut by the wire as it goes from PQ to $P'Q'$ is the amount contained within the area $P'PQ'Q$.

Equation 35.3 can be written thus:

$$B\,\Delta s\, l = \text{amount of flux cut by conductor}$$

and so:

$$B\frac{\Delta s}{\Delta t}l = \text{rate of cutting flux}$$

$$= Bvl$$

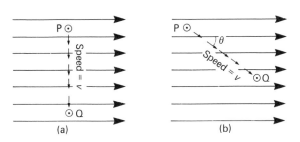

Fig. 35.5 Section through a straight conductor cutting lines of magnetic flux.

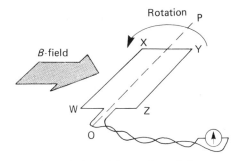

Fig. 35.6 The principle of the moving coil dynamo.

and as shown earlier (Eq. 35.2) this is the induced e.m.f. Thus we can write:

$$\text{induced e.m.f.} = \text{rate of cutting flux} \quad (35.4)$$

35.3 The e.m.f. induced in a rotating coil

Up to now we have considered electromagnetic induction in a conductor moving in a plane perpendicular to the direction of the magnetic flux. We shall now need to extend the ideas to the case of a conductor moving in a different plane.

Figure 35.5a represents a straight conductor which is perpendicular to the plane of the diagram and is moving across lines of magnetic flux, in a uniform magnetic field represented by a set of equally spaced parallel lines. Going from P to Q it cuts five lines. In Fig. 35.5b it goes the same distance but only cuts three lines. In both cases, the conductor is moving with the same speed, and experiments show that the e.m.f. induced is greatest in the case shown in Fig. 35.5a, and furthermore that the e.m.f. is proportional to $\sin \theta$, where θ is the angle between the direction of motion and the direction of the flux. If the conductor moves in a direction parallel to the flux, then $\theta = 0$ and the e.m.f. is zero. Clearly, therefore, it is the component of velocity perpendicular to the flux, $v \sin \theta$, which determines the magnitude of the induced e.m.f. Equation 35.4 becomes, for this situation:

$$\boxed{E = Bvl \sin \theta} \quad (35.5)$$

We are now equipped with the concepts needed to explain the action of a simple dynamo. Figure 35.6 shows the principle of a simple moving coil dynamo. A plane rectangular coil WXYZ rotates anti-clockwise about an axis OP which is set at right angles to a uniform magnetic field whose flux density is B. In the position shown in Fig. 35.6 the sides WX and ZY are moving at right angles to the flux, and in opposite directions. Thus the e.m.f.s induced in these sides reinforce each other and push electric charge one way around the circuit. The sides XY and WX are not cutting lines of flux, nor will they do so at any stage during rotation of the coil. Fig. 35.7 shows a section through the coil, viewed in a direction along its axis of rotation, at various stages during rotation.

During one half-revolution the current flows one way around the circuit, and it flows the opposite way during the next half-revolution. Consider the coil when the normal to its plane makes an angle θ with the direction of the field (Fig. 35.8). Let v be the speed of each of the sides WX and ZY. If the length of each of these sides is l then the e.m.f. induced in each side is given by Eq. 35.5:

$$E = Bvl \sin \theta$$

If the angular speed of the coil is ω, and if $\theta = 0$ when $t = 0$, then $\theta = \omega t$ and the linear speed of the sides WX and ZY is given by

$$v = \omega \frac{b}{2}$$

where b is the length of sides XY and WZ. Hence the total e.m.f. produced by WX and ZY together can be written:

$$E = 2 \times B\omega \frac{b}{2} l \sin \omega t$$

therefore

$$E = B\omega bl \sin \omega t$$

(a)

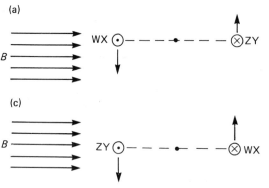

(b)

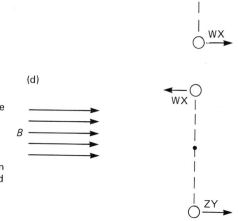

(c)

(d)

Fig. 35.7 Four glimpses, along the axis of rotation, of a section through the coil of a simple dynamo during one revolution. (a) Since the sides WX and ZY are cutting across the flux lines at right angles the induced e.m.f. is a maximum. The induced current flows away from us in side ZY and towards us in side WX. (b) The sides WX and ZY are now moving in a direction parallel to the flux; no e.m.f. is induced and no current flows. (c) Once again the two sides are cutting across the flux lines at right angles and the induced e.m.f. is a maximum; but the current now flows away from us in side WX and towards us in side ZY. (d) The sides are again moving parallel to the flux; no e.m.f. is induced and no current flows.

The area enclosed by the coil is bl, and we can call this A. Hence

$$E = BA\omega \sin \omega t$$

If the frequency of rotation (that is, the number of revolutions per unit time) is f, then $\omega = 2\pi f$, and so for a coil of N turns:

$$\boxed{E = 2\pi NBAf \sin 2\pi ft} \qquad (35.6)$$

This is a sinusoidally varying quantity whose amplitude is given by:

$$E_0 = 2\pi NBAf \qquad (35.7)$$

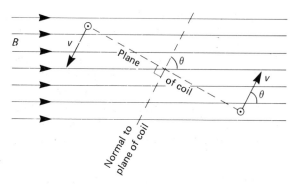

Fig. 35.8 Section through the dynamo coil, looking along the axis of rotation.

This expression shows that the *peak value* (or *amplitude*) of the e.m.f. is determined by the number of turns in the coil, the flux density of the field, the area enclosed by the coil, and the speed of rotation.

A dynamo designed exactly like that shown in Fig. 35.6 would be of little practical use, apart from demonstrating the principle. The following problem makes this evident.

Problem 35.1 A model moving coil dynamo as used in elementary physics courses is made of 20 turns of wire wrapped around a rectangular wooden former which has an area of about 20 cm². The average flux density of the magnetic field provided by a pair of ferrite magnets is about 0.02 Wb m^{-2}. Estimate the peak value of the e.m.f. generated when the coil is spun at 5 revolutions per second.

We use Equation 35.7 for the peak value of the e.m.f.

$$E_0 = 2\pi NBAf$$

The values to be substituted in this are as follows:

$$N = 20$$
$$B = 0.02 \text{ Wb m}^{-2}$$
$$A = 20 \times 10^{-4} \text{ m}^2$$
$$f = 5 \text{ s}^{-1}$$

Thus:

$$E_0 = 2 \times 20 \times 0.02 \times (20 \times 10^{-4}) \times 5$$
$$\approx 0.025 \text{ V}$$

This e.m.f. is far too small to light even the smallest filament lamp.

If a dynamo were to be constructed exactly as shown in Fig. 35.6 the coil could not be spun through more than a few turns without twisting the connecting wires too much. In some dynamos, including the model mentioned in Problem 35.1 above, the problem is solved by using *slip-rings* on the rotor, and *brushes* which press lightly against them to make electrical contact while allowing the rotor to spin unhindered. Alternatively, as in a bicycle dynamo (Fig. 35.1), it is the magnet which spins, while the coil (or coils) are stationary. As the permanent magnet rotor turns it magnetizes the iron of the *stator* (the stationary part) first in one direction, then in the opposite direction. This provides a magnetic flux through the coil which reverses twice during every revolution. In this way there is a continuous change of flux-linkage with the coil and so an e.m.f. is induced in it. The faster the rotation, the greater is the frequency and the peak value of the e.m.f.

35.4 Faraday's disc dynamo

The reasons for studying this here are not purely of historical interest. One important feature of it is that it is the simplest dynamo that can generate a steady e.m.f., and so provide a steady direct current in a circuit. It is sometimes used for demonstrating the laws of electromagnetic induction, although we have not chosen it for this purpose here. However, an understanding of its action will help you to understand how *induction motors* work, and the principles of *electromagnetic damping* and *braking*. These are discussed in Chapter 37.

Michael Faraday showed that if a conducting disc is spun in a magnetic field whose direction is perpendicular to the plane of the disc, a potential difference is created between the axis and the rim (Fig. 35.9). If electrical connections are made, by spring 'brush' contacts, to the axle and the rim and connected to a galvanometer, then a small induced current will flow, whose strength is directly proportional to the speed of rotation.

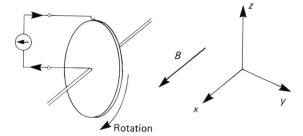

Fig. 35.9 The principle of the Faraday disc dynamo.

If the disc is rotating clockwise, as shown in Fig. 35.9, with its plane parallel to the zy plane, and the magnetic field parallel to the x-axis and in the positive x-direction, the direction of the induced current will be in the direction shown by the arrows on the connections to the galvanometer. Why is it in this direction? To answer this you should consider the disc as if it were made up of a large number of radial spokes, and then apply the same reasoning to one of these imaginary spokes as was applied to the wire PQ, in Fig. 35.4, as it cuts across the flux lines. Faraday himself said:

'If a single wire be moved like the spoke of a wheel near a magnetic pole, a current of electricity is determined through it from one end towards the other. If a wheel be imagined, constructed of a great number of these radii, and this revolved near the pole, in the manner of the copper disc, each radius will have a current produced in it as it passes by the pole.'

So, imagine a spoke of a wheel as it passes between the two brush contacts, and imagine positive charge-carriers inside this spoke, and reason out which way along the spoke they will be pushed by the magnetic force.

Now we shall obtain an expression for the e.m.f. induced between the axis and the rim of a disc dynamo, if the magnetic field is uniform over the *whole* area of the disc. We cannot simply use Eq. 35.2 because our imaginary spoke is *not* moving with uniform linear velocity; it is rotating, so that close to the axis its linear velocity is small, and near the rim it is larger. But we can use Eq. 35.4:

induced e.m.f. = rate of cutting flux

We obtain an expression for the amount of flux

which is cut by an imaginary spoke as it rotates. In one revolution of the disc, this spoke will cut all the flux which crosses the area of the whole disc, and this amount of flux is

$$\text{(flux density)} \times \text{(area)} = B\pi R^2$$

where B is the flux density of the field and R the radius of the disc. If the disc makes f revolutions per second, the flux cut per second is $fB\pi R^2$, and so this is the magnitude of the induced e.m.f.

35.5   Electromagnetic induction without physical motion

Any *transformer* provides an example of electromagnetic induction taking place without any observable movement, except perhaps for a small vibration which may produce an audible hum, in the case of transformers used with the 50 Hz a.c. mains.

So far we have considered examples in which an e.m.f. is induced in a conductor as a result of a relative movement of the conductor and a magnetic field. It is possible, however, to obtain a change of magnetic flux linkage with a circuit without any relative motion by simply altering the strength of the field. Faraday first studied this with the apparatus shown in Fig. 35.10, and wrote in 1831:

'A welded ring was made of soft round-bar iron ... the ring six inches in external diameter.'

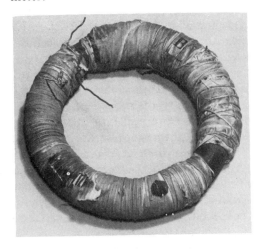

Fig. 35.10   Apparatus used by Faraday to investigate mutual induction. (Reproduced by permission of the Royal Institution of Great Britain.)

An insulated coil was wound around each half of the ring (Faraday used the word 'helix' for 'coil'). Then:

'The helix B was connected by copper wires with a galvanometer three feet from the ring.'

When 'helix' A was connected to a battery, he found that:

'The galvanometer was immediately affected ... but though the contact was continued, the effect was not permanent, for the needle soon came to rest in its natural position ... Upon breaking the contact with the battery, the needle was again powerfully deflected, but in the contrary direction to that induced in the first instance.'

In modern times, we would say that when the primary coil A was connected to the battery a sudden increase in magnetic flux occurred in the iron ring, and while this flux was changing, an e.m.f. was induced in the secondary coil B. Then, when A was disconnected from the battery, the flux rapidly decreased to zero, and this caused an e.m.f. to be induced in coil B, but in the opposite direction to the previous induced e.m.f. A sudden increase in *flux linkage*, followed by a sudden decrease in flux linkage.

To demonstrate the same phenomenon today we use basically the same arrangement, except that it is more convenient to have coils which can be quickly slipped on and off a magnetic core, as in Fig. 35.11. Provided that one chooses a pair of coils having a suitable number of turns, one can show the electromagnetic induction impressively by having a low-power, low-voltage lamp connected to one of the coils, and a battery and switch connected in series to the other coil. When the switch is closed, the lamp flashes briefly, and again when the switch is opened. With the switch closed, and thus a steady current flowing in the one coil, there is found to be no current in the other coil. Thus, an e.m.f. is generated, and current flows in the circuit, only while the magnetic flux produced by the primary coil is changing. One can refine the experiment of Fig. 35.11 a little by adding a variable resistor in series with the battery and one coil to vary the current, and hence also the magnetic flux, less suddenly than is done by a switch. With a suitable ammeter

(Fig. 35.3). We can vary the field by altering the current in the electromagnet; we shall refer to this current simply as the *magnet current*. We can alter the magnetic flux enclosed by the coil either, as was done before, by using a constant field and moving the coil into or out of the field region, or by setting the coil in a fixed position and altering the strength of the magnet current. The fact that these two methods are electrically equivalent can be shown by means of a game which could be entitled 'Unlink the Flux'.

The purpose of the game is to reduce the flux enclosed by the coil to zero in the shortest possible time, without the induced current exceeding a certain, agreed maximum value. Each player starts with the same amount of flux enclosed by the coil. The game begins with the coil of Fig. 35.3 pushed into a position that gives maximum flux through the coil and the electromagnet current set at a suitable, maximum safe value. One way to reduce the flux through the coil to zero is to leave the magnet current unaltered, and to withdraw the coil while watching the galvanometer to see that the induced current does not exceed the limit agreed (let us call this method A). Another way is to leave the coil where it is, and reduce the magnet current carefully, and finally snatch the coil out (let us call this method B). It is possible also to use a combination of methods A and B (we shall call this method C). After some practice, the competitors will discover that the operation cannot be performed in a time less than a certain definite minimum (provided, of course, that none of them exceeds the allowed limit for the induced current), regardless of whether they have used method A, B, or C. The 'perfect game' would be the one in which the induced current is kept at exactly the allowed limiting value throughout the operation – this is difficult to achieve.

What is the point of the game? If we imagine a 'perfect' game being played, using methods A, B and C, then it is reasonable to suppose that all three methods would require exactly the same time to reduce the flux in the coil to zero. Let us now see how this can be interpreted theoretically.

Let us return to Fig. 35.4. We use this example of a rectangular wire circuit moving into a region of magnetic field to develop the concept of an induced e.m.f. being caused by the side PQ *cutting the magnetic flux*. The rate of cutting flux was

Fig. 35.11 A modern version of Faraday's ring assembled from a pair of coils, two 'C' cores made from grain oriented silicon steel and a spring steel clip to hold the assembly together. (Courtesy Unilab Ltd.)

in series with the lamp it will then clearly be seen that there is current in the lamp circuit so long as the current in the primary coil is actually changing. It will also be seen that when the current in the primary coil is increasing, the current in the secondary foil flows in one direction; and with primary current decreasing, the secondary current flows in the opposite direction.

What would one have to do in order to keep a continuous, steady current flowing in the secondary coil? Presumably it would be necessary to make the current in the primary increase continually; or decrease continually. This, obviously, is not physically possible for any great length of time. But if the current in the primary is made to increase, then decrease, alternately, a current will flow in the secondary, continually reversing direction: in other words, an alternating current will flow in the secondary. This is the principle of the *transformer*, the action of which may be studied in detail later (see Section 38.2).

To study electromagnetic induction where there is no actual movement of the conductor we can use the same apparatus as was used previously

Blv. But another way of regarding the situation is to think of the *flux enclosed by the circuit* changing. Now

$$Blv = B \times \left(\begin{array}{c} \text{rate of change of area occupied} \\ \text{by flux within circuit} \end{array} \right)$$

and since *B* is constant here, this can be written

$$\text{rate of change of } \left(\begin{array}{c} B \times \text{ area occupied by} \\ \text{flux within circuit)} \end{array} \right)$$

which equals: rate of change of flux enclosed by circuit.

As the game 'Unlink the Flux' shows, we can equally well change the amount of flux enclosed within a circuit by keeping it stationary and altering the flux density of the field. Since

$$\text{flux} = \text{flux density} \times \text{area}$$

the amount of flux enclosed by a circuit can be altered by either:

1) keep the flux density constant, and alter the area occupied by the flux within the circuit, or
2) keep the area occupied by the flux constant, and alter the flux density,

or, of course, a combination of both procedures.

In all the situations mentioned above, if the coil has *N* turns, (all enclosing the same area), the e.m.f. induced in it is *N* times as great as the e.m.f. induced in a single turn coil of the same area. This leads to the concept of *flux linkage*.

$$\text{flux linkage} = \text{flux} \times \text{number of turns}$$

This can be applied in cases where every turn enclosed the same area. The concept is applied to cases where this is not so, however, and in such cases the flux linkage could be calculated by finding the value of the product (flux density × area enclosed by one turn) for each individual turn, and then adding all the products together.

Finally we can write an expression for induced e.m.f. which is applicable to all examples of electromagnetic induction:

$$\text{e.m.f.} = \text{rate of change of flux linkage}$$

In Section 35.1 we discussed Lenz's law, which concerns the *direction of the current* induced in circuit when electromagnetic induction occurs. We described there an experiment which demonstrates the truth of this law, and in which you can see the

effect of the force, due to the induced current, which opposes the motion. A similar experiment can be done using a *stationary* electromagnet instead of a *moving* magnet (Fig. 35.12). When the current in the electromagnet coil is switched on, the aluminium ring experiences a 'kick' away from the coil; this kick happens in the opposite direction when the current is switched off. When the experiment is repeated with an incomplete ring there is no observable kick. This suggests that the force on the ring must result from a current in the ring which can flow only if there is a complete conducting path. How does this experiment demonstrate Lenz's law? When the electromagnet is switched on, the magnetic flux through the aluminium ring increases, so inducing an e.m.f. and a current causing to flow in the ring. The force on the ring, owing to this induced current in the presence of the field of the electromagnet, pushes the ring away towards a region where the field is weaker. In this way the ring 'tries to avoid' the increase of flux through it. When the flux is reduced to zero by switching off the electro-magnet, the ring is pulled towards the coil, in a sense 'trying to avoid' the decrease of flux through the ring. In both instances the force on the ring is opposing the change of flux which causes the induced current.

We shall now look more closely into the question of the direction of the induced current, in situations where it is caused by a changing magnetic flux.

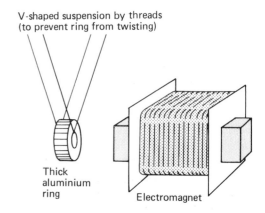

V-shaped suspension by threads (to prevent ring from twisting)

Thick aluminium ring

Electromagnet

Fig. 35.12 An experiment to demonstrate Lenz's law. When the electro magnet is switched on the ring is pushed away. When the magnet is switched off, the ring moves towards the magnet. The ring is suspended by V-shaped threads to prevent it from twisting.

Faraday's iron ring experiment showed that the induced current flows one way when the flux is increasing, and the opposite way when the flux is decreasing. We shall use the apparatus of Fig. 35.3. First of all we switch on the magnet current and use a small magnetic compass to find the direction of the field in the gap between the poles. We must now decide upon a positive direction for the flux – let us choose the positive y-direction (roughly left to right) in Fig. 35.3. If necessary we reverse the connections to the electromagnet so that we get the flux in this direction to begin with. Now we follow a similar procedure for the rectangular coil. When the induced current flows in this coil it produces its own magnetic flux, and we shall need to know the direction of this flux. The deflection of the galvanometer (to left or right of its centre zero) can be used to tell us this, provided we do a preliminary experiment to find out what deflection the galvanometer shows when the rectangular coil's flux is in the positive y-direction. This cannot be done with the electromagnet switched on because the electromagnet's flux is much stronger than the flux due to the induced current in the coil, and a compass placed close to the coil will simply indicate the direction of the electromagnet's flux. We need a way of getting a current to flow in the coil by some other means, with the electromagnet switched off.

The procedure is as follows. The electromagnet is switched off. The coil is withdrawn along its guide (but its orientation is not altered).

A simple shunt is connected across the galvanometer terminals so that a current of around 0.5 A can flow in the circuit without damaging the galvanometer, but nevertheless producing an observable deflection. The shunt will almost certainly have to have a very small resistance, much less than $1\,\Omega$, and a short length of fine copper wire is likely to suffice (see Fig. 35.13). A battery is inserted in series into the circuit. The magnetic flux produced by the coil should now be strong enough to have a marked effect on the compass placed near its centre. The results of the experiment will be easier to interpret if we arrange that the galvanometer shows a deflection to the right when the coil's flux is towards the right (positive y-direction) also. This may necessitate reversing the connections to the coil. We can now, therefore, tell from the galvanometer deflection the direction of the flux due to the current flowing in the coil. We then remove the battery and the shunt and set the apparatus once again, as shown in Fig. 35.3.

The experiment itself is quickly done. As we increase the magnet current we note which way the galvanometer pointer deflects. Then decrease the magnet current, and observe that the deflection is the opposite way. Then reverse the connections to the electromagnet (to reverse the direction of its flux), and increase the magnet current, noting the galvanometer deflection. Then finally decrease the magnet current while observing the deflection. The results are shown in Table 35a.

The symbol we use for magnetic flux is the Greek letter Φ (*phi*, pronounced 'fie'), and so the

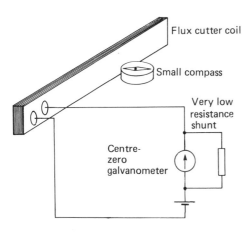

Fig. 35.13 Using the coil of the flux-cutter apparatus to find the relationship between the direction of the induced current and the flux it produces.

Table 35a

	(a)	(b)	(c)	(d)
Direction of magnet flux	→	→	←	←
Increasing or decreasing?	Increasing	Decreasing	Increasing	Decreasing
Sign for $d\Phi/dt$	+	–	–	+
Galvanometer deflection	←	→	→	←
Sign for flux due to induced current, and sign for induced e.m.f.	–	+	+	–

rate of change of flux can be represented by $d\Phi/dt$. Make sure that you understand why its value must be negative in case (c) and positive in case (d). If in doubt, refer back to Section 2.7.

If the coil has N turns, then the rate of change of flux linkage with the foil can be written:

$$N\frac{d\Phi}{dt}$$

So now, instead of writing:

induced e.m.f. = rate of change of flux linkage

We can write:

$$E = -N\frac{d\Phi}{dt} \qquad (35.8)$$

The minus sign is necessary because, as we see in Table 35a, the sign of the induced e.m.f. is opposite to the sign for the rate of change of flux, $d\Phi/dt$, in all four cases.

The result displayed in Table 35a is another instance of Lenz's law.

Consider case (a) in Table 35a. The magnet flux is directed towards the right and is increasing: the induced current in the coil flows in such a direction as to produce a flux directed towards the left, so it can be thought of as 'trying to prevent' the increase of magnet flux. In column (b), when the magnet flux is decreasing, the coil's flux 'tries to sustain' the decreasing magnet flux: in both cases it opposes the change in magnet flux. Similar arguments apply to columns (c) and (d).

Problem 35.2 In the game 'Unlink the flux' described in Section 35.5 typical values for the various quantities are as follows:

> Initial flux density between magnet poles: 0.25 Wb m^{-2}
> Area within the part of the coil through which the flux passes: 400 mm^2
> Number of turns in coil: 25
> Maximum allowed value for induced current: 2 μA
> Total resistance of galvanometer circuit: 50 Ω.

Calculate the minimum time required to reduce the flux-linkage with the coil to zero. Assume that it is possible to keep the induced current steady

at its maximum allowed value throughout the operation.

The amount of flux initially linking each turn of the coil is given by

$$0.25 \text{ Wb m}^{-2} \times (400 \times 10^{-6}) \text{ m}^2 = 10^{-4} \text{ Wb}$$

Let t be the time for the operation. Then the rate of change of flux linkage is given by:

$$N\frac{d\Phi}{dt} = -\left(25 \times \frac{10^{-4}\text{ Wb}}{t}\right)$$

(with a minus sign because the magnet flux, assumed to be in the positive direction, is decreasing). The induced e.m.f. is the product of the induced current and the circuit resistance, and so equation 35.8 gives:

$$(2 \times 10^{-6}\text{A} \times 50\,\Omega) = -\left[-\left(25 \times \frac{10^{-4}\text{ Wb}}{t}\right)\right]$$

Hence $t = 25$ s

35.6   Direct current motors

Figure 35.14 is a drawing of a model motor which can be quickly assembled from a kit of parts. Figure 35.15 is a simplified schematic diagram of the motor, to assist an explanation of its action. A rectangular coil of wire, QRSTUV, is free to rotate about an axis OO' between the poles of a pair of magnets. The ends of the coil are connected at P and W to a pair of almost semi-circular pieces of copper foil. Two brushes, B and B', make electrical contact with these pieces of foil without appreciably impeding the rotation. In the model (Fig. 35.14) the magnetic field is by no

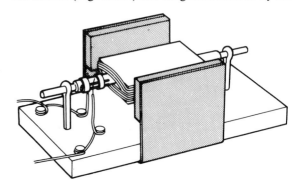

Fig. 35.14   A simple electric motor. (From J.M. Osborne, *Electromagnetism*, Longman, 1970. Reproduced by permission of the publisher.)

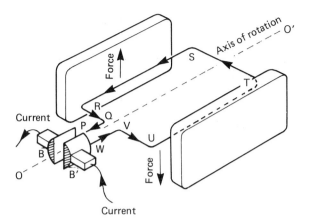

Fig. 35.15 Schematic diagram of a simple d.c. motor. The commutator segments shown are more robust than those in Fig. 35.14

means uniform, but for the purpose of our description here we shall assume that the field in Fig. 35.15 is uniform. When the coil is in the position shown the force on the side RS is directed vertically upwards, and the force on the side TU, vertically downwards. Since the parts of the coil ST, QR, UV are all parallel to the field, there is no force on them. The pair of equal forces on the sides RS and TU constitute a couple which rotates the coil clockwise. Consider now the situation when the coil has rotated through 45° from the position shown. The forces on each side of the coil will be the same as before, but their turning effect

is less because the lines of action of these two forces are closer to the axis than before. When the coil has rotated through another 45° the two forces simply pull against each other, because side RS is now vertically above side TU, and there is no turning effect. Assuming that the coil has acquired some energy of motion, we can see that it will overshoot this 'dead' position. As it does so, the gaps between the semicircular segments pass under the brushes, and an instant later the electrical connections to the coil are reversed. The segments and brushes constitute the *commutator*. A commutator is any kind of switch which reverses a pair of electrical connections. Thus the direction of current through the coil is reversed, and hence the directions of the forces on the sides RS and TU are reversed. The force on side RS now pulls it downwards, and the force on TU pulls upwards. This change-over process is shown in Fig. 35.16. The coil then makes a half revolution, the current is reversed once again, and hence the forces on the sides RS and TU are again reversed. In this way, the forces acting upon the sides of the coil are always in such directions as to make the coil rotate continuously in the same sense (clockwise, in our example).

Do forces on the other two sides on the coil have any effect? As stated earlier, when the sides RU (neglect the small gap Q–V) and ST are parallel to the field, as in Fig. 35.15, there is no

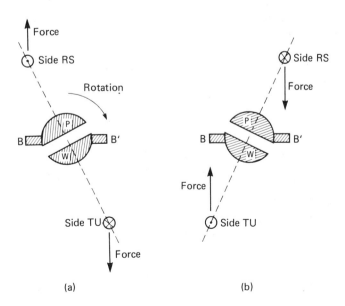

Fig. 35.16 A commutator (a) just before and (b) just after the reversal of the current.

(a)   (b)

force on them. When the plane of the coil is vertical (and you should argue this out for yourself), there are equal and opposite forces on RS and TU, but their lines of action lie along the axis, and thus exert no turning effect: they simply tug against each other (like one of the pairs of forces on the coil in Fig. 34.7). For any position of the coil, these two forces just oppose each other and make no contribution to the turning effect.

In Equation 34.4 we obtained an expression for the torque on a rectangular coil in a uniform magnetic field. We cannot use that expression for the rotating coils of electric motors because, neither in the simple model motor of Fig. 35.14, nor in a 'real life' motor, is the magnetic field ever like this. In the model motor the field is distinctly non-uniform, being strong close to the pole-faces and decreasing markedly with distance away from them. In a 'real life' motor the field is almost perfectly radial, like the field in a moving-coil galvanometer (see Fig. 34.10).

In a 'real life' motor the volume of space available in the field region is packed as densely as possible with soft magnetic alloy (to maximize the magnetic flux) and a number of overlapping coils, with their planes set at equal angles to one another (Fig. 35.17). The size of the air-gap between the *rotor* (that is, the rotating part: another name for it is *armature*) and the *stator* (the stationary part) is made as small as possible in order to maximize the flux density of the field affecting each coil. The way in which an air-gap influences magnetic flux will be studied in some detail in Section 37.2.

Each coil is connected to a pair of segments in the commutator (these can be seen in Fig. 35.17) in such a way that, whatever the position of the armature as it rotates, the currents in the coils are as shown in Fig. 35.18. Thus, in such a motor, the

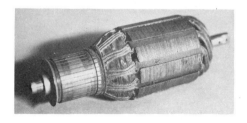

Fig. 35.17 The armature and commutator segments of a small d.c. motor.

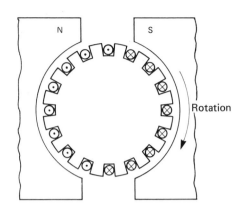

Fig. 35.18 Schematic diagram of the cross-section through the windings of a d.c. motor, showing the directions of the currents.

torque can be made to be nearly constant as the armature rotates.

Before discussing further the practical forms of motors, let us consider how a motor converts electrical energy into mechanical form. A simple experiment will open the discussion. Using, for convenience, a low-voltage d.c. motor, measure the current taken by the motor under different conditions. You will find two important facts:

a) When the motor is first switched on, the current is large; then, as the motor speeds up, the current decreases until the motor is running at a steady speed.

b) If the motor is forcibly slowed down (by gripping the shaft with a cloth, as a kind of crude friction-brake), the current is observed to rise, and it falls again when the motor is allowed to run freely.

A further experiment gives clues which help us to explain the facts just demonstrated. In the arrangement shown in Fig. 35.19 a motor has a flywheel attached to it so that, when the current is switched off, the armature keeps on spinning for a time. The switch S enables the motor to be disconnected from the battery and immediately connected to the lamp. The ammeter A must be a centre-zero type, because it is used here to indicate currents in both directions. Firstly, the switch is set to the left-hand position: the battery is thus connected to the motor which speeds up, and the current is shown on A. When the motor has reached its steady maximum speed the switch is flicked over to the right. Now acting as a dynamo

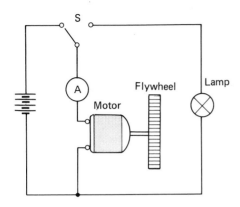

Fig. 35.19 A motor can also work as a dynamo.

while the flywheel is still spinning, the motor generates a current which lights the lamp. The important thing is that now the current registered by the ammeter is in the opposite direction to what it was before. Now, presumably, when the motor was connected to the battery it was nevertheless acting as a dynamo, generating an e.m.f. which opposed the current forced through it by the battery. This opposing e.m.f. is generally called a *back e.m.f.* The faster the armature spins, the greater is this back e.m.f., and hence the smaller the current flowing around the battery-motor circuit. This may become clearer when we construct a quantitative relationship, as follows. The net e.m.f. in the battery-motor circuit is the difference between the battery e.m.f. and the motor's back e.m.f. By the definition of resistance we can therefore write:

$$\frac{\text{total circuit}}{\text{resistance}} = \frac{(\text{battery e.m.f.}) - (\text{back e.m.f.})}{\text{current}}$$

or, in symbols,

$$R = \frac{V - E}{I}$$

Thus the current in the circuit is given by

$$I = \frac{V - E}{R} \qquad (35.9)$$

The back e.m.f., E, can be assumed to be proportional to the speed of the armature, just as the e.m.f. generated by a dynamo is proportional to the speed. Thus, the faster the armature spins, the greater the value of E and the smaller the value of

$(V - E)$ and hence the smaller the current.

We turn now to the question of the energy converted in the battery–motor circuit. The rate at which energy is converted (from chemical to electrical form) in the battery is VI. We rewrite Eq. 35.9 thus:

$$V = E + IR$$

and multiply both sides by I and obtain:

$$VI = EI + I^2R$$

This equation can be expressed in words as follows:

(power from battery)
   = (useful power) + (power lost at heat)

The term EI in the equation, the 'useful power' as we briefly call it, is the rate at which energy is used to keep the current I flowing against the opposing induced e.m.f., the 'back e.m.f.', and thus is the rate at which the motor does mechanical work. The term I^2R is the rate at which electrical energy is converted into heat in the resistance of the armature coil (or coils).

The existence of the back e.m.f., E, produced when a motor is running is made apparent by the experiment of Fig. 35.20. Here we have a voltmeter connected across the motor. When the switch is flicked across from left to right the voltmeter shows a voltage which, to start with, before the flywheel loses speed, is not much less than the battery voltage.

In the example above we assumed that the magnetic field of the stator was constant through-

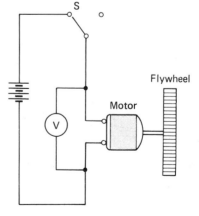

Fig. 35.20 Demonstration of the back e.m.f. generated by a motor.

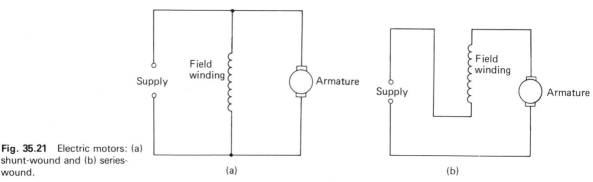

Fig. 35.21 Electric motors: (a) shunt-wound and (b) series-wound.

(a)   (b)

out. This is true for a *shunt-wound* motor in which the stator electromagnet, or *field winding*, is connected in parallel with the armature, as in Fig. 35.21a. In this arrangement the p.d. across the field winding is constant (assuming the supply voltage stays constant) and so the current flowing in it is constant. But the armature current, as the example above showed, depends upon the speed. When the motor is switched on from rest the armature current is first of all very high, and so the torque on the armature is large. As the motor speeds up, the back e.m.f. in the armature increases, the armature current decreases, and so therefore does the torque.

In a *series-wound* motor (Fig. 35.21b) the field winding and armature winding are connected in series to the supply. A motor of this kind does not have as many useful applications as a shunt-wound motor.

In an efficient motor running at its correct working speed the back e.m.f. is usually only slightly less than the battery (or other supply) voltage. For instance, a motor designed to work from a 12 V supply may generate a back e.m.f. of 11 V or even greater.

Problem 35.3 A certain d.c. motor, designed to work from a 12 V supply of negligible internal resistance, has an armature resistance of 0.25 Ω. When running freely, without any mechanical load, the current taken from the 12 V supply is 2 A. (The magnetic field of the stator can be assumed to be constant at all speeds.)

a) Calculate the back e.m.f. generated in the armature when the motor runs freely.

b) The motor is now 'harnessed' to some machinery, and its running speed becomes 80% of

the free running speed. Calculate the new back e.m.f., and the current taken from the supply,

c) If the armature is held so that it cannot rotate at all, what current is taken from the supply?

a) We can rewrite Eq. 35.9 thus:

$$E = V - IR$$

so the back e.m.f.

$$= 12 \text{ V} - (2 \text{ A} \times 0.25 \text{ Ω})$$
$$= 11.5 \text{ V}$$

b) Assuming that the back e.m.f. is proportional to the running speed, the new back e.m.f. will be:

$$\frac{80}{100} \times 11.5 \text{ V} = 9.2 \text{ V}$$

The current is given by:

$$I = \frac{V - E}{R}$$

so

$$I = \frac{12 \text{ V} - 9.2 \text{ V}}{0.25 \text{ Ω}}$$

$$= 11.2 \text{ A}$$

c) When the running speed is zero the back e.m.f. is zero, and so the current

$$I = \frac{V}{R} = 48 \text{ A}$$

35.7   Self induction

Here is a puzzling demonstration. A low-voltage a.c. supply (say 12 V) is used to light a lamp. It is

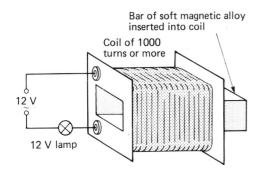

Fig. 35.22 Self-induction; the effect of a choke. A bar of soft magnetic alloy can slide into and out of the 1000 turn coil.

possible to dim the lamp without either altering the supply voltage, or altering the resistance of the circuit. Fig. 35.22 shows how this can be done. A coil consisting of several hundred turns (1000 is a suitable number) of thick copper wire, such as is used in transformer kits for school use, is connected in series with the supply and the lamp. When a bar of iron or mild steel, or, better still, a piece of laminated transformer core alloy, is pushed into the coil, the lamp dims. A coil, when used in this way to reduce the strength of an alternating current, is often called a *choke*. If the experiment is repeated using a steady smoothed, *d.c.* supply, the insertion of the magnetic material has no observable effect on the lamp's brightness.

The experiment suggests that it is the presence of alternating magnetic flux in the coil which is responsible for reducing the strength of the current. To understand how this comes about we need first of all to do some other simple experiments.

So far we have seen how electromagnetic induction can be brought about by the relative motion of a conductor and a magnetic field, and by changes in magnetic flux linking a circuit without any physical movement. It should not be surprising, therefore, that when a changing current flows in a conductor, the changing magnetic flux due to this current induces an e.m.f. in the self same conductor. This phenomenon is called *self induction*.

Figure 35.23 shows an experiment which demonstrates the effect of self-induction in a d.c. circuit. The circuit contains two parallel branches, each branch containing a similar lamp to give a rough indication of the strength of the current in the branch. L is a coil, of about 1000 turns of thick copper wire, of similar construction to the one used in the previous experiment, and which can be used with or without a core made of laminated transformer alloy. The coil's resistance is a few ohms only, and the variable resistor R is set so that the two lamps light to equal brightness. First of all the coil is used without the magnetic core. When the switch is closed the two lamps reach equal brightness simultaneously; but when the magnetic core is used with L the lamp in series with it is observed to light up later than the lamp in series with R. This suggests that it must be a magnetic effect in L which is causing a delay in the growth of the current in its branch of the circuit.

The growth of the current can be displayed on a c.r.o. by inserting a small resistance, 1 Ω for instance, in series with L, and connecting the input leads to the c.r.o. across this 'current sampling resistor', whose resistance is small enough to avoid appreciably altering the behaviour of the circuit. The p.d. across the current-sampling resistor is proportional to the current flowing through it, and so the c.r.o. can display the time-trace of the rising current, and this should appear similar to the curve of Fig. 35.24. The current increases rapidly at first, then less and less rapidly as it approaches the final steady value.

From now onwards we shall use the term *inductor* for a coil (with or without a magnetic core) in situations where its self-induction is significant. An inductor causes a delay in the

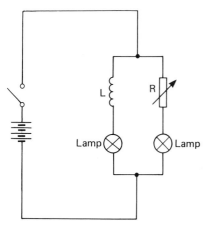

Fig. 35.23 Circuit to show that an inductor delays the growth of a current.

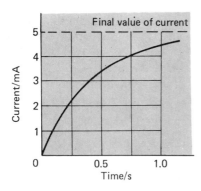

Fig. 35.24   The growth of current in an inductive circuit.

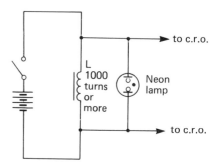

Fig. 35.25   When a current in an inductive circuit is switched off a large voltage can appear across the inductor. The straight line in the symbol for the inductor indicates that it has a magnetic core.

growth of current because the changing current produces a changing magnetic flux, and this changing flux induces an e.m.f. which (Lenz's law again) opposes the changing current. If we assume that the flux is proportional to the current, then the induced e.m.f. will be proportional to the rate of change of current. Thus the magnitude of this e.m.f. can be written:

$$L\frac{\mathrm{d}I}{\mathrm{d}t}$$

where L is a constant for the inductor and is called the *inductance* of the inductor. The unit for inductance is the *henry*, and the letter 'H' is used for it. Because the product: (inductance) × (rate of change of current) must be a *voltage*:

$$1\,\mathrm{H} = 1\,\mathrm{V\,A^{-1}\,s}$$

Because the e.m.f. induced in the inductor opposes the e.m.f. of the supply we can write an expression similar to the one for an electric motor (Eq. 35.9).

current

$$= \frac{(\text{e.m.f. of supply}) - (\text{e.m.f. of inductor})}{(\text{total circuit resistance})}$$

in symbols:

$$I = \frac{V - L\dfrac{\mathrm{d}I}{\mathrm{d}t}}{R}$$

hence

$$\frac{\mathrm{d}I}{\mathrm{d}t} = \frac{V - IR}{L} \qquad (35.10)$$

To find from this equation how the current varies with time one can use calculus to solve the

equation, but to avoid that, we shall choose suitable numerical values for V, R, and L, and carry out a step-by-step solution. At the instant when the switch is closed the time $t = 0$ and the current $I = 0$.

The numerical values we shall use are:

$$V = 2.5\,\mathrm{V}$$
$$R = 500\,\Omega$$
$$L = 250\,\mathrm{H}$$

At time $t = 0$ the term $IR = 0$ and so

$$\frac{\mathrm{d}I}{\mathrm{d}t} = \frac{2.5\,\mathrm{V}}{250\,\mathrm{H}} = 0.01\,\mathrm{A\,s^{-1}}$$

We assume, as a simplifying approximation, that the rate of rise of current remains constant at this value for a short interval of time Δt, and then we calculate the increment in current ΔI by the relation

$$\Delta I = \frac{\mathrm{d}I}{\mathrm{d}t}\Delta t$$

We shall choose $\Delta t = 0.1\,\mathrm{s}$ here. So, at a time $0.1\,\mathrm{s}$ after the instant of switching on, the current has increased by an amount

$$(10\,\mathrm{mA\,s^{-1}}) \times (0.1\,\mathrm{s}) = 1\,\mathrm{mA}$$

and so this is the current flowing at time $t = 0.1\,\mathrm{s}$. Then, using this value of current, I, we can compute a new value for the rate of change of current, $\mathrm{d}I/\mathrm{d}t$, hence a new value for ΔI, and thus obtain the current flowing at time $t = 0.2\,\mathrm{s}$. In this way we can compute the current at successive instants in time, and the results are shown in Table 35b and also plotted in Fig. 35.24.

After a very long time (we can write $t = \infty$) the current is changing no more, that is, $\mathrm{d}I/\mathrm{d}t = $

Table 35b Step-by-step calculation for growth of current in an inductive circuit.

$$R = 500\,\Omega,\ L = 250\,\text{H},\ V = 2.5\,\text{V}$$

$$\frac{\mathrm{d}I}{\mathrm{d}t} = \frac{V - IR}{L} \qquad\qquad \Delta t = 0.1\,\text{s}$$

t	I	IR	$V - IR$	$\dfrac{\mathrm{d}I}{\mathrm{d}t} = \dfrac{V - IR}{L}$	ΔI
/s	/A × 10^{-3}	/V	/V	/A s^{-1} × 10^{-3}	/A × 10^{-3}
0.0	0	0	2.5	10	1.0
0.1	1	0.5	2.0	8	0.8
0.2	1.8	0.9	1.6	6.4	0.64
0.3	2.44	1.22	1.28	5.12	0.512
0.4	2.952	1.476	1.024	4.096	0.410
0.5	3.362	1.681	0.819	3.276	0.328
0.6	3.690	1.845	0.655	2.620	0.262
0.7	3.952	1.976	0.524	2.096	0.210
0.8	4.162	2.081	0.419	1.676	0.168
0.9	4.330	2.165	0.335	1.340	0.134
1.0	4.464				

0, hence Eq. 35.10 simplifies to

$$I = \frac{V}{R}$$

showing that the final steady current is independent of the inductance and hence, with the numerical values used here, this gives $I = 5\,\text{mA}$. From the graph it can be seen that the time taken for the current to rise from zero to half the final value, namely 2.5 mA, is 0.31 s. After a further time interval of 0.31 s the current has risen to a value which is less than the final value by one quarter of the final value, namely 3.75 mA. After a further 0.31 s the current is less than the final value by one eighth of the final value, namely 4.375 mA. In other words, the difference between the instantaneous current and the final steady current is halved in successive equal intervals of time. In this respect the growth of current in an inductor resembles the growth of p.d. across a capacitor when being charged, and the decay of p.d. across a capacitor when being discharged (See Section 26.7.).

Finally we examine what happens in an inductive circuit when a current is switched off. We can use our mathematical model for an inductive circuit to predict what should happen. If a circuit is suddenly broken the current must decrease very rapidly, so the value of $\mathrm{d}I/\mathrm{d}t$ will be very large. Since the e.m.f. induced in an inductor is proportional to $\mathrm{d}I/\mathrm{d}t$, the e.m.f. should likewise be very large, although it will only be of very short duration. The experiment of Fig. 35.25 is used to test this prediction.

N is a neon lamp designed to work at mains supply voltage (that is, about 250 V r.m.s.). With a high-voltage d.c. supply, or a suitable battery, we can show that a glowing discharge in the neon lamp does not strike until the voltage is about 100 V. In the circuit of Fig. 35.25 we close the switch and find (not surprisingly) that the lamp does not glow, for the battery voltage is far too small to cause a discharge in the lamp. When the switch is opened a momentary flash is observed, indicating that for a very brief time the voltage is about a hundred volts or greater. The demonstration can be made more convincing if one has an oscilloscope in which connection can be made directly to the Y-deflector plates. If the two deflector plates are connected to the two terminals of the inductor, one can display the very large voltage 'spike' which occurs when the switch S is opened. If the sensitivity of the cathode ray tube is known (that is, the voltage needed to deflect the spot of light on the screen by one centimetre) it may be possible to measure the magnitude of this voltage spike. This very large but short-lived voltage is the e.m.f. induced in the inductor when the current decays very rapidly.

If any electrical circuit has a large inductance in it the problem of switching off a current in that circuit can be a serious one. It is perfectly possible for the spark which occurs between the switch contacts as they separate to persist and for there to be a continuous electric arc between the contacts, even when they are fully separated. In this condition the current will still be flowing: it will not have been switched off at all! Nearly all electrical machinery has considerable inductance, and so the design of suitable switches for use in the electrical supplies to factories and all industrial plant is of great importance. In some types of heavy-duty switchgear the electric arc which forms when the contacts separate is extinguished by a powerful blast of air, or quenched by having the whole switch mechanism immersed in a suitable oil insulating medium. Sometimes it is not possible to switch the current off rapidly: it has to be reduced gradually by switchgear of highly sophisticated design.

Chapter 36

ALTERNATING CURRENTS

36.1 Sinusoidal alternating currents and voltages

The importance of alternating currents and voltages can hardly be overestimated. In almost every country the national electricity supply uses alternating current, the voltages being 'stepped up' and 'stepped down' by transformers. In any radio communication system, the aerial of the radio transmitter emits electromagnetic waves as a result of alternating currents flowing in the antenna. In the electronic reproduction of sound, the vibrations of a sound wave are converted, by a microphone, into alternating voltages and currents in order that they may be electronically amplified and then converted, by means of loud-speakers, into sound vibrations of increased power.

To begin a study of alternating currents we need a supply which gives a very low frequency alternating current (a.c.). There are suitable low frequency electronic oscillators which are ideal, but a simple, mechanical means which can be operated by hand is shown in Fig. 36.1. It is a special kind of potentiometer, made from resistance wire wrapped closely around a rectangular card made of insulating material. When its spindle is rotated, sliding contacts A and B 'pick off' a p.d. which is proportional to the horizontal distance between them (by 'horizontal' here we mean in a direction parallel to the line joining P and Q). The terminals of P and Q are connected to a battery of small e.m.f., say 2.5 V. If the spindle is rotated at a constant speed the p.d. between A and B varies with time in an approximately sinusoidal fashion. The frequency of this alternating p.d. is

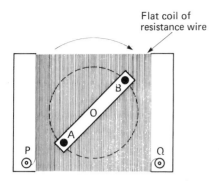

Fig. 36.1 A low frequency a.c. generator. Terminals P and Q are connected to a battery. A and B are sliding contacts mounted on an insulating, rotating arm. An alternating voltage appears between terminals A and B.

equal to the number of revolutions of the spindle per second.

Our first experiment compares an a.c. supply with a d.c. supply as a means of providing electrical power.

In the experiment we use the very low frequency source to light a lamp, as in Fig. 36.2 with the switch S in the left-hand position as shown. The lamp and ammeter are connected to a battery and variable resistor so that the lamp can be lit with d.c. instead of a.c. A is a centre-zero ammeter. When we rotate the shaft of the generator very slowly we observe that the lamp lights intermittently, reaching maximum brightness twice per revolution: once when the current flows one way, and again when the current flows the opposite way. If we spin the generator shaft faster we find that the lamp flickers more rapidly and that the variations in brightness become less; also, the ammeter cannot respond adequately to the variations of current, so it oscillates just a small distance either side of the centre-zero. We can run the generator at higher speed by using an electric motor to drive it, coupled by pulleys and a belt. Above a certain speed we find that the lamp

appears to be steadily lit, and the needle scarcely moves at all, registering zero all the time. We have the slightly paradoxical situation of an ammeter showing zero current in a circuit with a lamp showing clearly the presence of a current. The problem immediately arises, how does one measure the strength of the current in this situation? An ordinary moving-coil meter is useless but the brightness of the lamp gives a clue: perhaps one could use the heating effect of the current as a measure of its strength.

Suppose we arrange that the maximum current in the circuit (which, you will recall, is obtained at two instants during each revolution of the generator shaft) is 100 mA. We spin the shaft at high speed, and note the steady brightness of the lamp. Then we switch over to the battery and adjust the variable resistor until the lamp is at the same level of brightness, and we read the steady current on the ammeter. We will find that the ammeter reads about 70 mA. We deduce from these observations, therefore, that a sinusoidal alternating current whose peak value is 100 mA is equivalent to a steady d.c. of about 70 mA, if by equivalent we mean that the a.c. and the d.c. are carrying energy at the same average rate to the lamp bulb, where the energy is utilized as heat and light.

We now proceed to find a theoretical reason for this relationship between the peak value of a sinusoidal a.c. and the equivalent d.c. We can represent a sinusoidally varying a.c. by the expression

$$I = I_0 \sin \omega t \qquad (36.1)$$

where I is the instantaneous value of the current and I_0 is the amplitude or *peak value* of the current. In describing simple harmonic motion (Section 19.7) we found that $\omega = 2\pi f$, where f is the frequency of the oscillation.

The rate of conversion of electrical energy into heat in a resistance R when a current I flows is $I^2 R$, so the rate of conversion of energy by our sinusoidal a.c. at time t is given by

$$P = I_0^2 R \sin^2 \omega t$$

We are concerned with the average rate of conversion of energy by the a.c. because this is what determines the brightness of the lamp bulb.

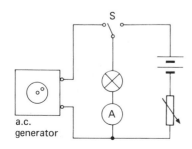

Fig. 36.2

To find this we need find only the average value of the term $\sin^2 \omega t$ because the term $I_0^2 R$ is constant. We find this average value by taking the value of $\sin^2 \omega t$ at equally spaced instants in time during one complete cycle of variation, that is, from $\omega t = 0$ to $\omega t = 2\pi$ radians, or $360°$. A suitable increment for ωt is $20°$, and Table 36a shows the values computed for one whole cycle. The whole cycle comprises 18 time intervals, so to find the mean value of $\sin^2 \omega t$ we divide the sum of all the values in the right-hand column by 18. This gives 0.500 as the mean value of $\sin^2 \omega t$. Thus the mean rate of conversion of energy per cycle can be written:

$$0.500 I_0^2 R, \quad \text{or} \quad \tfrac{1}{2} I_0^2 R$$

Figure 36.3 shows a graph of $\sin^2 \omega t$ plotted against t, and you can see from this that the average value of $\sin^2 \omega t$ over one complete cycle is $\frac{1}{2}$, because its value varies equally above and below the horizontal broken line, with the 'crests' above this line equal in size and shape to the 'troughs' below the line.

The average rate of energy conversion per cycle by the a.c. is $\tfrac{1}{2} I_0^2 R$. If we were to use a steady direct current to give the same rate of energy conversion in the same resistance R, then this current would have to be equal to $I_0/\sqrt{2}$, because

$$\left(\frac{I_0}{\sqrt{2}}\right)^2 R = \tfrac{1}{2} I_0^2 R$$

This 'equivalent steady current' is called the *root mean square* value of the alternating current. Here is its definition:

■ The root mean square (r.m.s.) value of an alternating current is the square root of the mean value of the square of the current, over one whole cycle. It is equal to that steady direct current which would dissipate energy at the same rate in a given resistance.

The numerical factor $\sqrt{2}$ applies only to a *sinusoidally* alternating current: and in such a case we can write the relationship between root mean square current and the amplitude, or *peak value*, thus:

$$\text{r.m.s. current} = \frac{\text{peak value of current}}{\sqrt{2}}$$

or, in symbols:

$$\boxed{I_{\text{r.m.s.}} = \frac{I_0}{\sqrt{2}}} \tag{36.2}$$

Table 36a

ωt	$\sin \omega t$	$\sin^2 \omega t$
0	0	0
20°	0.342	0.117
40°	0.643	0.413
60°	0.866	0.750
80°	0.985	0.970
100°	0.985	0.970
120°	0.866	0.750
140°	0.643	0.433
160°	0.342	0.117
180°	0	0
200°	−0.342	0.117
220°	−0.643	0.413
240°	−0.866	0.750
260°	−0.985	0.970
280°	−0.985	0.970
300°	−0.866	0.750
320°	−0.643	0.413
340°	−0.342	0.117
360°	0	0
Sum:		9.000
Mean value:		0.500

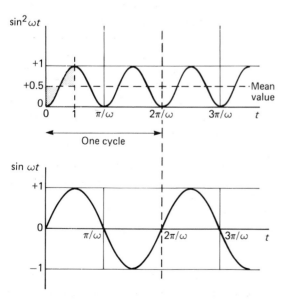

Fig. 36.3 The variation of $\sin^2 \omega t$ and $\sin \omega t$ with time t.

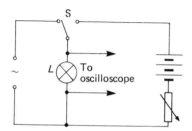

Fig. 36.4

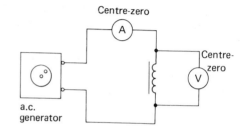

Fig. 36.5   a.c. generator

The r.m.s. and peak values of sinusoidally alternating *voltages* are related in the same way. In practice it is usually the r.m.s. value which is used to specify the magnitude of an alternating voltage. When we speak of the '240 volt a.c. mains' we mean that the r.m.s. voltage is 240 V: the peak value is therefore $\sqrt{2} \times 240$ V which equals 340 V.

An experiment which shows directly the relation between r.m.s. and peak voltage is represented in Fig. 36.4. Here an oscilloscope is used to measure the voltage; and the oscilloscope must be one which can respond to a d.c. input so that when a steady voltage is applied to the input terminals the spot of light on the screen is displayed vertically and stays displaced, the displacement being proportional to the voltage. When the switch is in the left-hand position the lamp L is fed with a.c. from the low voltage terminals of a step-down transformer, and the peak value of the voltage is measured by reading off the amplitude of the sinusoidal trace on the oscilloscope. The switch is then put to the other position, the variable resistor adjusted until the lamp has the same brightness, and then the steady voltage is measured from the displacement of the trace on the oscilloscope. One cannot achieve high accuracy of measurement here, but it will be found that the peak value of the voltage is about 1.4 times the steady voltage used ($\sqrt{2} = 1.414$). Testing for equality of brightness is best done by switching S back and forth rapidly.

36.2 Inductors in a.c. circuits

A demonstration of an inductor acting as a *choke* in an a.c. circuit was used to introduce Section 35.7. This showed that when a magnetic core was pushed into a coil in an a.c. circuit, the strength of the a.c. was reduced.

Before making a theoretical analysis of what happens when an alternating p.d. is applied to an inductor we can demonstrate the phase-difference between voltage and current which is observable in a circuit which contains a very large inductance, using the very low frequency generator (Fig. 36.5). The two meters are adjusted so that they have a centre-zero. When the generator shaft is rotated slowly the voltage is seen clearly to reach its maximum value at a different instant from the current. In fact the voltage will reach its maximum positive value before the current does, the phase-difference being usually less than, but never greater than, one quarter-cycle. The way in which this phase-difference arises can be appreciated if, having observed what happens when the generator shaft is rotated at steady speed, one rotates it in jerks, a quarter revolution at a time. The way in which the current lags behind the voltage will then become very apparent. That this phase-difference is not due to some difference in the action of the two meters can be demonstrated. This is done by using a resistor instead of the inductor: then it will be clearly seen that the current and voltage are in phase.

To understand how this phase-difference comes about we study the graphs of Fig. 36.6. Here the phase-difference is exactly one quarter cycle; this is the phase-difference we get whenever the resistance of the inductor is negligible.

Earlier we derived an expression (Eq. 35.10) for the current in a circuit containing a d.c. electrical supply of e.m.f. V, resistance R, and inductance L. If the resistance R is negligible the expression becomes

$$\frac{\mathrm{d}I}{\mathrm{d}t} = \frac{V}{L}$$

and we can now refer to V as the *applied voltage* (or p.d.) across the inductor. Suppose that $L = 3$ H and the peak value of the applied voltage is 6 V.

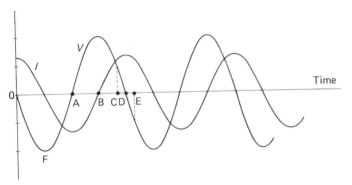

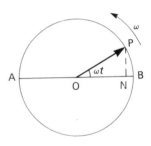

Fig. 36.6 The variation of current and voltage with time in a circuit containing pure inductance.

Fig. 36.7 Rotating vector, or *phasor* representation of a sinusoidally varying quantity.

Consider what is happening at point B on the time-scale of the graph in Fig. 36.6. At this instant $V = +6\,\text{V}$ and using the expression above we find that $\mathrm{d}I/\mathrm{d}t$ must be $+3\,\text{A s}^{-1}$: this is the rate at which the current is rising, and it is the maximum rate of rise because V is at its maximum at this point. If the applied voltage were to remain at a steady value of $+6\,\text{V}$ the current would have to continue rising at a rate of $+2\,\text{A s}^{-1}$. But the applied voltage decreases. At point C, $V = +3\,\text{V}$ and so $\mathrm{d}I/\mathrm{d}t = +1\,\text{A s}^{-1}$. At D, $V = 0$ and therefore $\mathrm{d}I/\mathrm{d}t = 0$; here the current is at its maximum value and its instantaneous rate of change is zero. At E we have $V = -3\,\text{V}$ and so $\mathrm{d}I/\mathrm{d}t = -1\,\text{A s}^{-1}$ which means that the current is decreasing at a rate of $1\,\text{A s}^{-1}$.

In order to develop the theory of a.c. circuits further we need to introduce a new concept.

A very convenient method of representing sinusoidally varying quantities, especially when there are phase-difference between them, is the rotating vector technique. Section 19.5 described how sinusoidal motion could be derived from uniform motion in a circle – in a similar way the variation of any sinusoidally varying quantity can be derived from a vector rotating with uniform speed. In this context these rotating vectors are often referred to as *phasors*.

In Fig. 36.7 imagine OP to be a rod, like the spoke of a wheel, rotating anticlockwise about O with uniform angular velocity ω. At time $t = 0$ the end P is at B. At the instant shown, therefore, the angle $\hat{PON}$ is ωt. The distance ON, which is the projection of OP upon the AOB, is OP $\cos \omega t$. Thus ON is a sinusoidally varying quantity whose

amplitude is equal to the radius of the circle. In this way we can represent two sinusoidally varying quantities with the same frequency which have a phase-difference of ϕ between them:

$$x_1 = A_1 \cos \omega t$$
$$x_2 = A_2 \cos (\omega t + \phi)$$

by the diagram of Fig. 36.8. If these two quantities are ones which can be added (for instance, the voltages across a pair of electrical components connected in series), then their sum at any instant is represented by the projection of OS upon the horizontal LOL′. OS is the diagonal of the parallelogram which has OP and OQ as two of its sides: in other words, **OS** is the vector sum of the two vectors **OP** and **OQ**.

Consider a circuit containing simply a resistor, R, and an inductor, L, connected in series with a supply of sinusoidally alternating voltage. The

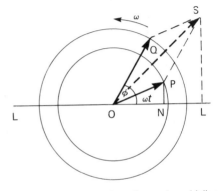

Fig. 36.8 Phasor representation of two sinusoidally varying quantities of different amplitude and phase.

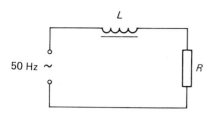

L = 100 H, with a resistance of about 300 Ω
R = 33 kΩ

Fig. 36.9

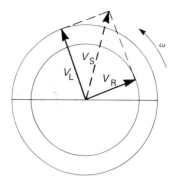

Fig. 36.10   Phasor diagram for the *LR* circuit of Fig. 36.9.

inductor, we shall assume, has negligible resistance. A circuit like this can easily be set up for
demonstration, but the demonstration is convincing only if the effect of the finite resistance of
the inductor *itself* is masked by the effect of its
inductance. Suitable values are suggested in Fig.
36.9. If an oscilloscope is connected in turn across
the terminals of the supply, across the inductor
alone, and across the resistor alone, the surprising
result is found that the sum of the voltage across
the inductor and the voltage across the resistor is
greater than the voltage across the terminals of the
supply! The vector diagram of Fig. 36.10 shows
how this comes about.

It must be remembered that the voltage across
a pure inductance leads the current flowing
through it with a phase-difference of one quarter-
cycle. The voltage across a pure resistance, on the
other hand, is in phase with the current through it.
Thus the voltage across the inductor, whose peak
value we shall call V_L, is a quarter-cycle ahead of
the voltage across the resistor, whose peak value
we call V_R. In the rotating vector diagram of Fig.
36.10 the vectors are rotating anticlockwise, and
the angle between V_L and V_R is $\pi/2$ radians, or
90°, all the time. The voltage of the supply is
represented by the vector V_S and this is the vector
sum of V_L and V_R. The lengths of these vectors are
in the ratio 4:3. Let us suppose that $V_L = 4$ V and
$V_R = 3$ V: then V_S is given by (Pythagoras'
theorem)

$$V_S^2 = V_L^2 + V_R^2$$

hence

$$V_S = 5 \text{ V}$$

So clearly, the arithmetical sum of V_L and V_R is
not equal to V_S: four plus three does not equal
five!

Calculations based on vector diagrams like
that of Fig. 36.10 do not require the presence of
the circles, or the horizontal diameter: so in future
we shall omit these parts of the diagram and
simply keep the vectors themselves, as in Fig.
36.14.

An inductor has the important property of
impeding the flow of an a.c. as our introductory
experiment showed. We shall now discuss this
property. The effect is convincingly shown in the
experiment depicted in Fig. 36.11. An audio signal
generator (with low impedance output) is used to
provide an alternating voltage whose frequency
can be varied as desired. The oscilloscope is used
to monitor the output of the signal generator,
simply to ensure that the peak value of the voltage
remains constant when the frequency is varied. A
is an a.c. ammeter of suitable range. L is an

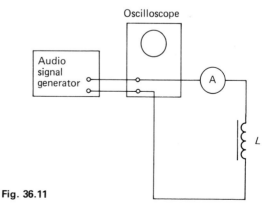

Fig. 36.11

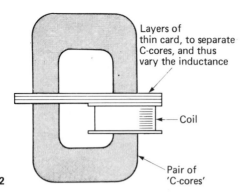

Layers of
thin card, to separate
C-cores, and thus
vary the inductance

— Coil

Pair of
'C-cores'

Fig. 36.12

inductor consisting of a coil wound on a former which can be slipped on a magnetic core, and the inductance can be varied by altering the distance between the two halves of the magnetic core, as shown in Fig. 36.12. The further apart the two halves, the smaller is the inductance. We find that, when the inductance is reduced, the current increases. When the frequency of the alternating voltage is increased the current is reduced, and a few measurements will suggest that, for a constant peak value of supply voltage, the current is inversely proportional to the frequency.

Why should the current (its peak value, let us say) be inversely proportional to the frequency? We can argue it in the following way, using imagined numerical values as we did in the argument of the last section. Referring back to that argument, suppose we start, as before, by considering the instant when the voltage applied to the inductor (assumed to have inductance only, and no resistance) is +6 V. The inductor has an

inductance of 3 H. At this instant the current must be changing at a rate of +2 A s⁻¹, because the value of $L\mathrm{d}i/\mathrm{d}t$ must equal the applied voltage. Suppose now that we double the frequency of the supply voltage, keeping its peak value unchanged. If the peak value of the current remained unchanged, then the rate of change of current at this instant would be +4 A s⁻¹, simply because the frequency has doubled and hence the rate of change of current at each point during a cycle is twice what it was before. But a rate of change of current of +4 A s⁻¹ would mean that the applied voltage would have to be +12 V: but in fact it is +6 V.

This must mean that the peak value of the current is half what it was before. Figure 36.13 shows the time-traces of the current at the old frequency and the current at the new frequency, for comparison. It follows, therefore, that if the supply voltage is kept constant, the peak value of current (hence also the r.m.s. value) must be inversely proportional to the frequency.

To summarize, we can write the relationship between the peak current and the peak voltage thus:

$$I_0 \propto \frac{1}{fL}$$

To find the precise form of the relationship we cannot avoid using calculus. Writing ω for $2\pi f$, we represent the instantaneous value of the current in the inductor thus:

$$I = I_0 \sin \omega t$$

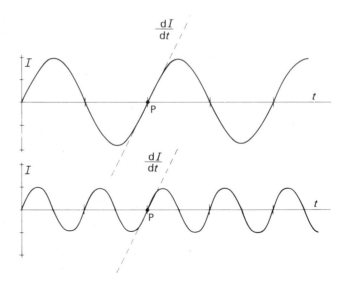

Fig. 36.13   The rate of change of current at point P is the same in both cases.

The applied voltage is given by

$$V = L\frac{dI}{dt}$$

Therefore, differentiating the expression for current with respect to t, we obtain

$$V = \omega LI_0\cos\omega t$$

and this can be written

$$V = V_0\cos\omega t$$

where

$$V_0 = \omega LI_0$$

and this is the peak value of the applied p.d. Thus

$$I_0 = \frac{V_0}{\omega L} = \frac{V_0}{2\pi fL}$$

This expression can be rearranged and written

$$\boxed{\frac{V_0}{I_0} = \omega L} \tag{36.3}$$

and here we see that it resembles the expression which defines the resistance of an electrical device, $V/I = R$. But the right-hand side of the expression, ωL, is not the resistance of the inductor (which we have assumed to have zero resistance anyway). It must be remembered that the voltage and current are quarter of a cycle out of phase. The quotient V/I only defines *resistance* when the voltage and current are in phase. Here this quotient defines a different quantity which is called the *reactance* of the inductor. It is, like resistance, measured in ohms but, unlike what happens in a resistance, no power is dissipated when a current flows in a pure inductance. It is not immediately obvious that the quantity ωL should be measurable in ohms, so let us see how this comes to be so. Frequency, f, has units: s^{-1}. Inductance has units: $V\,s\,A^{-1}$. So the product ωL has units: $(s^{-1}) \times (V\,s\,A^{-1}) = V\,A^{-1}$ or ohms.

So far our arguments have been based on the assumption that the inductor has no resistance. We are now equipped to understand the function of an inductor which does have appreciable resistance. Such an inductor can be treated simply as if it were a pure inductance in series with a pure resistance, as in Fig. 36.9 earlier. That phasor

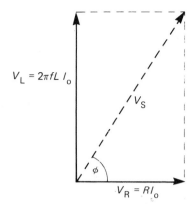

Fig. 36.14 Phasor diagram for an LR circuit.

diagram therefore applies to this situation, and we reproduce this diagram here in Fig. 36.14 without the circles, as we promised we would do. The peak value of the voltage across the series combination of inductance and resistance is the phasor V_S in the diagram. Now, the peak voltage across the inductance is V_L and we have already shown that V_0/I_0 for a pure inductance equals ωL, and so $V_L = \omega LI_0$. The peak voltage across the resistance R is given by $V_R = RI_0$. Hence, by Pythagoras' theorem,

$$V_S^2 = V_L^2 + V_R^2$$
$$= (\omega LI_0)^2 + (RI_0)^2$$

hence

$$\frac{V_S}{I_0} = \sqrt{(\omega L)^2 + R^2}$$

Now this quotient, voltage divided by current, is neither a reactance (where V and I are a quarter-cycle out of phase) nor a resistance (where V and I are in phase). We call it the *impedance*, and note that the voltage V_S across the inductor, and the current (which is in phase with V_R), have a phase-difference of ϕ between them. The impedance (Z) of an electrical device, or a whole circuit, is measured in ohms.

In the experiments, described earlier, demonstrating the properties of a pure inductance, we had to cheat a little by choosing suitable inductors and frequencies so that the effect of any resistance in the inductor itself was negligible, being masked by the effect of the inductance. What we did, in fact, was to choose values of L and f such that, in the phasor diagram of Fig. 36.14, V_L was much greater than V_R and so the phase-angle ϕ was very close to $\pi/2$ radians, or $90°$.

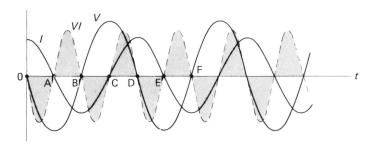

Fig. 36.15   Power in a purely inductive circuit.

It was said earlier, without any reasons given, that in a pure inductance no power is dissipated. This will now be explained.

We base our argument upon the graphs of voltage and current against time which were shown earlier, and are shown again here in Fig. 36.15. The shaded curve shows how the product of voltage and current varies with time.

Between points O and A on the time-scale, I is positive and V is negative, and so the product of VI is negative. Between A and B both I and V have negative values, thus their product is positive. Between B and C, V is positive but I is negative, and thus the product is negative. Between C and D the product is positive. When the product VI is positive, energy is being fed from the source into the inductor and being stored in the magnetic field of the inductor.

When VI is negative, energy is being returned from the inductor to the source. During one complete cycle, therefore, it is clear that there is no net energy transfer from source to inductor, or in the opposite direction.

36.3   Capacitors in a.c. circuits

One basic fact about capacitors and alternating currents is demonstrated strikingly by the experiment of Fig. 36.16. A capacitor whose capacitance is about $1000 \, \mu$F is connected in series with a 6 V battery and a 6 V 0.06 A lamp. (Care is taken to

connect the capacitor with the correct polarity.) The lamp does not light, although it lights perfectly well when connected directly to the battery. This is as expected. Then the battery is replaced by a 6 V r.m.s. a.c. supply, and the lamp is observed to light to almost full brightness. (Although we are using an electrolytic capacitor here, and it is being subjected to a voltage of about $6 \times \sqrt{2} = 8.5$ V, of the wrong polarity, every other half-cycle, it usually survives the treatment, provided that its working voltage is 15 V or greater.) The experiment shows that whereas the capacitor does not conduct a d.c., it does effectively conduct an a.c. The capacitor is, of course, being charged, discharged, and then charged with the opposite polarity, and discharged again, fifty times per second, and it is the charging current flowing in the wires of the circuit which lights the lamp.

To investigate the phase relationship between the voltage across a capacitor and the current in the circuit we can use the arrangement of Fig. 36.17. The voltage is found to lag behind the current (unlike the case of an inductor), reaching its maximum value about one quarter-cycle after the current reaches its maximum. The variations of voltage and current are shown in Fig. 36.18. To understand why the voltage and current vary in this way we must remember that current is the rate of flow of electric charge. At the instant A on the time-scale of the graph the voltage, and hence the

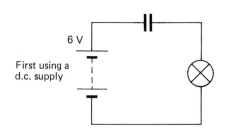

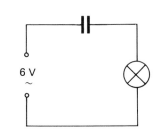

Fig. 36.16

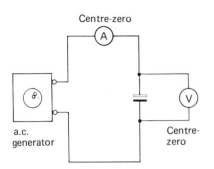

Centre-zero

a.c.
generator

Centre-
zero

Fig. 36.17

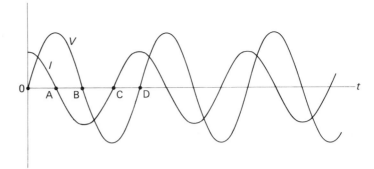

Fig. 36.18 Variation with time of current and voltage in a circuit containing pure capacitance.

amount of charge on the capacitor plates, is a maximum: the charge at this instant has ceased flowing into the plates and is about to flow away from them. Between A and B the voltage is decreasing which means that the amount of charge on the plates is decreasing also: the rate of decrease is a maximum at B, where the voltage curve has its greatest negative slope. At this instant therefore the current – the rate of flow of charge – has its greatest negative value. The rate of flow of charge then becomes less, as charge builds up on the capacitor plates, charging them with opposite polarity. At C the charge on the plates is again a maximum, and the rate of flow of charge, the current, is zero at this instant. You can continue the argument in this way, referring to the graphs.

Consider now a circuit containing capacitance and resistance, as in Fig. 36.19. (This experiment is similar to the one with an inductor shown in Fig. 36.9.) Suitable values for capacitance and resistance are shown. The peak value of the voltages across capacitance and resistance are measured with an oscilloscope and compared with the peak value of the voltage across the supply terminals.

The sum of the first two, $V_C + V_R$, is found to be greater than the supply voltage, V_S. The phasor diagram of Fig. 36.20 explains this situation. Remember that the vectors are imagined rotating anticlockwise: thus V_C is lagging behind V_R by one quarter-cycle (V_R is in phase with the current flowing through R, and we have shown already that the voltage across a capacitor lags behind the current). V_S is the vector sum of V_C and V_R and its magnitude is clearly less than the arithmetical sum of the magnitudes of V_C and V_R.

It is easy to get the phase relations between current and voltage for a capacitor confused with those for an inductor, and a useful memory-aid is the word *CIVIL*. Take the first three letters, *CIV*, and interpret them thus: in a capacitor (*C*) the current (*I*) leads the voltage (*V*). Take the last three letters, *VIL*, and interpret them: the voltage (*V*) leads the current (*I*) in an inductor (*L*).

When a capacitor is present in an a.c. circuit, no electrical energy is dissipated in the capacitor, unless its dielectric has finite resistance. In an ideal capacitor the dielectric has infinite resistance. Using an argument similar to that for an inductor, based on graphs of current and voltage as in Fig.

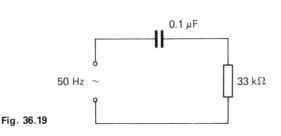

0.1 μF

50 Hz ~   33 kΩ

Fig. 36.19

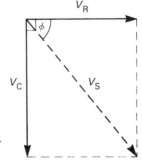

Fig. 36.20 Phasor diagram for an *RC* circuit.

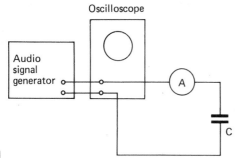

Fig. 36.21

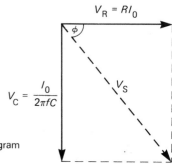

Fig. 36.22 Phasor diagram for an RC circuit.

36.18, you should show that, during one quarter-cycle, energy is fed from the source to the capacitor, and that this energy is returned to the source during the next quarter-cycle.

The reactance of a capacitor determines the relation between the magnitude of the voltage across it and the current in the circuit. The way in which the reactance depends on the capacitance and the frequency can be demonstrated by the experiment of Fig. 36.21. The capacitance can be varied by connecting other capacitors of the same value in parallel with C. In this way it is found that the current registered by the a.c. ammeter, A, is directly proportional to the capacitance: $I_0 \propto C$. If the frequency is varied, making sure that the output voltage of the signal generator is kept constant (this is what the oscilloscope is for), it is seen that the current is also directly proportional to the frequency: $I_0 \propto f$. Thus $I_0 \propto fC$. Also (although this is perhaps obvious and scarcely needs testing), if f and C are kept constant, the current is found to be directly proportional to the peak value of the supply voltage V_0. Thus the reactance, if we define it in the same way as for an inductor, is related to f and C thus:

$$\frac{V_0}{I_0} \propto \frac{1}{fC}$$

Using calculus one can show that the complete expression is:

$$\boxed{\frac{V_0}{I_0} = \frac{1}{2\pi fC}} \tag{36.4}$$

This is done as follows. We can write an expression for the applied voltage thus:

$$V = V_0 \sin \omega t$$

where V_0 is the peak value of this voltage. The charge on the capacitor is given by

$$Q = CV = CV_0 \sin \omega t$$

The current flowing in the circuit is

$$I = \frac{dQ}{dt}$$

substituting for Q in this expression, we obtain

$$I = \omega CV_0 \cos \omega t$$

which can be written

$$I = I_0 \cos \omega t$$

where I_0 is the peak value of the current, and

$$I_0 = \omega CV_0.$$

Hence

$$\frac{V_0}{I_0} = \frac{1}{\omega C}.$$

In Fig. 36.22 we see again the phasor diagram for a circuit containing capacitance and resistance. V_C, the peak value of the p.d. across the capacitor, is, as we have just explained, equal to $I_0/\omega C$. V_R, the p.d. across the resistor, is RI_0. Thus, by Pythagoras' theorem,

$$V_S^2 = V_C^2 + V_R^2$$

$$= \left(\frac{I_0}{\omega C}\right)^2 + (RI_0)^2,$$

hence

$$\frac{V_S}{I_0} = \sqrt{\left(\frac{1}{\omega C}\right)^2 + R^2}.$$

This quotient is the impedance of the circuit. Note that in this case the supply voltage, V_S, lags behind

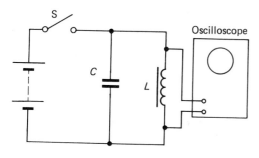

Fig. 36.23

the current in the circuit (which is in phase with V_R) by an amount ϕ. In an inductive circuit the voltage leads the current.

In the circuit of Fig. 36.22 power is dissipated only in the resistance, R. The peak value of the p.d. across R is V_R and

$$V_R = V_S \cos \phi$$

and if I_0 is the peak value of the current, then the peak power dissipation is the product $V_R I_0$. But the mean power dissipation is the product of the r.m.s. values of p.d. and current, and so the mean power equals

$$\frac{V_R}{\sqrt{2}} \times \frac{I_0}{\sqrt{2}}$$

which equals

$$\tfrac{1}{2} V_S I_0 \cos \phi. \tag{36.5}$$

The term $\cos \phi$ is often known as the *power factor* for the circuit. If the circuit were purely resistive ϕ would be zero and $\cos \phi$ would be unity. If the circuit were purely capacitive ϕ would be $\pi/2$ radians and $\cos \phi$ equal to zero, and, as we already know, no power is dissipated in a purely capacitance circuit. Expression (36.5) is applicable to any circuit containing resistance and reactance.

Finally, Table 36b summarizes concepts used in this section, showing their relationships and differences.

The concept of *resistance* is only applicable to an electrical device in an a.c. circuit if the voltage and current are in phase. The concept of *reactance* is applicable only to devices for which there is exactly one quarter-cycle phase difference (a phase angle of $\pm 90°$, or $\pm \pi/2$ radian) between the voltage across it and the current through it. The concept of *impedance* is used for devices which

Table 36b

	Resistance	Reactance	Impedance
Definition	$\dfrac{V_0}{I_0}$ or $\dfrac{V_{r.m.s.}}{I_{r.m.s.}}$	$\dfrac{V_0}{I_0}$ or $\dfrac{V_{r.m.s.}}{I_{r.m.s.}}$	$\dfrac{V_0}{I_0}$ or $\dfrac{V_{r.m.s.}}{I_{r.m.s.}}$
Phase angle between V and I (positive when V leads I)	$\phi = 0$	$\phi = \dfrac{\pi}{2}$ (inductive) $\phi = -\dfrac{\pi}{2}$ (capacitative)	ϕ may have any value in the range $-\dfrac{\pi}{2}$ to $+\dfrac{\pi}{2}$
Power dissipated	$V_{r.m.s.} \times I_{r.m.s.}$	Zero	$V_{r.m.s.} \times I_{r.m.s.} \cos \theta$

N.B. V_0 and I_0 are *peak values*; V and I, without suffixes, represent *instantaneous* values; and $V_{r.m.s.}$ and $I_{r.m.s.}$ are *root mean square* values.

have a *reactive component* and a *resistive component*, such as an inductor which has appreciable resistance, or a capacitor with 'leaky' dielectric; but its chief use is for circuits containing resistive and reactive devices. In general there will be a phase-difference between the voltage and current which is not precisely $\pi/2$ radian, nor precisely zero.

36.4 Oscillations in an *LC* circuit

The term '*LC* circuit', using the symbols for inductance and capacitance, is a brief way of saying 'circuit containing inductance and capacitance'. Circuits of this kind have widespread practical applications as *tuned circuits* or *resonant circuits*, chiefly in radio and television transmitters and receivers.

Let us start by speculating on what we might expect to happen if a capacitor is charged and then connected to an inductor. If there is no resistance in the circuit, then no power can be dissipated as heat, as has been explained in Sections 36.2 and 36.3. What would happen to the energy in the circuit? Would the capacitor discharge at all?

Figure 36.23 shows an experiment which can help us to answer the question. The oscilloscope, with its time-base set to give a very slow sweep rate (about 1 trace per second is suitable), records the voltage across the inductor and capacitor. Closing the switch S charges the capacitor. When the

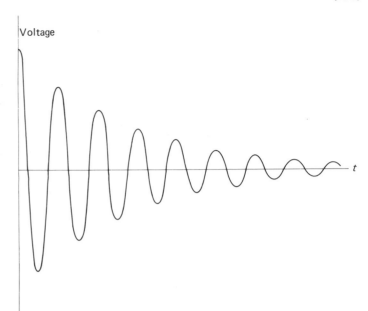

Fig. 36.24 The oscillations in the voltage across L and C in the circuit of Fig. 36.23.

switch is opened, a trace like that shown in Fig. 36.24 is seen on the screen. We have here an oscillating voltage which decays to zero after a few cycles.

If a resistor is connected across L and C the decay of oscillation is found to be more rapid: we have heavier damping. This suggests that the decay of oscillation may be due, in part at least, to the loss of energy being dissipated in the resistance in the circuit.

It is convenient, for further experiments, to arrange that a continuously repeating trace on the oscilloscope screen can be obtained. Figure 36.25 shows how this can be done. A sharp voltage pulse appears at the time-base output socket every time the spot flies back to the left-hand side, and this can be used to charge the capacitor which then discharges through the inductor as the spot traces from left to right.

We can now investigate experimentally what factors determine the frequency of the oscillations. The inductance can be varied by using the technique described in Section 36.2 and illustrated in Fig. 36.12. It is found that increasing the inductance reduces the frequency of oscillation. The capacitance can be altered by substituting other capacitors for C, or by adding further capacitors connected in parallel with C – this will show that increasing the capacitance also reduces

the frequency of oscillation. Thus we see that this kind of circuit has a characteristic frequency which depends upon the inductance and the capacitance in it. Suppose we have used an inductance of about 0.1 H and a capacitor of about 1 μF, we will find that this characteristic frequency is about 500 Hz.

A further experiment, using the same values of L and C as were used in the previous experiment, is shown in Fig. 36.26. The audio oscillator must be one which has a low-impedance output so that it can light a low-voltage lamp to full brightness even when its amplitude control is at less than its maximum setting. With the inductor alone, as in (a), the lamp lights at low frequencies but goes out when the frequency is increased: this is as

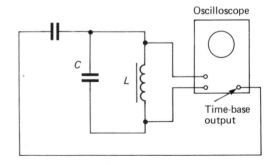

Fig. 36.25

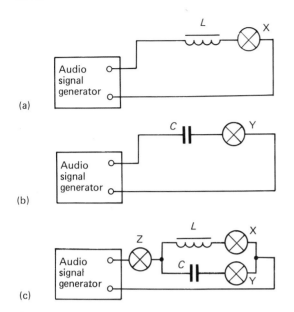

(a)

(b)

(c)

Fig. 36.26

expected. With the capacitor alone, as in (b), the reverse situation is found, and the lamp only lights when the frequency is high. This also is as expected.

In Fig. 36.26c we have combined the two previous circuits, and also inserted a third lamp, Z, to indicate the current flowing in one of the leads to the signal generator. At low frequencies X and Z glow but Y does not: thus it appears that most of the current is flowing by way of the branch containing *L* and X. At high frequencies the lamps Y and Z glow but X does not: so now most of the current must be flowing in the branch containing *C* and Y. But it is possible to set the frequency at an intermediate value such that lamps X and Y glow (although not necessarily at

full brightness) but lamp Z does not glow noticeably. Using a.c. ammeters, instead of lamps, at positions X, Y and Z makes these observations even more convincing.

At the characteristic frequency the current in the *L*-branch and the current in the *C*-branch of the circuit are both greater than the current drawn from the signal generator. This result seems a gross contradiction of Kirchhoff's first law (Section 23.1). But what we know about the phase-relationships of current and voltage in inductors and capacitors should resolve the paradox. We know that in a pure inductor the voltage across it leads the current by quarter of a cycle, and that in a pure capacitor the current leads the voltage, also by a quarter-cycle. Now, in our circuit, the capacitor and inductor are connected in parallel, so that the voltage across each must be the same in magnitude and phase. Thus the current in the inductor branch must be one half-cycle ahead of the current in the capacitor branch, that is, these two currents have opposite phase. The situation is shown in Fig. 36.27.

The way the currents go, as shown in the graph, implies that nearly all the current is surging around the *LC* circuit, first clockwise, then anticlockwise. It is the electric charge therefore that is in a state of oscillation, sloshing back and forth, as it were, like water in a bath. At the instant when the capacitor carries maximum charge (when the voltage is a maximum), the current is zero, hence the magnetic flux in the inductor is zero. This is the instant when all the energy in the oscillatory circuit is stored in the capacitor. A quarter of a cycle later the capacitor is uncharged, and the current is a maximum: this is the instant when the

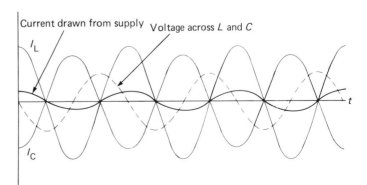

Fig. 36.27 The variation of current and voltage with time in a resonant circuit of negligible resistance.

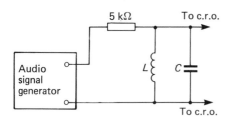

Fig. 36.28 Circuit to demonstrate resonance in an *LC* circuit.

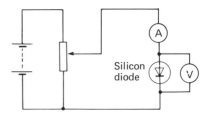

Fig. 36.29 Circuit to investigate conduction in a diode.

energy is all stored in the inductor. Thus, neglecting any energy dissipation in the resistance of the circuit, the energy is transferred back and forth from inductor to capacitor, from being associated with a magnetic field to being associated with an electric field, alternately.

We said earlier that the characteristic frequency of the *LC* circuit depended upon the inductance and the capacitance, and that increasing either of these quantities caused a reduction of the frequency. We shall now analyse the circuit theoretically to find precisely the relation between the frequency and *L* and *C*. Suppose the charge on the capacitor at some instant of time is *Q*, then the p.d. across the capacitor V_C is given by: $V_C = Q/C$. The p.d. across the inductor V_L is related to the rate of change of current through it by the expression (see Section 35.7)

$$V_L = L \frac{dI}{dt}$$

If we regard the capacitor as the electrical supply which drives current through the inductor, then the current is equal to the rate at which the capacitor *loses* charge, so we must write

$$I = -\frac{dQ}{dt}$$

Hence

$$\frac{dI}{dt} = \frac{d}{dt}\left(-\frac{dQ}{dt}\right) = -\frac{d^2Q}{dt^2}$$

Since the inductor and capacitor are connected in parallel, we must have $V_C = V_L$. Therefore we can write

$$\frac{Q}{C} = -L \frac{d^2Q}{dt^2}$$

hence

$$\frac{d^2Q}{dt^2} = -\frac{1}{LC} Q \qquad (36.6)$$

This equation is of the same form as the one describing simple harmonic motion. In Section 19.4 we found that, for such a motion, the acceleration d^2x/dt^2 was proportional to $(-x)$. In Section 19.7 this was written

$$\text{acceleration} = \frac{d^2x}{dt^2} = -\omega^2 x$$

And we found that the solution for this equation was

$$x = a_0 \cos \omega t$$

Equation 36.6 has the same form and so we may assume an analogous solution

$$Q = Q_0 \cos \omega t$$

implying that the charge on the capacitor varies sinusoidally, and the frequency of oscillation can be derived from the equation

$$\omega^2 = \frac{1}{LC}$$

hence

$$\boxed{f = \frac{1}{2\pi\sqrt{LC}}} \qquad (36.7)$$

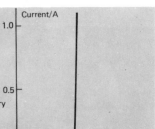

Fig. 36.30 Current/p.d. relationship for a silicon diode.

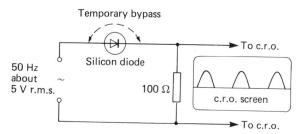

Fig. 36.31　Circuit to show half-wave rectification.

An experiment which demonstrates *resonance* in an *LC* circuit is shown in Fig. 36.28. The audio signal generator provides a sinusoidal alternating voltage whose peak value is in the order of one or two volts. A high resistance (about 5 kΩ is suitable) is put in series so that only a weak current is driven into the *LC* circuit. A c.r.o. is used to display the voltage across *L* and *C*. Provided that the resistance of the inductor is low (a few ohms at most) we find that, at one particular frequency, the voltage across *L* and *C* has a much greater peak value than at any other frequency. Here the *LC* circuit is *resonating* to the applied signal. If we already know the values of the inductance and capacitance, we can verify that this resonant frequency is equal to the value given by Eq. 36.7.

The principle demonstrated by the experiment above is used in radio receivers. A variable capacitor in the circuit enables one to 'tune in' to the radio station required: then the circuit gives a large signal voltage only at the frequency of that station. Broadcasting stations operating at any other frequencies should not be audible, because the voltage across *L* and *C* is too small at those frequencies. That is the principle used in the very simplest design of radio receiver: in practice, the frequencies of broadcasting stations are so close together that a single *LC* circuit is not *selective* enough to prevent unwanted stations from being heard. More complex circuits have to be used, usually incorporating the *superheterodyne* principle. To find out more you must refer to a textbook on radio.

36.5　Rectification of a.c.

All large-scale, national electrical supply systems use alternating current, chiefly for economic reasons, as is explained in Chapter 38. But almost all electronic devices, such as amplifiers, radio and television equipment, and computers, require a steady d.c. supply. Re-charging batteries requires d.c., but for this purpose it need not be a steady, or *smoothed* d.c. *Rectification* means the conversion of a.c. to d.c. A device which does this is called a *rectifier*.

The simplest type of rectifier is a single *diode*, and one of the commonest kinds of diode is made from two types of silicon, called *p-type* and *n-type* (see Section 32.7). We shall not attempt to explain its internal action here, but shall simply study its function. For our experiments we use any silicon rectifier diode which has a maximum current rating of about 1 A. Figure 36.29 shows a circuit for investigating the conduction of the diode. Note that the symbol for the diode is like an arrowhead: when it is connected in the circuit as shown a measurable current should flow; if it is removed, reversed, and put back in the circuit the current should then be negligibly small. This shows that it only allows current to flow one way through it: the way the arrow-head points. When it is 'the right way round' in the circuit, and the maximum allowable current is flowing, the p.d. across the diode is found to be about 0.7 V: this value is characteristic of all silicon diodes. The graph of Fig. 36.30 shows the current – p.d. relationship for a typical silicon diode. Note that it is not an ohmic conductor, and therefore does not have a constant resistance: its resistance becomes less as the p.d. across it is increased.

Figure 36.31 shows an experiment to demonstrate rectification of a.c. by a diode. A c.r.o. is used to display the p.d. across a resistor; this p.d. is proportional to the current in the diode at every instant and so the c.r.o. shows us how the current is varying. The diagram also shows the form of the trace observed on the c.r.o. This shows that the diode is allowing current to flow one way only, during every alternate half cycle of the a.c. This is called *half wave rectification*. If the diode is 'bridged' by a wire link, as shown in Fig. 36.31, the complete, unrectified 'waveform' is observed.

The term 'direct current' simply means current which flows in one direction only: it is not necessarily a *steady* current. In the experiment described we have produced a *pulsating d.c.* This would be quite satisfactory for re-charging a battery, but it would be useless as a d.c. supply for

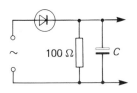
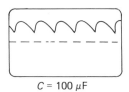
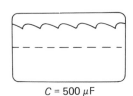

Fig. 36.32 Adding a smoothing capacitor to the half-wave rectifier circuit.

$C = 100 \, \mu F$ $C = 500 \, \mu F$

a radio receiver, for instance, unless the pulsations are smoothed out by some means. The simplest way of smoothing the d.c. is to use a capacitor, connected in parallel with the resistor, as shown in Fig. 36.32; here we also see the 'waveform' of the trace on the c.r.o. when two different values of capacitance are used, and clearly the greater capacitance gives the better smoothing. Why does the 'waveform' have precisely these shapes?

First you should recall the significance of the *time constant* of a circuit in which a capacitor is charging and discharging (see Section 26.7). We see, in Fig. 36.32, that when the p.d. across the resistor and capacitor is increasing its 'time trace' has the same shape as that of the alternating voltage of the supply. But when it is decreasing it does so more slowly than the supply voltage. When the capacitor is being charged the diode is conducting and has a low resistance. The supply also has a low resistance, and so the capacitor is being charged in a circuit of low total resistance, probably about 1 Ω, and the p.d. across the capacitor will 'follow' the varying p.d. of the supply with very little delay. But when the supply voltage falls the diode stops conducting, and the capacitor discharges through the resistor. In the experiment a 100 Ω resistor was used, so that, with a 100 μF capacitor, the time constant is given by:

time constant $= 100 \, \Omega \times (100 \times 10^{-6} \, F) = 0.01 \, s$

As explained in Chapter 26 this is approximately the time taken for the capacitor to lose half its charge (the precise time is 0.693 RC). Half a cycle of the 50 Hz supply has a duration of 0.01 s, and, as we see in Fig. 36.32, the p.d. across the capacitor is falling to about half the peak value of the supply when the diode is not conducting. The 500 μF capacitor increases the time constant of the

discharging circuit to 0.05 s and so the fall in p.d. is appreciably less. However, the smoothing provided by a 500 μF capacitor in this experiment would be quite inadequate in providing a supply for any amplifier, radio, or computer circuit.

Problem 36.1   A student wants to use the circuit of Fig. 36.32 to supply an electronic circuit whose voltage and current requirements make it behave as if it were simply a resistance of 100 Ω. She has been told that the 'ripple' (that is, the maximum variation of voltage) must not be greater than 1% of the mean supply voltage. What minimum value of smoothing capacitance must be used?

When the smoothing capacitor is discharging through a resistance the time taken for the p.d. across it to fall to half of its original value is approximately equal to the time constant RC. With $R = 100 \, \Omega$ and $C = 100 \, \mu F$ the time constant is 0.01 s. These values would cause a fall in p.d. of about 50% in the time of one half-cycle. To reduce the fall to less than 1% we need a time constant greater than 50 times the previous value, so a capacitance at least 50 times greater is needed: 5000 μF or more.

The bridge rectifier

The most widely used type of rectifier in electronic equipment is the *bridge rectifier*, so called because its circuit has the same configuration as the original Wheatstone bridge circuit (see Section 25.5) containing four resistors. Figure 36.33 shows a bridge rectifier in an experimental circuit. This type of rectifier has the advantage that it uses *both* half-cycles of the a.c. and gives *full wave rectifica-*

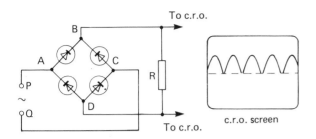

Fig. 36.33 Circuit to show full-wave rectification with a bridge rectifier.

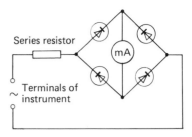

Fig. 36.34 Circuit of an a.c. voltmeter.

tion. During one half-cycle two of the diodes conduct, and during the other half-cycle the other two conduct. Consider the situation when terminal P of the a.c. supply is at a higher potential than Q (or, as one often says: 'P is positive with respect to Q'). The diode between A and B conducts because point A is at a higher potential than B; the diode between A and D does not conduct. Current flows through the resistor R, and the diode between D and C conducts because point D is at a higher potential than C (but less than A). The diode between B and C does not conduct. So current flows by way of P–A–B–R–D–C–Q. This current flows in a *downwards* direction through R, and so the top end of R is the high potential end. Similar reasoning can be applied to the other half-cycle; now the diodes between B and C, and between A and D, conduct, and the current path is Q–C–B–R–D–A–P. But once again the current flows *downwards* through R.

Clearly it requires less capacitance to smooth a full wave rectified supply than a half wave rectified one, because there is less time between the voltage maxima. For most everyday applications the four silicon diodes comprising the bridge are encapsulated into a single 'package' and provided with four terminals.

An important use for rectifiers is in a.c. meters. Figure 36.34 shows an example of a bridge rectifier linked to a moving coil milliammeter. This arrangement is often used in conjunction with a series resistor as the basis of an a.c. voltmeter, but its use is limited to situations in which the p.d. across each diode is negligible compared with the voltage being measured. The milliammeter indicates the *mean value* of the full wave rectified current, because the moving coil, pointer, etc., have sufficient inertia to prevent them from

any noticeable movement at the frequency of the a.c. being measured. It can be shown theoretically that the mean value of a full wave rectified sinusoidal current is $2/\pi$ times the peak value. As explained in Section 36.1 the peak value is $\sqrt{2}$ times the r.m.s. (root mean square) value. Thus the r.m.s. is $\pi/(2\sqrt{2})$ times the value indicated by the meter. This factor is 1.11, to two decimal places.

Problem 36.2   The arrangement of Fig. 36.34 is to be used to make an a.c. voltmeter reading up to 500 V r.m.s. The moving coil milliammeter has a coil resistance of $10\,\Omega$ and gives full-scale deflection for a current (d.c.) of 5 mA. What value of series resistance is required?

We can neglect the p.d. across each diode of the bridge, which cannot exceed about 0.7 V (as explained earlier). We want the current through the meter to be 5 mA when the r.m.s. voltage at the terminals is 500 V.

As explained above, when the meter reads 5 mA the r.m.s. current through it must be 1.11 times as great. Thus

$$\text{r.m.s. current} = 1.11 \times 5\,\text{mA} = 5.55\,\text{mA}$$

This is to be the current flowing via the resistor when the r.m.s. voltage at the terminals is 500 V, so the total resistance required is

$$\frac{500\,\text{V}}{5.55 \times 10^{-13}\,\text{A}} = 90\,090\,\Omega$$

The $10\,\Omega$ coil resistance is negligible in comparison with this. (Since moving coil meters are accurate only within about $\pm 2\%$, a resistance of $90\,\text{k}\Omega$ would suffice.)

Chapter 37

THE MAGNETIC FLUX DUE TO
ELECTRIC CURRENTS

37.1 Magnetic flux measurement

We shall need a technique for measuring amounts of magnetic flux, and flux density, in a variety of situations. The most suitable method for our purposes is to use a movable coil connected to an oscilloscope, and to arrange that the magnetic flux is alternating. When this alternating flux links the coil, an alternating e.m.f. is induced whose amplitude can be measured on the oscilloscope screen. Most of the measurements will be comparative ones: finding by what ratio the flux has increased or decreased. This technique enables such comparisons to be made very quickly.

Two distinct types of coil will be used (Fig 37.1) to investigate the magnetic flux produced by a long tubular *solenoid*. The larger coil, A, which we shall call a *girdle coil*, is used to measure the total flux produced in the solenoid. This girdle coil must fit closely so that it is not linked by any of the reverse, *external* flux from the solenoid. The small coil on an handle, B, is generally known as a

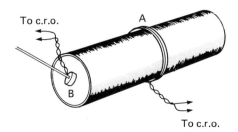

Fig. 37.1 A long solenoid fitted with a 'girdle' coil A for measuring the total flux. A search coil, B, is used for measuring the mean flux density at a point.

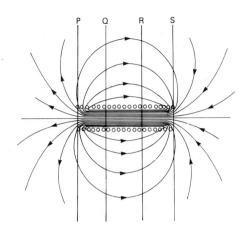

Fig. 37.2   The magnetic field of a solenoid; the ratio of length to diameter of the solenoid is 5:1.

search coil. It detects only the flux which links the very small area enclosed by it, and thus the e.m.f. induced in it gives an indication of the mean flux density in a direction parallel to its axis.

You should experiment with both types of coil, to explore thoroughly the flux produced by a long solenoid when an alternating current flows in it. To obtain a reasonably large induced e.m.f. in either coil one must use a frequency greater than the 50 Hz of the a.c. mains. This is a consequence of the law of electromagnetic induction: the e.m.f. induced in the coil is proportional to the rate of change of flux linkage, and so the higher the frequency, the greater will be the maximum rate of change of flux, and hence the greater will be the amplitude of the alternating induced e.m.f. A frequency of about 5 kHz is ideal for an air-cored solenoid which has a few hundred turns of wire on it. A frequency as high as this is needed so that the maximum rate of change of flux linkage with the girdle coil is great enough to produce an induced e.m.f. of sufficient amplitude to measure on the screen of a c.r.o.

The results of the investigation should match with the information given in Fig. 37.2 which shows a section through a solenoid and its magnetic field pattern. Inside the solenoid, between the transverse planes labelled Q and R, the field is very nearly uniform, and here very little flux 'leaks' out from the wall of the solenoid. This is demonstrated by the fact that if the girdle coil is moved back and forth between Q and R there is

very little change in the e.m.f. induced in it. When the search coil is used to probe the field inside the solenoid, between Q and R, keeping its axis parallel to the solenoid's axis, there is found to be very little variation in induced e.m.f., both when the search coil is moved back and forth parallel to the solenoid's axis, and from side to side across the axis. When the search coil is placed on the solenoid's axis, and level with the end P or S, the induced e.m.f. is half the value obtained previously, provided the solenoid is long in comparison with its diameter (see Table 37a for the significance of this).

The search coil can be used to find the direction of the flux at any point, simply by orienting it so that the induced e.m.f. is a maximum: this occurs when the flux is parallel to the search coil's axis.

For a full investigation of solenoids we require a set of solenoids having the same number of turns of wire per unit length but with different areas of cross-section, and ones having the same cross-section but different numbers of turns per unit length. With these it can be demonstrated that what determines the maximum flux density (in the inner regions of the solenoid, as between Q and R in Fig. 37.2) is simply the *total current circulating per unit length of solenoid*: neither the shape nor the area of cross-section affect it. Thus, a solenoid having 6 turns per centimetre and a current of 3 A gives the same maximum flux density as one having 9 turns per centimetre and a current of 2 A.

Sometimes solenoids are made from a roll of metal foil (aluminium foil is often used) wrapped around a hollow cylinder and not from wire. The foil has an insulated coating so that the current circulates round and round the cylinder. In the example of Fig. 37.3 the solenoid is 20 cm long and has 72 turns of foil around it. Electrical connections are made to the inner edge of the foil and to its outer edge. If a current flows uniformly, as suggested by the arrows in the figure, through the foil, then a current of 5 A in this solenoid would provide 72 × 5 A = 360 A circulating around it, so that the current circulating around every centimetre length would be 18 A, the same as in the previous example.

For some experiments, described later, we use a long, flexible solenoid which can be curved around to form a complete ring, as shown in Fig.

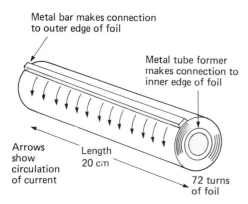

Fig. 37.3 A foil-wound solenoid.

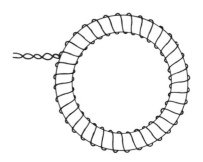

Fig. 37.4 A schematic diagram of a toroidal solenoid.

37.4. This is called a *toroidal* solenoid (from Latin *torus* meaning an anchor ring). Using a girdle coil we can show that the total flux produced by this flexible solenoid is unaltered if we change its shape from straight to toroidal form. This toroidal form has both practical and theoretical importance, chiefly because all of its flux is contained within it. This fact can be verified by using a search coil to hunt for flux leaking out from it. A girdle coil, on the other hand, is used to show that the flux is constant all the way around the toroid. As we have already seen, with a straight solenoid, if the flux density in the inner regions is B, then at each end, on the axis, the flux density is $B/2$. When the ends are joined, the fields at the two ends combine to produce a total flux density B. Figure 37.5 shows a graph of how the flux density on the axis of a long solenoid varies with distance, near an end; and how these two 'end-fields' add together to give a constant total when the two ends are joined.

The foregoing experiments suggest that the actual *length* of a solenoid is not a relevant factor in determining the flux density inside it, provided that it is long enough to make the field in its inner regions uniform. Strictly speaking this means that the solenoid must have infinite length, or else be toroidal so that it has no ends. In practice the flux density at the middle of a solenoid approximates well to that of a very long, or 'infinite solenoid' value if its length is a few times greater than its diameter. Table 37a shows how good the approximation is.

So far we have concentrated attention on the flux *density* in the region in and around a

Table 37a

Ratio $\dfrac{\text{length}}{\text{diameter}}$	Flux density at centre, as a percentage of 'infinite solenoid' value/%
5	98.0
7	99.0
10	99.5
20	99.9

solenoid, and have established that in a long solenoid the flux density deep inside it depends only upon the current circulating per unit length of solenoid. Soon we shall be mainly concerned with the *total flux* produced by coils.

To help you to start thinking about measuring total flux, look at Fig. 37.6. A set of three long solenoids, all having the same number of turns per unit length, is connected in series to an a.c. supply. Each one has around its middle a ten-turn girdle coil which can be connected to an oscilloscope to measure the e.m.f. induced in it. A search coil also is used to probe the field inside the solenoids. The areas of cross-section of the solenoids are as shown. When the search coil is placed near the middle of solenoid X the e.m.f. induced in it is 10.0 mV. The e.m.f. induced in the girdle coil around it is 7.1 mV. Can you predict the search coil e.m.f. and the girdle coil e.m.f. for the other two solenoids, Y and Z? The flux density shown by the search coil will be the same in each because the current circulating per unit length is the same. The total flux, therefore, will be proportional to the area of cross-section, and so will be the e.m.f. induced in the girdle coil.

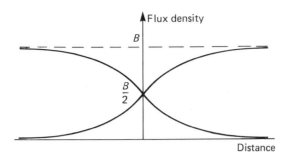

Fig. 37.5 Graph showing the variation of flux density with distance at each end of a flexible solenoid. When the two ends of the solenoid are joined together the total flux density throughout the coil is *B*.

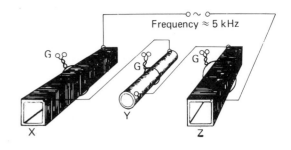

Fig. 37.6 Measuring the flux in a set of three solenoids. Each solenoid has a ten-turn girdle coil around its middle. All solenoids have the same length but the areas of cross-section are 20 cm^2 (X), 10 cm^2 (Y) and 40 cm^2 (Z).

37.2   The magnetic circuit concept

How do we design an electromagnet to produce the strongest possible field in the air-gap between its poles? The concept of the *magnetic circuit* helps to solve this problem, as well as enabling us to find the exact relationship between magnetic flux and the currents which produce the flux. From the studies completed so far you should be able to offer some suggestions about what is needed for a strong electromagnet, for instance:

1) use the strongest possible current
2) use the greatest number of turns in the coils
3) make the air-gap as small as possible.

We cannot proceed further than this without a deeper understanding of the way magnetic flux behaves. To attain this we require an experimental system in which the magnetic flux is completely contained within a closed, loop-like path. The apparatus of Fig. 37.7 can be used for this purpose. The flux-path is provided by a ring made up of thin strips of overlapping magnetic alloy – used for making the cores of transformers – and bound together with adhesive tape. This alloy is a magnetically very 'soft' material: that is to say, it is easily magnetized and demagnetized, but cannot retain any permanent magnetism. It would be useless for making permanent magnets. The length of the ring in Fig. 37.7 and its area of cross-section, can be altered by building it from different numbers of strips, although this is a tedious operation. Unfortunately this is the only way of ensuring adequate magnetic contact between the separate parts which go to make up the ring. (This

is fully explained later in this Section). Exactly the same principle is applied to the building of transformer cores. Indeed, the apparatus of Fig. 37.7 is a sort of transformer, but here we are using the secondary coil as a means of measuring the flux in the magnetic ring, and not as an electrical supply.

The movable girdle coil is connected to an oscilloscope which must be sensitive enough to measure voltages of a few millivolts. The primary coil is connected to a low voltage a.c. supply (audio signal generator) with a frequency of about 1 kHz to 5 kHz. An a.c. ammeter is used to measure the current in the primary coil: this current will not be greater than about 0.1 A r.m.s., for reasons which will become apparent later.

The first experiment is to investigate how the flux in the magnetic ring depends upon the current and the number of turns in the primary coil. Keeping the same primary coil and varying the current, it is found that the e.m.f. induced in the girdle coil (the peak-to-peak e.m.f. is measured on

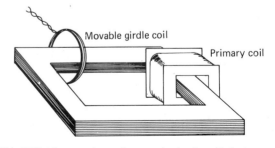

Fig. 37.7 An experimental magnetic circuit made by overlapping strips of transformer alloy bound tightly together.

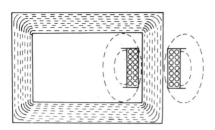

Fig. 37.8 Nearly all the flux is contained by the alloy, but a little leaks out into the air.

the oscilloscope screen) increases when the current is increased. Substituting coils having different numbers of turns, but using the same current in each, we find that we get more flux with a larger number of turns. In fact it is the product: (current) × (number of turns) which determines the amount of flux. This product is often named 'ampere turns'.

The fact that almost all the flux is contained within the metal ring is quickly checked by moving the girdle coil to different positions and observing that there is very little change in the induced e.m.f. An increase will probably be observed close to the fixed coil because a little of the flux takes a path through the air, instead of through the metal, as shown in Fig. 37.8.

The next experiment is to construct a ring of greater total length, but keeping the same area of cross-section, and using the same number of ampere turns. This operation will reduce the inductance of the coil and the a.c. supply voltage will need to be reduced in order to restore the same current in the coil as before. Then it will be found that the flux has been reduced by lengthening the ring. By adding more alloy strips on to the core to thicken it, we find that this increases the flux for the same number of ampere turns.

Finally we can construct a ring of non-uniform cross section area and demonstrate that, in spite of this, the flux is the same at all places around the ring.

These experiments show that the magnetic flux in a ring of suitable magnetic alloy behaves like an electric current in an electrical circuit. We shall sum up the points of similarity in Table 37b, using the term *magnetic circuit* for the apparatus with which we have experimented.

The 'proper' name for 'battery voltage' in this context is of course *electromotive force*, or

'e.m.f.'. By analogy with this, the term *magneto-motive force*, or *m.m.f.*, is used to signify what we have so far called 'ampere turns'. More precise experiments than we have described show that the flux in a magnetic circuit can be described mathematically by expressions which are exact analogues of the expressions used in electrical circuits. Thus, for a magnetic circuit consisting of a ring of one material of uniform cross-section, we can write:

$$\Phi = \mu M \frac{A}{l} \qquad (37.1)$$

where μ depends upon the flux density within the material, but within strict limits may be *approximately* constant (this point will be taken up later:) and where Φ is the flux, M is the magnetomotive force (the 'ampere turns' product for the coil), A is the area of cross-section, and l is the length of the magnetic circuit. For an electric circuit comprising simply a battery of negligible internal resistance of e.m.f. E, and a uniform resistance wire of cross-sectional area A and length l, the current would be given by the expression:

$$I = \sigma E \frac{A}{l} \qquad (37.2)$$

Here σ is the *conductivity* of the material of the wire. Conductivity is the reciprocal of resistivity:

Table 37b

Magnetic circuit	Electric circuit
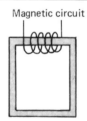	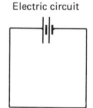
The flux depends on the 'ampere turns'	The flow of charge depends on the 'battery voltage'
If length increases; flux decreases	If length increases; the current decreases
If cross-sectional area increases, the flux increases	If cross-sectional area increases, the current increases
The flux is constant all around the circuit regardless of variations in cross-sectional area.	The current is constant all around the circuit regardless of variations in cross-sectional area.

in symbols, $\sigma = 1/\varrho$ (see Section 23.4). The conductivity of most conducting solid materials is constant, at constant temperature, for a very large range of current densities in the material. In this way it differs markedly from its magnetic analogue: permeability (μ).

Rearranging the two Eqs. 37.1 and 37.2, and writing them so that they can be compared, we have:

$$\frac{M}{\Phi} = \frac{1}{\mu}\frac{l}{A} \qquad (37.3)$$

and

$$\frac{E}{I} = \frac{1}{\sigma}\frac{l}{A}\left[= \varrho\frac{l}{A}\right] \quad (37.4)$$

The electrical circuit expression is shown in terms of resistivity, ϱ, as well as conductivity, σ: and it will be recognized as the expression for the electrical resistance of the circuit. By analogy the quotient M/Φ is defined as the magnetic *reluctance* of the magnetic circuit. The unit of reluctance is: $A\,Wb^{-1}$. This concept of reluctance can be applied to any one *part* of a magnetic circuit also, and the total reluctance of the entire circuit is equal to the sum of the reluctances of all the parts: a similar rule applies to the electrical resistances of the parts of any electrical circuit, in which all the parts are in series.

Before we go on to use these concepts quantitatively in problems, for instance that of the design of an electromagnet, we must emphasize the limitations of the analogy between magnetic and electric circuits. The first important difference concerns *energy*. In an electric circuit which has resistance, energy must be continuously supplied to maintain a steady current in the circuit. This is not so in a magnetic circuit: energy is not used in *maintaining* a steady magnetic flux, although energy is needed to 'build up' a magnetic flux from zero, and energy will be converted into heat in the magnetizing coil unless it is made of a super-conductor. Another point, and related to the first, is that whereas our theoretical model of an electric current in a conductor involves the idea of moving charged particles of matter, for which there is plenty of evidence (see Chapter 33), there is no evidence whatsoever that magnetic flux is a flow of any material substance (although the Latin word *flux* means 'flow'). Furthermore, the

electrical conductivity of a vacuum is precisely zero, but the magnetic permeability of a vacuum has a finite value. This value is determined by the definition of the ampere, as will be explained later in this chapter (Section 37.3).

The fundamental difficulty in measuring the magnetic permeability of a medium is in ensuring that all the flux is kept within the medium. To adapt the experiment just described to measure the permeability of transformer alloy, or iron, the ideal way would be to fill the toroidal solenoid with the material in question. However, we can obtain good estimates, for transformer alloy, using the apparatus of Fig. 37.7 and applying the theory of the experiment we have just described. Question 9.29, at the end of this book, concerns such a measurement. But we must emphasize, at this point, that the permeability of magnetic materials is never constant, unlike the conductivity of any metal at constant temperature. At best, permeability is approximately constant over a limited range of flux density within the material; and the relationship: flux $\propto$ m.m.f. is an approximate one, valid within certain limits, but nevertheless useful in spite of the limitations. This should explain why, in the experiment described in Section 37.2, the current was kept small.

Problem 37.1 This example uses the concept of the reluctance of a magnetic circuit, and shows how large an effect a small air-gap can have on the flux.

A pair of 'C'-shaped transformer half-cores join perfectly to form a magnetic circuit. The alloy of which they are made has a mean permeability of $3 \times 10^{-3}\,Wb\,A^{-1}\,m^{-1}$ for small flux densities. The mean length of the flux-path in the complete ring is 200 mm, and the area of cross-section is 400 mm². Suppose now that an air-gap of width 0.04 mm is introduced between both pairs of joining faces. By what factors is the reluctance of the magnetic circuit increased?

We use Eq. 37.4 for the reluctance of a uniform magnetic circuit:

$$\text{Reluctance} = \frac{M}{\Phi} = \frac{1}{\mu}\frac{l}{A}$$

and we shall also use it for individual parts of a complete circuit.

Thus the reluctance of the pair of half-cores (without air-gap) is

$$\frac{1}{3 \times 10^{-3}\,\text{Wb A}^{-1}\text{m}^{-1}} \times \frac{200 \times 10^{-3}\,\text{m}}{400 \times 10^{-6}\,\text{m}^2}$$

$$= 1.7 \times 10^5\,\text{A Wb}^{-1}$$

Taking the permeability of air as being the same as that of a vacuum, $4\pi \times 10^{-7}\,\text{Wb A}^{-1}\text{m}^{-1}$, and assuming that the flux crosses the air-gap without spreading over a larger area of cross-section, we obtain, for the reluctance of the two air-gaps (total width 0.08 mm):

$$\frac{1}{4\pi \times 10^{-7}\,\text{Wb A}^{-1}\text{m}^{-1}} \times \frac{0.08 \times 10^{-3}\,\text{m}}{400 \times 10^{-6}\,\text{m}^2}$$

$$= 1.6 \times 10^5\,\text{A Wb}^{-1}$$

This adds to the reluctance of the half-cores making a total of $3.3 \times 10^5\,\text{A Wb}^{-1}$, just about double the original reluctance. This means that, for the same coil current, the flux in the core will be approximately halved by introducing the small air-gaps. Gaps of this magnitude can easily arise from ill-fitting, rusty, or damaged faces to the half-cores.

The problem should help you to appreciate why it was so important, in the experiments of Fig. 37.7, to avoid having 'butt joints' with an air-space between. Figure 37.9 should make this clear. The overlapping construction is used in transformer cores: transformers are described in Chapter 38.

Now, in answer to the question which we posed at the beginning of this section, about electromagnet design, you should work through Question 9.30 (see end of book) which applies the theory we have developed.

Finally, we use the theory to derive an expression for the inductance of a magnetic-alloy cored inductor. Suppose this consists of a coil of N turns wound upon a magnetic core, forming a closed magnetic circuit, of reluctance S. The flux in the core is given by

$$\Phi = \frac{\text{m.m.f.}}{\text{reluctance}} = \frac{NI}{S}$$

Assuming that the reluctance is constant, then if the current changes at a rate $\mathrm{d}I/\mathrm{d}t$, the flux will change at a rate given by

$$\frac{\mathrm{d}\Phi}{\mathrm{d}t} = \frac{N}{S}\frac{\mathrm{d}I}{\mathrm{d}t}$$

This changing flux will induce an e.m.f. in the coil E, where

$$E = -N\frac{\mathrm{d}\Phi}{\mathrm{d}t}$$

if we assume that all the flux links *every single turn* of the coil. Hence

$$E = -\frac{N^2}{S}\frac{\mathrm{d}I}{\mathrm{d}t}$$

But inductance, L, is defined by the relation (see Section 35.7):

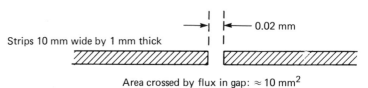

Strips 10 mm wide by 1 mm thick |←— 0.02 mm

Area crossed by flux in gap: $\approx 10\,\text{mm}^2$

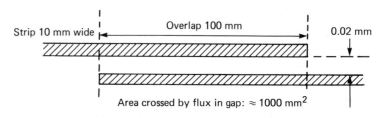

Strip 10 mm wide |←— Overlap 100 mm —→| 0.02 mm

Area crossed by flux in gap: $\approx 1000\,\text{mm}^2$

Fig. 37.9 Why an overlap joint makes better magnetic contact than a butt joint.

Reluctance of air-space reduced by about 100 times

$$E = -L \frac{\mathrm{d}I}{\mathrm{d}t}$$

therefore

$$L = \frac{N^2}{S} \qquad (37.5)$$

One type of inductor used frequently in school laboratories has a coil of 1100 turns which can be fitted on to a core of the type described in Problem 37.1. We estimated there that the reluctance of this core for small flux densities, was about 1.7×10^5 A Wb^{-1}. Substituting these values in Eq. 37.5, we get

$$L = \frac{1100^2}{1.7 \times 10^5} \approx 7 \text{ H}$$

Note the unit, the henry. Equation 37.5 shows that the unit for inductance, which we already know to be the henry, must also be the reciprocal of the unit for reluctance (N, the number of turns, has no units). Thus inductance could also be expressed in terms of the unit: Wb A^{-1}.

As shown in Problem 37.1 an air-gap of 0.08 mm total width would just about double the reluctance of the core. Since, as Eq. 37.5 indicates, the inductance is inversely proportional to the reluctance, this air-gap would approximately halve the inductance.

37.3 Formulae for flux density

We now come to a fundamental theoretical problem: how can we predict the magnetic flux density at points in the field of any conductor with a current flowing it it? In the previous section we have solved the problem for the field inside a long solenoid, but that is just one particular example of a current-carrying conductor: we now proceed to consider the magnetic fields of straight conductors and of flat coils.

A long straight wire

First we shall study the field around a very long straight wire with a current in it, taking up this topic from where we left it at the end of Section 34.2. In that section we established that the magnetic force between two parallel straight conductors of length separated by a distance a was given by the expression (34.1):

$$F = \left(2 \times 10^{-7} \text{N A}^{-2} \right) \frac{I_1 I_2 l}{a}$$

where I_1, I_2 are currents in the two conductors, expressed in amperes; l and a are expressed in metres, and the force is in newtons. Now, by the definition of magnetic field strength, (which we prefer to call *flux density*), the flux density at any point whose distance is a from the wire carrying the current I_1 is given by

$$B = \frac{F}{I_2 l}$$

and substituting for F, we obtain

$$B = \left(2 \times 10^{-7} \text{N A}^{-2} \right) \frac{I_1}{a} \qquad (37.6)$$

As you already know, the pattern made by the flux lines around a long straight wire is one of concentric circles; and furthermore the pattern made by flux lines around any conductor, or set of conductors, carrying current is a pattern of closed loops, although these loops are not, in general, circular in shape.

The French physicist Ampère established a very important law which applies to these flux patterns, whatever their shape, and whatever the configuration of conductors carrying the current. To avoid too mathematical a discussion, we need to go back a little into the history of electromagnetic theory. For more than a century after Ampère's time (the 1820s) the strength of a magnetic field was expressed, not as now, in terms of the force on a current-carrying conductor, but in terms of the force on an imaginary single magnetic pole. This idea has survived in the convention we use for defining the direction of magnetic field: this was originally defined as the direction of the force on a 'free' north-seeking magnetic pole. As you know, magnetic poles cannot exist singly in isolation, but only in pairs (N and S). However, it is possible to imagine an isolated pole and how it would perform in a magnetic field. Consider the field of a long straight wire, as shown in Fig. 37.10: the current is in a direction at right angles to the plane of the diagram. If a free N pole were released at point P the force on it would push it around the circle (we assume that this imaginary pole has no inertia, no mass, of course!). For every revolution of the pole

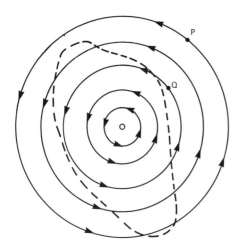

Fig. 37.10 The magnetic field of a long straight wire carrying a current perpendicular to the plane of the paper.

a certain amount of work would be done on it. If the N pole were released at Q the force on it would be greater, but it would not travel so far during each revolution. In all such cases:

work done on pole
     ∝ field strength × distance around circle

hence

$$\text{work done} \propto \frac{I}{a} \times 2\pi a$$

therefore

$$\text{work done} \propto I$$

Thus the work done is independent of the distance from the wire: it depends only upon the current 'encircled'. It can be proved mathematically that the result is exactly the same for an imaginary path, in the form of a loop, of *any* shape, surrounding the current; but the proof is beyond the scope of this book.

A solenoid

We can show that the same rule is true for the field inside a toroidal solenoid. The total flux inside the solenoid (see Section 37.2) is given by the expression

$$\Phi = \mu M \frac{A}{l}$$

The magnetomotive force M is the product of the number of turns in the solenoid, N, and the current, I. Assuming there is no material inside the toroid we can therefore write

$$\Phi = \mu_0 N I \frac{A}{l}$$

and so the mean flux density (or field strength, if we choose to call it so) is given by

$$B = \frac{\Phi}{A} = \frac{\mu_0 N I}{l} \qquad (37.7)$$

We say 'mean flux density' because the field is not strictly uniform: it is stronger close to the inner edge of the toroid because here the wires are closer together than at the outer edge. We can neglect this non-uniformity if we imagine the radius of the toroid to be large compared with the radius of the hollow tube itself: then the total length of the toroid, measured along its axis, is given by

$$l = 2\pi r$$

As before, we imagine a free north pole making one revolution in the field inside the solenoid, along its axis. Then, since work done on the pole is proportional to the field strength × distance around circle
we have:

$$\text{work done} \propto \mu_0 \frac{N I}{2\pi r} \times 2\pi r$$

$$\propto N I$$

The total current 'encircled' by the pole as it goes once round is NI, so once again it is true that the work done on the pole is proportional to the current 'encircled' by the pole as it makes one revolution.

We stated earlier, without any proof, that the rule is true for a looped path of any shape. We can give one instance of this by considering the field in part of a long straight solenoid. Figure 37.11

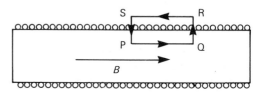

Fig. 37.11 The application of Ampère's circuital law to part of a long solenoid. PQRS is the imaginary path considered.

shows a section through the solenoid: we have 'sliced through' the wires and the current is coming towards us along the wires on the upper side. The field inside the solenoid is thus towards the right. We imagine moving a free N pole along the path PQRS which is partly inside the solenoid and partly outside. From P to Q the force on the pole is along the direction PQ, so the work done in moving it from P to Q is proportional to Bd. Going from Q to R the pole is first of all travelling in a direction perpendicular to the field, so no work is done on it by the field. When the pole is outside the solenoid, going along the upper part of QR, the field is zero, so again no work is done. From R to S there is no field, so no work is done. From S to P the same arguments apply as to the section QR, and no work is done. Thus the total work is simply $\propto Bd$. Now, if as before N is the number of turns in the solenoid whose length is l,

$$B = \mu_0 \frac{N}{l} I$$

and so the work done

$$\propto m_0 N \frac{d}{l} I$$

but $N(d/l)$ is the number of turns in a distance d along the solenoid, and so once again we have the result:

work done $\propto$ current 'encircled' by path of pole

By now we hope that you are convinced of the general truth of this rule. Finally, we can make a statement of it, in modern terms, but avoiding advanced mathematical ideas, and avoiding use of the concept of a free magnetic pole.

Ampère's circuital law

Imagine a closed path of any shape in a magnetic field in a vacuum, such as the broken line in Fig. 37.10. Imagine this path divided up into infinitesimally small portions, each of length ds. For each portion consider the product:

(component of B-field parallel to ds) × ds

The sum of all such products equals:
μ_0 (net current enclosed by path).

You should check back for yourself at this point that the law is correct for the toroidal solenoid for which B is given by Eq. 37.7.

Ampère's law and a long straight wire

Consider once again the expression we obtained for the flux density at a distance a from a long straight wire carrying current I:

$$B = \left(2 \times 10^{-7} \text{N A}^{-2} \right) \frac{I}{a}$$

We apply Ampère's circuital law to this, for a circular path of radius a around the wire. Because B is parallel to the path at every point the sum of products:

(component of B-field parallel to ds) × ds

is simply

$$B \times \text{(length of path)}$$

which is

$$B \times 2\pi a \qquad \text{and this must equal } \mu_0 I$$

therefore

$$B = \frac{\mu_0}{2\pi} \frac{I}{a} \qquad (37.8)$$

comparing this with the earlier expression, a few lines above, we see that

$$\mu_0 = 4\pi \times 10^{-7} \text{N A}^{-2}$$

Thus the numerical value of the constant μ_0 is determined by the number '2×10^{-7}' which appears in the definition of the unit of current, the ampere.

The symbol μ, without a suffix, is used for the permeability of any medium other than a vacuum, and it is often convenient in practice to use the concept of *relative permeability* for magnetic materials. This is defined as follows:

$$\text{relative permeability of material} = \frac{\text{absolute permeability of material}}{\text{absolute permeability of free space}}$$

By 'absolute permeability' we mean the permeability expressed in its proper units. These proper units can be N A^{-2}, as used above, or other alternatives: see Question 9.37 at end of this book.

A circular coil

The final example of a current-carrying conductor and its magnetic field for us to study is a plane, circular coil (Fig. 37.12). This is more difficult to analyse theoretically than the previous examples (solenoid, and straight wire), but it is important because in many experiments a magnetic field is provided by one or more ring-shaped coils.

Using the search coil technique we can discover what variable factors determine the flux density at the centre of a plane circular coil. The coil can be made from thin, flexible insulated wire wound around a former made of non-magnetic pegs stuck into a large piece of pegboard. The number of turns, N, used to make the coil, and the radius of the coil, r, can be varied, as well as the current in the coil. Altering only one of the variables at a time we find that:

$$B \propto N \qquad B \propto \frac{1}{r} \qquad B \propto I$$

These three expressions can be combined into one:

$$B \propto \frac{NI}{r}$$

To determine theoretically the constant of proportionality in this expression is the awkward bit. It cannot be done by applying Ampère's circuital law, because the magnetic flux pattern, as shown in Fig. 37.12, is geometrically complex, and it is not possible to find an imaginary path enclosing the current for which the flux density is constant, as we were able to do for a solenoid and a straight wire.

In order to avoid advanced mathematics we adopt an experimental approach to the problem of determining the constant in the expression for the flux density at the centre of a plane circular coil. Figure 37.13 shows a long solenoid made of plane

Fig. 37.12 The magnetic field pattern of a current in a plane circular coil. (From D.S. Heath *et al.*, PSSC Physics (1965). Reproduced by permission of the publisher.)

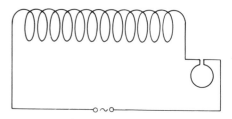

Low voltage supply, frequency about 5 kHz

Fig. 37.13 A long solenoid made of plane circular coils of radius r, separated by distance $r/2$. The flux density near the middle is four times the flux density at the centre of an isolated plane coil of a single turn having the same radius and carrying the same current.

circular coils, of radius r, which are separated by a distance $r/2$. Although this is a very 'open-wound' solenoid, the flux density on its axis is very nearly constant, as the use of a search coil will show. Near the middle of the solenoid, where, as for any other long solenoid, the flux density approximates closely to that of an infinite solenoid having the same number of turns per unit length, the mean flux density is found to be exactly *four times* the flux density at the centre of a single coil of radius r with the same current flowing in it. This 4:1 ratio is easily demonstrated by means of a search coil connected to an oscilloscope. Now, the flux density in a long solenoid is

$$\mu_0 \frac{NI}{l}$$

where N is the number of turns and l the length. If we call the flux density at the centre of the single plane coil B, then

$$4B = \mu_0 \frac{NI}{l}$$

But since the turns in the solenoid are spaced by a distance $r/2$,

$$N\frac{r}{2} = l$$

and so

$$N = \frac{2l}{r}$$

and substituting this in the expression above, we get

$$B = \frac{\mu_0}{2}\frac{I}{r}$$

For a flat coil of N turns

$$\boxed{B = \frac{\mu_0}{2}\frac{NI}{r}} \qquad (37.9)$$

Chapter 38

SOME APPLICATIONS OF ELECTROMAGNETIC THEORY

38.1 Transformers

All national, large-scale electrical supply systems use alternating current. The chief reason for this is one of economics. It is cheaper to send electrical power over large distances, by cables (usually overhead), if very high voltages are used. This was mentioned in Section 24.3, with a numerical problem (24.3) which showed the very great reduction of power loss (as heat in the cables) achieved when a high voltage is used. The British system uses voltages of 132 kV, 275 kV, and 400 kV. For obvious reasons these voltages would not be suitable for the mains supply to your house. In practice, *transformers* are used to 'step down' the voltage by stages to 240 V r.m.s. for ordinary domestic supplies. Also, at the other end of the system – at the power generating station – 'step up' transformers are used to increase the voltage of the supply, from the generators, from 25 kV to 132 kV, 275 kV, and 400 kV.

Transformers are used in many domestic appliances. In the audio amplifier of a mains-operated cassette recorder, for instance, a step-down transformer is used to reduce the voltage from 240 V r.m.s. to about 10 V r.m.s.: this supply is then rectified and smoothed (see Section 36.5) to give a steady d.c. supply at, typically, 9 V. On the other hand, in a colour television receiver, a step-up transformer is needed to provide the very high voltage (about 20 kV) needed to accelerate the electrons in the picture tube: again, the supply from the transformer is rectified and smoothed.

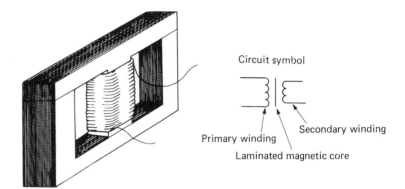

Fig. 38.1 A common type of transformer, with only a few turns of one winding shown. The magnetic alloy core is made up of interleaved 'T'-shaped and 'E'-shaped laminations. The circuit symbol for a transformer is shown inset.

A transformer works by electromagnetic induction. The very first transformer ever made was probably Faraday's iron ring apparatus which we saw in Section 35.4 (Fig. 35.10).

The two coils, or windings, of a simple transformer are generally wound concentrically on a common core made of soft magnetic alloy. For reasons which will be explained later this core is usually built up of thin layers of metal, called *laminations*, which are electrically insulated from each other by very thin layers of paper. Figure 38.1 shows part of a common type of transformer and also the conventional circuit-symbol for a transformer. The form of the magnetic core is such that it provides two closed paths for the magnetic flux produced by currents in the windings: in a well-designed transformer there is negligible leakage of magnetic flux outside the core. Having the two windings concentric upon a common core ensures good flux linkage between the two windings, that is to say, nearly all the flux produced by current in the primary winding threads through, or links, the secondary winding.

Before attempting to analyse the action of a transformer theoretically we will look at the results of some simple experiments with transformers. For the first experiment we can use a 'do-it-yourself' transformer in which we make our own secondary winding. Figure 38.2 shows two kinds of design suitable for this purpose. In each of these the primary winding has a fixed number of turns and can be connected directly to the 250 V a.c. mains. We take a length of insulated copper wire and wind first one, then two, then three, and more, turns around the transformer core. For each number of turns used we connect the ends of the wire to a suitable low-voltage lamp with an a.c. voltmeter connected in parallel with it to measure the secondary voltage of the transformer. The arrangement is thus as shown in Fig. 38.3.

When we plot a graph of secondary voltage against N_s, the number of turns in the secondary winding, we find that it is a straight line. Thus, in principle, it is a simple matter to find how many turns we need to obtain any other chosen value of secondary voltage. For the primary winding the

Fig. 38.2 Examples of laboratory transformers. The one on the left is a *toroidal* transformer. That on the right has a removable 'yoke' as the top part of its magnetic core, making it easy to interchange the coils. See also Fig. 35.11.

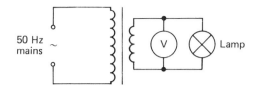

Fig. 38.3 Circuit for investigating the effect of changing the numbers of turns on the secondary coil.

Table 38a

Primary voltage V_p/V	Primary current I_p	Secondary voltage V_s/V	Secondary current I_s
250	0.100	13.0	0
250	0.175	12.9	1.76
250	0.250	12.7	3.30
250	0.388	12.5	6.20
250	0.490	12.4	8.20

number of turns is determined by the primary voltage and, for a transformer with good flux-linkage between the two windings, the relationship

$$\frac{\text{primary voltage}}{\text{secondary voltage}} = \frac{\text{number of primary turns}}{\text{number of secondary turns}}$$

is very nearly obeyed. Thus, by choosing suitable numbers of turns for the two windings we can convert the voltage of an alternating supply into a higher voltage (as in a step-up transformer) or into a lower voltage (step-down transformer).

Now we take a closer look at the voltages and currents in a transformer when it is working. The circuit of Fig. 38.4 is used to apply different loads to a transformer, and to measure voltage and current in primary and secondary.

The results of such an experiment are shown in Table 38a. The first set of readings was taken with the secondary on open circuit (that is, with nothing at all connected to it apart from the voltmeter which we can assume takes negligible current). The different loads were simply different numbers of 12 V lamps connected in parallel. Also, the resistance of the primary and secondary windings are measured with an ohm-meter. The results were: primary 9.0 Ω, secondary 0.2 Ω.

Table 38a needs to be studied carefully. Consider, first, the situation when there is no load. The primary current is 0.100 A. (This is shown not to be due to the presence of the voltmeter connected to the secondary simply by temporarily disconnecting that voltmeter.) The resistance of the primary winding was found to be

9.0 Ω, so if we apply the relation $V = IR$ to the primary, we should get, for the primary current:

$$I_p = \frac{250}{9.0} = 28 \text{ A}$$

approximately. But we found the primary current to be 0.100 A. Why the huge discrepancy? Obviously, because the primary has appreciable inductance (in fact, the inductance is approximately 8 H as you can verify, if you wish, from the data). The inductive reactance (see Section 36.2) is very much greater than the resistance, and so the phasor diagram for the primary winding, relating the applied voltage with the p.d. across the inductive part and the p.d. across the resistive part would be as shown in Fig. 38.5. The phase-angle ϕ, which, of course, is the phase-difference between the applied p.d. and the current, is very nearly $\pi/2$ radians (90°), and the applied p.d. very nearly equal in magnitude to the p.d. across the inductive part of the winding.

Thus the primary behaves very nearly as a pure inductance, and, as explained in Section 36.2, no power is dissipated in a pure inductance. But, as we see from the table above, when a load is connected to the secondary and a current is drawn from it, the primary current increases in magnitude. Since power is now being fed from the secondary to the load, power must be going from the mains into the primary (if the principle of conservation of energy applies, as obviously it must). Let us consider the case in which a purely resistive load is connected to the secondary. The

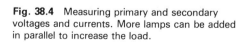

Fig. 38.4 Measuring primary and secondary voltages and currents. More lamps can be added in parallel to increase the load.

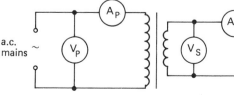

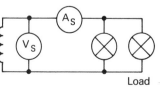

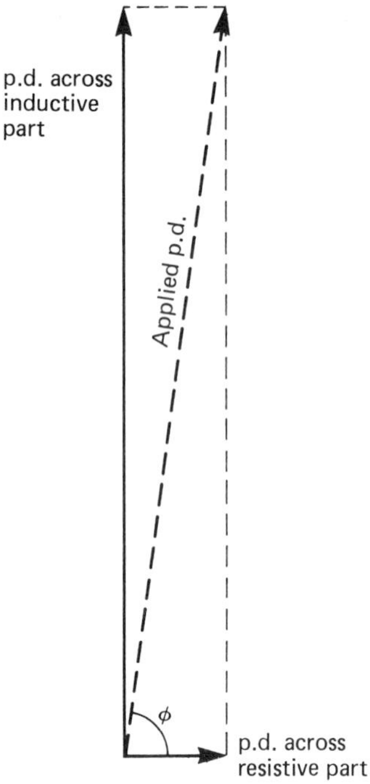

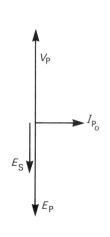

Fig. **38.5** Phasor diagram for the primary of a transformer.

Fig. 38.6 Phasor diagram for the secondary of a transformer (no load).

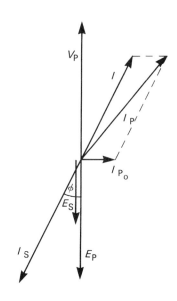

Fig. 38.7 Phasor diagram for a transformer supplying a load.

secondary current flowing will be in phase with the secondary voltage. This current creates a magnetic flux which tends to reduce the total flux in the core due to the primary current, and if this flux is reduced then the back e.m.f. in the primary is reduced, and if so then the primary current increases. To enable you to understand this better it is helpful to build up a phasor diagram which includes voltages and currents. Figure 38.6 shows the situation where there is no load connected to the secondary. Assuming that the primary has negligible resistance, then the back e.m.f. E_p in the primary is equal and opposite to the applied voltage, V_p. The primary current in this condition is I_{p_0}, lagging behind V_p with a phase-angle of $\pi/2$ radians. The e.m.f. induced in the secondary is E_s, and has the same phase as the back e.m.f. in the primary (because they are both e.m.f.s induced in coils by the same changing magnetic flux). Since there is no load connected to the secondary, no secondary current flows, and therefore the figure is a complete representation of the situation. But when a load is connected to the secondary, current flows. We call this current I_s. Since, even with a purely resistive load, the secondary circuit is partly inductive (simply because the secondary winding has self-inductance), the secondary current will not be in phase with the secondary voltage, but will lag behind it, with a phase-angle ϕ, as shown in Fig. 38.7. As has been said already, this current tends to reduce the magnetic flux in the core, and thus the primary current increases by an amount I. I is the current needed to produce a magnetic flux equal to the reverse flux produced by I_s. The current drawn from the supply is I_p, the vector sum of I and I_{p_0}.

In order to calculate correctly the power input to the primary and the power output from the secondary one has to take into account the phase-relationships of V_p, I_p, etc. To do this in detail is beyond the scope of this book, but these facts explain why you cannot simply take the values of voltage and current in Table 38a (which are r.m.s. values, of course) and multiply V_p by I_p to get the power input, and V_s by I_s to get the power output. As a rough rule, however, when a transformer is working with the load for which it is designed, one can use the following approximate relation:

$$V_p I_p \approx V_s I_s$$

$$\boxed{\therefore \quad \frac{I_p}{I_s} \approx \frac{V_s}{V_p} \approx \frac{N_s}{N_p}} \qquad (38.1)$$

where N_s and N_p are the numbers of turns in the primary and secondary windings, respectively.

As has been already explained, if more current is taken from the secondary, the primary current immediately increases. Presumably, there must be a limit to the strength of secondary current which can be taken. It can be seen in Table 38a that the effect of taking more current from the secondary is that the secondary voltage decreases slightly. This happens because the secondary winding has resistance, and of course heat is dissipated in this resistance. Likewise, some heat is dissipated in the primary when any current is flowing. This dissipation of heat in the windings is what electrical engineers often refer to as *copper losses* and these determine the maximum current which can be safely drawn from a transformer.

The laminated form of the transformer's magnetic core, mentioned earlier, is designed to minimize eddy-current losses. If the core were made of a single solid piece of magnetic alloy the current induced in it by the changing magnetic flux would be sufficient to cause excessive heating of the metal. The paths of these eddy currents are in planes perpendicular to the magnetic flux so that if the laminations (which, remember, are insulated from each other) are parallel to the magnetic flux, the eddy-current paths are interrupted by the layers of insulator. It is found that the rate of energy loss due to eddy-currents is roughly proportional to the square of the lamination thickness, and it is easy in practice to use laminations thin enough to reduce eddy-current loss to a negligible amount. Although this method of transformer core construction is adequate at low frequencies (especially the frequency of the a.c. mains, 50 Hz), at higher frequencies, in the megahertz region, for instance, magnetic alloys are not suitable and ferrites are used instead nowadays. Ferrites are ceramic materials, not metals, and have extremely high resistivities so that they can be classed as insulating materials, and eddy-current loss in them can be negligibly small at high frequencies.

At very high frequencies, around 100 MHz and above, it is often not necessary for a transformer to have a magnetic core at all. In such air-cored transformers the magnetic flux linkage between a pair of coils wound concentrically, one inside the other, can often be quite enough for efficient power-transfer from primary to secondary.

38.2 Eddy currents

We have already discussed the problem of eddy currents in the magnetic cores of transformers, and how they can cause power loss through heating. Now we shall investigate some of the useful applications of this class of induced currents.

The heating effect of strong eddy currents is used in several industrial processes where a piece of metal has to be heated in a vacuum or in an atmosphere of inert gases, in order to join it to another piece by the process of *brazing* (like soldering but at a higher temperature). Figure 38.8 shows a schematic diagram of a high frequency *induction furnace*. The large coil is fed with alternating current at a very high frequency, typically around 50 kHz. The alternating magnetic flux at the centre of the coil induces currents in the workpiece strong enough to heat it to bright red heat and melt the brazing metal which flows into the joint. This technique can obviously only be used for heating materials which have an electrical resistivity low enough to allow large induced currents to flow.

Electromagnetic damping and *braking* are examples of Lenz's law (see Section 18.7) in action. Figure 38.9 shows apparatus used for demonstrating these effects. The flat pendulum (a)

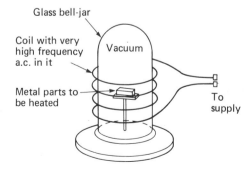

Fig. 38.8 Schematic diagram to show the principle of induction heating.

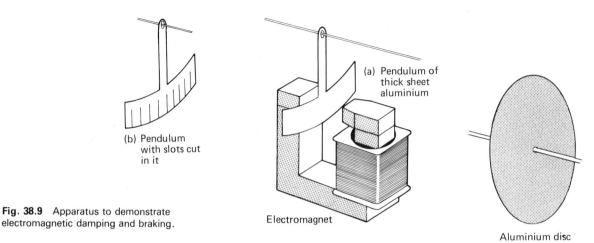

Fig. 38.9 Apparatus to demonstrate electromagnetic damping and braking.

(b) Pendulum with slots cut in it

(a) Pendulum of thick sheet aluminium

Electromagnet

Aluminium disc

make of thick sheet aluminium is set up so that it can swing freely in its own vertical plane. It is set swinging between the poles of a large electromagnet, as shown, first of all with the electromagnet switched off. When the electromagnet is switched on the oscillation of the pendulum is strongly damped and it soon comes to rest. The interpretation is as follows: when the metal is moving in relation to the magnetic field, currents are induced in it, and the force of interaction between these currents and the field opposes the motion (Lenz's law).

If pendulum (b) is now substituted for (a) we find that the damping is very much less. The fine cuts in the aluminium reduce the effect of the eddy-current considerably, just as *laminating* the core of a transformer does: this was explained in the previous section.

The aluminium disc in Fig. 38.9 can be set to spin in a vertical plane, with its edge between the poles of the electromagnet. If it spins freely with the electromagnet switched off, then it will quickly be stopped when the electromagnet is switched on. We have here a situation similar to the Faraday disc dynamo, as described in Section 35.4. When the disc is spinning an e.m.f. is induced which acts in a radial direction, in the region of magnetic field, just as it did in the disc dynamo. But in Fig. 38.9 only a small part of the disc is in a region of magnetic field. Suppose that, in this part, the induced e.m.f. is driving current radially from the rim towards the centre: this current must find a return path through other parts of the disc which are *not* in the magnetic field. The force which

opposes motion is the result of interaction of the induced current in the magnetic field region with the magnetic field of the electromagnet.

Electromagnetic braking has many applications in mechanical engineering. It makes possible an electrically controlled brake in which there is no physical contact, no friction, and therefore no wear of material. Another advantage is that the braking force or torque is exactly proportional to the speed of the moving part.

In all the demonstrations described above the force produced opposes the motion. Suppose, with the pendulum (a) of Fig. 38.9, we start with the pendulum at rest but move the electromagnet (switched on already) in a direction parallel to the plane of the pendulum. We find that the pendulum moves with the magnet a little: the magnet is 'trying to' drag the metal along with it. This dragging effect can be studied further using the apparatus of Fig. 38.10 which also shows the principle of a car *speedometer* of traditional type. A turntable with very free-running bearing has a thick aluminium disc fixed to it. Just above the disc and very close to it is another turntable, fixed upside down, with several strong, flat magnets stuck on its surface. When the upper disc is spun we find that the lower one begins to rotate in the same direction as the upper one. The force of interaction between the eddy currents induced in the aluminium disc and the magnets on the upper disc opposes the *relative motion* of the two, so the aluminium disc 'tries to' follow the magnets. To get some indication of the torque produced by this force we wrap a thin elastic cord around the

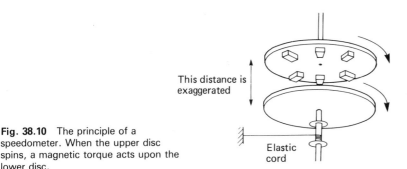

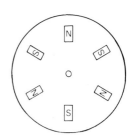

Fig. 38.10 The principle of a speedometer. When the upper disc spins, a magnetic torque acts upon the lower disc.

Elastic cord

How the magnets are arranged on the upper disc

spindle of the lower turntable, with the outer end of the cord tied to a fixed support. The upper disc is spun at a steady speed, being driven by an electric motor. The faster the upper disc spins, the greater is the torque on the lower one. To prove that the effect is not due to moving air between the two discs, a sheet of cardboard is interposed between the two.

In one type of speedometer used in cars a flexible drive cable links the propeller shaft to a flat magnet mounted on a spindle in the instrument. This magnet spins very close to an aluminium disc mounted on another spindle and restrained by a fine spiral hair-spring. The pointer is attached to this disc. The magnetic torque acting upon the disc is directly proportional to the speed of the spinning magnet, and so the angular deflection of the pointer is proportional to the speed. (We assume that the hair-spring obeys Hooke's law, giving an opposing torque which is proportional to the angular deflection.)

These demonstrations lead us to study how the force of interaction between an induced current and an *alternating* magnetic field can be employed in various kinds of electric motor which operate from an a.c. supply.

38.3   Induction machines

We begin with an experiment which shows the principle of an *induction motor*. Figure 38.11 shows a pair of small electromagnets, each made from a single C-core (of transformer alloy) and a coil of about 200 turns. A large capacitance is connected in series with one of the coils: this is made by connecting two electrolytic capacitors 'back-to-back' so that the alternating voltage across them will not damage them (remember that electrolytic capacitors are designed to be used with the correct polarity). The purpose of the capacitor is to alter the *phase* of the a.c. in one of the electromagnet coils, and hence to alter the phase of the alternating magnetic field of this electromagnet. This phase-difference is demonstrated by means of a double-beam c.r.o. and two search coils, each search coil being connected the same

Fig. 38.11 The principle of the induction motor. The circuit diagram shows how the two electromagnet coils are connected to the a.c. supply.

Aluminium disc

Rotation

5000 μF

5000 μF

A

B

240 turns

240 turns

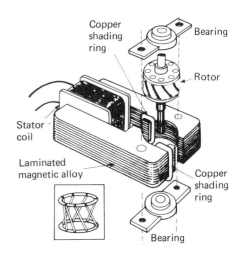

Fig. 38.12 A simple shaded-pole induction motor. The rotor is made of copper rods in a 'squirrel-cage' pattern (inset) which is embedded in solid magnetic alloy.

way round to each of the two inputs to the c.r.o. The two electromagnets are set side by side, one pole of each being very close to the disc, but not touching it. When the supply is switched on the disc begins to rotate. Use of the search coils and c.r.o. shows that the alternating field above magnet B reaches a maximum slightly *later* than the field above magnet A; and we observe that the edge of the disc above them moves in the direction A to B. Connecting the capacitor in series with B instead of A causes the rotation to reverse. It is as if the aluminium dics were being 'tricked into believing' that we were actually moving a magnet beneath the disc in the direction A to B. A theoretical analysis of the effect is beyond the scope of this book, but at least you should be able to appreciate that it is possible to set up an alternating magnetic field whose maximum moves from one place to another and so behaves as if a permanent magnet were present and actually moving along.

Instead of using the large capacitors to provide a phase-difference between the alternating magnetic fields we can use the *shaded pole* principle. Although, with the apparatus of Fig. 38.11, this will not produce such a great torque as the previous method, it should give enough to turn the disc, provided there is very little friction in the axle bearings of the disc. 'Shading' one pole of one of the electromagnet is simply done by putting a thick copper ring around it, or by placing a small piece of thick copper or aluminium sheet above

the pole. The torque on the disc is found to be in the direction from the unshaded to the shaded pole; and using the pair of search coils and double-beam c.r.o., as before, one can show that the alternating magnetic field above the shaded pole reaches its maximum slightly later than the field above the unshaded pole. Thus the shading ring (or plate) *retards* the phase of the field. In the capacitor method, the presence of the capacitors in series with electromagnet A *advanced* the phase of its field.

In the foregoing experiments the torque on the disc was very weak. How can the principle be used in such a way as to produce enough torque to do a useful job? In fact there is one important application of the simple disc motor, and that is in the ordinary domestic *kilowatt hour meter* which records how much electrical energy the household has used from the a.c. mains. There is not space here to explain how it works except to say that the speed at which the disc rotates is proportional to the product of the r.m.s. voltage, the r.m.s. current, and to the cosine of the phase-angle between them. This product, as mentioned at the end of Section 36.3, is the *power* used in an a.c. circuit. The torque driving the disc has only to be enough to turn the small wheels which display the numbers of units.

Figure 38.12 shows a type of shaded-pole induction motor often used in record players, fans, and other low-power domestic appliances. Notice that there are two shading rings, one on

Fig. 38.13 A 500 MW turbo-generator in the turbine hall at the coal-fired 1000 MW Ironbridge B power station in Shropshire. (Courtesy C.E.G.B.)

one half of each pole of the electromagnet. The rotor is made of thick copper sections joined in a 'squirrel-cage' construction as shown in the lower part of the diagram. This structure is embedded in magnetic alloy to form a solid cylinder. The air-gap between rotor and stator is generally less than a millimetre in order to reduce the reluctance of the magnetic circuit (stator–rotor–air-gap) to a minimum so that the greatest possible magnetic flux is produced and hence the greatest possible torque. The construction is simple, robust, and relatively cheap.

More complicated designs are needed for a.c. motors which provide greater power than the type we have discussed, and in factories large machinery is driven by a.c. motors which operate from a *three-phase* supply. But the same fundamental principle applies: an alternating magnetic field is set up in which the position of maximum field strength moves along, from one pole-face of the stator to the next. In a *linear induction motor* the stator pole-faces lie in a plane, rather than around a cylinder as in all rotary machines.

38.4   The generation of a.c. power

The industrial generator or *alternator* is usually driven by a steam turbine (see Fig. 38.13). The steam is raised in a conventional boiler (usually coal-burning) or in a nuclear 'boiler' (see Section 49.1). In Britain the turbine shaft usually rotates at 3000 revolutions per minute (50 Hz).

Within the alternator, a two-pole d.c. electromagnet or rotor is carried on the shaft so that it rotates between three symmetrically arranged pairs of coils (stator). (See Figs. 38.14 and 38.15).

Fig. 38.14 A stator assembly for the Littlebrook D power station prior to the threading of the rotor. (Courtesy C.E.G.B.)

Fig. 38.15 Checking the clearance as the rotor is installed in the 500 MW unit at the oil-fired 2000MW Pembroke power station in South Wales.

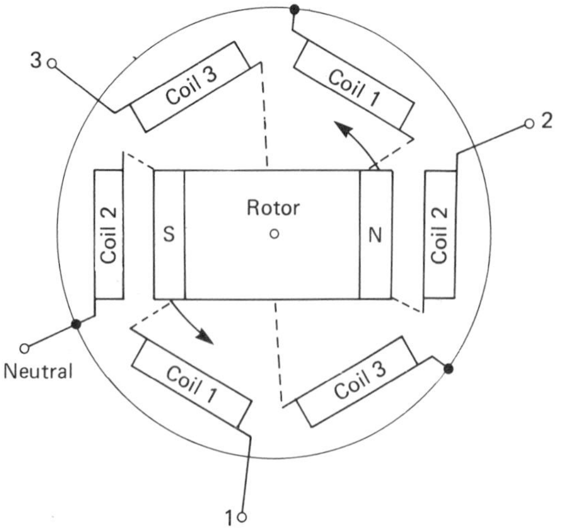

Fig. 38.16 Schematic diagram to show how the bi-polar rotor rotates between the three pairs of stator windings. The three separate output voltages appear at the terminals 1, 2 and 3; the common neutral wire is also shown.

The rotating field induces an alternating e.m.f. in each pair of stator coils in turn. Consequently the phases of these three e.m.f.s are separated by 120° or one-third of a cycle. The r.m.s. value of each of the three e.m.f.s is about 25 kV.

Figure 38.16 shows how the e.m.f.s occur in each of the three stator windings – the maximum in winding 1 leading that in winding 2 by 120° and leading that in winding 3 by 240°.

The three stator coils share a common (neutral) wire, marked N in Fig. 38.16. The three separate phases appear at the terminals marked 1, 2 and 3. These three output voltages are fed to a suitable three-phase, step-up transformer which raises the output voltage to that required for the grid system (132 kV, 275 kV or 400 kV). (See Fig. 38.17)

Only four wires are required for the distribution of the energy through the grid; one for each of the three phases (the line) and the common return (neutral) (see Fig. 38.18). This confers a

Fig. 38.17 A 132/275 kV Supergrid transformer at a substation near Stourport-on-Severn, Worcestershire. (Courtesy C.E.G.B.)

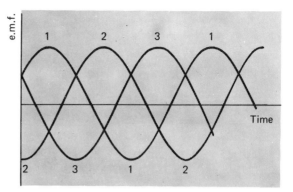

Fig. 38.19   Variation of the e.m.f. of each of the three phases of a three-phase generator with time.

achieved, the current in the common return wire is zero – as can be seen either from Figs. 38.19 and 38.20 which shows the phasors (rotating vectors).

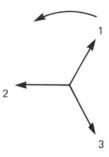

Fig. 38.20   Phasor diagram for a three-phase supply.

Fig. 38.18   Suspension towers carrying two 400 kV three-phase transmission lines from the nuclear station at Sizewell, Suffolk. (Courtesy C.E.G.B.)

great economic advantage especially as the neutral wire is of light construction and does not need to be insulated.

This arises because every effort is made to ensure that the loads on each of the three phases are very nearly the same. If exact balance is

In practice, exact balance is rarely achieved; but a sufficiently close approximation can be attained. In the case of domestic supplies, which use one of the phases only, houses are grouped together and each group receives its energy from one of the three phases.

Chapter 39

ELECTROMAGNETIC WAVES IN SPACE

39.1 The propagation of *E* and *B* fields in space

In Unit 5 we saw that the behaviour which characterizes the transmission of energy by waves includes the phenomena of diffraction, interference, polarization, refraction and reflection. We found too that light and the other members of the family of electromagnetic waves travel in free space with a speed of $3 \times 10^8 \, \mathrm{m \, s^{-1}}$.

Our present concern is with the make-up of these electromagnetic waves and how it is that they share this speed. We shall first consider a problem which arose about a century and a half ago. This was the question of the speed with which electrical signals are propagated in very long cables. Modern equipment makes the measurement of this speed a simple one.

If one takes a 200 m length of co-axial cable (that is a cable in which an inner conductor is sheathed in a dielectric which is itself surrounded with an outer conductor, often of braided copper, the whole being coated with a protective layer of a plastic) one can send a train of pulses into one end and receive them again at the other. Figure 39.1 shows the circuit which one might use.

The double-beam c.r.o. displays the signal entering the cable in one beam and the signal received from the cable on the other beam. The signal consists of a train of pulses from a 200 kHz pulse generator. The time base of the c.r.o. must be set at a suitable high speed; about 1 cm per μs.

With such a cable the time delay between the entry of a pulse into the 'near' end and its arrival at the 'remote' end is in the order of $1 \, \mu$s which corresponds to a pulse speed in the cable of some $2 \times 10^8 \, \mathrm{m \, s^{-1}}$.

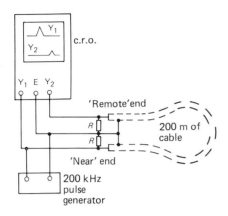

Fig. 39.1 Circuit used to measure the speed of a pulse along a length of coaxial cable.

The two resistors shown, each of about 70 Ω and of equal impedance to the cable itself, serve to prevent the reflection of the pulse from the ends of the cable. Without them, each pulse would arrive at the remote end, be reflected back to the near end, reflected back again, and so on. Since the cable has an impedance of about 70 Ω, the resistors will have just the right value to absorb these reflected pulses (see Section 24.4).

We note that such a cable is a form of capacitor: two electrodes separated by a dielectric. And to get some understanding of what is going on in the cable we shall consider an extended parallel plate capacitor; that is, one which is very long in comparison with its width.

Although it is not quite so obvious, the plates also have some inductance. As the charge is being established, the charging current produces magnetic flux and the flux linkage gives rise to inductance (alongside the capacitance). It is this dual role of the extended system, in which it has capacitance and inductance at the same time, that holds the key to understanding its behaviour.

Figure 39.2a shows the parallel plate capacitor to be considered. The width of the plates is *a* and the separation is *b*. The plates can be connected to a battery by means of a switch.

At time $t = 0$, the switch is closed and after a time *t* charge will have spread along the two plates as shown in Fig. 39.2b. The upper plate will have received negative charge and the lower will have lost negative charge. These charges will spread along the two plates from left to right with speed *v* as current *I* flows in the circuit. We shall assume

that the advancing edge of the sheet of charge is sharp.

At time *t* the front edges of the charge sheets will have reached a distance *vt* from the left-hand end.

After time *t* a negative charge $Q = -It$ will have spread on to the upper plate and an equal positive charge on to the lower plate.

The resulting charge density (the charge per unit area, ϱ) will be

$$\frac{Q}{a\,vt} = \frac{It}{a\,vt} = \frac{I}{a\,v} \tag{39.1}$$

One consequence of this flow of charge is that an electric field *E* is established within the volume *ab vt* (Fig. 39.2b).

If we combine Eq. 26.1

$$C = Q/V$$

with Eq. 26.2

$$C = \epsilon_0 \frac{A}{d}$$

for a parallel-plate capacitor, we find that

$$\frac{V}{d} = \frac{Q}{A} \times \frac{1}{\epsilon_0}$$

where *V/d* is the field intensity, *E*, and *Q/A* is the charge density, ϱ. So we have

$$E = \frac{\varrho}{\epsilon_0}$$

In the volume *ab vt* the field *E*

$$= \frac{1}{\epsilon_0} \frac{I}{a\,v}$$

$$\therefore \frac{I}{a} = \epsilon_0 Ev \tag{39.2}$$

As the charge sweeps along from left to right it also establishes a magnetic field *B* within the volume *ab vt*.

The pair of plates can be thought of as a flat, elongated solenoid with a single turn (Figs. 39.2c and d).

For a solenoid (from Eq. 37.7) we have flux density

$$B = \mu_0 \frac{NI}{a}$$

Since, in this case, $N = 1$, the flux density

$$B = \mu_0 \frac{I}{a}$$

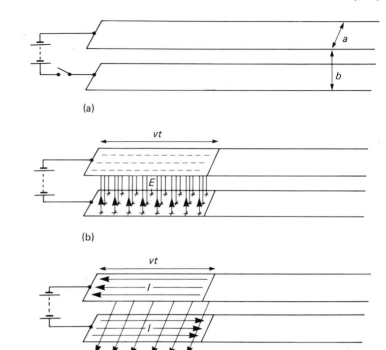

Fig. 39.2 (a) An extended (very long) parallel plate capacitor. When the circuit is completed charge flows on to the plates and (b) an electric field E is established and (c) a magnetic field B develops. (d) Shows a flat elongated single-turn solenoid which is equivalent to the system shown in (c).

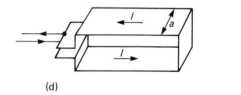

Substituting for I/a from Eq. 39.2, we have

$$B = \mu_0 \epsilon_0 Ev \qquad (39.3)$$

The two fields are extending themselves into an increasing volume of space between the plates.

In time t, the flux in this new volume is (flux density) × (the area normal to the direction of the flux) = $B b vt$

From the laws of a electromagnetic induction (see Chapter 35) we deduce that to keep a magnetic flux Φ increasing we must maintain an applied voltage which is given by $V = d\Phi/dt$, $d\Phi/dt$ being the rate of a change of flux.

In this case we have

$$V = \frac{d\Phi}{dt} = B b v \qquad (39.4)$$

The existence of such a potential difference between the charged regions of the plates neces-

sarily implies the existence of an electric field such that

$$E = \frac{V}{b} \qquad (39.5)$$

Combining the two equations 39.4 and 39.5, gives

$$E = \frac{B b v}{b} = Bv \qquad (39.6)$$

From Eq. 39.3 above

$$B = \mu_0 \epsilon_0 Ev$$

Combining these last two equations we see that

$$v^2 = \frac{1}{\mu_0 \epsilon_0}$$

so that

$$v = \frac{1}{\sqrt{\mu_0 \epsilon_0}} \qquad (39.7)$$

The speed with which the two fields are propagated in the space between the plates is related to two fundamental constants of free space – its permeability (μ_0) and its permittivity (ϵ_0).

How large is this speed? By definition (see Section 37.3)

$$\mu_0 = 4\pi \times 10^{-7}\,\mathrm{N\,A^{-2}}$$

and as we have seen in Section 26.3

$$\epsilon_0 = \frac{1}{36\pi} \times 10^{-9}\,\mathrm{F\,m^{-1}}$$

Hence,

$$\mu_0\epsilon_0 = \tfrac{1}{9} \times 10^{-16}\,\mathrm{m\,s^{-1}}$$

So

$$v = 3 \times 10^8\,\mathrm{m\,s^{-1}}$$

This we recognize as the speed of light.

Before seeking some experimental support for this, we will make one further check on our analysis which, it must be admitted is founded on reasonable assumptions rather than on exact deductions. We have seen that the unit in which μ_0 (the permeability of a vacuum) is measured is the newton ampere^{-2} (see Section 37.3). We have also seen that the unit in which ϵ_0 (the permittivity of a vacuum) is measured is the joule^{-1} coulomb^{-2} metre^{-1}. This latter unit is equivalent to the coulomb^2 newton^{-1} metre^{-1} (see Section 30.15). Consequently the units in which $\mu_0\epsilon_0$ is quoted are:

$$\frac{\text{newton}}{\text{ampere}^2}\;\frac{\text{coulomb}^2}{\text{newton metre}^2}$$

Since the coulomb represents the quantity of electricity passing round a circuit when 1 ampere flows for one second, we can write

$$\text{coulomb}^2 = \text{ampere}^2\ \text{second}^2$$

Thus, the units of $\mu_0\epsilon_0$ are second^2 metre^{-2}, and the units of $1/\sqrt{(\mu_0\epsilon_0)}$ are metre second^{-1}, the units of speed, or velocity. This gives us some check on the result we have derived.

Our analysis has examined the case of the spreading of continuous sheets of charge along the two plates. Had we connected a pulse generator to the two plates instead of a battery we should have had the same situation except that the region of charge would have had an end as well as a beginning. The graph of potential difference between the plates against position would have had the form shown in Fig. 39.3b rather than that in Fig. 39.3a.

We have assumed that the trailing edge of the pulse travels with the same speed as the leading edge: and that conforms to our experience with the travelling pulses in the coaxial cable; even though, in practice, the leading and the trailing edges of such a pulse are not the sharp discontinuities we have assumed.

Indeed the situation within the coaxial cable when the pulses are travelling along it is very similar in form to the case of the two parallel plates we have analysed. One difference is that there is a dielectric other than a vacuum in the space between the two conductors and so we might expect the speed to be lower because the permittivity of any dielectric is greater than that of a vacuum. Another is the difference in geometry, but it can be shown theoretically that the speed of the pulses along a pair of conductors in a vacuum is independent of their geometry, and is 3×10^8 m s^{-1}.

The configuration of the two long parallel conductors we have examined is known as a *wave-guide*; and the problem now is how it is that the pair of fields which accompanies the moving charges can, and does, leave the ends of the wave-guide. We have already seen that the fields do so when we used the 2.8 cm wave transmitter (Section 18.7) which is equipped with a flared wave-guide.

The charges themselves must stop at the ends of the plates, and the pair of plates shown in Fig. 39.2 then reverts to an ordinary parallel plate capacitor. One would expect that there would be·

Fig. 39.3 (a) Graph of p.d. between the plates against position when a steady d.c. supply is used. (b) Graph of p.d. between the plates against position when a single pulse is used.

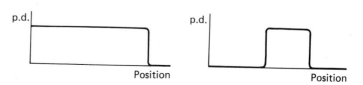

some reflection of the charge pulse at the end (and indeed the experiment with the co-axial cable can be adapted to show this) but, after a short period of time, the situation will become static. There will be two sheets of charge, a uniform electric field between them and the usual non-uniformity at the edges. The magnetic field will no longer be present because the charges are no longer moving.

Had we used a sinusoidal alternating voltage to provide the potential difference between the plates and the charges, both the E and the B fields would persist. They too would alternate sinusoidally. Such a pair of oscillating fields can certainly travel out from the plates but the mechanism by which this happens is beyond the scope of this book.

We may describe the electromagnetic wave as compounded of two fields: an E field and a B field. They are at right angles to each other and to the direction of propagation of the wave. In empty space the wave travels with speed $c = 3 \times 10^8$ m s^{-1}. The variations in the two fields are in phase and we may represent them as in Fig. 39.4 which shows the variations of the E- and the B-fields in a sinusoidal, plane polarized electromagnetic wave.

In Section 18.6, we examined some of the properties of such radiation. In one set of experiments the radiation was transmitted into space from a pair of metal rods (a dipole) with a frequency of 1 GHz (and a wavelength of 30 cm); in another set, a flared wave-guide was used to transmit the electromagnetic waves with a frequency of just over 10 GHz and a wavelength of just less than 3 cm.

In particular, we observed that both sets of waves were polarized with the implication that the electric vector (which is the one detected in each case) is transverse in nature. It follows that the magnetic vector must also be transverse.

39.2   The special theory of relativity

There is much experimental evidence to suggest that the speed of light measured in a vacuum is independent of the motion of either the source or the observer. That it is independent of the motion of the source is characteristic of a wave motion. For example, the speed of sound is independent of the motion of its source. That the speed of light is

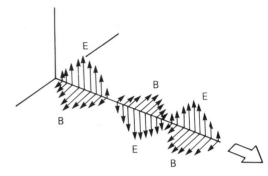

Fig. 39.4   Variations in the B and E fields in a sinusoidal plane polarized electromagnetic wave.

independent of the speed of the observer or apparatus used to measure it came as a great surprise.

As if this were not enough, evidence now exists to suggest that the speed of light in a vacuum is a limiting speed for the movement of all material objects.

The consequences of this experimental fact are far reaching. Our concepts of force and energy lead us to believe that the transfer of energy can be measured in terms of forces which act during this transfer and of the distance moved by their point of application. This leads us to equate the kinetic energy of a moving body with $\frac{1}{2}mv^2$.

An electron accelerated in a suitable linear accelerator can attain a speed of 3×10^8 m s^{-1} (or as close to this as makes no difference). Are we to assume that, after reaching this speed, it gains no more energy? Experiment suggests not. The energy transferred to the moving electrons can be measured in terms of their charges and the potential difference through which they pass. The energy they possess as a result of the acceleration can be converted into heat by allowing the electrons to hit a metal target. As the potential difference is raised, so does the energy attained by the electrons rise – and it continues to rise long after their speed has apparently reached 3×10^8 m s^{-1}.

We have defined force as that which changes the motion of a body. Is this basic definition of force (and therefore of energy) wrong? We prefer to think not. Instead we accept other consequences. If the kinetic energy is still to be described in terms of the work done on a body, we find we must assume that energy itself has mass. The

Fig. 39.5 A 'laser-clock'

total mass (i.e. the sum of the rest mass and the mass of the kinetic energy) of the body is no longer unchanging. It is a function of the speed of the body *relative to the person making the measurements*. We refer to this person as 'the observer'.

This need to describe events in relation to particular observers instead of in relation to some ill-defined *absolute* system is the central idea of the *theory of special relativity*.

Galileo and Newton recognized that the laws of mechanics were independent of the uniform motion of one observer relative to another. In other words, Newton's laws of motion are equally true whether they are tested in a laboratory fixed on the earth, in an aircraft moving at a steady speed relative to the earth, or in a spaceship which is not accelerating relative to the earth. This truth is often referred to as the *principle of Galilean relativity*.

Einstein recognized that the constancy of the speed of light merely represented the extension of the principle of Galilean relativity to the whole of physics. We have seen that the speed of light in a vacuum is related to two fundamental quantities,

$$c = \frac{1}{\sqrt{\epsilon_0 \mu_0}}$$

This is a law of electromagnetism. We know that electromagnetic fields are self-propagating when they move at this speed relative to the observer. Electromagnetic laws as well as mechanical laws are independent of the state of motion of an observer provided he is not accelerating relative to the earth.

It might appear that the earth is a specially chosen reference system. In a way it is, since all our fundamental experiments have been made in laboratories fixed to it. Now we recognize that even the earth suffers small accelerations (as it moves around the sun for instance). Scientists find it useful to visualize a frame of reference in which all our assumed laws of physics are exactly true: this is called an *inertial frame of reference*. For most purposes a frame fixed relative to the earth can be regarded as an inertial frame. An observer in a frame of reference which is in uniform motion relative to an *inertial* frame will also (by Einstein's reasoning) be in an inertial frame.

We cannot explore in this book all the consequences of the principle of relativity, but we will look very simply at two of the most important.

First of all we shall consider how two different observers in uniform motion relative to each other measure *time*. To do this, we shall suppose that we are one of the observers and that another is moving steadily past us (on a balloon, in a train, observer who moves past us with a special type of clock – a *laser clock*. We choose such a clock to illustrate the consequence of the constancy of the speed of light. It can be shown, however, that what is true for *this* clock is true for *all* clocks.

In this laser clock, we imagine a pulse of light bouncing up and down between mirrors at either end which are distance *l* apart (see Fig. 39.5). We will suppose that every time the pulse hits end A the clock gives out a tick, or moves some clock hands. The clock will have a natural period (to the observer moving with the clock) of

$$t_0 = \frac{2l}{c}$$

where t_0 is a rather special time interval; that recorded by the clock between each pair of ticks according to the observer moving along with it. We call t_0 the *proper time*.

Now we will see how the clock appears to behave from *our* point of view. As the light pulse bounces up and down, so the clock moves steadily to the right at speed *v* relative to us. The light pulse seems to traverse the line AB as it moves from the bottom mirror to the top (Fig. 39.6).

Now AB:AC :: *c*:*v*, since the light pulse moves along AB with speed *c according to us*.

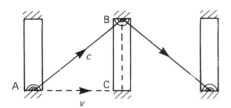

Fig. 39.6 A stationary observer's view of a laser clock moving from left to right.

Hence

$$\frac{AB}{BC} = \frac{c}{\sqrt{(c^2 - v^2)}}$$

and

$$AB = \frac{c}{\sqrt{(c^2 - v^2)}} l$$

As the light pulse travels from A to B and returns to A, it seems to us to travel a distance 2AB or

$$\frac{2lc}{\sqrt{(c^2 - v^2)}}$$

But it does this at speed c and thus takes a time

$$t = \frac{2lc}{\sqrt{(c^2 - v^2)}} \frac{1}{c}$$

$$= \frac{2l}{\sqrt{(c^2 - v^2)}}$$

Thus, whereas the observer moving with the clock says it ticks with a period

$$t_0 = \frac{2l}{c}$$

we say it ticks with a period

$$t = \frac{2l}{\sqrt{(c^2 - v^2)}}$$

so:

$$\frac{t}{t_0} = \frac{c}{\sqrt{(c^2 - v^2)}}$$

$$= \frac{1}{\sqrt{(1 - v^2/c^2)}}$$

$$\therefore t = \frac{t_0}{\sqrt{(1 - v^2/c^2)}}$$

Let us suppose that the laser clock ticks seconds according to the observer moving with the clock.

We will further suppose by way of example that the clock and observer move at the high speed relative to us of $0.6\,c$ (that is $1.8 \times 10^8\,\mathrm{m\,s^{-1}}$).
Then

$$v^2 = 0.36\,c^2$$

and

$$\frac{v^2}{c^2} = 0.36$$

so that

$$\left(1 - \frac{v^2}{c^2}\right) = 0.64$$

and

$$\sqrt{\left(1 - \frac{v^2}{c^2}\right)} = 0.8$$

$$\therefore t = \frac{t_0}{0.8}$$

$$= 1.25\,t_0$$

Thus in one minute of *our* time, the moving clock only appears to make $60/1.25 = 48$ ticks. Since each tick interval is called one second by the observer accompanying the clock, *we say* that the moving clock is running slow. It records only 48 s in the time our clock records 60 s.

There is a good deal of experimental evidence for this apparent running slow of a moving clock: an effect referred to as *time dilatation* or *dilation*. Unstable particles called μ-mesons, or more simply muons have, according to observations made when they are at rest with respect to us, a half-life of only $2 \times 10^{-6}\,\mathrm{s}$. But travelling towards us from the top of the atmosphere with a speed of $0.99\,c$, they appear to live much longer. During the passage of 1 microsecond of time measured by the moving muon clock, we shall record the elapse of more than 7 microseconds. Without this idea we cannot, in fact, account for the number of muons reaching the lower atmosphere without decaying first. It is on the basis of such arguments and experiments that it has been proposed that a space traveller, setting out from earth and then returning will take a shorter time over the journey according to *his* clock than we record on *ours*. Consequently it is believed that the traveller will have *aged* less than his earth-bound friends. The arguments supporting this are more complex than we can give here but many believe this outcome to be true.

Another consequence of the principle of relativity is to be found in the measurement of length. A body moving past is recorded as having a shorter length *in the direction of travel* (though not sideways!) than an observer moving *with* the body says it has. Again we can use the laser clock to see why this is so (see Fig. 39.7).

A pulse of light starts out from A, is reflected at B and returns to A′. According to an observer moving along with the clock this takes a time

$$t_0 = \frac{2l_0}{c}$$

Consequently

$$l_0 = \frac{t_0 c}{2}$$

where l_0 is the length of a body according to an observer moving along with it and is called the *proper length*. To ourselves, the pulse takes a time t_1 to move from A to B such that

$$ct_1 = l + vt_1$$

$$t_1 = \frac{l}{(c - v)}$$

It takes a further time t_2 to go from B back to A

$$ct_2 = l - vt_2$$

$$t_2 = \frac{l}{(c + v)}$$

So the time t the pulse takes to move from one end and back again is

$$t = t_1 + t_2$$

$$= \frac{l}{(c + v)} + \frac{l}{(c - v)}$$

$$= \frac{2lc}{(c^2 - v^2)}$$

But we have seen already that

$$t = \frac{t_0}{\sqrt{(1 - v^2/c^2)}}$$

So

$$\frac{t_0}{\sqrt{(1 - v^2/c^2)}} = \frac{2lc}{(c^2 - v^2)}$$

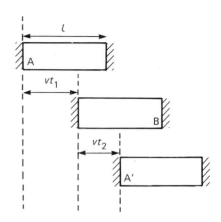

Fig. 39.7

Since

$$t_0 = \frac{2l_0}{c}$$

$$\frac{2l_0/c}{\sqrt{(1 - v^2/c^2)}} = \frac{2lc}{(c^2 - v^2)}$$

So

$$\frac{l_0}{\sqrt{(1 - v^2/c^2)}} = \frac{lc^2}{(c^2 - v^2)}$$

$$= \frac{l}{(1 - v^2/c^2)}$$

$$\therefore l_0 = \frac{l}{\sqrt{(1 - v^2/c^2)}}$$

Since

$$\sqrt{1 - \frac{v^2}{c^2}} \leq 1, \qquad l_0 \geq l$$

Thus a moving length always appears to be shorter than an observer moving *with* the length would say.

For an apparent contraction of 1% of the proper length:

$$\frac{l}{l_0} = \frac{99}{100}$$

Consequently,

$$\sqrt{1 - \frac{v^2}{c^2}} = \frac{99}{100}$$

$$1 - \frac{v^2}{c^2} = (0.99)^2 = 0.980$$

$$\therefore \frac{v^2}{c^2} = 0.020$$

$$v^2 = 0.02\, c^2$$

and

$$v = 0.14\, c$$

The moving length must be travelling at a speed of 14% of the speed of light relative to the observer making the measurements – about $4 \times 10^7 \, \text{m s}^{-1}$. A spacecraft, escaping from the earth's gravitational field travels at only $1.1 \times 10^4 \, \text{m s}^{-1}$. For all normal speeds length contraction effects are very small indeed.

Despite these effects of length contraction and time dilation, two observers moving past each other will always agree about their relative speed. If we form one set of observers, we measure the moving observer's speed by comparing the time he takes (according to *our* clocks) to move a length measured in *our* frame of reference. If the same measurements are made by the moving observer, using times and lengths in his frame of reference, we shall claim that his measured lengths are too long, but his measured time intervals are too short. But as you can see by looking at the equations for time dilation and length contraction, these two effects *just cancel each other*, so both of us record the same speed of relative motion.

Why should the speed of light be the ultimate speed? This is a direct consequence of measuring speeds – we are compelled to make all measurements in our own frame of reference. Thus we assess a body's speed in terms of the distance it travels in our frame (as *proper length*) and the time it takes in *our frame of reference*. This latter is not a *proper time*: we know now that our clock appears to run progressively faster compared with a clock carried by the moving body, the faster that body moves relative to us. As relative speeds approach c, the time dilation factor approaches infinity. We conclude that c is the ultimate speed.

We started by observing the surprising consequences this limiting speed c has on our ideas about inertial mass. Derivation of the consequent relationship between mass and relative speed is beyond the scope of this book. It is from the relationship between mass and relative speed that Einstein was able to show a link between the energy carried by a moving body and its apparent mass, now celebrated in the famous equation $E = mc^2$. A full discussion of the implications of equation is beyond us here, but one observation might form a valuable conclusion to this brief discussion of the special theory.

From all that we have said so far, it might appear that the relationship between measured

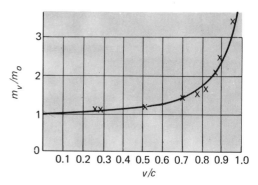

Fig. 39.8 Some experimental results showing the variation in the inertial mass of electrons with speed.

masses, lengths and time intervals and the relative speeds of observers is an artificial contrivance, designed by physicists in order to preserve a few well-known laws. We have already been able to distinguish some special lengths and time intervals called *proper length* and *proper time* which are measurements made by observers stationary relative to the length measured, or clock used. In a similar way, we are able to define a *rest mass* of an object, as the mass of the object relative to which the observer is stationary.

Fig. 39.8 brings together a number of experimental results showing the increase of inertial mass with speed for high-speed electrons. Only a few of the experimental results are marked.

It was Einstein who first proposed that energy itself has mass, and also that the extra mass of a body moving relative to an observer could be extended to some of the mass of the matter making up the body. Mass is a measure of inertia; it is both conserved and additive. Rest mass is the sum of the inertial mass of the matter in the body and of the energies of the particles which make up the body. That this was true was violently demonstrated by the release of nuclear energy. The energy which can be released from the breakdown of a nucleus is related to the change in mass observed by the relativity equation $E = mc^2$.

It is this fact more than any other which has served both to demonstrate that the special theory of relativity is something more than an exercise in mass and energy book-keeping and to turn what may seem an esoteric exercise into something which is the 'bread and butter' of a particle-physicist's life.

UNIT TEN

Electronics

Chapter 40

DIGITAL ELECTRONICS

Developments in electronics are occurring at a great rate. At the domestic level, high quality sound-reproducing equipment and television have been with us for a long time, and their technology has been steadily improved over the years. But now there is a spate of more advanced electronic equipment available to the domestic user: video recording equipment, microcomputers, microprocessors for controlling home-heating systems, cookers – not to mention the ever more sophisticated electronic intruder alarm systems which house-owners install to protect their houses full of expensive electronics!

Other spectacular developments are too many to mention here, but one area which may be familiar to some readers is that of electromechanical aids for disabled people. A home computer, with suitably designed *interfacing* (that is, the means by which the user can communicate with the machine), can enable a person who has almost no muscular power whatsoever to write letters, make design drawings, do complicated mathematics: many of the functions needed to carry on a professional career.

In a few years' time more and more people will be profoundly influenced by the 'information explosion', in which information of all kinds – verbal, written, pictorial, numerical – will become almost instantly communicable from one place to any other. All these data are processed by electronics. In order for this to happen the data must be converted into a sequence of electrical pulses: often the number of pulses per second is many millions. These pulses are used to represent digits, and hence the term *digital electronics* is used for the branch of electronics which concerns the processing of data in digital form.

This chapter introduces some of the fundamental 'building blocks' of digital electronics: what they do, and how they are used.

40.1 Encoding data in digital form

We begin with one example of how a piece of data, in this case a single letter of the alphabet, can be represented by a series of digits. The digits of binary arithmetic are used: 0 and 1. First of all a suitable *digital code* is needed. A set of eight binary digits allows 256 symbols to be encoded ($2^8 = 256$). The word 'bit' is often used instead of 'binary digit'. An eight-bit code is sufficient for all the letters, numbers, and symbols found on a typewriter keyboard, as well as for a considerable number of other functions.

One eight-bit code in common use is the ASCII system (the letters stand for: American Standard Code for Information Interchange). In this system a capital letter 'Q', for example, is represented by the sequence of eight bits: 01010001. A set of eight bits is often referred to as a 'byte', although in some contexts a 'byte' is taken to mean sixteen bits.

When data are processed by electronic circuits, every byte has to be converted into electrical form. The usual way of doing this is to use voltages to represent the binary digits. Typically, in a computer circuit, the bit '0' is represented by a low voltage near to 0 V, and the bit '1' by a high voltage about $+5$ V. If data are being sent along a single pair of conductors, then the p.d. between these conductors will vary with time when the letter 'Q' is being transmitted as shown in Fig. 40.1. Here we have an example of data being transmitted in *serial* form (a series of bits, one after the other). Digital data are also transmitted in *parallel* form: in our example a set of eight conductors would be used to transmit the eight bits simultaneously.

In Fig. 40.1 the data are being transmitted serially at a rate of 9600 bits per second, or 9600 *baud*: thus the duration of each bit is 1/9600 second, except for the *stop bit* which is always longer than the standard bit, and serves as an instruction to the data receiving device that a byte of data is about to follow. (Note that, in our example, the *least significant bit* of the byte is sent out *first*.)

What has been said so far raises a number of questions, for instance: how can the pressing of a key labelled 'Q' on a keyboard produce a set of eight voltage pulses, to represent the eight binary digits? And how can this set of pulses be made to cause a letter 'Q' to be displayed on the screen of a television receiver? And how can the byte 01010001 representing the letter be stored in a computer's 'memory'? Certainly we shall not be able to answer fully any of these questions in this book, but after studying this chapter you will have gained some knowledge of the fundamental 'building blocks' of computers and other digital systems.

The symbols on a typewriter keyboard are by no means the only kind of data which can be

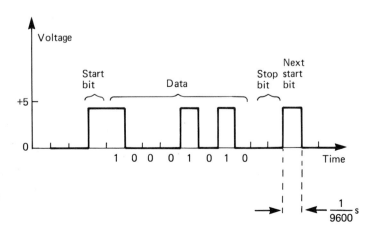

Fig. 40.1 Sequence of pulses representing the letter 'Q' in the ASCII code (and using positive logic).

Fig. 40.2 The logical NOR function.

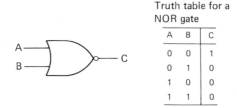

Truth table for a NOR gate

A	B	C
0	0	1
0	1	0
1	0	0
1	1	0

(a) The logic symbol and the truth table for a NOR gate.

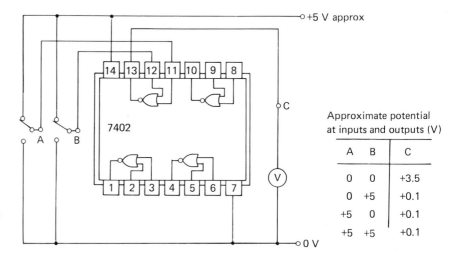

Approximate potential at inputs and outputs (V)

A	B	C
0	0	+3.5
0	+5	+0.1
+5	0	+0.1
+5	+5	+0.1

(b) Using one of the four NOR gates of the IC type number 7402.

represented by a digital code. Pictorial symbols, musical notes and sounds, any physical measurement, the phonetic elements of human speech; even entire pictures, complete musical performances, and human utterance in any language – all these can be digitally encoded, for the purpose of transmitting the data from one place to another, for recording and for reproducing.

Before proceeding we should point out that it is purely a matter of convention and convenience whether the binary digit '1' is represented by the presence of a voltage, or by the absence of a voltage. In some systems, termed *negative logic* systems, a '1' is represented by a zero voltage and a '0' by a non-zero voltage. Furthermore, the answers to logical questions: 'yes' and 'no', 'true' and 'false', are represented in digital systems by zero and non-zero voltages. In summary, we can say that a *bit* is the smallest unit of information that can be represented by means of a voltage (or current, in some cases) which can be either zero or non-zero in value.

Just as letters of the alphabet are strung together to form words, so bits are strung together to make binary 'words'. In any digital electronic system the number of bits thus strung together (typically eight or sixteen, and termed a *byte*) is constant.

In the experiments which follow we shall not get as far as dealing with complete bytes: we shall be concerned with the principles of processing individual data bits. For further study you must refer to a textbook in which the principles of microprocessors and computers are introduced.

40.2   Logic gates

At the end of this chapter you will see how to make a circuit which will add together two numbers. The numbers are expressed in binary digital form, and the circuit is built up from elements known as *logic gates*. Logic gates are used to perform the functions of *combinational logic*: in Section 40.5 you will see examples of circuits which perform some operations of *sequential logic*.

Fig. 40.2 (continued)

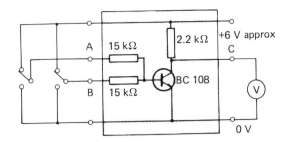

A	B	C
0	0	+5.5
0	+6	+0.5
+6	0	+0.5
+6	+6	+0.5

Approximate potentials at inputs and outputs/V

(c) Using a module made of resistors and a transistor.

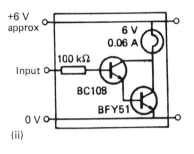

(ii)

(d) Lamp indicator modules for use in experiments on digital electronics:
(i) A quadruple lamp indicator module, using *light-emitting diodes* (LEDs) and the IC type number 7404.
(ii) A single lamp indicator using a filament lamp.

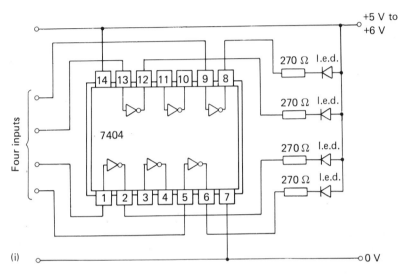

(i)

The first logic gate to be studied is known as a NOR gate, and its symbol (using the American convention) is shown in Fig. 40.2a. For experiments we can either use the integrated circuit (IC) shown in Fig. 40.2b (type number 7402 and known as a 'quad 2-input NOR gate'), or the module shown in Fig. 40.2c and which is made from one transistor and three resistors (and which can be used for a large variety of other experiments). At this stage we are treating these devices as 'black boxes' with no attempt to explain their internal construction and *how* they work: we are concerned for the time being with *what* they do.

Look at Fig. 40.2b or 40.2c. A voltmeter and/or a suitable *indicator lamp* is connected between the output terminal, C, and the negative supply 'rail'. (Suitable indicator lamp modules are shown in Fig. 40.2d.) A pair of two-way switches enables each of the inputs A and B to be held at either the positive rail potential or at zero potential. It will be found that if either input A or input B are held at high potential the output voltage is low – nearly zero. If both inputs are simultaneously held at a

high potential, then the output voltage is low once again. The output voltage goes high only when both inputs are simultaneously at a low potential (when they are both connected to the negative 'rail').

The function of the NOR gate is summarized by the tables in Fig. 40.2a. The digit '0' corresponds to 'low potential' and '1' corresponds to 'high potential'. These are examples of *truth tables*. The function described by these tables is termed the *logical NOR* function because the output of the gate is 'high' only when neither input A *nor* input B are 'high'.

In Fig. 40.3 the NOR gate is shown along with three other logic gates. Each of the gates can be made from transistors and resistors, but also each gate can be made from combinations of other gates, as you will find.

The simplest logic function of all is that of *inversion*. The symbol for an *inverter* is shown in Fig. 40.3. It has only one input: when this is 'high' the output is 'low', and vice versa. In binary digital notation, the output is '1' when the input is

	AND	OR	NAND	NOR	Inverter
	A–⟍ ⟍–C B–⟋ ⟋	A–⟍ ⟍–C B–⟋ ⟋	A–⟍ ⟍o–C B–⟋ ⟋	A–⟍ ⟍o–C B–⟋ ⟋	A–▷o– $\overline{A}$

Inputs A B	C	C	C	C	A	$\overline{A}$
					0	1
0 0	0	0	1	1	1	0
0 1	0	1	1	0		
1 0	0	1	1	0		
1 1	1	1	0	0		

Fig. 40.3 The principal logic gates and their functions.

'0', and the output is '0' when the input is '1'. A NOR gate can be used as an inverter by having one of its inputs kept at 'low' potential. In the symbolism of binary logic, if the input to an inverter is X (where X can be 0 or 1) the output can be written 'NOT X' or '$\overline{X}$'. In Fig. 40.2d an IC containing six inverters is being used.

Figure 40.4 shows the logic symbol for an AND gate, with circuits for experimenting. As in Fig. 40.2 a pair of two-way switches is used to provide the inputs, and either a voltmeter or a lamp indicator module, or both, are used to show the state of the output. The truth-table in Fig. 40.3 shows why the name AND is appropriate: the output C is 'high' only when A *and* B are 'high' together. An AND gate can be constructed from a

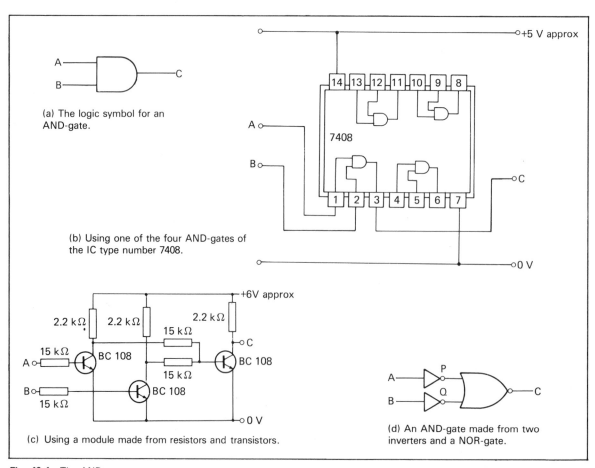

(a) The logic symbol for an AND-gate.

(b) Using one of the four AND-gates of the IC type number 7408.

(c) Using a module made from resistors and transistors.

(d) An AND-gate made from two inverters and a NOR-gate.

Fig. 40.4 The AND-gate.

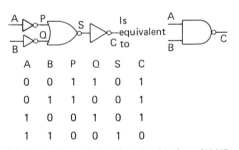

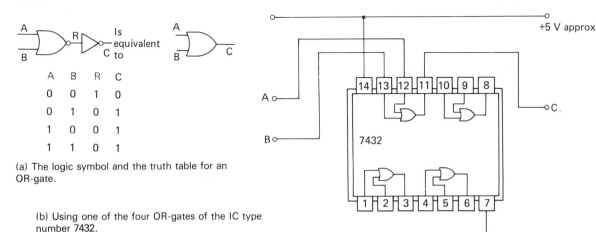

(a) The logic symbol and the truth table for an OR-gate.

A	B	R	C
0	0	1	0
0	1	0	1
1	0	0	1
1	1	0	1

(b) Using one of the four OR-gates of the IC type number 7432.

Fig. 40.5 The OR-gate.

pair of inverters and a NOR gate, as shown in Fig. 40.4d: and the circuit of Fig. 40.4c is exactly this, with the two transistors on the left, each with a pair of resistors, constituting a pair of inverters. The transistor on the right with its three resistors forms the NOR gate (compare Fig. 40.2c).

In Fig. 40.5 we have an OR gate. With this the output is 'high' when either input A *or* B are high, or when both are high together. This can be made from a NOR gate followed by an inverter, as shown.

Figure 40.6 shows a NAND gate (the name is a contraction of 'NOT' and 'AND'). This could, if necessary, be made from three inverters and a NOR gate.

Other combinations of gates, providing functions different from the ones already described, will be found in the Questions for this unit at the end of the book.

40.3 Storing digital data

There are many systems in use for storing data in digital form. Data bits can be stored on magnetic tape or disc, for instance. A typical 'mini-floppy' magnetic disc can hold more than 200 000 bytes (200 kilobytes) on one side: this is a total of 1 600 000 bits. (One kilobyte is actually 1024 bytes. Binary arithmetic dominates the world of digital electronics, and $1024 = 2^{10}$. The prefix 'kilo-' is used because 1000 is close enough to 1024 for practical purposes.) Punched paper tape and cards are used in older computer systems, but magnetic discs provide a much more compact

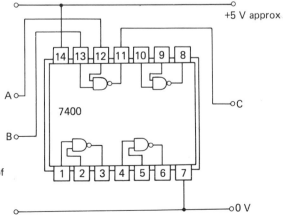

(a) The logic symbol and truth table for a NAND-gate.

A	B	P	Q	S	C
0	0	1	1	0	1
0	1	1	0	0	1
1	0	0	1	0	1
1	1	0	0	1	0

(b) Using one of the four NAND-gates of the IC type number 7400.

Fig. 40.6 The NAND-gate.

means of storage. As well as these there are several electronic circuit elements which can be used to store digital data. Nowadays a variety of semiconductor memory devices is used. A device of this kind is microscopically small and constructed on a single chip of semiconductor material, usually silicon. These devices are incorporated into the circuitry of the computer, for temporary or permanent storage of data.

Here we shall explain how one type of temporary data storage device works. The basic memory 'cell' of this is called a *bistable circuit*, or '*flip-flop*' (it is also sometimes called a *binary* circuit).

A *bistable circuit* is a circuit which has two stable states: that is to say, it can have two, but only two, distinctly different patterns of currents and voltages existing in it. One of these patterns will persist until some external influence 'flips' the circuit and establishes the other pattern. The two distinct conducting states are used to represent the bits 1 and 0.

Figure 40.7 shows one example of a bistable circuit, both in its logical symbolic form, and as a full circuit diagram. It can be constructed by linking a pair of transistor NOR-gate modules, as shown. When the power is first switched on, it is a matter of pure chance whether the output Q is low

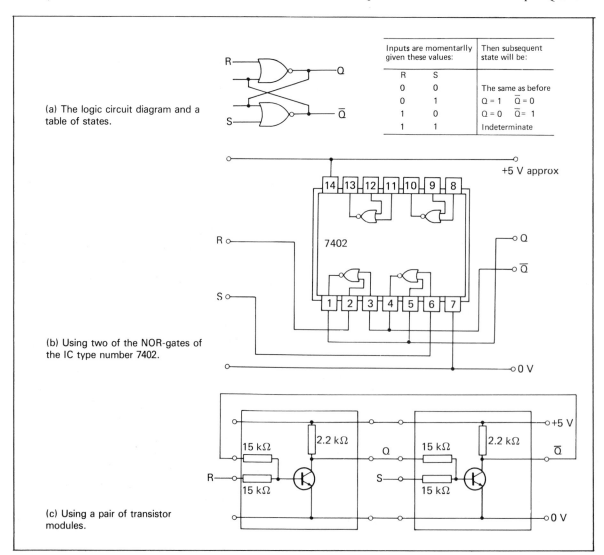

(a) The logic circuit diagram and a table of states.

Inputs are momentarily given these values:		Then subsequent state will be:
R	S	
0	0	The same as before
0	1	Q = 1 $\overline{Q}$ = 0
1	0	Q = 0 $\overline{Q}$ = 1
1	1	Indeterminate

(b) Using two of the NOR-gates of the IC type number 7402.

(c) Using a pair of transistor modules.

Fig. 40.7   One kind of bistable circuit, made from a pair of cross-linked NOR-gates.

and $\overline{Q}$ high, or Q high and $\overline{Q}$ low. The symbol $\overline{Q}$ means 'NOT Q'. But if the positive rail potential is applied to the input R (or, as we say, we make R *high*), then the output Q must go low, as a result of the NOR-function of the module – this will happen regardless of the state (high or low potential) of the other input to the upper module. And when Q goes low, $\overline{Q}$ must go high, because of the inverting function of the module (the input S is 'floating' that is to say, not connected to anything at present). The high potential of $\overline{Q}$ is fed back to one input of the upper module, thus holding it high. If input R is then disconnected the circuit will remain in this state, with Q low and $\overline{Q}$ high. In order to make the circuit 'flip' to the opposite state (Q high, $\overline{Q}$ low) the input S must be made high momentarily.

The symbols R, S, Q, and $\overline{Q}$ are the conventional ones used in the context of bistable devices. The S stands for 'set' and the R for 'reset', and, by convention, 'setting' the bistable means making Q = 1 and $\overline{Q}$ = 0 (in binary digital notation).

The example of a bistable circuit just described is not in fact the one must commonly used in semiconductor memory devices. It was chosen because it is the simplest one to construct using standard transistor modules that are used in many other experiments in elementary electronics. A commoner type of flip-flop is made from a pair of cross-coupled NAND-gates (Question 10.5 at the end of this book concerns the action of this type).

For full details of the various types of flip-flop in present day use you must refer to a textbook on digital electronics. The type of flip-flop described above is shown as an *RS flip-flop*.

Bistable circuits of all types are made so small that 16,384 of them can be built as an integrated circuit on a single silicon chip measuring 2 mm × 2 mm. This can store 16 384 data bits. A set of eight of these would be used to store 16 kilobytes of data.

One important feature of bistable memory devices like the ones mentioned above is that all the stored data is lost as soon as the power supply is switched off. When the power is switched on again it is a matter of chance into which of the two bistable states the circuit will settle. Thus, if the memory device is 'read' it will be found to contain 'rubbish': a random sequence of '0's and '1's.

Memory devices such as these are used for temporary storage of data while the computer, or calculator, is operating. There are other types of memory device which do not lose their data when the power is switched off. For instance, an EPROM (erasable programmable read-only memory) will store digital data for as long as is wished. The data can be erased only by exposing the chip to ultraviolet radiation for a certain minimum length of time. EPROMs are used in pocket calculators, for instance, to store the numerical values of mathematical constants, such as π, in binary digital form.

40.4 Pulse generating circuits

In the introduction to this chapter we explained how digital data is often encoded into a sequence of *voltage pulses*. By 'pulse' in this context we mean a voltage which changes abruptly from one steady value to another value, remains at that new value for a certain length of time, then returns abruptly to its previous value. An example is shown in Fig. 40.8.

In any digital circuit, all 'signals' are in the form of voltage pulses – not only the data itself. All computers have an internal *clock*: an oscillator which generates a precisely regular sequence of pulses at a high frequency (typically 8 MHz). These pulses are used to control and synchronize the operations of all the different parts of the computer circuit. Also, in a computer, all instructions sent from one part of the circuit to another are voltage pulses.

Figure 40.8 shows an experiment using a timer IC, type number 555, to produce a single pulse of any chosen duration. When the switch S is closed the potential at the output terminal (connected to pin 3 of the 8-pin package) rises sharply to nearly + 5 V, and remains at this potential for a time determined by the capacitance C and resistance R; then the potential falls abruptly to zero.

To be able to use this timer IC effectively you need only know the essential details of its function, and these are illustrated by the experiments of Fig. 40.8 and Fig. 40.11.

This is what happens in Fig. 40.8. When switch S is closed, pin 2 (Trigger) is 'grounded' – that is, taken to zero potential – then immediately the internal logic circuitry causes the output (pin 3) to

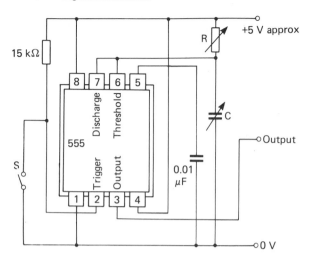

Fig. 40.8 (a) A monostable circuit using IC type number 555. A single pulse is produced when the switch S is closed.

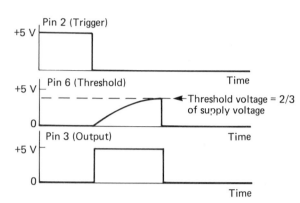

(b) Graphs showing the sequence of changes of potential.

go to high potential. Until now, pin 7 (Discharge) has been 'grounded' (held at zero potential), and so capacitor C is uncharged at the start. But at this instant pin 7 becomes disconnected from 'ground', by means of the internal logic circuitry of the device. Immediately C begins to be charged via resistor R. (The currents in the connections to pins 6 and 7 are negligible.) When the potential at pin 6 (Threshold) reaches 2/3 of the supply voltage, the internal logic causes the output (pin 3) to go back to zero potential, and also causes pin 7 (Discharge) to be grounded and so C discharges almost instantaneously. If, at any stage during this sequence of events, switch S is opened, it will have no effect on the events: the Trigger (pin 6) is activated only by a *decreasing* potential, as the experiment of Fig. 40.11 will show.

If you experiment with the circuit of Fig. 40.8 you should try the effect of varying the value of C over a range of about 1 μF to 10 μF (a capacitance substitution box is ideal for this); and of varying R from about 1 kΩ to 1 MΩ. You will then see that this device can be used to produce a pulse of very short or very long duration. If the pulse has very short duration it may not succeed in lighting the lamp indicator: in this case a c.r.o. should be used to display the pulse. The duration of the pulse should be found to equal 1.1 RC. As explained in Chapter 26, the product RC is the *time constant* for the capacitor and resistor combination; and using the theory of capacitor discharge given there, it is possible to show that the time required

for a capacitance C to be charged via a resistance R to a p.d. equal to 2/3 of the supply voltage is (ln 3) RC which equals 1.099 RC.

A circuit, like that of Fig. 40.8, which produces a single pulse then returns to its resting, or *stable*, state, is called a *monostable* circuit. A cruder but simpler monostable circuit, using a transistor module, is shown in Fig. 40.9. Note that the upper input terminal is connected to the positive rail, thus making the input to the module 'high'. The output therefore is low: this is the same *inverting* function that we met in Section 40.2. When the transistor module is in this state the potential of its base (labelled 'b' in Fig. 40.9) is approximately +0.7 V.

With the switch in the 'up' position, the left-hand 'plate' of the capacitor is held at a potential of +6 V and its right-hand plate is at the potential of the base of the transistor, about +0.7 V. Thus the p.d. across the capacitor is about 5.3 V, and the capacitor is charged. When the switch is put to the 'down' position, the p.d. across the capacitor will initially be 5.3 V, and then this p.d. will decrease as the capacitor discharges. But at the very instant when the switch is operated, making the left-hand plate of the capacitor equal to zero, the right-hand plate must go to a potential of −5.3 V, and this has the same effect as making the input to the module 'low', and so the output goes high. The capacitor immediately starts to discharge through the 15 kΩ resistor in Fig. 40.10a. It continues to do so until fully discharged and then it becomes charged with the opposite polarity. With its right-hand plate reaches a potential of about $+0.7$ V (as it was originally) the output of the module goes high. The graphs in Fig. 40.9 show the changes in potential at the lower input terminal of the module, and at the output terminal.

The duration of the pulse is determined almost entirely by the time required for the capacitor to discharge through the 15 kΩ resistor. Although a small current does flow via the base 'b' of the transistor, it is negligible in this case, being much smaller than the current flowing via the 15 kΩ resistor.

The duration of the pulse can be increased by adding extra resistance in series between the upper input terminal and the positive rail, as shown in Fig. 40.10a.

As explained in Chapter 26, the rate at which a capacitor discharges through a resistor depends upon the *time constant*, which is the product of the resistance and the capacitance. In this case we have:

$$RC = 15 \times 10^{-3}\,\Omega \times 25 \times 10^{-6}\,\text{F}$$
$$\simeq 0.4\,\text{s}$$

It was also explained that the *half-life* for capacitor discharge (the time required for the charge on the capacitor, and hence also the p.d. across it, to decrease to half of its original value) is $\ln 2\,RC$. Now $\ln 2 \simeq 0.7$, so in our case the half-life is approximately $0.7 \times 0.4 \simeq 0.3$ s. But, in the circuit of Fig. 40.10a, the capacitor has to lose considerably more than half of its charge before the output of the module returns to its original high potential. Because of this, and because the

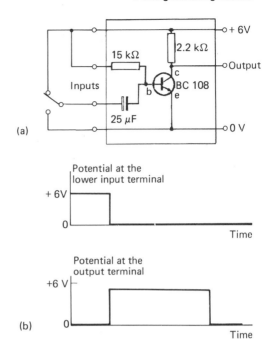

Fig. 40.9 (a) A monostable circuit using a transistor module. (b) Graphs showing the sequence of changes of potential.

actual capacitance of any electrolytic capacitor is nearly always appreciably greater than its marked value, the pulse duration obtained with this module is likely to be nearer to 0.5 s or possibly even greater. Nevertheless, the product RC gives a rough estimate of the pulse duration to be expected.

Using the circuit of Fig. 40.10a you can increase the pulse duration, as already suggested, by adding extra resistance in series between the upper input terminal and the positive rail; but you

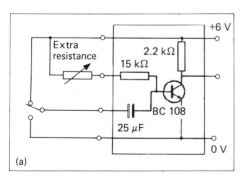

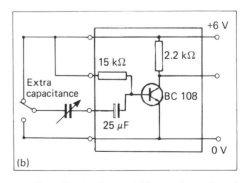

Fig. 40.10 A monostable circuit with (a) extra resistance added to increase the pulse duration and (b) capacitance in series to decrease the pulse duration.

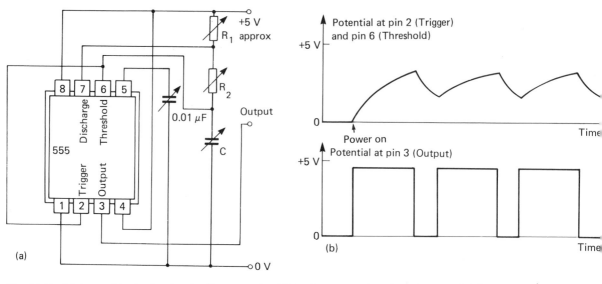

Fig. 40.11 (a) An astable circuit using the IC type number 555; pulses are produced continuously. (b) Graphs showing the sequence of changes of potential.

will find that if the resistance is too great, no pulses can be produced, and that when the switch is set to the 'down' position the output of the module goes high and stays high. The cause of this is the *leakage resistance* of the electrolytic capacitor. With the switch in the 'down' position the capacitor acts like a high resistance connected between the base of the transistor and the negative rail, and if this high resistance is of similar magnitude to the resistance between the base of the transistor and the positive rail, the input to the module will be effectively held 'low' and so its output will stay high. The maximum extra resistance that can be used to produce pulses may vary from as little as $10\,\mathrm{k}\Omega$ to greater than $100\,\mathrm{k}\Omega$, depending upon how 'leaky' the capacitor is.

The pulse duration can be reduced by adding another capacitor in series with the internal capacitor of the module, as shown in Fig. 40.10b, where the extra capacitor is connected in series with the lower input terminal of the module. If you do this, you should try to predict what pulse duration will be obtained, using first of all the formula for the capacitance of two capacitors connected in series (see Eq. 26.9):

$$\frac{1}{C} = \frac{1}{C_1} + \frac{1}{C_2}$$

and then calculating the time constant RC.

Figure 40.11 shows a circuit which produces pulses continuously: an *astable* circuit. Here you will see that pin 6 of the timer IC, the Threshold pin, is linked to pin 2, the Trigger pin. The sequence of events is as follows:

1. When the power supply is first switched on, the capacitor C is uncharged, and pin 7 (Discharge) is not grounded.
2. The capacitor C charges up via resistors R_1 and R_2; while this is happening the output (pin 3) stays at high potential.
3. When the potential at pin 6 (Threshold) reaches 2/3 of the supply voltage, pin 7 (Discharge) is grounded. As this instant the internal logic makes the output (pin 3) go to zero potential.
4. The capacitor now discharges via R_2.
   When the potential at pin 6 (Threshold) falls to $\frac{1}{3}$ of the supply voltage, so also does pin 2 (Trigger) because it is connected to pin 6; this causes the output to 'go high', disconnects pin 7 from ground, and the cycle repeats from stage 2.

The graphs in Fig. 40.11b show the relevant changes of potential during this sequence.

When experimenting with the circuit of Fig. 40.11a you should study the effect of varying each of the resistances R_1 and R_2, as well as the

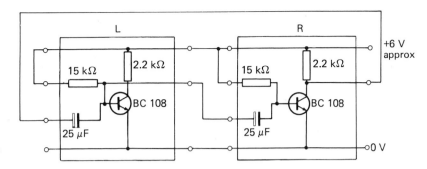

Fig. 40.12 An astable circuit made from two transistor modules.

capacitance C. From the theory developed in Section 26.5, it can be shown that the time taken for a capacitor to lose half of its original charge (so that the p.d. across it also halves) is $\ln 2\, RC$ which equals $0.693\, RC$. This is also the time needed for it to charge up from a p.d. equal to *one*-third of the supply voltage to a p.d. equal to *two*-thirds of the supply voltage. In the circuit of Fig. 40.11a, C is being charged via resistors R_1 and R_2 in series, then being discharged via R_2 alone (when pin 7 is grounded). Thus, theoretically, the duration of the positive output pulse, t_1, and the interval between pulses, t_2, should be given by:

$$t_1 = 0.693\,(R_1 + R_2)\,C$$
$$t_2 = 0.693\,R_2 C$$

In practice t_1 and t_2 are slightly less than this, partly because small currents flow via the connection to pin 6.

Figure 40.12 shows an astable circuit made from two transistor modules. With the values of resistance and capacitance already built into these modules, as shown, a lamp indicator connected to either output terminal will blink on and off repeatedly. As was done in the experiment of Fig. 40.10, extra resistance or capacitance can be added externally to alter the duration of the pulses produced.

With either of the astable circuits described above, a c.r.o. should be used to display the output pulses if they are occurring rapidly. Also, if an earphone, or an audio frequency amplifier with loudspeaker is available, it should be connected to the output of the astable circuit. With a slow pulse rate clicks will be heard: faster pulse rates will produce harsh-sounding musical notes.

40.5   Some other digital circuits

(i) A circuit which adds two binary digits

This is an example of *combinational logic*. The circuit performs the simplest operation of binary arithmetic: adding two binary digits to produce a 'sum' digit and a 'carry' digit. Figure 40.13 shows the arrangement of logic modules needed to achieve this, and the truth table which describes its action.

You should check for yourself that the truth table is correct for the circuit of Fig. 40.13 by writing an intermediate truth table to show the logical states (0 or 1) of points C and D in the circuit, for each of the four possible combinations of input states A and B.

The circuit is called a 'half adder' because when two binary numbers are being added (each number consisting of several digits), two half adders are needed to add together not only one pair of digits but also the 'carry' digit from the previous addition operation. Figure 40.14 shows a circuit which does this.

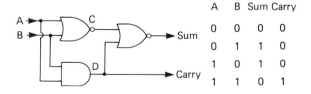

A	B	Sum	Carry
0	0	0	0
0	1	1	0
1	0	1	0
1	1	0	1

Fig. 40.13   A half-adder circuit with its truth table.

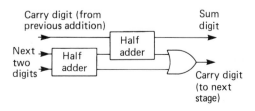

Fig. 40.14   A whole-adder circuit.

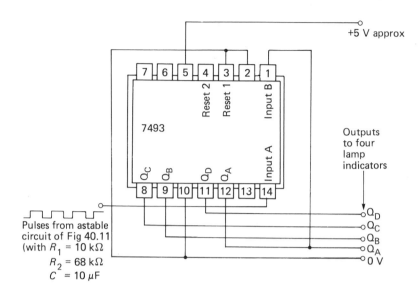

Fig. 40.15 Using the 4-bit binary counter IC type number 7493.

Pulses from astable circuit of Fig 40.11 (with R_1 = 10 kΩ R_2 = 68 kΩ C = 10 µF

+5 V approx

7493

Reset 2
Reset 1
Input B

Outputs to four lamp indicators

Input A

Q_C Q_B Q_D Q_A

Q_D
Q_C
Q_B
Q_A
0 V

(ii) Using a counter IC

This provides an example of a *sequential logic* circuit. The *4-bit binary counter* IC, type number 7493, counts voltage pulses and outputs the count as a 4-bit binary number. In the circuit of Fig. 40.15 the pulses are provided by the astable circuit, previously seen in Fig. 40.11, but here with values of resistance and capacitance which will produce pulses at a rate of about one per second. A set of four lamp indicators should be connected to the four output terminals of the counter, and arranged on the bench in the order suggested by the diagram. Q_A gives the *least significant bit* and Q_D the *most significant bit* of the binary number. The lamps will then show that the counting sequence goes: 0000, 0001, 0010, 0011, etc., up to 1111, then starts again at 0000.

Although we make no attempt to explain the internal structure of this counter IC, we can point out that it is built around several bistable (flip-flop) circuits. Four of these are arranged in a sequence, so that the output from one triggers the next one. In this way each of the bistables operates as a scale-of-two counter, and four of them constitute a circuit which has $2^4 = 16$ stable states.

(iii) Gating an audio frequency signal

This circuit produces a single audible 'bleep' of chosen duration, by using an AND-gate to allow the audio frequency signal via one of the inputs to

the AND-gate when, and only when, the other input to the AND-gate is held at high potential. Figure 40.16 shows a schematic diagram of the circuit. Here the 'box' represents the monostable circuit of Fig. 40.8. The Z-shaped symbol above the input line to the monostable is a concise way of indicating that the monostable circuit is 'triggered' when the input voltage goes from high to low. The symbol is in effect a portion of a graph of voltage against time. The same kind of symbolism is used to represent the output pulse from the monostable; the audio frequency signal is also shown in this way (with no attempt to represent the frequency realistically. however).

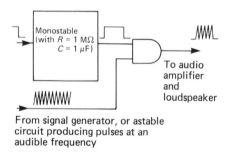

Monostable
(with R = 1 MΩ
C = 1 µF)

To audio amplifier and loudspeaker

From signal generator, or astable circuit producing pulses at an audible frequency

Fig. 40.16 Gating an audio-frequency signal.

In the circuit of Fig. 40.16 we see that the AND-gate functions here as a gate which is 'opened' by making one of its inputs high.

Chapter 41

AMPLIFICATION

The kind of amplifier with which most of you are likely to be familiar is often termed an *audio amplifier*, as used in any sound reproducing system, such as a record player or tape-cassette player. In a tape-cassette player, for example, the magnetic play-back 'head' converts the changes of magnetization in the magnetic tape into an alternating e.m.f. of very small amplitude, typically about one millivolt or less. This alternating e.m.f. is called the *signal*, and in this example it is an *audio frequency* signal, or *a.f. signal*. This means that the frequency of the alternating e.m.f. lies within the range of frequencies which are audible to the human ear. This signal is fed to the input terminals of an amplifier, amplified, and the output signal from the amplifier is fed to one or more loudspeakers or earphones. When loudspeakers are used the output voltage will usually have an amplitude in the order of tens of volts, when loud sounds are being reproduced, and the amplitude of the current may be about one ampere.

The example described illustrates the point that *amplification* involves increasing voltages, or currents, or both. Although an audio amplifier may be the kind of amplifier most familiar to you, it is by no means the simplest kind. The faithful reproduction of sounds does not simply require that all audio signals should be amplified by a constant amount: the amplification, as a rule, has to be different for different frequencies. There are many reasons for this. For example, with some types of player for disc recordings, the pick-up cartridge emphasizes the high frequencies and to compensate for this the amplifier provides less amplification at high frequencies than at lower frequencies. Also, with disc recordings, it is standard practice to record low frequency signals with a smaller amplitude than high frequency signals: thus, when the recording is being played back there must be more amplification at these low frequencies than at the higher frequencies. This point is taken up further in Section 42.4.

The number of different kinds of amplifier, for different purposes, is huge. Amplifiers are essential in radio transmitters and receivers, television equipment, electronic measuring instruments, radar, missile guidance systems, automatic control equipment, and hosts of other applications.

Amplifier circuits can be classified in various ways. There are, however, two main categories: *active* and *passive* amplifiers. An example of a *passive* amplifier is a step-up transformer, which produces an increase in voltage, but no increase in power: the output power from the secondary of the transformer can never be greater than the power input to the primary. In an *active* amplifier the power output is greater than the power input, the extra power being taken from the battery or from some other *power supply*. Nearly always, if the word 'amplifier' is used on its own, it is an active amplifier that is meant.

In this chapter we shall be concerned entirely with examples of active amplifiers, and the power supply can be either a battery or a low voltage, smoothed d.c. supply.

41.1 An amplifier module for experiments

Our study of amplification begins with a series of experiments using a simple amplifier *module*. The same module (see Fig. 40.2c) which was used for the experiments in Chapter 40, can be used as an amplifier. Alternatively, the module shown in Fig. 42.5, which incorporates a *linear IC* amplifier, can be used. Either module performs much the same *function*, and it is chiefly the function that concerns us at present.

The module of Fig. 41.1 needs an external d.c. power supply giving a voltage of about 5 V to 6 V; the precise value is not critical. Dry batteries or re-chargeable cells giving about 6 V are ideal. The maximum current taken by the module from the supply is less than 3 mA, so that several modules can be supplied from the same batteries. In the circuit diagrams the power supply is not shown:

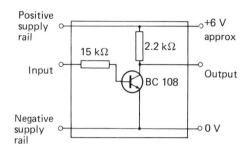

Fig. 41.1 A junction transistor amplifier module for use in experimental work.

instead the *power rails* are simply labelled ' + 6 V' and '0 V'.

In order to perform the experiments which follow it is not necessary to understand *how* the transistor in the module of Fig. 41.1 functions, but we give here a brief description of *what* happens.

The particular type of transistor used in the module of Fig. 41.1 is known as a *bipolar junction transistor*, of the *npn* type, made of silicon. To find out more about this you will have to refer to a textbook which includes elementary semiconductor physics. The three terminals of this kind of transistor are known as the *base* (b), *emitter* (e), and *collector* (c). The base is used, as a rule, as a control electrode, with only very small current flowing into or out of it, while a considerably greater current flows via the emitter and the collector. Whatever these currents may be individually, they must always obey Kirchhoff's first law, so that the total current flowing into the transistor must equal the total current flowing out.

We can assemble the circuit shown in Fig. 41.2a, to find out how the currents in the three connections to the transistor are related. The 5 kΩ potentiometer is used to vary the potential at the input terminal and hence the current flowing into the base of the transistor. When this base current is increased we find that the collector current also increases (that is, the current flowing via the 2.2 kΩ resistor into the collector of the transistor). This collector current is much greater than the base current, and is very nearly directly proportional to the base current over a large range of currents. Typical results are shown in the graph of Fig. 41.2b.

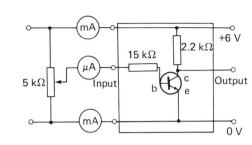

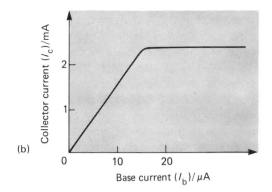

Fig. 41.2 (a) Measuring the currents in the three connections to the transistor. (b) Typical graph of collector current against base current.

According to Kirchhoff's first law the current flowing out from the emitter should equal the sum of the base current and the collector current. In symbols we can write:

$$I_e = I_b + I_c$$

In practice I_e and I_c will probably appear to be exactly equal, because I_b is much less than either of them.

Consider now what happens to the potential of the output terminal when we increase the potential of the input terminal of the module. An increase in potential at the input terminal causes an increase in the base current to the transistor. This causes an increase in the collector current, and this causes an increase in the p.d. across the 2.2 kΩ resistor. Because the top end of this resistor is held at the potential of the positive rail, about +6 V, the potential of its lower end, connected to the collector, must decrease. Thus an increase of input potential results in a decrease of output potential.

This accounts for the inverting effect which we met in Chapter 40. We now proceed to study the relationship between input and output voltage (using the everyday term 'voltage' to mean 'potential with respect to the negative supply rail' in this context).

41.2 Measuring input and output voltages

Figure 41.3 shows the circuit for making these measurements. The potentiometer across the supply rails provides an input voltage which can be varied from 0 V to about +6 V. The results of a typical experiment are shown in the graph of Fig. 41.3. The output voltage drops very suddenly from nearly +6 V to almost zero when the input voltage V_i is increased by a very small amount, from about 0.5 V to 0.7 V. Within this range of input voltages a small change in V_i causes a much greater change in V_o: this behaviour enables the module to be used for amplification.

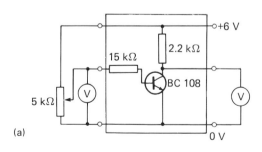

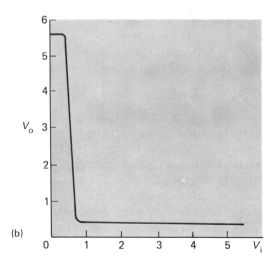

Fig. 41.3 (a) Measuring input and output voltages on the amplifier module. (b) The voltage transfer characteristic for the amplifier module.

The graph of Fig. 41.3 can be used to predict the *voltage gain* (or *voltage amplification factor*) obtainable with this module. Over the range V_i = 0.5 V to V_i = 0.7 V the graph is very nearly linear. The difference between these two values, for which the symbol ΔV_i is used, is 0.2 V. When the input voltage increases by this amount the output voltage decreases from 5.5 V to 0.5 V. For the change in output voltage we write:

$$\Delta V_o = -5.0 \text{ V}$$

using a minus sign because the change in V_o is a decrease. The voltage gain A is defined as the ratio

$$\frac{\Delta V_o}{\Delta V_i}$$

and so, with our experimental results:

$$\text{voltage gain} = \frac{-5.0}{0.2} = -25$$

This result can be expressed in words by saying that we have an *inverting amplifier* with a voltage gain of 25. As will be shown later, the fact that the amplifier inverts the voltage is no disadvantage: on the contrary, in many applications the inverting function is essential.

41.3   Amplifying an audio frequency signal

To use the module to amplify an alternating voltage, which we call the 'signal', we must arrange that this signal will cause the output voltage V_o to vary within the limits $V_o = +1$ V and $V_o = +5$ V. In the absence of any signal, therefore, V_o must be set at about $+3$ V, midway between the two limits. This will be the *quiescent value* of the output voltage. When a signal is fed to the amplifier the output voltage will vary above and below this quiescent value. The point on the graph (Fig. 41.3b) corresponding to the quiescent value of V_o is called the *operating point* (or *working point*). To set this operating point correctly we have to apply *bias* to the amplifier input. There are several ways of providing this bias, but for the sake of convenience we shall use a potentiometer.

Figure 41.4 shows the complete circuit for the simple a.f. (audio frequency) amplifier. To set the correct bias we connect a voltmeter between terminal A and the negative rail and then adjust

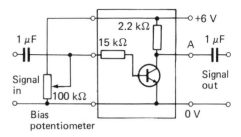

Fig. 41.4   The experimental audio amplifier.

the potentiometer until the voltmeter reads $+3$ V. The voltmeter should then be removed. The input signal is fed to the amplifier module via the 1 μF capacitor which allows a.c. to flow, but does not provide any d.c. connection between the signal source and the amplifier, so that the d.c. bias is undisturbed. The value of capacitance is not critical provided that its reactance, at the signal frequency used, is not greater than a few tens of ohms. The reasons for this will be clarified in Section 41.6

A suitable frequency for this experiment, and those which follow, is about 5 kHz. An audio signal generator is used to provide a sinusoidal alternating voltage at this frequency, with amplitude initially set at about ± 0.1 V, and is connected to the terminals labelled 'signal in' in Fig. 41.4. A c.r.o. is connected to the terminals labelled 'signal out'. Care must be taken to link the earthed terminal of the signal generator (usually coloured black) with the earthed input terminal of the c.r.o. via the negative power rail connection (labelled '0 V') in the circuit, as shown in the diagram.

The amplitude of the signal is gradually adjusted until the output signal appearing on the c.r.o. screen has an amplitude of about ± 2 V. It cannot be any greater than this without 'clipping' of the waveform occurring. The sensitivity (or *gain*) of the c.r.o. input should be adjusted so that the amplified signal waveform nearly fills the screen, with ideally about three or four complete cycles displayed, as in Fig. 41.5b. The c.r.o. 'live' (or positive) input connection is then transferred to the upper 'signal in' terminal, and a trace of much smaller amplitude appears, as in Fig. 41.5a.

The voltage gain can be estimated using this relationship:

$$\text{voltage gain} = \frac{\text{amplitude of output voltage}}{\text{amplitude of input voltage}}$$

Fig. 41.5 Signal waveforms obtained with the audio amplifier circuit, as displayed on a c.r.o. with its gain controls unaltered during the experiment.
(a) Input signal.
(b) Output signal showing little distortion.
(c) 'Clipped' output signal which results from overloading the amplifier.

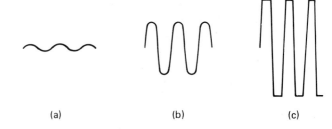

(a)   (b)   (c)

When using a c.r.o. to make the voltage measurements it is best to measure the double amplitude of each voltage (the 'peak-to-peak' value). If a double-beam c.r.o. is available the input and output signals can be displayed simultaneously. The voltage gain measured in this way should agree approximately with the value predicted in the previous experiment. Typical values range from about 20 to about 40 (depending upon the characteristics of the transistor itself). The same transistor could be used in a circuit to provide greater values of voltage gain, but values in the range 20 to 40 are best suited to this series of experiments.

The experiment just described (Fig. 41.4) can be used to show the effect of *overloading* the input of the amplifier. This will not harm the module, provided that the input signal voltage amplitude does not exceed a few volts. This means that an audio signal generator can be turned up to full output without any harm being done. If the amplitude of the signal going into the amplifier is gradually increased while the output signal is being displayed on the c.r.o., then sooner or later the waveform of the output signal will be 'clipped', as shown in Fig. 41.5c. This clipping happens because the output voltage cannot go beyond the limits set by the power rail voltages, + 6 V and 0 V in this case. This clipping is one form of *distortion* which can occur in amplifiers.

When it occurs in a sound reproducing system it causes a very unpleasant distortion of the sound. In this experiment the effect can be made audible by connecting a high impedance earphone across the 'signal out' terminals. A loudspeaker is not suitable because its impedance is too low for use with this experimental amplifier: this point will be clarified in the next section.

41.4 The output part of the amplifier

For further experiments on the principles of amplification we add a resistive load to the output of the amplifier. Up to now the amplifier's output has been effectively 'on open circuit', because the only item connected to it has been the c.r.o., and the resistance between the input terminals of a c.r.o. is very high: about 1 MΩ.

Adding a load to the amplifier output simulates more closely the conditions under which any single stage of amplification works in practice. Figure 41.6 shows the complete circuit including the load. A resistance box providing values of resistance from around 100 Ω up to about 1 MΩ is suitable for the load. The output coupling capacitor allows the alternating signal current to flow in the load, but does not provide any d.c. connection, so there is no risk of disturbing the quiescent value of the output voltage.

The first experiment to do with the load

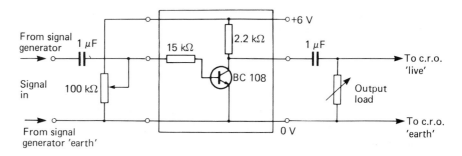

Fig. 41.6 Circuit for investigating the effect of loading the amplifier output.

Table 41a Performance of amplifier on load

Data from experiment			calculated from data		
Load resistance/kΩ	Output voltage amplitude/V	Output current amplitude/mA	'Lost volts'/V	Output resistance/kΩ	Output power/mW
R_L	V_o	$I_o = \dfrac{V_o}{R_L}$	$= 1.40\,V - V_o$	$= \dfrac{\text{lost volts}}{I_o}$	$I_o^2 R_L$
Infinite	1.40	0.00	0.00	–	0.00
10	1.25	0.13	0.15	1.2	0.08
4.7	1.1	0.23	0.30	1.3	0.13
2.2	0.85	0.39	0.55	1.4	0.16
1.5	0.75	0.50	0.65	1.3	0.19
1.0	0.60	0.60	0.80	1.3	0.18
470 Ω	0.35	0.74	1.05	1.4	0.13

connected is to measure what effect altering the load resistance has upon the output signal voltage amplitude. In this experiment we are not concerned with measuring the input voltage amplitude: this will be kept constant. We should, however, check that this input signal voltage does in practice stay very nearly constant, regardless of alternations in the output load.

A typical set of results is shown in Table 41a. The data from the experiment are shown in the first two columns. Reducing the load resistance, thus drawing more current from the amplifier, causes a drop in the output voltage. This is similar to what happens with any electrical supply when progressively more and more current is taken from it as a result of reducing the resistance of whatever load is connected to it.

For each value of load resistance the current flowing in the load has been calculated by dividing the output voltage V_o by the load resistance R_L. Very nearly the same kind of variation of voltage with current would be shown by an electrical supply having constant e.m.f. and constant internal resistance. For such a supply, as we saw in Chapter 23, there is the relationship:

$$\text{internal resistance} = \frac{\text{'lost volts'}}{\text{current}}$$

Using this relationship to calculate values for the effective internal resistance of the amplifier's output yields the results shown in the fifth column of Table 41a. These are near enough to being constant for us to suppose that the relationship is valid for the amplifier output, bearing in mind that there is a sizeable range of uncertainty in the experimental data. The resistors in the inexpensive resistance box had a tolerance of $\pm 10\%$, and one must allow at least $\pm 5\%$ uncertainty in the measurements made with the c.r.o.

From the foregoing results we can devise a very useful theoretical model for the amplifier. The correct name for this model is *small signal*

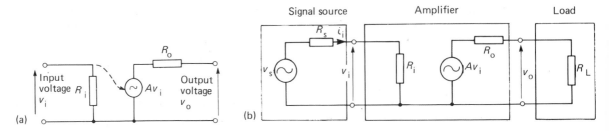

Fig. 41.7 (a) The simplest equivalent circuit model for the audio amplifier. R_i is the input resistance, R_o is the output resistance and Av_i is the output e.m.f. *A* is the voltage amplification factor or voltage gain. (b) The equivalent circuit for the complete amplification system.

a.c. equivalent circuit. Figure 41.7a shows the simplest version of such an equivalent circuit, and this simple version will enable us to develop the theory of amplification much further.

The term *output resistance* is the name given to the effective internal resistance of the output part of the amplifier. From the figures in the fifth column of Table 41a we see that the output resistance of our amplifier is approximately 1.3 kΩ. The symbol R_o will be used for this. The power figures in the last column of the table show that the power output is greatest when the load resistance is about 1.5 kΩ. This is consistent with the theory developed in Chapter 23 where it was shown that a source of electrical energy delivers maximum power to a load when the load resistance equals the internal resistance of the source.

41.5   The input part of the amplifier

As shown in Fig. 41.7, the input part of the amplifier can be represented by a single resistance, R_i. This drastic simplification can be made only if a number of conditions are fulfilled: these are discussed in the next section. For the present we shall assume that these conditions are satisfied.

The most direct method of measuring R_i would be to make simultaneous measurements of voltage and current at the input of the amplifier, but in practice these are usually so small that they would be very difficult to measure without sensitive instruments.

A much simpler, although indirect, method is illustrated in Fig. 41.8. It must be emphasized, however, that this simple method can be used only if the internal resistance of the signal source, R_s, is small compared with the input resistance of the amplifier, R_i. As Fig. 41.8 shows, an extra resistance R_x is added in series with the signal input. This resistance is increased until the amplitude of the output signal decreases to half of its original value. Assuming that the amplifier keeps constant gain when this is done, we can deduce that the value of v_i must have decreased to half of its original value: and this occurs when $R_x = R_i$. For the amplifier circuit of Fig. 41.4 the value of R_x is found to be in the range of about 3 kΩ to 6 kΩ. Differing results are obtained because different transistors, even though they have the same type number, have different *current gains*.

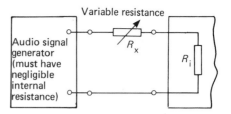

Fig. 41.8 Arrangement for finding the input resistance R_i of the amplifier.

41.6   A simple equivalent circuit for the amplifier

The amplifier, at its output, behaves like a generator whose e.m.f. is Av_i and internal resistance is R_o. This e.m.f. is controlled by the input to the amplifier. The input behaves like a resistance R_i. The whole amplifier can be represented by the *equivalent circuit* shown in Fig. 41.7a. The broken line in the diagram indicates the control that the input has over the 'generator' of e.m.f. Av_i. An analogy is the control that the handle of a tap has over the flow of water through the tap. In practice this broken line is omitted, as in Fig. 41.7b.

The simple equivalent circuit of Fig. 41.7a represents a remarkable simplification of a circuit which in fact contains six components: a transistor, two fixed resistors, one potentiometer, and two capacitors. But we must be cautious in using the equivalent circuit to make predictions about the amplifier's performance. During our development of the ideas leading up to this simple equivalent circuit we have made, explicitly or implicitly, several bold assumptions, approximations, and simplifications. These can be summed up as follows:

1. All signal voltages referred to, and appearing in all formulae and calculations relating to the equivalent circuit, are the *amplitudes of alternating voltages*. The amplifier would perform entirely differently if steady d.c. voltages were applied to its input (remember the input coupling capacitor, for instance).

2. The instantaneous values of signal voltages must never go beyond the linear portion of the input–output voltage characteristic. Signals which fulfil these conditions are called *small signals*, and in all formulae and expressions we use lower-case letter symbols for them.

3. All coupling capacitors, and any other capacitance and inductance within the amplifier circuit and within the components themselves, are assumed to have *negligible reactance*: this is so that all passive elements in the circuit can be assumed to behave as pure resistances.

4. The pure resistances of (3) above are assumed to have constant values. This is, in fact, only an approximation to the truth.

5. The amplifier is assumed to have only *forward transfer* effects and no *reverse transfer* effects. That is, altering voltages or currents at the input will produce an effect upon the output signal, but altering whatever is connected to the output will have no effect on any voltages and currents at the input. This, again, is not strictly true for our experimental amplifier: a careful experimenter will notice that there are slight reverse transfer effects.

Because of these limitations the theoretical equivalent circuit which we have developed should be termed, in full, a *non-reactive small-signal a.c. equivalent circuit*.

A little more needs to be said about the simplifying assumptions which have been made. Considering point (3) above: are we justified in ignoring any reactive effects in the amplifier circuit? Can we allowably 'forget' the presence of the two 1 μF coupling capacitors, for instance? As shown in Section 37.3 the reactance of a capacitor in an a.c. circuit is given by

$$X = \frac{1}{2\pi f C}$$

If $f = 5$ kHz and C is 1 μF, this gives $X = 32\,\Omega$. Compared with the input resistance of the amplifier which, as we have seen, is about 6 kΩ, this 32 Ω is very small. This means that the amplitude of the alternating voltage across the capacitor will be a negligible fraction of the input voltage, and also that the phase-difference between voltage and current in the input circuit will likewise be negligible. However, if the signal frequency were much lower, since $X \propto 1/f$, this would no longer be true, and the 1 μF capacitor would 'drop' an appreciable fraction of the ingoing signal voltage, thus reducing the overall gain of the amplifier.

A similar argument applies to the 1 μF output coupling capacitor. As was shown, the output resistance of the amplifier is about 1.3 kΩ, and compared with this the 32 Ω reactance of the capacitor is negligible.

41.7   Using the equivalent circuit model

Provided that we keep within the limitations set out in the previous section, we can proceed to use the equivalent circuit to develop the theory of amplification further. The chief use of the simple equivalent circuit is to predict the values of input voltage and current and of output voltage and current, and hence voltage gain, current gain, and power gain. To do this we have to consider the complete amplification system, which can be represented as in Fig. 41.7b

At the input of the amplifier the values of input voltage and current do not depend only upon the e.m.f. of the signal source: the total input circuit resistance has to be taken into account. As shown, the equivalent circuit model is used once again to represent the signal source, which behaves as if it were simply a generator of an alternating e.m.f. of amplitude v_s in series with a resistance R_s. The total input circuit resistance is $R_s + R_i$, and so the input current is given by

$$i_i = \frac{v_s}{R_s + R_i}$$

The input circuit, in our theoretical model, has the same structure as the output circuit, and similar rules will therefore apply to it. It has already been shown (Section 41.4) that, at the output, maximum power is delivered to the load when $R_L = R_o$. Similarly, at the input, maximum power will be transferred from the signal source to the amplifier input when $R_s = R_i$. Thus, if it is required to obtain the greatest possible power gain with this amplification system, the source should be *matched* to the amplifier input and the load matched to the amplifier output. This is often what is required in r.f. (radio frequency) amplification systems: but you should not suppose that this is always the most desirable arrangement. In many applications the designer makes R_i much greater than R_s, so that when the amplifier input is connected to the signal source it imposes negligible load upon the source. In a similar way it is often best to arrange that R_o is much less than R_L. This is usually the case with the final stage of an audio amplifier feeding a loudspeaker (the *power amplifier* section). A typical loudspeaker unit has an impedance of $4\,\Omega$ to $8\,\Omega$. The impedance includes a reactance which varies with frequency as well as resistance: nevertheless the same broad principles are applicable. If R_o is about $0.1\,\Omega$ or less, as is often the case, then the amplifier's performance will be very little affected by altering the load, provided that the impedance of the load is never less than a few ohms.

For most applications the 'ideal amplifier' is one which has infinite input impedance and zero output impedance. Using integrated circuit operational amplifiers enables one to approximate very closely to this ideal quite easily: much more easily than by using a circuit built around separate transistors, although a single *field effect transistor* (FET) can be used as the basis of a simple amplifier which has extremely high input impedance.

41.8 The frequency bandwidth of the amplifier

Having developed the simple equivalent circuit model for the experimental amplifier, and having made some use of the model, we straight away go on to break the bounds of its usefulness! If the equivalent circuit of Fig. 41.7 were the *actual* circuit for the amplifier, then it ought to perform in the same way at all frequencies, because pure resistance in a circuit causes a constant relationship between voltage and current, regardless of frequency. The experiment now to be described shows how the gain of the amplifier falls off at low frequencies and at high frequencies. This leads to the concept of the frequency *bandwidth* of the amplifier.

For this experiment we use the same circuit as for the previous experiment, Fig. 41.6. The load resistance is set so that it is approximately matched to the amplifier output resistance. In our example, a value in the range $1\,\text{k}\Omega$ to $1.5\,\text{k}\Omega$ is satisfactory. Under these conditions we shall get most of the available output power transferred to the load.

The method of expressing power gain in decibel (dB) units will be explained in Section 41.9. As we saw in Section 21.1 a decrease in power of 3 dB means that the power has been

halved. The frequency bandwidth of an amplifier is usually understood as being the *half power bandwidth*. This concept is not applicable to every type of audio amplifier; only to those which are designed to have an approximately 'flat' response over a range of frequencies, or to *tuned* amplifiers which have a single, sharp 'peak' in the graph of gain against frequency.

Our experimental amplifier has an approximately flat response over much of the audible range of frequencies, as shown in Fig. 41.9. Note that this is a graph of voltage gain, not power gain, against frequency: but this is no drawback because the half-power points are easily located. The power delivered to the load is proportional to V^2, and so to halve the power the voltage must be reduced to $V/\sqrt{2}$. Thus the half-power points are where the output voltage is a fraction $1/\sqrt{2}$ of its maximum value. $1/\sqrt{2}$ is almost exactly 0.7. In Fig. 41.9 it will be seen that these half-power points occur at frequencies of about 15 Hz and 20 kHz: so these are the limits of the frequency bandwidth. Note that the scale of frequency in Fig. 41.9 is a logarithmic scale. This is usual practice for response graphs of this kind.

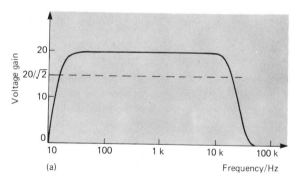

(a)

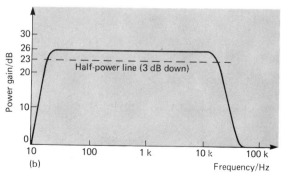

(b)

Fig. 41.9 The frequency response of the amplifier shown in Fig. 41.6 (a) in terms of voltage gain and (b) in terms of power gain using decibels.

If our experimental amplifier were to be used as one amplifying stage in a complete audio amplification system, it would not be too bad, in that its bandwidth covers nearly the whole range of audible frequencies. However, if you study the circuit diagrams of good audio amplifiers, you will find that the circuit of each stage of amplification is nearly always considerably more complicated than the circuit of our experimental amplifier. The reasons for this chiefly concern the problem of *stability*: ensuring that the amplifier's performance will not alter as a result of changes in temperature (to which all semiconductor devices are sensitive); of changes in the characteristics of components due to ageing; and of variations in the characteristics of individual components within each manufactured batch. Obviously, when mass-producing electronic equipment, it is economically undesirable to have to alter the circuit design to suit each individual transistor, for instance. Within one manufactured 'run' of transistors having the same type number there are inevitably variations in characteristics. As long as these variations are within certain (often quite wide) limits, correct circuit design ensures that these variations do not cause appreciable variations in the performance of the circuit. These practical considerations will be taken up in the next chapter, concerning *feedback*.

Problem 41.1   A certain audio power amplifier, which has a flat response curve down to a frequency of 20 Hz, has a negligibly small output impedance. It is to be coupled to a load via a capacitor, C, as shown in Fig. 41.10. The load can be regarded as a pure resistance of 8 Ω. What minimum capacitance must C have, if the power to the load is not to fall below half of its mid-range value at a frequency of 20 Hz?

We assume what at mid-range frequencies the reactance of C is negligible, so that the voltage

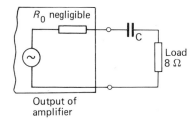

Fig. 41.10   Output of amplifier

across the load is equal to the output voltage of the amplifier. The power in the load will fall to half of its mid-range value when the reactance of the capacitor has the same magnitude as the load resistance, because then the voltage across C and the voltage across R will have the same amplitude, and since these two voltages are 90° out of phase, the amplitude of each will be $1/\sqrt{2}$ of the output voltage of the amplifier. (See Section 37.3.) The power in the load is proportional to the square of the voltage across it.

Thus, we have to find what value of C gives it a reactance of 8 Ω at a frequency of 20 Hz.

$$\frac{1}{2\pi \times 20\, C} = 8\,\Omega$$

therefore

$$C = \frac{1}{2\pi \times 20 \times 8}\, F$$

$$\approx 1000\, \mu F$$

41.9   Decibels

The *decibel* is a unit which can be used to express the *power gain* of an amplifier, as has already been mentioned in Section 41.8. It can also be used to express the *power attenuation* of a circuit which reduces the power of a signal. We shall now explain how and why the decibel unit is used.

The *decibel* originated as a unit for expressing the loudness of sounds. You should refer to Section 21.1, where this is discussed, before proceeding. The abbreviation for decibel is dB.

The power gain of an amplifier can be expressed thus:

$$\begin{matrix}\text{power gain} \\ \text{(in decibel)}\end{matrix} = 10 \log_{10} \left(\frac{\text{power output}}{\text{power input}} \right)$$

Problem 41.2   A certain audio frequency amplifier provides an output power of 100 W to a loudspeaker when the input power to the amplifier is 0.25 mW. What is the power gain, expressed in decibel?

$$\text{power gain} = 10 \log_{10} \left(\frac{100}{0.25 \times 10^{-3}} \right)$$

$$= 10 \log_{10} (4 \times 10^5)$$

$$= 10 \times 5.60$$

$$= 56.0\,\text{dB}$$

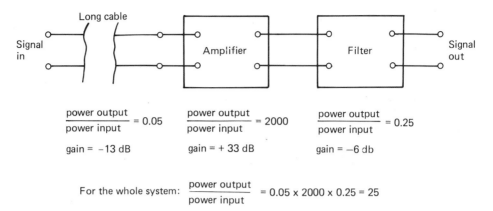

$$\frac{\text{power output}}{\text{power input}} = 0.05 \qquad \frac{\text{power output}}{\text{power input}} = 2000 \qquad \frac{\text{power output}}{\text{power input}} = 0.25$$

gain = –13 dB   gain = + 33 dB   gain = –6 db

For the whole system: $\dfrac{\text{power output}}{\text{power input}}$ = 0.05 x 2000 x 0.25 = 25

Fig. 41.11   Or, in decibels: power gain = –13 dB + 33 dB –6 dB = + 14 dB

For a circuit which reduces the power of a signal the *power attenuation* can likewise be expressed in decibel. The word 'attenuation' comes from Latin 'attenuare' meaning 'to make thin'. We use exactly the same expression as the one given above for power gain.

Problem 41.3   A certain tone-control filter circuit reduces the power of an audio signal by a factor of 100 at one particular frequency. What is its power attenuation, expressed in decibel units?

$$\begin{aligned}
\frac{\text{power gain}}{\text{(in decibel)}} &= 10\log_{10}\frac{1}{100} \\[4pt]
&= 10\log_{10}0.01 \\
&= 10 \times -2 \\
&= -20\,\text{dB}
\end{aligned}$$

It is indeed correct to say that this filter provides a *power gain* of − 20 dB. The minus sign indicates that the power output is less than the power input, and that the circuit therefore provides power attenuation. We can express the result in two equally valid ways:

'− 20 dB of gain'   or   '20 dB of attenuation'

The decibel system can be useful for expressing the overall gain of a system in which a signal is passing through a number of distinct 'stages' arranged one following another. Figure 41.11 shows an example of this. In this example there are three stages in the system through which the

signal passes: the first stage is a long length of cable which reduces the power of the signal, the second stage is an amplifier, and the third stage is a filter circuit which reduces the power. For each stage the ratio: power output/power input is given below. The value of this ratio for the whole system can be found by multiplying together the ratios for the three stages; but if the power gain for each stage is expressed in decibel units then we simply *add* the gains in order to find the overall power gain. Thus:

For the three stages in fig. 41.10 we have:

$\dfrac{\text{power output}}{\text{power input}}$	0.05	2000	0.25
gain in dB	− 13	+ 33	− 6

for the whole system:

$$\frac{\text{power output}}{\text{power input}} = 0.05 \times 2000 \times 0.25 = 25$$

or, in decibel:

$$\begin{aligned}
\text{power gain} &= -13\,\text{dB} + 33\,\text{dB} - 6\,\text{dB} \\
&= +14\,\text{dB}
\end{aligned}$$

In the example of Fig. 41.11 note that the power gain in decibel has, for each stage, been calculated to the nearest whole number. For instance, $\log_{10}0.05$ is $\overline{2}.6990$ to four places of decimals, and

$$\overline{2}.6990 = -2 + 0.6990 = -1.3010$$

so that $10 \log_{10} 0.05 = -13.010$, to three places of decimals. In practice a two significant figure value for power gain in decibel is usually accurate enough, and often the figure can be 'rounded' to the nearest whole number without appreciable loss of accuracy. Thus, if an amplifier doubles the power of a signal, we say that the power gain is $+3$ dB. Quadrupling the power means a gain of $+6$ dB. Increasing the power by a factor of 5 times means a gain of $+7$ dB. Likewise halving the power means a gain of -3 dB, and so on.

Refer back now to Fig. 41.9b, where you see an example of the generally accepted way of displaying the frequency response of an amplifier. In the language of electronic engineers, we can say that the output signal power is '3 dB down' at frequencies of about 15 Hz and 20 kHz, meaning that these are the frequencies at which the output power falls to half of its mid-range value.

It is common, although not strictly correct, practice in the world of electronics to use decibels for expressing *voltage* gain and attenuation. Suppose that in a certain amplifier the r.m.s. value of the input voltage is v_i and the input resistance of the amplifier is R_i: then the power input to the amplifier is given by:

$$p_i = \frac{v_i^2}{R_i}$$

Similarly, if the r.m.s. output voltage is v_o and the load resistance is R_L, then the power output from the amplifier to the load is given by:

$$p_o = \frac{v_o^2}{R_L}$$

If, and *only if*, $R_i = R_L$ (and this is by no means always the case), we can write:

$$\text{power gain} = 10 \log_{10} \left(\frac{v_o^2}{v_i^2} \right) \text{dB}$$

$$= 20 \log_{10} \left(\frac{v_o}{v_i} \right) \text{dB}$$

In systems where a signal passes through a succession of stages whose inputs and outputs are matched to each other, the expression given above can be correctly used. In such a context we could say that, for example, a certain stage has a 'voltage gain of 60 dB'. What we really mean is that the *power* gain is 60 dB, but that the voltage gain is related to it by the expression above. So in this example:

$$20 \log_{10} \left(\frac{v_o}{v_i} \right) = 60$$

hence

$$\log_{10} \left(\frac{v_o}{v_i} \right) = 3$$

and so

$$\frac{v_o}{v_i} = 1000$$

In other words, this particular stage amplifies the voltage by a factor of 1000.

It has become common practice to express voltage gains (and attenuations) in decibel, regardless of the values of input resistance and load resistance, and this can be misleading.

Chapter 42

FEEDBACK

42.1 Positive and negative feedback

Feedback, in the context of electronics, means taking a portion of the output signal from an amplifier and 'feeding it back' to the input. If the fed-back portion of the output signal is added to the input signal to produce a larger input signal, it is called *positive feedback*. With a.c. signals this means that the fed-back signal is in phase with the input signal.

Negative feedback is the result of arranging that the fed-back portion of the output signal is subtracted from the input signal, thus diminishing it. With a.c. signals this is done by arranging that the fed-back signal is in antiphase with the input signal.

Positive feedback, when applied to an amplifier, increases the gain: negative feedback decreases the gain. In the very early days of radio, when thermionic valves were being used to amplify the signal, positive feedback was sometimes used to obtain increased amplification, and greater sharpness of tuning (greater *selectivity*). The danger in doing this was that a little too much positive feedback caused distortion of the signal, and a bit more feedback produced ear-splitting oscillation. This unpleasant effect can easily be produced with any audio amplification system having a microphone and a loudspeaker: simply bring the microphone near to the loudspeaker and sooner or later a very loud whistle or shrieking sound will result.

If negative feedback reduces an amplifier's gain, then it might be supposed that this can never be of any practical value. But this is the exact opposite of the truth, as we shall see.

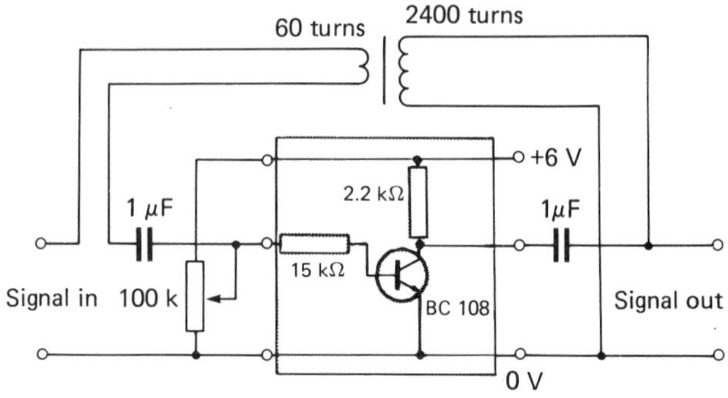

Fig. 42.1 Circuit for investigating positive and negative feedback.

42.2   An experiment to demonstrate feedback

Figure 42.1 shows a system which is used to study the effects of positive and negative feedback. It is built around the same audio amplifier circuit as was used in the previous experiments. (We should emphasize that the circuit has no practical use apart from demonstrating the *principles* of feedback.)

A step-down transformer is connected so that a fraction of the alternating output voltage from the amplifier is fed back to the input. Previously, with no load connected to the output, the voltage gain of the amplifier was in the range of about 15 to 20. In the system of Fig. 42.1 the turns ratio of the transformer is 1:40. In one version of the experiment this was achieved by using a standard type of small transformer kit with interchangeable coils and a magnetic alloy core. A 2400 turn coil and a 60 turn coil were used. The voltage across the 60 turn coil is thus 1/40 of the voltage across the 2400 turn coil. In this way 1/40 of the amplifier's output voltage is fed back and 'inserted' in series with the input, as shown.

According to which way round we make the connections to the 60 turn coil, we can obtain positive or negative feedback. In the experiment this is most easily done by trial and error. A c.r.o. is used, as in the earlier experiments, to compare the amplitude of the output signal with that of the input signal. It is found that, with the connections to the 60 turn coil one way round, the output signal is increased as a result of feedback, and reduced when the connections are reversed. The former case, giving positive feedback, may cause the circuit to oscillate and produce large signals with distorted waveform, but this will happen only in the unlikely event of the amplifier's voltage gain

being appreciably greater than 40. Positive feedback and oscillatory circuits are dealt with in Section 42.6.

Having shown that the feedback does affect the overall voltage gain, we can go on to make some measurements. Firstly the 60 turn coil is disconnected, and the input signal is fed straight to the amplifier, and the voltage gain without any feedback is measured. Typical results are shown in Table 42a. Throughout the experiment we keep the transformer primary connected to the output of the amplifier because it does to some extent load the amplifier output, and we must keep this load constant. Note that we have used a new symbol for the input voltage: v_g. This is in order to distinguish it from the voltage across the input terminals of the amplifier itself, for which we continue to use the symbol v_i. The letter 'g' in v_g can be thought of as standing for 'generator' in this context.

The results of Table 42a were obtained using a frequency of 5 kHz, but very nearly the same results were obtained at other frequencies in the range 1 kHz to 10 kHz: thus the frequency used is not critical. Within this frequency range the transformer 'behaves' satisfactorily in that its resistive and reactive components do not complicate the situation by introducing any appreciable loadings or any appreciable phase-shifts.

Table 42a   Effect of feedback on amplifier gain

Feedback	v_g	v_o	Overall gain v_g/v_o
None	0.12	2.0	17
Positive	0.10	2.5	25
Negative	0.14	1.6	11

42.3   The theory of feedback

An expression will be derived for the overall voltage gain of an amplifier with feedback, such as the experimental one shown in Fig. 42.1.

The symbol A is used for the voltage gain of the amplifier *on its own*, just as it has been used in the previous chapter. The *feedback fraction* is B and is provided by the feedback circuit, as shown in Fig. 42.2. Thus the feedback voltage is Bv_o. This is added to the voltage across the 'signal in' terminals, v_g; so we can write

$$v_g + Bv_o = v_i$$

but

$$v_o = Av_i$$

and the overall gain of the system is given by

$$\frac{v_o}{v_g} = \frac{Av_i}{v_i - Bv_o}$$

If we divide top and bottom of the expression by v_i we obtain

$$\frac{v_o}{v_g} = \frac{A}{1 - AB} \qquad (42.1)$$

Equation 42.1 is a general one for the voltage gain of any amplifier with feedback. But we must give a warning at this point: the writers of textbooks and article on electronics are not universally agreed about how positive and negative signs should be used in the algebraic expressions employed in the theory of feedback. You may think that we have introduced too many 'minus' signs in the following paragraphs: but we prefer to be rigorous in the matter of signs, as illustrated in the examples in Table 42b.

In Eq. 42.1, if the term AB (the *loop gain*) is positive, but less than unity, then $(1 - AB)$ is less than unity and so

$$\frac{A}{1 - AB} > A$$

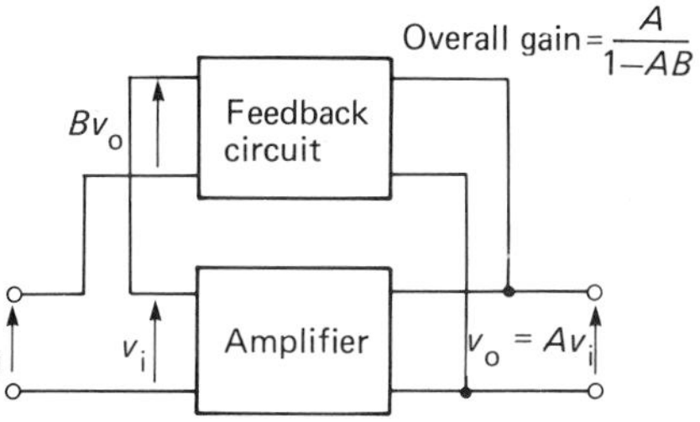

Table 42b   Examples of values of A and B, as used in the theory of feedback

$A = +100$	A non-inverting amplifier with a voltage gain of 100
$A = -25$	An inverting amplifier with a voltage gain of 25
$B = +0.05$	A feedback circuit which does not reverse the phase of the signal, providing a feedback fraction of 1/20
$B = -0.025$	A feedback circuit which reverses the phase of the signal, providing a feedback fraction of 1/40

This means that the overall gain is now greater than the gain of the amplifier without feedback: here we have positive feedback.

If AB has a negative value, then $(1 - AB) > 1$ and

$$\frac{A}{1 - AB} < A$$

here we have negative feedback. For the amplifier of Fig. 42.1 A is negative because we are using an *inverting* amplifier, as explained in Section 41.2. To obtain positive feedback, therefore, we had to ensure that B was negative also, so that the product AB was positive. This meant that the transformer was connected in such a way that the fed-back voltage was of opposite phase to the output voltage from the amplifier. We did this simply by trial and error, finding which way round the connections to the 60-turn coil of the transformer had to be made in order to cause an *increase* in the output voltage from the amplifier.

Does the theory predict correctly the experimental results shown in Table 42a? The measured value of A was -17, and the feedback fraction was 1/40: so $B = -0.025$ for positive feedback and $+0.025$ for negative feedback. So we calculate the overall gain as follows:
with positive feedback

$$\frac{A}{1 - AB} = \frac{-17}{1 - (-17 \times -0.025)} = -30$$

with negative feedback

$$\frac{A}{1 - AB} = \frac{-17}{1 - (-17 \times +0.025)} = -12$$

These give acceptable agreement with experi-

Fig. 42.2   Simplified diagram for amplifier with feedback.

mental results, if we allow for an uncertainty of about $\pm 10\%$ in the measurement of the input signal voltage amplitude, this being the least precise measurement in the experiment. (If you doubt this last statement, try calculating the overall gain with $A = -15$ instead of $A = -17$.)

What would happen if AB were to equal $+1$? This would make $(1 - AB) = 0$, and the overall gain equal to infinity: we shall discuss the effect of doing this in Section 42.6.

42.4   Negative feedback: why it is used

Let us consider an example which illustrates the usefulness of negative feedback. The example concerns the type of integrated circuit (IC), constructed on a silicon chip, which is used for *operational amplifiers* (see Section 42.5). These have, as a rule, a very high *intrinsic gain*. By 'intrinsic gain' we mean the gain of the device without any external circuitry added to it. The intrinsic gain of such an 'op amp IC', as well as being very large, is likely to vary appreciably from one individual in a manufactured batch to another. Negative feedback is almost always used to reduce the gain to a precisely determined value.

Suppose that one such 'op amp' has an intrinsic voltage gain of 10 000 (*not* an inverting amplifier in this case, so that A is positive). A circuit is constructed around it to provide negative feedback so that $B = -1/100$. The overall gain is given by the expression used above:

$$\frac{A}{1 - AB}$$

Then suppose that we substitute an op amp which has a higher intrinsic gain, say 100 000 into the circuit. How does the overall gain compare in these two cases? Here are the calculations, side by side for direct comparison.

$A = 10\,000$	$A = 100\,000$
$B = -\dfrac{1}{100}$	$B = -\dfrac{1}{100}$
Overall gain $= \dfrac{10\,000}{101}$	Overall gain $= \dfrac{100\,000}{1001}$
$= 99.0$	$= 99.9$

This shows that a tenfold difference in the intrinsic gain of the op amp has resulted in only a minute difference in the overall gain. Furthermore, the overall gain is determined almost entirely by the value of B, the feedback fraction, and the value of A has very little effect upon the result. This arises because, in these examples, $AB \gg 1$, and so

$$1 - AB \approx -AB$$

and so

$$\frac{A}{1 - AB} \approx \frac{A}{-AB} \approx -\frac{1}{B}$$

In our example

$$B = -\frac{1}{100} \quad \text{and so} \quad -\frac{1}{B} = 100$$

The overall gain, in either case dealt with in our example, does not differ from this value by more than 1%.

Nowadays there is scarcely any single commercially produced amplifier, of any type, which does not employ negative feedback in one or more stages of amplification. An example of the value of negative feedback has already been given: minimizing the effect of variations in the intrinsic gain of an operational amplifier IC. Three of the chief reasons for using negative feedback are concerned with:

i)   Stability of gain.
ii)   Frequency bandwidth.
iii) Frequency-dependent gain.

i) *Stability of gain.* With IC amplifiers, such as the one in the example above, and with transistors, the characteristics can vary considerably from one individual to another in a manufactured batch. Also the characteristics can alter with age and with temperature. Negative feedback prevents variations in the intrinsic gain of these devices from altering the overall gain of the amplifying circuit.

ii) *Frequency bandwidth.* If negative feedback is applied by means of circuitry which is made up of pure resistances (or it could be provided by an 'ideal' transformer), then it is bound to increase the frequency bandwidth of the amplifier. As was demonstrated in the experiment of Section 41.8, the gain of an audio amplifier falls off at low and

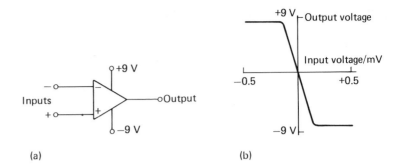

Fig. 42.3   (a) Symbol for an operational amplifier. (b) Input and output voltages for a typical operational amplifier.

(a)

(b)

high frequencies. In other words, the value of A decreases at high and low frequencies. If the feedback circuitry is not frequency dependent, then B will be constant regardless of frequency. Then if, as in (i) above, the value of the product AB is large compared with unity, the overall gain will be approximately $-1/B$ and thus very nearly independent of frequency.

iii) *Frequency-dependent gain*. Suppose that we wished to design an audio amplifier whose gain decreases in a regular fashion when the signal frequency is increased. An example of such an amplifier is any pre-amplifier designed for use with a magnetic pick-up cartridge for reproducing sound from disc recordings. This type of pre-amplifier is designed so that the power gain decreases by about 4 dB for every octave (that is, the power gain is reduced to a little less than half for every doubling of the frequency). This is most easily achieved by using a semiconductor amplifying device which has very high intrinsic gain, and a feedback circuit, made up of resistors and capacitors, whose attenuation increases by 4 dB per octave increase in frequency. Thus the value of B increases when the signal frequency increases, and the overall gain which is approximately $-1/B$ decreases. To find out the reasons why this kind of frequency-dependent gain is needed the reader must refer to a textbook which deals with audio amplification in detail. The chief reason for having an amplifier gain which becomes less as the frequency increases concerns the fact that the amplitude of the e.m.f. generated by a magnetic pick-up cartridge, when the stylus oscillates in the groove on the disc, increases with frequency: this increase has to be 'cancelled' by the amplifier.

42.5   Integrated circuit operational amplifiers

Here we shall introduce a type of amplifier, constructed as an integrated circuit on a single silicon chip, which is used in a great range of everyday applications. The term *operational amplifier* was originally used to describe the type of amplifier used in circuits to perform mathematical operations, in a certain class of computers called *analogue computers*. The integrated circuit amplifiers that were developed for this particular application were found to be well suited to many other applications, but the name 'operational amplifier' (or 'op amp') has been kept.

The internal circuit of an op amp is complex, comprising many transistors and resistors, and we do not need to know the details of this in order to be able to use it correctly. The most usual symbol for it is shown in Fig. 42.3. It requires a d.c. power supply giving positive *and* negative voltages (typically $+9\,V$ and $-9\,V$). It has two input connections, the upper one (labelled '$-$') being the *inverting input* and the lower one ('$+$') the *non-inverting input*. Which of the two inputs is used depends upon the application and in some applications both inputs are used. Figure 42.3 shows also a graph of the output voltage against an input voltage applied between the two input connections. From this you will see that the mean output voltage is zero: this allows the operating point to be set at zero when the op amp is used to amplify a.c. signals, unlike the single transistor amplifier (see Section 41.3) for which the operating point was about $+3\,V$.

The intrinsic gain of an op amp is extremely high, although it decreases as the signal frequency increases, as the graph of Fig. 42.4 shows. When

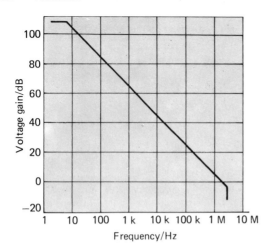

Fig. 42.4 Data for the 741 operational amplifier (courtesy RS Components Ltd.).

interpreting this graph, note that the frequency scale is logarithmic; note also that the voltage gain is expressed in decibels. As explained in Section 41.9, decibels are used primarily for expressing *power* ratios, but the decibel system is extended (rather loosely, as a rule) to expressing voltage ratios also. Refer back to that section to see exactly how that was done. This particular op amp serves as a good example of how negative feedback is used to stabilize the gain of an amplifier. To apply a constant amount of negative feedback we use two resistors, as shown in Fig. 42.5. The magnitude of the feedback factor, B, here is simply the ratio of the resistance R_a to the resistance R_f (the *feedback resistance*). Thus

$$B = \frac{R_a}{R_f} = \frac{4.7 \text{ k}\Omega}{120 \text{ k}\Omega} \approx \frac{1}{25}$$

As explained in Section 42.4, if the magnitude of the product $AB \gg 1$, where A is the intrinsic voltage gain of the amplifier, then the overall gain

of the amplifier with feedback $\approx -1/B$, and so in this example it is approximately -25.

This amplifier (Fig. 42.5) is an inverting amplifier with a voltage gain of about 25, and thus is similar to the single transistor amplifier module of Fig. 41.1.

The input resistance of an amplifier with feedback, as in Fig. 42.5, is approximately equal to R_a, and so in this case it is about 4.7 kΩ. The output resistance, however, is negligibly small. This makes it useful for applications in which a low resistance load is connected to its output: this principle was explained in Section 41.7.

Figure 42.6 shows an audio amplifier, suitable for the experiments of Chapter 41. An extra resistance of 1.2 kΩ has been added in series with the output to make the output resistance of the amplifier equal to about 1.2 kΩ, so that this, and its input resistance, are both similar to the values for the amplifier circuit of Fig. 41.4.

We have described only one application of an integrated circuit operational amplifier. The many other applications are described in most standard textbooks on electronics. Among these applications are: the voltage comparator, voltage follower, and analogue computer circuits for summing, differentiating, and integrating.

42.6   Positive feedback and oscillators

It has already been mentioned that almost any amplifier, given sufficient positive feedback is applied to it, may spontaneously break into oscillation. The experiment of Fig. 42.1 can be used to demonstrate this if the transformer step-down ratio can be reduced, preferably by using a secondary transformer having a larger number of turns. With most amplifier modules of the kind

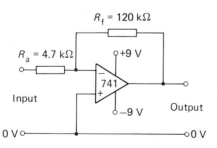

Fig. 42.5 Operational amplifier with negative feedback.

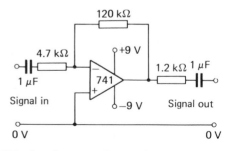

Fig. 42.6 Experimental audio amplifier using an operational amplifier.

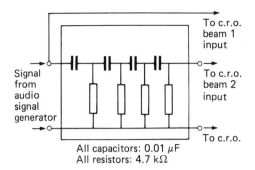

To c.r.o.
beam 1
input

Signal
from
audio
signal
generator

To c.r.o.
beam 2
input

To c.r.o.

All capacitors: 0.01 μF
All resistors: 4.7 kΩ

Fig. 42.7 Investigating the function of a phase-shift module.

used in that experiment, a transformer voltage ratio of 20:1 should cause the system to oscillate. A c.r.o. used to display the oscillating voltage at the output of the amplifier is likely to show a waveform which is not sinusoidal, probably a 'sawtooth' waveform of some kind. The frequency of oscillation may be anything from a few hundred hertz up to tens of kilohertz: this is not at all a stable or well-behaved oscillator and would certainly not be used by an electronic designer as a means of producing an alternating voltage.

It is instructive to carry out an experiment in which positive feedback is applied in a controlled way, and the frequency of oscillation is predictable. For this purpose we introduce a new four-terminal module which houses a *phase-shift network*. Figure 42.7 shows the circuit diagram. Before putting the oscillator together one should find out by experiment what the function of this phase-shift module is. Using an audio signal generator and a c.r.o., as shown in Fig. 42.7, the input and output voltage amplitudes are compared at several different frequencies over a range of about 1 kHz to about 20 kHz. Not just the amplitudes are compared, but the relative phase of the input and output voltage: for this a double beam c.r.o. is essential.

It will be found that the input and output voltages do not have the same phase, and that at one particular frequency the output voltage is in precise antiphase with the input voltage. At this frequency, which we shall call f_o, the module causes a phase-shift of 180° (π radian). At low frequencies the output voltage is much smaller in amplitude than the input voltage; in other words there is a large *attenuation*. This attenuation

becomes progressively less at higher frequencies. It can be shown theoretically that, at the phase-reversal frequency f_o, the ratio

$$\frac{\text{output voltage}}{\text{input voltage}} = -\frac{49}{901} \approx -\frac{1}{18}$$

In the circuit of Fig. 42.8 the output signal from the amplifier is fed back to its input via the phase-shift module: thus, for any signal present in the circuit at the frequency f_o, the feedback fraction $B = -1/18$. If the voltage gain of the amplifier by itself, A, is -25, then the loop gain $AB = 1.4$. Let us suppose that, when the power supply is switched on, there is present at the amplifier input a very weak signal at the frequency f_o. Every time this signal 'goes round the loop' its amplitude will increase by 1.4. Eventually the amplitude of the output signal from the amplifier will reach its maximum limit (which is about half the supply voltage, as was explained in Section 41.3), even though the input signal amplitude is still increasing. This means that the ratio

$$\frac{\text{output signal amplitude}}{\text{input signal amplitude}}$$

is decreasing: in other words, the value of A is decreasing. This process continues until $AB = 1$: then oscillation is maintained with a steady amplitude.

Although the theory is beyond the scope of this book, we can quote the formula which gives the phase-reversal frequency for the four-section phase-shift network of Fig. 42.7:

$$f_o = \frac{1}{2\pi} \sqrt{\frac{7}{10}} \frac{1}{RC} = \frac{0.133}{RC}$$

This expression should predict fairly closely the oscillation frequency of the circuit in Fig. 42.8.

The conditions under which oscillation will be maintained at a frequency f_o in an amplifier with feedback can be summarized as follows:

1) The loop gain $AB = 1$ at frequency f_o.
2) The overall phase-shift around the loop including amplifier inversion must be 360° (2π radian) or a whole number multiple of 360°, at frequency f_o

In the experiment of Fig. 42.8 one nearly always finds that the waveform of the output voltage is somewhat distorted from a pure

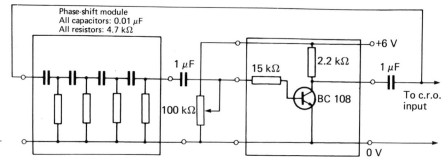

Fig. 42.8 Experimental phase-shift oscillator.

sinusoidal shape, because the intrinsic gain of the amplifier is too great. This can be reduced by introducing extra resistance in series with the input to the amplifier (next to the 1 μF capacitor), using, for instance, a 100 kΩ variable resistor, or a resistance substitution box.

Finally, one particular point of difficulty must be dealt with. In the circuit of Fig. 42.8, how can an oscillation ever 'get going'? Our argument above assumed that, at the instant of switching on the power, there was a weak signal at frequency f_o present in the circuit. But how can there be? To understand this we have to use a concept that was introduced in Unit 5: the concept embodied in

Fourier analysis. This is the idea that *any* time-varying quantity can be regarded as being composed of purely sinusoidal components, of different frequencies and amplitudes. This concept can be applied to the changing voltages which inevitably appear in any circuit when its power supply is first switched on: it is just as if there were countless sinusoidal signals, all of different frequencies, present. All these signals will die away rapidly except for the one at the frequency f_o: this one alone will grow in amplitude, as described earlier.

Oscillatory circuits of other types will be found in any standard textbook on electronics.

The Atom, Electrons and Radiation

Chapter 43

EXCITATION AND IONIZATION

43.1 Introduction

In Unit 4 we discussed a model for ideal gases in which particles collided with one another or with the walls of the containing vessel without the dissipation of energy. The forces involved were conservative (see Section 8.16) and the collisions were assumed to be perfectly elastic. The argument can be extended to real monatomic gases. In the case of the diatomic and other more complex molecules, energies other than the translational kinetic energy are involved. Nevertheless the effect, overall, is the same; the consequences of the many inelastic and superelastic collisions balancing out. (A superelastic collision is one in which translational kinetic energy is gained during the interaction. Although rarely met with in laboratory experiments on dynamics, such collisions are common in the microphysical world. Should, for example, two vibrating polyatomic molecules collide, they may show an increase of translational kinetic energy at the expense of their vibrational energy. It is such collisions which we have in mind when we say that collisions between the molecules of a gas can, *on average*, be taken to be elastic (see Section 10.2).) Under different circumstances, for example in a fine beam tube or a cloud chamber, interactions between particles cause changes which involve forces which are dissipative and the collisions are inelastic. A collision between an alpha particle and an atom, or between an electron and an atom may excite or even ionize the atom.

It is legitimate to enquire what the conditions are which lead to gaseous excitation and ionization. Controlled experiments are hardly possible in atmospheric air as we find it in the laboratory; but we can exercise control if we enclose some gas in a tube at low pressure and examine its properties when it is bombarded with a stream of electrons. The method has the great merit that we can determine the energies of the bombarding electrons by using a voltmeter. For we know that the energy of an electron (which has a charge of -1.6×10^{-19} C) passing across a potential difference of, say, 5 V is 5 eV or $5 \times 1.6 \times 10^{-19}$ J. That is 8×10^{-19} J.

With justice, it has been said that twentieth-century physicists have been great bombardiers! To understand the significance of the experiments they have performed we must return to the conditions which apply to elastic and to inelastic collisions (see Chapter 3).

43.2 Energy transfer in collision processes

a) Inelastic collisions

Consider a particle of mass m_1, moving with velocity u which collides inelastically with a stationary particle of mass m_2, making a direct hit and sticking to it. The two move off together with velocity v in the same direction as u (Fig. 43.1).

Momentum is conserved and so

$$m_1 u = (m_1 + m_2) v$$

But kinetic energy is not; for dissipative forces are at work. After the collision, the kinetic energy is

$$\begin{aligned} &\tfrac{1}{2}(m_1 + m_2) v^2 \\ &= \tfrac{1}{2}(m_1 + m_2)\left(\frac{m_1 u}{m_1 + m_2}\right)^2 \\ &= \tfrac{1}{2} m_1^2 u^2 \frac{1}{m_1 + m_2} \\ &= \tfrac{1}{2} m_1 u^2 \frac{m_1}{m_1 + m_2} \\ &= (\text{original K.E. of } m_1)\frac{m_1}{m_1 + m_2} \end{aligned}$$

The kinetic energy remaining is dependent upon the value of

$$\frac{m_1}{m_1 + m_2}$$

			Momentum	Kinetic energy
Before	m_1 $\rightarrow u$	m_2	$m_1 u$	$\frac{1}{2} m_1 u^2$
After		$m_1 + m_2$ $\rightarrow v$	$(m_1 + m_2)v$	$\frac{1}{2}(m_1 + m_2)v^2$

Fig. 43.1

This we see is

$$\frac{1}{1 + m_2/m_1}$$

So it is the ratio of the two masses which is important.

Let us take a few cases:

a) If $m_1 \ll m_2$ (for example a meteorite colliding with the moon) the ratio is very small. Almost all the kinetic energy of m_1 disappears.
b) If $m_1 = m_2$ (as when two dynamics troileys of equal mass collide and stick together) the ratio is one half. Half the kinetic energy of m_1 is lost.
c) If $m_1 \gg m_2$ (for example, when a moving motor-car strikes a moth) the ratio is nearly unity. The car loses very little of its kinetic energy.

b) Elastic collisions

Consider an elastic collision between the two particles m_1 and m_2. Before the collision, m_1 has velocity u and m_2 is at rest. After it, the masses have velocities v_1 and v_2 as in Fig. 43.2. Momentum is conserved so

$$m_1 u = m_1 v_1 + m_2 v_2$$

or

$$m_1(u - v_1) = m_2 v_2 \qquad (43.1)$$

			Momentum	Kinetic energy
Before	m_1 $\rightarrow u$	m_2	$m_1 u$	$\frac{1}{2} m_1 u^2$
After	m_1 $\rightarrow v_1$	m_2 $\rightarrow v_1$	$m_1 v_1 + m_2 v_2$	$\frac{1}{2} m_1 v_1^2 + \frac{1}{2} m_2 v_2^2$

Fig. 43.2

Since the collision is elastic we also know (see Chapter 3) that the

$$\left(\begin{array}{c}\text{relative velocity}\\\text{before collision}\end{array}\right) = -\left(\begin{array}{c}\text{relative velocity}\\\text{after collision}\end{array}\right)$$

$$(u - 0) = -(v_1 - v_2)$$

$$u + v_1 = v_2 \qquad (43.2)$$

Multiplying Eqs. 43.1 and 43.2 together gives:

$$m_1(u - v_1)(u + v_1) = m_2 v_2^2$$

$$m_1 u^2 - m_1 v_1^2 = m_2 v_2^2$$

$$\tfrac{1}{2}m_1 u^2 - \tfrac{1}{2}m_1 v_1^2 = \tfrac{1}{2}m_2 v_2^2$$

Thus the kinetic energy is conserved and is shared between the particles.

In order to relate v_2 and u, use Eq. 43.2 to eliminate v_1 in Eq. 43.1:

$$m_2 v_2 = m_1(u + u - v_2)$$

$$v_2(m_1 + m_2) = 2m_1 u$$

$$v_2 = \frac{2m_1 u}{(m_1 + m_2)}$$

$$v_2 = \frac{2u}{[1 + (m_2/m_1)]} \qquad (43.3)$$

Before discussing this equation, it is a small step to obtain a relationship between v_1 and u by combining Eqs. 43.2 and 43.3. Thus,

$$v_2 = u + v_1 = \frac{2u}{[1 + (m_2/m_1)]}$$

It is left as an exercise in algebra to show that this leads to:

$$v_1 = u\left[\frac{1 - (m_2/m_1)}{1 + (m_2/m_1)}\right] \qquad (43.4)$$

Considering some examples:

a) If $m_1 \ll m_2$ (for example, an electron striking an atom), the ratio m_2/m_1 is large and v_2 is small. v_1 is very little different from u but is reversed in direction. The electron bounces away having lost very little energy to the atom. This is rather like a rubber ball bouncing back from a wall against which it has been thrown.

b) If $m_1 = m_2$ (for example, an alpha particle striking a helium atom) the ratio of the two masses is unity and we see that the alpha particle stops, and all its energy is transferred to the atom (cf. Experiment B in Chapter 3).

c) If $m_1 \gg m_2$ (for example, an alpha particle striking an electron) m_2/m_1 is small and $v_1 \approx u$ and $v_2 \approx 2u$. The velocity of the alpha particle is little changed whilst the electron acquires almost twice the velocity of the alpha particle.

43.3 The experiment of Franck and Hertz

Among the pioneer bombardiers were J. Franck and G. Hertz (the nephew of H. Hertz). Their apparatus is represented diagrammatically in Fig. 43.3.

In the experiment, described in 1914, mercury vapour was used at a pressure of about $100\ \text{N m}^{-2}$ (normal atmospheric pressure is $10^5\ \text{N m}^{-2}$). The tube was contained in an oil bath at about 380 K.

The filament was heated by a current and produced thermionic electrons. These could be attracted to the grid by the application of a small variable potential difference. Electrons moving under the influence of this electric field would, of course, collide frequently with mercury atoms, and many would pass through the grid to fall on the anode. The anode was connected to a sensitive galvanometer. The reading of this galvanometer provided a measure of the number of electrons arriving at the anode each second.

By applying a small reverse voltage to the anode (relative to the grid) electrons passing through the grid could only reach the anode if they had sufficient energy as they entered this reverse

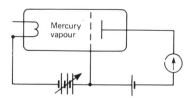

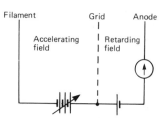

Fig. 43.3 Schematic representation of the Franck-Hertz experiment and of the electric fields within the Franck-Hertz tube.

Table 43a

Accelerating potential (filament to grid)/V; and electron energy at mesh/eV	Retarding potential (grid to anode)/V	Electron energy at the anode/eV	Current
0	−1	No electrons	No current
1	−1	No electrons	No current
2	−1	1	Current flow
3	−1	2	Current flow
4	−1	3	Current flow
5	−1	No electrons	No current
6	−1	1	Current flow

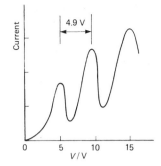

Fig. 43.4 Variation of current with potential for a mercury-filled tube.

field (see Fig. 43.3). If, for example, the reverse potential difference was 1 V only electrons with energies greater than 1 eV would be able to reach the anode and so contribute to the current registered by the galvanometer. Table 43a shows some typical situations.

In the experiment itself the retarding potential difference between the grid and the anode was kept at a small fixed value and the accelerating potential difference between the filament and the grid was raised slowly. The current flowing from the anode was recorded. As long as the accelerating potential was smaller than the retarding potential, the current was zero. As the accelerating potential rose above this value, so the current rose in just the same way as the current rises in a vacuum tube. The collisions being elastic, the electrons transferred negligible amounts of energy when in collision with the massive mercury atoms. It was just as though the atoms were not there. But when the accelerating potential reached about 5 V the current through the galvanometer fell almost to zero. The electrons had insufficient energy to cross the gap between the grid and the anode against the retarding field.

Franck and Hertz suggested that the electrons acquired just the right energy to excite the mercury atoms in collisions near to the grid. Their energy was all transferred to the excited atoms leaving far too little to overcome the retarding potential.

If the accelerating potential was then raised, the excitation zone in the tube would move away from the grid towards the filament, so that an electron which had lost its energy in a collision would gain some further energy in the field, pass the grid and possibly reach the anode. The current would rise once more (see Fig. 43.4).

Figure 43.4 reveals that the total current flowing reached a larger maximum than before. More electrons were being collected. Later research showed that these additional electrons originated in the ionization of some of the mercury atoms which entered a metastable state allowing ionization at an accelerating potential of 5.7 V. At an accelerating potential of about 10 V, the current fell for the second time, when the excitation zone occurred twice in the space between the filament and the grid.

Excitation occurred in two zones in the space between the filament and the grid; the electrons suffered two inelastic collisions and the sequence of phenomena was seen to recur again and again as the accelerating potential was further raised. Franck and Hertz concluded that an electron required an energy of 4.9 eV or 7.8×10^{-19} J if it was to excite a mercury atom. Or, to look at it from another point of view, a mercury atom can accept energy in a 'packet' of 7.8×10^{-19} J.

This experiment by Franck and Hertz was the first of a series performed by many workers. In the case of mercury, the excitation at 7.8×10^{-19} J (4.9 eV) and at multiples of this energy which was so marked in the original experiment obscured the complexities of the phenomenon. Using another method devised by Hertz in 1923 and developed by J. C. Morris in 1928, a whole series of *excitation levels* for mercury was revealed. This series includes *ionization* (i.e., the removal of an electron from the atom) at an energy of 16.5×10^{-19} J (10.3 eV).

Subsequently the *critical potentials* of many other gases have been studied. In every case there is a discrete energy packet which atoms can accept and which causes ionization. See Table 43b.

Table 43b   Ionization energies

Hydrogen	13.6 eV	or	21.8×10^{-19} J
Helium	24.6 eV	or	39.2×10^{-19} J
Nitrogen	14.5 eV	or	23.2×10^{-19} J
Oxygen	13.6 eV	or	21.8×10^{-19} J
Neon	21.6 eV	or	34.4×10^{-19} J
Argon	15.8 eV	or	25.3×10^{-19} J
Xenon	12.1 eV	or	19.3×10^{-19} J

Problem 43.1 Molecules of the gases which make up the air do not appear to ionize one another although we assume that collisions are frequent. Why is this?

The energy of a typical molecule at room temperature (about 300 K) is $3/2\,kT$, where k is Boltzmann's constant. So energy

$$= 3/2 \times 1.38 \times 10^{-23} \times 300 \text{ J}$$
$$\approx 6 \times 10^{-21} \text{ J (about } 4 \times 10^{-2} \text{ eV)}$$

Now, typically it requires about 20×10^{-19} J to ionize a gas particle. The mean energy of the gas molecules is about 300 times too small at room temperature and it is unlikely that any molecules in the energy distribution will have enough energy to succeed in causing ionization during a collision.

As we have seen in Section 10.4, the energies of the molecules are grouped in a distribution pattern such that, even when the mean energy is as low as this, a few molecules will have considerably higher energies. If the gas is heated appreciably, by contact with a hot wire, for example, ionization can be detected. A few molecules acquire sufficient energy to ionize others with which they may collide. See Problem 33.2.

Chapter 44

MATTER, ELECTRICITY AND LIGHT

44.1 Photoelectricity

It will be recalled that electrons can be produced within a vacuum tube by heating a filament. Under the right conditions, however, light can cause electrons to be ejected from metal and other surfaces. The effect was first observed by H. R. Hertz and can readily be demonstrated in the laboratory (Fig. 44.1).

The clean, zinc plate attached to the negatively charged cap of the electroscope is illuminated with ultraviolet light through the wire mesh shown. Very soon the gold leaf will be seen to fall. If the experiment is repeated with a glass sheet between the light source and the zinc plate, no effect is observed. Evidently there are frequencies of light which are absorbed by glass and these have the ability to drive carriers of negative charge from the zinc.

The simplest hypothesis is that the negative charges are associated with electrons.

It was P. E. A. Lenard who made a systematic study of this photoelectric effect.

u.v. light →

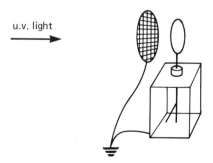

Fig. 44.1 Simple demonstration of the photo electric effect.

The facts which Lenard uncovered are listed below. Note that the light is referred to by its frequency (f) rather than by its wavelength (λ). Remembering that $f\lambda = c = 3 \times 10^8 \, \text{m s}^{-1}$, then red light with a long wavelength has a lower frequency than violet light with a short wavelength.

a) The emission of electrons does not occur for light of all frequencies. If below a certain value, called the *threshold frequency*, f_0, then emission will not occur *no matter how intense the light*.

b) For a given frequency f (where $f > f_0$) the kinetic energy of the emitted electrons has a spread of values from zero up to a maximum value, E_{max}, which is proportional to $(f - f_0)$. For any chosen emitter E_{max} has the same value, *no matter how weak or intense the light*.

c) If emission occurs, it does so as soon as the light reaches it (in modern terms, within 10^{-9} s), *no matter how weak the light*.

d) If emission occurs, the number of electrons emitted per second (i.e., the electron current) is proportional to the light intensity.

e) The value of the threshold frequency, f_0, depends upon the material of the surface being illuminated.

Points (a), (b), and (c) are surprising. There is emission if the frequency f is only a little above the threshold f_0 even if the light is very dim; and making the light brighter does not increase the maximum energy of the electrons. It only increases the number emitted. On the other hand, even with the most intense source of light which is available no emission is possible if its frequency is below the threshold, no matter by how little.

In an earlier problem connected with the emission of radiation from hot bodies, Max Planck, the great German physicist, had suggested (somewhat reluctantly) that the energy might be radiated not in a continuous stream but in tiny packets which he called 'quanta of action' or, as we would say, *quanta of energy*. These quanta of light energy (or *photons*) had a magnitude given by

$$E = hf \qquad (44.1)$$

where f is the frequency of the radiation and h is a constant now known as *Planck's constant*. This constant is very small and has the value of 6.63×10^{-34} J s (to 3 significant figures). Table 44a illustrates the very tiny sizes of photons of various types of radiation.

Figure 44.2 is a cloud chamber photograph of the track of a photoelectron from a copper plate which has been irradiated with a very weak beam of X-rays (Cu K radiation). The length of the track of the photoelectron is what would be expected if a photon of the X-ray radiation had been absorbed (8 keV). The very short track to the right is that left by an electron which has been released from an atom by the photon which was emitted when the excited copper atom regained another electron. (C. T. R. Wilson, *Proc. Roy. Soc.*, London (A) **104**, 1 (1923).)

In 1905, Albert Einstein related the photoelectric phenomena studied by Lenard to this quantum model of Planck. He supposed that the light quanta (or photons) could 'penetrate into the surface layer of the body and their energy is transformed into kinetic energy of the electrons. The simplest way to imagine this is that a light quantum delivers its entire energy to a single electron.' If the electron is at the surface (i.e., is ejected) it will have kinetic energy equal to that of the absorbed photon less any energy (ϕ) necessary to remove it from the field within the metal, that is

$$(\tfrac{1}{2}mv^2)_{max} = hf - \phi \qquad (44.2)$$

Table 44a

Photon:	Radio	TV	Red	Ultra violet	X-rays	γ-rays
Wavelength/m	1500	1	6×10^{-7}	10^{-7}	10^{-10}	10^{-12}
Frequency/Hz	2×10^5	3×10^8	5×10^{14}	3×10^{15}	3×10^{18}	3×10^{20}
Energy/J	10^{-28}	10^{-25}	3×10^{-19}	2×10^{-18}	2×10^{-15}	2×10^{-13}
Energy/eV	10^{-9}	10^{-6}	2	12.5	10^4	10^6

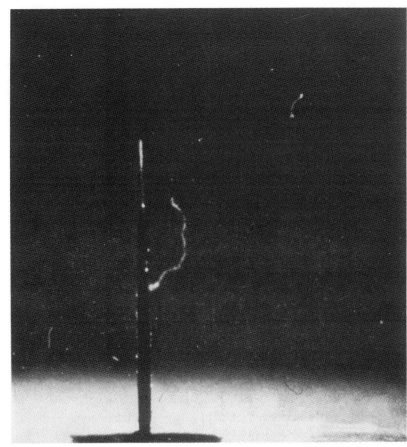

Fig. 44.2 A cloud-chamber photograph of the track of a photoelectron from a copper plate which has been irradiated with a very weak beam of X-rays (Cu K radiation). The length of the track of the photoelectron is what would be expected if a photon of the X-radiation had been absorbed. The very short track to the right is that left by an electron which has been released from an atom by the photon which was emitted when the excited copper atom regained an electron. (Photograph by C.T.R. Wilson, *Proc. R. Soc. London* (A), **104**, 1, 1923).

The term ϕ is known as the *work function*.

The model accounts satisfactorily for the experimental results: more photons cause the emission of more electrons; the greater the frequency of the photon, the greater the energy of the emitted electrons. And if $hf < \phi$, no electrons are emitted.

This must make us reconsider the Franck–Hertz experiment where we observed that an atom could accept energy from an electron provided that there was enough. In the case of mercury, the appropriate energy was 4.9 eV or 7.8×10^{-19} J. The excited atom must eventually return to the unexcited state. If the extra energy is emitted as a photon of light, that photon might have a frequency given by the relationship $E = hf$, that is

$$f = \frac{7.8 \times 10^{-19}}{6.6 \times 10^{-34}} \text{ Hz}$$

$$= 1.18 \times 10^{15} \text{ Hz}$$

which corresponds to a wavelength of 2.5 ×

10^{-7} m. This lies in the ultraviolet and reference to tables of line spectra confirms the existence of that line.

44.2 Energy levels

The experiments show that, in this excitation, a mercury atom accepts just a single quantum (7.8×10^{-19} J) of energy and that the line spectrum of mercury includes a line of such a frequency that the photon involved carries exactly the same quantum of energy. In ionization, the mercury atom accepts a single larger quantum (16.5×10^{-19} J) and the line spectrum includes a line of such a frequency that the photon involved carries exactly this quantum of energy. The atomic model must accommodate these observations.

Let us assume that the neutral atom accepts 16.5×10^{-19} J from an electron which collides with it and that, as a result, an electron is removed

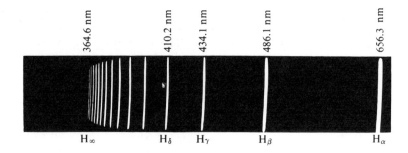

Fig. 44.3 The Balmer series of atomic hydrogen. (From G. Herzberg, *Atomic Spectra and Atomic Structures*, Prentice-Hall, 1937. Reproduced by permission of the publisher.)

from the atom to leave it ionized. When recombination occurs between this positive ion and an electron, the energy of the incoming electron is just 16.5×10^{-19} J and this appears as a photon of electromagnetic radiation.

What of the other line? Indeed, the line spectrum of mercury contains many other lines; photons of many different energies are emitted by excited mercury atoms. It appears that the electrons in the unexcited atom can accept various discrete energies as a result of collisions and subsequently re-emit that energy in the form of a photon. This has led to the idea of the *energy level*. In the case considered, an electron can exist in the atom at certain levels of energy. Subsequently, it may return to a lower energy state and in so doing the atom emits the appropriate photon.

As we have indicated the line spectrum of mercury is a very complex one. So, in accordance with usual practice, let us turn to something simpler – hydrogen. This too is found to possess a line spectrum which was extensively studied in the 1880s by J. J. Balmer. Balmer was a mathematics teacher in Switzerland with an interest in mathematical puzzles rather than physics.

Figure 44.3 shows a photograph of the lines with which Balmer worked and some of the wavelengths are given. This series, which lies almost entirely in the visible spectrum, shows some remarkable regularities. The lines are spaced closer and closer together as we move from the red (long wavelength) to the violet end of the spectrum. And there seems to be a limiting value of wavelength within the series. Balmer developed an empirical relationship to describe this hydrogen series:

$$\lambda = b \left[\frac{n^2}{n^2 - 2^2} \right] \qquad (44.3)$$

where λ = wavelength in nanometres, b = 364.6 nm, and n is a positive integer. Each integer, starting with $n = 3$ gives a different line of the series. For the first line (labelled α in Fig. 44.3):

$$\lambda_\alpha = 364.6 \left[\frac{9}{9 - 4} \right] = 656.2 \text{ nm}$$

For the second line (labelled β), $n = 4$ and so the wavelength is:

$$\lambda_\beta = 364.6 \left[\frac{16}{16 - 4} \right] = 486.1 \text{ nm}$$

Table 44b shows the results obtained by calculating the wavelengths of seven of these lines up to $n = 9$, and also the values of the wavelengths as measured by experiment. These values are those which Balmer himself reported in the *Annalen der Physik und Chemie* in 1885, and the agreement between the calculated and the experimental values is striking.

This remarkable piece of work provides good reason for naming this series of lines in the hydrogen spectrum after Balmer. We can see how the series tends towards a limiting value which, when $n = \infty$, is 364.6 nm.

Table 44b Balmer's Series in the spectrum of hydrogen

Name of line	n	Wavelength (nm)	
		Calculated	Observed
α	3	656.2	656.2
β	4	486.1	486.1
γ	5	434.1	434.1
δ	6	410.0	410.0
ϵ	7	396.9	396.8
ζ	8	388.8	388.7
η	9	383.5	383.4

Table 44c   The Hydrogen spectrum

Lyman series frequency $\times 10^{14}$ Hz	Balmer series frequency $\times 10^{14}$ Hz	Paschen series frequency $\times 10^{14}$ Hz
24.66	4.57	1.60
29.23	6.16	2.34
30.82	6.90	2.74
31.56	7.31	2.98
31.97	7.55	3.14
32.21	7.71	
32.36		
–		
–		
Limit 32.88	8.24	3.66

Table 44d   The Hydrogen spectrum photon energies

Lyman series energy $\times 10^{-19}$ J	Balmer series energy $\times 10^{-19}$ J	Paschen series energy $\times 10^{-19}$ J
16.3	3.02	1.06
19.4	4.10	1.55
20.4	4.57	1.82
20.9	4.85	1.97
21.1	5.00	2.08
21.3	5.11	
21.4		
–		
Limit 21.80	5.46	2.42

Balmer wrote that he was initially aware only of the existence of the first four lines. Imagine his excitement when a friend told him that more lines were known in the spectra of the stars and his satisfaction on discovering that their wavelengths agreed with his prediction. Encouraged by the success of his skill in numerology, Balmer speculated on the existence of hitherto undiscovered lines that would be given by his formula when the 2^2 in the denominator is replaced by such numbers as 1^2, 3^2, 4^2, etc.

Balmer's work inspired others to search for additional lines in the hydrogen spectrum.

It was not until the first decade of the twentieth century that Balmer's speculation about further series of lines within the spectrum of hydrogen bore fruit. C. S. Lyman found lines in the far ultraviolet and F. Paschen reported a series in the near infrared. A total of six such series is now known.

Let us now consider the Lyman, Balmer and Paschen series of lines in the hydrogen spectrum in the light of the suggestion that each line represents a different photon with an energy which is associated with the energy of an electron within the atom. Table 44c tabulates the frequencies of some of the lines in these series.

If we assume that we may calculate photon energies for each line by applying Planck's formula $E = hf$ we find that these lines correspond to photon energies as given in Table 44d.

Inspection reveals one familiar number: 21.8×10^{-19} J. This was the energy quoted for the process of ionization by collision in Franck–Hertz type experiments with hydrogen (see Table 43b).

There is a remarkable observation to be made about the numbers in Table 44d which may best be displayed in diagrammatic form. Figure 44.4 shows on the left the position of electron energies corresponding to the lines on a linear scale of energy. Energies between 6×10^{-19} and 16×10^{-19} J have been omitted since no hydrogen lines fall within that region.

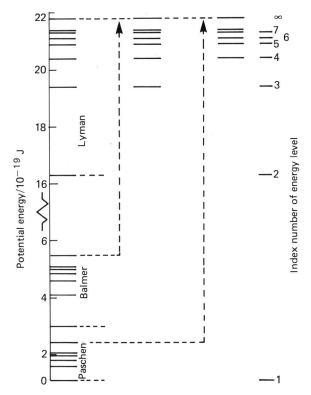

Fig. 44.4   Photon energies in the Lyman, Balmer and Paschen series for hydrogen.

The two remaining columns show the result of repositioning the Balmer and Paschen series of lines alongside the Lyman series, with the three limiting energies placed level with one another. The correspondence between the energies of the photons is quite remarkable.

To interpret these observations, let us assume that, within an hydrogen atom, an electron can exist with energies, above some unknown zero of energy, which are given by the photon energies of the Lyman series. We shall number these excess energies successively from level 1 (zero excess energy) through level 2 (16.3 × 10¹⁹ J), level 3 (19.4 × 10⁻¹⁹ J) and so on. These 'energy levels' are shown to the right of Fig. 44.3. We see that the energy steps get closer and closer together as we approach the limit and ionization. And we see how each of the Lyman lines can be associated with the transition of an excited electron to the ground state.

What is likely to happen if the transitions occur to the level for which $n = 2$? Presumably the photons emitted will have energies which correspond to transitions from

$n = 3$ to $n = 2$

   19.4 × 10⁻¹⁹ to 16.3 × 10⁻¹⁹ J

= 3.1 × 10⁻¹⁹ J

$n = 4$ to $n = 2$

   20.4 × 10⁻¹⁹ to 16.3 × 10⁻¹⁹ J

= 4.1 × 10⁻¹⁹ J

$n = 5$ to $n = 2$

   20.9 × 10⁻¹⁹ to 16.3 × 10⁻¹⁹ J

= 4.6 × 10⁻¹⁹ J

These energies correspond very closely to those for the photons of the Balmer series. The model can accommodate this series as well as the Lyman series.

Problem 44.1 Use the energy level model to account for the Paschen series of lines in the hydrogen spectrum (Fig. 44.4).

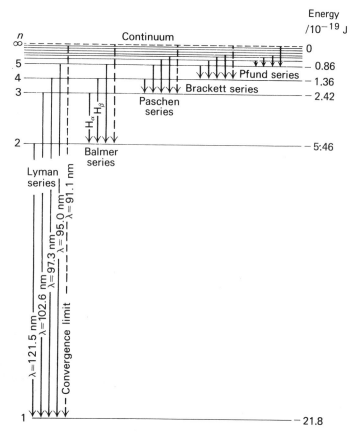

Fig. 44.5 Electron transitions in hydrogen. The zero for energy is at an infinite distance. (Adapted from M. Alonso and E.J. Finn, *Physics*, Addison-Wesley, 1970.)

44.3   Summary

A reasonable model for an atom allows a bound electron to exist in a number of *states*. In an unexcited atom, the electron is bound to the nucleus and exists in the *ground state* or level 1. When the atom is excited, the electron is still bound to the nucleus but exists in one of a large number of *excited states* or energy levels. As the atom gets nearer to being ionized, the gaps between these levels get closer together. Once the atom is ionized, the electron is free of the nucleus and can exist with any energy.

These energy levels are very well defined and simple numerical relationships exist between them. They are often referred to as *stationary* states.

The emission line spectra of atoms are associated with transitions of electrons from energy level to energy level, the energy difference determining the energy (and therefore the frequency and wavelength) of the radiation emitted (Fig. 44.5).

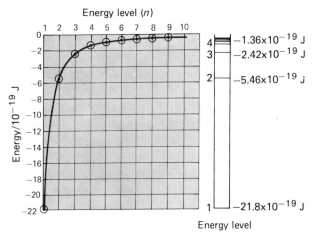

Fig. 44.7   The energy levels of hydrogen plotted against the index numbers of the levels, with, on the right, the sequence of energy levels.

where n is the number of the energy level considered. Figure 44.7 shows a graph of the energies associated with the energy levels for hydrogen plotted against the number of the level concerned. In accordance with a practice which we first introduced in Section 13.1, the zero of energy chosen here is the energy of a free electron at rest; so an electron in the ground state has energy -21.8×10^{-19} J (-13.6 eV). The graph shows clearly how it is that the energy levels crowd closer and closer together as the limit is approached.

We may ask why it is that an electron in a hydrogen atom is subject to the $1/n^2$ law and why there are discrete energy levels anyway. This is a matter to which we must return later.

Problem 44.2   Figure 44.6 shows what happens when white light is passed through sodium vapour and then examined in a spectroscope. Dark lines appear on a continuous spectrum and the wavelengths are identical with those appearing in the emission spectrum of sodium. Account for this absorption spectrum in terms of the idea of energy levels.

44.4   Balmer's $1/n^2$ law

There is an alternative and very useful way of presenting the energy level view for the hydrogen atom. It derives from Balmer's rule. In energy level terms, each level can be written in the form

$$E_n = -\frac{21.8 \times 10^{-19}}{n^2}$$

44.5   A problem experiment

In 1909, G. I. Taylor performed an experiment which raises in the clearest possible way one of the most surprising of physical concepts. In present-day terms, consider a small electric lamp; perhaps one from a torch emitting about 0.3 W. Most of that is heat, but it is safe to assume that about 10^{-2} W is visible light. If we accept that photons

Fig. 44.6   Comparison of the emission and absorption spectra of sodium. (From A.B. Arons, *Development of the Concepts of Physics*, Addison-Wesley, 1965.)

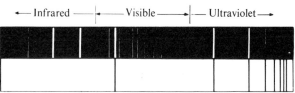

exist, we may ask how many are emitted in each second by this lamp. Taking the frequency of the light as around 6×10^{14} Hz, we find the photon energy is around 4×10^{-19} J. So the number emitted per second is around 2.5×10^{16}. Standing about 30 cm from that lamp, about 1 photon in 50 000 will enter the eye. So the number entering the eye in each second is around 10^{12}. That is one photon every 10^{-12} s. Since light travels at 3×10^{8} m s^{-1}, these will be separated from one another by, on average, $3 \times 10^{8} \times 10^{-12}$ m which is 3×10^{-4} m.

They are rather close together. If now the intensity of the lamp is reduced to about 10^{-4} of its original value by placing a very dark filter such as a piece of fogged film in the beam, the number of photons received at the eye will be reduced and their separation in space increased by 10^4 times. That is about 3 m – ten times the distance of the lamp from the eye. The chance that there are two or more photons in the space between the eye and the filter is small. Nevertheless the dark-adapted eye will continue to see the lamp quite clearly.

If now a diffraction grating is held just in front of the eye, a diffraction pattern will still be seen. Here we are dealing with individual photons, no one of which is likely to pass through more than one of the slits of the grating.

Our explanation of diffraction phenomena has assumed a wave model for light. Here we are working with a particle model – the photons being the particles. It seems that both the wave and the particle models for light are necessary to explain G. I. Taylor's experiment.

We may note too that our explanation of the photoelectric effect employed both models. We thought of the light as arriving as a wave but delivering its energy to the ejected photoelectron in a packet.

The two models for light are inseparable.

44.6   Some applications

a) The photoelectric effect

Tubes using this effect were once commonplace. In one form, a curved plate of the photosensitive material (often caesium) was made the cathode of a vacuum tube. The anode, a loop of wire, served to collect any electrons ejected from the cathode when radiation of the appropriate frequency fell

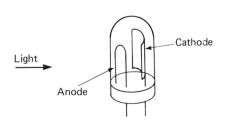

Fig. 44.8   A vacuum photo-electric cell.

upon it (Fig. 44.8). A potential difference was maintained between the electrodes and a suitable meter measured the current. Since the response was a linear one, the meter could be calibrated to give readings of the illuminance (or illumination) directly.

Today, semiconductor materials which are sensitive to light are used instead. Such photo-diodes, used in conjunction with a suitable integrated circuit can be used to control circuits carrying quite heavy currents – for example, street lighting circuits – and for counting inter-ruptions (in, for example, the light pen used to read the bar-codes on items sold in supermarkets).

b)   A measurement of the Planck constant

A vacuum photocell (Fig. 44.8) can be used to provide an estimate of Planck's constant. Figure 44.9 shows the circuit.

When light of a known frequency f (greater than the threshold frequency for the material of the cathode) falls on the cathode, photoelectrons are emitted with a range of energies extending up to a maximum value $(\frac{1}{2}mv^2)_{\text{max}}$. If a variable reverse potential difference is applied between the anode and the cathode, it will be possible to find a

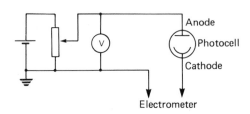

Fig. 44.9   Using a vacuum photocell in a determination of the Planck constant.

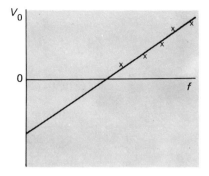

Fig. 44.10 Graph of the stopping voltage (Vo) against frequency.

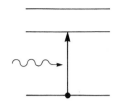

Fig. 44.11 Excitation.

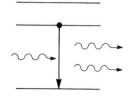

Fig. 44.12 Stimulated emission.

value of that p.d. (V_o) which will just prevent photoelectric emission. Then

$$(\tfrac{1}{2}mv^2)_{\text{max}} = eV_o \qquad \text{and also } = hf - \phi$$

where ϕ is the work function.

This may be written

$$V_o = \frac{h}{e}f - \frac{\phi}{e}$$

Further values of the 'stopping voltage' V_o can be obtained for light of different frequencies. If then the results are plotted as in Fig. 44.10 the slope of the graph is h/e and the intercept on the axis of V_o (corresponding to $f = 0$) is $-\phi/e$. So both the work function and a value for Planck's constant can be found.

c) Solar cells

Solar cells are a more recent application of the photoelectric effect. Cells are available which develop an e.m.f. of about 0.45 V and which can drive a current of about 80 mA through a circuit. But such cells remain an expensive energy source. Nevertheless they are an attractive proposition in view of the enormous demand for energy which has already been discussed in Chapter 8. Research is concentrated on the production of low-cost solar cells. Even so, it has been estimated that if a cell with an efficiency of about 20% were available at a reasonable price, it would require 1% of the total land area of the United States to be covered with such cells if the 'solar cell farm' were to supply the needs of the whole country for electrical energy. That is indeed a formidable undertaking with serious environmental consequences. And, even then, such a farm would only operate by day!

Nevertheless solar cells are particularly useful for supplying energy in orbiting satellites and in space probes. They also find an application in some pocket calculators and watches.

d) The stimulated emission of radiation

Figure 44.6 shows what happens when white light passes through sodium vapour. Light of a frequency corresponding to the normal emission frequencies of sodium is absorbed. A radiation quantum corresponding to a frequency in the sodium emission spectrum carries just the right energy to raise an electron in the sodium atom from its ground state to an excited state (Figure 44.11). Consequently radiation of this frequency is absorbed and dark lines appear in the continuous white light spectrum. Of course, photons of this frequency are emitted when the electron returns to the unexcited state but such emission will take place in all directions so the intensity of the beam in the original direction is reduced.

If, by chance, some of the sodium atoms within the radiation field are already at this higher energy level, interaction with the incident photons can still take place. But now the atom falls to a lower energy level, itself emitting a quantum of energy of the same size as that of the incident photon (see Fig. 44.12). This process is called *stimulated emission*. The interesting feature of this process is that the stimulated quantum of radiation is precisely in phase with the original quantum. Consequently, the on-going radiation is enhanced in energy but remains coherent.

Normally energetic atoms fall to lower energy levels of their own accord and at random. This process of spontaneous emission leads to the production of quanta which bear random phase relationships with each other and the radiation is non-coherent.

Einstein showed that, if there were equal numbers of atoms in the ground state and an excited state, an incident photon of the right energy has the same chance of stimulating emission as of being absorbed. If a situation could be produced in which there were more atoms in a high excited state than in a low one, stimulated emission would be more probable than absorption and a beam of highly coherent radiation could be produced. However, this is an improbable state of affairs under normal circumstances and this is fortunate for highly coherent light radiation can be dangerous to the sight.

In normal equilibrium when a beam of light radiation is passing through a suitable medium, more atoms will be in low energy states than in high ones. The reversed state to this is associated with a population inversion and is utilized in the laser. The word *laser* is derived from the initials of the expression 'light amplification by the stimulated emission of radiation'.

e) The helium-neon laser

This is one of the commonest forms of laser. The neon gas is the source of the stimulated emission. The helium gas is the agency which produces a population inversion in the neon.

By chance, one energy level in helium is identical with an energy level in neon and it is also 'metastable'. This means that excited atoms in this energy level tend to stay in it rather longer than is normal before they return to the ground state.

Helium atoms are excited into high energy states by continuous bombardment by electrons in a low pressure gas discharge (step 1 in Fig. 44.13). These atoms are continuously in collision with one another and with neon atoms. Those which are in the metastable state will, on colliding with a neon atom, have a high probability of transferring this

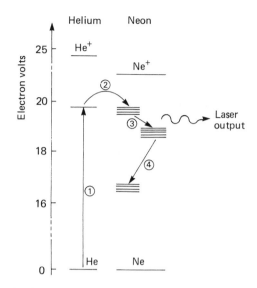

Fig. 44.13 Energy level systems of helium and neon atoms. Successive energy transitions are indicated by the figures in circles. From D.J.E. Ingram, *Contemporary Physics*, **3**, 435 (1966). (Reproduced by permission of the publisher.)

energy to the neon atom so raising it to a high energy level (step 2 in Fig. 44.13). By suitably adjusting the proportion of helium to neon in the tube it is possible to have more neon atoms in this high energy state than in a lower one. This population inversion will then be maintained in dynamic equilibrium. Radiation quanta, produced by spontaneous emission from neon atoms (step 3 in Fig. 44.13) will then stimulate others into emission and a beam of highly coherent stimulated radiation will be produced.

In order to enhance the interaction of the emitted radiation with the energetic neon atoms, the discharge is contained within a tube bounded at one end by a highly reflecting mirror and at the other by a partially reflecting mirror. Thus most quanta traverse the gas many times before escaping in the emitted radiation.

Chapter 45

THE ATOM AND ITS ELECTRONS

45.1   Matter waves

'The universe', said J. B. S. Haldane, 'is not only queerer than we suppose, but queerer than we can suppose.'

We have seen that light, whose macroscopic properties are so well described by a wave theory, also demonstrates microscopic behaviour which is characteristic of a particle. Maybe electrons, whose macroscopic properties in, for example, an electron beam, are so well described by particle dynamics may also reveal wave behaviour.

Figure 14.19 shows a photograph of the X-ray diffraction pattern obtained when X-rays pass through an alum crystal. The behaviour is characteristic of a wave. Fig. 45.1 compares a remarkably similar pattern on the screen of an electron tube with that obtained with X-rays. There is a series of light and dark rings on the screen. The electrons are not special – they are produced by thermionic emission from a hot cathode, accelerated in a field and then strike a target made of thin aluminium foil. Evidently the rings are the result of the electrons exhibiting the phenomenon of crystal diffraction.

If the electrons are given additional energy by increasing the accelerating field, the ring diameters decrease. Now, when we considered diffraction, decreasing the angle through which the waves were diffracted was associated with decreasing wavelength. So it appears that there must be some relationship between the kinetic energy of the electrons and the wavelength with which they are associated. At the time of the discovery of electron diffraction (1927), with which the names of C. J. Davisson and L. Germer in the U.S.A. and of G. P. Thomson in the U.K.

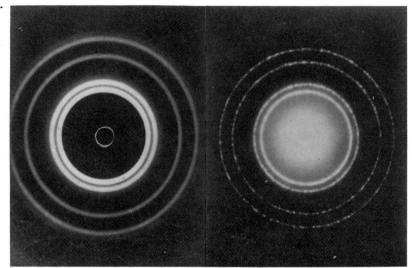

Fig. 45.1 Similar patterns produced by a beam of X-rays (left) and a beam of electrons (right) when passed through the same aluminium foil. (From the P.S.S.C. film 'Matter Waves', Educational Development Center.)

will always be associated, there was in existence a suggestion by the French physicist Louis de Broglie that the wavelength which he believed to be associated with an electron was given by the equation

$$\lambda = \frac{h}{p} = \frac{h}{m_0 v} \qquad (45.1)$$

where p was the momentum of the electron (which at slow speeds was m_0 (the rest mass) $\times$ the velocity.)

As de Broglie pointed out in his Nobel Prize address (given in 1929): 'to obtain electrons with the same velocity they are subject to the same potential difference V. Then $\frac{1}{2} m_0 v^2 = eV$.'

If we accept the de Broglie equation for the wavelength we need to find the momentum in terms of the accelerating voltage V. In place of $\frac{1}{2} m_0 v^2$ we may write

$$\frac{(m_0 v)^2}{2m_0}$$

and then

$$\frac{(m_0 v)^2}{2m_0} = eV$$

and

$$m_0 v = \sqrt{2m_0 eV}$$

Consequently

$$\lambda = \frac{h}{\sqrt{2m_0 eV}}$$

Numerically

$$\lambda = 1.226 \times 10^{-9}/\sqrt{V} \dots \text{in metres}$$

de Broglie continued: 'As we can only use electrons that have fallen through a potential difference of at least some tens of volts, it follows that the wavelength λ, assumed by the theory, is at most of the order of 10^{-8} cm. This is also the order of the magnitude of the wavelength of X-rays.'

'The length of the electron wave being thus of the same order of magnitude as that of X-rays, we may fairly expect to obtain a scattering of this wave by crystals in complete analogy with the Laue phenomenon.'

First published in 1924, de Broglie's theory aroused considerable astonishment at the time but is today recognized as the beginning of the study of wave mechanics.

The work of Davisson and Germer and of G. P. Thomson brought experimental support to de Broglie's theory. The graph (Fig. 45.2) taken from a paper by Davisson and Germer shows that the wavelength of the electron was directly proportional to $1/\sqrt{V}$.

de Broglie's reasoning which led him to suggest the relationship between wavelength and momentum started with the assumption that: '... it is necessary to introduce the particle concept and the wave concept at the same time. The existence of particles accompanied by waves has to be assumed in all cases'. And then he reasoned thus:

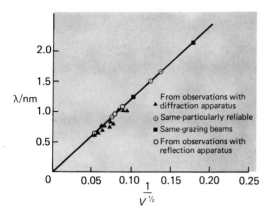

Fig. 45.2 Graph of electron wavelength against $1/\sqrt{V}$ (From C.J. Davisson, Are Electrons Waves? *Franklin Institute Journal*, 205, 1928.)

According to Planck's quantum theory
the energy of a photon $= hf$

According to Einstein
the energy of the photon $= mc^2$

So, for the photon,

$$hf = mc^2$$

and its momentum, p is

$$mc = hf/c$$

For a light wave

$$c = f\lambda$$

$$\therefore \frac{f}{c} = \frac{1}{\lambda}$$

and the photon has momentum

$$= \frac{h}{\lambda}$$

Assuming that this result applies in all cases (as de Broglie did) the wavelength of the electron will be given by

$$\lambda = \frac{h}{p}$$

Matter waves are part of the fabric of physics. It is unprofitable to ask whether an electron is a particle. To get an answer an experiment will be done which examines particle behaviour. It is equally unprofitable to ask if an electron is a wave. The experiment will examine wave behaviour

alone. We have to accept that in the quantum world particle and wave behaviour characterize the same entities. One has to concentrate one's thoughts on the behaviour of the entity – whether photon, electron, proton, neutron. The entity is real enough and associated with it is a wave which appears to have the function of determining where the particle may be. Today it is accepted that the chance of arrival of a particle at a particular place is proportional to the (wave amplitude)^2.

Problem 45.1 With what speed will an electron travel which has been accelerated in a field of 1 kV, as in a small cathode ray oscilloscope? With what wavelength will this electron be associated?

The speed is obtained by applying the equation $\frac{1}{2}mv^2 = eV$ and is in the order of $10^7 \, \text{m s}^{-1}$.

The wavelength is found from

$$\begin{aligned}
\lambda &= h/mv \\
&= 6.6 \times 10^{-34}/(9 \times 10^{-31} \times 10^7) \, \text{m} \\
&= 7 \times 10^{-11} \, \text{m}
\end{aligned}$$

Problem 45.2 By making estimates for the mass and velocity of a tennis ball in ordinary play, estimate the wavelength associated with the ball!

You will find that the wavelength is vastly smaller than that associated with the electron of Problem 45.1. Indeed it is far smaller (by some 23 magnitudes) than the 'size' of an atom! So far as we are concerned we can ignore the wave behaviour of a tennis ball since it is impossible to imagine a suitable diffraction grating!

45.2   Electrons in atoms

a) The ground state

There is a host of questions to be asked about electrons within atoms. We have already wondered why there are discrete levels of energy at which such electrons exist and why Balmer's $1/n^2$ law supplies to those levels. We might even enquire why it is that a neutral hydrogen atom does not collapse under the attraction of the Coulomb law forces between the proton and the electron. Hydrogen atoms certainly do *not* collapse, even though the forces involved are very large indeed in relation to the masses involved.

The discovery of the wave behaviour of the electron provides the clue to the answers to these and many other questions. In Section 20.2 we examined the properties of progressive waves and also of standing or stationary waves in strings. Such waves are, one might say, trapped within nodes at the anchored ends of the string in which they form. And the possible wavelengths for this to happen are restricted to those obeying the relationship

$$\lambda = \frac{2l}{n}$$

where l is the length of the string and n is 1, 2, 3, etc. Eigen frequencies were described – for each string a fundamental frequency and a set of higher harmonics. The values of these frequencies depended on the conditions.

Are we, perhaps, dealing with some comparable phenomenon for the electron bound within the atom? If so, we must remember that although the waves will not themselves have a physical reality, their amplitudes provide a measure of the probability of locating the electron. Where the amplitude is large, there is a high chance of finding the electron. With that idea in mind, we should proceed to examine the conditions which will apply.

b) Electron waves in boxes: why is the size of the hydrogen atom in the order of 10^{-10} m?

Let us perform a thought-experiment. We shall consider first an electron which is confined within a one-dimensional box, bounded by two rigid walls where the wave amplitude is zero. Then, as for the stationary wave on a string, we shall have

$$\text{wavelength, } \lambda = \frac{2l}{n} \qquad (45.2)$$

where l is the distance between the walls and $n = 1, 2, 3$, etc.

The first two possible states for the matter wave associated with this electron are shown in Fig. 45.3.

Assuming the conclusion, already mentioned, that the chance of finding a particle at a particular place is proportional to (wave amplitude)^2, the corresponding graphs representing the probabilities of the electron being at a position r will take the forms shown in Fig. 45.4.

The momentum of the electron is given by $mv = h/\lambda$ and, since its kinetic energy (E_k) is $(mv)^2/2m$ it follows that

$$E_k = \frac{h^2}{\lambda^2} \times \frac{1}{2m}$$

Substituting the possible values given by Eq. 45.2, we have

$$E_k = \frac{h^2}{2m} \left(\frac{n^2}{4l^2}\right) = n^2 \left(\frac{h^2}{8ml^2}\right) \qquad (45.3)$$

where $n = 1, 2, 3$, etc.

The term $h^2/8ml^2$ is constant for our particular thought-experiment and so we see that the kinetic energies which the electron may possess are proportional to n^2. These then are permitted energy levels. The lowest of these is $1^2(h^2/8ml^2)$; it corresponds to the ground state.

It is important to note that the minimum kinetic energy of the electron is *not* zero.

We may now turn to the question we asked about the minimum size of such a box into which an electron may be confined.

There is evidence to suggest that the hydrogen atom is about 2×10^{-10} m across. So let us assume that we have a one-dimensional box of that length and that we wish to confine an electron within it. If the available space (l) is 2×10^{-10} m, the wave-

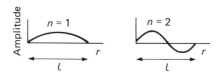

Fig. 45.3   The first two states of an electron confined in a potential box.

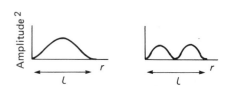

Fig. 45.4   The probabilities of finding an electron at a particular position in a potential box for the first two states.

length of the electron in the ground state will be 4×10^{-10} m. From Eq. 45.3 we have

$$E_k = 1^2 \frac{h^2}{8m(2 \times 10^{-10})^2} \cdots \text{ in J}$$

$$= 1 \frac{(6.6 \times 10^{-34})^2}{8 \times 9.1 \times 10^{-31} \times 4 \times 10^{-20}}$$

$$= 14.9 \times 10^{-19} \text{ J}$$

This is in the right order. It is less than the measured ionization energy $(-21.8 \times 10^{-19} \text{ J})$. It is also less than the potential energy of an electron at a distance of 10^{-10} m from the proton which is

$$-\frac{1}{4\pi\epsilon_0} \frac{e^2}{r} = -(9 \times 10^9) \frac{(1.6 \times 10^{-19})^2}{10^{-10}}$$

$$= -23 \times 10^{-19} \text{ J}$$

This value of the potential energy is negative because we take the zero of potential energy as corresponding to the value of the potential energy when the electron is remote from the proton.

Suppose now that we attempted to confine this electron within a pair of walls separated by half the diameter of the hydrogen atom. The available space is now 10^{-10} m: and the wavelength in the ground state is 2×10^{-10} m.

With l reduced by a factor of 0.5, l^2 is reduced by a factor of 0.25. Inspection of the calculation carried out above shows that E_k must be larger by a factor of $1/0.25$, that is by 4. So E_k is now 59.6×10^{-19} J.

Not only is this much larger than the measured ionization energy but it is also larger than E_p for an electron at a distance of 0.5×10^{-10} m from a proton which is exactly twice that at 10^{-10} m; this is 46×10^{-19} J.

This smaller box is quite incapable of holding an electron within its walls; the electron has so much kinetic energy that it can escape without difficulty.

This offers a crude understanding of why the hydrogen atom is the size that it is. The kinetic energy of the electron in the ground state is insufficient to allow it to escape from the electrostatic field. Radii smaller than about 10^{-10} m are not possible ones.

No wonder, as we noted in Unit 4, that solids resist compression so very strongly. Any attempt to compress a solid into a smaller space is also an attempt to squeeze its electrons into a smaller

space; that means a shorter wavelength, larger momentum and larger kinetic energy. And they won't do it! Even very large compressive forces have but little effect.

c) Electrons in $1/r$ shaped boxes

The potential field in the vicinity of a proton cannot possibly lead to the box with rectangular sides we have already used in our thought-experiment. The potential energy E_p of an electron at distance r from a proton (charge $+1.6 \times 10^{-19}$ C) is given by

$$E_p = \frac{1}{4\pi\epsilon_0} \times \frac{e^2}{r} = (9 \times 10^9) \frac{(1.6 \times 10^{-19})^2}{r}$$

$$= 2.3 \times 10^{-28} \times \frac{1}{r} \cdots \text{ in J}$$

Table 45a gives a set of values.

Table 45a

$r/10^{-10}$ m	0.2	0.4	0.5	0.8	0.9	1.0	1.2	1.5	1.8
$V/10^{-19}$ J	115	57.6	46.1	28.8	25.6	23.0	19.2	15.4	12.8

Figure 45.5 is a graph of the potential energy of an electron at various distances from the proton – a graph of the 'potential well' of the proton. (Compare also Section 29.6).

Now let us imagine that the electron is confined within a box of this shape. If it should 'fall' towards the proton, its kinetic energy will increase, its momentum will increase and the associated wavelength will decrease since, as before

$$\lambda = \frac{h}{mv} = \frac{h}{\sqrt{2m E_k}} \tag{45.4}$$

The total energy (i.e., the sum of the potential and the kinetic energies) of this proton–electron system is constant.

$$E_k = E - E_p \tag{45.5}$$

where E is the total energy.

We have already seen that, for this system,

$$E_p = -\frac{1}{4\pi\epsilon_0} \times \frac{e^2}{r}$$

Substituting in the Eq. 45.4 we have

$$\lambda = \frac{h}{\sqrt{2m\left(E + \frac{1}{4\pi\epsilon_0} \times \frac{e^2}{r}\right)}} \tag{45.6}$$

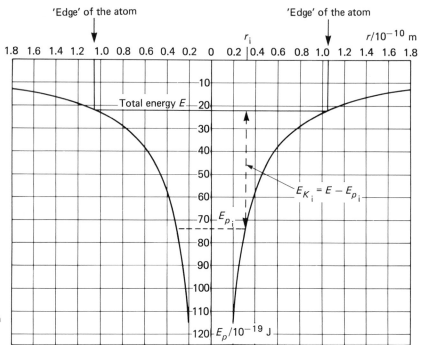

'Edge' of the atom

'Edge' of the atom

Fig. 45.5 The variation of the potential energy of an electron with distance from a proton; this is often referred to as the 'potential well' of the proton.

As r decreases the wavelength also decreases since the only variables is this last equation are λ and r.

It would appear, then, that the stationary wave associated with the electron has, in this case, a varying wavelength.

Such standing waves do exist. For example, if you take a length of light chain, hold it up by one end and then oscillate that end from side to side, a stationary wave will develop. As the tension in the chain diminishes towards the lower end, so does the wavelength (see Fig. 45.6).

We can, if we wish, read the kinetic energy associated with an electron at various distances from the central proton off the graph (Fig. 45.5). An electron in the ground state must be given the ionization energy of 21.8×10^{-19} J to escape from the field of the proton. The line marked 'total energy E' on the graph represents this. It intersects the curve of E_p against r at a distance of 1.1×10^{-10} m and we shall take this to be the edge of this one-dimensional atom. Now let us assume, as we did in the first thought-experiment with the rectangular box, that the kinetic energy of the electron is zero at this distance (i.e., that the wave has a node at this distance).

If the electron moves in towards the proton, its kinetic energy increases at the expense of its potential energy – but the sum of the two energies remains E. At a distance r_i, where the potential energy is E_{p_i} the kinetic energy E_{k_i} is $(E - E_{p_i})$, as shown in Fig. 45.5.

d) Matter waves in three dimensions

The two thought-experiments just considered were confined to one dimension. The argument of the first can be extended to a rectangular membrane in two dimensions for which the wavelength of the wave is given by

$$\frac{1}{\lambda^2} = \left(\frac{m}{2b}\right)^2 + \left(\frac{n}{2l}\right)^2$$

where l is the length, b the breadth and both m and n are 1, 2, 3, etc. We recall that for the one dimension we had

$$\frac{1}{\lambda^2} = \left(\frac{n}{2l}\right)^2$$

Fig. 45.6 Using a light chain to demonstrate a stationary wave of varying wave-length.

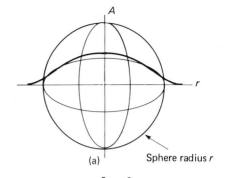

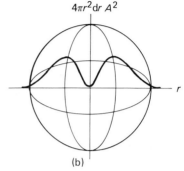

Fig. 45.7 (a) A stationary electron wave with a varying wavelength in the ground state of a spherical hydrogen atom. (b) The probability of finding such an electron at a particular distance from the proton in the hydrogen atom.

Further extension to a box of dimensions l, b and h leads to a very similar expression for the wavelength

$$\frac{1}{\lambda^2} = \left(\frac{l}{2h}\right)^2 + \left(\frac{m}{2b}\right)^2 + \left(\frac{n}{2l}\right)^2$$

We may assume that the arguments can be extended to the case of an electron confined within a spherical box by the electrostatic force of the proton to make a hydrogen atom.

We have already assumed that the amplitude of the matter wave at a point tells us the chance of finding the electron near to that point. In fact, the chance of finding an electron within a distance dr of its distance r from the proton is $4\pi r^2\, dr\, A^2$, where A is the appropriate wave amplitude.

When r is zero (i.e., at the centre) the chance of finding the electron is vanishingly small. For although A^2 may be large, the volume $4\pi r^2\, dr$ of the shell of radius r and thickness dr in which to search for it is zero. As r increases the volume of the shell increases (proportionally to the square of its radius) whilst the value of A^2 decreases.

We have to imagine how we might fit a standing wave of variable wavelength (as suggested by the second thought-experiment within a $1/r$ shaped box) into a spherical atom. At the 'edge' of the atom, there is a small chance of finding the electron. There is zero chance at the centre. Somewhere in between there will be a maximum chance.

The wave is to have a varying wavelength – small near to the centre and increasing towards the edge.

Figure 45.7a shows a possible shape for the wave and Fig. 45.7b a possible shape for the chance of finding the electron at a particular distance from the centre. Figure 45.8 is an attempt to show the distribution of the possible location of the electron in the ground state of the hydrogen atom. The one electron appears as an 'electron cloud'.

Our next concern must be to extract whatever further information we can from this extremely crude model of the hydrogen atom. We might start by guessing where the maximum chance will be found. Perhaps half way between the proton and the atom edge. That's not very probable

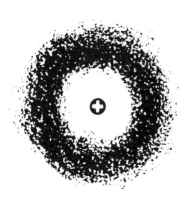

Fig. 45.8 An impression of the electron cloud; that is, of the possible locations of the electron in the ground state of the hydrogen atom.

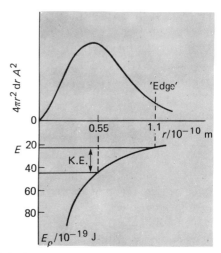

Fig. 45.9 The potential energy diagram for an electron in the ground state of a hydrogen atom and the graph showing the probability of finding the electron at a particular distance from the proton.

because the need to take account of the way the wavelength varies is likely to pull the maximum a little nearer to the proton than to the 'edge'. But it won't be too far wrong.

At the half-way position $r = 0.55 \times 10^{-10}$ m. The potential here must be half of the potential at the 'edge' because the potential is $\propto 1/r$. The potential energy at the edge is -21.8×10^{-19} J. So the potential at $r = 0.55 \times 10^{-10}$ m must be -43.6×10^{-19} J. Applying the same technique as we used in Section 45.2(c), this suggests that the electron at this position has $E_k = 21.8 \times 10^{-19}$ J (see Fig. 45.9).

The corresponding momentum is $\sqrt{2m\,E_k}$

$$= \sqrt{2(9.1 \times 10^{-31})(21.8 \times 10^{-19})}$$
$$= 2 \times 10^{-24} \text{ in kg m s}^{-1}$$

The associated wavelength is

$$h/mv = 6.6 \times 10^{-34}/2 \times 10^{-24}$$
$$= 3.4 \times 10^{-10} \text{ m}$$

A stationary wave with such a wavelength will have 1.7×10^{-10} m between nodes. A little large for our atom – the wave would extend somewhat beyond its 'edge'. But not unsatisfactory in view of all the assumptions made.

The electron in the ground state in the hydrogen atom appears to have a potential energy of -21.8×10^{-19} J; the associated matter wave

has a single loop (the minimum number) which extends, but with decreasing amplitude beyond the 'edge' of the atom. The maximum chance of finding the electron is at a location at a distance of about 0.5×10^{-10} m from the proton. And, incidentally, we may note that this coincides with the Bohr radius which is usually taken to be 0.53×10^{-10} m.

45.3   Extending the argument to the excited states; the Balmer rule

You will recall that Balmer gave a rule which, in energy level terms, was written

$$E_n = -\frac{21.8 \times 10^{-19}}{n^2}$$

(See Section 44.4.)

If the ground state of the electron in the hydrogen atom is associated with a stationary wave with only one loop, maybe the first excited state is associated with two loop waves, the third with three, and so on.

Using Balmer's rule the energy in the first excited state will be

$$E_2 = -\frac{21.8 \times 10^{-19}}{2^2} = -5.45 \times 10^{-19} \text{ J}$$

Figure 45.10 shows the potential energy diagram relevant to this energy. Since the curve is that for $E_p \propto 1/r$, the edge of the excited atom has been moved out from a distance r to a new distance $4r$. Into this increased space the two loops of the standing wave have to be fitted. The total energy having been reduced by one quarter, the mean momentum is reduced by $1/\sqrt{4}$, that is, by one half. So the wavelength is doubled. And it can be fitted in since there is now four times the space available for it (remember that two loops implies four times the wavelength).

Figure 45.11 shows the chance of finding the electron at various distances from the proton when it is in this first excited state. Remembering that this is determined by $4\pi r^2 dr A^2$, we see that as r increases, the $4\pi r^2 dr$ term increases rather rapidly and this accounts for the higher probability of finding the excited electron in the outer of the two loops.

The argument applies to the other excited states. And we now realize that Balmer's rule works because electrons in hydrogen atoms are associated with standing waves – each of which may have only a fixed number of loops – never less than one, of course.

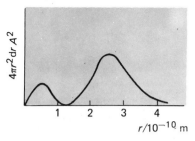

Fig. 45.11 Graph showing the probability of finding an electron in the first excited state at a particular distance from the proton.

45.4   A comment on the argument given

All the later work in this chapter has been in the 'back of an envelope' style. Calculations have been concerned with orders of magnitude rather than with exact values. Rough assumptions have been accepted. But the work is nevertheless of great importance. Remembering that the product $4\pi r^2 dr A^2$ for an electron wave tells us the chance of locating an electron and noting that the first excited state offers two maxima for this rather than one; we would expect to find the excited electron at one or other of these locations for much of the time but with a definite preference for the outer one. The hydrogen atom would appear to be a very 'fuzzy' thing – a proton surrounded by an electron cloud within which the one electron may be anywhere except at the proton and with a high probability of it being somewhere around some specified distance from the proton when in the ground state. Excite that electron by providing the right energy quantum to the first excited state (the second energy level) and it is most likely to be found in one or two locations. For the third level there will be three such regions. These are the 'orbitals' referred to by chemists (see, e.g., Kneen, Rogers and Simpson, *Chemistry: facts, patterns and principles,* Section 2.3.6. Longman, 1972).

The theory, of which the above account is the merest (and the crudest) outline was developed by Heisenberg, Born, Schrödinger, Dirac and others; known as wave or quantum mechanics, its further development is beyond our present scope.

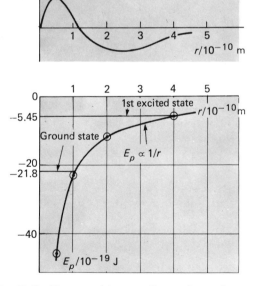

Fig. 45.10 The potential energy diagram for an electron in the first excited state of a hydrogen atom and the graph showing the variation of the amplitude of the wave with distance from the proton.

The Nuclear Atom

Chapter 46

RADIOACTIVITY

An atom consists of a number of electrons existing within a confined space; it is therefore reasonable to assume that they are held within that space by the attraction of an equal positive charge making the whole atom electrically neutral. The seat of that positive charge is called the nucleus. Since the mass of the electron is very small (9×10^{-31} kg) compared with the mass of even a hydrogen atom (1.7×10^{-27} kg) we see that by far the greater part of the mass of the atom is associated with the positive charge. This unit is concerned with the evidence for this nuclear model of the atom and we shall start with the phenomenon of radio-activity.

46.1 The discovery of radioactivity

The discovery of radioactivity resulted from research into phosphorescence by Henri Becquerel of Paris. He was using a salt of uranium which he deemed to be phosphorescent after exposure to sunlight. He wrote

> 'A photographic plate was wrapped with two sheets of thick black paper, so thick that the plate was not clouded by exposure to the sun for a whole day. Externally, over the paper sheet, was placed a piece of the phosphorescent substance (a uranium salt), and all were exposed to the sun for many hours. Upon developing the photographic plate I recognized the silhouette of the phosphorescent substance in black on the negative.'

Becquerel then proceeded to show that the radiations could pass through sheets of glass, aluminium, or copper, placed between the photographic plate and the uranium salt. He continues:

> 'Some of the preceding experiments were prepared during Wednesday the 26th and Thursday the 27th of February (1896 – a leap year), and since on those days the sun appeared only intermittently, I stopped all experiments and left them in readiness by placing the wrapped plates in a drawer of a cabinet, leaving in place the uranium salts. The sun did not appear on the following days and I developed the plates on March 1st (Sunday), expecting to find only very faint images. The silhouettes appeared, on the contrary, with great intensity.'

Becquerel was becoming aware that the radiations were not, after all, induced by sunlight. However, being a careful experimenter he wanted to make certain and so kept the uranium salts in an opaque box for two months. When these were tested in a darkroom on a photographic plate he found that the radiations were emitted without noticeable decrease.

46.2 Subsequent developments

The subsequent study of the Becquerel rays by Marie Curie, her husband Pierre, and Ernest Rutherford, will not be covered in detail here. Let it be sufficient to say that the Curies discovered the radioactive elements polonium and radium, and Rutherford investigated the nature of the radiations.

By 1909, research, mostly guided by Rutherford, had established the following facts: three types of radiation can be emitted from radioactive substances, alpha (α) particles; beta (β) particles; gamma (γ) rays. The properties of each of these are listed in Table 46a.

It was also known that when an atom of a radioactive element decays, it changes into a different element according to the following *displacement* laws.

a) When the atom disintegrates by the emission of an alpha particle, it turns into an element with chemical properties similar to those of an element two places lower in the periodic table. For example, if radium decays, it does so by alpha emission and becomes radon.

b) When an atom disintegrates by the emission of a beta particle it turns into an element with chemical properties similar to those of an element one place higher in the periodic table. For example, if an atom of a radioactive potassium decays, it emits a beta particle and the resulting atom is calcium.

Table 46a

Alpha particles	Beta particles	Gamma radiation
Easily absorbed by matter. e.g., 3–10 cm of air at atmospheric pressure, or 10^{-2} mm of aluminium, thick paper, etc.	Penetrate matter more easily than α-particles, and will pass through 1 m of air or 3 mm of aluminium	Very penetrating. γ-Rays will pass through a few centimetres of lead
Cause intense ionization, e.g., 10^5 ion pairs in 3–10 cm of air	Cause some ionization, e.g., about 100 ion pairs per cm of air	Cause relatively little ionization per centimetre of air
Deflected by magnetic and electric fields with difficulty	Deflected by magnetic and electric fields easily	Cannot be deflected by magnetic or electric fields
Emitted with velocities up to 10^7 m s^{-1}, or 0.1 c	Emitted with velocities up to 0.99 c	Electromagnetic radiation. Travels at c, or 3×10^8 m s^{-1}
Carry a charge $= +2e$	Carry a charge $= -e$	Carries no charge
Mass $= 4 \times$ mass of hydrogen atom	Mass $= 1/1840 \times$ mass of hydrogen atom $=$ mass of electron	Has no mass as ordinarily understood
Are helium nuclei	Are electrons	Are photons. Higher frequency than X-rays

46.3 Detection of the radiations

Becquerel's original discovery of radioactivity depended on the photochemical effects of the radiation as these fell on to photographic emulsions. Most of the work of Rutherford depended on two additional tools: the ionization chamber with an electrometer and a method based on the scintillations which occurred when alpha particles fell on to screens coated with zinc sulphide.

Later techniques involve the use of spark detectors, of cloud chambers, of Geiger–Muller (GM) tubes and, more recently, bubble chambers and solid-state devices. Nevertheless, methods utilizing the ionization chamber and the scintillation method remain in use. All these techniques require that an ionizing event should occur from which the behaviour of the original radiation can be inferred. We shall consider those detectors which are readily available in the context of an experiment which might be performed using the radioactive sources available to teaching establishments. These sources are all very weak.

The *activity* of a source of radioactive radiation is the number of spontaneous nuclear disintegrations which occur in unit time. One gram of radium undergoes 3.7×10^{10} disintegrations per second. Since 1 disintegration per second is known as a *becquerel* (Bq), the activity of 1 g of radium is 3.7×10^{10} Bq. At one time this was known as a curie (Ci).

Sources used in schools in Britain have activities of around 0.18 MBq (180×10^3 Bq which is about 5μCi in the older unit). They include:

Nuclide	Final emission	Activity/Bq	Note
^{226}Ra	α, β, γ	180×10^3	
^{90}Sr	β	180×10^3	
^{60}Co	γ	180×10^3	β emission is absorbed by an aluminium shield over the source
^{241}Am	α + little γ	180×10^3	

Although very weak, these sources are handled with great care. They are designed to be lifted from the lead-lined boxes in which they are kept using long tongs or a special source holder. Strict

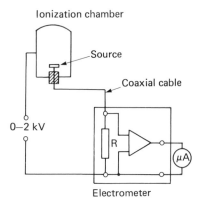

Fig. 46.1 An ionisation chamber in use with an electrometer.

procedures are laid down to ensure that users of these sources do not receive harmful dose of radiation. See also Section 49.4.

Experiment A: Ionization by alpha particles using an ionization chamber

Essentially an ionization chamber is a conducting box in which ionizing events can occur in the presence of an electric field. Figure 46.1 shows such a chamber in which a well-insulated plug carries a conductor on to which one of the radioactive sources may be fitted. A suitably high voltage (in the order of 1 kV) is maintained between the source and the chamber wall.

As alpha particles are emitted, they will ionize some of the air molecules in their path and these ions will move in the electric field within the box. There will, in fact, be a current within the box; but it will be very small – in the order of 10^{-9} A. It will need to be measured by using an electronic electrometer with an input resistance of 10^9 ohm. Electrometers of this sort measure the voltage developed by the current across the high resistance (R). In a typical example, a voltage of 1 V across the 10^9 ohm resistor will produce an output current of 100μA from the amplifier within the electrometer. This output current is large enough to give a full scale deflection on a moving coil meter. It can easily be shown that the meter reading is directly proportional to the input voltage and hence to the very small ionization current.

Since the charges on the ions within the chamber are in the order of the charge on the electron, it is possible to estimate the number of

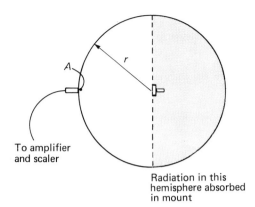

To amplifier
and scaler

Radiation in this
hemisphere absorbed
in mount

Fig. 46.2   The geometry of counting from a radioactive
source.

ions which the source of the alpha particles is
producing each second.

Obviously the voltage applied across the
chamber must be large enough to ensure that all
the ions do, in fact, drift across to the electrodes
before recombination occurs. The chamber itself
must be large enough to accommodate the full
range of the alpha particles in air.

**Experiment B: Counting alpha particles using a
solid-state detector**

A solid-state detector is a semiconductor which
produces a current pulse whenever the sensitive
layer is subject to ionizing radiation. The amplitude of this pulse depends on the extra electron–
positive hole pairs produced. This means that the
device can be used to discriminate between the
various radiations. Alpha particles produce the
largest pulses but these still require amplification
before they can be counted by a suitable pulse
counting device (scaler).

The sensitive area (A) of the detector is usually
small: about 3 mm in diameter. It is placed
directly in front of the source S at a known
distance (r) – say 20 mm (Fig. 46.2).

The source (S) emits radiation in all directions
– but much of this will be absorbed in the source
and its mount. If the number of particles entering
the detector in unit time is C, the total number of
particles emitted in the forward direction is in the
order of $2\pi r^2 C/A$ (see Fig. 46.4).

Problem 46.1   A source of α-particles is placed in
an ionization chamber which is connected to an

electrometer. The resulting ionization current is
found to be 10^{-8} A.

Held at a distance of 20 mm from a solid-state
detector of effective area 7 mm² (7×10^{-6} m²), the
count rate is 515 s⁻¹.

What is the activity of the source (i.e., how
many disintegrations occur per second)? And
what is the energy of the α-particles on emission?

The current of 10^{-8} A results from the ion-pairs
produced by all the α-particles emitted by the
source in the forward direction. It follows that the
charge produced in each second is 10^{-8} C.

Each ionizing event produces two ions which
we may assume to carry charges of $+ 1.6 \times 10^{-19}$ C
and $- 1.6 \times 10^{-19}$ C. The number of ionizing
events will be

$$\frac{10^{-8}}{2 \times 1.6 \times 10^{-19}} \text{s}^{-1}$$

$$= 3.1 \times 10^{10}\,\text{s}^{-1}$$

Only the fraction

$$\frac{\text{area of detector}}{\text{area of hemisphere}} = \frac{7 \times 10^{-6}}{2\pi (0.02)^2}$$

of the α-particles enter the detector in the second
experiment and they produce a count rate of
515 s⁻¹.

The total number of α-particles emitted in the
forward direction then is

$$\frac{2\pi (0.02)^2}{7 \times 10^{-6}} \times 515 = 14.8 \times 10^5\,\text{s}^{-1}$$

This is the activity of the source: 1.84×10^5 Bq
(about 5 μCi).

1.85×10^5 α-particles cause 3.1×10^{10}
ionizing events every second. Each particle is
responsible for 1.7×10^5 of these.

Not all the collisions in the ionization chamber
will cause ionization; some will involve the
transfer of energy by excitation. Consequently the
average energy transferred by an α-particle for
each ionizing event it causes is likely to be greater
than the ionization energy of the gas in the
chamber. This is mostly nitrogen and, as we have
seen in Section 42.3, the ionization energy of
nitrogen is 23×10^{-19} J. Experiment has shown
that an α-particle passing through nitrogen gas
transfers about 48×10^{-19} J for each ionizing
event it causes.

Applying this result, the average initial energy of the α-particles emitted by the source is $(1.7 \times 10^5) \times (48 \times 10^{-19}\,\text{J})$
$$= 8.2 \times 10^{-13}\,\text{J}$$
$$= 5.1\,\text{MeV}$$

Problem 46.2 At what speed will an α-particle with energy 5 MeV travel on emission from the surface of a source?

If we assume that the whole of the energy $(8 \times 10^{-13}\,\text{J})$ is transferred to kinetic energy we have
$$\tfrac{1}{2}mv^2 = 8 \times 10^{-13}\,\text{J}$$

An α-particle has mass $6.8 \times 10^{-27}\,\text{kg}$ so
$$v^2 = \frac{2 \times 8 \times 10^{-13}}{6.8 \times 10^{-27}}$$

whence
$$v = 1.5 \times 10^7\,\text{m s}^{-1}$$

The energy possessed by such an α-particle (which is typical of those emitted by radium or by americium) is very large in proportion to its mass as these calculations show.

Experiment C: The range of alpha particles in air using a spark detector

The detector consists of a grid of taut, fine wires held on an insulating frame about 2 mm in front of, and parallel to, a smooth brass plate. The grid, which is at the front of the device (Fig. 46.3), is held at earth potential and the plate at around 4 kV. There is then a large electric field between the wire grid and the plate. When an α-particle

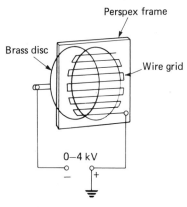

Perspex frame

Brass disc

Wire grid

0–4 kV

Fig. 46.3 A spark counter.

ionizes a molecule of the air within that field, the ions are accelerated rapidly and acquire sufficient energy to ionize further molecules with which they may collide. The result is an 'avalanche' of ions – seen as a spark.

When the detector is placed in front of and near to an americium or a radium source, large numbers of sparks will be seen; they occur over a wide area of the grid and apparently at random. If so desired, they can be counted using a suitable counting device. If now the source is moved away from the grid, the moment will come when the sparks suddenly cease. At that point the distance from the source to the grid of the detector is the range of the alpha particles in air.

The range is related to the energy of the particles. Those from americium (^{241}Am) have energies of about 5.4 MeV and a range of about 37 mm. Alpha particles from radium itself have a range of about 80 mm.

If, whilst the sparks are appearing, a thin sheet of paper is inserted between the detector and the source, the sparking will cease immediately. Shielding from alpha radiation is a simple matter – 10 cm of air, a thin sheet of paper, the skin itself, will suffice.

Experiment D: The range of beta particles in aluminium using a Geiger–Muller tube and scaler

The Geiger–Muller (GM) tube is a development of the ionization chamber. Those in common use consist of a metal cylinder, one end of which is sealed with a thin (and therefore delicate) mica window. The other end is sealed into the base and carries two electrode connection pins. The metal cylinder constitutes the cathode whilst the anode is a thin coaxial rod which extends the length of the tube. A strong electric field develops in the space near to the anode when the working voltage of about 400 V is applied. See Fig. 46.4.

In the intense electric field near to the anode, a single ion-pair can develop into an avalanche of electrons (and positive ions). The current pulse which ensues develops a voltage across the resistor (R) in series with it. Such voltage pulses can be counted by a suitable electronic counting circuit (scaler).

A typical tube will operate at about 400 V with a dead-time of 100 μs. This means that it takes 100 μs for the tube to recover from the passage of

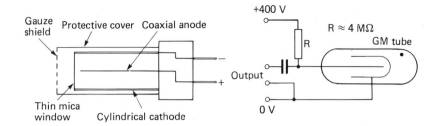

Fig. 46.4 A Geiger-Müller tube.

a pulse. So it cannot discriminate between pulses which arrive at rates in excess of $10\,000\ \text{s}^{-1}$.

Such a tube will respond to all three types of radiation. But the thickness of the end-window severely limits its use as a detector of alpha particles.

Background count; randomness
One cannot fail to notice that pulses will be counted once the tube is at the operating voltage whether or not there is a source nearby. This is owing partly to radioactive substances which are always present in the earth, in buildings and in the atmosphere, and partly to cosmic radiation entering the atmosphere from space. With a typical GM tube these pulses occur at around 30 counts per minute. Before doing any experiments with such detectors as solid-state detectors or GM tubes, this background count should be measured over a suitable period of time (say 500 s) and the mean rate deducted from later measurements made.

Observing the background count draws attention to the way in which the ionizing events occur at unpredictable intervals of time. In consequence, the count rates recorded will inevitably show random variations. For example, in an experiment with a GM tube and a radium source, 200 different counts of duration 1 second produced the following results for the count rate:

Range of counts	1100–1199	1200–1299	1300–1399	1400–1499	1500–1599	1600–1699
No of occurences	11	88	61	29	7	4

In statistical work of this sort, the *standard deviation* provides a good measure of the confidence with which the mean number of counts is quoted. If N counts are taken the standard

deviation is $\sqrt{N}$. So if, for example, 100 counts were taken we would express the result as $100 \pm \sqrt{100} = 100 \pm 10$. If 1000 counts were taken the result would be expressed as $1000 \pm \sqrt{1000}$ or 1000 ± 31. The more counts you take the more confidence you will have in your measurement since the fractional uncertainty decreases.

The experiment
In the experiment the GM tube is set to face the source of β-radiation and a series of thin aluminium sheets is placed between the two. The count rate is recorded for various thicknessess of aluminium (from about 1 to 5 mm), due allowance being made for the background count.

Since the β-particles have to penetrate the mica window of the GM tube and some centimetres of air as well as the aluminium, one cannot expect to arrive at an accurate measure of the range which, as in the case of the α-particle, is dependent on the energy of the particle.

The energy of a β-particle from a ^{90}Sr source is about 0.5 MeV. Bearing in mind the small mass of the β-particle (9.1×10^{-31} kg compared with 6.8×10^{-27} kg for an α-particle), it will be seen that the maximum velocity of a β-particle is likely to be a substantial fraction of that of a light itself (see Problem 46.3).

Unlike α-particles, which for a given nuclide are monoenergetic, β-particles exhibit a continuous energy spectrum. The maxima are characteristic of the nuclide involved. The figure quoted for the strontium source is the maximum energy.

Problem 46.3 What is the maximum speed of a β-particle which is emitted from protactinium with energy 2.3 MeV?

The mass of the particle is 9.1×10^{-31} kg; its energy of 2.3 MeV is equivalent to $2.3 \times 10^{6} \times 1.6 \times 10^{-19}$ J which is 3.7×10^{-13} J. *Relying on* *the $\frac{1}{2}mv^2$ expression* to give us the kinetic energy we have

$$\frac{1}{2} \times 9.1 \times 10^{-31} \times v^2 = 3.7 \times 10^{-13}$$

whence

$$v^2 = 8.1 \times 10^{17}$$

and

$$v = 9 \times 10^8 \, \text{m s}^{-1}$$

Which is absurd! The speed of light is 3×10^8 m s^{-1} which is the ultimate speed! It is quite impossible for the particle to have the speed computed. Evidently our application of the expression $\frac{1}{2}mv^2$ for kinetic energy was wrong. The mass of the particle itself is not constant but rises as the speed of the particle approaches that of light. In this case the speed of the β-particle is over 80% of the speed of light.

One may now wish to question the validity of the solution to Problem 46.2. The speed of the α-particle was there shown to be only 5% of the speed of light. The calculated speed is not very far out! (See Sections 33.7 and 39.2.)

Experiment E: The range of gamma radiation in lead

In this case the GM tube is placed in front of the source of gamma radiation and lead absorbers replace the aluminium ones used in Experiment D. These absorbers will need to build up to a thickness of at least 20 mm. In this case it is best to find the thickness of lead which will reduce the count rate by one half.

The energy of the gamma photon from ^{60}Co is about 1.3 MeV.

Experiment F: The behaviour of β-particles in a magnetic field

Experiment D may be modified to show the effect of a magnetic field on the motion of β-particles by interposing such a field between the source and the GM tube. Figure 46.5 illustrates the arrangement. The experiment will provide evidence for the sign of the charge on the particles.

Unfortunately, to show the behaviour of alpha particles in such a field requires the use of a solid-state detector with the whole apparatus contained in a vacuum (but see cloud chamber photographs on pages 532–533).

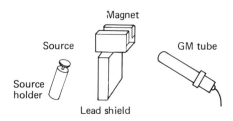

Fig. 46.5 The deflection of β-particles in a magnetic field.

Experiment G: Cloud chamber tracks

It was C. T. R. Wilson who applied the known ability of ions to act as nuclei for the condensation of water (or alcohol) vapour to the examination of the behaviour of the radiations from radioactive substances. The first cloud chambers were of the expansion type: that is a chamber containing a saturated vapour was allowed to enlarge (by, e.g., lowering a piston) so that the gas within it expanded and cooled. Ions within the chamber during that expansion would act as condensation nuclei and minute liquid droplets would gather about each one.

A disadvantage of the expansion chamber is that it is only sensitive for about 0.2 s after the expansion and then it takes several seconds for it to be recycled before another expansion can take place. A type of chamber that permits continuous observation of particle tracks is the diffusion cloud chamber, first suggested by Langsdorf in 1939. It relies on solid CO_2 to provide cooling, rather than on the expansion of a gas. A simple version of a continuous cloud chamber is shown in Fig. 46.6. The transparent case is divided horizontally by a thin black metal plate that forms the floor of the chamber. In the lower half is placed solid CO_2 that is pressed into contact with the underside of the floor by a piece of foam plastic sponge. The top half of the case is the chamber proper and near the perspex viewing window is a ring of felt soaked with alcohol. While the top of the chamber is at room temperature, say 20 °C, the floor is at −78 °C, and alcohol vapour evaporating from the felt diffuses downwards. As it does so the gas and vapour are cooled and therefore the degree of saturation of the gas in the chamber increases. At about 1 cm from the floor the saturation is sufficient for condensation to occur on the ions left in the wake of the radiation

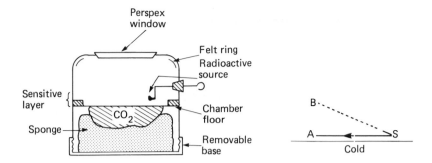

Fig. 46.6   A simple diffusion cloud chamber.

from a radioactive source, and tracks can be seen in this sensitive layer.

A vertical electric field must be established (by charging the perspex with a wool cloth) but its function is not that of a clearing field. It is needed to pull ions quickly into the sensitive layer before they have had time to diffuse. Consider first an α-particle that travels horizontally along SA through the sensitive layer; each of the ions it produces will immediately be surrounded by a droplet of alcohol (there are about 10^5 ions per track). Each droplet is of considerable mass compared with the mass of the ion on which it is formed. It falls, with neighbouring droplets, under gravity to the floor of the chamber; there is no relative motion of the droplets because of their great inertias. Thus tracks in the sensitive layer are thin and well defined. However, if an α-particle moves at, say, 30° to the floor along SB, it will first travel up through the sensitive layer where it leaves a sharp track of droplets, but above the layer, a line of ions will be formed that have not yet collected droplets. Because they are 'bare' ions they have little mass and can separate easily from one another, consequently, what was a thin line of ions soon becomes wide and diffuse. If an electric field pulls freshly formed ions down rapidly into the sensitive layer before spreading occurs, a sharp track will result. It must be remembered that the consequence of this is that an observed track length is the projection of the true length on to the horizontal plane of the chamber floor.

Bubble chambers

The cloud chambers have the grave disadvantage that the density of the gas inside is not large enough to allow high-energy particles a reasonable number of interactions with the gas molecules as they cross the chamber. Very few, if any, ions are

produced. So for work with such high-energy particles bubble chambers are preferred. These contain a liquid (usually liquid hydrogen at about 20 K) which is at such a pressure and temperature that it is just below its boiling point. When the pressure is decreased suddenly, the liquid becomes 'super-heated', i.e. its temperature is now above the boiling point at the new pressure. This, like the super-saturation of the vapour within the cloud chamber, is an unstable state. Any ions which happen to be present in the liquid will act as 'evaporation nuclei' (compare the condensation nuclei referred to in Section 15.2) and small bubbles will form around them. So the track of the particle which caused the ionization will be revealed as a line of bubbles which can be photographed. Such photographs usually record a number of events within a relatively small space. Fig. 46.7 is an example.

The sharpness of the tracks shown in the bubble chamber photograph is significant in itself. Although the tracks are caused by the interactions of the high-speed particles with the hydrogen atoms, deviations from the paths are very rare, even though the density of the liquid is high. This simple observation suggests strongly that the hydrogen atoms must be largely empty.

A small collection of photographs or tracks arising from ionizing events in cloud chambers are shown on the following pages.

Photograph A. The gas in the chamber was air. The track was one of the first ever photographed and the sharpness of it is a tribute to the skill of the inventor of the cloud chamber, C. T. R. Wilson.

A ^{222}Rn atom, at the left of the photograph, has ejected an α-particle. The residual ^{218}Po atom recoils and it too produces some ionization; hence

Fig. 46.7 Tracks of sub-atomic particles in a bubble chamber. (Courtesy of the Cavendish laboratory, Cambridge.)

the tiny dot at the start of the track. The number of ions, and therefore the number of water droplets produced per cm, increases as the α-particle slows down towards the right of the photograph and spends longer in the vicinity of gas atoms. The increase of interaction time as the speed falls also explains why α-particles so often undergo a large deflection towards the end of their tracks. By looking, in the plane of the page, along the track, two smaller deviations can be seen.

Photograph B. α-particle tracks from a small source. Note that there are particles with two different energies and, therefore, two different ranges. *J. Chadwick, Cavendish Laboratory, University of Cambridge.*

Photograph C. α-particle tracks in a magnetic field. The particles enter the field of about 4.5 T from the left. This, the first photograph showing magnetic deflection of α-particles, was taken by *P. Kapitza, 1924.*

Photograph D. Track of a β-particle in air. The radioactive source is to the right of the picture. The track shows how easily β-particles are

deflected by the electric fields of electrons in the atoms they are passing. *C. T. R. Wilson, 1912.*

Photograph E. β-particle tracks. One fast particle crossed the tracks of several slow ones which have been released in the gas by X-rays. *C. T. R. Wilson, 1912.*

Photograph F. Photo and recoil electrons from hard X-rays in air. The longer tracks are made by photoelectrons; the short tracks are made by recoil electrons from the Compton effect. *C. T. R. Wilson, 1912.*

Photograph G. The absorption of Cu K radiation in argon. A narrow pencil of monochromatic X-rays enters the chamber from the left. Photoelectrons are emitted from the argon and so the X-rays are absorbed causing an exponential decrease in the number of photoelectrons in the beam. See Section 47.1. *E. J. Williams.*

Photograph H. α-particles have passed through hydrogen. One of the particles has collided with a hydrogen nucleus and the knock-on proton leaves a long track showing less ionization than the α-particle track. The angles made by the branches of the fork are consistent with the ratio (mass of α-particle/mass of proton) being equal to 4. *Blackett and Lees, 1932.*

Photograph I. α-particle tracks in helium. One collided with a helium nucleus. After the collision the two tracks make 90° with each other. The photograph is a lucky choice of a case when the fork occurred in a plane parallel to the plane of the plate in the camera. *N. Feather, 1933.*

Photograph J. α-particle tracks in nitrogen. One collided with a nitrogen nucleus and moved to the right. The recoiling nucleus was responsible for the other track. *P. M. S. Blackett, 1925.*

Photograph K. Below the chamber there is a polonium source which is surrounded with beryllium. Within the chamber there is a paraffin target seen as a horizontal straight line. A neutron from the beryllium, invisible in the chamber, releases a long range proton by collision in the paraffin. *I. Curie and F. Joliot, 1932.*

Photograph L. The transmutation of nitrogen by α-particles. An α-particle is captured when it strikes the nitrogen nucleus, and the long branch is an example of an ejected proton travelling backwards, i.e., with a velocity component towards the source. The short branch of the fork is caused by the recoiling oxygen nucleus which results from the transmutation. Using a fully automatic chamber, Blackett found only eight examples of such transmutations in 400 000 tracks. *P. M. S. Blackett and D. S. Lees, 1932.*

Photograph M. A neutron which came from a source below the chamber and left no track, struck a nitrogen nucleus and was absorbed. The resulting nucleus ejected an α-particle, and the remainder recoiled. *N. Feather, 1932.*

Photograph N. This illustrates neutron-induced transmutation of a heavy nucleus and, because the nucleus splits into two roughly equal parts, the process is called nuclear fission. The chamber contains air and has across its middle a screen of gold foil coated with uranium. Neutrons bombarding the foil have caused a uranium nucleus to undergo fission. The total energy released is large, being of the order of 200 MeV which appears mostly as kinetic energy of the fission products. The two 'heavy' fragments are charged and leave tracks that can be seen travelling in opposite directions from the sides of the foil. As each carries as many as 50 proton charges the ionization density is very high and the tracks' ends show branches indicating collisions with nuclei of the chamber gas. At the moment of fission, 2 or 3 neutrons are emitted and if at least one of these can be induced to produce fission of another nucleus, a chain reaction results which is self-sustaining. This has not occurred in N, the other tracks are the result of recoil nuclei and transmutations caused by neutron bombardment of the gas in the chamber. *Bøggild, 1940.*

Photographs A, C, D, E, F, H, I J, L and M and Fig. 44.2 are reproduced, with permission, from the proceedings of the Royal Society.

46.4   Radioactive decay

Photographs O, P and Q were taken at 30 s intervals and show the tracks left by alpha particles originating from the decay of the radioactive gas radon in a simple diffusion cloud chamber. In the 60 s between the first and the last picture, the activity has diminished significantly. The radon is decaying rapidly. This phenomenon deserves closer study.

Experiment I: The decay and recovery of protactinium

When uranium decays it initiates a chain of events which we might represent

^{238}U ⟶ ^{234}Th ⟶ ^{234}Pa ⟶

| emits an α-particle | emits a low energy β-particle | emits an energetic β-particle |

and so on.

The uranium decays very slowly; the thorium quite slowly and the protactinium relatively quickly. This makes it a very suitable element for the study of radioactive decay. Moreover, the protactinium salt can be separated from the thorium parent in a suitable organic solvent, leaving the uranium grandparent and the thorium parent behind in the original aqueous solution.

The experiment is performed in a small plastic bottle about half full of an aqueous solution of a uranium salt. The rest of the space is filled with the organic solvent. The bottle, which is carefully

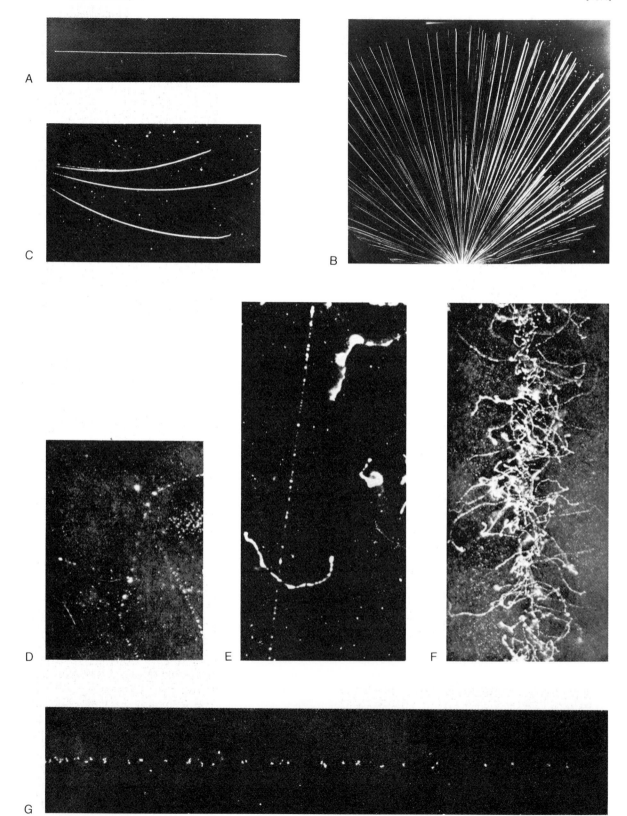

H

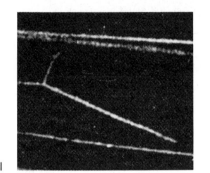

I

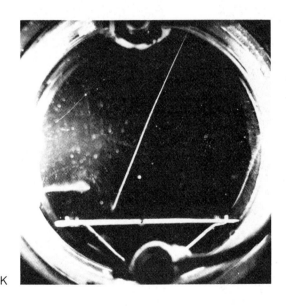

K

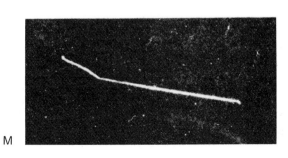

M

J

L

N

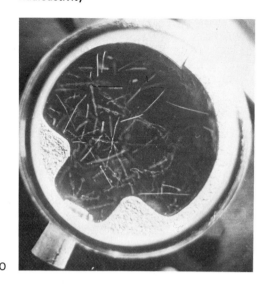

O

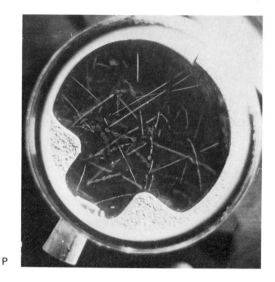

P

Q

sealed, rests on a tray lined with absorbent tissue as a precaution against spillage. It is shaken for at least 15 s and returned to the tray. The protactinium salt transfers to the solvent so that, after the two layers have separated out, the upper half of the bottle contains the protactinium. A GM tube is placed close to the shoulder of the bottle and starts to count the energetic beta particles from the protactinium (see Fig. 46.8). The count is noted every 10 s. Figure 46.9 is a graph of a typical set of results showing how the count rate changes with time.

A second GM tube and counter can be used to monitor the recovery of the protactinium in the aqueous layer as the parent thorium continues to decay.

A very similar graph to that shown in Fig. 46.8 was analysed in Section 26.7. A characteristic feature of this and all other graphs of exponential decay is the constancy of the *half-life* of the quantity which is decaying. Six values taken from the graph of this quantity for protactinium are shown below the graph; their mean is 71 s.

In each period of 71 s the activity of the protactinium was reduced to one half of its value at the beginning of the period.

Some typical half-lives of radioactive elements are shown in Table 46b.

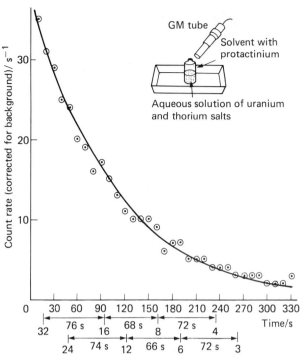

GM tube

Solvent with protactinium

Aqueous solution of uranium and thorium salts

Fig. 46.8 Determining the half-life of protactinium with a graph of some results.

Table 46b Half lives of a selection of radioactive nuclides

These are the 1st six elements in the ^{238}U series		These, with radium, are the sources referred to in Section 46.3	
$^{238}_{92}$U	4.5×10^9 years	$^{241}_{95}$Am	460 years
$^{234}_{90}$Th	24.1 days	$^{90}_{38}$Sr	28.8 years
$^{234}_{91}$Pa	71 s	$^{60}_{27}$Co	5.24 years
$^{234}_{92}$U	2.48×10^5 years		
$^{230}_{90}$Th	8×10^4 years		
$^{226}_{88}$Ra	1600 years		

The mathematics of the decay process
The equation for the discharge of a capacitor took the form (see Eq. 26.10)

$$\frac{\Delta Q}{\Delta t} = -\frac{1}{RC} Q$$

The solution was given by Eq. 26.11

$$Q = Q_0 e^{-t/RC}$$

The half-life of the decaying charge was given by setting $Q = Q_0/2$ and solving for $t_{1/2}$ (Eq. 26.12).

$$t_{1/2} = RC \ln 2 = 0.693 \, RC$$

The activity of the protactinium is sampled by the Geiger counter and reported in counts per second. These diminish according to an exponential law as the ability to find a half-life shows. We may therefore assume that the activity A also obeys an exponential law and so we assume

$$\frac{dA}{dt} = -\lambda A$$

where λ is the *decay constant*.

Now the activity depends directly on the number (N) of atoms involved so we may also write

$$\frac{dN}{dt} = -\lambda N \qquad (46.1)$$

By analogy with the case of the exponential decay of charge on the capacitor already quoted we see that the solution to this equation will be

$$N = N_0 e^{-\lambda t} \qquad (46.2)$$

where N_0 is the number of atoms at the commencement of the time t.

The half-life will be given by

$$t_{1/2} = \frac{1}{\lambda} \ln 2 \qquad (46.3)$$

If we take logarithms, Eq. 46.2 becomes

$$\ln N = \ln N_0 - \lambda t$$

If a straight line, a graph of $\ln N$ against t will confirm the exponential nature of the observations and the slope of the graph will give λ and hence $t_{1/2}$ using Eq. 46.3.

All our experience with radioactive decay has shown that the process is a random one; even the points plotted on Fig. 46.8 show this. And yet the decay curve which smooths out the random fluctuations in the counting obeys the exponential law. The randomness is real; but so too is the law governing the decay process.

Indeed we can find further examples in which processes which we recognize to be governed by chance nevertheless obey strict laws. Even an analysis of the behaviour of dice when they are thrown leads to an exponential function. In a simple game with 500 dice, all the sixes thrown were discarded after each throw. The figures obtained were:

Throw: 1 2 3 4 5 6 7 8 9 10 11 12 13 14 15 16 17 18 19 20 21 22 23 24 25 26 27 28 29 30

Sixes
discarded: 80 71 55 42 36 37 23 31 13 15 10 8 9 11 10 10 5 2 4 3 2 1 2 1 3 1 2 1 1 2

Problem 46.4   Plot a histogram of these figures. Draw a smooth curve to represent the 'decay' of the sixes and then make four different estimates of the half-life of a die.

If N dice are thrown and λN 'decay' at each throw, we expect the rate of decay per throw $\Delta N/\Delta t$ to be $-(1/6)N$ since a die has six faces. There will be random fluctuations as your histogram shows so clearly, but this is the most probable decay rate.

Let us now take the general case:

$$\frac{dN}{dt} = -\lambda N$$

$$\int \frac{dN}{N} = \int (-\lambda)\, dt$$

$$\ln N = -\lambda t + \text{constant}$$

When $t = 0$

$$\ln N_0 = \text{constant}$$

therefore

$$\ln N = -\lambda t + \ln N_0$$

whence

$$N = N_0 e^{-\lambda t} \qquad (46.4)$$

When

$$N = \frac{N_0}{2}, \qquad t = t_{1/2}$$

and

$$\frac{N_0}{2} = N_0 e^{-\lambda t_{1/2}}$$

$$e^{-\lambda t_{1/2}} = \tfrac{1}{2}$$

Or

$$e^{\lambda t_{1/2}} = 2$$

$$t_{1/2} = \frac{1}{\lambda} \ln_e 2 \qquad (46.5)$$

For the dice, λ is, in fact, 0.182 and so

$$t_{1/2} = \frac{1}{0.182} \ln 2 \ .$$

$$= 5.49 \times 0.693$$
$$= 3.8 \text{ throws}$$

You should compare this value with your own obtained from Problem 46.4.

It may seem that we have wandered far away from the case of radioactive decay. But we now realize that the mathematical law of exponential decay applies to the behaviour of dice which we recognize to be governed by chance. It should come as no surprise then if we assume that the success of the exponential law when applied to the decay of protactinium and other radioactive substances is also due to the operation of chance. The observations suffered random variations; but given long enough time intervals – and recognizing that vast numbers of atoms are involved – we find that the behaviour of the radioactive nuclides is determined by chance and described by the exponential law. We must be strongly reminded of an earlier analysis (Chapter 12) of the flow of heat in an Einstein solid (see especially Fig. 15b).

46.5   Growth and decay

As one nuclide decays another grows. Figure 46.9 shows the now familiar exponential decay of one nuclide (M) and the necessary growth of the daughter nuclide (N) assuming that the latter is stable. But N may itself be radioactive; if so, it will decay at a rate which is proportional to the number of atoms present. So, as the number of atoms of the daughter product N grows, so will the rate of decay. Figure 46.10 illustrates this situation.

Problem 46.5   The half lives of ^{234}Th and ^{234}Pa are 24.1 days and 71 s respectively. A bottle as used in Experiment I is left undisturbed for some time and the nuclei are in equilibrium – that is, as many protactinium atoms are produced in unit time as decay. If there are 10^{12} atoms of thorium in

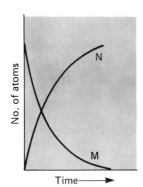

Fig. 46.9 The growth of a stable daughter nuclide (N) and the decay of the parent nuclide (M).

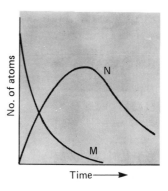

Fig. 46.10 The growth of an unstable daughter nuclide (N) and the decay of the parent nuclide (M).

the bottle, how many atoms of protactinium are there and what is their mass?

For thorium

$$t_{\frac{1}{2}} = 24.1 \times 24 \times 60 \times 60 = \frac{1}{\lambda_{Th}} \times 0.693 \text{ s}$$

whence

$$\lambda_{Th} = 3.3 \times 10^{-7} \text{s}^{-1}$$

For protactinium

$$t_{\frac{1}{2}} = 71 = \frac{1}{\lambda_{Pa}} \times 0.693 \text{ s}$$

whence

$$\lambda_{Pa} = 9.8 \times 10^{-3} \text{s}^{-1}$$

Since the solution is in equilibrium, the rate of decay of the thorium is equal to the rate of production of protactinium.

Hence

$$\lambda_{Th} N_{Th} = \lambda_{Pa} N_{Pa}$$

$$N_{Pa} = \frac{\lambda_{Th}}{\lambda_{Pa}} N_{Th} = \frac{3.3 \times 10^{-7}}{9.8 \times 10^{-3}} 10^{12}$$

$$= 3.4 \times 10^7 \text{ atoms}$$

The Avogadro number is 6.02×10^{23} mol^{-1}. Therefore 234 g of protactinium contain 6.02×10^{23} atoms.

It follows that the mass of protactinium present is

$$\frac{3.4 \times 10^7 \times 234}{6.02 \times 10^{23}}$$

$$= 1.3 \times 10^{-14} \text{g}$$

46.6   Very long half-lives

Table 46b includes several radioactive nuclides with half lives in the order of a million years. Such half-lives can hardly be determined directly! But, as we have seen, the half-life is related to the decay constant (Eq. 46.3). So measurements of λ, the decay constant, can lead to the calculation of these enormously long half-lives.

Problem 46.6   A ^{226}Ra source commonly used in schools has an activity of 18.5×10^4 Bq (or 5 μCi). The mass of radium in the source is 5.1×10^{-6} g. What is the half-life of Radium-226?

226 g of radium contain 6.02×10^{23} atoms (Avogadro number)

5.1×10^{-6} g of radium contain

$$\frac{6.02 \times 10^{23} \times 5.1 \times 10^{-6}}{226} \text{ atoms}$$

$$= 1.36 \times 10^{16} \text{ atoms}$$

The decay constant, λ

$$= \text{activity/number of atoms}$$

$$= \frac{18.5 \times 10^4}{1.36 \times 10^{16}}$$

$$= 1.36 \times 10^{-11} \text{s}^{-1}$$

And $t_{\frac{1}{2}} = 0.693/\lambda$

$$= \frac{0.693}{1.36 \times 10^{-11}}$$

$$= 5.1 \times 10^{10} \text{s} = 1620 \text{ years}$$

Chapter 47

THE RUTHERFORD MODEL

We have already seen in Unit 11 that all atoms contain electrons; for whatever technique is used for removing these electrons (thermionic emission, irradiation by ultraviolet light, passing electric discharges) all have the same specific charge.

The discovery and exploration of radioactivity, with the identification of the α-particle as a charged helium ion, of the β-particle as a negative electron and of gamma radiation as photons of very high frequency, provided physicists with some evidence for an internal structure for an atom as well as a tool for exploring that structure. The α-particle, which is both massive and charged, became that tool.

That these particles were helium ions was first demonstrated by Rutherford and Royds in 1909. Essentially the experiment consisted in collecting 'spent' α-particles which had come from radon (a radioactive gas), and had passed through the very thin wall of the glass tube containing the gas. After about a week had passed, the gas in the outer chamber was compressed, by admitting mercury, into a small discharge tube (see Fig. 47.1). The spectrum obtained was that of helium.

Confirmation of this identification came from an examination of cloud chamber pictures of the tracks made when α-particles pass through helium gas. (See Photograph I – Chapter 46). The right angle fork is characteristic of a two-dimensional collision between bodies of equal mass (See Section 8.17).

Since the helium atom is some 7000 times as massive as an electron it is hardly surprising that the demonstration of the magnetic deflection of α-particles eluded earlier experimenters. See Photograph C – Chapter 46.

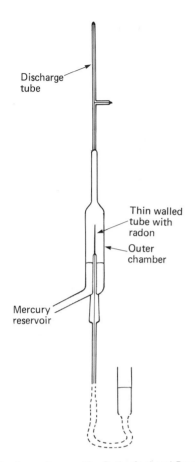

Fig. 47.1 Apparatus used by Rutherford and Royds in the experiment which confirmed the identification of α-particles with the element helium.

The identification of the β-particle with the electron was a much simpler task. It was enough to measure the specific charge. And magnetic deflection, was, as you have seen, easy to achieve.

The identity between the behaviour of gamma radiation and X-rays was sufficient evidence for the electromagnetic nature of this radiation. They share the speed 3×10^8 m s^{-1} with light and radio waves.

47.1 The scattering of alpha particles

We have seen that, in a typical case, an α-particle has energy in the order of several MeV (10^{-13} J) which is very large indeed for a particle of atomic dimensions. As a positively charged, massive and highly energetic particle it made the ideal projectile for early probing of the structure of atoms.

As we have seen the very sharpness of the

tracks in bubble chambers suggests that the atoms themselves must be largely empty space.

And yet there must be some centre of mass and of positive charge around which the electrons cluster.

Earlier investigations by J. J. Thomson into the scattering of X-rays by gases and light elements had led him to deduce that the number of electrons in an atom was *in the same order* as its *mass number* (which, for reasons which will appear later, we now call the nucleon number A). In 1911 C. Barkla, an English physicist, also using X-ray scattering, found that the number of electrons was more nearly equal to half the mass number.

When metal foils were subjected to a beam of α-particles, the facts remain much the same. The majority of the particles pass straight through such foils which may have as many as 10 000 atoms across their thickness. But occasionally, a particle will be turned; and in a very, very few cases (about 1 in 1800) will even be turned through an angle larger than a right angle. This scattering of α-particles was investigated by Geiger and Marsden under the direction of Rutherford in the Cavendish laboratory of the University of Cambridge in 1909 to 1911.

The observation of large angle scattering led Rutherford to postulate that most of the mass of an atom was concentrated in a minute positively charged *nucleus* at the centre of the volume of space in which the cloud of electrons existed. Close to such a nucleus, the α-particle will be repelled by the electric force existing between it and the nucleus. Assuming the inverse square law of force still to apply at such short distances, the force when the particle approaches close to the nucleus might be large enough to produce the large angle scattering observed. Rutherford first

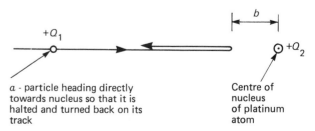

a - particle heading directly towards nucleus so that it is halted and turned back on its track

Centre of nucleus of platinum atom

Fig. 47.2 A head-on collision between an α-particle and a nucleus.

calculated how close an α-particle could approach a positive nucleus if it was going to make a head-on collision. Figure 47.2 illustrates the situation.

The α-particle initially has a kinetic energy of $\frac{1}{2}mu^2$, and at the point where it is momentarily at rest, all this energy is converted into electrical potential energy. If at this point, the centres of the two charges are separated by a distance b, and the inverse square law applies, then

$$\frac{1}{2}mu^2 = \frac{Q_1 Q_2}{4\pi\epsilon_o b}$$

Whence

$$b = \frac{1}{4\pi\epsilon_o} \frac{Q_1 Q_2}{\frac{1}{2}mu^2}$$

Since for an α-particle, m is 6.7×10^{-27} kg, Q_1 is $2(1.6 \times 10^{-19})$ C, and u is about 2×10^7 m s^{-1} it will be possible to calculate b in terms of Q_2.

For the charge on a platinum atom, Rutherford accepted Barkla's estimate that the number of electrons, and therefore positive charges, is approximately equal to half the mass number. Thus for a platinum atom,

$$Q_2 \approx 100 \times (1.6 \times 10^{-19}) \, C$$

As you can check, with a little arithmetic, the formula gives

$$b = 3.4 \times 10^{-14} \, m$$

This value cannot be quoted as the size of the nucleus. All that can be said is that the nucleus cannot have a radius greater than 3.4×10^{-14} m, if an inverse square law of force can be assumed to operate at such minute distances. If, in the calculation, a higher speed of $u = 3 \times 10^7$ m s^{-1} is used then the corresponding value of b becomes 1.5×10^{-14} m.

It is now clear, however, that the nucleus is very small indeed compared with the radius of the atom itself, for this is in the order of 10^{-10} m. (See calculation of the distance between atoms in Section 14.5).

It is hard to picture a structure in which the outermost fringe has a radius 10 000 times greater than that of the nucleus at its centre.

It is difficult to draw a scale diagram of an atom unless a very large piece of paper is used. Inside the square of Fig. 47.3 is printed a very small dot having a radius of about 0.05 mm. If it

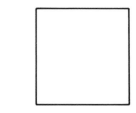

Fig. 47.3

represents the nucleus of an atom, then the edge of the atom, drawn to scale, will be 0.5 m away, and the nucleus of the next atom will be a whole metre away.

Once it has been realized that the nucleus is so small compared with the open space around it, one is not so surprised that α-particles can pass right through thin foils of metal and only very occasionally suffer a direct collision with a nucleus. What now becomes difficult to understand is that most of the mass of an atom is concentrated in such a tiny volume as a nucleus.

In his paper of 1911, Rutherford showed that if an α-particle did not head directly towards the nucleus then it would undergo a glancing collision and move in a hyperbolic trajectory. In Fig. 47.4 particle B is an example, and had it been able to continue on its original course, it would have missed the centre of the nucleus by a distance p. However, it is repelled and finally travels in a direction making an angle ϕ with its original path. It should be noted that if the angle between the asymptotes of the hyperbola (which is 180–ϕ in Fig. 47.4) is bisected, then the bisector will pass through the nucleus.

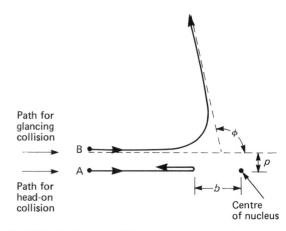

Path for glancing collision

Path for head-on collision

Fig. 47.4 A glancing collision between an α-particle and a nucleus.

Rutherford derived an equation relating b, p and ϕ, namely,

$$\cot \tfrac{1}{2}\phi = \frac{2p}{b} \qquad (47.1)$$

As has already been pointed out, Rutherford assumed that the inverse square law applied to the forces between the α-particle and the nucleus – with the implication that both particles act as point charges. Had the α-particle entered the nucleus, this could no longer apply since, as we have seen in Unit 7, a spherical distribution of charge acts like a point charge only outside itself.

Table 47a

$p =$	100b	10b	5b	2b	b	$\tfrac{1}{2}b$	$\tfrac{1}{4}b$	$\tfrac{1}{8}b$	0
$\phi =$	0.6°	6°	11°	28°	53°	90°	127°	152°	180°

Table 47a was obtained by putting values in Eq. 47.1. This gives us an idea of the scattering angles that can be expected for various 'miss-distances' or 'aiming errors', p, which have been expressed as multiples of the nuclear radius. Some of these values have been used to draw Fig. 47.5 which emphasizes that ϕ is a large angle if, and only if, p is less than b.

The deflection is certainly negligible when $p = 100b$ and, to return to the scale diagram of Fig. 47.3 this means that an α-particle will hardly be deflected unless it falls within 5 mm of that tiny dot. Recall that the radius of the atom on this scale is 500 mm and you will see that there is

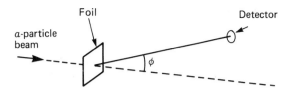
Fig. 47.6 The principle of the Geiger and Marsden experiment.

plenty of opportunity for an α-particle to remain undeflected.

Subsequently (1911), Rutherford developed an equation which related the numbers, ΔN, of α-particles scattered in particular directions to measurable quantities:

$$\Delta N = \frac{b^2 N n t}{16 R^2 \sin^4 \tfrac{\phi}{2}}$$

where, as before,

$$b = \frac{1}{4\pi\epsilon_0} \frac{Q_1 Q_2}{\tfrac{1}{2}mu^2}$$

ΔN is the number of particles falling on unit area of the detecting screen per second, N is the number of particles falling per second on the foil, n is the number of atoms per unit volume of the foil, t is the foil thickness, R is the distance from the point of incidence of the particle beam on the foil to the detector, ϕ is the scattering angle, Q_1 and Q_2 are the charges on the α-particle and the nucleus respectively and $\tfrac{1}{2}mu^2$ is the initial kinetic energy of the α-particle (see Fig. 47.6).

It was this equation which was applied to the results which Geiger and Marsden got using the apparatus shown in Fig. 47.7: 'The apparatus mainly consisted of a strong cylindrical metal box B, which contained the source of particles R, the scattering foil F and a microscope M to which the zinc sulphide screen was rigidly attached. The box was fastened down to a graduated circular plat-form A, which could be rotated by means of a conical airtight joint C. By rotating the platform the box and microscope moved with it, whilst the scattering foil and the radiating source remained in position, being attached to the tube T, which was fastened to the standard L. The box B was closed by the ground-glass plate P, and could be exhausted through the tube T.'

The variables, given the specified source of α-particles and a specified scattering foil, are ΔN and $\sin^4 \phi/2$. Table 47b is taken from the original

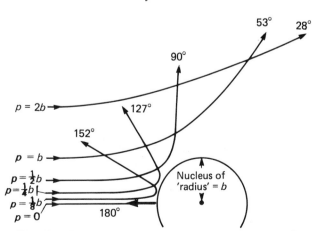
Fig. 47.5 Trajectories followed by α-particles for a range of miss-distances.

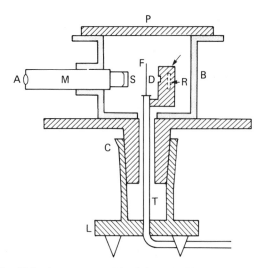

Fig. 47.7 Apparatus used by Geiger and Marsden to investigate the scattering of α-particles.

paper and relates to gold foil. The real test of the Rutherford model is the constancy of the product $\Delta N \sin^4 \phi/2$.

'It will be seen that the values of $\Delta N \sin^4 \phi/2$ are approximately constant. The deviations are somewhat systematic, the ratio increasing with decreasing angle. However, any slight asymetry in the apparatus and other causes would affect the results in a systematic way so that the deviations from constancy are probably well within the experimental error.'

This remarkable experiment not only confirmed the scattering prediction for a nuclear atom but also justified the assumption that the inverse square law of force between electric charges was

valid right down to atomic dimensions.

But that was not all. The experiment was performed with foils of several different elements – gold, platinum, tin, silver, copper and aluminium as well as carbon – and Geiger and Marsden concluded that the nuclear charge *was* equal to half the atomic mass. However, a Dutch physicist, A. van der Broek, pointed out that their data showed better proportionality if the number of alpha particles scattered by the foils was compared with the square of what came to be known as the atomic number rather than with the square of the mass number. He suggested that the nuclear charge was the product of this number and the numerical value of the electronic charge. The atom of gold then would consist of a nucleus carrying a charge of $+79\,e$ with 79 electrons in the nearby space.

Confirmation of this suggestion came from several quarters; from the radiochemical evidence developed by F. Soddy and A. Fleck and from the work which was being carried out with X-rays notably by C. G. Barkla and H. G. Moseley. The former had found that when X-rays produced at one target strike a second target, the second target can emit a *characteristic radiation* with a definite wavelength or series of wavelengths as well as scatter the original rays. This characteristic radiation could also be excited directly from the target electrode if the energy of the incident electrons was high enough and it proved to be characteristic of the metal used (see Fig. 47.8). This was, evidently, the X-ray analogue of the optical line spectrum. But Moseley revealed that it was a surprisingly simple spectrum.

Table 47b

Angle of deflection ϕ	$\dfrac{1}{\sin^4 \phi/2}$	No. of scintillations ΔN	$\Delta N \sin^4 \phi/2$
150°	1.15	33.1	28.8
135°	1.38	43.0	31.2
120°	1.79	51.9	29.0
105°	2.53	69.5	27.5
75°	7.25	211	29.1
60°	16.0	477	29.8
45°	46.6	1435	30.8
37.5°	93.7	3300	35.3
30°	223	7800	35.0
22.5°	690	27300	39.6
15°	3445	132000	38.4

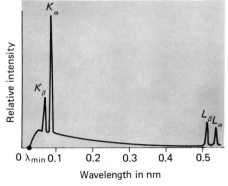

Fig. 47.8 Graph of the intensity of X-rays against wavelength for a solid target.

Table 47c Wavelengths of the K_α and K_β lines in certain X-ray spectra

Element	Proton number Z	Relative atomic mass A	Wavelength/nm	
			K_α	K_β
Manganese	25	54.93	0.210	0.191
Iron	26	55.85	0.194	0.175
Cobalt	27	58.94	0.179	0.162
Nickel	28	58.69	0.166	0.148
Copper	29	63.54	0.154	0.139

Moseley found that the wavelengths of the lines became progressively shorter as the atomic (or proton) number of the target increased. Table 47c shows this and also indicates that the observed regularities must be related to the atomic number and not to the relative atomic mass. Consider, for example, the cases of cobalt and nickel; the latter has the higher atomic number but the lower relative mass. The evidence of the wavelengths supports the view that it is the atomic number which is the fundamental measure. Moseley found that the frequency of the K line was given by the empirical relationship

$$f = 2.48 \times 10^{15} (Z - 1)^2 \, \text{Hz}$$

where Z is the atomic (or proton) number.

47.2   The nuclear model

According to this model, the atom comprises a nucleus which is not only positively charged but which also carries by far the greater proportion of the mass. Around this minute nucleus (which is smaller than 10^{-14} m across) there is a cloud of electrons whose number is equal to the number of positive charges (of size 1.6×10^{-19} C) on the nucleus. The atom itself is 4 orders of magnitude larger than its nucleus. As we saw in Unit 11, the extra-nuclear electrons exist in that space with definite energies and they can receive and emit extra energy provided that this is supplied in packets of the right size.

There can be little doubt that the radioactive radiations are emitted by the nuclei. The events, which in nature, are confined to a relatively small group of elements, occur spontaneously and at random. They are quite unaffected by external conditions as Rutherford himself tested by subjecting radioactive material to a severe but controlled explosion within an iron container.

The naturally-occurring radioactive elements emit α-particles, β-particles and γ-radiation. One is tempted to ask: what the charge on the nucleus is; what is the relationship between that charge and the serial numbers of the elements in the periodic table; what happens as a result of radioactive decay and so on.

Chapter 48

SOME APPLICATIONS OF
THE NUCLEAR MODEL

48.1   The proton

Marsden extended his work on the scattering of α-particles to some of the elements at the lower end of the periodic table. In particular, direct hits on hydrogen nuclei caused the recoil of the latter at speeds which could be estimated from Eq. 43.3

$$v_2 = \frac{2u}{(1 + m_2/m_1)}$$

where m_1 and u represent the mass and initial speed of the α-particle and m_2 and v_2 represent the mass and recoil speed of the hydrogen nucleus. Since the ratio m_2/m_1 is 1/4,

$$v_2 = \frac{8}{5}u = 1.6u$$

The hydrogen nucleus leaves the scene of the collision 1.6 times faster than the incident speed of the α-particle. Cloud chamber Photograph H (Chapter 46) shows such a collision. The particle knocked on produces a long track with less ionization per unit length than the incident α-particle. Although its speed is so high, it has only half the charge and one quarter of the mass of the α-particle and does not ionize as effectively. So, by 1914, a third particle had been identified and given the name of 'proton'. It is positively charged with a charge of $+e$ and has a mass of 1.7×10^{-27} kg.

48.2   Transmutation

When, after the end of the First World War, Rutherford returned to his studies of radioactivity at Cambridge, he observed that protons appeared to be obtained when he exposed the gas nitrogen to α-particle radiation. He wrote:

'If this be the case, we must conclude that the nitrogen atom is disintegrated under the intense forces developed in a close collision with a swift α-particle, and that the hydrogen atom which is liberated formed a constituent part of the nitrogen nucleus.' (Rutherford, 1919.)

It was clear that protons form a constituent part of a nitrogen nucleus, and that nitrogen nucleii can be disintegrated by being bombarded with α-particles. But what does the nitrogen atom turn into? This was the question that next had to be answered.

While thought was being given to this problem, Rutherford bombarded other light elements, and in conjunction with Chadwick found that disintegration protons could be emitted by all elements from boron to potassium with the exception of carbon and oxygen. A surprise was that some of these protons were emitted with *more* energy than that possessed by the incoming α-particle. This, and the fact that the protons were emitted in all directions and not just *forwards* indicated that the mechanism was not that of an elastic collision, but a superelastic one (see Section 43.1). (In, for instance, an elastic collision between billiard balls, a formerly stationary ball will always be knocked forwards or at most, sideways; it can never be knocked backwards in the direction from which the moving ball approached. However, if the stationary ball is a small bomb that detonates as soon as it is struck, a fragment is likely to go in any direction, and the total kinetic energy of the system will be increased.)

The process behind these nuclear disintegrations was made clear in 1925 by P. M. S. Blackett, who suggested that the process was one of alpha capture, i.e., an α-particle collides with a nitrogen nucleus to form an unstable nucleus of fluorine which decays, ejecting a proton and leaving a recoiling oxygen nucleus.

$$^4_2\text{He} + {}^{14}_7\text{N} \rightarrow {}^{18}_9\text{F} \rightarrow {}^1_1\text{H} + {}^{17}_8\text{O}$$

A simpler notation for this last reaction is

$$^{14}\text{N}\,(\alpha,\,p)\,^{17}\text{O}$$

in accordance with the convention that nuclear reactions should be written in the form

$$\text{initial nuclide} \begin{pmatrix} \text{incoming} & \text{outgoing} \\ \text{particle(s),} & \text{particle(s)} \\ \text{or quanta} & \text{or quanta} \end{pmatrix} \text{final nuclide}$$

Cloud chamber photograph L (Chapter 46) is one of Blackett's illustrating this phenomenon. Here was confirmation that Rutherford had achieved the first artificial transmutation of an atom in his 1919 experiment.

Natural transmutation has, of course, been occurring since the earth emerged as a planet. Radon, ^{220}Rn, is one of the long chain of naturally occurring elements which are involved in the thorium decay series. Photographs O, P and Q (Chapter 46), which illustrate the decay of ^{220}Rn, also show events linked with the daughter product ^{216}Po. This also decays with the emission of an α-particle and a short half-life of 0.15 s. There is a high probability that this second disintegration will occur very soon after the first, so leaving a V-shaped pair of tracks. Several of these can be seen in Photographs P and Q.

Three radioactive decay chains are known in nature. The parent nuclei are $^{232}_{90}\text{Th}$, $^{238}_{92}\text{U}$ and $^{235}_{92}\text{U}$; the final stable nuclei are $^{208}_{82}\text{Pb}$, $^{206}_{82}\text{Pb}$ and $^{207}_{82}\text{Pb}$. Figure 48.1 shows one of these chains – the uranium 238 series – in detail, and relates the neutron number (see below) to the proton number (a Segre chart).

48.3 The neutron

The existence of a neutral particle had been foreseen by Rutherford, at the Cavendish Laboratory, for such a particle would provide the solution to the problem posed by the difference between the number of positive charges on nuclei and their masses. Why, for example, did the nitrogen nucleus have a charge of $+7e$ (which suggested the presence of 7 protons) but a mass which was 14 times that of the proton? The idea of a neutral particle with the mass of a proton was an attractive one. Its existence was eventually confirmed by J. Chadwick at Cambridge in 1932.

But there was an alternative: the nucleus might be made up of protons and some electrons. The nitrogen nucleus might have 14 protons, so conferring the requisite mass, and 7 electrons, so that

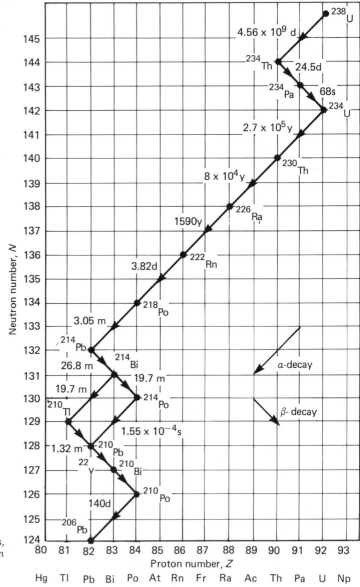

Fig. 48.1 Chart showing the U-238 decay series, terminating with the stable nuclide Pb-206. (From F.W. Sears, M.W. Zemansky, H.D. Young, *College Physics*, Addison-Wesley, 1980.)

the total charge was again +7e. But this is not really a tenable proposition. When examining the atom of hydrogen in Section 45.2 we found that an electron with energy appropriate to confinement within a space 10^{-10} m across would have enough energy to escape from the field of the proton. Now the nucleus is four orders of magnitude smaller than this. No electron could be contained within that space – its immense kinetic energy would cause it to burst out immediately.

The story of the neutron began in 1930 when

W. G. Bothe and G. F. Becker in Germany found that a very penetrating radiation appeared when α-particles bombarded the lighter elements. The effect, which was very pronounced when the target was beryllium, was assumed to be yet another example of a knock-on proton. In 1932 the French husband and wife team of the Joliot-Curies (Irene Joliot-Curie was the daughter of Mme Curie, the discoverer of radium) used a very powerful polonium source with beryllium to provide an intense beam of the 'beryllium radiation'.

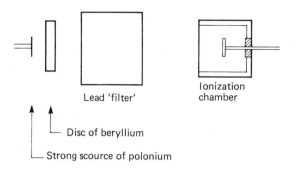

Lead 'filter'

Ionization chamber

└─ Disc of beryllium

└─ Strong scource of polonium

Fig. 48.2 The observation of neutrons.

This beam passed through 15 mm of lead with no appreciable reduction in intensity. They were surprised to find that the intensity of the radiation, as measured in an ionization chamber, *increased* if a thin sheet of paraffin wax was placed between the lead and the chamber (see Fig. 48.2). This increase was quickly attributed to knock-on protons from the wax (see Photograph K in Chapter 46). They concluded that the original 'beryllium radiation' must consist of very high-energy gamma photons.

Chadwick repeated these experiments just as soon as he heard about them with the belief that the increase in the ionization current was not caused by gamma radiation knocking on protons in the wax, but by the long-sought neutral particle. He was able to confirm this view by careful analysis and measurement of cloud chamber tracks of the events. He was able also to show that the new particle had a mass which was very slightly larger than that of the proton. It is $1.674\,8 \times 10^{-27}$ kg compared with $1.672\,5 \times 10^{-27}$ kg.

The nuclear reaction is

$$^9\text{Be}\,(\alpha,\,n)^{12}\text{C}$$

48.4 Nuclear terms

Rutherford's problem about the make-up of the nitrogen nucleus was now solved. That nucleus comprised 7 protons (so conferring the electric charge of $+7e$) and 7 neutrons.

The sum of the numbers of protons and neutrons in a nucleus is known as the *nucleon number A* (or the mass number). The charge (and the serial number in the periodic table) is given by the number of protons or the *proton number Z* (or the atomic number).

The properties of the element are determined by the proton number (which is the same as the number of extra-nuclear electrons); but the number of neutrons is not necessarily the same in all atoms of the same element. For example, uranium exists in nature in the forms $^{238}_{92}\text{U}$, $^{235}_{92}\text{U}$ and $^{234}_{92}\text{U}$. Each of these *nuclides* contains 92 protons, but the number of neutrons is 146, 143 or 142. Chemically these are indistinguishable. But they behave differently as they decay:

^{238}U emits an α-particle with energy 4.2 MeV and half life 4.5×10^9 years.
^{235}U emits an α-particle with energy between 4.18 and 4.60 MeV and half life 7×10^8 years.
^{234}U emits an α-particle with energy 4.77 MeV and half-life 2.5×10^8 years.

These are isotopic nuclides or *isotopes*. Although they cannot be distinguished chemically they may be sorted out in a mass spectrometer. This device can compare the masses of the nuclei to one part in 10^7 or better.

48.5 The mass spectrometer

Deriving from the pioneering work of J. J. Thomson and of F. W. Aston the modern mass spectrometer takes many forms, one of which is shown in Fig. 48.3. In such a machine the positive

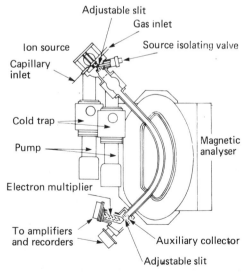

Fig. 48.3 Schematic diagram of the AEI MS12 industrial mass spectrometer. The instrument uses an electron multiplier with a gain of 10^6 to measure the ion current. (Courtesy AEI Scientific Apparatus Ltd.)

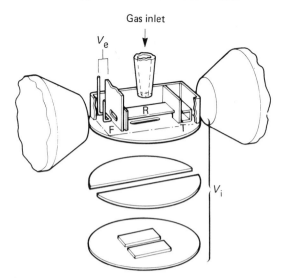

Gas inlet

V_e

R

F   T

V_i

Fig: 48.4 Schematic diagram of an ion source. (From G.P. Barnard, *Modern Mass Spectrometry*, Chapman and Hall for the Institute of Physics, 1953. Reproduced by permission).

ions are produced in an ion source by electron bombardment, as shown in Fig. 48.4.

Electrons from the heated filament F are accelerated towards an anode by a voltage V_e which is in the order of 100 V. Those which pass through the hole in this anode will move across the ionization chamber towards the trap T under the guidance of the external magnet system. The sample of material to be investigated, which is in gaseous form, enters the ionization chamber from the top and encounters the stream of electrons. Some of these atoms will be ionized by collision (see Chapter 43) and will pass with low speeds through a slit in the lower plate of the chamber.

Now they come under the influence of the potential V_i and accelerate towards the lower plate through a series of slits which serves to narrow the ion beam.

The emergent beam, containing a mixture of ions of differing speeds, differing charges, and differing masses enters the field of the electromagnet (magnetic analyser – Fig. 48.3). In this uniform field, the ions will follow a curved path. The radius of this path will depend upon the mass to charge ratio of the ion.

The accelerating voltage V_i determines the energy of the ion in the field and the ratio of mass to charge Q is given by Eq. 33.6

$$\frac{m}{Q} = \frac{B^2 r^2}{2V_i}$$

In the case of the mass spectrometer, r is the radius of the path of the ion which is steered by the magnetic field B into the slit of the collector. This radius is, of course, determined in manufacture. The beam of ions can be scanned across the slit by changing either the value of the field B or of the accelerating voltage V_i. The ion current passing the collector slit is amplified before being recorded.

48.6   Alpha particle emission

Our model of the nucleus allows explanation of the displacement laws relating to natural transmutation. If a nucleus of $^{226}_{88}$Ra emits an α-particle, then since this is a helium nucleus, the radium must lose 2 protons and 2 neutrons, i.e., it loses a charge of 2 and a mass of 4. Consequently, a new element is formed having a nucleon number 222 and proton number 86. Element 86 is radon and so the *daughter* product is $^{222}_{86}$Rn, or radon-222. This process can be expressed by the equation:

$$^{226}_{88}\text{Ra} \rightarrow \, ^{222}_{86}\text{Rn} + \, ^{4}_{2}\text{He}$$

48.7   Beta particle emission

We were reminded in Section 48.3 that it is not possible for an electron to be confined within the volume occupied by a nucleus. The same objection does not apply to a proton because of the much greater mass of the latter. But it is still true that nuclei do emit electrons as β-particles!

Unlike α-particles which are emitted from the parent nucleus with a definite speed, the β-particles are emitted with a continuous range of speeds (and therefore energies) from zero up to a maximum. This maximum varies with the nucleus involved. The existence of an energy spectrum is difficult to reconcile with the principles of conservation of energy and momentum.

Maybe an explanation could be found in the suggestion that a neutron can itself decay into a proton and an electron which, becoming the emitted β-particle, carries away $-e$ charge units. In this case the extra proton within the nucleus causes the proton number of the daughter product to be one greater than the parent and so the daughter has the chemical properties of an element one place higher in the periodic table. The nucleon number remains the same, for all that is lost is the tiny mass of the ejected electron. A typical decay might then be written

$$^{214}_{82}Pb \rightarrow \, ^{214}_{83}Bi + e^-$$

or in general terms

$$^{A}_{Z}X \rightarrow \, _{Z+1}^{A}Y + e^-$$

where X and Y represent the two atoms involved, A is the nucleon number and Z the proton number of the parent nucleus.

This simple explanation suggests that the emitted electrons, all of which result from the same process, should have identical energies. But, as we have seen, this is not so. It appears that beta decay is a process in which momentum and energy may not be conserved.

In a bold effort to solve this second problem and, at the same time, to retain the laws of conservation of momentum and of energy. W. Pauli suggested that another particle must be involved. To fit the observations, this hypothetical particle must have no charge, very little or even zero rest mass and the ability to carry energy. It was Enrico Fermi who named it 'the little neutral one' or *neutrino*. When a low energy β-particle is emitted, the neutrino carries a large amount of energy; a high energy β-particle is accompanied by a low energy neutrino. For a given nuclide the total energy is constant, the neutrino having 'balanced the books'.

Today we know that such particles exist in two forms – neutrinos, v_e, and antineutrinos, $\tilde{v}_e$. With such properties they are extremely difficult to detect experimentally but this was achieved by C. Cowan and F. Reines in 1956, twenty-five years after Pauli had suggested their existence.

The equation for the decay of the neutron may now be written

$$n \rightarrow p + e^- + \tilde{v}_e$$

and that for the decay of the nuclide becomes

$$^{214}_{82}Pb \rightarrow \, ^{214}_{83}Pb + e^- + \tilde{v}_e$$

or in general terms

$$^{A}_{Z}X \rightarrow \, _{Z+1}^{A}Y + e^- + \tilde{v}_e$$

where e^- is the electron (β-particle) and $\tilde{v}_e$ the neutrino. In fact the particle involved is the anti-neutrino and this is indicated by the symbol $\sim$

above the symbol for the particle. In these equations the neutrino ensures that sufficient energy and momentum are carried away for the conservation laws to apply. Historically this incident provides an unusual example of a successful attempt to save a physical law: an attempt which remained an act of faith for 25 years.

48.8   Gamma radiation

Gamma radiation, it must be stressed, does not consist of particles but of electromagnetic photons of nuclear origin. The wavelength is very short, being in the range 10^{-10} m to 10^{-13} m.

Many models of the nucleus have been proposed; all agree that the structure cannot be a rigid one and that it may suffer excitations, and oscillations or changes of shape rather like a blob of mercury that vibrates after being disturbed. Within the nucleus it appears that there are energy levels comparable to the energy levels available to the electrons in an atom. For example, the α-particles emitted by $^{226}_{88}Ra$ fall into two groups, one with kinetic energy 4.879 MeV (7.806×10^{-13} J) and one with 4.695 MeV (7.512×10^{-13} J). When a radium nucleus emits an α-particle of the lower energy, the daughter nucleus, $^{222}_{86}Rn$, has excess energy and is in an excited state. When this nucleus changes to the stable state it emits a γ-photon with energy $(4.879 - 4.695)$ MeV, that is 0.184 MeV. The measured energy of the γ-photon is in close agreement with this. If the radium nucleus emits an α-particle with the higher energy, the daughter radon nucleus is already in the stable ground state and no γ-photon is emitted. A similar process is observed with the emission of β-particles.

It may be thought that if gamma radiation can remove energy during β-emission, then the introduction of the neutrino is unnecessary. However, this is not the case since some β-emitters produce particles with a range of energies and yet do not radiate γ-rays at all. Also if the neutrino did not exist, then when gamma rays are emitted they would have a large energy when the β-energy was small, and vice versa; since β-particles have a wide range of energy then γ-rays should have a wide range too. However, such a continuous spectrum of gamma energy is not observed during β-emission.

One other important process which leads to the emission of gamma radiation is known as 'K-electron capture'. In this the positive nucleus captures one of the orbital electrons from the innermost, K, shell. Consequently a proton is converted into a neutron in the nucleus. This causes the atomic number to decrease by one, and may leave the resulting nucleus in an excited state. It then drops to the ground state with the emission of a gamma ray. When the vacant place in the K-shell is filled by another electron X-rays are also emitted. These X-rays may be recognized as they have wavelengths which are characteristic of the element that has been formed.

48.9   Other particles

The neutrino has already been introduced. There is an extensive range of other particles – far too extensive to discuss here. One, however deserves a mention. This is the positive electron or *positron*.

These particles were first observed when a γ-ray photon of very high energy interacted with a nucleus, giving rise to a pair of oppositely charged particles. One of these was an electron; the other the positron.

If energy is to produce matter, the mass–energy relationship $E = mc^2$ suggests that the minimum energy for pair-production of this sort will be $2m_0c^2$, where m_0 is the rest mass of the electron and of the positron. This is 1.02 MeV. Experiment shows that it is only γ-ray photons with energies in excess of this figure which are able to produce the electron/positron pair.

The positron lives but briefly, soon joining with an electron and emitting two or three γ-ray photons of total energy 1.02 MeV.

48.10   An early example of the conversion of mass to energy

The mass–energy relationship just mentioned is a logical consequence of Einstein's relativity theory. An early confirmation of the validity of the relationship came from the experiment of J. D. Cockcroft and E. T. S. Walton who, in the Cavendish Laboratory in 1932, achieved the first artificial transmutation. Using a particle accelerator of their own design (it cost about £1000, a substantial sum at the time) they fired protons at a target. Using lithium as the target they observed that high speed α-particles were produced.

$$^1_1\text{H} + ^7_3\text{Li} \rightarrow ^4_2\text{He} + ^4_2\text{He} \qquad \text{or} \qquad ^7_3\text{Li}\,(p, 2\alpha)$$

The event was shown to proceed in two stages: first, capture of the proton by the lithium nucleus to form a highly unstable nucleus ^8_4Be, and then secondly the disintegration of this latter nucleus into two α-particles.

The accelerator, which, by modern standards, was puny, gave 0.6 MeV (9.6×10^{-14} J) to each proton. The kinetic energy of the two α-particles was 27×10^{-13} J. That is twenty-eight times larger.

The mass of a proton is
$$1.673 \times 10^{-27} \text{ kg}$$
The mass of the ^{7}Li nucleus is
$$11.647 \times 10^{-27} \text{ kg}$$
The total mass of the proton and the nucleus is
$$\underline{13.320 \times 10^{-27} \text{ kg}}$$
The mass of two α-particles is
$$\underline{13.288 \times 10^{-27} \text{ kg}}$$
The difference is
$$0.032 \times 10^{-27} \text{ kg}$$
Energy equivalent using $E = mc^2$ is

$$0.032 \times 10^{-27} \times (3 \times 10^8)^2 \text{ J} = 29 \times 10^{-13} \text{ J}$$

And this is close to the value obtained from the experiment.

Modern accelerators, for example the one at CERN near Geneva (see Fig. 48.5), can provide energies of 300 GeV or even more. That is five orders of magnitude larger than the one built by Cockcroft and Walton.

48.11   Holding the nucleus together

The Cockcroft and Walton experiment raises the question of the stability of nuclei. If ^7_3Li is a stable nucleus formed from 3 protons and 4 neutrons, why is ^8_4Be (4 protons and 4 neutrons) highly unstable?

Certainly the three protons which are crammed into the nucleus of the lithium atom must exert very powerful forces of repulsion on one another. Somehow these are contained. There must be some other force which is presumably supplied by the presence of the neutrons to counteract this electric force. This is the *strong nuclear interaction* which appears to be effective over the very short ranges which apply – something in the order of a few nucleon diameters (a proton has a radius of about 10^{-15} m).

This implies that the nucleons which are within the fields of other nucleons possess potential energy just as the extra nuclear electrons possess potential energy in the electric field of the positively charge nucleus.

Problem 48.1   Applying the techniques of Section 45.2 find the least kinetic energy of a proton contained within a nucleus 10^{-14} m across.

The wavelength associated with such a proton will not be greater than 10^{-14} m.

The minimum momentum associated with this proton is h/λ

$$= 6.6 \times 10^{-34}/10^{-14}\,\text{kg m s}^{-1}$$
$$= 6.6 \times 10^{-20}\,\text{kg m s}^{-1}$$

The kinetic energy will then be $(mv)^2/2m$

$$= (6.6 \times 10^{-20})^2/2 \times 1.67 \times 10^{-27}$$
$$\approx 10^{-12}\,\text{J (or about 10 MeV)}$$

Problem 48.1 suggests that the potential energy of a nucleon within the nucleus is in the order of 10 MeV.

Nuclear arithmetic of the sort explained in Section 48.10 shows that the mass of the nucleus is always less than the mass of the component nucleons. Let us consider, for example, the helium nucleus. This has two protons and two neutrons.

Mass of two protons
  $= 3.346 \times 10^{-27}\,\text{kg}$
Mass of two neutrons
  $= 3.350 \times 10^{-27}\,\text{kg}$
Total mass of the nucleons
  $= 6.696 \times 10^{-27}\,\text{kg}$
Mass of a helium nucleus
  $= 6.644 \times 10^{-27}\,\text{kg}$
Mass deficit or defect
  $= 0.052 \times 10^{-27}\,\text{kg}$

The energy associated with this is

$$0.052 \times 10^{-27} \times (3 \times 10^8)^2\,\text{J}$$
$$= 4.68 \times 10^{-12}\,\text{J}$$
$$= 29.2\,\text{MeV}$$

This is the amount of energy which has to be supplied to break the helium nucleus. It is the

binding energy of the nucleus and is often quoted per nucleon. In the case of helium, the binding energy per nucleon is 1.17×10^{-12} J or 7.3 MeV.

The binding energy per nucleon in a nucleus is a measure of the stability of that nucleus.

A plot of the binding energies per nucleon for the elements in the periodic table reveals a pattern of great significance. With only one nucleon, the hydrogen nucleus can hardly be said to have binding energy. The next element, helium, has a remarkably high binding energy – the combination of two protons and two neutrons has great stability. This is the α-particle which plays such an important part in radioactive decay and which does itself not decay. Beyond helium, the graph is rather smooth but with minor peaks corresponding to $^{12}_{6}\text{C}$ and $^{16}_{8}\text{O}$. (See Fig. 48.6.) It climbs to a peak at ^{56}Fe and then falls steadily down towards uranium.

The first twenty elements have about equal numbers of protons and neutrons within their nuclei. As the number of nucleons increases, the proportion of neutrons to protons also increases. Up to element 83 all but two of the elements have stable isotopes as well as unstable ones. But elements with more than 83 protons are all unstable and radioactive.

It appears that as more and more nucleons come together, stability is conferred by adding more neutrons in proportion to protons up to a total of about 60 nucleons, but beyond that the ability of the additional neutrons to hold the nucleus together diminishes. It seems that the nuclear force is effective over very short ranges, acting only between nearest neighbours. On the other hand, the electrostatic repulsion between the protons acts between *all* the protons within the nucleus and not merely between near neighbours. The cumulative effect is that, in the long run, the electrostatic repulsion can counter-balance the nuclear force.

A nucleus can remain stable if there are small numbers of protons to repel one another; this explains why the heavy nuclei have a dearth of protons (see Fig. 48.7). Beyond $^{209}_{83}\text{Bi}$ complete stability is not possible.

Figure 48.7 which like Fig. 48.1 is known as a Segre chart, shows the neutron and proton numbers for the stable nuclei.

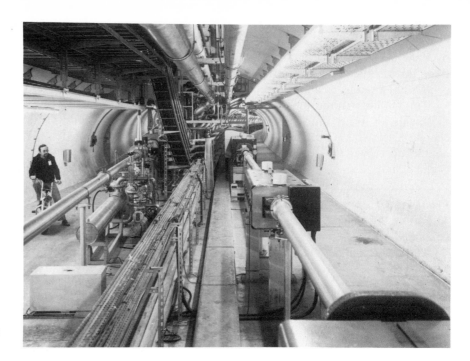

Fig. 48.5 The super proton synchronton.

The graph (Fig. 48.6) implies that energy can be released from nuclei on either side of $^{56}_{26}$Fe if they can be moved along the curve towards that element. If two light nuclei can be fused together the resulting nucleus will have less binding energy per nucleon. And the excess energy will be released. Consider, for example, hydrogen and its isotope deuterium.

The mass of a proton $= 1.6725 \times 10^{-27}$ kg
The mass of a neutron $= 1.6748 \times 10^{-27}$ kg
Together the mass $= 3.3473 \times 10^{-27}$ kg

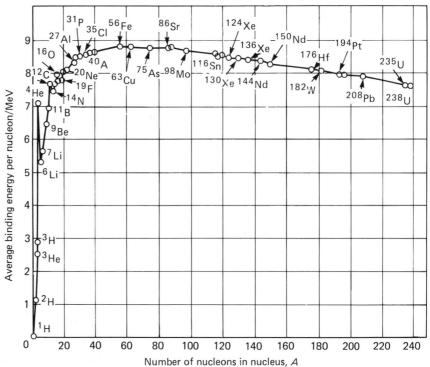

Fig. 48.6 The variation of the binding energy per nucleon with nucleon number for the elements.

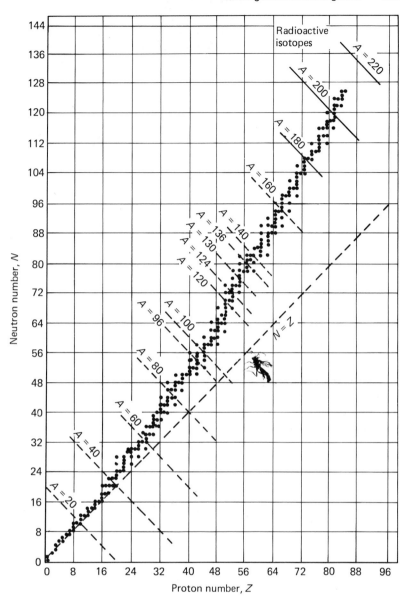

Fig. 48.7 The variation of neutron number with proton number for the stable nuclei. (From F.W. Sears, M.W. Zemansky, H.D. Young, *College Physics*, Addison-Wesley, 1980.)

But the mass of the nucleus of deuterium $_1^2$H is 3.3433×10^{-27} kg.

The difference (which is shown as the *mass defect*) is 0.0040×10^{-27} kg. In energy terms this is 3.6×10^{-13} J or 2.25 MeV. This is the binding energy of the deuterium nucleus; the binding energy per nucleon is just half of this.

The process of bringing light nuclei together to form more massive ones with a consequential release of energy is termed *fusion*. It is the

mechanism which is believed to operate in the stars (including our own sun). Under the extreme temperature prevailing, the kinetic energies of the nuclei (all stripped of their electrons) are easily capable of allowing nuclei to come close enough to one another to fuse and to emit energy as they do so. Several reaction chains have been identified – all leading to the building up of helium from hydrogen. Four protons have a total mass of 6.6900×10^{-27} kg; one helium nucleus has a mass of 6.6442×10^{-27} kg. The mass deficit is 0.0458×10^{-27} kg and this is equivalent to an energy release of 4.122×10^{-12} J or 25.8 MeV.

If *all* the hydrogen nuclei present in the sun could be converted to helium, the present rate of emission of energy could continue for 100×10^9 years; the earth is thought to be 4.6×10^9 years old!

At the 'heavy' end of the periodic table, energy is released when the radioactive elements decay towards elements of lower nucleon number. This source of energy is certainly responsible for maintaining the temperature of the earth's core – which is not less than 5000 K (almost as hot as the 'surface' of the sun).

48.12   Fission

Once available, the neutron offered the bombardiers of physics a new and very effective tool. Being uncharged, it was far more likely to reach a target nucleus than an α-particle or a proton. A pioneer in this exploration was the Italian, Enrico Fermi. He soon realized that slow neutrons were more effective as projectiles than fast ones; and that fast ones could be slowed down by allowing them to collide with particles of the same mass – hydrogen nuclei which were abundantly provided in paraffin wax.

Uranium showed unexpected behaviour. The nuclei, after bombardment, emitted β-particles. Fermi, in 1934, proposed that the uranium nucleus absorbed the bombarding neutron, becoming unstable in the process. It then emitted an electron and so moved one place higher in the periodic table. It became a transuranic nucleus. In fact this does happen; but the majority of the β-particles do not arise in this way.

It wasn't until Christmas 1938 that Otto Hahn and Fritz Strassman realized that the result of the interaction of slow neutrons was the creation of an isotope of the relatively light element $^{138}_{56}$Ba with a considerable release of energy, since this involved a considerable movement along the binding energy curve (Fig. 48.5) towards iron. It appeared that the additional neutron drove the uranium nucleus to split into two parts, one of which was a barium nucleus.

Uranium, with 82 protons, is the most massive nucleus to occur naturally. Lise Meitner and her nephew Otto Frisch (both physicists of high reputation and refugees from the Nazi regime in Germany) advanced the hypothesis that the acceptance of a neutron by a uranium nucleus produced a highly unstable nucleus which promptly broke down into approximate halves – which repelled each other under the action of the considerable electric forces which were now predominant. The fragment nuclei would be less massive than the parent by about 0.1% and the release of energy would be very large. Frisch named the process *fission*. See Photograph N (Chapter 46).

Further research revealed that the slow neutrons were causing fission in $^{235}_{92}$U rather than in $^{238}_{92}$U (naturally occurring uranium contains about 0.7% of the 235 isotope). The reaction varies but is of the form:

$$^{235}_{92}\text{U} + {}^{1}_{0}\text{n} \rightarrow {}^{236}_{92}\text{U} \rightarrow {}^{138}_{56}\text{Ba} + {}^{95}_{36}\text{Kr} + 3\,{}^{1}_{0}\text{n}$$

$$\underset{\text{fragments}}{\text{Fission}}$$

and the energy release is in the order of 200 MeV (3.2×10^{-11} J).

48.13   The chain reaction

The sample equation above shows the appearance of three neutrons. If one of these was to enter a second ^{235}U nucleus the reaction could be sustained. If two, the rate of the reaction could grow. These would be *chain reactions*. The first chain is controlled and steady; the second can grow. The sequence is 2, 4, 8, 16, 32, 64, 128, ..., 2^n. After a mere 17 steps the number involved is over a million. If done quickly enough, the reaction is explosive.

Control of such reactions can be exercised by deliberately absorbing some of the neutrons. Some will leak away; some will be ineffective in causing fission by involvement with ^{238}U nuclei and others can be absorbed in rods of such elements as cadmium. There exists a critical mass for the chain reaction; there must be sufficient of the fissile material to ensure that, allowing for leakages and absorption, each single fission produces enough effective neutrons to keep the reaction going.

The first controlled chain reaction was achieved by Fermi and his associates at Chicago in

1942. Their reactor was made of a pile of 40 000 large graphite blocks – alternate layers being made of blocks into which pellets of uranium oxide and uranium metal had been inserted. The graphite's function was to slow down the fast neutrons which resulted from the reaction so that they would be effective in causing further fission. Control rods of cadmium (a very effective absorber of neutrons) could be moved in and out of the 'pile' to control the chain reaction.

That primitive 'pile' was the ancestor of the modern nuclear power station.

Chapter 49

SOME APPLICATIONS OF
NUCLEAR KNOWLEDGE

49.1 Nuclear reactors

We shall first consider the Advanced Gas-Cooled Reactors (AGR) now producing electrical energy for the British Grid system (See Section 38.4). These were developed from the Magnox type which first went into service supplying energy to the Grid as long ago as 1956. In 1982 some 12% of Britain's electrical power came from the nuclear reactors of the Central Electricity Generating Board (CEGB). This figure is likely to rise to 20% when the reactors now building come 'on stream'. A similar figure applies to the United States but in France as much as 60% of electrical energy is derived from nuclear plants. In several other Western European countries the figure lies between 20 and 30%.

A reactor is no more than a sophisticated form of boiler whose purpose is to raise steam for conventional turbine-driven generators. Figure 49.1 shows a section through such a unit. The fuel is uranium dioxide, the uranium having been enriched so that it contains 2.3% of ^{235}U. Natural uranium contains 0.7% of the fissile ^{235}U and is used as the fuel in the earlier Magnox stations.

The reactor is moderated with graphite and the heat is extracted by forcing carbon dioxide gas over the fuel rods. The gas leaves the reactor core at a temperature of about 650 °C and then passes through a heat exchanger. There it surrenders some of its energy to water; steam is raised and used to drive the turbine-generator sets in the usual way. The steam emerging from the turbines is condensed and recycled through the heat exchanger.

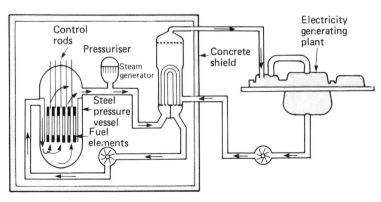

Fig. 49.1 Advanced Gas-cooled Reactor (AGR). Designed to give an output of 660 MW, these reactors used enriched uranium (2.3% U-235) as fuel and graphite as the moderator. The energy is extracted from the core by carbon dioxide gas which leaves the core at a temperature of 650°C. The fuel elements are placed in vertical holes in the graphite. (See also Fig. 49.3.) (Courtesy U.K.A.E.A.)

Control rods of boron-steel, an effective neutron absorber, can shut the reactor down very quickly.

A typical AGR has an output of 1320 MW and produces electrical energy at a cost which is significantly lower than that produced in a conventional coal or oil-burning station.

Pressurized Water Reactors (PWR) use enriched uranium dioxide fuel (containing 3.2% of ^{235}U) and ordinary water as the moderator. This water, which is under high pressure (150×10^5 N m^{-2}), carries the energy to a heat exchanger where water in a separate circuit is converted to steam. Figure 49.2 shows a diagrammatic section through such a station. The nominal output of these units is about 700 MW. The design originated in the USA and stations have been built all over the world. There are plans to build at least one such station in Britain.

This design achieved world-wide notoriety after the so-called Three Mile Island incident in Pennsylvania; a combination of a violation of the operating procedures and a misunderstanding of the situation within the reactor led to gross over-heating of the reactor core. Substantial amounts of gaseous fission products escaped into the building which had been designed to contain them. This containment system effectively limited the release of radioactive materials to the atmosphere. Regrettably, the same could not be said of the hybrid (i.e. graphite-moderated but water-cooled) Russian RBMK reactor which caused the Chernobyl disaster in 1986.

The most efficient users of nuclear fuels are the fast reactors which have been developed in Britain, France and the USSR. These reactors use a fuel mixture of plutonium and uranium dioxides. They run very hot and are cooled by liquid sodium metal; no moderator is required. Britain's prototype reactor operated at Dounreay in Caithness for 17 years before being closed in 1977. It was replaced by a larger version producing 250 MW in that year. The significant difference between these reactors and those referred to above is that fast neutrons are used (and hence the name 'fast reactor'). The ^{239}Pu fuel, which is derived from the thermal reactors using enriched uranium, is placed at the centre of the reactor core and

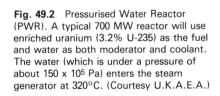

Fig. 49.2 Pressurised Water Reactor (PWR). A typical 700 MW reactor will use enriched uranium (3.2% U-235) as the fuel and water as both moderator and coolant. The water (which is under a pressure of about 150×10^5 Pa) enters the steam generator at 320°C. (Courtesy U.K.A.E.A.)

Fig. 49.3 The machine used to charge and discharge the nuclear reactors at the AGR Hinkley Point B power station in position on one of the two pile caps (bottom of picture). (Courtesy C.E.G.B.)

surrounded with a 'blanket' of ^{238}U. The plutonium is fissile to fast neutrons; moreover, when ^{238}U absorbs fast neutrons, ^{239}Pu is produced. So this reactor generates energy and produces more fuel for future use. They are sometimes referred to as fast or 'breeder reactors'. One designed to develop 1200 MW of electrical energy is currently (1982) being built in France.

49.2 Re-processing fuel and the storage problem

Large scale generation of electrical energy is inevitably accompanied by pollution of one sort or another. All power stations, of whatever type or design, discharge huge quantities of heat to their surroundings. Coal and oil-fired stations produce large quantities of solid ash and gaseous pollutants such as carbon dioxide and sulphur dioxide. They even discharge some radioactive nuclei!

Nuclear stations yield considerable quantities of radioactive waste products. Many of these nuclides have short half-lives; others long. It is these latter which give the industry its major problem.

The Magnox and AGR stations utilize 50 to 60% of the fissile ^{235}U fuel and none of the ^{238}U. Some of the latter is, however, transmuted to ^{239}Pu which is itself fissile. At all the stations there is a continuous programme of fuel element replacement (see Fig. 49.3). When first removed from the reactor the spent fuel cans are, radioactively speaking, 'hot'. But a major part of this activity is due to nuclides with quite short half-lives. A spell of several months under water in a 'cooling pond' allows that activity to die down to reasonably low levels.

The material is then taken to a re-processing plant at which chemical methods are used to separate the relatively small volumes of waste products from the bulk of the uranium and the plutonium (which can be used as fuel). The separated radioactive waste made up of a mixture of nuclides with quite short and rather long half-lives has then to be stored in some way until its activity is comparable with, say, that of naturally occurring uranium. This takes around 500 years. Currently these waste products are stored as liquids in double-walled stainless steel tanks within concrete vaults which are themselves lined with stainless steel. Cooling is provided and the mixture is stirred continuously. By 1978 the waste from the entire United Kingdom nuclear programme, which was then 25 years old, amounted to a volume of only 800 m^3 – roughly the volume of a four-bedroomed house.

Research and development is being concentrated on the possibility of solidifying this liquid waste into glass blocks. This would certainly make the storage problem an easier one than it is at present.

49.3 Nuclear weapons

In Section 48.13 we noted the possibility of an uncontrolled nuclear chain reaction. The typical application is the nuclear bomb. The earliest were fission bombs in which two masses of fissile material were forced together so that the critical mass was reached. This allowed the chain reaction to develop explosively, that is more neutrons were produced than were lost by wastage and leakage. These early bombs were equivalent in explosive terms to 14 kilotonnes of TNT; but their effect was far more damaging because of the intense burst of radiation which accompanied that blast.

The 1950s saw the development of the hydrogen or fusion bomb. The earliest of these were equivalent in explosive terms to 10 megatonnes of TNT. As we have seen, the fusion process requires very high temperatures and these exist within a fission bomb. So the hydrogen bomb system comprises a fission bomb which provides the necessary energy and also produces the necessary neutrons. The fusion material is deuterium ($^2_1$H). The reaction also requires tritium ($^3_1$H):

$$^2_1H + ^3_1H \rightarrow ^4_2He + ^1_0n + 28.2 \times 10^{-13} \text{ J}$$

The tritium can be obtained from lithium by bombardment with neutrons:

$$^6_3Li + ^1_0n \rightarrow ^4_2He + ^3_1H + \text{energy}$$

It follows that the compound lithium deuteride ^{6}LiD can provide both the tritium (from the lithium) and the deuterium.

49.4 Energy from nuclear fusion?

Energy from fusion would be a great prize – for the basic reaction involves only light nuclei and is therefore unlikely to produce radioactive waste on anything like the scale of either volume or activity as the fission reactors of today. But the problems are immense. The necessary condition is a temperature in the order of 10^8K in a sample of adequate density for an appropriate time. In spite of promising beginnings with plasmas contained by magnetic fields in machines known as Tokamaks and, latterly, with intense laser beams, the industrial production of energy from fusion must be regarded as remaining well in the future.

49.5 The biological effects of radioactive radiations

As we have seen, all radiation from radioactive substances causes ionizing events – and any ionizing event which may occur within a living cell can damage the chemicals found in that cell. There appears to be no threshold below which radiation damage does not take place. Nevertheless, living things have *always* been exposed to such radiation. And this ever-present background of ionizing events, which arises from the incidence of cosmic radiation, from the decay of the radioactive nuclides which are naturally present in the soil, in the air, and even in our own bodies (where naturally occurring ^{40}K and ^{14}C both emit β-particles), forms and always has formed part of the environment in which living things, both plant and animal, have evolved. Indeed there are those who suggest that it may even have contributed to the process of evolution itself.

High levels of exposure to radiation are certainly damaging and we must protect ourselves from these. Fortunately, the techniques of doing this are well understood. The Health and Safety Commission and the National Radiological Protection Board, which are responsible for the implementation of safety standards in the United Kingdom, publish standards and codes of practice which must be adhered to wherever radioactive isotopes are used: in hospitals, in industry, in teaching institutions and so on. The *dose of absorbed radiation* (which is the ratio of the energy imparted by the ionizing radiation to the mass of living material involved) is measured in $J\,kg^{-1}$ or *gray* (Gy). The earlier unit still seen in literature was the *rad* and 100 rad = 1 Gy.

Once determined, the absorbed dose has to be modified for its biological effectiveness – and the *dose equivalent* is then quoted in *sievert* (Sv). The earlier unit was the *rem* and 100 rem = 1 Sv.

For the ordinary population not subject to the health and safety monitoring insisted upon in industry, hospitals and so on, we may take as a yardstick the normal, unavoidable background radiation already referred to. Although this varies from place to place across the world (with no significant differences to the health of the populations involved), the average level in Britain is 100

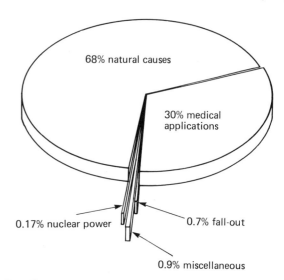

Fig. 49.4   Where the background radiation comes from.

millirem (1 mSv) per person per year. That compares with the 500 millirem (5 mSv) which the regulations of the European Economic Community (EEC) specify as the upper dose limit for whole body radiation for a member of the public per year.

To the 100 millirem of natural radiation we must add 50 millirem derived from man-made sources. By far the largest part of this (about 45 millirem) comes from medical applications such as X-rays. The remainder is the result of fall-out from weapons testing, from miscellaneous sources and from the nuclear power industry. Table 49a shows the details, which are also displayed in Fig. 49.4.

Man cannot avoid exposure to radiation; nevertheless it is very important to limit the amount to which people are exposed. As we have seen the EEC specifies the upper dose limit for an ordinary member of the public as 500 millirem (5 mSv) per year – that is over three times the amount to which we are currently exposed. Risk calculations show that an additional 1000 millirem of radiation per person per year (which is twice the EEC limit and 4000 times the amount due to the nuclear power industry) increases the risk of contracting cancer in an individual by 1 in 10 000. To place that figure in context, remember that the risk of death through a motor car accident is about 1 in 4000 per year, and through a fall is also about 1 in 10 000 per year.

Table 49a   Background radiation

Source	Dose equivalent per person per year	Percentage of total (approx.)
Natural sources	100 millirem (1 mSv)	68
Medical applications	45 millirem (0.45 mSv)	30
Fall-out from weapons testing	1 millirem (0.01 mSv)	0.7
Miscellaneous sources	1.4 millirem (0.014 mSv)	0.9
Nuclear power stations	0.25 millirem (0.0025 mSv)	0.17
Total	About 150 millirem (1.5 mSv)	

49.6 The nuclear power debate

People *are* concerned about the development of the nuclear power industry which, as we have seen, provides some 12 to 20% of our need for electrical energy. That concern stems from the fear of the effects of radioactive radiations inevitably associated with the industry. The properties of these radiations are but little understood by a public which is more familiar with the effects of nuclear bombs.

Now a textbook of physics is hardly an appropriate place to examine the case for and against a nuclear power programme. In the preceding pages we have tried to state the facts; but clearly not all the relevant facts can be offered. One must also consider the safety record of the nuclear power industry, now over 25 years old in Britain; the safety of the conventional energy raising industry (from coal mine through to the electrical output); the likely demand for electrical energy in the future; the problem of waste disposal whether of radioactive isotopes or of such gases as SO_2 into the atmosphere; the contribution which can be and is being made by improved methods of energy usage and energy saving; the future development of alternative, renewable energy resources and the impact that these new resources will themselves have on the environment.

The matter is far less simple than many people assume. If you are interested, then look at the opposing points of view as dispassionately as you can – remembering that this subject arouses strong emotions and that it is very prone to rhetoric and to the pronouncement of half-truths *on both sides*. In particular, you should remember what was pointed out in Chapter 8; that all estimates as to how long a particular reserve of fuel will last *must* be accompanied by statements of the rate of usage assumed and of the rate of change of that usage. Without that information, such estimates are totally without meaning.

49.7 Other applications

Radioactive isotopes as tracers

Living organisms take up many elements from their food and from their surroundings. But the function of these elements is not always clear. Once a stable isotope has entered a living system its identity is lost. But an isotope of that element which is radioactive can be traced by following its activity: and very low concentrations are all that are required given the sensitivity of the detecting devices used. The technique is now a standard practice in the diagnosis of certain diseases and malfunctions, and also in treatment.

The technique of labelling atoms also finds industrial application.

Radiometric dating

This is a simple direct application of radioactivity to the understanding of problems in geology, palaeontology and archaeology.

Living plants (and therefore animals which feed upon those plants) contain large numbers of carbon atoms with the stable nucleus of $^{12}_{6}C$. But the radioactive isotope $^{14}_{6}C$ is also present. These nuclei result from a reaction between atmospheric nitrogen and secondary neutrons which are the result of cosmic ray interactions

$$^{14}N + {}^{1}n \rightarrow {}^{14}C + p$$

This isotope of carbon has a half-life of 5.76×10^3 years.

In living tissue the ratio of ^{14}C to ^{12}C nuclei is constant. As some ^{14}C nuclei decay, others are absorbed. But, in dead tissue, no renewal of the decayed nucleus can take place. So the ratio of ^{14}C to ^{12}C found in dead tissue is a measure of the time elapsed since death.

The activity caused by ^{14}C is very small and that means that there is an effective limit to the carbon-dating process of around 100 000 years.

Geologists require a longer time scale than that. Such time scales can be provided by measurements on the ratio of ^{238}U to ^{206}Pb in a rock (see Fig. 48.1) or by the ratio of ^{87}Rb to ^{87}Sr or by the ratio of ^{40}K to ^{40}Ar. The half-lives of the nuclides involved are all long enough to be geologically significant.

Consider a sample of rock which is found to contain n atoms of ^{238}U and n_1 atoms of ^{206}Pb. If we can assume that no uranium or lead atoms have been lost since the rock was laid down, the number of ^{238}U atoms present when that rock formed time t ago was $(n + n_1)$. Let us call the original number of ^{238}U atoms n_0.

$$n_0 = n + n_1$$

From Eq. 46.5 we know that

$$n = n_0 e^{-\lambda t}$$

So

$$\ln \frac{n_0}{n} = \lambda t$$

The half-life

$$t_{\frac{1}{2}} = \frac{\ln 2}{\lambda}$$

Hence

$$t = t_{\frac{1}{2}} \frac{\ln (n_0/n)}{\ln 2} = t_{\frac{1}{2}} \frac{\ln (n_1 + n)/n}{\ln 2} \qquad (49.1)$$

The method has, of course, been applied to the samples of the moon rocks brought back by the Apollo missions and gives an age for the moon of 4.5×10^9 years. This, interestingly enough, is ten times older than the figure similarly obtained for the rocks we know on earth. One must conclude that, since the two bodies must be very nearly the same age, most rocks forming the crust of the earth were not part of its original surface. They emerged from the interior and solidified at a later date. There are rare examples of earth rocks whose age is placed at 3.6×10^9 years.

Problem 49.1   Fossils from a bed of sandstone contain both ^{87}Rb and ^{87}Sr in the ratio of 1 to 0.004 by mass. Estimate the age of the fossils given that the half-life of ^{87}Rb is 47×10^9 years.

The ratio n_1/n is 0.004/1
    Applying Eq, 49.1, we get

$$t = 47 \times 10^9 \frac{\ln 1.004}{\ln 2} \text{ years}$$

$$= 2.7 \times 10^8 \text{ years}$$

QUESTIONS

Unit One

1.1 The model of a gas as an assemblage of elastic particles moving about in all directions can give an adequate explanation of how a real gas, enclosed in a container, can exert a pressure. Explain how this model can also account for the pressure of the atmosphere, which is not in a closed container.

1.2 Descartes proposed a very simple model to explain the phenomenon of seeing. He suggested that the eyes shot out rays which, on touching an object, give rise to the sensation of sight.

a) Describe briefly (half a page) what you understand by the use of the word *model* when it is used in this context.

b) Suggest one experiment or observation you could make which could support or refute this model.

c) It has been suggested that all that experimental observation can do is to show if a model is *not* correct. Experiments can never provide absolute confirmation of the correctness of a model. Give a reason for accepting this view.

1.3 This is a question about describing different kinds of scientific statements. Here are some possible descriptions of such statements:

'states an experimental fact'
'makes an hypothesis'
'quotes a scientific law'
'is a rough estimate'
'is a deduction from earlier statements'.

You are asked to give a brief description of each of the numbered statements in the passage below. Your descriptions should use phrases like those given above; you may use, combine or adapt the phrases above or invent others of your own.

1) If we assume that in a gas the atoms have a radius of about 10^{-10} m and a mean separation of about ten atomic diameters . . .

2) ... it is clear that an alpha particle in traversing several centimetres of the gas must encounter some thousands of atoms of gas.

3) Only a minute fraction of such encounters, however, produce any appreciable deflection of the alpha particle.

4) It is difficult to avoid the conclusion that the greater part of the atomic volume is effectively empty.

Oxford and Cambridge Schools Examination Board
Nuffield A-level Physics (1970)

1.4 Sir Humphry Davy, a famous scientist who worked at the Royal Institution, London, during the early years of the nineteenth century, once said 'Nothing leads to the advancement of knowledge as the application of a new instrument.'

There are many instances in the history of science where this is true. Find out, and write a short note on, how

a) the telescope influenced the development of the present day model of the solar system,

b) heat engines profoundly influenced the theory of heat,

c) electron microscopes have influenced models of the structure of matter, both physical and biological.

1.5 The mass of the earth is 5.98×10^{24} kg; its radius is 6.37×10^6 m. Calculate its density, taking care to quote the correct number of significant figures and to use standard form.

1.6 The mass of an electron is 9.1×10^{-31} kg and its charge is 1.602×10^{-19} coulomb. (a) What is the ratio of charge to mass? Use standard form and quote the appropriate number of significant figures. (b) What is the mass of a mole of electrons?

1.7 Use the method of dimensions to decide which, if any, of the following statements is/are invalid?

a) The area of the ring contained between two concentric circles of radius r_1 and r_2 is $\pi(r_1 + r_2)(r_1 - r_2)$.

b) The volume of a sphere is $\frac{4}{3}\pi R^3$, where R is the radius.

c) The volume of a cone of height h and radius of base r is $\frac{1}{3}\pi r^2 h$.

d) The area cut off on a sphere of radius R by a pair of parallel planes distance h apart is $2\pi R^2 h$.

1.8 An aircraft flies 20 km in a direction 30° East of North, then 30 km due East and finally 10 km due North. Show by both a scale drawing and by a calculation that the aircraft reached a point just over 48 km from its starting point. What is the bearing of this point (from the direction of North) from the starting point?

1.9 A block of concrete rests on an inclined plane which is inclined to the horizontal at 20°. The pull of the earth on the concrete block is 200 N in the vertical direction. Draw an appropriate diagram showing this force and the components of the force parallel to and normal to the inclined plane. Calculate these two components.

1.10 During the testing of an Inter-City 125 train along a straight track these measurements were obtained:

Time/s	0	100	200	300	400	500	600	700	800
Displacement/km	0	1.12	4.37	8.62	13.5	18.7	24.0	29.5	34.9
Speedometer reading/ms^{-1}	0	24.3	39.5	45.6	50.5	53.0	54.4	55.0	55.0

a) Plot graphs of displacement against time and of speed against time.

b) Explain the differences between the shapes of the two graphs.

c) The measurements show that the train travelled 3.25 km in the 100 s to 200 s time interval. This suggests a speed of 32.5 ms^{-1}. What does this speed represent? Why does it differ from the measured speeds at 100 s and 200 s?

d) Find the gradient of the displacement/time graph at 150 s. Comment on your value in relation to your answers to (c).

e) Find the gradient of the speed/time graph at 150 s. What physical interpretation can you give to this gradient?

1.11 During the car journey discussed in Section 2.7, the temperature of the water in the cooling system rose during the first five minutes of the journey thus:

time/min	0	1	2	3	4	5
temperature/K	293	295.4	302.5	314.6	331.4	353

a) Plot a graph of these measurements.

b) Find the average rate of change of temperature $\Delta T / \Delta t$.

Given that the equation to the graph is $T = 2.4t^2$ find the average rate of change of temperature over each of the following time intervals (in minutes):

c) 3 to 4,   d) 3 to 3.1,   e) 3 to 3.01,

f) 3 to 3.001,   g) 3 to 3.0001.

h) What is your best estimate of the instantaneous rate of change of temperature with time at $t = 3$ min?

i) Draw, as carefully as you can, the tangent to the curve at $t = 3$ min and find its gradient. Compare this with your answer to (h).

Unit Two

2.1 A rifle of mass 5 kg is used to fire a bullet of mass 0.15 kg with a muzzle velocity of 600 m s^{-1}. Calculate the velocity with which the rifle starts to recoil. If this

Fig. Q2.1

recoil is to be reduced to zero in a distance of 5 cm, what is the average force required?

2.2 An astronaut is engaged in some activity some distance from his orbiting spacecraft to which he is attached by a line. How could he return to the spacecraft without pulling on the line?

2.3 In a toy known as Newton's cradle, a small number (say 5) of similar steel balls is suspended, each from a pair of strings so that they are in line and each ball is in contact with its neighbour. See Fig. Q2.1. Use your knowledge of momentum to explain why, when one ball is allowed to swing into the end of the line, only one ball leaves the other end of the line. You may assume that the spheres are almost elastic. What will happen if the end balls are lifted away by the same distance and then released simultaneously?

2.4 A golf ball (mass 45 g) leaves the club which struck it at $70 \, \text{m s}^{-1}$. If the impact lasted 0.5 s, find the mean force applied to the ball.

2.5 Two students stand 5 m apart at opposite ends of a flat-bottomed boat floating on a river. One throws a 2 kg ball at $8 \, \text{m s}^{-1}$ to the other who catches it and holds it still. Explain what happens to the boat. You may assume a mass of 200 kg for the boat and the students.

2.6 A car of mass 1000 kg is travelling at $15 \, \text{m s}^{-1}$ along a straight road. If the application of the brakes causes an average retarding force of 4000 N to act on the car, how far does the car travel before coming to rest? Also calculate the stopping distance if the car had been travelling at $30 \, \text{m s}^{-1}$. What moral can be drawn from your answers?

2.7 The table following gives the 'stopping distances' for cars travelling at given speeds.

a) What do you understand by 'thinking distance' and why is it directly proportional to speed?

Speed/ m s^{-1}	Thinking distance/m	Braking distance/m	Overall stopping distance/m
10	7	7	14
20	14	28	42
30	21	63	84

b) Use your knowledge of physics to explain the figures in the braking distance column.

2.8 A supertanker has a mass of 409 600 tonnes, its engines have been stopped but there is a slight forward motion of only 1 m in 2 s. In order to stop, the anchor is dropped and, once it starts to hold, the ship comes to rest in 20 m. Estimate the average tension in the anchor chain. If the tensile strength of the chain's steel is $8 \times 10^8 \, \text{N m}^{-2}$, what is the minimum area of cross-section of one side of a link? How would your answer be modified if the ship had been moving at an initial speed of $1 \, \text{m s}^{-1}$. You may assume the same stopping distance.

2.9 a) The equation $F = kma$ is simplified by measuring F in newtons and making k unity. What units does k have?

b) The choice of k is arbitrary. Make $k = 2$, name the resulting unit of force after yourself and write out its definition and its dimensions.

c) Is your unit of force bigger or smaller than the newton? By how much?

2.10 The gravitational field of the earth is $9.8 \, \text{N kg}^{-1}$.

a) What is the force of attraction on 2 kg?

b) What is the force of attraction on x kg?

c) The 2 kg mass is allowed to fall. What is its acceleration?

d) The x kg mass is allowed to fall. What is its acceleration?

e) What can be concluded about freely falling masses near to the earth's surface?

2.11 A spring balance supports a 4 kg mass in a lift which is accelerating upwards at $2 \, \text{m s}^{-2}$. What will the balance read? How will the balance reading change as the lift reaches a maximum uniform upward speed and then slows down before coming to rest? What will the balance read if the lift now starts to accelerate downwards at $2 \, \text{m s}^{-2}$?

2.12 A man descending by parachute, and having nothing to do for a minute or two, passes the time by thinking about Newton's third law of motion. He reasons that his speed is constant because his weight is equal but opposite to the force the parachute exerts on him and so he experiences no resultant force and hence he has zero acceleration.

a) Is this reasoning correct?
b) Is he correct in applying Newton III to the forces mentioned?
c) Name four other forces involved in this situation. (Consider the air, parachute, man and earth.)
d) Group all six forces into three pairs to which Newton III applies.

2.13 Explain why Newton III applies when you press your hand against a brick wall, but not when you are holding a brick and equating the weight of the brick to the upward force exerted by your hand.

2.14 A rocket has an initial mass M and its exhaust gases pass through a jet of effective area A with a uniform velocity v. If the density of the gases is d, calculate the minimum value of v which will allow the rocket to take off vertically. Calculate also the acceleration with which the rocket starts to rise, and discuss qualitatively how you would expect the acceleration to vary during the vertical ascent of the rocket.

2.15 At the moment of lift-off a rocket has a total mass of 3×10^6 kg and the exhaust gases have a velocity of $1500 \, \mathrm{m \, s^{-1}}$ relative to the rocket. Calculate the minimum rate, in $\mathrm{kg \, s^{-1}}$, of fuel consumption as this rocket takes off vertically.

2.16 An old-fashioned toy is a clockwork model of a steam engine. As it travels forward, a ball is shot vertically from the funnel at the front of the engine.

a) If the forward speed of the toy is $2 \, \mathrm{m \, s^{-1}}$ and the ball is ejected with a speed of $5 \, \mathrm{m \, s^{-1}}$, where would the ball land? You may ignore air resistance.
b) Describe the motion of the ball as seen by an observer travelling on the engine and also by an observer standing to one side.

2.17 A hunter takes careful aim straight at a monkey which is hanging from a branch of a tree. What would a wise monkey do: hang on or drop from the branch the moment it sees the flash from the gun? Explain fully. Would it make any difference if the monkey was level rather than above the hunter?

2.18 A car (mass 1000 kg) moving in a straight line at a steady $9 \, \mathrm{m \, s^{-1}}$ is acted upon by the viscous drag of the air and other frictional forces, but the engine maintains the car's speed by causing the driving wheels to turn against the frictional forces between the tyres and the road. The car then enters a bend with a radius of 30 m. Calculate the force which the car must experience in order to undergo this change of direction. How does this force arise?

2.19 Taking the radius of the earth to be 6400 km, estimate the speed of a satellite travelling relatively close to the earth's surface. Assume that $g = 10 \, \mathrm{N \, kg^{-1}}$. How long would this satellite take to complete the orbit?

2.20 A synchronous communications satellite moves in a circular orbit in the equatorial plane of the earth and remains directly above the same point on the earth at all times. If the radius of orbit is 42 400 km, calculate the gravitational field strength at this distance.

Is your answer consistent with the hypothesis that the mass of the earth may be considered as acting as though it were concentrated at a point at the earth's centre so that Newton's law of gravitational attraction can be applied? Take the earth's radius to be 6400 km and g at its surface to be $9.8 \, \mathrm{N \, kg^{-1}}$.

2.21 Car maintenance manuals specify the torque to be applied when the nuts on the cylinder head of a car engine are tightened up. Why not specify the force to be applied to the handle of the wrench instead?

2.22 What determines whether a tall, massive object – a refrigerator, for example – slides or tips when one tries to push it across a rough floor?

2.23 A light truck has a distance of 2.7 m between the front and the rear axles (the wheelbase). Of the weight of the truck 60% is supported by the front wheels. Where is the centre of gravity of the truck?

2.24 A door 0.8 m wide and 2 m tall weighs 150 N. The two hinges, which are secured 25 cm from the top and bottom of the door respectively, each support half its weight. Assuming that the centre of gravity of the door is at its centre, find (a) the vertical and horizontal components of the force exerted on the door by each of the hinges, and (b) the magnitude and the direction of the force exerted by each hinge.

2.25 A television receiving aerial is fixed to a vertical mast which is itself secured to the chimney of a house by a bracket. The bracket is bolted to the chimney by two bolts, one above the other. The bolts are 25 cm apart and the mast, which is 3 m tall, is held 15 cm away from the chimney by the bracket. The mast has a mass of 2 kg. The aerial is 1 m long and has a mass of 2 kg. See Fig. Q2.2.

Describe the forces (in magnitude and direction) exerted by each of the two bolts on the bracket. You may assume that the weight of the mast and aerial is shared equally between the two bolts and that the mass of the bracket is negligible.

Unit Three

3.1 The engine and transmission of a car deliver 50 b.h.p. to the wheels (b.h.p. means 'brake horsepower', i.e., the actual measured power rather than a calculated

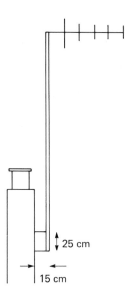

25 cm

15 cm

Fig. Q2.2

value; 1 b.h.p. = 0.746 kW). If it is travelling along a level road at 144 km h^{-1}, calculate the total wind and road resistance to the motion.

3.2 A car of mass 1 tonne climbs a hill, 300 m high, along a road 2.1 km in length. The total resistance to motion owing to air and road friction is on average, 600 N. Assume that $g = 9.8$ N kg^{-1}.

a) Calculate the energy required by the car to climb the hill.
b) Calculate the component, parallel to the road, of the average force exerted by the tyres while the car is climbing the hill.
c) If the car stops on the hill, what force must be exerted parallel to the road to prevent the car from running backwards?
d) What assumptions have you made about the hill in parts (b) and (c)?

3.3 A car and driver with a total mass of 1800 kg are travelling at 24 m s^{-1}. The driver sees an obstruction ahead, brakes sharply and strikes the obstruction at a speed of 8 m s^{-1}. During the collision the forward motion of the driver is brought to rest in 0.1 s by his tightly fitting seat-belt. The mass of the driver himself is 60 kg.

a) What is the change in kinetic energy of the car during the braking?
b) What happens to this energy?
c) What is the mean acceleration of the driver during the impact?
d) What force does the seat-belt exert?
e) If the seat-belt had been fitted loosely, what

differences would be apparent in the motion of the driver?

3.4 An archer's bow is drawn back using a spring balance. Readings of the force (F) and the displacement (d) are:

F/N	0	50	100	160	250	400
d/m	0	0.2	0.3	0.4	0.5	0.6

a) Plot a graph of these values and use it to determine the energy transferred to the bow when it is fully drawn.
b) An arrow of mass 400 g is fired with this bow. Calculate the maximum velocity of the arrow.

3.5 A trolley of mass 40 kg has free-running wheels and is initially at rest on a horizontal surface (see Fig. Q3.1). A light rope is attached to the trolley, passes over a well-oiled pulley, and its other end is tied to a mass of 10 kg. Assume that g is 10 N kg^{-1} and that both friction and the mass of the string can be neglected.

a) What is the horizontal force on the trolley?
b) What is the total mass in motion?
c) What is the acceleration of the masses?

If the trolley moves a distance of 2 m from rest:

d) What is the potential energy transformed by the system?
e) What kinetic energy is acquired by the trolley?
f) What is the speed of the trolley?

3.6 An alpha particle of mass (or nucleon) number 4 travelling at 1.8×10^7 m s^{-1} has a head-on collision with a gas atom. The gas atom moves forwards at 0.8×10^7 m s^{-1} and the alpha particle rebounds at 1.0×10^7 m s^{-1}. Calculate the mass or nucleon number of the atom. State clearly what assumption you have made about the collision.

3.7 A method that can be used to determine the speed of a bullet is to fire the bullet into a suspended box of sand (see Fig. Q3.2). The bullet embeds itself in the

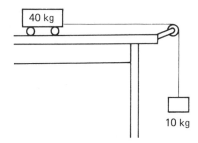

40 kg

10 kg

Fig. Q3.1

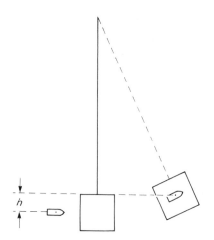

Fig. Q3.2

sand; the box swings to the right and rises through a vertical distance h as shown. Calculate the speed of the bullet given that its mass is 0.2 kg, the mass of the box of sand is 29.8 kg and h is 0.8 m. Why is it not correct to equate the kinetic energy of the bullet to the rise of potential energy of the bullet and the box?

3.8 A bullet of mass m travelling with speed v passes through a stationary pendulum bob of mass M and emerges with with speed $\frac{1}{2}v$. The pendulum bob is at the end of a string of length l. What is the minimum value of v such that the bob will swing through a complete circle?

3.9 Two equal lumps of putty are suspended side by side from two long strings so that they are just touching. One is drawn aside so that its centre of gravity rises a vertical distance h. It is released and then collides inelastically with the other one. Calculate the vertical distance risen by the centre of gravity of the combination.

3.10 When driving his car a driver regrets that, on applying the brakes, the kinetic energy of the car is mostly converted to heat and wastefully discarded. He invents a mechanical system which enables him to connect the wheels of the car through a system of gears to a flywheel whenever he wants to stop. As the flywheel accelerates it stores some of the kinetic energy lost by the car in slowing down. The energy in the rotating flywheel is then available to set the car in motion once again. Supposing that the mechanical difficulties could be overcome, is this a viable idea? To help you decide, calculate the mass of a suitable flywheel assuming it to be a disk of radius 0.5 m with a maximum speed of 3000 revolutions per minute. Take the mass of the car to be 1 tonne and its initial speed as 20 m s^{-1}.

3.11 a) What are the differences between a tangential acceleration and a radial acceleration?
b) Consider a point P on the rim of a flywheel which is rotating with constant angular velocity. Does P have a tangential acceleration, a radial acceleration or both? Explain.
c) Consider the same point P if the flywheel is given a constant angular acceleration. Does P have a tangential acceleration, a radial acceleration, or both? Explain.

3.12 A small metal block of mass 50 g is tied to a string about 1 m long. The end of the string is passed through a hollow cylindrical handle and a knot is tied in it so that it cannot slip back. The block is then swung round in a circle which is almost horizontal. The radius of the circle is 0.8 m when the speed of the block is 3 radians per second.

a) What is the tension in the string?

The cord is then pulled from below so that the radius of the circle is reduced to 0.4 m. Considering the block as a point mass:

b) What is the new angular velocity?
c) What is the change in the kinetic energy of the block?

3.13 A constant-volume gas thermometer has a pressure of 1.50×10^4 Pa at the triple point of water. What is the pressure at an indicated temperature of 373 K?

3.14 In Joule's experiments with a paddle wheel immersed in water and driven by a pair of falling masses, each of about 14 kg, twenty successive falls of about 2 m were arranged between each reading of the thermometer in the water. The total mass of water was about 7 kg. Estimate the temperature rise. (Specific heat capacity of water is 4200 J kg^{-1} K^{-1}.)

3.15 An air-gun pellet moving at 100 m s^{-1} becomes embedded in a massive wooden block. Estimate the rise in temperature of the pellet noting all the assumptions you make.

3.16 Estimate the minimum velocity of the air-gun pellet of Question 3.15, if it is to melt on becoming embedded in the block. (Melting point of lead of 328 °C.)

3.17 Hot coffee is sealed up in a vacuum flask and then shaken vigorously. Assuming that the coffee is an isolated system, would you expect its temperature to rise? Explain.

3.18 An electric kettle working at a rate of 1.5 kW is used to raise the temperature of 2 kg of water from 293 K to the boiling point. The mass of the kettle is

0.5 kg and its mean specific heat capacity 900 J kg^{-1} K^{-1}. How long will this process take? You may assume that 2% of the energy is wasted.

3.19 Devise an experiment to investigate how the temperature change in a body to which energy is transferred by a heater is related to the mass of the body and to the material of which the body is made.

3.20 If work is done on a gas by compressing it into a cylinder, the temperature of the gas rises. After a time, the temperature falls to that of the surroundings and yet the compressed gas can still transfer energy by expanding. Where does this energy come from?

3.21 The door of a refrigerator is left open. What effect will this have on the temperature of the room in which the refrigerator is kept? Explain.

3.22 A '30 gallon' tank containing 140 litres of water at 10 °C is left in the loft of a house under conditions in which it loses energy at an average rate of 100 W. How long will it take for the water temperature to reach 0 °C and how long after that for the water to freeze completely?

The density of water changes as follows:

Temperature/K	Ice at 0 °C	Water at 0 °C	Water at 4 °C	Water at 10 °C
Density/kg m^{-3}	920.0	999.9	1000	999.7

How will the loss of energy affect the behaviour of the water in the tank as it first cools and then freezes?

The specific heat capacity of water is 4200 J kg^{-1} K^{-1}; the latent heat of fusion of ice is 330 kJ kg^{-1} at 273 K.

Unit Four

Some molecular data:

Avogadro constant (N_A)	6.02×10^{23} mol^{-1}
Molar gas constant (R)	8.3 J mol^{-1}K^{-1}
Boltzmann constant (k)	1.38×10^{-23} J K^{-1}
The mass of a hydrogen atom	1.67×10^{-27} kg
Atomic mass unit	1.66×10^{-27} kg
The mass of an oxygen molecule	$32 \times 1.66 \times 10^{-27}$ kg
The mass of a nitrogen molecule	$28 \times 1.66 \times 10^{-27}$ kg

4.1 Estimate

a) the number of moles in a glass of water (say 300 g).

b) the volume of a mole of an ideal gas at a temperature of 273 K and a pressure of 101 kPa.

c) the total random translational kinetic energy of the molecules in a mole of helium at temperatures of 300 K and 301 K and, hence, the molar heat capacity of helium.

4.2 A motor car tyre is stated to be correctly inflated at a pressure of 26 lb/in^2. In this system the atmospheric pressure is about 15 lb/in^2, which in the S.I. is about 10^5 N m^{-2}.

a) What is the absolute pressure within the tyre, in S.I. units?

b) A pump is full of air at normal atmospheric pressure and the length of the stroke is 15 cm. At what point in the stroke does air begin to enter this tyre? Assume the compression to be isothermal.

4.3 A motor car tyre is correctly inflated to an absolute pressure of 2.7×10^5 N m^{-2}.

a) What is the reading of a tyre gauge connected to it if the atmospheric pressure is 10^5 N m^{-2}?

b) If the volume of the tyre is 0.035 m^3 and the temperature of the air is 20 °C, how many moles of air does the tyre contain?

c) If the tyre supports one quarter of the weight of the car (mass 1500 kg) what area of surface of the tyre must be in contact with the road?

d) If the car is driven at high speed in hot weather what changes will occur to the air in the tyre? What steps would you advise the driver to take?

4.4 Compute a value for the r.m.s. speed of molecules in the air at a pressure of 10^5 N m^{-2} given that the density of air is 1.2 kg m^{-3}.

A tightly-fitting cork is pulled from a bottle of ammonia at the far end of a draught-free corridor. How does the kinetic theory of gases account for

a) the time taken for the sound to reach the other end of the corridor,

b) the time taken for the smell of the ammonia to reach the same point?

Account for the wide disparity in these times.

Would these times be increased or decreased if the experiment were performed in a building on a mountain at such a height that the pressure was reduced to one-half? Explain your reasoning carefully. You may assume that the air temperature has remained constant (velocity of sound in air is 340 m s^{-1}).

4.5 How high would the atmosphere be if the density of the air were maintained at the value it has at s.t.p.?

Calculate the velocity of a particle which falls from rest through this height assuming that no retarding forces act.

4.6 a) Plot a histogram to display the distribution of speed among the molecules of nitrogen at 0 °C from the information given in Table Q4.1

b) Calculate the r.m.s. speed of nitrogen at this temperature. Does this value agree with the value you would expect to get by considering the graph? Explain.

4.7 a) Estimate the r.m.s. speeds of oxygen and of hydrogen molecules at a temperature of 300 K.

b) The 'escape' speed, i.e., the speed at which a particle can escape from the gravitational field of the earth is 11 km s^{-1}. Estimate the temperatures at which (i) a molecule of oxygen and (ii) a molecule of hydrogen would have sufficient energy to escape from the earth. Use your estimates to explain why so little hydrogen is to be found in the earth's atmosphere.

4.8 a) At the pumped storage system at Ffestiniog (see Fig. 8.4) the minimum water level in the reservoir is 483 m OD and the level of the pumping station is 182 m OD. What is the minimum gauge pressure at the pumps (i.e., the difference between the absolute pressure and the atmospheric pressure) when water is being pumped into the reservoir?

b) If the barometric pressure is 1013 millibars, what is the absolute pressure provided by the pumps? 1 bar is a meteorological unit of pressure and it is equal to a pressure of 10^5 N m^{-2}.

c) What is the height of the mercury column in a barometer which is reading 1013 mb? Take g to be 9.8 N kg^{-1}.

4.9 a) An ice-floe of constant thickness and derived from a freshwater glacier floats on sea-water with 1 m of ice appearing above the water level. What is the total thickness of the ice? (The densities of ice and sea water in kg m^{-3} are 920 and 1030.)

b) An explorer with his stores (total mass 1200 kg) finds himself marooned on the ice-floe. He estimates the surface area of the flow to be 20 m^2. What is the height of the upper surface of the flow above the level of the water now?

4.10 When a car is passed at speed by a large lorry on a motorway, the car driver senses that there is a force on

the car towards the lorry. Explain in terms of the Bernoulli effect.

4.11 a) A ball of radius r and density ϱ_1 falling through a viscous liquid (viscosity η and density ϱ_2) obeys Stokes' law (see Section 31.6). State the forces acting on the ball, write down an expression for each and then find an equation for the terminal speed of the ball.

b) Estimate the terminal speed of a steel ball-bearing (radius 3 mm and density 7700 kg m^{-3}) which is falling through a large tank of glycerol (for which density is 1260 kg m^{-3} and the viscosity is 1.50 N s m^{-2} at the appropriate temperature).

4.12 A patient receives a blood transfusion from a bottle set so that the level of the blood in it is 1.2 m above the needle. The needle is 3 cm long and has a bore (diameter) of 0.4 mm. Given that blood has a density of 1020 kg m^{-3} and a viscosity of 2.4 × 10^{-3} N s m^{-2}, at what rate is blood being supplied through the needle?

4.13 a) A student is told that the air in a room is at a temperature of 20 °C and the air outside is at 0 °C. The window has an area of 2 m^2 and the glass, which is 1.5 mm thick, has a thermal conductivity of 1.0 J m^{-1} s^{-1} K^{-1}. He computes the rate of loss of energy and finds it to be 27 kW! How did he arrive at this figure? Discuss whether or not this is a probable rate of loss of energy?

b) When a measurement was made of the temperatures at the surfaces of the glass, the temperature drop was found to be only 0.2 °C. At what rate must energy be supplied to the air in the room to maintain this temperature difference assuming that all the heat losses are confined to the window? Is this answer any more realistic than that proposed in part (a)? Comment.

4.14 An aluminium window frame experiences over the year a range of temperature from −5 °C to 30 °C. If its internal dimensions at 20 °C are 80 cm × 40 cm what are its maximum and minimum sizes? What advice would you offer a glazier who is to replace the glass during the summer? The expansivity of aluminium is 24 × 10^{-6} K^{-1} and of glass is 8.5 × 10^{-6} K^{-1}.

4.15 A simple heat engine can be made by placing two

Table Q4.1

Speed range/ m s^{-1}	0–99	100–199	200–299	300–399	400–499	500–599	600–699	700–799	800–899	900–999	1000–1099	1100–1199
% of molecules within the range	1	5	14	20	21	16	10	6	3	2	1	<1

metal tanks in good thermal contact with a semi-conductor thermopile. This unit will generate an e.m.f. if subject to a temperature difference. The e.m.f. can be used to run a small electric motor.

As energy is transferred from the hot tank to the cold tank through the thermopile, the motor will run and can be made to do a small amount of work (say by raising a small load on a thread).

a) Why must there be good thermal contact between the tanks and the thermopile?

b) Why does the motor cease to operate when the temperature of the two tanks becomes the same?

c) Explain why the motor can be re-started by *either* adding hot water *or* ice to one of the tanks.

d) If the temperature of the hot water is initially 90 °C while that of the cold water is 20 °C calculate the maximum efficiency of the heat engine. Would you expect to find this efficiency in practice?

4.16 Suppose 1 mole of an ideal gas, contained initially in one half of a vessel, is allowed to expand at constant temperature into the previously evacuated other half (Fig. Q4.1). A molecule can now be in either half of the vessel whereas before it was in the left-hand half. The number of possible arrangements is doubled.

a) What is the increase in the number of ways of arranging all N_A molecules?

b) What is the increase in the entropy of the gas?

c) Why is the gas overwhelmingly most likely to found evenly distributed in the vessel.

4.17 A liquid will evaporate if its molecules acquire some energy E great enough to overcome the potential energy due to the attractions of neighbouring molecules. At moderate temperatures, $E \gg kT$, and the number of molecules with energy E or greater is proportional to $e^{-E/kT}$. The energy needed to evaporate water is 2.3 MJ kg^{-1}. Estimate the temperatures at which an airing cupboard should be maintained in order to dry clothes 10 times faster than it would at 27°C.

What simplifying assumptions have you made in obtaining your estimate?

Closely based upon a question in Cambridge Colleges Examination (Awards and Entrance) 1977

4.18 Particles may react chemically if they interact with an energy greater than some critical amount, E. If $E \gg kT$, the number of particles with this energy is proportional to $e^{-E/kT}$. As a rule of thumb many chemical reactions double in rate for each 10 °C rise in temperature at temperatures for which $E \gg kT$.

a) Explain why such small increases in temperature have such a considerable effect on the rate of reaction.

b) Assuming $T = 300$ K initially, estimate a value for the ratio E/kT from the above data.

4.19 The bar chart (Fig. Q4.2) shows the distribution of energy quanta with numbers of atoms for 1 kg of a particular solid at 300 K, assuming the simple Einstein model developed in the text.
Sketch similar bar charts to show the distribution of energy quanta with numbers of atoms for

a) 2 kg of the same solid at 300 K

b) 1 kg of the same solid at 100 K

c) 2 kg of the same solid at 600 K

4.20 A 3.5 m length of aluminium wire of diameter 2.4 mm was suspended from a beam with a load of 20 N hanging from it in order to keep the wire taut. A scale was set up so that the position of the 3 m mark on the wire could be kept under observation and the wire was then loaded up. The following readings were obtained:

Additional load/N	0	40	100	160	220	280	340	350	355
Scale reading/mm	28.29	28.67	29.23	29.80	30.37	30.95	31.52	31.95	32.35

a) Plot a suitable graph and describe the behaviour of the wire during the process of loading.

b) Fearing that he had over-loaded the wire, the experimenter removed the additional load (355 N) and found that the 3 m mark was now alongside the 28.90 mm mark on the scale. Discuss what had happened.

c) Use your graph to find: (i) Young modulus for aluminium; and (ii) the elastic limit for aluminium.

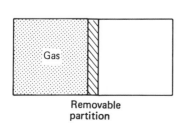

Fig. Q4.1

Removable partition

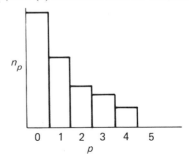

Fig. Q4.2

4.21 a) Two wires, one of steel and one of brass, each 3 m long and with cross-sectional area 0.20×10^{-6} m^2, are used to support masses of 5 kg. By how much will each wire stretch? The Young modulus for steel is 200×10^9 N m^{-2} and for brass is 100×10^9 N m^{-2}.

b) If the two wires are then joined end to end, what load will produce a total extension of 0.02 m?

4.22 The elastic limit of a steel cable used in a lift or elevator is 200×10^6 N m^{-2}. If the stress is not to exceed 25% of this figure what is the maximum upward acceleration which can be given to the car of total mass 1000 kg? The cable has a cross-sectional area of 3 cm^2.

4.23 A steel rod is used to secure a vessel to a motor so that the vessel and its contents can be whirled around in a horizontal circle. The rod has a cross sectional area of 1.5 cm^2 and the stress must not exceed 25% of the elastic limit for the steel (200×10^6 N m^{-2}). What is the maximum speed of rotation which can be reached safely if the total mass of the vessel is 0.5 kg and the centre of mass of the system is 20 cm from the axis of rotation?

4.24 The tensile strength of a material depends not only upon the nature of the material but also upon its cross-sectional area.

The legs of animals have sufficient strength to withstand the weight of the animal.

Use these facts to explain why the legs of a mouse are relatively much thinner than the legs of an elephant.

An explorer of the planet Jupiter reports seeing ant-like creatures 1 m high. Would you accept this? Give a reason for your answer.

4.25 The linear dimensions of a simple pendulum made up of a spherical bob and a steel wire are scaled up by a factor of three. By what factor do each of the following change

a) the volume of the bob,
b) the mass of the bob,
c) the surface area of the bob,
d) the volume of the wire,
e) the mass of the wire,
f) the surface area of the wire,
g) the tensile strength of the wire,
h) the period of the pendulum?

4.26 The wire supporting the bob of Question 4.25 will be stretched. What change will occur in the stretch of the wire when the pendulum is scaled up by the factor of 3? ($E = 200 \times 10^9$ N m^{-2}.)

4.27 Figure Q4.3 shows the unit cell of a face centred cubic crystal, each sphere representing an atom. See also Figs. 14.7 and 14.8.

If the radius of each sphere is r find:

a) The length of the side of the unit cell (a).
b) The volume of the unit cell.
c) How many atoms constitute the unit cell shown best in Fig. 14.8?
d) How many corner atoms in the unit cell?
e) How many face atoms?
f) Among how many unit cells is one corner atom shared?
g) Among how many unit cells is one face atom shared?
h) How many corner and how many face atoms may be thought of as belonging to a single unit cell? This is the number of effective atoms in the cell.
i) What is the total volume of the effective atoms in a cell?
j) What is the ratio of the volume of the effective atoms in a unit cell to the volume of the cell? This is the *atomic packing factor* for a face centred cubic structure.

4.28 Copper has a f.c.c. structure (c.f. Question 4.27) and an atom radius of 0.1278 nm.

a) What is the length of side of a unit cell of copper?
b) How many atoms per unit cell?
c) Given that the molar mass of copper is 0.0635 kg, find the density of copper.

4.29 X-rays with a wavelength of 0.58×10^{-10} m were used to investigate the structure of a crystal. Reflections were observed at angles of 6.45°, 9.15° and 13.0°. What interplanar spacings do these angles represent? Comment on your answers.

4.30 a) Sap can obviously climb to the top of a tree. Giant redwoods reach to heights of 100 m or even more. Estimate the diameter of the capillaries in the wood on the assumption that surface tension provides an explanation for the rise. You may assume that the surface tension is the same as that of water, and that the angle of contact is zero.

b) The diameter of the smallest capillaries is, in fact, in the order of 0.01 mm. Comment on the reasonableness or otherwise of the assumption suggested.

c) What is the minimum pressure that has to be generated in order for the sap to reach the top of the 100 m high tree?

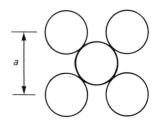

Fig. Q4.3

4.31 A film of soap solution ($\gamma = 25 \times 10^{-3}\,\mathrm{N\,m^{-1}}$) is formed across the end of a glass tube which is 0.5 cm in internal diameter. Calculate the excess pressures required to blow the film into bubbles of diameters 1, 2, 4, 6, 8, 10 cm. Plot a suitable graph to display your results and describe how the pressure changes during the process. Use your graph to describe what you would expect to see if a bubble of diameter 10 cm was connected to a bubble of 5 cm diameter.

4.32

	Young modulus/ $10^9\,\mathrm{N\,m^{-2}}$	Tensile strength/ $10^6\,\mathrm{N\,m^{-2}}$	Compressive strength/ $10^6\,\mathrm{N\,m^{-2}}$
Aluminium	70	100	100
Wood across the grain	0.6	—	—
Wood along the grain	12	18	15

a) The table shows that the Young modulus for wood across the grain is less than that along the grain. In which direction is it easiest to stretch timber? Explain.
b) What light does this throw on the construction of plywood, which is essentially a sandwich of thin sheets of wood arranged so that alternate layers have their grain directions at right angles?
c) Would you expect the tensile strength (i.e., the stress at which the wood will break) across the grain to be larger, smaller or about the same as the tensile strength along the grain? Explain.
d) Why do the tensile and compressive strengths of aluminium have the same values whereas those for wood do not?

4.33 The average male has a mass of 70 kg and the minimum cross-sectional area of the bone in each leg is approximately $5.0 \times 10^{-4}\,\mathrm{m^2}$. The compressive breaking stress of bone is approximately $1.0 \times 10^7\,\mathrm{N\,m^2}$.

a) Assuming that a man is standing with his weight equally shared by each leg, calculate (i) the maximum stress in his leg bones, (ii) the ratio (maximum stress in bones)/(breaking stress).
b) Now suppose that all the linear dimensions of the human body are increased by a factor of nine. (i) What is the new value of the ratio found in (a)(ii)? Show and explain how you reach your answer. (ii) Explain what would happen to this 'giant' if he attempted to chase a normal human being.

Oxford and Cambridge Schools Examinations Board
Nuffield A-level Physics (1980)

Unit Five

5.1 a) Calculate the frequency of sound waves that travel through air with a speed of $330\,\mathrm{m\,s^{-1}}$ and a wavelength of 0.30 m.
b) If these sound waves enter another medium in which their wavelength is found to be 0.20 m, calculate the speed of the sound waves in this second medium.

5.2 Place the following waves (a) in order of wavelength and (b) in order of frequency, in each case starting with the largest.
Waves: light, sound corresponding to normal speech, gamma rays, VHF radio, infrared, water waves produced by an incoming tide.

5.3 Explain the meaning behind the following statement: 'If it were not for diffraction, the Young's double slit method of producing interference patterns would not work!'

5.4 The map (Fig. Q5.1) shows part of a coastline, with two land-based radio navigation stations A and B. Both stations transmit continuous sinusoidal radio waves with the same amplitude and wavelength (200 m).

A ship X, midway between A and B, detects a signal whose amplitude is twice that of either station alone.

a) What can be said about the signals from A and B?
b) The ship X travels to a new position by sailing 100 m in the direction shown by the arrow. What signal will it now detect? Explain this.
c) A ship Y also starts at a position equidistant from A and B and travels in the direction shown by the arrow. Exactly the same changes to the signal received were observed as in the case of ship X in (b).

Explain whether Y has sailed 100 m, more than 100 m, or less than 100 m.

Oxford and Cambridge Examinations Schools Board
Nuffield A-level Physics (1972)

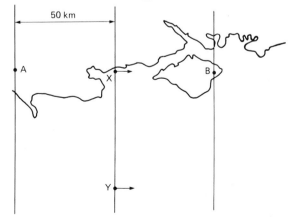

Fig. Q5.1

5.5 The following passage describes the setting up of three different interference patterns. Explain and comment on why the particular arrangements are essential in each case.

a) An interference pattern in a ripple-tank can be clearly observed using a pair of sources (mechanical dippers) about 5 cm apart. A barrier containing two slits 5 cm apart does also produce an interference pattern when a plane wave falls on it, but the pattern is by no means clear.

b) An interference pattern using sound waves can be readily observed using two loudspeakers, about half a metre apart, connected in parallel to a signal generator set at a frequency of about 2 kHz. One loudspeaker placed behind a barrier containing two slits will not be successful in producing an interference pattern.

c) An interference pattern using light can only be observed if a narrow source is placed some distance (one or two metres) behind two slits less than a millimetre apart. Two separate sources of light never seem to produce an interference pattern.

5.6 A heavy rope, whose mass per unit length is 0.1 kg per metre, is stretched nearly horizontally with a tension of 40 N. What is the speed at which a transverse sinusoidal wave will travel along the rope?

If this rope is hanging freely and vertically from a fixed support, and a tranverse wave-pulse is sent up the rope from the bottom end, will the pulse travel at constant speed? Explain your answer.

5.7 The top string (E) of a steel-string guitar is tuned to vibrate, in its fundamental mode, at a frequency of 330 Hz. The length of the string is 64 cm. If the mass per unit length of the string is 0.5 g m^{-1}, calculate the tension (in newtons) needed to make it vibrate at this pitch. You may need to refer to Section 20.2 (on stationary waves) and to Section 17.2 (speed of waves on a stretched string).

5.8 Here is an idealized sketch of a pulse travelling along a taut, narrow spring from left to right at a speed of 0.5 m s^{-1} (Fig. Q5.2).

a) Sketch a graph of the velocity of point D against

time, taking the time at the instant shown in the sketch as $t = 0$. Mark a suitable scale, with units, on both axes.

b) Indicate the place on your graph where point D has a large acceleration and the place where it has a large deceleration.

c) Use the answer to (b) to say why the sketch and your graph cannot represent exactly what happens to a real spring.

Oxford and Cambridge Schools Examinations Board
Nuffield A-level Physics (1971)

5.9 The expression for the speed of longitudinal waves along a rod or wire of elastic material, derived in Section 17.4 can be used to find the speed of longitudinal waves along a 'column' of gas enclosed in a tube if, instead of the Young modulus, E (which is applicable only to solid materials), one substitutes the bulk modulus of elasticity of the gas. Given that the bulk modulus for air is about 1.4×10^5 N m^{-2}, and the density about 1.25 kg m^{-3}, at ordinary room temperature, estimate the length of a pipe which, open at both ends, will have a fundamental frequency of 260 Hz (about middle C). (The fundamental mode of oscillation of the air-column in a pipe open at both ends occurs when the length of the pipe is approximately equal to half the wavelength of the stationary wave.)

5.10 A particular slinky spring has an unstretched length of 0.10 m and a total mass of 0.30 kg. The spring is stretched until it it 5.0 m long, when the tension in the spring is measured to be 1.0 N.

a) Calculate the speed of a small transverse pulse travelling along the spring.

b) The length of the spring is now *doubled* to 10.0 m. Assuming the spring obeys Hooke's law over this range of extension, calculate the speed of a small transverse pulse travelling along the longer spring.

c) From the result calculated in (b), explain carefully why the time taken for a transverse pulse to travel along a slinky spring is almost independent of its length. Under what circumstances is this not true?

5.11 A friend says to you that he cannot see the point in saying that light consists of waves. 'After all', he says, 'the obvious facts that light travels in straight lines, that it is reflected and that it changes direction if its speed changes are very easily explained by thinking of light as a stream of particles. The only thing you have shown me that you say requires *waves* to explain it is when light passes through two slits so close together it takes a microscope to see them – and then you had to spend more time explaining why the experiment didn't work too well than you did anything else!' Try to explain convincingly to your friend:

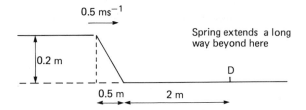

0.5 ms^{-1}

Spring extends a long way beyond here

0.2 m

D

0.5 m   2 m

Fig. Q5.2

a) That all *his* observations are in fact better explained using a wave model.

b) How only the wave model has enabled us to obtain an adequate understanding of things like colour, polarization and the similarities between all the members of the electromagnetic spectrum.

5.12 Plane waves travelling over the surface of water in a ripple-tank are 'frozen' using a stroboscope and the separation between one wave front (counted as '0') and the tenth ahead of it is found to be 15 cm. Using the same stroboscope, the frequency of vibration of the wave source is measured to be 14 Hz.

a) Calculate the speed of the waves in the ripple tank.

b) The waves now pass over a shallow region in the tank (produced by placing a glass plate in the bottom of the tank) (Fig. Q5.3). In this region the wave fronts are found to be 1.0 cm apart. What is the speed of the waves in the shallow region of the tank?

c) Waves approach the shallow region so that the wave direction in the deeper region makes an angle of 30° to the normal to the line separating shallow and deep regions. What angle does the wave direction in the shallower region make with this line?

d) Suppose the waves are generated in the shallow region and cross the boundary into the deeper region. What is the *maximum* angle the wave direction can make with the normal to the line separating deep and shallow regions, if waves are to enter the deeper part?

e) The angle you have found in (d) is called the *critical angle*. What will happen to the waves if the angle between their direction and the normal to the dividing line is greater than the critical angle? Find out how this phenomenon enables 'light pipes' to be constructed which will pass light with little loss down fine glass fibres. This has led to a new branch of optics called *fibre optics*.

5.13 a) Figure 18.1b shows a record of earthquake waves received at Cambridge, Massachusetts. The speed of the first waves to arrive (P-waves) is given by the expression $v_p = \sqrt{E/\varrho}$, while that of the second or S-waves is given by $v_S = \sqrt{\mu/\varrho}$. Using the data given below estimate a value for the distance of the earthquake epicentre from the receiving station. (The average density of rocks on the earth's surface (ϱ) is $2.9 \times 10^3 \, \text{kg m}^{-3}$. The elastic modulus $E = 9.4 \times 10^{10} \, \text{N m}^{-2}$ while the rigidity modulus, $\mu = 3.16 \times 10^{10} \, \text{N m}^{-2}$.)

b) Other measurements suggest that the epicentre was in fact very much farther away than the distance calculated in (a). Remembering that these earthquake waves have passed through the earth, suggest an explanation for the difference between the observed and calculated results.

c) If the epicentre had been on the opposite side of the earth, experience shows that *no* S-waves would have been detected. Explain why evidence such as this suggests that the earth has a liquid core.

5.14 Light and X-rays are both said to be electromagnetic waves.

a) What kind of experimental results suggest that both are wave motions?

b) What arguments, based on experimental results, can you give to suggest that they have very different wavelengths?

c) Mention any evidence that might suggest that light and X-rays are the same kind of radiation.

Oxford and Cambridge Schools Examinations Board
Nuffield A-level Physics (1970)

5.15 Here are two phenomena you may have observed, or perhaps you will observe one day, which you should try to explain by applying your understanding of what happens when waves pass the edge of a barrier or go through an aperture.

a) Suppose a band of musicians is playing in the main street in a town and you are in a side street, so placed that you can hear the music but cannot see the band. The sound will seem muffled, and the most prominent sound is likely to be that of the bass drum, and other low-pitched instruments. You may not hear the high-pitched piccolos at all until you are actually in sight of the band.

b) When the doors of a concert hall are left open, a person outside and some distance away from the hall hears the bass notes much more clearly than the treble notes.

5.16 The speed of sound in air can be measured by means of the apparatus sketched in Fig. Q5.4. The loudspeaker, L, is fed with the signal from an audio signal generator so that the diaphragm vibrates at a constant frequency. Fine dust (for instance, lycopodium powder)

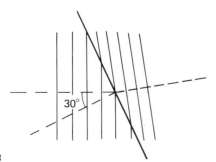

Fig. Q5.3

30°

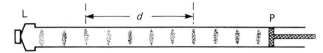

Fig. Q5.4

is put in the tube. This dust is whipped up in places where the air is vibrating strongly, and the small whirls of dust locate the antinodes of the longitudinal ·stationary wave which is set up when the tube is tuned to resonate at the frequency of the loudspeaker by adjusting the position of the piston, P.

Suppose that, in a particular experiment, the antinodes are as shown in the diagram. The distance d is found to be 1.65 m when the frequency is 500 Hz. Calculate the speed of travelling sound waves in air from this information.

5.17   In an experiment designed to illustrate the interference of sound, two loudspeakers were placed outdoors 0.25 m apart and connected in parallel to a signal generator, set to a frequency of 600 Hz.

a) What is the significance in placing the loudspeakers outdoors?
b) Unfortunately no interference maxima or minima were detected. Describe *two* changes that could be made to this arrangement in order that interference maxima might be detected.
c) For each of these changes, give the quantitative details necessary to ensure that the listeners can pick up at least *five* interference maxima.
d) On another occasion the same experiment was set up to ensure that maxima and minima were heard, but this time the point midway between the speakers was found to be a minimum instead of a maximum. Explain how this could have come about.

5.18   A white light source is placed behind a narrow slit 2.0 m from a double slit system. The double slits are separated by 0.40 mm. Light passes through the double slits and falls on a screen a further 2.0 m away. Six light and dark bands are observed on the screen each separated from its neighbour by 2.5 mm.

a) Calculate the average wavelength of the light.
b) Why is the word 'average' used in (a)?

If a red filter is placed across the light source, many more bright and dark lines are observed. However, removing the narrow slit from in front of the lamp causes the interference pattern to vanish.

c) Explain why placing a red filter across the light source enables many more bright and dark lines to be seen.
d) Explain why the removal of the first, single slit from in front of the light source causes the pattern to vanish.

5.19   a) If a receiver of 3 cm electromagnetic waves is arranged so that its dipole is in the same plane, but at right angles to the direction of the transmitting dipole, no radiation is picked up, although the transmitter is working normally. Why is this?
b) If a grid of parallel wires is placed between the transmitter and the receiver so that the plane of wires is parallel to the plane containing the transmitting and receiving dipoles, a signal may be picked up by the receiver. This signal reaches a maximum if the wires are at an angle of 45° to the direction of the transmitting and receiving dipoles. Explain why this is so and show that the maximum intensity of the received signal is approximately one quarter of that which will be picked up by the receiver when its dipole is parallel to the transmitting dipole.

5.20   Calcite is a crystalline form of calcium carbonate. Large crystals transparent to light are easily obtainable. If a black dot is made on a sheet of paper and a crystal placed over it, the dot will seem to have become two. If the crystal is rotated, one dot will appear to revolve round the other.

If instead of looking at the dot directly though the crystal, a piece of polaroid film is placed between the crystal and the eye and the crystal rotated, each dot will seem periodically to disappear from view. During each complete revolution each image of the dot disappears twice, alternately with the other.

Using a diagram of the wave fronts passing through the crystal, try to explain these observations in as much detail as you can. If possible repeat the observations for yourself.

5.21   Two transmitting aerials A and B, 200 m apart are radiating equal powers at the same frequency of 100 MHz. A straight road running parallel to AB passes within 4 km of A and B at its nearest point 0. The signal power received in a car travelling along the road is found to fluctuate in intensity at a frequency of 0.4 Hz when passing 0.

Explain this and calculate the speed of the car.

The Colleges of Oxford Joint
Scholarship and Entrance Examination 1976

5.22   Each of the sketches in Fig. Q5.5 represents a frictionless track made up of straight sections, upon which a particle can slide. At the junctions between the straight portions the angle can be imagined to be 'radiused', but the curved bits can be assumed negligibly small compared with the straight bits.

Sketch the velocity–time graph for a particle released from the right-hand end of each track. Take velocities to the right as positive, and velocities to the

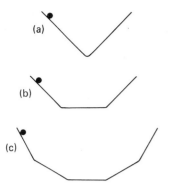

(a)

(b)

(c)

Fig. Q5.5

left as negative.

It should become quite clear to you, after doing this, that only a track without discontinuities can give rise to a velocity–time graph without discontinuities.

5.23 Figure Q5.6 shows one cycle of the motion of a simple harmonic oscillator. Using the arbitrary time-scale, answer the following questions.

a) At what times is the speed zero?
b) At what times is the speed greatest?
c) At what times is the acceleration zero?
d) At what times is the acceleration greatest?
e) At what times do you estimate the speed will be half its maximum value?

5.24 A trolley of mass 0.80 kg is attached to a pair of supports by two springs, and it is set into oscillation between them. Describe as quantitatively as possible what changes will take place in (i) the period, (ii) the total energy, and (iii) the amplitude of the oscillation if another mass of 0.80 kg is dopped on to and sticks to the trolley;

a) when the trolley is momentarily at rest at one end of its travel
b) when the trolley is passing through its normal rest position.

You should ignore the effect of friction on the motion. The combined spring constant of the springs is 10 N m^{-1}, and the initial amplitude of the motion is 10 cm.

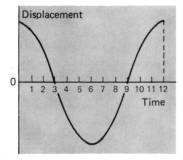

Fig. Q5.6

5.25 If the amplitude of a simple harmonic oscillator is halved, what change, if any, occurs in:

a) the period,
b) the maximum velocity,
c) the total energy,
d) the maximum acceleration?

5.26 A simple harmonic oscillator moves from its central equilibrium position to its maximum displacement of 5 cm in a time of 1 second. Calculate:

a) the period,
b) the angular frequency, ω,
c) the maximum velocity,
d) the maximum acceleration.

5.27 An object whose mass is 2 kg is attached to the lower end of a spring and held at rest in such a position that the spring is straight but, as yet, unstretched. When released from this position the object oscillates vertically up and down with an amplitude of 30 cm about the equilibrium position (in which it finally comes to rest). To answer the following questions, neglect damping, and use the principle of conservation of energy.

a) Calculate the amount of gravitational potential energy lost by the object when it descends from its initial position to the equilibrium position.
b) Calculate the increase in elastic potential energy stored in the spring during the same descent.
c) Account for the difference between the amounts of energy in (a) and (b).
d) What is the total energy of the oscillator?

5.28 In this exercise you will plot a sequence of four wave-profiles, each with an amplitude of 1 cm, on a sheet of graph paper.

In Section 19.10 it was explained that a travelling sinusoidal wave can be described by the expression

$$y = a_0 \cos \omega \left(t - \frac{x}{v} \right).$$

Suppose we have such a wave whose amplitude is 1 cm, wavelength 2 cm, and frequency 0.25 s.

a) Calculate the transverse displacement, y, at time $t = 0$, for values of x in the range $x = 0$ to $x = 4$ cm, increasing by intervals of 0.25 cm. Plot these displacements against x near the top of a sheet of graph paper.
b) Repeat the calculations for the same values of x but at times $t = 0.5$ s, 1 s, and 2 s. Plot the three wave-profiles below the previous one.
c) Use the graphs to verify the relationship: frequency × wavelength = wave velocity.

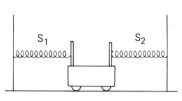

Fig. Q5.7

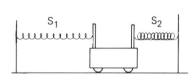

Fig. Q5.8

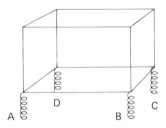

Fig. Q5.9

d) Describe (but do not draw) in what way the results would be different if the expression for displacement

$$y = a_0 \cos \omega \left(t + \frac{x}{v} \right)$$

were used instead.

5.29 A trolley of mass 0.8 kg is held in equilibrium between two fixed supports by identical springs (S_1 and S_2) as shown in Fig. Q5.7; each spring has an extension of 0.10 m.

In Fig. Q5.8 the trolley is shown moved to the right a distance of 0.05 m.

The relation between force (F) in newton and the extension (x) in metre for each spring is given by $F = 20x$.

a) What is the *change* in force exerted by spring S_1 caused by moving the trolley to the right as shown in Fig. Q5.8?
b) If the trolley is now released what will be the magnitudes of:
   (i) the resultant force acting on the trolley at the moment of release?
   (ii) the initial acceleration of the trolley?
c) Showing the steps in your calculation, determine the total energy stored in S_1 and S_2 when the springs are stretched: (i) as in Fig. Q5.7; (ii) as in Fig. Q5.8.
d) What is the kinetic energy of the trolley as it passes through the equilibrium position?

Oxford and Cambridge Schools Examinations Board
Nuffield A-level Physics (1981)

5.30 Any experienced walker knows that the style of walking which uses the least energy concerns choosing the correct rhythm of leg movements, and thinking always of swinging one's leg from the hip. Sit on the edge of a table or other surface which is high enough for you to dangle one leg freely. Let it swing like a pendulum, from the hip: ideally get someone else to keep your leg swinging by repeated light pushes while you keep your leg completely relaxed. Note the approximate period of swing. Why, do you think, is it a good idea to move your legs with this rhythm when walking? Do you walk this way?

5.31 A simple pendulum can be forced into oscillation by moving its point of support very slightly back and forth horizontally with a regular rhythm.

Describe how you would expect the pendulum to behave when the frequency of the movement of the support is varied from a low to a high value. Consider, in your discussion, the frequency, amplitude, and relative phase, of the motions of the support and the pendulum bob. You can get a feel for the situation by tying a small heavy object to a length of cotton (about 30 cm is suitable) and using your hand as the support.

How is it possible, in this situation, for the pendulum to oscillate at frequencies which are different from its natural frequency?

5.32 Figure Q5.9 shows a massive rectangular box, which can be assumed to be perfectly rigid, supported on four similar springs. How many modes of vibration are possible in this system? The mode that probably first comes to mind is the one in which the box simply bobs up and down, all springs being compressed simultaneously and relaxed simultaneously. But the box can also oscillate in a pitching mode, with springs A and B being compressed while C and D are being relaxed, and vice versa. What other modes do you think are possible? Attempt to complete a table on the lines suggested below.

Description of mode	What the springs are doing at one instant			
	A	B	C	D
Bobbing	All being compressed			
Pitching	Compressed		Relaxed	

5.33 This can be done as a home experiment or, alternatively, reasoned out in your imagination.

Two exactly similar simple pendulums are set up side by side and linked, not far from the top, by a light rigid strut (a straw, or thin strip of wood, for example). The system has two distinct normal modes of oscillation (i.e., modes in which each pendulum oscillates with simple harmonic motion of constant amplitude, if one neglects damping). What are these normal modes? How does the frequency of each normal mode depend upon the length *l* and the distance *a*?

If one of the pendulums is stationary and the other is drawn aside and released, what is the subsequent motion of the system? Illustrate your answer by sketch-graphs of displacement and time.

(*Note*: The motion of the two pendulums is assumed to be entirely in one plane: the plane of the diagram in Fig. Q5.10.)

5.34   a) A 440 Hz tuning fork is held over the end of a long narrow vertical tube filled with water. A tap at the bottom allows water to flow slowly out of the tube. A sound of loud intensity is heard when the water level reaches 19 cm and 58 cm from the top of the tube.

Explain why the loud sound occurs and calculate a value for the speed of sound.

b) A loudspeaker whose aperture is approximately twice as wide as the tube, is connected to a signal generator to produce a note of the same frequency as the tuning fork. The loudspeaker is mounted about 1 cm above the top of the tube. The experiment is repeated. As the water level falls the sound heard falls to a *minimum* as the water level reaches 19 cm and 58 cm from the top of the tube. Suggest a reason for this.

5.35   The frequency of the note emitted by a wire under tension is that corresponding to the longest wavelength standing wave that can be set up on the wire. The wave speed is given by $\sqrt{(\text{tension}/\text{mass per unit length})}$. A violin carries four such wires, or 'strings', which are tuned in fifths – that is to say, the frequency of each is 50% higher than the string immediately below it in pitch.

Suppose that the tension of all four strings is the same and that they are all made of the same material. If

the diameter of the string of lowest pitch is 0.80 mm, what is the diameter of the other three strings?

If you can, compare your answers with the diameters found in practice.

5.36   Engineers have a long narrow flat-bottomed tank, containing water to a depth *d* of 1.6 m, in which they want to test the effects of waves on a new 18 m boat of the type rowed by eight oarsmen and with a cox.

A side view of the boat in the tank might look as shown in Fig. Q5.11.

a) (i)   Sketch a situation in which the wavelength of the waves is such as to put the greatest strain on the boat. Explain your choice of wavelength.
   (ii)   What is the approximate wavelength of the waves you have chosen?
b) If the speed of the waves *c* is given by $c = \sqrt{(gd)}$, calculate the frequency at which the wave generator would have to operate in order to make the waves you chose in (a) (i).
c) To test the effect of standing waves on the boat the engineers placed barriers in the tank and obtained a standing wave as shown in Fig. Q5.12. Now sketch what the standing wave would have been like if the depth of water had varied as shown in Fig. Q5.13. Why have you drawn it this way?

Oxford and Cambridge Schools Examinations Board
Nuffield A-level Physics (1978)

5.37   Two identical loudspeakers were connected to a signal generator and a sound-level meter, calibrated in decibels (dB), was placed some distance away and at a point midway between them. The sound level meter recorded a reading of 66 dB.

One loudspeaker was then disconnected from the signal generator and the sound level meter now recorded a reading of 60 dB.

Fig. Q5.10

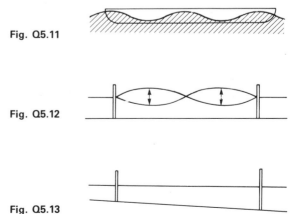

Fig. Q5.11

Fig. Q5.12

Fig. Q5.13

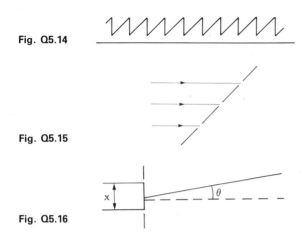

Fig. Q5.14

Fig. Q5.15

Fig. Q5.16

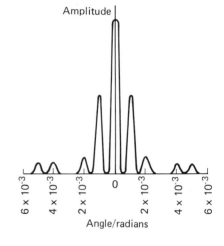

Fig. Q5.17   Angle/radians

a) Calculate the change in sound intensity which was brought about by disconnecting one loudspeaker.
b) A cathode ray oscilloscope connected to a microphone indicated that the amplitude of the sound was halved when one loudspeaker was disconnected. Compare this result with that found in part (a) and explain the difference.

5.38 A diffraction grating, 25 mm wide, is ruled with 5000 lines. When it is set up in a spectrometer to examine light from a sodium lamp, the bright yellow part of the spectrum is seen clearly to consist of two closely spaced lines. Yet if the same grating is held up to the eye to observe the sodium lamp, the two lines appear as one.

Explain these observations, supporting your explanation with appropriate calculations. (The wavelengths of the two lines are 589.0 nm and 589.6 nm; the diameter of the pupil of the eye is about 3 mm.)

5.39 Spectacular claims have been made for the ability of 'spy' satellites to record details on the earth's surface. Let us assume that such a satellite is in a circular orbit 100 km above the earth's surface.

a) What diameter lens would a camera on the satellite need in order to resolve the details of a car number-plate? (Assume $\lambda_{light} = 500$ nm.)
b) Would an astronaut using a normal 35 mm film camera with a lens diameter of 2 cm be able to resolve two houses 50 m apart on the earth's surface?
c) Give one advantage and one disadvantage of using infrared light instead of visible light for earth surveillance from such a satellite.

5.40 Some diffraction gratings are ruled so that their surfaces take up the profile found in Fig. Q5.14.

a) When light is transmitted by such a grating, the diffracted light tends to be concentrated into one or two orders on one side of the central maximum. Explain why the profile of the grating makes this so.
b) What advantage could be offered by such a grating?
c) If red light of wavelength 650 nm forms a second order interference maximum at an angle of 60° to the direction of the incident light, calculate the separation between adjacent slits in the grating.
d) Draw a diagram to show how such a grating can be used as a *reflection grating*, reflecting the incident light from its surface.
e) Under what circumstances might it be helpful to use a reflection grating rather than a transmission grating?

5.41 a) The largest optical telescope so far built has an aperture of 5.0 m. Can such a telescope resolve two stars 10 light-years apart in our nearest-neighbour galaxy – the one in Andromeda – into separate sources? (The Andromeda galaxy is 1.5×10^6 light-years from the earth.)
b) It has been suggested that there would be great advantage for astronomy if a large optical telescope could be mounted on an orbiting space laboratory. In view of the difficulty in making and mounting the mirror of the present 5.0 m telescope on Mt. Palomar it seems unlikely that such a telescope could have so large an aperture. What then can be the advantages? (1 light-year = distance travelled by light in one year, i.e., 9.5×10^{15} m.)

5.42 A narrow parallel beam of light of wavelength, 5.0×10^{-7} m falls normally on to a diffraction grating (12.5×10^{-7} m spacing): Diffracted beams are observed on the other side of the grating from the light source.

a) Calculate the angle between the first-order diffracted beam and a line perpendicular to the grating. Show how you arrive at your answer.

b) Calculate the angle between the second-order diffracted beam and a line perpendicular to the grating. Show how you arrive at your answer.

c) Explain why no third and subsequent orders of interference are produced when the grating is set up as described above.

d) A third-order diffracted beam can be produced by tilting the grating so that the incident beam is no longer normal to the grating. Copy and complete the diagram (Fig. Q5.15) and add whatever further written explanation is necessary to explain how the tilted grating can produce a third-order diffracted beam.

Oxford and Cambridge Schools Examinations Board
Nuffield A-level Physics (1976)

5.43 A parallel beam of red light of wavelength 6.0×10^{-7} m falls normally on the double slits shown in Fig. Q5.16 (which is not drawn to scale).
The graph (Fig. Q5.17) shows how the amplitude of the transmitted light varies with the angle θ.

a) Sketch a graph to show the way the amplitude would vary with angle θ if one of the slits was covered up.

b) Use the graph in Fig. Q5.17 to calculate the width of each single slit in Fig. Q5.16. Explain how you make your calculations.

c) Sketch a graph to show how the amplitude would vary with the angle θ if the central section X were removed from the double slit to make a single slit.

Oxford and Cambridge Schools Examinations Board
Nuffield A-level Physics (1974)

5.44 When infrared radiation is passed through hydrogen chloride gas it is found that radiation of wavelength 3.3×10^{-6} m is strongly absorbed.

Hydrogen chloride gas consists of molecules each composed of a hydrogen atom attached to a much more massive chlorine atom. The hydrogen atom behaves like a mass on the end of a spring.

a) Explain why radiation of one particular wavelength is strongly absorbed by the gas.

b) What becomes of the energy absorbed from the radiation?

c) Estimate a value for the spring constant of the bond joining the hydrogen and chlorine atoms.

Unit Six

6.1 a) A certain American electric toaster has a power-

rating of 1 kW and is designed to work on a 110 V supply (as in the USA and Canada). It is brought by a visitor to Britain who fits it with a 13 A fused plug and connects it to the 240 V mains. Discuss whether the toaster can be expected to work satisfactorily or not. If not, what will be the consequences?

b) A British electric kettle, rated at 2.75 kW and designed for use on the 240 V mains is taken to the USA and plugged into the mains there. Discuss whether the kettle will work satisfactorily in this situation.

c) Suppose you are arranging strings of electric lamps to provide lighting for a garden party at night. Calculate how many 240 V 60 W lamps you can safely connect to the 240 V mains using a plug containing a 13 A fuse.

d) A British fast-boiling kettle, rated at 2.75 kW, is found on a certain day to take 4 min 40 s to heat 2 litres of water (mass: 2 kg) from 15 °C to boiling point (100 °C). Assuming that the kettle actually is working at its rated power, calculate the percentage of electrical energy which is converted into heat to boil the water. (It takes 4200 J to raise the temperature of 1 kg of water by 1 °C.)

6.2 a) Given three resistors, each 1.2 kΩ, what values of resistance can you obtain by using them in combination, either two of them, or all three together? You may connect them together any way you wish. Sketch each combination and label it with its overall resistance.

b) For a particular function, in a certain electronic circuit, a resistance of exactly 100 Ω is required. Suppose you have a carbon resistor marked with its nominal value of 100 Ω, but you have found that its actual resistance is a little too high, say 102 Ω. (This is perfectly possible if the resistor's tolerance is ± 10%). You decide to reduce its resistance slightly by connecting another resistor, of high value, in parallel with it. Among your collection of resistors you have the following preferred values: 470 Ω, 820 Ω, 1 kΩ, 4.7 kΩ, 8.2 kΩ, 10 kΩ, 47 kΩ, 100 kΩ. Which one of these, in parallel with the 102 Ω resistor, will bring you nearest to the desired value? If you want the resistance to be within ± 1% of 100 Ω, will this be satisfactory?

6.3 A certain bicycle dynamo has an internal resistance of 0.1 Ω. When running at normal speed it lights the bicycle's front and rear lamps, each containing a bulb rated at 6 V 6 W, to normal brightness. The lamps are connected in parallel to the dynamo.

a) Assuming the p.d. across the dynamo terminals is 6 V exactly, calculate the e.m.f. of the dynamo.

b) Of the electrical energy generated by the dynamo,

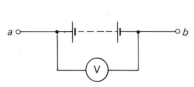

Fig. Q6.1

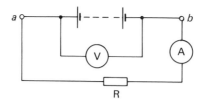

Fig. Q6.2

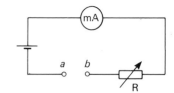

Fig. Q6.3

what percentage is dissipated as heat in its internal resistance?

6.4 The battery shown in Fig. Q6.1 provides an e.m.f. of 12 V and has an internal resistance of 2 ohm.

a) If the voltmeter has infinite resistance, what is the p.d. between the terminals a and b?

A resistor R of 4 ohm resistance and an ammeter of negligible resistance are connected in series to the terminals *a* and *b* as shown in Fig. Q6.2.

b) What is the reading of the voltmeter now?
c) What is the current in R?
d) What is the rate of energy conversion in the battery?
e) What is the rate of energy dissipation in the battery?
f) What is the useful power output of the battery?
g) What is the rate of energy dissipation in the resistor?

R is inadvertently short-circuited.

h) What are the new values of the ammeter and voltmeter readings?
i) What is the rate of energy conversion in the battery?
j) What is the rate of energy dissipation in the battery?
k) What is the useful power output of the battery?

6.5 A moving-coil meter may be used as an ohm-meter. The circuit is shown in Fig. Q6.3.

In a typical case the meter has a resistance of 100 Ω and a full scale deflection of 1 mA, the cell provides 1.5 V and the variable resistor, R, is used to 'zero' the instrument whilst the terminals *a* and *b* are 'shorted'. In this condition the instrument reads full-scale.

a) What is the value of the resistance R?
b) What current reading would indicate a resistance between the terminals *a* and *b* of 600 ohm?
c) To what resistance would a current reading of 0.1 mA correspond?

6.6 As a safety precaution, a typical school 5 kV supply has a rather large internal resistance. Explain why this is a sensible precaution. Assuming that the current output is limited to 3 mA, what is the internal resistance of the supply?

6.7 A typical moving-coil bench meter has a basic range of 3–0–15 mA and 15–0–75 mV.

a) What is the resistance between the terminals?
b) If this meter is 'shunted' with a suitable resistor it may be used as an ammeter in the range

0.6–0–3 A. What is the value of the shunt resistor?
c) If a resistor of 995 ohm is connected in series with the meter, what p.d. across the meter and resistor would give full-scale deflection?

6.8 Four 'black boxes' labelled H, J, K and L, a 6 V battery, a high resistance voltmeter and a milliammeter are connected in turn into the circuits shown. The meter readings were as indicated in the Fig. Q6.4a–d.

Make sketches to show what you would expect to find in each of the boxes and justify each of them.

6.9 The following tests were made on a 'black box', with results as described. Try to deduce from these what might be inside the box. You can be assured that every component inside the box obeys Ohm's law (Fig. Q6.5).

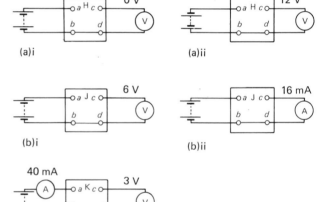

(a)i

(a)ii

(b)i

(b)ii

(c)

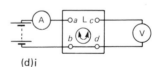

(d)i

As the knob was turned the voltmeter reading increased at a steady rate from 0 to 6 V. The ammeter reading remained at 1.2 mA throughout.

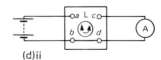

(d)ii

As the knob was turned the ammeter reading increased at a steady rate from 0 to 80 mA

Fig. Q6.4

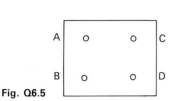

Fig. Q6.5

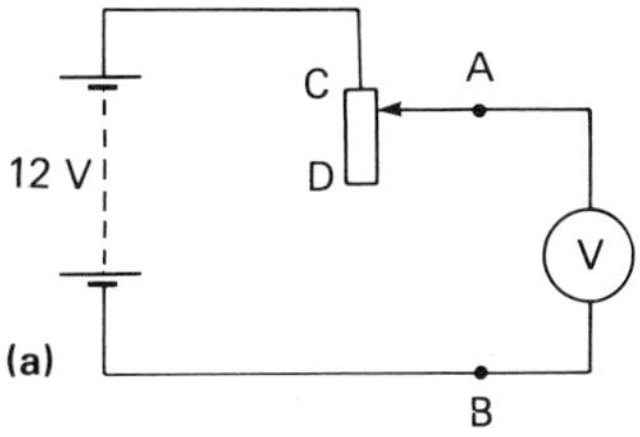

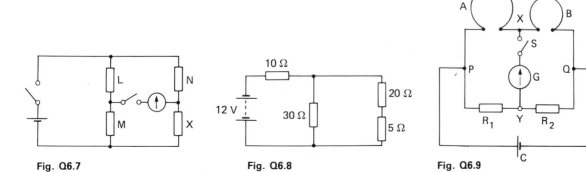

Fig. Q6.6 (a) 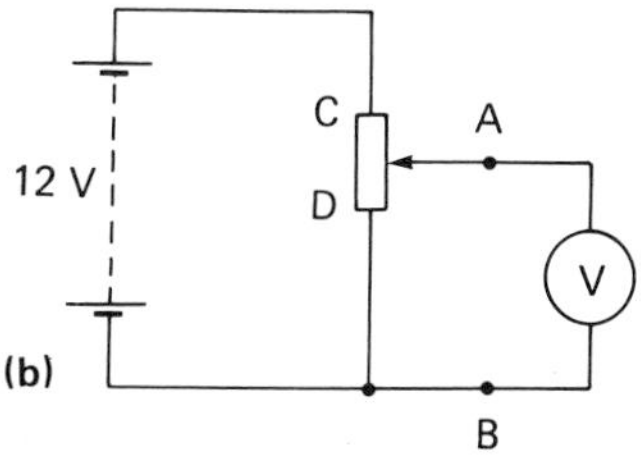   (b)

a) When a voltage V is applied to the terminals AB, the voltage between the terminals C and D is always found to be equal to V.

b) When a voltage V is applied to C and D, the voltage appearing between A and B is always found to be $\frac{3}{4} V$.

c) Whatever voltage is applied between the terminals A and C, the voltage between B and D is zero.

d) The resistance between terminals C and D is 800 Ω.

e) The resistance between A and B is 600 Ω.

6.10   The current, in amperes, through a certain type of non-linear resistor is given by $I = 0.20 V^3$, where V is the p.d. across the resistor. This resistor is connected in series with an ordinary carbon resistor (which obeys Ohm's law, that is, it is linear) to a constant d.c. voltage source of 6.0 V. What value of resistance should the carbon resistor have so that the current in the circuit is 0.40 A?

6.11   A student wanted to light a lamp labelled 3 V 0.2 A but had available only a 12 V battery of negligible internal resistance. In order to reduce the battery voltage he connected up the circuit shown in Fig. Q6.6a. He included a voltmeter – using it rather stupidly – so that he could check the voltage before connecting the lamp between A and B. The maximum value of the resistance of the rheostat CD was 1000 ohm.

a) He found that, when the sliding contact of the rheostat was moved down from C to D, the voltmeter reading dropped from 12 V to 11 V. What was the resistance of the voltmeter?

b) He modified his circuit as shown in Fig. Q6.6b using the rheostat as a potentiometer, and was now able

to adjust the rheostat to give a meter reading of 3 V. What current would now flow through the voltmeter?

c) Assuming that this current is negligible compared with the current through the rheostat, how far down from C would the sliding contact have moved?

d) The student then removed the voltmeter and connected the lamp in its place, but it did not light. How would you explain this? (The lamp itself was not defective.)

Oxford and Cambridge Schools Examinations Board
Nuffield A-Level Physics (1979)

6.12   The circuit (Fig. Q6.7) shows a balanced Wheatstone bridge in which the values of the resistors L, M and N are 10.00 ohm, 1000 ohm and 43.72 ohm.

a) What is the resistance of the resistor X?

b) What difference would it make to the balance of the bridge if the battery deteriorated during the measurement? Explain.

6.13   Use the rules due to Kirchhoff to find (a) the current in each of the resistors and (b) the p.d. across each resistor in the circuit shown in Fig. Q6.8. You may assume that the internal resistance of the battery can be neglected.

6.14   Two resistance wires A and B, made of different materials, are connected into a circuit with identical resistors R_1 and R_2 ($R_1 = R_2$), a sensitive high-resistance galvanometer G, a cell C and a switch S, as shown in Fig. Q6.9.

Fig. Q6.7   **Fig. Q6.8**   **Fig. Q6.9**

a) If A and B have equal resistance, no current will flow through G when the switch S is closed even though the cell is still delivering a current. Explain why this is so.
b) The diameter of A is twice that of B and the resistivity of the material of which B is made is $6 \times 10^{-6}\,\Omega\text{m}$. It is found that for zero current through G the length of A has to be three times that of B. Calculate, showing your working, the resistivity of the material of which A is made.
c) If the length of wire B is now reduced by a small amount so that the current through G is no longer zero, say which way the current will flow through G and explain why.

Oxford and Cambridge Schools Examinations Board
Nuffield A-Level Physics (1981)

6.15 Capacitors of small values in a micro-circuit are often made of a thin layer of gold deposited on the flat substrate, a thin layer of silicon monoxide dielectric deposited on the gold, and another thin layer of gold. Assuming that this construction constitutes a parallel-plate capacitor, estimate the capacitance per square millimetre of surface which can be obtained, if the relative permittivity of silicon monoxide is 4.5 and its thickness is 400 nm.

For higher values of capacitance TaO_5 can be used as a dielectric: its relative permittivity is about 22. Assuming that a 400 nm thickness of this dielectric is used, deduce whether or not it would be practicable to make a $1\,\mu\text{F}$ capacitor by this technique in a micro-circuit.

6.16 In various electronic circuits one requires a voltage which increases at a steady rate, for instance, in an analogue computer where voltages are used to represent numerical quantities. A way of achieving this is to use a large capacitor in conjunction with a large resistance.

Suppose we have a $5000\,\mu\text{F}$ capacitor connected in series with a $1\,\text{M}\Omega$ resistor, and that this combination is connected to a 12 V d.c. supply. Initially the capacitor is uncharged (Fig. Q6.10).

a) Calculate the initial current flowing in the resistor,

when the p.d. across the capacitor is zero.
b) Assuming that this current gives the rate of flow of charge into the capacitor for the first second, by how much will the p.d. across the capacitor increase in the first second?
c) Using the result of (b), write down the p.d. across the resistor after the first second.
d) What is the time constant for this circuit?
e) How long will it take for the capacitor to acquire half its final charge?
f) What is the value of the final charge?

6.17 In an electronic flash-gun, for use with a camera, a large capacitor is charged to a high voltage and then discharged through a gas discharge lamp, usually containing Xenon, which gives a very intense flash of white light of extremely short duration.

In a typical flash-gun, a $500\,\mu\text{F}$ capacitor is charged up to 400 V. After the flash has been fired the capacitor is not completely discharged: there is a p.d. across it, usually about 50 V. Use this information to calculate how much energy is obtained from the capacitor in this process.

6.18 In the circuit shown in Fig. Q6.11 the neon lamp is found to flash at regular intervals. A typical neon lamp does not conduct any current until the voltage across it rises to 110 V, the striking voltage, and then it keeps glowing until the voltage across it falls to 80 V. When it is glowing its resistance can be considered to be negligibly small in comparison with R.

a) Explain why the neon flashes repeatedly.
b) What would happen to the rate of flashing if R were increased?
c) What would happen to the rate of flashing if C were increased?

6.19 The circuit shown in Fig. Q6.12 might be used to study the storage of electric charge on a pair of parallel plates. The meter is a centre zero galvanometer. Describe with a short explanation in each case what would happen to the meter needle if

a) the top plate is moved sideways so that the overlap is reduced but the separation remains constant,
b) the top plate is returned to its original position and

Fig. Q6.10

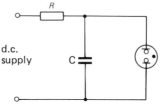

Fig. Q6.11

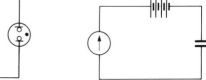

Fig. Q6.12

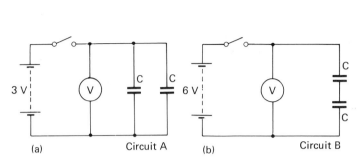

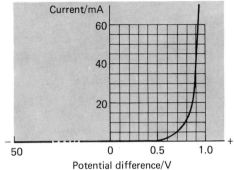

Fig. Q6.13

Fig. Q6.14

the plates pushed closer together,

c) the p.d. applied to the plates is increased after they have been restored to their original positions,

d) a sheet of polythene is slipped between the plates.

6.20 Two *identical* capacitors, each of capacitance C, are connected to a battery of e.m.f. 3 V, a switch and a high resistance voltmeter, as in circuit A (Fig. Q6.13a). Two more capacitors, *identical* with those in circuit A, are connected to a battery of e.m.f. 6 V, a switch and a high resistance voltmeter, as in circuit B (Fig. Q6.13b). The two voltmeters used are *identical*.

a) At first both switches are closed. (i) Is the electric field between the plates of the capacitors greater in circuit A, greater in circuit B, or the same in A and B? Justify your answer. (ii) Is the total energy stored in the capacitors greater in circuit A, greater in circuit B, or the same in A and B? Justify your answer.

b) Both switches are now opened. (i) State briefly why the voltmeter readings do not fall *steadily* with time. (ii) In which circuit will the voltmeter reading drop more quickly to half its initial value? Why is this?

Oxford and Cambridge Schools Examinations Board
Nuffield A-Level Physics (1981)

6.21 Figure Q6.14 shows approximately how the current through a silicon diode varies as the p.d. across the diode is altered over the range of −50 V to +1.0 V.

a) Use the graph to estimate the resistance of the silicon diode at p.ds. of −50 V, +0.5 V, +0.7 V and +0.9 V.

b) Two such diodes connected as shown in Fig. Q6.15 may be used to protect a sensitive galvanometer against electrical overload. Explain qualitatively how this arrangement provides protection.

c) Suppose that a galvanometer of resistance 25 ohm, protected in this way is used in a circuit as shown in Fig. Q6.16. A reed switch S operates to charge the capacitor C, of capacitance 10^{-9} F, to 50 V and discharge it through the galvanometer 100 times in each second. Calculate the average value of the current through the branch XY of the circuit.

d) In fact, the average value of the current indicated by G would be considerably below the value calculated in (c). How do you account for this?

e) Connecting a 10 kΩ resistor in series with the galvanometer restores the reading of G to the value calculated in (c). Why is this?

Oxford and Cambridge Schools Examinations Board
Nuffield A-Level Physics (1977)

Unit Seven

In answering the questions on gravitational fields you may need the following information:

The radius of the earth: 6.37×10^6 m
. The mass of the earth: 6×10^{24} kg

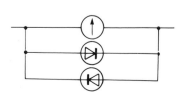

Fig. Q6.15

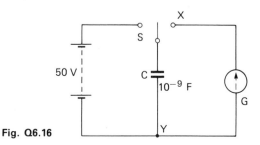

Fig. Q6.16

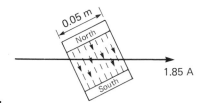

Fig. Q7.1

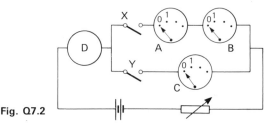

Fig. Q7.2

The gravitational constant, G, 6.67×10^{-11} $N\,m^2\,kg^{-2}$

7.1 a) A horizontal magnetic field of flux density 0.02 T acts at right-angles to a horizontal conductor 40 mm long that carries a current of 2.5 A. Figure 27.4 shows such an arrangement. What is the magnitude of the force on the conductor and state its general direction?

b) The conductor is next turned through 35° in a horizontal plane. What is the force now acting and state its direction?

7.2 A current balance (see Fig. 27.3) has a horizontal conductor 200 mm long lying E–W and carrying a current of 3 A. The horizontal component of the earth's magnetic field has a flux density of 20 μT. Calculate the force acting on the conductor. Does this force act in such a direction that deflection would occur if the balance was sensitive enough? Explain your reasoning. Would your answers be different if the conductor ran N–S?

7.3 A horizontal conductor 0.5 m long is carrying a current of 4 A in the magnetic field of the earth at a place where the flux density is 52 μT inclined at 70° to the horizontal. The conductor is turned so that it remains in a horizontal plane. State the direction (N–S, E–W, etc.) of the *conductor* when it experiences a maximum force, calculate the size of this force and give its direction measured from the horizontal.

Find also the magnitude and direction of the minimum force.

7.4 A current balance, initially in equilibrium, is placed between the poles of a magnet as in Fig. 27.3. When a current of 1.85 A passes, 50 cm of paper tape had to be hung over the conductor to restore equilibrium. Five metres of the tape was found to have a mass of 4.23 g, and 0.05 m of the conductor was estimated to be in the field of the magnet. Calculate the magnetic flux density.

Without altering the current flowing, the magnet is turned so that the field is no longer at 90° to the wire, as shown in Fig. Q7.1. This has no effect on the equilibrium of the balance although it is very sensitive. Explain why this is so.

7.5 A student has read Section 27.6 on giving the scale of an ammeter, D, marks that are proportional to 1, 2, and 3 units of current. He reconnects the circuit as shown (Fig. Q7.2) and explains how to place marks of $\frac{1}{2}$ and $1\frac{1}{2}$ units. Explain his method. (Remember, all the meters are identical and so have the same resistance.)

7.6 Calculate the gravitational attraction between the earth and a 1 kg mass on its surface.

7.7 Calculate the gravitational attaction between two 1 kg masses placed with their centres of mass 8 cm apart.

7.8 Two spheres each of radius R and density ϱ are just touching. Deduce a formula for their mutual attraction in terms of R, ϱ, and G.

7.9 The mass of the earth is 81 times that of the moon. Their centres are separated by a distance of 3.844×10^8 m and there is a point on the line joining them where the attraction of the earth is equal but opposite to the attraction of the moon. How far is this point from the centre of the moon?

7.10 A 5 kg mass weighs 49 N on the surface of the earth. Calculate a value for the mean density of the earth.

Newton was unable to do this calculation; what information did he lack?

7.11 A satellite is in a circular orbit 300 km above the earth's surface. What is its speed, and how long does it take to complete one orbit of 360°?

7.12 Suppose the satellite of Question 7.11 is orbiting in the equatorial plane and is travelling east. What time elapses between two successive passes over a town on the equator?

7.13 Find the height and speed of a synchronous communications satellite that remains over the same point on the equator at all times. Assume that the orbit is circular and in the equatorial plane. The radius of the earth is 6.37×10^6 m and g on its surface is 9.8 N kg^{-1}.

7.14 As well as the earth with its moon, other planets in our Solar system have satellites, some having several. Data on a few of these are tabulated (Table Q1). Which of these are likely to belong to the same planet?

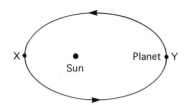

Fig. Q7.3

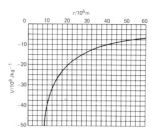

Fig. Q7.4

Table Q1

Name of satellite	Mean radius of orbit/10^8 m	Period of orbit/10^5 s
Ariel	1.90	2.16
Europa	6.67	3.06
Phoebe	130.0	476.0
Titania	4.36	7.49
Umbriel	2.68	3.56
Oberon	5.82	11.2

7.15 Calculate the mass of the planet around which Ariel is orbiting (see Question 7.14). Find also the value of g on the surface of the planet if it has a radius of 2.4×10^7 m.

7.16 Find the mass of the earth if the moon is 3.844×10^8 m away and the length of the mean sidereal month is 27.3 days.

7.17 Find the relationship giving the speed (v) of a satellite if it is to maintain a circular orbit of radius (R) around the earth. The only other information available is the radius (r) of the earth and the gravitational field strength (g) on its surface.

7.18 What happens to the loss of gravitational potential energy if a satellite is (a) falling in outer space towards the earth, or (b) falling at its terminal velocity in air.

7.19 A mass of 4 kg is raised through 3 m. What is

a) its change of potential energy,
b) the change of potential.

Assume that $g = 10 \, \text{N kg}^{-1}$ and is constant.

7.20 A planet moves around the sun in an ellipse, as shown in Fig. Q7.3. It is closest to the sun at X, and farthest from the sun at Y. Which one of the following quantities is *greater* at Y than at X?

A the gravitational force on the planet caused by the sun
B the gravitational potential energy of the planet–sun system
C the kinetic energy of the planet in its orbit
D the acceleration of the planet towards the sun
E the speed of the planet in its orbit.
  Explain your answer.

7.21 Taking the zero of gravitational potential to be at a very great distance, calculate the gravitational potential (a) at the earth's surface, and (b) at a point 3630 km above the surface. What is the change of potential if a spacecraft travels from this point to the surface? If the spacecraft has a mass of 6 tonnes, how much potential energy has it lost in falling from this height of 3630 km.

To find this loss of potential energy, a student used the formula P.E. lost = mgh. Comment.

7.22 The graph (Fig. Q7.4) shows the variation on the earth's gravitational potential (V) with distance (r) from the centre of the earth.

a) (i) What is the potential at an infinite distance from the earth?
   (ii) Why is the potential on the graph negative?
b) What is the value of the gravitational field at 25×10^6 m from the centre of the earth? Explain how you arrive at your answer.
c) How much energy is required to raise a mass of 70 kg from 10×10^6 m to 50×10^6 m above the earth's centre?

7.23 A very small sphere carrying a charge of 6×10^{-8} C is placed centrally between a pair of large parallel conducting plates separated by a distance of 40 mm. If the p.d. between the conductors is 1.5 kV, calculate:

a) the field between the plates;
b) the force acting on the sphere due to the electric field;
c) the energy required to move the sphere from one plate to the other;
d) the ratio of the energy required to the charge carried.

7.24 A small sphere of mass 0.1 g hangs by a thread between two vertical plates 50 mm apart. The charge on the sphere is 6 nC. What p.d. between the plates will cause the thread to make an angle of 30° with the vertical.

7.25 Two small metal spheres, each of mass 10 g are attached to nylon threads 1 m long and hung from a common point. When the spheres are given equal quantities of positive charge each thread makes an angle of 4° with the vertical.

a) Draw a diagram showing all of the forces acting on a sphere.
b) Calculate the charge on each sphere.

7.26 A particle carries a charge of 3 nC and is held in a uniform electric field. It is then released. After moving a distance of 50 mm its kinetic energy is found to be 4.5×10^{-5} J.

a) How much work has been done by the electric force?
b) What is the difference in potential between the starting point and the endpoint?
c) What is the magnitude of the electric field?

7.27 A charge of 8 nC is distributed uniformly over the surface of a sphere of radius 0.1 m. If the potential approaches zero at a very large, what is the potential at a point on the sphere? Calculate also the p.d. between the sphere and a point 0.15 m away from its surface.

7.28 In a flame probe experiment the sphere has a radius of 60 mm and is kept at a negative potential of 1.5 kV. By how much does the electroscope reading change when the flame is moved from a point 0.4 m to a point 0.1 m from the centre of the sphere? Is it a rise or a fall of potential?

7.29 A uniformly charged spherical balloon is isolated (from all electrical supplies) and initially it has a radius of 100 mm. A flame probe is used to measure the potential at a point 0.2 m from the centre of the balloon and a reading of 800 V is obtained. The balloon has a leak and its radius decreases to 50 mm.

a) Is the total charge on the balloon more, the same, or less than it was initially?
b) What is the final potential reading if the flame and the centre of the balloon remain fixed?
c) How does the potential of the surface of the balloon now compare with its initial value?

Explain your answers.

7.30 The potential at a certain distance from a point charge is 600 V and the electric field there is $200 \, \text{N C}^{-1}$.

a) What is the distance to the point charge?
b) What is the size of the point charge?

7.31 In an experiment for displaying electric fields the two electrodes had the shape as shown in Fig. Q7.5. Make a copy of this arrangement with electrodes 4 cm long and a gap of 1 cm. On your diagram draw:

a) electric field lines (solid);
b) equipotential lines (dotted);

State two important characteristics of these lines that your diagram illustrates.

Unit Eight

In answering some of these questions you may need the following information:

The charge of the electron (e): 1.6×10^{-19} C
The mass of the electron (m_e): 9.1×10^{-31} kg
The specific charge of an
electron (e/m_e): 1.76×10^{11} C kg^{-1}

8.1 A very small conducting ball on a long insulating thread shuttles to and fro between two large, closely-spaced metal plates which are connected to a high voltage supply. The p.d. between the plates is V; the charge on the ball as it moves Q; the plate separation d; the time taken by the ball to go from one plate to the other t; the capacitance of the plates C. Write expressions for:

a) The force on the ball mid-way between the plates.
b) The energy taken from the supply as the ball moves from one plate to the other.
c) The average current taken from the supply.
d) The charge on the plates if the ball is removed.

8.2 a) Using the data given in section 31.2 calculate the field strength between the plates.
b) If the radius of the ball is 19 mm and it is charged to a potential of 5 kV, what is the charge on the ball? ($\epsilon_0 = 8.854 \times 10^{-12} \, \text{Fm}^{-1}$)
c) Estimate the force acting on the ball and causing motion.
d) Why are your answers to questions (b) and (c), estimates?
e) If the ball hits each plate 10 times per second what current will register on the galvanometer.
f) How can the frequency of oscillation of the ball be increased?

8.3 Five envelopes contain different numbers of identical coins. Their masses are 65.1 g, 93.0 g, 83.7 g, 37.2 g, 46.5 g. What is the possible mass of a single coin and the total number of coins?

8.4 A number of differently charged drops but all with the same mass are used in a Millikan experiment. They were kept stationary when the following voltages were

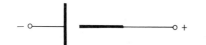

Fig. Q7.5

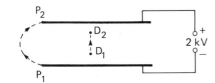

Fig. Q8.1

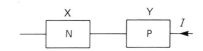

Fig. Q8.2

applied to the plates: 200, 100, 125, 330, 250, 500. Calculate a possible number of units of electric charge on each of these drops.

8.5 In a Millikan experiment, a charged oil drop of mass 6.78×10^{-15} kg is stationary between parallel plates 5 mm apart when the p.d. between them is 300 V. What is the charge on the drop? Estimate the number of electron charges that this represents.

8.6 Calculate the electric field strength needed to keep an oil drop of mass 8×10^{-15} kg stationary if it carries two surplus electrons. What is the direction of the field, and if it is provided by a pair of plates, what is the polarity of the lower one?

8.7 A charged oil drop of mass 4×10^{-15} kg falls under gravity at a terminal velocity of 0.04 cm s^{-1}. When an electric field is applied the drop rises steadily through 5 mm in 12.5 s. If the field is provided by parallel plates 1 cm apart having a p.d. of 700 V between them, calculate the charge on the drop.

Is your answer consistent with the notion that all charges are multiples of the charge on an electron?

8.8 What is meant by the statement that charge is quantised? May we also say that mass is quantised?

8.9 The diagram (Fig. Q8.1) shows a pair of flat, wide conducting plates. They are parallel, 5 mm apart, and are connected to a steady p.d. of 2 kV. An oil drop between the plates moves from D_1 and D_2 along a straight line at right angles to the plates and between their centres.

a) If the drop carries five electron charges, each 1.6×10^{-19} C, and moves 2.5 mm, how much electrical energy is transformed?
b) State what you can about the amount of electrical energy transformed if the drop were to move along the curved path P_1P_2 from one plate to the other, outside the plates. Explain your statement.

8.10 A raindrop carries a negative charge and it falls from a cloud that has a positive charge with respect to the earth. The electric field in the space under the cloud is 230 V m^{-1} and g there is 9.8 N kg^{-1}.

If the drop falls through 20 m, what is:

a) the change of gravitational potential of the drop.
b) the change of electrical potential of the drop.

Pay particular attention to signs and explain the

convention used.

If the mass of the raindrop is 34 mg and it carries a charge of 240 nC, what is the overall change of energy? Is it a rise or a fall?

8.11 Hall effect voltages are much larger for relatively poor conductors like silicon, than for good conductors like copper, when comparable currents, fields and dimensions are used. Why is this? What steps would you take in an experiment on the Hall effect in silver to make the measured Hall voltage as large as possible?

8.12 Figure Q8.2 shows two flat slices of semi-conductor connected in series, with a current I flowing through them. N and P indicate which sample has negative and which has positive charge carriers. What are the polarities of edges X and Y when a magnetic field is applied perpendicular to the plane of the diagram and directed out of the paper?

8.13 A strip, 0.1 mm thick, 6 mm wide is made of a metal that has 8×10^{28} free electrons per m^3. A current of 15 A flows along the strip, and a magnetic field of 0.4 T acts at right angles to its surface.

a) What is the drift speed of an electron along the strip?
b) What force is acting on an electron due to the magnetic field?
c) What force is acting on all the electrons in a 50 mm length of strip?
d) Electrons will move to one edge of the strip. What is the resulting electric field strength across the width of the strip?
e) What is the p.d. between opposite edges due to the Hall effect?
f) Check your answer to (c) by an alternative calculation.

8.14 Experiments involving the Hall effect often use conductors in the form of foils, thin slices, flat strips or ribbons. Explain why this is so.

Discuss in detail the possibility that the Hall effect might be observed with ordinary wires of circular cross-section. You should mention the exact positions on the wire where you would place contacts for detecting any Hall effect that might exist.

8.15 Two students wish to investigate experimentally the Hall effect in wires of circular cross-section. Student A says that a thick wire is better because the Hall voltage is proportional to the current flowing and thick wires can carry large currents. Student B argues that thin foils give large Hall voltages and so a thin wire would be better. What diameter would you choose? Explain your reasoning carefully.

8.16 A sample of aluminium foil is 10 mm wide, 0.02 mm thick and it carries a current of 4 A. It is to be used in a Hall effect experiment and contacts are placed on opposite edges. The contacts were misaligned by 0.1 mm instead of being exactly opposite each other. Calculate the p.d. that exists between the contacts when there is zero B-field. Take the resistivity of aluminium to be $2.7 \times 10^{-8} \Omega$ m. Is this p.d. going to cause problems? To help you decide, calculate the Hall voltage when a B-field of 1.9 T is applied, perpendicular to the surface of the foil. Take the number of electrons per unit volume in aluminium to be 2×10^{29} m^{-3}.

8.17 In a magnetic field directed vertically downward, a particle initially moving north is deflected towards the east. What is the sign of the charge on the particle? Explain your reasoning.

8.18 A straight electron path is observed in a fine-beam tube. Can you be certain that no magnetic field is acting? Explain your answer.

8.19 Electrons are accelerated through a p.d. of 250 V in a fine beam tube. They are then deflected into a circle of diameter 96 mm by a magnetic field perpendicular to the electron path. Calculate the speed of the electrons and the flux density of the field.

8.20 A beam of electrons is undeflected when it passes simultaneously through an electric field of 2.4×10^4 V m^{-1} and a magnetic field of 1.5×10^{-3} T. Both fields are perpendicular to the beam and to each other. What is the speed of the electrons?

When the electric field is removed the radius of the electron orbit is 6 cm; calculate a value for the ratio e/m.

8.21 Give a reason why the spot on the screen of a cathode ray tube can never be a perfect point.

8.22 In an electron gun, electrons are emitted by a heated cathode and accelerated to an anode 6 cm away by a p.d. of 600 V. Another electron gun also operates with a p.d. of 600 V but has its anode only 2 cm away from the cathode. By taking the electrodes to be parallel plates, compare the final electron speeds produced by each gun. (To do this you should consider the force acting on an electron, the acceleration experienced by its mass, and the resulting speed reached in the space available.) What conclusion can you draw from your answers about the effect of gun size on speed? Would your conclusion be different if the electrodes were not parallel plates?

8.23 Assuming that the energy of a particle is all in the form of kinetic energy, calculate the speed of:

a) a 3 keV electron;
b) 3 keV alpha particle (mass $= 6.7 \times 10^{-27}$ kg);
c) an alpha particle that has been accelerated through a p.d. of 3 kV.

8.24 An electron is projected along the axis of a cathode ray tube midway between a pair of deflecting plates 2 cm apart and 4 cm long measured along the axis. The electron has an initial velocity of 2×10^7 m s^{-1} and there is a p.d. of 400 V between the plates.

a) How long does the electron spend between the plates?
b) How far away from the axis has the electron moved when it reaches the end of the plates?

8.25 The tube of a colour TV receiver operates with an accelerating p.d. of 25 kV and the total beam current is 1.6 mA.

a) What is the energy in joules gained by a single electron?
b) How many electrons are hitting the screen per second?
c) What power is being delivered to the screen?

8.26 An X-ray tube operating at 150 kV takes a current of 10 mA. If only 0.5% of the energy of the electrons is converted into X-radiation,

a) estimate the rate at which heat is dissipated from the anode.
b) what is the X-ray power output?

8.27 A current of 0.80 A is passed for 30 minutes through a voltameter consisting of two platinum electrodes dipping into dilute sulphuric acid. 200 cm^3 of hydrogen are collected at a pressure of 1000 mb and a temperature of 17°C. Given that the density of hydrogen at s.t.p. is 8.99×10^{-5} g cm^{-3}, calculate a value for the Faraday.

If a copper voltameter (made of two copper plates dipping into a solution of copper sulphate) had been connected in series with the water voltameter in the experiment, how much copper would have been deposited on the cathode? 1 mole of copper has a mass of 63.5 g and a copper ion carries 2 positive charges.

8.28 Electrical conductivity is affected by changes in the temperature of the material. Write a short paragraph accounting as far as you can for each of the following.

a) The electrical conductivity of copper falls as the temperature rises.
b) The electrical conductivity of copper sulphate solution rises as the temperature rises.
c) The electrical conductivity of common salt is almost zero at room temperature but rises rapidly when it

melts.

d) The electrical conductivity of paraffin wax, which is almost zero at room temperature, is unchanged by heating and melting.

e) The electrical conductivity of air is very low unless either the pressure is reduced, or the potential gradient is raised to $20\,000\ \mathrm{V\,m^{-1}}$.

Unit Nine

9.1 The three-core flex (flexible cable) connecting an electric heater to the mains supply contains three conductors whose spacing is 2 mm between centres. Estimate the magnetic force between the two conductors which carry the current, per metre of their length, at an instant when the current in each is 10 A. Comment on the answer.

9.2 In a piece of apparatus commonly used to demonstrate the fact that there is a force between two current-carrying conductors, strips of very thin aluminium foil are used, hanging vertically side by side. A force of 1 mN produces a clearly visible movement of the strips. If the strips are 0.5 m long and 20 mm apart, how much current (roughly) must flow in each strip?

9.3 A student makes a model of a moving-coil meter by winding 20 turns of insulated wire around a rectangular former of square shape measuring 40 mm by 40 mm, mounted on a spindle. He sets this up with the plane of the coil horizontal and between two magnet poles which give a horizontal field whose strength is about 0.1 T in the region of the coil. Estimate the magnetic torque on the coil when a current of 2 A flows in it.

9.4 In a certain moving-coil galvanometer the pointer turns through an angle of 2 radian in going from zero to full-scale deflection. The magnetic field strength in the air-gap between the magnet poles is 0.5 T. The coil has 100 turns and encloses an area of $64\ \mathrm{mm^2}$. The pair of hairsprings provides torque of $8 \times 10^{-7}\ \mathrm{N\,m}$ per radian of twist. Calculate the current which produces full-scale deflection.

9.5 In Fig. Q9.1 two wires with currents flowing in them are in the same plane but are not parallel to each other. The arrowheads show the direction of current.

a) In what direction is the field at point Q due to the current in wire X?

b) In what direction is the field at point P due to the current in wire Y?

c) Copy the diagram and draw an arrow at Q to show the direction of the force on wire Y due to the field of wire X. (Fleming left hand rule.)

d) Draw an arrow at P to show the direction of the force on wire X due to the field of wire Y.

e) If you have done (c) and (d) correctly you will see that the situation appears to disobey Newton's third law. But certain factors have not been taken into account: what could these be?

9.6 Figure Q9.2 shows a long straight wire with a current flowing in the direction of the arrow, and in the same horizontal plane is a flat square wire frame with a current flowing in the direction shown (the current flows in and out of the square via a pair of flexible leads whose magnetic field can be ignored). The direction of North is shown in the figure.

a) In what direction is the force, due to the field of the current in PQ, acting upon each side of the square: WX, XY, YZ, ZW?

b) Two of the forces on the sides of the square are equal and opposite to each other: which two?

c) The other two forces produce a resultant force: what is the direction of this resultant? Give your reasoning.

d) If wire PQ is fixed and the frame WXYZ is free to move until it finds equilibrium, where will WXYZ finally settle?

9.7 Estimate the maximum magnetic field strength produced by the deflector coils in a television picture tube, using the following data:

Radius of curved path in magnetic field, for maximum deflection: 5 cm.
Speed of electrons in electron beam: $1 \times 10^8\ \mathrm{m\,s^{-1}}$.
Mass of electron (at this speed): $9.7 \times 10^{-31}\ \mathrm{kg}$.
Charge on electron: $1.6 \times 10^{-19}\ \mathrm{C}$.

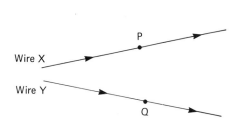

Fig. Q9.1

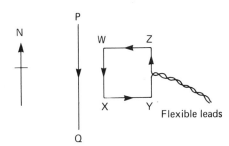

Fig. Q9.2

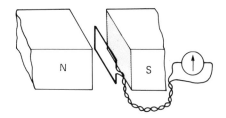

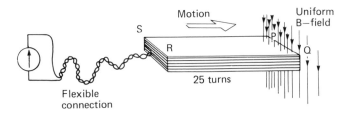

Fig. Q9.3   **Fig. Q9.4**

9.8 The speed of electrons in the electron beam of a colour television picture tube is very nearly 1×10^8 m s^{-1}. The vertical component of the earth's magnetic field in Britain is about 5×10^{-5} T. The charge on an electron is 1.6×10^{-19} C and its mass, travelling at this speed, is 9.7×10^{-31} kg.

The vertical component of the earth's field will give the beam curvature in a horizontal plane and the theory of Section 34.5 can be used to estimate the radius of this curvature, just as if it were a plane circle.

a) Estimate the radius of curvature of the electron beam in the horizontal plane.
b) Estimate the distance between the electron gun and the screen, in a large television receiver.
c) Estimate how much displacement of the point where the beam hits the screen is due to this effect of the earth's field.

9.9 a) Particles of two different types are entering a uniform magnetic field in a plane perpendicular to the field. Particles of type X have mass m_X, charge Q_X, and speed v_X. Particles of type Y have mass m_Y, charge Q_Y, and speed v_Y. Derive an expression for the ratio of the radii of their circular paths in the field, r_X/r_Y.

b) A magnetic field is often used to guide charged particles of different types into separate paths. One instance of this is the separation of alpha particles and beta particles from a radioactive source. Here is some data for these: the mass of an alpha particle is about 1840 times the mass of a beta particle. The (positive) charge of an alpha particle is twice that of the charge (negative mostly; sometimes positive) of a beta particle. The speed of beta particles is about ten times that of alpha particles.

Use the data to find the approximate ratio of the radii of their paths when moving in a plane at right angles to a uniform magnetic field.

9.10 Figure Q9.3 shows a rectangular wire frame connected by flexible leads to a centre-zero galavano-meter. When the frame is pulled out of the space between the magnet poles (towards us) the meter shows a momentary deflection to the right. Describe what sort of deflection, if any, you would expect the meter to show when:

a) the frame is pushed back into the field-region;
b) the frame is moved vertically upwards and out of the field-region;
c) the frame is moved vertically downwards back into the field;
d) the frame is rotated clockwise through 90° to make its plane horizontal;
e) keeping the plane horizontal, the frame is moved vertically upwards out of the field, and then returned to its former position;
f) still with its plane horizontal, the frame is pulled out horizontally towards us.

9.11 Figure Q9.4 shows a rectangular coil of wire with 25 turns. The side PQ which is 10 mm long is in a region of uniform magnetic field of strength 0.25 T. The plane of the rectangle is perpendicular to the field. The coil is moved at a steady speed of 1 mm s^{-1} in the direction shown.

a) Calculate the e.m.f. induced in one wire of the side PQ.
b) Why is no e.m.f. induced in the sides PS and QR, although they are partly in the field region?
c) The resistance of the galvanometer is 12.5 Ω and the resistance of the rest of the circuit is negligible. Calculate the current through the galvanometer.

9.12 Figure Q9.5 shows schematically a piece of apparatus which can be used to demonstrate the laws of electromagnetic induction. An e.m.f. is induced in the spinning copper disc which is set with its plane perpendicular to the uniform magnetic field inside a large solenoid. Connection is made to the spindle and the rim of the disc by 'brushes'.

The radius of the disc is 10 mm. Calculate the e.m.f. induced between axis and rim when the disc makes 20 revolutions per second in a field of 5 mT.

9.13 a) A wire of length l is horizontal and oriented north–south. It moves east with a velocity v through the earth's magnetic field which has a downward vertical component of flux density B. Write down

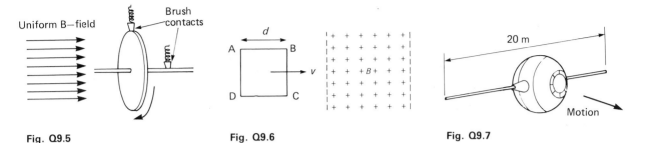

Fig. Q9.5   Fig. Q9.6   Fig. Q9.7

an expression for the potential difference between the two ends of the wire. Which end of the wire is at the more positive potential?

b) A horizontal square wire frame ABCD, of side d, moves with velocity v parallel to sides AB, DC from a field-free region into a region of uniform magnetic field of flux density B (Fig. Q9.6). The boundaries of the field are parallel to the sides BC, AD of the frame and the field is directed vertically downward. Write down expressions for the electromotive force induced in the frame

(i)   when side BC has entered the field but side AD has not,

(ii) when the frame is entirely within the field region,

(iii) when side BC has left the field but side AD has not.

For each position derive an expression for the magnitude and direction of the current in the frame and the resultant force acting on the frame due to the current. The total resistance of the wire frame is R, and its self inductance may be neglected.

Oxford and Cambridge Schools Examinations Board
A-level Physics (1978)

9.14 A satellite orbiting the earth is travelling at $8000 \, \mathrm{m \, s^{-1}}$ at a height of a few hundred kilometres above the surface, and in a region where the vertical component of the earth's magnetic field is $5 \times 10^{-5} \, \mathrm{T}$.

Two metal poles stick out on either side as shown in Fig. Q9.7 (these are supports for radio aerials). The distance between the tips of the these poles is 20 m, and they are on a line perpendicular to the satellite's direction of motion.

a) Calculate the p.d. between the ends of the poles due to electromagnetic induction in the earth's field.

b) Suppose we wish to make the poles part of a complete circuit and use the induced e.m.f. as a source of electrical energy for equipment in the satellite. Imagine having a wire leading from each pole-end, into the satellite, and connected to a resistor to complete the circuit. No current would flow: why not?

9.15 A certain electromagnet (Fig. Q9.8) produces a field which can be considered as approximately uniform over an area of $500 \, \mathrm{mm^2}$. Around it is a coil of 50 turns connected to a millivoltmeter.

a) When the mean field strength is 0.5 T, what is the total flux of the magnetic field?

b) Calculate the e.m.f. induced if the field strength is reduced to zero at a constant rate in 5 second.

9.16 A certain small d.c. motor, designed to work with a supply of 12 V, has a permanent magnet to provide a radial field of strength of 0.5 T, in the region of the armature coil. The armature carries a single coil of 200 turns having the rectangular shape shown in Fig. Q9.9. The motor can also be used as a dynamo.

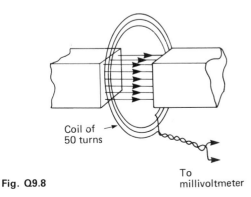

Coil of
50 turns

To
millivoltmeter

Fig. Q9.8

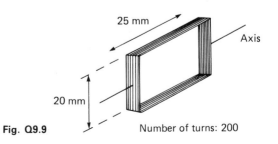

25 mm

Axis

20 mm

Fig. Q9.9   Number of turns: 200

a) If the armature spins at 10 rev s^{-1}, what is the speed of each long side of the rectangular coil?
b) Use the answer to (a) to calculate the e.m.f. induced in one 25 mm length of wire on one side of the coil.
c) Calculate the e.m.f. induced in the whole coil.
d) The motor is now connected to a 12 V d.c. supply, and allowed to run freely, and is found to spin at 35 rev s^{-1}. Given the resistance of the armature coil is 5 Ω, what current is taken from the supply at this speed?
e) The mechanical power is calculated by multiplying the 'back' e.m.f. induced in the coil by the current. When spinning freely, this mechanical power is being used to overcome friction in the motor itself. Calculate this power for the speed of 35 rev s^{-1}.
f) Do the other calculations needed to write numerical values for the terms in this expression, for the motor running freely:

$$\begin{pmatrix} \text{power taken} \\ \text{from supply} \end{pmatrix} = \begin{matrix} \text{power} \\ \text{dissipated} \\ \text{as heat in} \\ \text{armature} \\ \text{resistance} \end{matrix} + \begin{pmatrix} \text{power used} \\ \text{to overcome} \\ \text{friction} \end{pmatrix}$$

g) The motor is now used to drive a piece of model machinery, and the armature spins at 30 rev s^{-1}. Calculate the total mechanical power now produced. If the power needed to overcome friction is the same as in (f), what is the *useful* mechanical power now being produced?
h) At what running speed would the motor produce the greatest total mechanical power of which it is capable? What is the back e.m.f. at this speed? Comment upon the answer.

9.17 Discuss the principle of operation of a d.c. electric motor. Explain what is meant by the term *back e.m.f.*

a) A simple d.c. motor consists of a plane coil XY of area A having N turns. It is free to rotate about an axis through O perpendicular to a uniform magnetic field B of flux density B. The coil has a resistance R and its self-inductance may be neglected.

It is connected to a battery of e.m.f. V and negligible internal resistance via a split-ring commutator with brushes arranged so that the current through the coil is reversed whenever the plane of the coil is perpendicular to the direction of B. The coil rotates with constant angular velocity ω. In Fig. Q9.10 θ ($\theta = \omega t$) is the angle between the plane of the coil and B.

Sketch a graph showing the variation of back e.m.f. with time t.

Deduce expressions for:

(i) the maximum value of the back e.m.f. generated, stating the angles θ for which this occurs.
(ii) the maximum current passing through the coil.
(iii) the torque exerted by the coil when $\theta = 0$.

b) A d.c. electric motor is connected to a 200 V supply having negligible internal resistance. The motor coils have a resistance of 2.0 Ω. At a particular rotational speed, the average current flowing is 5.0 A. Calculate the average back e.m.f. generated. Neglecting friction, calculate the mechanical power being supplied by the motor and the power being dissipated in the coil resistance.

Oxford and Cambridge Schools Examinations Board
A-level Physics (1981)

9.18 Four electronic oscillators generate alternating e.m.f.s which vary with time as shown by the four graphs in Fig. Q9.11. Each of them has negligible internal resistance, and each is connected to a load resistor of 50 Ω. For each oscillator consider one cycle and calculate:

a) the mean value of the e.m.f.;
b) the mean power delivered to the load;
c) the root mean square (r.m.s.) e.m.f.

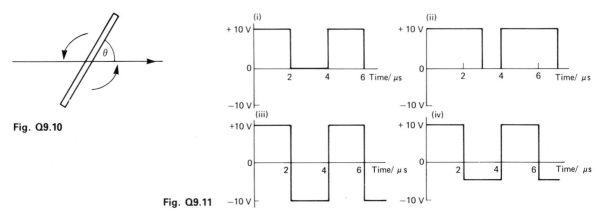

Fig. Q9.10

Fig. Q9.11

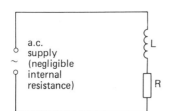

Fig. Q9.12

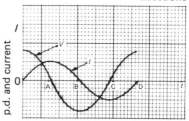

Fig. Q9.13

9.19 A certain electric kettle is labelled, on its underside: 'Volts 230/250 Watts 2750'. Assuming that it is operating from a mains supply voltage of 240 V r.m.s., what is the peak value of the current flowing in the connecting lead? Is the usual 13 A fuse in the plug suitable for this?

9.20 A *v.h.f. suppressor choke* is a small inductor which is connected in series with electrical supply leads to prevent *v.h.f. interference*, that is, to prevent stray alternating currents of very high frequency from flowing in the leads. They are used in television equipment in which the signal frequency is in the order of 100 MHz, or greater. What is the reactance of a choke, whose inductance is 5 μH, at this frequency?

9.21 A student is experimenting with a bicycle dynamo. He finds that the r.m.s. value of the generated e.m.f., as indicated by an a.c. voltmeter, appears to be directly proportional to the speed of rotation. He then tests the dynamo 'on load', using a 10 Ω wire-wound resistor as the load, with an a.c. ammeter in series with it to measure the current, at various speeds of rotation. He finds that at one particular speed the current is 0.5 A r.m.s., but to his surprise, at double this speed the current is only 0.6 A r.m.s. What could account for the fact that the current does not increase in proportion to the speed? (It may help to study the diagram, Fig. 35.1.)

9.22 The circuit shown in Fig. Q9.12 is set up and an a.c. voltmeter (which draws negligible current) is used to measure the r.m.s. voltage across the terminals of the supply, V_S, across the inductor, V_L, and across the resistor, V_R. These results are obtained:

$$V_S = 5 \text{ V}, \quad V_L = 4 \text{ V}, \quad V_R = 3 \text{ V}$$

If the circuit were a d.c. one we should expect $V_S = V_L + V_R$. Explain why this does not apply in this case.

If the resistor was 100 Ω and the frequency of the supply was 1 kHz, what was the inductance of L?

9.23 Figure Q9.13 shows how the alternating potential difference applied to the ends of a coil made from thick copper wire, and the current I through the coil, vary with time t.

a) Why is the applied potential difference zero when the current has a maximum value?

b) During which of the periods of time OA, AB, BC, and CD would the source of power be (i) supplying energy, (ii) receiving energy?

How did you decide on these answers?

When the source of power *receives* energy where does the energy come from?

Oxford and Cambridge Schools Examinations Board
Nuffield A-Level Physics (1978)

9.24 A certain metal can be made to have no resistance at all (i.e., superconducting) by keeping it at a very low temperature. It will remain superconducting at this low temperature, provided it is not in a magnetic field greater than a certain magnitude; if this value is exceeded, the metal will then have a resistance.

A coil of wire, made from this metal and having sufficient number of turns to give it an appreciable self-inductance, is placed in a low temperature bath and then connected to a steady d.c. voltage source. The graph, Fig. Q9.14, shows how the current I changes with time t since switching on the voltage.

Explain, in as much detail as you can, what is happening during each of the three stages of the graph labelled A, B and C.

Oxford and Cambridge Schools Examinations Board
Nuffield A-Level Physics (1977)

9.25 In Section 38.3 a model induction motor is described in which a capacitor is used to alter the phase of the a.c. in one part of a circuit. The frequency of the a.c. is 50 Hz and the capacitance is 2500 μF. If the current in the connections to the capacitor is 5 A r.m.s., what is the r.m.s. voltage across it?

9.26 In electronic circuits a *decoupling* capacitor is often connected between the positive and negative power supply lines to prevent high frequency currents getting from one part of the circuit, via the power supply

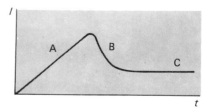

Fig. Q9.14

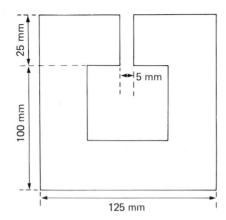

Fig. Q9.15
$\vdash\!\!\!-\!\!\!-\!\!\!-\!\!\!-\!\!\!-\!\!\!-\!\!\!-$ 125 mm $-\!\!\!-\!\!\!-\!\!\!-\!\!\!-\!\!\!\dashv$

lines, to another part. The decoupling capacitor must have a very low reactance at the frequency in question. For example, in a certain amplifier circuit, we require a decoupling capacitor which will have a reactance of not more than 0.1 Ω at a frequency of 1 kHz. What is the minimum capacitance required?

9.27 Suppose you are given three 'black boxes', each having a pair of terminals. You are told that one contains a resistor, one contains an inductor of negligible resistance, and one contains a capacitor. You have available an *audio signal generator* to provide an alternating e.m.f. at any desired frequency, an a.c. milliammeter, and an a.c. voltmeter.

a) Explain how you would find out, without opening the boxes, which box contained which component.
b) Explain how you would determine the resistance, the inductance, and the capacitance.

9.28 The tuned circuit of a radio receiver often consists of a fixed-value inductor connected in parallel with a variable capacitor. Inevitably there is stray capacitance in the circuit which is effectively an extra capacitance connected in parallel with the variable capacitor. In a typical tuned circuit the stray capacitance amounts to about 50 pF. Suppose such a circuit has an inductance of 120 μH, and is to be tunable over a frequency range from 0.6 MHz to 1.5 MHz (the medium wave band). What must be the maximum and minimum values of the variable capacitor?

9.29 The apparatus of Fig. 37.7 can be used to find an approximate value for the magnetic permeability of the transformer alloy used to form the ring. Here is the data from an experiment:

  Number of turns in the large coil: 240
  Frequency of sinusoidal a.c. supply: 1000 Hz
  Current in large coil: 10 mA r.m.s.
  Length of alloy ring: 400 mm
  Area of cross-section of ring: 100 mm²

  Number of turns in 'girdle' coil: 10
  e.m.f. induced in this coil: 0.14 V peak value

a) Calculate the amplitude (peak value) of the flux in the core, from the e.m.f. induced in the 'girdle' coil (see Section 37.2).
b) Calculate the amplitude of the m.m.f. produced by the coil (remember that, for any sinusoidally varying quantity, peak value = $\sqrt{2}$ × r.m.s. value).
c) Hence calculate the reluctance of the alloy ring.
d) Hence calculate the permeability, μ, of the alloy, and finally its *relative permeability*, which is the ratio μ/μ_0.

9.30 Figure Q9.15 shows the core of an electromagnet made from sections of transformer core alloy. Its coil has 300 turns (not shown). Estimate the current required to provide a magnetic field whose mean flux density is 0.1 T in the air-gap. All the alloy sections have a total area of cross section of 625 mm². We make the simplifying assumption that the electromagnet can be treated as a magnetic circuit, partly alloy and partly air, of constant cross-sectional area (that is, neglecting the 'bulging out' of the magnetic flux in the air-gap).

Data: permeability of free space (and air also): $4\pi \times 10^{-7}$ Wb A^{-1} m^{-1}. Relative permeability of the alloy, for a flux density of 0.1 T: 4000

a) Using the expression for reluctance,

$$S = \frac{1}{\mu}\frac{L}{A}$$

estimate the reluctance of the alloy part of the circuit, and of the air-gap.
b) Calculate the flux required to produce a flux density of 0.1 T throughout the circuit.
c) Use the relationship: m.m.f. = flux × reluctance to find the magnetomotive force required.
d) Calculate the current needed in the 300 turn coil to provide the m.m.f.

9.31 The graph (Fig. Q9.16) shows the variation with distance of the magnetic field strength, B, near a long straight wire carrying a current of 10 A. The small circle with a cross in it to the right of the graph represents the wire carrying the current perpendicularly into the paper.

a) Show on a drawing, using the symbol shown in the diagram, the shape and direction of the magnetic field near the wire. Does your drawing also represent the variation of field strength shown by the graph? Say, either how your drawing does this, or, how it could be modified to do so.
b) Using information from the graph, determine (i) the strength of the magnetic field at a distance of 0.005 m from the wire when the current is 10 A, and (ii) the current which, at a distance of 0.02 m from

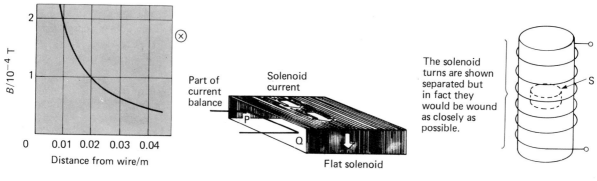

Fig. Q9.16

Fig. Q9.17

Fig. Q9.18

the wire, would give a value of B equal to 2×10^{-5} T. In each case show how you obtain your answer.

c) A second wire of length 0.3 m, which also carries a current of 10 A, is placed parallel to the original wire and at a distance of 0.01 m from it. What is the magnitude of the force on this second wire due to the current in the original wire? Show the steps of your calculation.

Oxford and Cambridge Schools Examinations Board Nuffield A-Level Physics (1980)

9.32 Figure Q9.17 shows the principle of a simple current balance which can be used to make an absolute measurement of a current (see Section 34.2). It is easier to use than a parallel-wire ampere balance because it is more sensitive and because the field deep inside a solenoid is almost uniform and so the position of the balance wire is not critical.

The flat solenoid in the diagram has 250 turns of wire in a single layer over a length of 0.2 m. The balance wire is 0.2 m long.

a) Use Eq. 37.7 for the flux density inside a long solenoid to calculate the flux density near the middle of this solenoid when a current of 1 A flows in it.

b) The direction of current around the solenoid is as shown in the diagram. If the current in the balance wire PQ flows from P to Q, what is the direction of the magnetic force on the balance wire?

c) Calculate the force on the balance wire when a current of 1 A flows in it as well as in the solenoid.

d) How great (approximately) will the force be if the balance-wire is in the same vertical plane as the end of the solenoid winding, instead of being at the middle of the solenoid?

9.33 A biologist wishes to expose a small living organism to a steady magnetic field of different magnitudes for quite long periods of time, during which he intends to monitor the temperature inside the glass

container, S, holding the specimen near the middle of a long solenoid, as in Fig. Q9.18.

a) What information will he need in order to calculate the value of the magnetic field (the B-field) inside the solenoid?

b) The length and diameter of the solenoid, and the fact that it was to be wound with a single layer of turns touching one another, had already been decided, but he had to choose between using 1 mm or 0.5 mm diameter copper wire for the windings. How would his choice affect: (i) the current to give a particular magnetic field; and (ii) the heat produced by that current in the coil? (Show the steps in your argument.)

c) What could he do to ensure that any change in temperature observed was in fact not caused by the heating effect of the current in the coil?

Oxford and Cambridge Schools Examinations Board Nuffield A-Level Physics (1979)

9.34 In certain experiments to demonstrate the magnetic deflection of an electron beam ('cathode rays') large, open, plane circular coils are used. A field strength of the order of 1 mT at the centre of each coil is required, and the coil diameter must be about 150 mm. Calculate how many turns such a coil must have to give the required field strength with a current of 1 A.

9.35 In a type of galvanometer, now obsolete, called the *tangent galvanometer*, a plane circular coil is mounted in a vertical plane, and this plane is set in the earth's magnetic meridian (that is, a vertical plane passing through the earth's magnetic North and South poles). A magnetic compass at the centre of the coil is deflected towards East or West when a current flows in the coil.

a) The strength of the current in the coil is proportional to tan θ, where θ is the angle through which the compass needle is deflected (from its original

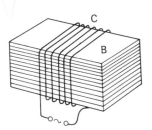

Fig. Q9.19

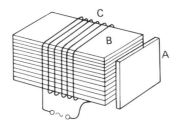

Fig. Q9.20

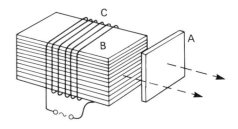

Fig. Q9.21

northerly direction). Why is this? (This, of course, is why it is called a *tangent* galvanometer.)

b) To deflect the compass needle through 45° the strength of the coil's field at the centre must be equal in magnitude to the horizontal component of the earth's field. If this deflection is produced by a coil of 20 turns having a radius of 60 mm, when a current of 86 mA flows in the coil, what is the strength of the horizontal component of the earth's magnetic field?

9.36 Figure Q9.19 shows a laminated iron bar B with a coil C of many turns wound around it, the ends of the coil being connected to an a.c. supply.

a) Why would dividing the r.m.s. value of the supply voltage by the resistance of the coil give quite a wrong value for the r.m.s. current in the coil?

b) A thick aluminium plate A is placed squarely over one end of the bar as shown in Fig. Q9.20. Copy Fig. Q9.20 and add to it an indication of the pattern of any current flow in the plate, and then explain your sketch.

c) The plate is now moved to cover half the end face of the bar B as shown in Fig. Q9.21. Explain any differences you think there may be now between the alternating magnetic fields at X (in front of the uncovered bar) and at Y (in front of the part covered by the plate).

d) With the plate in place as in (b), is the alternating current in the coil likely to be the same, be larger, or be smaller than when the plate is not present? Give a reason.

Oxford and Cambridge Schools Examinations Board
Nuffield A-Level Physics (1977)

9.37 This question concerns the unit for μ_0, the magnetic permeability of free space. The usual unit is henry per metre (H m^{-1}). In Chapter 37 the alternatives Wb A^{-1} m^{-1} and N A^{-2} also appeared. Prove that these three forms are in fact the same. The following may help.

The *henry* (H) is the unit of inductance, and inductance is defined in Section 35.7.

The *weber* (Wb), the unit of magnetic flux, first appears in Section 35.2.

The *volt*, (V), unit of potential difference, is defined in Section 23.2

9.38 In an induction motor of the *shaded-pole* type, as described in Chapter 38, the presence of the thick copper 'shading' ring in front of part of a magnet pole causes a slight phase-delay in the alternating magnetic field. Why is this?

a) Suppose the alternating magnetic flux, due to the magnet, which passes through the copper ring, is represented by the expression

$$\Phi = \Phi_0 \sin \omega t$$

Write an expression for the e.m.f. induced in the ring (refer back to Eq. 358.)

b) The e.m.f. induced in the ring causes a current to flow in the ring, and because the self-inductance of the ring is negligible, this current will be in phase with the e.m.f. This induced current gives rise to a magnetic flux, Φ', directly proportional to the induced current. What is the phase-relation between this induced flux, Φ', and the original flux, Φ, due to the magnet?

c) The induced flux Φ' has a much smaller amplitude than the original flux Φ due to the magnet. Draw a graph of Φ and Φ' against time, and on the same axes show how the total flux $(\Phi + \Phi')$, varies with time.

9.39 Figure Q9.22 shows a length of coaxial cable, made of an outer sheath of conductor wrapped round a long roll of insulator, inside which is a conducting wire running along the axis of the whole cable. The capacitance of each metre of cable is 200×10^{-12} F. Some time before the instant shown, the switch S was switched to the 1.5 V battery for 10^{-9} s and then back to the position shown. At the time shown, in the region BC the inner wire carries positive charge and the outer conductor carries negative charge. The regions AB, CD are uncharged.

a) Give a reason why no electricity has yet reached the distant place E.

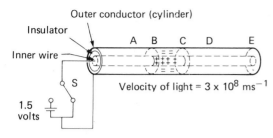

Fig. Q9.22

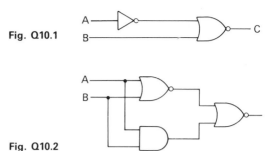

Fig. Q10.1

Fig. Q10.2

b) Estimate roughly the length of the charged region *BC* (the insulator does actually reduce the velocity, but you may ignore this.)

c) Make a sketch showing the directions of the electric field around the central wire within the region *BC*.

d) Explain why the charge on each conductor in the region *BC* is about 90×10^{-12} C (again neglect any effect of the insulator).

e) What do you think will happen when the charged region reaches the end of the cable beyond *B* if the conductors end abruptly and are not connected to anything else or to each other?

Oxford and Cambridge Schools Examinations Board
Nuffield A-Level Physics (1970)

Unit Ten

10.1 With inexpensive microcomputers, digital data are often recorded on magnetic (cassette) tape at a rate of 300 baud. There is an interval between one data byte and the next, and this interval is of $2\frac{1}{2}$ bits' duration. How many bytes (to the nearest whole number) are recorded per second?

10.2 Write a truth table for the system shown in Fig. Q10.1, to show how the output C depends upon the inputs A and B.

10.3 Draw a diagram showing how one inverter and one NAND-gate can be combined to produce a function having this truth-table:

A	B	C
0	0	1
0	1	1
1	0	0
1	1	1

10.4 Write a truth table for the function of the circuit shown in Fig. Q10.2. In this circuit the two inputs A and B are fed to a NOR-gate and also to an AND-gate: the outputs of these two gates are then fed to another NOR-gate. This function is known as the EXCLUSIVE-OR function. Explain why it has this name.

10.5 Figure Q10.3 shows two NAND-gates cross-coupled to make a bistable circuit ('flip-flop').

a) If both inputs, R and S, are in logic state 0, what are the possible states for the outputs X and Y? (Consider what happens if each of the other inputs to the NAND-gates are 0 or 1.)

b) If R = 0 and S = 1, there is only one possible state for the outputs X and Y. What is it?

c) If the circuit is in the stable state with R = 0 and S = 1, what happens when R is made equal to 1 momentarily?

d) If R = 1 and S = 1, what states can X and Y have?

10.6 Design a circuit, using AND-gates and inverters, which has two inputs X and Y, and four outputs ABCD, and which will function according to this truth-table:

X	Y	A	B	C	D
0	0	1	0	0	0
0	1	0	1	0	0
1	0	0	0	1	0
1	1	0	0	0	1

The circuit is known as a *two-line to four-line binary decoder*, and is a kind of circuit commonly used in computers. Decoder circuits of this kind are usually made as integrated circuits on a single silicon chip.

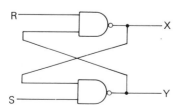

Fig. Q10.3

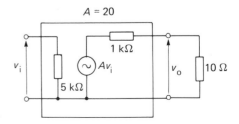

Fig. Q10.4

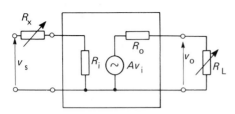

Fig. Q10.5

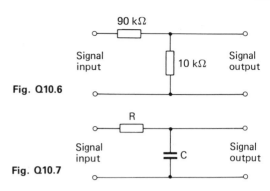

Fig. Q10.6

Fig. Q10.7

10.7 A certain a.c. voltage amplifier has these small-signal parameters: voltage gain, A, = 20; output resistance, R_o, = 3 kΩ. If a resistive load of 1 kΩ is connected to its output, what voltage will be developed across the load when the input voltage is 3 mV?

10.8 Figure Q10.4 shows an a.c. voltage amplifier connected to a load of 10 Ω resistance. The voltage gain, A, of the amplifier without load is 20. Calculate the ratio v_o/v_i when the 10 Ω load is connected. Why would this amplification system probably be unsatisfactory?

10.9 For the circuit of Fig. Q10.4 calculate the ratio: output current/input current. Use this result and the answer to Question 10.8 to decide whether or not the system provides any power gain.

10.10 Figure Q10.5 is the equivalent circuit, for small signals of an a.f. amplifier. An audio signal generator of negligible internal resistance provides a signal of constant voltage amplitude (v_s = 0.05 V pk-to-pk). The output voltage amplitude v_o is measured using a c.r.o. When R_x, a variable resistor in series with the input and R_L, a variable output load resistor have the values shown, the output voltage v_o is as given below:

R_x/kΩ	0	5	5
R_L/kΩ	∞	∞	0.5
v_o/V pk-to-pk	2.0	1.0	0.5

Find the voltage gain A, the input series resistance R_i and the output resistance R_o.

10.11 Figure Q10.6 shows the circuit for a simple *attenuator*. It is simply the *voltage divider* circuit we first met in Section 25.2.

a) If the voltage across its input terminals is v_i, what is the voltage across its output terminals (assuming that no load is connected to the output)?

b) Using the expression which appears near the end of Section 25.2, find the *voltage gain* of this attenuator and express it in decibel.

c) If a load of resistance 10 kΩ were connected across the output terminals the output voltage would become a smaller fraction of the input voltage than

it was before. What is the voltage gain now, in decibels (to the nearest whole number)?

10.12 Figure Q10.7 shows a simple *low-pass filter* circuit, so called because it allows signals of low frequencies to pass, but progressively attenuates signals of higher and higher frequencies. The *cutoff frequency* is defined as the frequency at which the reactance of the capacitor equals the resistance of the resistor. At this frequency the peak voltage across the capacitor equals the peak voltage across the resistor, but because there is a $\frac{1}{4}$ cycle phase–difference between these voltages, each voltage is $1/\sqrt{2}$ of the input voltage (see Section 36.3).

a) What is the voltage gain, in decibels, of this filter at the cutoff frequency?

b) If R = 4.7 kΩ, what value must C have to give a cutoff frequency of 10 kHz?

10.13 Suppose the circuit of Fig. 42.6 is to be used as an a.c. signal amplifier, for signals having frequencies in the range 10 Hz to 100 kHz.

a) Use the data of Fig. 42.4 to estimate the voltage gain *expressed as a ratio* (see Section 41.9.) at the limiting frequencies of 10 Hz and 100 kHz.

b) As explained in Section 42.5, the feedback ratio for the amplifier circuit of Fig. 42.6 is approximately 1/25. Use the theory developed in Section 42.4 on feedback, and your answers to (a), to estimate the magnitude of the overall voltage gain of the amplifier at 10 Hz and 100 kHz. Give the answers both as ratios, and in decibels.

c) If the overall gain of the amplifier is not to vary by more than 3 dB from its value over most of the frequency range, what is the approximate upper frequency limit for the amplifier? Would the amplifier be suitable for use as an audio frequency amplifier?

10.14 A certain non-inverting amplifier has a voltage gain of 60, before any feedback is applied. A feedback circuit is then added which reverses the sign of the

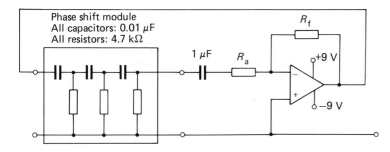

Fig. Q10.8

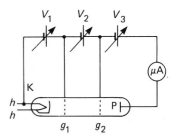

Fig. Q11.1

voltage to provide negative feedback, and provides a feedback factor of magnitude 0.05. Calculate the overall voltage gain of the amplifier with feedback.

(*Hint*: For a non-inverting amplifier A has a positive numerical value; for a feedback circuit which reverses the sign of the voltage B has a negative value.)

10.15 A certain inverting amplifier has a voltage gain of -100. Negative feedback is to be added in order to reduce the overall gain to -5. What must the feedback factor be?

10.16 A phase-shift oscillator is being made, using the circuit of Fig. Q10.8. The three-section phase-shift module gives an output voltage, when no load is connected to its output, which is 1/29 of the input voltage to it, at the frequency which causes phase reversal (and this will be the frequency of oscillation). To avoid appreciable loading of the output of the phase-shift module we should make the input resistance of the amplifier at least five times as large as the final resistor in the phase-shift module. What are suitable values for R_a and R_f, so that positive feedback, and hence oscillation, will occur?

Unit Eleven

11.1 A truck, A, of mass 3000 kg is shunted along a railway line at 8 m s^{-1} when it collides, and links with, truck B of mass 5000 kg that is travelling in the same direction at 4 m s^{-1}. What is their final common velocity? Also, calculate the energy lost in the process.

11.2 Repeat Question 11.1, but reverse the direction of B so that it has a head-on collision with A. If, as before, they link together calculate their final common velocity, stating carefully its direction of course. Also, what is the energy lost this time?

11.3 An air track vehicle of mass 0.2 kg is moving at 5 m s^{-1} when it collides elastically with a stationary vehicle of mass 0.5 kg. Calculate the velocities of the vehicles after the interaction. Calculate the fraction of

the original energy that is transferred to the second vehicle.

11.4 An air track vehicle, A, of mass 0.3 kg is travelling at 2 m s^{-1} towards another vehicle, B, of mass 0.2 kg, which is moving at 5 m s^{-1}. As a result of the collision the velocity of A is reversed. What is the final velocity of B and calculate the loss of energy.

11.5 A gas-filled tube or thyratron produced commercially for the demonstration of the Franck–Hertz experiment with mercury vapour is shown schematically in Fig. Q11.1.

As V_2 is raised from 0 to 20 V, the sensitive ammeter A shows three distinct current minima at intervals of 4.9 V.

a) Explain this.
b) Consider the case when V_1 is 1 V, V_2 is 10 V and V_3 is -1 V. In what region of the space in the tube can the electrons give up energies of 4.9 eV in inelastic collisions? Sketch the graph of current against V_2 in the voltage range 0 to 10 V.
c) Consider now the case when the three voltages are 1 V, 20 V, and -1 V. Sketch the graph which you would expect to obtain in this case.

11.6 Collisions between gas molecules in the air rarely cause ionization. Use the histogram of Question 4.6 to explain why this is so.

What is the minimum relative velocity of two nitrogen molecules which, in collision, might lead to the ionization of one of them?

11.7 Calculate the energy associated with each of the following photons:

a) *X*-radiation at a wavelength of 0.15 nm,
b) ultra-violet radiation at 121 nm,
c) a television transmission at 200 MHz,
d) a radio transmission at 2×10^5 Hz.

11.8 A molecule of silver bromide, a light sensitive compound which is to be found in many photographic films, can be decomposed by about 1.6×10^{-19} J. What is the minimum frequency of the photon which would decomposed a molecule of silver bromide?

11.9   Photographic paper of a type known as bromide paper can be handled quite safely in red light but not in white light. How do you account for this?

11.10   A simple laboratory method for obtaining almost monochromatic light is to heat a sodium salt in a bunsen flame. The yellow light emitted comprises, in fact, two discrete wavelengths at 589.0 and 589.6 nm. We shall assume a wavelength of 590 nm and that light of this wavelength is a consequence of the return of electrons from the first excited state to the ground state.

a) What is the photon energy involved?
b) What is the temperature of a gas whose molecules have an average kinetic energy equal to that of the photon energy?
c) The average gas kinetic energy in the gas flame may be taken as about 2000 K. How then is it possible for excitation to occur at all?

11.11   In an experiment with a photocell it was found that electron emission could be prevented by the application of a suitable reverse voltage. This voltage varied with the wavelength of the light used. From the results given in Table Q11.1 plot a graph of 'stopping potential' against frequency and use it to determine (a) the threshold frequency and (b) the work-function of the material of the cell cathode. Find a value of Planck's constant.

Table Q11.1

Wavelength/nm	365	436	470	530	589	605
Reverse voltage/V	2.10	1.60	1.32	1.05	0.80	0.70

11.12   A metal plate of material with a work function of about 5 eV is placed at a distance of 2 m from a weak source of monochromatic light of wavelength about 500 nm. The light output of the light source is 10^{-2} W. Assuming that light behaves as a wave, estimate how long it would take for a single photoelectron to be emitted. You may assume that an individual photoelectron collects its energy over a circular area of the plate centred on the atom from which it will come. This area may be assumed to have a radius of about 10 atomic diameters – say 10^{-9} m. Comment on your estimate.

11.13   In 1909 G. I. Taylor devised an experiment to test whether light photons could be responsible for the diffraction patterns which are usually regarded as a characteristic feature of the wave model for light. This question is concerned with a similar situation.

a) The retina of the human eye can readily detect energy arriving at a rate of about 10^{-12} W. How many photons per second does the retina receive under these conditions if the wavelength of the light used is 600 nm.
b) On average, how far apart will these photons be?
c) Light of this intensity nevertheless produces clear diffraction effects. State the problem which G. I. Taylor was considering.

11.14   According to the wave theory the intensity of the light emitted from a point source varies inversely as the square of the distance from the source. If photons are emitted in random directions from a point source of radiation, show that the number passing through unit area (i.e., the intensity) must also be governed by an inverse square law.

11.15   The spectrum of singly ionized helium is very similar to that of hydrogen. It includes lines with wavelengths of 303, 256, 243, and 237 nm. Compare this series with the corresponding series for hydrogen. If the ionization energy for hydrogen is 21.8×10^{-19} J, how much energy would you expect to supply if you wished completely to ionize this helium?

11.16   What are the differences and the similarities between a photon and an electron?

11.17   The H_α line of hydrogen is emitted during an electron transition from the energy level $n = 3$ to $n = 2$. Calculate the wavelength of the H_α line. Is this in the visible spectrum?

The first of the Lyman series corresponds to an electron transition from $n = 2$ to $n = 1$. Calculate the wavelength. In which part of the spectrum does this line occur?

11.18   What is the de Broglie wavelength associated with an electron accelerated through 20 kV, which is a typical voltage for the tubes used in colour television sets?

11.19   What is the de Broglie wavelength associated with an electron which has been accelerated in an electric field of 100 V?

How would this electron behave if it were to meet a slit which was 1 mm across?

The following three questions extend the arguments given in Chapter 45 to atoms with proton numbers in excess of 1.

11.20   a) How is the electrical potential energy (E) of an electron at distance r_0 from a nucleus related to the proton number (Z) of the nucleus?
b) How does the wavelength (λ) associated with the electron depend on the distance (r_0) at which all the

energy of the electron is potential?

c) Recalling that the kinetic energy and the momentum of a particle are related by the equation $E_k = p^2/2m$ find a relationship between the de Broglie wavelength of an electron and its mean kinetic energy (E_k).

d) Assuming, as before, that the mean kinetic energy is approximately equal to the potential energy when at distance r_o, find how the mean kinetic energy of an electron is related to the proton number Z.

11.21 a) Using the result derived in the last question, estimate the ionization energy for helium ($Z = 2$).

b) In fact the ionization energy is 24.6 eV. Considering the numbers of charged particles involved, explain the discrepancy.

11.22 a) The next element in the periodic table to He is Li for which $Z = 3$. Estimate its ionization energy.

b) It is, in fact, 5.4 eV. Advance an explanation for this discrepancy.

c) What light does this throw on the marked differences in chemical behaviour between helium and lithium?

11.23 What is the kinetic energy of an electron confined within a spherical box which has a radius of 10^{-11} m? Compare this value with the potential energy provided by the electrostatic field of a proton at this distance. Which is the larger and by how much? What would be the consequence of this confinement?

Unit Twelve

12.1 A 6 MeV α-particle has a range in air (at s.t.p.) of 4.7 cm. In this distance it loses its initial energy by ionizing atoms in its path. If it loses 30 eV at each encounter, how many atoms does it ionize before coming to rest?

Explain why an α-particle travels further if:

a) the pressure of the air is reduced;

b) a gas such as hydrogen is used instead of air.

12.2 A 0.18 MBq (5 μCi) alpha source is placed so that all the particles it emits pass into an ionization chamber containing air which brings the particles to rest. The ionization current, measured with a d.c. amplifier, is 2.96×10^{-9} A. Calculate the number of ions being produced per second, and the number of ions produced by one α-particle. If the ionization potential is 30 eV, calculate the initial energy of an α-particle.

12.3 A radioactive substance, labelled R, is set up in front of a G–M tube connected to a scaler. The scaler is set counting at the same moment that a stopwatch is started. Every time the second hand reaches a minute, a boy records the reading of the scaler without stopping it counting. He tabulates the readings as shown below:

Time	0	1	2	3	4	5	6	7	8
Scaler reading	0	6015	8026	9016	9401	9541	9802	9636	9673

a) One of the readings is suspicious. Which one? What did the boy probably mean to record?

b) Having made this correction, work out from the readings what was the count rate in each successive minute, giving your answers in number of counts per minute.

c) What do these readings suggest is happening?

d) Plot the count rates on a suitable graph and deduce a value for the half-life of the radioactive substance R. (The mean of at least two values is required.)

e) Jill says that a better result would be obtained if the readings were extended over a further eight minutes and that the boy stopped too soon. Is this a good idea or a bad one? Give reasons.

f) Jack says that there are random fluctuations in radioactive experiments. He says a better result would be obtained if more counts were taken and that it would be better to count at five minute intervals, not one minute intervals.

Give your reasons why this is a good or a bad idea.

g) There is a slight increase in count rate during the eighth minute. Is this significant? Give the reason for your answer.

h) Suggest any way in which you think the experiment might be improved.

Nuffield O-level examination; reproduced by permission of the Oxford and Cambridge Schools Examination Board

12.4 a) The success of the provision of medical care for mothers and babies during this century is shown by the fall in the death rate of babies before and soon after birth. The table shows the decrease in infant mortality in the U.K. since 1901.

Year	1901	1911	1921	1931	1941	1951	1961	1971
Mortality per 1000 live births	144	111	81	60	44	32.5	22.5	16

Is this an example of an exponential decay? Give details of your argument.

b) The following table shows how the population of the world has changed over a period of time

Year	1800 1850 1900 1950 1970 1978
Population/10^6 persons	910 1130 1600 2510 3375 4219

Is this an example of exponential growth? Give details of your argument.

12.5   Radon-220 is a radioactive gas with a half-life of 54 s. What fraction of the mass of a sample of this gas remains as Radon after

a) 108 s;
b) 216 s;
c) 540 s?
d) A radioactive isotope, ^{124}Sb, having an initial activity of 2 m Ci (7.4×10^7 Bq) has a half-life of 60 days. If it decays for a year, what activity remains?

12.6   The activity of carbon found in living specimens is 2.6×10^2 Bq per kg, due to the ^{14}C present. The charcoal taken from the fire pit of an Indian camp site has an activity of 1.78×10^2 Bq kg^{-1}. The half-life of ^{14}C is 5760 years. Calculate the year the camp site was last used.

12.7   Deduce the relationship between half-life and decay constant.

Calculate the half-life of ^{226}Ra if 1 kg of radium emits 3.7×10^{13} α-particles per second and the mass of the radium atom is 3.8×10^{-25} kg.

12.8   Polonium $^{215}_{84}$Po decays with the emission of an α-particle with energy 7.38 MeV.

a) What is the daughter nuclide?
b) What is the speed of emission of the α-particle?
c) The daughter nuclide emits a β-particle; what nuclide results?

12.9   Imagine you are standing at the edge of an atom (where the outermost electrons are) and are looking in towards the nucleus. Would it appear to be about the same size as an apple held, (a) 3 m; (b) 30 m; (c) 300 m; (d) 3000 m away?

12.10   The radius of a nucleus is taken to be the distance from the centre of the nucleus to the point at which the nuclear density is reduced by one-half. The radii of nuclei are approximately equal to $1.4 \times 10^{-15} \times A^{1/3}$ in metre. A is the nucleon number.

a) Find the proton and nucleon numbers for a platinum nucleus.

Calculate:

b) the 'radius' of this nucleus;
c) the volume of this nucleus;
d) the mass in kg of this nucleus;
e) the density of the nuclear material.

If the density of a sample of platinum is 21×10^3 kg m^{-3}, calculate:

f) the volume of a platinum atom;
g) the radius of a platinum atom.

Compare your answers to (b) and (g), and to (c) and (f) and comment on them.

h) Use the equation for the radius of a nucleus to show that the density of nuclear material is the same for all nuclei.

12.11   A piece of foil measures 0.01 m $\times$ 0.01 m and is 2×10^{-6} m thick. It is made of a metal each of whose atoms can be contained in a cube of side 2×10^{-10} m. Calculate:

a) the number of atoms in the surface layer;
b) the number of layers that make up the foil's thickness;
c) the number of nuclei in the foil.

When α-particles bombard the foil, it is found that 1 in 2500 are deflected through an angle greater than 53°. What does this tell you about:

d) the aiming error, p, for these deflected particles;
e) the ratio: $\dfrac{\text{area of the foil}}{\text{total target area of the nuclei}}$,

(Assume that the nuclei are not one behind the other in the line of fire of the particles.) Hence calculate:

f) the target area of a nucleus;
g) an estimate of the 'radius' of a nucleus.

12.12   In an α-particle scattering experiment, 1 in 10 000 are scattered through more than 20°. How many would be scattered through more than 20° if:

a) the thickness of the foil was trebled;
b) the original thickness was used but the α-particles have twice the velocity;
c) again the original foil is used but protons bombard the foil that have the same velocity as the original α-particles?

12.13   An experiment is performed with a beam of α-particles that can bombard each of three foil target in turn. When the gold foil is used, 1400 particles are scattered through 120° in 10^4 s. For the same angle and time, foil X scatters 500 particles, and Y, 40. The three foils are of such thicknesses that, for each foil, the beam passes the same number of atoms per unit surface area. Calculate the proton numbers of X and Y, *and suggest a material from which each could be made.*

12.14   a) Calculate the gravitational attraction between an α-particle and a platinum nucleus when their centres are separated by 10^{-14} m. G = 6.7×10^{-11} N m^2 kg^{-2}.

Show that the ratio: $\dfrac{\text{coulomb force}}{\text{gravitational force}}$

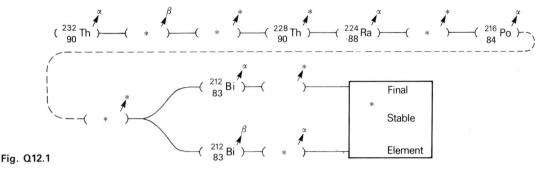

Fig. Q12.1

has a value of 2.5×10^{35} irrespective of separation.
b) Calculate the coulomb force acting when the centres of an α-particle and a platinum nucleus are separated by 10^{-14} m. Compare the answers you obtain with the weights of familiar objects, e.g., a grain of sand, a book, a bicycle, a car.

12.15 A proton and a deuteron each have the same energy and are on paths that will lead to a head-on collision with a silver nucleus. Compare their closest distances of approach to the nucleus.

12.16 An α-particle with a speed of 10^7 m s^{-1} undergoes elastic collisions with these stationary particles:

a) an electron;
b) the nucleus of a hydrogen atom;
c) the nucleus of a nitrogen atom.

What is the maximum possible speed of the struck particle in each case, and (d) what percentage of the α-particle's original energy is transferred?

12.17 How much energy does a neutron lose in a head-on collision with:

a) a hydrogen nucleus;
b) a nitrogen nucleus;
c) the nucleus of a lead atom?

What are the desirable properties of a substance to be used as a shield for a beam of neutrons? Explain.

12.18 Supply the missing data (indicated in Fig. Q12.1) in the thorium series which is one of the three radioactive decay series that occur in nature. Note that there are alternative paths towards the end; 36% of the atoms take the upper one. The final stable element is the same whichever path is taken. The upward arrows indicate the particles ejected.

12.19 Make a copy of the following table:

	^{1}H	^{2}H	^{4}He	^{9}Be	^{12}C	^{27}Al	^{206}Pb	^{226}Ra	^{238}U
A									
Z									
N									

Fill in the nucleon number A, the proton number Z, and the neutron number N.

12.20 Complete the following nuclear reactions in which ? = missing data, e = electron; p = proton, n = neutron, d = deuteron, α = alpha particle, γ = photon, $\bar{\nu}_e$ = neutrino.

a) $^{27}_{13}$Al + $\alpha \rightarrow$ p + ?
b) $^{24}_{12}$Mg + ? $\rightarrow$ $^{24}_{11}$Na + p
c) $^{238}_{92}$U + n $\rightarrow$? + γ
d) n $\rightarrow$? + e + $\bar{\nu}_e$
e) ^{14}N(α, p)?
f) ^{10}B(n, ?)^{7}Li
g) ^{198}Hg(n, d)?

12.21 a) In a helium nucleus there are two protons and two neutrons; if the protons are separated by 10^{-14} m, what is the electric force of repulsion between them?
b) Calculate the binding energy per nucleon if the mass of an atom of helium $^4_2$He is 4.00260 u.

12.22 Using Fig. 48.6 which gives binding energies *per nucleon*, find the binding energies for ^{235}U, ^{141}Ba, and ^{92}Kr. Use these binding energies to show that when the fission of ^{235}U occurs, about 200 MeV of energy is released.

12.23 In a mass spectrometer, hydrogen ions or protons are given an energy of 10 keV and they then enter a uniform magnetic field of 0.5 T in a direction perpendicular to the field. Calculate the radius of the path of these protons in this field.

The hydrogen in the ion source will also contain deuterium which will also become ionized. Calculate the radius of the path of the deuterons. (Mass of a proton = 1.66×10^{-27} kg, charge on a proton = 1.6×10^{-19} C.)

ANSWERS TO QUESTIONS

Unit One

1.5 $5.52 \times 10^3 \, \text{kg m}^{-3}$.

1.6   (a) 1.8×10^{11} coulomb per kilogram (or C kg^{-1}); (b) $5.5 \times 10^{-7} \, \text{kg}$.

1.7   (d).

1.8   48.4 km; bearing 55.7° East of North.

1.9   68 N along the plane; 188 N normal to the plane.

1.10   (c) The speed of $32.5 \, \text{m s}^{-1}$ is an average over the 100 s interval. Speeds measured by the speedometer are instantaneous speeds. (d) About 31 m s^{-1}. This is the instantaneous speed at $t = 150$ s. (e) About $0.13 \, \text{m s}^{-2}$. This is an acceleration.

1.11   (b) $12 \, \text{K min}^{-1}$; (c) $16.6 \, \text{K min}^{-1}$; (d) $14.6 \, \text{K min}^{-1}$; (e) $14.4 \, \text{K min}^{-1}$; (f) $14.4 \, \text{K min}^{-1}$; (g) $14.4 \, \text{K min}^{-1}$; (h) $14.4 \, \text{K min}^{-1}$.

Unit Two

2.1 $18 \, \text{m s}^{-1}$; 16 200 N.

2.2   By, e.g., throwing some object away in a direction away from the spacecraft – taking care not to produce rotation.

2.4   6.3 N.

2.5   The boat moves at $0.08 \, \text{m s}^{-1}$ in a direction opposite to that of the ball. When the second student catches the ball, the boat also comes to rest.

2.6   28 m; 112 m.

2.7   (a) The reaction time (thinking time) is the same in each case.

(b) An average constant force is assumed which gives, in each case, an acceleration of -7.1 m s^{-2}.

2.8 $2.56 \times 10^6 \, \text{N}$; $16 \, \text{cm}^2$; the force is increased by 4 times.

2.9 (a) None, k is a dimensionless constant.

(b) The unit of force will be such as to give an acceleration of $0.5 \, \text{m s}^{-2}$ to a mass of 1 kg. It has dimensions M L T^{-2}.

(c) Smaller by a factor of 0.5.

2.10 (a) 19.6 N; (b) $9.8x$ N; (c) $9.8 \, \text{m s}^{-2}$; (d) 9.8 m s^{-2}; (e) they have the same acceleration.

2.11 48 N; 32 N.

2.14 $\sqrt{Mg/Ad}$; $Adv^2/M - g$.

2.15 $2 \times 10^4 \, \text{kg s}^{-1}$

2.16 (a) Back in the funnel.

(b) The observer on the train sees vertical rise and fall; the observer by the track sees a parabolic curve.

2.17 In both cases the wise monkey will hang on.

2.18 2700 N. The force is the resultant, inward, of the viscous air drag and the tyre–road friction.

2.19 $8000 \, \text{m s}^{-1}$; 84 minutes.

2.20 $0.22 \, \text{N kg}^{-1}$; yes.

2.21 The 'tightness' of the nut on the bolt is determined by the torque applied and not by the force alone.

2.22 Consider the point of action of the force relative to the centre of gravity.

2.23 1.08 m behind the front axle.

2.24 (a) The vertical components of the force on each hinge are each 75 N; the horizontal components at the upper and lower hinges are -40 N and $+40$ N respectively;

(b) 85 N at 62° and at 118° to the horizontal.

2.25 The horizontal component of the force *on* the upper bracket is 64 N to the left; on the lower bracket is 64 N to the right. The vertical components are each 20 N upwards.

Unit Three

3.1 932.5 N.

3.2 (a) 4.2 MJ; (b) 2000 N; (c) 1400 N; (d) Constant gradient.

3.3 4.6×10^5 J; (b) Heat in brakes; (c) $-80 \, \text{m s}^{-2}$; (d) 4800 N.

3.4 (a) About 77 J; (b) About 19 m s^{-1}.

3.5 (a) 100 N; (b) 50 kg; (c) 2 m s^{-2}; (d) 200 J; (e) 160 J; (f) 2.8 m s^{-1}.

3.6 14.

3.7 $600 \, \text{m s}^{-1}$.

3.8 $\dfrac{4M}{m}\sqrt{lg}$.

3.9 $h/4$.

3.10 32 kg.

3.11 (b) P has radial acceleration; (c) P has both radial and tangential acceleration.

3.12 (a) 0.36 N; (b) 12 rad s^{-1}; (c) 0.432 J.

3.13 2.05×10^4 Pa.

3.14 0.4 K.

3.15 Assuming that all kinetic energy is available to the pellet, the temperature rise of the pellet of lead is about 40 K.

3.16 $275 \, \text{m s}^{-1}$ (on same assumption as previous question).

3.18 8 minutes.

3.22 16.3 hours; 128 hours. As the water cools to 4 K a convection current takes the denser water to the bottom of the tank. Thereafter water which is cooled below 4 K will collect at the top of the tank so that the freezing will start at the top. This freezing will exert forces on the tank itself as the water turns to ice and expands.

Unit Four

4.1 (a) 16.7 mol; (b) 0.0224 m³; (c) 3738 J, 3750 J, 12 J mol^{-1} K^{-1}.

4.2 (a) 2.7×10^5 N m^{-2}; (b) 5.5 cm.

4.3 (a) 1.7×10^5 N m^{-2}; (b) 3.9 mol; (c) 0.022 m².

4.4 500 m s^{-1}. Remember that the r.m.s. speed does not depend on pressure but the mean free path does.

4.5 8 km; 400 m s^{-1}.

4.6 (b) 490 m s^{-1}.

4.7 (a) 484 m s^{-1} and 1930 m s^{-1}; (b) 1.6×10^5 K, 9.7×10^3 K.

The distribution of speeds among the molecules is such that the hydrogen molecules have a high probability of eventual escape. At the 'top' of the atmosphere (say 200+ km) the temperature of the gases is in the order of 1000 K which greatly increases the chance of escape for the faster hydrogen molecules.

4.8 (a) 3.01×10^6 Pa; (b) 3.11×10^6 Pa; (c) 0.76 m.

4.9 (a) 9.4 m; (b) 0.95 m.

4.10 The air pressure between the vehicles falls below atmospheric.

4.11 (a) $\eta v = \frac{2}{9} r^3 g(\varrho_1 - \varrho_2)$; (b) 8.6 cm s^{-1}.

4.12 0.11 cm³ s^{-1} (6.6 cm³ per minute).

4.13 (a) The student wrongly assumed that the temperature difference between the glass surfaces was 20 °C; (b) 270 W.

4.14 The maximum and minimum sizes are 800.19 $\times$ 400.09 mm and 799.52 $\times$ 399.76 mm; so cut the glass a little under-size.

4.15 (a) To allow energy transfer through the thermopile;

(b) No *net* transfer of energy;

(c) Any temperature difference will cause energy flow;

(d) 19.3%. Other energy losses (which?) prevent it.

4.16 (a) 2^{N_A}; (b) $k \ln W = 5.8 \, \text{J K}^{-1}$; (c) Many more ways of achieving this distribution than any other.

4.17 348 K.

4.18 (a) The change in the reaction rate depends on the ratio $e^{-(E/kT_1)}/e^{-(E/kT_2)}$. If E/kT is itself large (say 20), the ratio can be quite large for quite small differences of temperature.
(b) 8.9×10^{-20} J/particle.

4.19 (a) The ratio of the lengths remains the same (0.6) but all the lengths are doubled;
(b) Steeper but with the sum total of the bar lengths unchanged;
(c) Shallower but with the sum total of the bar lengths doubled.

4.20 (a) Hooke's law obeyed up to the elastic limit (340 N);
(b) The wire had acquired a permanent set having yielded;
(c) (i) $6.8 \times 10^{10} \, \text{N m}^{-2}$; (ii) $75 \times 10^6 \, \text{N m}^{-2}$.

4.21 (a) 3.75 mm, 7.5 mm; (b) 8.9 kg.

4.22 $5 \, \text{m s}^{-2}$.

4.23 2610 rev min^{-1}.

4.24 Remember that Jupiter is by far the most massive planet.

4.25 (a) 27; (b) 27; (c) 9; (d) 27; (e) 27; (f) 9; (g) 9; (h) $\sqrt{3}$;

4.26 9 times larger.

4.27 (a) $2\sqrt{2}r$; (b) $8 \times 2^{3/2} \times r^3$; (c) 14; (d) 8; (e) 6; (f) 8; (g) 2; (h) 1, 3; (i) $\frac{16}{3}\pi r^3$; (j) 0.74.

4.28 (a) 36.1 nm; (b) 4; (c) 8940 kg m^{-3}.

4.29 25.6 nm, 18.2 nm, 12.9 nm. Since the last spacing is twice the first, it represents a second order reflection with $n = 2$.

4.30 (a) 0.0003 mm;
(b) There must be some other mechanism since the capillary rise in a tube of diameter 0.01 mm is only 3 m;
(c) 10^6 Pa.

4.31 (a) The excess pressure decreases with increasing radius;
(b) the smaller bubble transfers air to the larger.

4.32 (a) Across the grain;
(b) Plywood, unlike a sheet of timber, will have similar properties however the sheet is stressed;
(c) smaller;
(d) unlike aluminium, wood is not homogeneous.

4.33 (a) (i) $700 \times 10^3 \, \text{N m}^{-2}$, (ii) 0.07 to 1; (b) (i) 0.63 to 1, (ii) disaster.

Unit Five

5.1 (a) 1100 Hz.
(b) $220 \, \text{m s}^{-1}$.

5.2 (a) Water waves; VHF radio; sound; infrared; light; gamma.
(b) Gamma; light; infrared; VHF radio; sound; water.

5.3 The diffraction of the waves at each slit leads to the waves spreading out after passing through the slits; the waves can thus overlap and interfere.

5.4 (a) They must be being transmitted in phase with each other and thus interfere at X to give an enhanced signal.
(b) The signal will again have an amplitude twice that from either station on its own.
(c) More than 100 m. $\text{AY} - \text{BY} = 200 \, \text{m}$ – think about the geometry of the situation.

5.5 (a) The dippers produce waves that are spatially and temporally coherent; the limited diffraction at the slots in the barrier produces an inferior pattern.
(b) A similar situation occurs with the two loudspeakers; in this case the two slots would be too far apart to produce an interference pattern that can be heard.
(c) Two sources of light are never spatially or temporally coherent. The two slits must be close due to the short wavelength.

5.6 $20 \, \text{m s}^{-1}$; no, such a pulse will travel faster as it moves up the rope. Can you think of a reason why?

5.7 89.2 N.

5.8 (a) Your graph should be a 'square wave' shape spanning the two times 4 to 5 s, and of a height of $0.2 \, \text{m s}^{-1}$.
(b) Such a graph assumes infinitely large accelerations – why is this impossible to achieve?

5.9 0.64 m.

5.10 (a) $4.1 \, \text{m s}^{-1}$;
(b) $8.2 \, \text{m s}^{-1}$;
(c) The pulse covers double the length at double the speed. One would not expect to find this to be true unless the extension of the spring was many, many times the original length.

5.11 (a) Here are some points you might make: if the aperture is small enough light in fact does not travel in straight lines; refraction using the particle model gives the wrong result for the change in speed; interference effects may be difficult to show but they *do* exist.
(b) Colour depends on forced oscillations and resonance amongst the atoms of the coloured material; polarization can only be explained by assuming light is a transverse wave – the particle model fails completely. Members of the electromagnetic spectrum are linked by a common speed; differences can be related to differences in wavelength and frequency.

5.12 (a) 21 cm s^{-1};   (b) 14 cm s^{-1}; (c) 20°; (d) 42°;
(e) Waves are internally reflected.

5.13 4710 km.

5.14 (a) You ought to mention interference somewhere here.
(b) This essentially depends on the difference in 'slit spacing' to produce patterns of comparable size.
(c) Common speed; possibly their transverse character.

5.15 Both your answers ought to have mentioned diffraction and the way the extent of the diffraction depends on the size of the wavelength of the wave.

5.16 330 m s^{-1}.

5.17 (a) Absence of reflected waves.
(b) Increase frequency or speaker separation.
(c) In either case the speakers must be *at least* 2 wavelengths apart.
(d) The speakers have been connected in antiphase.

5.18 (a) 5×10^{-7} m.
(b) White light consists of a range of wavelengths.
(c) Think about the different positions of the maxima from the different wavelengths in the white light.
(d) Different parts of the light source produce independent, overlapping interference patterns.

5.19 (a) The transmitted waves are polarized in a plane at right angles to the plane of the receiving dipole.
(b) The amplitude of the received wave is $A \cos^2 45$, where A is the amplitude of the original wave. Thus the intensity received is one quarter the transmitted intensity.

5.20 There must be *two* refracted waves, polarized in planes at right angles to each other. (By arranging a prism of such an angle that *one* of these waves is internally reflected, very precise polarizers and analysers can be constructed. They are called Nicol prisms.)

5.21 24 m s^{-1}.

5.22 The graphs are in fact an upside down picture of the tracks!

5.23 (a) 0, 6, 12 units; (b) 3, 9 units; (c) 3, 9 units; (d) 0, 6, 12 units; (e) 1, 5, 7 and 11 units.

5.24 (a) Period = 2.5 s; amplitude and energy are unchanged.
(b) Period = 2.5 s; energy is halved and amplitude reduced by $1/\sqrt{2}$.

5.25 (a) None; (b) Halved; (c) Reduced to 1/4; (d) Halved.

5.26 (a) 4 s; (b) 1.57 s^{-1}; (c) 7.85 cm s^{-1}; (d) 24.7 cm s^{-2}.

5.27 (a) 6 J; (b) 3 J; (c) 3 J.

5.28 This is a matter of following instructions.

5.29 (a) 1 N, (b) (i) 2 N, (ii) 2.5 ms^{-2}; (c) (i) 0.2 J, (ii) 0.25 J; (d) 0.05 J.

5.30 Walking with the natural rhythm of the swing of your leg will enable you to make best use of the energy of the swing – as in resonance.

5.31 This is essentially a discussion of forced oscillations and resonance, and your answer should cover all the main features.

5.32 You ought to have identified another pitching mode and also pitching modes across the diagonal. There is also the possibility of a rolling motion as each spring compresses in turn.

5.33 This is a matter for observation – it would spoil the experiment to tell you what will happen!

5.34 (a) 343 m s^{-1};
(b) Using a loudspeaker, the directly radiated energy is as intense as that reflected up the tube. If you think about it carefully you will find that these two waves are 180° out of phase and thus interfere destructively.

5.35 0.53 mm; 0.36 mm; 0.24 mm. You may find that this does not quite match what is found in practice because the different springs are rather differently constructed and may not in fact be of the same tension.

5.36 (a) (i) The wavelength should be a little less than the boat's length. (ii) A little less than 18 m.
(b) $c = 4$ m s^{-1}, so $f = 0.22$ Hz, if $\lambda = 18$ m.
(c) The shallower water should have a wave of shorter wavelength than the deeper water has, because c increases with depth.

5.37 (a) Sound intensity is reduced to $\frac{1}{4}$ of its previous value.
(b) Sound intensity is proportional to (amplitude)^2.

5.38 The resolving powers of the grating are different in the two cases. In the first case, if light passes through the whole grating, two wavelengths 0.12 nm apart will be resolved. In the second it will just resolve wavelengths 1 nm apart.

5.39 (a) Assuming he wishes to resolve detail 1 cm across, the lens will have a diameter of 5 m.
(b) Yes, he would be able to do so.
(c) It could be conducted at night, but resolving power could be less.

5.40 (a) Think about the *refraction* of light by such a grating.
(b) A great deal of light energy is concentrated into one side of the grating.
(c) 1.5×10^{-6} m.
(d) Your diagram should show light being reflected off the oblique faces.
(e) It would be useful for infrared radiation

which cannot be transmitted by the glass.

5.41 (a) The resolving power required is 6.7×10^{-6}; the resolving power of the telescope is 10^{-7} so it should be able to resolve such a pair of stars. (The effective resolving power may be less than this owing to atmospheric conditions.)

(b) The absence of atmospheric turbulence, etc., will mean that any instrument will approach its theoretical resolving power in use. Earth-based instruments hardly ever do.

5.42 (a) $23.6°$; (b) $53.1°$;

(c) $\sin \theta > 1$ for 3rd and subsequent orders.

(d) With extreme tilting the path difference between adjacent waves can approach *twice* the separation of the silts.

5.43 (a) Your sketch should follow the envelope of the curve drawn.

(b) 2×10^{-4} m.

(c) The total slit width is 0.8 mm so the minima should occur at $\pm 0.75 \times 10^{-3}$ radians; $\pm 1.5 \times 10^{-3}$ radians; $\pm 2.25 \times 10^{-3}$ radians.

5.44 (a) The frequency of the waves corresponds to the natural frequency of vibration of the HCl molecule.

(b) The gas warms up.

(c) 546 N m^{-1}.

Unit Six

6.1 (a) Current is 20 A and the fuse blows;

(b) Kettle works at well below the rated power at a current of 5.2 A;

(c) 52;

(d) 93%.

6.2 (a) 0.4, 0.6, 1.8, 2.4 and 3.6 KΩ, (b) 4.7 kΩ is satisfactory.

6.3 (a) 6.2 V; (b) 3.2%.

6.4 (a) 12 V; (b) 8 V; (c) 2 A; (d) 24 W; (e) 8 W; (f) 16 W; (g) 16 W; (h) 6 A, 0 V; (i) 72 W; (j) 72 W; (k) zero.

6.5 (a) 1400 Ω; (b) 0.71 mA; (c) 13.5 kΩ.

6.6 1.67 MΩ.

6.7 (a) 5 Ω; (b) 0.02513 Ω; (c) 15 V.

6.8 H contains a 6 V battery; J contains a resistor equivalent to 375 Ω; K contains a 75 Ω resistor between a and c and another between c and d; L contains a 5 kΩ linear potentiometer between a and b with the sliding contact connected to c through a 75 Ω resistor.

6.9 A to B is 600 Ω; A to C is 200 Ω: B and D are connected together; C and D are not.

6.10 11.9 Ω.

6.11 (a) 11 kΩ; (b) 270 μA; (c) 75%; (d) *Hint*: the assumption made in (c) no longer applies.

6.12 (a) 4372 Ω; (b) None.

6.13 (a) $I_{10} = 0.51$ A; $I_{30} = 0.23$ A; $I_{20} = L_5 = 0.28$ A; (b) $V_{10} = 5.1$ V; $V_{30} = 6.9$ V; $V_{20} = 5.6$ V; $V_5 = 1.4$ V.

6.14 (a) The p.d. between X and Y is zero; (b) $8 \times 10^{-6} \Omega$m; (c) From X to Y.

6.15 99 pF mm^{-2}; 21 cm^2; No.

6.16 (a) 12 μA; (b) 0.0024 V; (c) 11.9976 V; (d) 5000 s; (e) 58 min; (f) 0.06 C.

6.17 39.4 J.

6.18 (b) Rate decreases; (c) Rate decreases.

6.19 (a) Charge flows from the capacitor. (b,c,d) Charge flows to the capacitor.

6.20 (a)(i) The same; (ii) The same (*Hint*: use $\frac{1}{2}CV^2$); (b)(ii) Circuit B (*Hint*: compare the time constants).

6.21 (a) The resistances are respectively very large, very large, 140 Ω and 25.7 Ω; (c) 5 μA; (d) Consider the p.d. across the diodes; (e) 10 kΩ in series prevents the p.d. across G and the diodes from reaching a value which will allow a significant current to flow through the latter.

Unit Seven

7.1 2×10^{-3} N (vertical); 1.6×10^{-3} N (vertical).

7.2 12 μN (deflection possible); no vertical force and no deflection.

7.3 E−W. 104 μN at 20° to horizontal; 97.7 μN horizontally E−W.

7.4 45 mT.

7.6 9.86 N.

7.7 10^{-8} N.

7.8 $\dfrac{4\pi^2 GR^4 \varrho^2}{9}$.

7.9 3.844×10^7 m.

7.10 5506 kg m^{-3}; Newton lacked G.

7.11 7.75 km s^{-1}; 90 min.

7.12 96 min.

7.13 3.58×10^7 m above surface; 3069 m s^{-1}.

7.14 Ariel, Titania, Umbriel, Oberon are all moons of Uranus.

7.15 8.7×10^{25} kg; 10.1 N kg^{-1}.

7.16 6×10^{24} kg.

7.17 $v^2 = \dfrac{r^2 g}{R}$.

7.18 P.E. → K.E.; P.E. → heat, etc.

7.19 120 J; 30 J kg^{-1}.

7.20 B.

7.21 -6.28×10^7 J kg^{-1}; -4×10^7 J kg^{-1}; -2.28×10^7 J kg^{-1}; 1.37×10^{11} J; Student assumed that g is constant.

7.22 (a) zero; (b) 0.64 N kg^{-1}; (c) 2.24 × 10^9 J.

7.23 (a) 3.75 × 10^4 N C^{-1}; (b) 2.25 mN; (c) 9 × 10^{-5} J; (d) 1500 V or J C^{-1}.

7.24 4715 V.

7.25 61 nC.

7.26 4.5 × 10^{-5} J; 15 kV; 3 × 10^5 V m^{-1}.

7.27 720 V; 432 V.

7.28 675 V; potential has decreased.

7.29 (a) Same; (b) 800 V; (c) Potential is doubled.

7.30 (a) 3 m; (b) 2 × 10^{-7} C.

Unit Eight

8.1 (a) VQ/d; (b) VQ; (c) Q/t; (d) CV.

8.2 (a) 5 × 10^4 N C^{-1}; (b) 1.05 × 10^{-8} C; (c) 5.285 × 10^{-4} N; (e) 0.21 μA; (f) The frequency of oscillation of the ball can be increased by reducing the distance between the plates. The effect of doing this is two-fold: The field strength and therefore the force on the ball is increased and not only does the ball travel faster but it has a shorter distance to travel between hits. Another method of increasing the frequency is to raise the voltage of the E.H.T. supply which will result in a larger field strength without changing the plate separation.

8.3 9.3 g; 35.

8.4 5, 10, 8, 3, 4, 2.

8.5 1.11 × 10^{-18} C; 7.

8.6 2.45 × 10^5 V m^{-1}; downwards; negative.

8.7 1.12 × 10^{-18} C; yes, it carries 7e.

8.9 (a) 8 × 10^{-16} J; (b) 16 × 10^{-16} J.

8.10 (a) − 196 J kg^{-1}; (b) + 4600 J C^{-1}; − 5.56 × 10^{-3} J (fall).

8.12 X is − , Y is + .

8.13 (a) 1.95 mm s^{-1}; (b) 1.25 × 10^{-22} N; (c) 0.3 N; (d) 7.8 × 10^{-4} V m^{-1}; (e) 4.7 μV; (f) $F = BIL$ = 0.3 N.

8.16 54 μV; 12 μV.

8.17 Negative.

8.18 No! Consider all possible field directions in three dimensions.

8.19 9.4 × 10^6 m s^{-1}; 1.1 mT.

8.20 1.6 × 10^7 m s^{-1}; 1.78 × 10^{11} C kg^{-1}.

8.21 Electrons carry like charges.

8.23 (a) 3.25 × 10^7 m s^{-1}; (b) 3.78 × 10^5 m s^{-1}; (c) 5.35 × 10^5 m s^{-1}.

8.24 (a) 2 ns; (b) 7 mm.

8.25 (a) 4 × 10^{-15} J per electron; (b) 10^{16} s^{-1}; (c) 40 W.

8.26 (a) 1.5 kW; (b) 7.5 W.

8.27 96 000 C; 0.475 g.

Unit Nine

9.1 0.01 N. This is too small to strain the insulation or to produce noticeable vibration.

9.2 About 14 A.

9.3 6.4 × 10^{-3} N m.

9.4 500 μA.

9.5 (a) Normal to the plane of the diagram and down into it; (b) normal to the plane of the diagram and up out of it; (c) perpendicular to wire Y and towards X; (d) perpendicular to wire X and towards Y; (e) each wire must be part of a complete circuit. Any effects due to the rest of these circuits have been ignored. It can be shown, both theoretically and experimentally, that two circuits carrying current, whatever their shape, exert equal and opposite forces and torques upon each other.

9.6 (a) WX due west, XY due south; YZ due east, ZW due north; (b) the forces on WZ and XY; (c) due west. The force on WX is greater than that on YZ because WX is closer to PQ than YZ; (d) with WX in contact with PQ and *not*, as might be guessed, with WX and YZ on either side of PQ and equidistant from PQ. Although this would be symmetrical geometrically, the square in this position would experience a resultant force eastwards.

9.7 0.012 T.

9.8 (a) 12 m; (b) 0.2 m; (c) 1.7 mm.

9.9 (a) $\dfrac{m_x v_x Q_y}{m_y v_y Q_x}$;

(b) $\dfrac{\text{radius of } \alpha\text{-particle track}}{\text{radius of } \beta\text{-particle track}} = 92$.

9.10 (a) deflection left and back to zero; (b) deflection right and back to zero; (c) deflection left and back to zero; (d) deflection right and back to zero; (e) no deflection if the frame is held symmetrically between the poles throughout the motion; the e.m.f.s in each side then exactly cancel; (f) no deflection because no flux lines are cut.

9.11 (a) 2.5 μV; (b) Sides PS and QR are not cutting flux lines; (c) 5 μA.

9.12 31.4 μV.

9.13 (a) p.d. is Blv; the north end is more positive; (b) e.m.f. in the frame: (i) Bdv, (ii) zero, (iii) Bdv. Current in the frame: (i) Bdv/R anticlockwise, (ii) zero, (iii) Bdv/R clockwise. Force on frame: (i) $B^2 d^2 v/R$ towards the left, (ii) zero, (iii) $B^2 d^2 v/R$ towards the left.

9.14 (a) 8 V; (b) an equal and opposite e.m.f. is induced in the connecting wires.

9.15 (a) 250 μWb; (b) 2.5 mV.

9.16 (a) 0.628 m s^{-1}; (b) 7.85 mV; (c) 3.14 V; (d) 0.202 A; (e) 2.22 W; (f) Power from supply =

2.424 W, power as heat in armature resistance = 0.204 W; (g) 4.86 W; useful mechanical power = 2.64 W; (h) 19.1 rev s^{-1}; at this speed the back e.m.f. equals half the supply voltage.

9.17 (a)(i) $BAN\omega$, which is a maximum when θ is 0° or 180°; (ii) V/R; (iii) $BAN\left(\dfrac{V - BAN\omega}{R}\right)$; (b) 190 V; 950 W; 50 W.

9.18 (a) 5 V, 7.5 V, 0 V, 2.5 V; (b) 1 W, 1.5 W, 2 W, 1.25 W; (c) $5\sqrt{2}$ V, $5\sqrt{3}$ V, 10 V, $5\sqrt{5/2}$ V.

9.19 16.2 A. The fuse rating is r.m.s. current which is 11.5 A.

9.20 3140 Ω.

9.21 The stator has inductance and its reactance increases with frequency.

9.22 0.021 H.

9.25 6.37 V.

9.26 1590 μF.

9.28 536 pF and 94 pF.

9.29 (a) 2.2×10^{-6} Wb; (b) 3.4 A; (c) 1.5×10^{6} A Wb^{-1}; (d) 2.7×10^{-3} Wb A^{-1} m^{-1}, 2100.

9.30 (a) 1.26×10^{5} A Wb^{-1}, 6.37×10^{6} A Wb^{-1}; (b) 6.25×10^{-5} Wb; (c) 406 A; (d) 1.35 A.

9.31 (b)(i) 4×10^{-4} T; (ii) 2 A; (c) 6×10^{-4} N.

9.32 (a) 1.57×10^{-3} Wb m^{-2}; (b) Vertically upwards; (c) 3.14×10^{-4} N; (d) Half of the previous value; about 1.6×10^{-4} N.

9.34 120 turns.

9.35 1.8×10^{-5} T.

9.39 (b) 0.3 m.

Unit Ten

10.1 29 bytes s^{-1}.

10.2

A	B	C
0	0	0
0	1	0
1	0	1
1	1	0

10.3

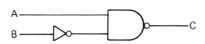

10.4

A	B	S
0	0	0
0	1	1
1	0	1
1	1	0

10.5 (a) Only one state possible, with X = 1 and Y = 1; (b) X = 1, Y = 0; (c) No change in the outputs; (d) X = 0 and Y = 1, or X = 1 and Y = 0.

10.6

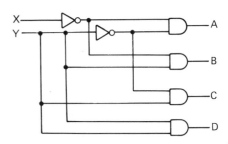

10.7 15 mV.

10.8 0.20.

10.9 99; power gain = 20 (13 dB).

10.10 40; 5 kΩ; 500 Ω.

10.11 (a) $0.1 v_i$; (b) −20 dB; (c) −26 dB.

10.12 (a) −3 dB; (b) 3.4 nF.

10.13 (a) 100 000 at 10 Hz, 10 at 100 kHz; (b) 25 (or 28 dB) at 10 Hz, 7.1 (or 17 dB) at 100 kHz; (c) Overall gain 25 dB at 16 kHz. This would not be a high enough upper frequency limit for a high quality a.f. amplifier (but 25 kHz would be acceptable).

10.14 +15.

10.15 +0.19.

10.16 26 kΩ and 1 MΩ.

Unit Eleven

11.1 5.5 m s^{-1}; 15 000 J.

11.2 0.5 m s^{-1} in the direction initially followed by the 3000 kg truck; 135 000 J.

11.3 −2.14 m s^{-1}; 2.86 m s^{-1}; 0.82.

11.4 1 m s^{-1} (reversed); 2.4 J.

11.6 10^{4} m s^{-1}.

11.7 (a) 1.3×10^{-15} J; (b) 1.64×10^{-18} J; (c) 1.3×10^{-25} J; (d) 1.3×10^{-28} J.

11.8 2.4×10^{14} Hz.

11.10 (a) 3.4×10^{-19} J; (b) 16 000 K.

11.11 (a) 3.2×10^{14} Hz; (b) 2.2×10^{-19} J or 1.4 eV; $h = 6.7 \times 10^{-34}$ J s.

11.12 1300 s (about 21 min).

11.13 (a) 3×10^{6} (b) 100 m.

11.15 Compare the Lyman series; 87.2×10^{-19} J.

11.17 (a) 656 nm; (b) 121 nm.

11.18 8.7 pm.

11.19 0.12 nm; particle behaviour dominant.

11.20 (a) $E \propto Z/r_o$; (b) $\lambda \propto r_o$; (c) $\lambda = h/\sqrt{2mE_k}$; (d) $E \propto Z^2$.

11.21 (a) 54 eV; (b) Each electron shields the other from the full attractive force of the nucleus.

11.22 (a) 122 eV; (b) Two of the electrons have a major shielding effect. The electron involved in the ionization is not in the ground state.

11.23 E_k; $E_p = 6:1$.

Unit Twelve

12.1 2×10^5; (a) fewer encounters per cm; (b) lower ionization energy.

12.2 1.85×10^{10} s^{-1}; 1.03×10^5; 3 MeV.

12.3 (a) 9802; (d) about 40 s.

12.4 In each case an appropriate graph on semi-log paper is straight, confirming that the changes are (a) exponential decay and (b) exponential growth.

12.5 (a) $\frac{1}{4}$; (b) $\frac{1}{16}$; (c) $(\frac{1}{2})^{10}$; (d) 1.1×10^6 Bq.

12.6 3160 years ago.

12.7 1562 years.

12.8 (a) $^{211}_{82}$Pb; (b) 1.9×10^7 m s^{-1}; (c) $^{211}_{83}$Bi.

12.9 300 m.

12.10 (a) 78, 195; (b) 8.1×10^{-15} m; (c) 2.2×10^{-42} m^3;

(d) 3.24×10^{-25} kg;

(e) 1.4×10^{17} kg m^{-3};

(f) 1.55×10^{-29} m^3;

(g) 1.6×10^{-10} m.

12.11 (a) 25×10^{14}; (b) 10^4; (c) 25×10^{18}; (d) 3.4×10^{-14} m; (f) 4×10^{-8} m^2; (g) 2×10^{-14} m.

12.12 (a) 3 in 10^4; (b) 1 in 16×10^4; (c) 4 in 10^4.

12.13 47, 13, silver, aluminium.

12.14 (a) 1.44×10^{-33} N; (b) 359 N.

12.15 The same.

12.16 (a) 2×10^7 m s^{-1}; (b) 1.6×10^7 m s^{-1}; (c) 4.45×10^6 m s^{-1}; (d) 0.05%, 64%, 69%.

12.17 (a) 100%; (b) 25%; (c) 2%.

12.20 (a) $^{30}_{14}$Si; (b) n; (c) ^{239}U; (d) p; (e) ^{17}O; (f) α; (g) ^{197}Au.

12.21 (a) 2.3 N; (b) 7.1 MeV.

12.23 28.9 mm; 40.75 mm.

APPENDIX 1

The International System of Units (SI)

1.1 Base units

length	metre	m
mass	kilogram	kg
time	second	s
current	ampere	A
thermodynamic temperature	kelvin	K
luminous intensity	candela	cd
amount of substance	mole	mol

1.2 Supplementary units

plane angle	radian	rad
solid angle	steradian	sr

1.3 Some derived units

area	square metre		m^2
volume	cubic metre		m^3
frequency	hertz	Hz	s^{-1}
density	kilogram per cubic metre		$kg\,m^{-3}$
speed, velocity	metre per second		$m\,s^{-1}$
angular velocity	radian per second		$rad\,s^{-1}$
acceleration	metre per second squared		$m\,s^{-2}$
angular acceleration	radian per second squared		$rad\,s^{-2}$
force	newton	N	$kg\,m\,s^{-2}$
pressure, stress	pascal	Pa	$N\,m^{-2}$
torque, moment of a force	newton metre		$N\,m$
momentum	kilogram metre per second		$kg\,m\,s^{-1}$
	or newton second		$N\,s$
viscosity	newton second per square metre		$N\,s\,m^{-2}$
energy, heat, work	joule	J	$N\,m$
entropy	joule per kelvin		$J\,K^{-1}$
power	watt	W	$J\,s^{-1}$
electric charge	coulomb	C	$A\,s$
electric potential difference electromotive force	volt	V	$J\,C^{-1}$
electric field intensity	newton per coulomb		$N\,C^{-1}$
	or volt per metre		$V\,m^{-1}$

electric resistance	ohm		Ω	$V A^{-1}$
electric conductance	siemens		S	$A V^{-1}$
electric capacitance	farad		F	$C V^{-1}$
magnetic flux	weber		Wb	V s
magnetic flux density	tesla		T	$Wb\ m^{-2}$
electric inductance	henry		H	$Wb\ A^{-1}$
magnetomotive force	ampere turns		A	A
luminous flux	lumen		lm	cd sr
illumination	lux		lx	$lm\ m^{-2}$
thermal conductivity				$W\ m^{-1} K^{-1}$
activity of a radioactive source	becquerel		Bq	s^{-1}
absorbed dose of ionising radiation	gray		Gy	$J\ kg^{-1}$
dose equivalent of ionising radiation	sievert		Sv	$J\ kg^{-1}$

Prefixes for SI units

Sub-multiple	Prefix	Symbol	Multiple	Prefix	Symbol
10^{-1}	deci	d	10^{1}	deca	da
10^{-2}	centi	c	10^{2}	hecto	h
10^{-3}	milli	m	10^{3}	kilo	k
10^{-6}	micro	μ	10^{6}	mega	M
10^{-9}	nano	n	10^{9}	giga	G
10^{-12}	pico	p	10^{12}	tera	T
10^{-15}	femto	f	10^{15}	peta	P
10^{-18}	atto	a	10^{18}	exa	E

(Note: The 'billion' is avoided since, in Britain, it is equivalent to 10^{12}, but in the United States it is equivalent to 10^{9}.)

Some other units

a) *Units exactly defined in terms of SI units.*

time	minute	min	60 s
	hour	h	3600 s
	day	d	86 400 s
angle	degree	°	$\pi/180$ rad
	minute	′	$\pi/10\,800$ rad
	second	″	$\pi/648\,000$ rad
volume	litre	l, L	$10^{-3}\,m^3$
mass	tonne	t	$10^3\,kg$
pressure	bar	bar	$10^5\,Pa$

b) *Units defined in terms of physical constants.*

mass	unified atomic mass unit	u	$1.660 \times 10^{-27}\,kg$
energy	electronvolt	eV	$1.602 \times 10^{-19}\,J$

Definitions

metre   The metre is the length equal to 1 650 763.73 wavelengths in vacuum of the radiation corresponding to the transition between the levels $2p_{10}$ and $5d_5$ of the krypton-86 atom.

kilogram   The kilogram is equal to the mass of the international prototype of the kilogram.

second	The second is the duration of 9 192 631 770 periods of the radiation corresponding to the transition between the two hyperfine levels of the ground state of the caesium-133 atom.
ampere	The ampere is that constant current which, if maintained in two straight parallel conductors of infinite length, of negligible cross-section, and placed 1 metre apart in vacuum, would produce between these conductors a force equal to 2×10^{-7} newton per metre of length.
kelvin	The kelvin is the fraction 1/273.16 of the thermodynamic temperature of the triple point of water.
candela	The candela is the luminous intensity, in a given direction, of a source that emits monochromatic radiation of frequency 540×10^{12} hertz that has a radiant intensity in that direction of 1/683 watt per steradian.
mole	The mole is the amount of substance of a system which contains as many elementary entities as there are atoms in 0.012 kilogram of carbon-12. *Note*: when the mole is used, the elementary entities must be specified and may be atoms, molecules, ions, electrons, other particles or specified groups of such particles.
radian	The radian is the plane angle between two radii of a circle which cut off on the circumference an arc equal in length to the radius.
steradian	The steradian is the solid angle which, having its vertex in the centre of a sphere, cuts off an area of the surface of the sphere equal to that of a square with sides of length equal to the radius of the sphere.
newton	The newton is that force which gives to a mass of 1 kilogram an acceleration of 1 metre per second squared.
joule	The joule is the work done when the point of application of a force of 1 newton is displaced a distance of 1 metre in the direction of the force.
watt	The watt is the rate of transformation of energy of 1 joule per second.
coulomb	The coulomb is the quantity of electric charge transported in 1 second by a current of 1 ampere.
volt	The volt is the electric potential difference between two points when the energy transformed is 1 joule for each coulomb which passes.
ohm	The ohm is the electrical resistance between two points of a conductor when a constant potential difference of 1 volt applied between those points produces in the conductor a current of 1 ampere, the conductor not being the source of any electromotive force.

APPENDIX 2

Constants and Formulae

1. Fundamental physical constants

Speed of light in a vacuum	c	$2.998 \times 10^8 \, \text{m s}^{-1}$
Charge of electron	e	$1.602 \times 10^{-19} \, \text{C}$
Faraday constant	F	$9.649 \times 10^4 \, \text{C mol}^{-1}$
Gravitational constant	G	$6.673 \times 10^{-11} \, \text{N m}^2 \, \text{kg}^{-2}$
Planck constant	h	$6.626 \times 10^{-34} \, \text{J s}$
Boltzmann constant	k	$1.381 \times 10^{-23} \, \text{J K}^{-1}$
Avogadro constant	N_A, L	$6.022 \times 10^{23} \, \text{mol}^{-1}$
Molar gas constant	R	$8.314 \, \text{J mol}^{-1} \, \text{K}^{-1}$
Rest mass of electron	m_e	$9.109 \times 10^{-31} \, \text{kg}$
Rest mass of neutron	m_n	$1.675 \times 10^{-27} \, \text{kg}$
Rest mass of proton	m_p	$1.673 \times 10^{-27} \, \text{kg}$
Permittivity of free space	ϵ_0	$8.854 \times 10^{-12} \, \text{F m}^{-1}$
	$\dfrac{1}{4\pi\epsilon_0}$	$8.987 \times 10^9 \, \text{N m}^2 \, \text{C}^{-2}$
Permeability of free space	μ_0	$4\pi \times 10^{-7} \, \text{H m}^{-1}$

2. Some other constants

Standard atmospheric pressure	$1.013 \times 10^5 \, \text{Pa}$ (1.013 bar or 1013 millibar)
Absolute zero of temperature	$-273.15 \, °\text{C}$ (0 K)
Molar volume of ideal gas at s.t.p.	$2.24 \times 10^{-2} \, \text{m}^3$
Standard gravitational field strength	$9.807 \, \text{N kg}^{-1} \, (\text{m s}^{-2})$

3. Mathematical constants

$\pi = 3.142$ $\quad$ $e = 2.718$ $\quad$ $\ln 2 = 0.693$
$\pi^2 = 9.870$ $\quad$ $\log_{10}e = 0.4343$

4. Some mathematical equations and formulae

Geometrical: Circle: circumference $2\pi r$ $\quad$ area πr^2
$\qquad\qquad$ Sphere: surface area $4\pi r^2$ $\quad$ volume $\frac{4}{3}\pi r^3$

Algebra:

$$ax^2 + bx + c = 0 \Rightarrow x = \frac{-b \pm \sqrt{b^2 - 4ac}}{2a}$$

$$a^2 - b^2 = (a + b)(a - b)$$
$$(a \pm b)^2 = a^2 \pm 2ab + b^2$$

Coordinate geometry:
Straight line $\quad\quad y = mx + c$
$$y - y_0 = m(x - x_0) \text{ through } x_0, y_0$$

Trigonometry:

$$\sin \theta = \frac{y}{r} = \frac{1}{\operatorname{cosec} \theta}$$

$$\cos \theta = \frac{x}{r} = \frac{1}{\sec \theta}$$

$$\tan \theta = \frac{y}{x} = \frac{1}{\cot \theta} = \frac{\sin \theta}{\cos \theta}$$

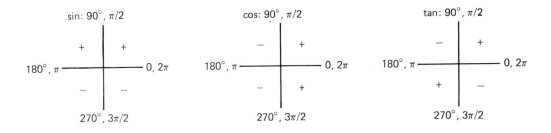

$$\cos^2 \theta + \sin^2 \theta = 1; \ \sec^2 \theta = 1 + \tan^2 \theta; \ \operatorname{cosec}^2 \theta = 1 + \cot^2 \theta$$

For any triangle: $\quad a^2 = b^2 + c^2 - 2bc \cos A$, (cosine rule)

$$\frac{a}{\sin A} = \frac{b}{\sin B} = \frac{d}{\sin C} \text{ (sine rule)}$$

Differentiation:

If $y = ax^n$ $\quad\quad \dfrac{dy}{dx} = anx^{n-1}$

$y = uv$ $\quad\quad \dfrac{dy}{dx} = v\dfrac{du}{dx} + u\dfrac{dv}{dx}$

$y = \text{constant}$ $\quad \dfrac{dy}{dx} = 0$

$y = \sin x$ $\quad\quad \dfrac{dy}{dx} = \cos x.$ If $y = \sin \omega t$ $\quad \dfrac{dy}{dt} = \omega \cos \omega t$

$y = \cos x$ $\quad\quad \dfrac{dy}{dx} = -\sin x.$ If $y = \cos \omega t$ $\quad \dfrac{dy}{dt} = -\omega \sin \omega t$

$y = \ln x$ $\quad\quad \dfrac{dy}{dx} = \dfrac{1}{x}$

$y = e^{ax}$ $\quad\quad \dfrac{dy}{dx} = ae^{ax}$

5. Selected physics equations

Kinematics, linear motion:

$$v = \frac{ds}{dt}$$ Section 2.7

$$a = \frac{d^2s}{dt^2} = \frac{dv}{dt}$$ Section 5.3

For constant acceleration:

$$v_t = v_0 + at$$ Eq. 5.9
$$v_t^2 = v_0^2 + 2as$$ Eq. 5.12

circular motion:

$$v = r\omega$$ Section 7.3

$$a = \frac{v^2}{r}$$ Section 7.2

Dynamics, linear motion: $F = ma$ Eq. 5.6
 impulse $F\Delta t = m\Delta v$ Eq. 5.4
 work $W = F\Delta s$ Eq. 8.1
 kinetic energy $E_k = \frac{1}{2}mv^2$ Eq. 8.3

inelastic collisions: $\dfrac{E_k \text{ remaining}}{\text{initial } E_k} = \dfrac{m_1}{m_1 + m_2}$ Section 43.2

circular motion:

 central force $F = \dfrac{mv^2}{r}$ Section 7.2

 torque $T = Fr$ Section 6.8
 moment of inertia $I = \Sigma mr^2 = Mk^2$ Section 8.19
 angular momentum $= I\omega$ Section 8.19

$$T = \frac{2\pi}{\omega}$$ Section 7.3

$$\omega = 2\pi f$$ Section 7.3
simple harmonic motion: $a = -\omega^2 x$ Eq. 19.4
$$x = a_0 \cos(\omega t + \phi)$$ Eq. 19.3
$$E_p = \frac{1}{2}kx^2$$ Eq. 19.7
$$E_k = \frac{1}{2}k(a_0^2 - x^2)$$ Eq. 19.8
Vectors: $\boldsymbol{F} = \boldsymbol{F}_x + \boldsymbol{F}_y = F\sin\theta + F\cos\theta$ Section 2.13

Pressure: $$p = \frac{F}{A}$$ Eq. 11.1

$$\Delta p = -\varrho g\Delta h$$ Eq. 11.2

Elasticity: $\text{Stress} = \dfrac{F}{A}$; $\text{strain} = \dfrac{\Delta l}{l_0}$ Section 13.2

 Young modulus $E = \dfrac{F/A}{\Delta l/l_0}$ Section 13.2

Strain energy per unit volume	$= \frac{1}{2}\dfrac{(\text{stress})^2}{E}$	Eq. 15.2
Springs etc., obeying Hooke's law	$F = kx$	Section 15.1
energy stored	$= \frac{1}{2}kx^2$	

Thermal physics:

Temperature	$T_x = 273.16\,\dfrac{p_x}{p_{tr}}$	Section 9.1
Conduction of heat	$\dfrac{\Delta Q}{\Delta t} = \lambda A\,\dfrac{\Delta T}{l}$	Eq. 11.5
Linear expansivity	$\alpha = \dfrac{\Delta l}{l_0 \Delta T}$	Section 15.6
Specific heat capacity,	$c = \dfrac{1}{m}\dfrac{\Delta Q}{\Delta T}$	Eq. 9.3
For gases	$V = V_0(1 + \alpha \Delta T)$	Eq. 10.6
	$p = p_0(1 + \beta \Delta T)$	Eq. 10.7
1st law of thermodynamics	$\Delta Q = \Delta U + \Delta W$	Eq. 9.1

Kinetic model:

	$pV = \frac{1}{3}Nmc^2$	Eq. 10.5
	$pV_m = RT$	Eq. 10.8
	$\sqrt{\bar{c}^2} = \sqrt{3k\dfrac{T}{m}}$	Eq. 10.10
Internal energy per mole of ideal gas	$= \frac{3}{2}N_A kT = \frac{3}{2}RT$	Eq. 10.14

Viscosity:

Poiseuille	$\dfrac{\Delta V}{\Delta t} = \dfrac{a^2}{8\eta}A\,\dfrac{\Delta p}{l} = \dfrac{\pi a^4}{8\eta}\dfrac{\Delta p}{l}$	Eq. 11.4
Stokes	$F = 6\pi\eta va$	Eq. 31.2

Statistical physics:

	$\dfrac{W^*}{W} = 1 + \dfrac{N}{q}$	Eq. 12.1
	$\dfrac{N}{q} = \dfrac{\epsilon}{kT}$	Eq. 12.5
	$n_E = n_0 e^{-E/kT}$	Eq. 12.6
Entropy change	$\Delta S = \dfrac{\Delta Q}{T}$	Eq. 12.9

Surface effects:

Surface energy per unit area	$\gamma = F/2l$	Eq. 15.5
Excess pressure in a drop	$\Delta p = 2\gamma/r$	Eq. 15.6
Capillary rise	$= \dfrac{2\gamma\cos\theta}{\varrho gr}$	Eq. 15.8

Waves:

	$c = f\lambda$	Eq. 16.1
Interference	$n\lambda = s\sin\theta$	Eq. 16.2

Speed of transverse pulse in elastic string	$= \sqrt{F/\mu}$	Eq. 17.4
Speed of longitudinal pulse in solid bar	$= \sqrt{E/\varrho}$	Eq. 17.6
Bragg diffraction	$2d \sin \phi = n\lambda$	Eq. 14.6
Refractive index	$_a n_b = \dfrac{\text{speed of wave in medium a}}{\text{speed of wave in medium b}}$	Eq. 18.1
Wave equation	$x = a_0 \cos \dfrac{2\pi}{\lambda}(ct - z)$	Eq. 19.9
Stationary waves	$\lambda = 2l/n$	Eq. 20.5
	$f = \dfrac{n}{2l}\sqrt{\dfrac{F}{\mu}}$	Eq. 20.6
Diffraction – single slit	$\sin \theta = \dfrac{n\lambda}{d}$	Section 21.4
circular aperture	$\sin \theta = 1.22 \, \lambda / b$	Eq. 21.2
grating	$n\lambda = s \sin \theta$	Eq. 21.1
Rayleigh's criterion	$\theta = 1.22 \, \lambda / b$	Section 21.5

Electricity:

Kirchhoff's 1st law for series circuit:	$I_0 = I_1 + I_2 + I_3$	Section 23.1
	$Q = It$	Eq. 23.1
	$W = VIt$	Eq. 23.2
	$R = V/I$	Eq. 23.3
	$R = \varrho l/A$	Eq. 23.4
	$\dfrac{R_\theta - R_0}{R_0} = \alpha\theta$	Eq. 23.5
	$P = VI$	Eq. 24.1
Kirchhoff's 2nd law	$V_{AD} = V_{AB} + V_{BC} + V_{CD}$	Section 24.5
Series resistors	$R = R_1 + R_2 + R_3$	Eq. 25.2
Parallel resistors	$\dfrac{1}{R} = \dfrac{1}{R_1} + \dfrac{1}{R_2} + \dfrac{1}{R_3}$	Eq. 25.4
Wheatstone bridge	$\dfrac{R_1}{R_2} = \dfrac{R_3}{R_4}$	Eq. 25.7
Capacitance	$C = \dfrac{Q}{V}$	Eq. 26.1
Parallel plate capacitor	$C = \epsilon \dfrac{A}{d} = \epsilon_r \epsilon_0 \dfrac{A}{d}$	Eq. 26.2
Series capacitors	$\dfrac{1}{C} = \dfrac{1}{C_1} + \dfrac{1}{C_2} + \dfrac{1}{C_3}$	Eq. 26.9
Parallel capacitors	$C = C_1 + C_2 + C_3$	Eq. 26.6
Capacitance of a sphere	$= 4\pi\epsilon_0 r$	Eq. 30.10
Energy stored in a capacitor	$= \tfrac{1}{2}QV = \tfrac{1}{2}CV^2 = \tfrac{1}{2}Q^2/C$	Section 26.6
	$\dfrac{\Delta Q}{\Delta t} = -\dfrac{Q}{RC}$	Eq. 26.10

$$Q = Q_0 \, e^{-t/RC}$$ Eq. 26.11

$$t_{\frac{1}{2}} = RC \ln 2$$ Eq. 26.12

$$\text{Time constant} \ = RC$$ Eq. 26.12

Fields:

Gravitational field strength $\qquad g = F/m = GM/r^2$ Eq. 27.1 and 28.9

Gravitational force between masses $\quad F = G\dfrac{m_1 m_2}{r^2}$ Eq. 28.1

Gravitational potential $\qquad V = -G\dfrac{M}{r}$ Eq. 29.3

Gravitational potential gradient $\qquad g = -\dfrac{dV}{dr}$ Eq. 29.2

Potential energy difference near earth's surface $\approx mgh$ Eq. 8.2

Electric field strength $\qquad E = F/Q = -\dfrac{dV}{dr}$ Eq. 27.2 and 30.2

Uniform electric field $\qquad E = V/d$ Eq. 30.1

Electric field, point charge $\qquad E = \dfrac{1}{4\pi\epsilon_0}\dfrac{Q}{r^2}$ Eq. 30.6

Electric potential, point charge $\qquad V = \dfrac{1}{4\pi\epsilon_0}\dfrac{Q}{r}$ Eq. 30.5

Electric potential gradient $\qquad E = -\dfrac{dV}{dr}$ Eq. 30.2

Magnetic field strength $\qquad B = F/Il$ Eq. 27.3

Force on a conductor $\qquad F = BIl \sin\theta$ Eq. 27.4

Conduction:

Faraday's law of electrolysis $\qquad m \propto It$ Section 32.9

Charge carrier flow $\qquad I = nAve$ Eq. 32.2

Hall voltage $\qquad V = IB/ned$ Eq. 32.4

Speed of electron from electron gun $\quad v = \sqrt{\dfrac{2Ve}{m}}$ Eq. 33.5

Specific charge of electron $\qquad \dfrac{e}{m} = \dfrac{2V}{B^2 r^2}$ Eq. 33.6

Force between 2 currents $\qquad F = (2 \times 10^{-7}\,\mathrm{N\,A^{-2}})\dfrac{I_1 I_2 l}{a}$ Eq. 34.1

Force on a charge in a B field $\qquad F = BQv \sin\theta$ Section 34.5

Induced e.m.f.s:

Induced e.m.f. $\qquad E = Bvl \sin\theta$ Eq. 35.5

e.m.f. induced in rotating coil $\qquad E = E_0 \sin 2\pi ft$ Eq. 35.6

where $\qquad E_0 = 2\pi NBAf$

Induced e.m.f. $\qquad E = -N\dfrac{d\Phi}{dt}$ Eq. 35.8

Alternating currents:

$$\omega = 2\pi f$$

$$I = I_0 \sin \omega t$$ Eq. 36.1

$$I_{\text{r.m.s.}} = \frac{I_0}{\sqrt{2}}$$

Eq. 36.2

Reactance of inductor	$= V_0/I_0 = 2\pi fL = \omega L$	Eq. 36.3
Reactance of capacitor	$= V_0/I_0 = 1/2\pi fC = 1/\omega C$	Eq. 36.4
Power dissipated in circuit with impedance	$= V_{\text{r.m.s.}} \times I_{\text{r.m.s.}} \cos \phi$	Table 36b

Resonant frequency $\qquad f = \dfrac{1}{2\pi \sqrt{LC}}$

Eq. 36.7

For ideal transformer $\qquad \dfrac{V_s}{V_p} = \dfrac{N_s}{N_p}$

Eq. 38.1

Flux density:

solenoid $\qquad B = \dfrac{\Phi}{A} = \dfrac{\mu_0 NI}{l}$

Eq. 37.7

plane circular coil $\qquad B = \dfrac{\mu_0 NI}{2r}$

Eq. 37.9

long straight wire $\qquad B = \dfrac{\mu_0}{2\pi}\dfrac{I}{a}$

Eq. 37.8

Speed of an electromagnetic wave in vacuum: $\quad c = 1/\sqrt{\mu_0 \epsilon_0}$

Eq. 39.7

Radiation and matter: $\qquad E = hf = \dfrac{hc}{\lambda}$

Eq. 44.1

Photoelectric effect $\qquad (\tfrac{1}{2}mv^2)_{\max} = hf - \phi$

Eq. 44.2

de Broglie wavelength, $\qquad \lambda = h/p = h/m_0 v$

Eq. 45.1

Atoms and nuclei:

Radioactive decay $\qquad \mathrm{d}A/\mathrm{d}t = -\lambda A \text{ or } \mathrm{d}N/\mathrm{d}t = -\lambda N$

Eq. 46.1

$$N = N_0 e^{-\lambda t}$$

Eq. 46.2

$$t_{\frac{1}{2}} = \frac{1}{\lambda}\ln 2$$

Eq. 46.3

Mass and energy $\qquad E = mc^2$

Section 39.2

INDEX